# ORGANIC CHEMISTRY OF SULFUR

# ORGANIC CHEMISTRY OF SULFUR

*Edited by*

## S. Oae

*Department of Chemistry*
*University of Tsukuba*
*Ibaraki, Japan*

PLENUM PRESS • NEW YORK AND LONDON

Library of Congress Cataloging in Publication Data
Main entry under title:

Organic chemistry of sulfur.

Includes bibliographical references and index.
1. Organosulphur compounds—Addresses, essays, lectures. I. Oae, Shigeru.
QD412.S1067                                    547'.06                              75-20028
ISBN 0-306-30740-5

©1977 Plenum Press, New York
A Division of Plenum Publishing Corporation
227 West 17th Street, New York, N.Y. 10011

Printed in the United States of America

# Contributors

**H.A. Bent**
Department of Chemistry, North Carolina State University, Raleigh, North Carolina 27607, USA.

**W.W. Brand**
Department of Chemistry, Purdue University, West Lafayette, Indiana 47907, USA.

**L. Field**
Department of Chemistry, Vanderbilt University, Nashville, Tennessee 37203, USA.

**E.T. Kaiser**
Department of Chemistry, University of Chicago, Chicago, Illinois 60637, USA.

**T.C. Klingler**
Department of Chemistry, Purdue University, West Lafayette, Indiana 47907, USA.

**N. Kunieda**
Department of Applied Chemistry, Faculty of Engineering, Osaka City University, Sumiyoshi-ku, Osaka, Japan.

**R. Mayer**
Department of Chemistry, Technical University of Dresden, Germany (GDR)

**S. Oae**
Department of Chemistry, University of Tsukuba, Sakura-mura, Niihari-gun Ibaraki-ken, Japan.

**A. Ohno**
Institute for Chemical Research, Kyoto University, Uji, Kyoto 611, Japan.

**M. Porter**
The Natural Rubber Producers' Research Association, Welwyn Garden City, Hertfordshire, England.

**C.J.M. Stirling**
Department of Chemistry, University College of North Wales, Bangor LL57 2UW.

**W. Tagaki**
Department of Chemistry, Faculty of Engineering, Gumma University, Kiryu, Gumma-Ken, Japan.

**W.E. Truce**
Department of Chemistry, Purdue University, West Lafayette, Indiana 47907, USA.

# Preface

In recent years organic sulfur chemistry has been growing at an even faster pace than the very rapid development in other fields of chemistry. This phenomenal growth is undoubtedly a reflection of industrial and public demands: not only was sulfur recently in overall surplus for the first time in the history of the chemical industry but it has now become a principal environmental hazard in the form of sulfur dioxide, sulfuric acid and hydrogen sulfide. Another reason, discernible in the last fifteen years, has been the desire, on the part of individual chemists and all types of research managers, to move away from the established chemistry of carbon into the less well understood and sometimes virgin chemistries of the other elements which form covalent bonds.

As a result of this movement the last decade has seen the development of sulfur chemistry into a well-organized and now much better understood branch of organic chemistry. Enough of the detail has become clear to see mechanistic interrelationships between previously unconnected reactions and with this clarification the whole subject has in turn become systematized and subdivided. The divalent sulfur chemistry of thiols, monosulfides, disulfides and polysulfides is a large area in itself, much of it devoted to oxidation—reduction and the breakage and formation of sulfur—sulfur bonds, although interesting discoveries are now being made about the reactivity of certain sulfur—carbon bonds. Of course, this area has its own massive biochemical branch involving enzymes and proteins.

The chemistry of sulfur in higher oxidation states evolved at first in an analogous way around sulfur—oxygen compounds — sulfoxides, sulfones, and sulfur oxyacids and their thiol esters. There has since been an expansion into sulfur compounds containing other hetero atoms and into the recently characterized tetracovalent sulfuranes. A third area with many ramifications is that of sulfur heterocycles. Much of the recent interest here has been engendered by pseudoaromatic ring systems and structures exhibiting "no bond" resonance, such as the thiathiophthens.

A number of guide books have appeared for this fast-advancing frontier of chemistry. Professor Charles C. Price and I wrote one of the pioneering books than ten years ago and I wrote another in Japanese six years ago. However, the growth of the subject in recent years has been so rapid that when I was asked to consider the preparation of a new guide book I readily agreed. However, the subject is now so broad that I felt I should not assume sole authorship but rather enlist the help of intimate colleagues, the top experts in their respective fields. Thus this book was born.

The text offers a comprehensive survey of most of the basic aspects of organosulfur chemistry. Since it is intended to be used primarily as a textbook for graduate courses, the subjects of each of the twelve chapters, written by a well-known authority, are carefully chosen and are explained in the light of the most modern concepts in physical organic chemistry. Therefore, I hope that not only students but also many of those actively engaged in industrial and biological chemistry, agriculture and medicine will find this book useful and stimulating.

As readers will realize, this book is a synthesis of international cooperation. There are one German, two British, four Japanese and five American authors. Although the cooperation has been harmonious and genuine, readers will note considerable differences in the styles of the chapters. This was to some extent inevitable in any case but, as editor, I have taken the view that each individual author will have chosen the wording he considers best for expression and to revise that wording was to risk destroying the nuances of meaning and flavors of the sentences. I have therefore preserved the authors' wording as much as possible.

It is a pleasure to acknowledge the real cooperation of all the authors who have willingly devoted their time in the pursuit of this international undertaking. In addition, I wish to thank many young colleagues in our laboratories at Osaka City University who have rendered assistance in various ways, among them Drs. T. Numata, A. Nakanishi, K. Fujimori, T. Aida and F. Yamamoto. Of course, I alone am responsible for all errors.

Shigeru Oae

# Contents

*Chapter 1*

# SULFUR BONDING

Henry A. Bent

*Department of Chemistry, North Carolina State University*
*Raleigh, North Carolina 27607, USA*

## 1.1 INTRODUCTION

### 1.1.1 The *d*-Orbital Problem

Does sulfur use *d* orbitals? That has been the central issue in theoretical studies of bonds to sulfur [1-4] — and other second-row elements [5,6] — from the time Pauling employed, in qualitative descriptions of octahedrally co-ordinated atoms, $sp^3d^2$ hybrid orbitals [7,8].

1

The question as to whether or not sulfur atoms use $d$ orbitals arises chiefly with regard to bonds to sulfur —

1) From such atoms as oxygen, nitrogen, and carbon (in carbanions) that have, at least nominally, unshared valence-shell electrons; and

2) In such molecules as $SF_6$, where the classical valence of sulfur — less the overall charge (if any) on the sulfur-containing species (+1 for sulfonium compounds, Chapter 9; −1 for $RS^-$; −2 for $S^{2-}$) — is greater than 2. (No compounds with a sulfur-valence-less-ionic-charge number *less* than 2 are discussed in this volume.)

Some investigators have asserted that sulfur's $3d$ orbitals are *too diffuse* to be used effectively in chemical bonds [6]. In atomic sulfur, *e.g.*, the excited state configuration $3s^2 3p^3 3d$ lies, in energy, over eighty percent of the way up from ground state S to ground state $S^+$ plus an electron in the most diffuse of all orbitals, the *continuum* [9].

Other investigators have stressed that in the intense field of the positively charged kernel of a highly electronegative ligand, such as F (in $SF_6$), the *effective* orbital exponent $\alpha$ of a Slater-type sulfur $3d$ orbital, $Nr^2 \exp(-\alpha r)$, is significantly *increased* over its value in atomic sulfur, leading to a marked *contraction* of the sulfur $3d$ orbital, whose maximum probability for the radial function occurs at $r = 3/\alpha$ [1,10,11].

Still other investigators have stressed, however, that, in molecular orbital theory, to rationalize the existence of such compounds as $SF_6$ (rare gas halides, polyhalides and, generally, "hypervalent" compounds), it is not necessary to use any $d$ orbitals at all [12-14]. An explicit, if qualitative, set of low-lying molecular orbitals adequate in number to accommodate all valence-shell electrons can be set down solely in terms of ligand orbitals and sulfur $p$ orbitals. In these applications of molecular orbital theory, as in the earliest orbital theories of valence, an attempt is made to account for the broad features of molecular geometry — bond angles, particularly — purely with $p$ orbitals.

### 1.1.2  Importance of the Exclusion Principle

If sulfur can form a hexafluoride without the use of $d$ orbitals, one might wonder, why cannot oxygen? Why, essentially, is sulfur so different, physically and chemically, from oxygen?

Insight into this sulfur/oxygen, second-row/third-row, $p$-only/contracted-$d$ orbital question can be achieved by examining a model of electronic structure that exhibits *directly*, in visualizable, physical terms, the outcome of the operation in chemical systems of the fundamental principles of electron physics.

To understand the distinctive chemistries of the elements, a model is needed that will enable one to sense intuitively what the outcome will be, in complex systems, of the joint operation of three physical factors: coulombic forces, the wave-like character of electrons, and the Exclusion Principle.

Coulombic forces have long been recognized as important in chemistry. Berzelius's early models of polar attractions have been steadily refined, most notably, for homopolar compounds, by Gilbert N. Lewis [15,16], in the shared-electron-pair bond, and, for heteropolar compounds, by Born [17], Bragg [18], Goldschmidt [19], and Pauling [8], in the ion-packing model of crystals.

During the last half-century, most work in theoretical chemistry has been focused on the problem of examining the implications for chemistry of the wave-like character of electrons. The perceived problem has been to find analytical expressions, or numerical values, for a wave function $\Psi$ that is an eigenfunction of Schrödinger's electrostatic Hamiltonian. Relatively little work has been expended on examining directly the role of the Exclusion Principle in chemistry.

Yet, as Lennard-Jones emphasized in 1954, *the effect of the Exclusion Principle is most powerful, much more powerful than electrostatic forces.* "It does more to determine the shapes and properties of molecules," wrote Lennard-Jones, "than any other single factor. It is the Exclusion Principle which plays the dominant role in chemistry" [20].

How important the Exclusion Principle is can be seen in the simplest case: the ionization potentials of the first three elements, hydrogen, helium, and lithium. The ionization potential (in ev) of hydrogen is 13.6; of $He^+$ $2^2$ x 13.6 = 54.4; of He, only 24.6, owing to electrostatic replusion — or "screening of the nucleus" — of one electron for — or by — the other. From those numbers, one might suppose, correctly, that, without any significant modification in procedure, a direct application of Schrödinger's equation to lithium would yield for lithium's first ionization potential 30-50 ev. In fact, to a zeroth approximation, the ionization potential of lithium is almost zero; it is only 5.39 ev. The effective positive field seen by the most easily removed electron of lithium is remarkably small. It is smaller than the field seen by either electron in a helium atom. It is smaller, even, than the positive field seen by the single electron in an hydrogen atom, a most remarkable conclusion. Two electrons in lithium somehow screen the third electron from the field of the nucleus more effectively than they would were they localized entirely within the atom's nucleus. The effect of the Exclusion Principle is indeed most powerful, much more powerful than electrostatic forces.

### 1.1.3 Pauli Mechanics

At the present time, it would appear that any fundamental model of electronic structure should emphasize explicitly the prominent, even dominant role played by the Exclusion Principle in any close confederation of particles that contains more than two electrons.

Indeed, for centuries chemists have been articulating, slowly but surely, and most explicitly, in the classical structural model of homopolar compounds, in the ionic model of heteropolar compounds, in the doctrine of co-ordination, and in such concepts as directed valence, steric effects, electron sharing, inner-shell electrons, unshared valence-shell electrons, and hybrid orbitals, a practical, working mechanics of the Exclusion Principle as it applies to complex, chemical systems [21,22].

Unlike classical mechanics, unlike even pure quantum mechanics without the Exclusion Principle, this chemical mechanics − this *Pauli Mechanics* − has the peculiar, but to chemists familiar feature that it produces stable, *complexly articulated structures*. The salient features of these Pauli mechanical structures can be usefully, if not yet perfectly described with simple packing models that incorporate and display perspicuously nearly all the accumulated insights into chemical bonding of classical structural theory [22].

### 1.1.4 Representation-Independent Terms

One of the principal functions of a model is to supply a terminology. Physical models are said to be useful if they account for a large number of facts in simple ways and if (consequently) they suggest terms and a nomenclature eventually deemed essential for a proper description of the phenomena [23]. Such functions are achieved, emphasized Schrödinger, through the *deliberate suppression of quantitative details* [24].

Suppression of details, continues Schrödinger, may yield results more interesting than a full treatment. Most importantly, it may suggest new concepts. Pure quantum mechanics alone, in all its detail, cannot supply a definition of, *e.g.*, an acid or a base, or a double bond [25].

Unlike conventional valence-bond and molecular orbital models, Pauli mechanical packing models supply concepts and terms that, through a deliberate suppression of quantitative details, are, appropriately, largely representation-independent [25]. Hence, they permit one to see, as will be shown, that the bonding in, *e.g.*, $SF_6$ can be described without the use of sulfur $3d$ or sulfur $3p$ orbitals. They permit one to see, also, that the $p$-only and contracted-$d$ orbital models are merely different ways of describing, qualitatively, the same thing: the potential energy space about the kernel of a medium-sized atom. Most importantly, perhaps, Pauli packing models permit one to see, at a glance, how operation of the Exclusion Principle conjointly with coulombic forces and the wave-like character of electrons makes the chemistry of sulfur, so amply illustrated in this volume, so different from that of its Group VI congener oxygen.

## 1.2 BONDING IN SF$_6$

### 1.2.1 Conventional Molecular Orbital Model

Symmetry-determined combinations of fluorine $2p$ orbitals for construction of molecular orbitals for SF$_6$, assumed octahedral, are given in Table 1.1, following Mitchell [1]. Orbital axes are shown in Fig. 1.1.

The atomic core of sulfur has been taken to be S$^{6+}$ ($1s^2 2s^2 2p^6$), that of a fluorine atom F$^{5+}$ ($1s^2 2s^2$), leaving 36 electrons, 6 from sulfur, 5 from each of 6 fluorine ligands, to be placed in valence-shell molecular orbitals.

Following Rundle [12], one ignores, initially, participation by sulfur $3d$ orbitals — and additional sulfur orbitals of still higher energies with representations $t_{1g}$ and $t_{2u}$. The $e_g$ and $t_{2g}$ molecular orbitals, like the $t_{1g}$ and $t_{2u}$ orbitals, are then completely determined by symmetry; together, the 11 orbitals can accommodate 22 electrons. Three relatively weak "back-bonding type" molecular orbitals of $t_{1u}$ symmetry formed between sulfur $3p$ orbitals and $\pi$-type combinations of fluorine $2p$ orbitals can accommodate 6 more electrons, leaving 8 electrons for the four strongly bonding $a_{1g}$ and $t_{1u}$ molecular orbitals [1].

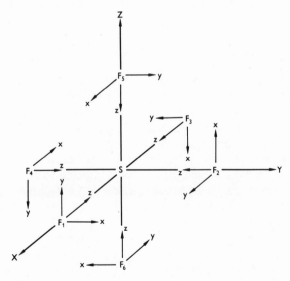

**Fig. 1.1.** Axes used in the description, Table 1.1, of the symmetry orbitals for octahedral SF$_6$.

From the molecular orbital description of $SF_6$ given above (and from similar m.o. descriptions of other hypervalent compounds), it is often said to follow that "the stability of $SF_6$ [and hypervalent compounds, generally] can be *interpreted* without recourse to $3d$-orbital participation"[1] [emphasis added].

TABLE 1.1

Symmetry Orbitals for Octahedral $SF_6$

| Repre-sentation | Sulfur orbital | Fluorine orbitals | |
| --- | --- | --- | --- |
| | | $\sigma$ | $\pi$ |
| $a_{1g}$ | $3s$ | $z_1 + z_2 + z_3 +$ $z_4 + z_5 + z_6$ | |
| $t_{1u}$ | $3p_x$ | $z_1 - z_3$ | $y_2 + x_5 - x_4 - y_6$ |
| | $3p_y$ | $z_2 - z_4$ | $x_1 + y_5 - y_3 - x_6$ |
| | $3p_z$ | $z_5 - z_6$ | $y_1 + x_2 - x_3 - y_4$ |
| $e_g$ | $3d_{x^2-y^2}$ | $z_1 - z_2 + z_3 - z_4$ | |
| | $3d_{z^2}$ | $2z_5 + 2z_6 - z_1 -$ $z_2 - z_3 - z_4$ | |
| $t_{2g}$ | $3d_{xy}$ | | $x_1 + y_2 + y_3 + x_4$ |
| | $3d_{xz}$ | | $y_1 + x_5 + x_3 + y_6$ |
| | $3d_{yz}$ | | $x_2 + y_5 + y_4 + x_6$ |
| $t_{1g}$ | | | $x_1 - y_2 + y_3 - x_4$ |
| | | | $y_1 - x_5 + x_3 - y_6$ |
| | | | $x_2 - y_5 + y_4 - x_6$ |
| $t_{2u}$ | | | $y_2 - x_5 - x_4 + y_6$ |
| | | | $x_1 - y_5 - y_3 + x_6$ |
| | | | $y_1 - x_2 - x_3 + y_4$ |

It is well known, however, that simple m.o. theory suggests stability for molecular species that have not so far been synthesized. Change, for example, the principal quantum number for the sulfur orbitals in Table 1.1 from 3 to 2, and the m.o. description given above for $SF_6$ applies, *mutatis mutandis*, to octahedral oxygen hexafluoride.

It is possible to give a "satisfactory", qualitative molecular orbital description for $OF_6$ — and $NF_5$. From reasoning based on the $p$-only molecular orbital model, Rundle was, indeed, led to suppose that it should be possible to synthesize a compound in which a nitrogen atom has a ligancy of five[12].

Rundle's $p$-orbital model can, in fact, be taken a step further. The $p$-orbitals themselves can be omitted. The valence-shell electrons of $SF_6$ can be described, qualitatively, solely in terms of ligand orbitals. One can omit completely explicit reference to all sulfur extra-kernel orbitals, $3d$ and $3p$.

For the essential physical feature of qualitative molecular orbital descriptions of $SF_6$ lies in the assignment of a relatively low potential energy for electrons in the region of space immediately surrounding the sulfur kernel (the "bonding region"), into which project, separately and collectively, the ligand's sigma orbitals. It is recognition of the field of the kernel $S^{6+}$ — together (see later) with unstated assumptions regarding stable occupancy of that region by electrons — that leads one to assert that the $a_{1g}$ and three $t_{1u}$ molecular orbitals (Table 1.1) are "strongly bonding". It is the existence of the field of $S^{6+}$, not some representation-dependent expression of that field (local or non-local) in terms of spherical harmonics ($s$, $p$, $d$, ...), spherical gaussians, or whatever, that lies at the basis of all interpretations of the stability of $SF_6$. Since, however, the field about an $O^{6+}$ kernel is even greater than the field about an $S^{6+}$ kernel, the conventional molecular orbital theory summarized in Table 1.1 leaves unanswered the previous question: If $SF_6$ is stable, why is not $OF_6$ stable?

### 1.2.2 Orbitals and Wave Functions

#### 1.2.2.1 Antisymmetrization and Spin-Exclusion

The key to understanding, from a quantum mechanical point of view, the extraordinary stability of $SF_6$ and the non-existence of $OF_6$, lies in examining the relation between orbitals and wave functions.

In electronic interpretations of chemistry, the properties of a wave function are often assumed to be, in essence, the sum of the properties of the wave function's component orbitals. Orbital properties are assumed to carry over to the wave functions. Moreover, it is often assumed, implicity, but incorrectly, that *all* chemically significant properties of a wave function are contained in, and adequately revealed by a study of, the properties of a set of delocalized molecular orbitals $\phi_1, \phi_2, \phi_3, \ldots$.

A wave function constructed from delocalized molecular orbitals is not, however, merely a simple Hartree product $\phi_1(1)\phi_2(2)\phi_3(3)\ldots$.

A wave function for a many-electron system has properties — for chemistry, enormously important properties — not possessed by delocalized molecular orbitals or a simple product formed therefrom.

From reasoning based on simple Hartree products, for example, one might suppose that, for a wave function constructed from a set of conventional, symmetry-adapted, mutually overlapping molecular orbitals, such as those exhibited in Table 1.1, the *configuration of maximum probability* occurs when each electron is assigned to the place in space when its orbital has a maximum absolute value. To illustrate with a simpler case [26], one might suppose, incorrectly, that the most probable location of two spin-paralleled electrons in an olefinic carbon-carbon double bond is for one electron, say 1, to be on the internuclear axis at the center of a σ orbital and for the other electron, 2, to be off the axis in one lobe of a π orbital, thereby maximizing the product

$$\sigma(1)\pi(2) \tag{1.1}$$

In fact, it is a fundamental postulate of electron physics that *electrons are indistinguishable*. A change in electron-labels should not change the absolute value of the wave function. Generally, however, scrambling the labels in a single Hartree product changes the absolute value of the product. In the present example, $\sigma(1)\pi(2) \neq \sigma(2)\pi(1) = 0$.

On physical grounds, *a Hartree product is a profoundly unsatisfactory wave function*. It does not satisfy a fundamental guiding principle that is applicable to all wave functions and that governs all reasoning about atomic and molecular systems [27]. This guiding principle was stated in 1926 by Heisenberg. It is a generalization of Pauli's initial statement of 1925 of the Exclusion Principle [28]. Curiously, Pauli's preliminary statement — that no more than two electrons may be placed in the same spatial orbital — is still cited more frequently in chemistry textbooks than is Heisenberg's more fundamental statement, that electronic wave functions must be antisymmetric.

Pauli's restricted statement of the Exclusion Principle cannot be applied to highly precise, many-electron wave functions — ones that, to allow for electron-electron repulsions, contain explicitly inter-electronic co-ordinates $r_{ij}$. Pauli's statement is restricted to wave functions that are expressed in terms of physically fictitious, if mathematically convenient electron orbitals. Heisenberg's statement, on the other hand, is applicable to *all* wave functions.

Heisenberg's statement rigorously excludes, as a simple Hartree-product function and Pauli's statement do not, configurations in which two electrons of parallel spin are at the same point in space at the same time [29]. *Electrons of like spin tend to avoid each other* [20]. The geometrical implications of this *spin-exclusion*, so manifest in a comparison of the ionization potentials of hydrogen, helium, and lithium, are not fully or adequately expressed by Pauli's statement of the Exclusion Principle. For, although Pauli's familiar statement requires that electrons of like spin be placed in different spatial orbitals, most single- and multi-centered orbitals are not spatially mutually exclusive. Most delocalized, valence-shell molecular orbitals, though perhaps mutually orthogonal, penetrate into similar, if not identical, regions of space.

In the $\sigma$-$\pi$ description of a double bond, for example, Heisenberg's statement requires that the wave function for two spin-parallel electrons be written as this antisymmetric, linear combination of Hartree products:

$$\sigma(1)\pi(2) - \sigma(2)\pi(1) \qquad (1.2)$$

The mathematical properties of the antisymmetric function (1.2) are profoundly different from those of the Hartree-product function (1.1). Unlike (1.1), expression (1.2) always changes its sign, but never its absolute value, when the labels on the electrons are interchanged. Consequently, whenever the co-ordinates of the two spin-parallel electrons are the same, expression (1.2), unlike (1.1), vanishes.

This spatial, spin-exclusion, which does more to determine the shapes and properties of molecules than any other single factor, is a property of the entire, properly antisymmetrized wave function. It is not a property of individual spectroscopic orbitals. Its implications for chemistry cannot be gleaned from an examination, however careful, of the properties of delocalized molecular orbitals.

Because the molecular orbital model is, in essence, an *independent-particle* model, *the most significant property of a wave function for chemistry is not contained in, or revealed by, a study of the properties of delocalized molecular orbitals.* Spin-exclusion, the physical basis − or explanation − for the existence of complexly articulated chemical structures is, in essence, a property of a *collective-particle* model.

Fortunately, the chief conceptual advantages of an orbital, independent-particle model can be largely retained − and, with regard to the transferability of structural parameters, greatly extended − in a simple, collective-particle model that captures for chemistry the essence of Heisenberg's statement of the Exclusion Principle. Owing to a mathematical property of antisymmetrized orbital products, it is possible to express in terms of mutually dependent, yet transferable orbitals − in vivid, visual, geometrical terms − the dominant role in structural chemistry of spin-exclusion[22].

### 1.2.2.2 Determinantal Wave Functions and Chemical Orbitals

In 1929 Slater pointed out that an antisymmetrized Hartree product, such as expression (1.2), can be written as a determinant[30],

$$\begin{vmatrix} \sigma(1) & \pi(1) \\ \sigma(2) & \pi(2) \end{vmatrix} \qquad (1.3)$$

Determinants have the property that columns can be added to each other without changing the value of the determinant. The expression

$$-\frac{1}{2} \begin{vmatrix} \sigma(1) + \pi(1) & \sigma(1) - \pi(1) \\ \sigma(2) + \pi(2) & \sigma(2) - \pi(2) \end{vmatrix} \tag{1.4}$$

yields identically, on expansion, expression (1.2), as does (1.3). Mathematically, the two expressions, (1.3) and (1.4), are equivalent.

It is convenient to introduce abbreviations for the functions $(\sigma + \pi)$ and $(\sigma - \pi)$ that appear in expression (1.4). Let

$$\eta_1 \equiv \frac{1}{\sqrt{2}} (\sigma - \pi)$$

$$\tag{1.5}$$

$$\eta_2 \equiv \frac{1}{\sqrt{2}} (\sigma + \pi)$$

The "hybrid" orbitals $\eta_1$ and $\eta_2$ are called (in this instance) *equivalent orbitals* [31,32]. Like, *e.g.*, $sp^3$ hybrid orbitals, they are identical to each other except for their orientation in space: $\eta_1$ is concentrated below the molecular plane, $\eta_2$ above it. They correspond closely to the bent- or banana-bonds of classical structural theory. For that reason, they have been called *chemical orbitals.* Being relatively localized, spatially, they have been called, also, *localized molecular orbitals* [33]. Substitution from (1.5) into (1.4) yields

$$\begin{vmatrix} \eta_1(1) & \eta_2(1) \\ \eta_1(2) & \eta_2(2) \end{vmatrix} \tag{1.6}$$

Mathematically, expressions (1.6) and (1.3) are equivalent, for all values of the electron co-ordinates. Physically, however, the two descriptions of a double bond — the $\sigma$-$\pi$ description [expression (1.3)] and the bent- or banana-bond description [expression (1.6)] — do not, at first glance, look the same. The apparent difference between descriptions of bonding in terms of a chemist's localized orbitals and a physicist's delocalized orbitals has, in fact, been likened to the difference between day and night — or night and day.

Localized and delocalized orbital descriptions of bonding do lead to different mental pictures of electronic structure — when the working model one has of the link between orbitals and wave functions corresponds, perhaps unwittingly, to a simple Hartree product. Properly antisymmetrized, however, products of localized orbitals and products of linearly related delocalized orbitals lead to identical conclusions. Particularly, and importantly, they yield identical configurations of maximum probability.

### 1.2.2.3 Localized Molecular Orbitals and Configurations of Maximum Probability

The set of values of nuclear and electronic co-ordinates that maximizes a molecular wave function has been called the molecule's *configuration of maximum probability*, CMP [34]. Although configurations other than the CMP contribute to molecular expectation values, the configuration of maximum probability is exceedingly useful in predicting, or rationalizing, from the standpoint of quantum physics, a molecule's structure and chemical properties.

The question arises: How, from a set of molecular orbitals, is a system's most probable, mutual disposition of electrons determined? Here localized, mutually exclusive molecular orbitals [35] are more useful than delocalized, mutually overlapping molecular orbitals. For two spin-parallel electrons of a double bond, for example, the CMP is not one electron on the internuclear axis, one off it (Sect. 1.2.2.1). That configuration maximizes the absolute value of one of the products in expression (1.2), but minimizes the absolute value of the other product. Expression (1.2) has a larger value if the two electrons are placed symmetrically about the internuclear axis.

The CMP for spin-parallel electrons of a double bond (and other electronic systems) is not immediately apparent from a casual inspection of the form of the individual, delocalized, mutually overlapping orbitals ($\sigma$ and $\pi$) used in the conventional molecular orbital description of the system. Consider, however, the mathematically equivalent, localized orbital, bent- or banana-bond representation of a double bond. Expansion of expression (1.6) yields

$$\eta_1 (1)\eta_2 (2) - \eta_1 (2)\eta_2 (1) \qquad (1.7)$$

Always, in an exclusive orbital representation of a system, for any distribution of electrons, at most only one of the products of a Slater determinantal wave function is finite, and then only if, in accordance with Pauli's statement, there is one but no more than one spin-parallel electron in each orbital.

In an exclusive orbital representation of an N-electron system, the entire N!-term determinantal wave function reduces, in essence, to a single Hartree product. With exclusive orbitals, orbital properties do carry over to the wave function; a single Hartree product is a satisfactory mathematical model of the wave function; Pauli's statement of the Exclusion Principle does lead to a useful, geometrical representation of spin-exclusion.

Many, perhaps most of the properties of a wave function important in structural chemistry are contained in, and revealed by a study of, the properties of a set of *localized* molecular orbitals.

The most important *occupied*, localized orbitals in molecular systems not containing transition metals generally correspond closely to Gilbert N. Lewis's unshared, inner-shell, "kernel" or "atomic core" electrons; to his valence-shell, unshared, "lone-pair" electrons; and to his valence-shell, shared, bonding electron pairs [33,36-38].

Two sets of occupied, localized orbitals, one for each set of electron spins ($\alpha$ and $\beta$), are sometimes useful in describing the structures and properties of systems in which spatial pairing of electrons of opposite spins would increase repulsions between electrons of opposite spin without any attendant, counter-balancing lowering of energy through, chiefly, enhanced nuclear-electron attractions [22,39]. "Anticoincident spin-sets" [40], "non-spatially paired orbitals" (NSPO), and "different orbitals for electrons of different spin" (DODS) [41] have been used in discussions of the electronic structures of helium [42]; the alkali metals [43]; oxygen and ozone [40]; the oxides of nitrogen [40]; benzene (generally, any system in which, classically, "resonance" occurs between two equivalent structures [40]); anions in lattices with large, weak-field cations [22,44]; and the kernel of the sulfur atom in $H_2S$ and $CH_3SH$ [45].

Chemically, the most important (nominally) *vacant* localized orbitals in molecular systems — the "pockets" in an atomic core's electron-pair co-ordination polyhedron [46] — correspond closely to the lobes of $\sigma$ antibonding orbitals off the more electropositive atoms of the corresponding bonding orbitals, in tetrahedral co-ordination (*e.g.*, the backside of a carbon atom of a carbon-halogen bond [44]), and to the individual lobes of classical $d$ and $f$ orbitals about kernels of large-core atoms that are co-ordinatively unsaturated [22].

### 1.2.3 Localized Orbitals for $SF_6$

A simple and clear picture of the bonding in $SF_6$ can readily be obtained from an inspection of the system's CMP for the valence-shell electrons. To obtain the system's CMP, localized orbitals may be created by taking linear combinations of the delocalized, spectroscopic orbitals listed in Table 1.1.

Several criteria have been given for obtaining "best" localized orbitals [33,47]. For the present purposes, it is sufficient to note that, for example, the sum of the listed fluorine $\pi$ orbitals containing $x_1$ is simply four times $x_1$, a localized, atomic $2p$ orbital on fluorine atom 1. Mathematically, in a determinantal wave function, the 12 fluorine delocalized $\pi$ orbitals are essentially equivalent to 12 localized, largely non-bonding atomic $p$ orbitals on fluorine, into which may be placed 24 (largely unshared) electrons (12 Lewis "lone pairs").

To obtain localized orbitals for the remaining 12 valence-shell electrons, let $\phi_{1-6}$ represent the six listed delocalized fluorine $\sigma$ orbitals

$$
\begin{aligned}
\phi_1 &\equiv z_1 + z_2 + z_3 + z_4 + z_5 + z_6 \\
\phi_2 &\equiv z_1 - z_3 \\
\phi_3 &\equiv z_2 - z_4 \\
\phi_4 &\equiv z_5 - z_6 \\
\phi_5 &\equiv z_1 - z_2 + z_3 - z_4 \\
\phi_6 &\equiv 2z_5 + 2z_6 - z_1 - z_2 - z_3 - z_4
\end{aligned}
\tag{1.8}
$$

Solving the expressions in (1.8) for $z_{1-6}$, one finds that

$$
z_1 = \frac{1}{12}(2\phi_1 + 6\phi_2 + 3\phi_5 - \phi_6)
$$

$$
z_2 = \frac{1}{12}(2\phi_1 + 6\phi_3 - 3\phi_5 - \phi_6)
$$

$$
z_3 = \frac{1}{12}(2\phi_1 - 6\phi_2 + 3\phi_5 - \phi_6)
$$

$$
z_4 = \frac{1}{12}(2\phi_1 - 6\phi_2 - 3\phi_5 - \phi_6)
\tag{1.9}
$$

$$
z_5 = \frac{1}{12}(2\phi_1 + 6\phi_4 + 2\phi_6)
$$

$$
z_6 = \frac{1}{12}(2\phi_1 - 6\phi_4 + 2\phi_6)
$$

From expressions (1.8) and (1.9), and from earlier remarks concerning determinantal wave functions, one sees that, mathematically, in an antisymmetrized wave function, the 6 fluorine delocalized $\sigma$ orbitals listed in Table 1.1 are essentially equivalent to 6 atomic $p$ orbitals, each localized on a different fluorine atom and pointing toward the sulfur kernel. Into those 6 localized orbitals may be placed 12 (shared) electrons (6 Lewis bonding pairs). In sulfur hexafluoride's CMP, the sulfur kernel is surrounded octahedrally by 6 electron pairs.

Often the question is asked, "Does the sulfur atom in $SF_6$ violate the

Octet Rule [12]?" If by that question one means, "To give an approximate description of the orbitals of an approximate wave function for the molecule's valence-shell electrons, is it necessary to use more than four orbitals centered at sulfur?", the answer is No. To describe, approximately, a set of orbitals for the valence-shell electrons of $SF_6$ it is not necessary to use *any* sulfur orbitals.

Nor is it necessary in a description of $SF_6$ to use any fluorine orbitals. A Lewis-basis-set of orbitals for the valence-shell electrons of $SF_6$ could be described semi-quantitatively with floating spherical gaussian functions [48,49]. On the other hand, to describe accurately by configuration interaction the wave function of $SF_6$ in terms of atom-centered spherical harmonics would require, in an efficient calculation, more than merely four, or six, functions from sulfur.

If, by the question, "Does sulfur in $SF_6$ violate the Octet Rule?", one means, however, "Are there in the near vicinity of the sulfur kernel more than 8 electrons?", the answer is probably Yes [16,44]. There are probably 12 electrons in the sulfur atom's valence shell, however one chooses to describe that space.

In co-ordinatively saturated $SF_6$, the sulfur atom's valence shell can be described, usefully if not perfectly, solely in terms of six, electronegative-ligand orbitals. Six such orbitals just fill it. About a sulfur kernel there appears to be room, though just barely (see below), for 6 localized, fluorine-like orbitals [22], each of which, owing to the strong field of $S^{6+}$, can be doubly occupied.

Evidently the Octet Rule is used currently in two senses [44]: in an older sense, by Lewis [15,16] and Langmuir [50], with reference to the number of electrons in an atom's valence shell; and in a newer sense, in molecular orbital theory, with reference to the number of atomic orbitals deemed necessary to account for molecular shapes. Many times it has been pointed out, however, that, with the use of ion and electron-pair packing models, it is possible to account for the shapes of such molecules as, *e.g.*, $ML_n$, M a central atom, L a ligand, without using any directional atomic orbitals from the central atom [21,22,29,45,51-53].

### 1.2.4 Ionic Model of $SF_6$

Although the structures of both homopolar and heteropolar compounds can be understood in terms of ion packing models [22], the "ions" in the former instance being individual valence-shell electron pairs ("electride ions" [54,55], or occupied, localized molecular orbitals) and atomic cores (small cations), it is usually asserted, for reasons based on the data in Table 1.2, that the bonding in homopolar compounds is largely covalent.

**TABLE 1.2**

"Lattice Energy" of $SF_6$

| Reaction | $\Delta H$ (kcal) |
|---|---|
| $SF_6(g) = S(c) + 3F_2(g)$ | 262 |
| $S(c) \quad = S(g)$ | 53 |
| $3F_2(g) = 6F(g)$ | 110 |
| $S(g) \quad = S^{6+}(g) + 6e$ | 6380 |
| $6F(g) + 6e = 6F^-(g)$ | $-477$ |
| $SF_6(g) = S^{6+}(g) + 6F^-(g)$ | 6322 |

In Table 1.2 the last reaction, the gas-phase ionization of $SF_6$, is the sum of the five preceding reactions. Its enthalpy change, the "lattice energy" of ionic (?) $SF_6$, is the sum of the five preceding, experimental $\Delta H$ values. A conventional ionic model for $SF_6$ does not yield an enthalpy of ionization as large as 6322 kcal per mole of $SF_6$ ionized. For a micro-crystal composed of a spherically symmetric, hard-sphere +6 cation surrounded octahedrally by 6 similar, univalent anions, center-to-center distance 1.57 Å (the S-F distance in $SF_6$), the calculated enthalpy of ionization, by Coulomb's law is, per mole, only 5500 kcal. Inclusion of a Born-repulsive term might lower that value 15~25 percent [8,17]. It would appear that the bonding in $SF_6$ is significantly stronger than that expected for purely ionic bonds.

However, the assumption that, at the small separation of 1.57 Å, a fluoride ion may be treated as a simple, spherical charge distribution is unjustified. It has long been known that the spherical-ion assumption does not adequately account for one of the most important facts of chemistry: the degree of chemical combination necessary to saturate the affinity of an element [22,56]. To account for the maximum co-ordination numbers of the elements, it is necessary to take a more intimate view of the distribution of charge in a tightly co-ordinated fluoride ion [22].

A stereochemically useful model of a fluoride ion is a +7 kernel, Pauling radius 0.07 Å [8], surrounded tetrahedrally by four doubly occupied, localized, approximately spherical electron orbitals, radius 0.61 Å [22,54]. For such a model, all charge domains treated as "hard", Coulomb's law yields for the lattice energy of an ionic model of $SF_6$ 9130 kcal/mole [57].

Allowance for the fact that in the ionic model of $SF_6$ the "valence state" of a fluoride ion (one valence-shell, bonding pair coincident; three valence-shell, unshared pairs trigonally anticoincident [44]) is higher in energy than the state of a free, uncoordinated, gaseous fluoride ion (all four valence-shell, unshared pairs tetrahedrally anticoincident [39,40]) would bring the calculated lattice energy down to approximately 8370 kcal/mole, for uniformly charged electron-domains. Inclusion of a Born repulsive term would lower still further the calculated lattice energy, probably another 15~25 percent, the more so the softer the repulsive interaction. (Bartell has suggested that the repulsive potential between individual electron pairs is relatively soft [58].)

With refinements, it would appear that a relatively good Born-Haber type calculation could be made for $SF_6$ based upon a description of the compound's electronic structure in terms of an ionic model. Of course, a structure can be partitioned in many ways. By the Principle of Local Electrical Neutrality [8], partitions can usually, perhaps always, be drawn so as to include each atomic kernel in a neutral, atom-like domain. What the calculations described above imply is that the distribution of charge suggested by the $S^{6+}(F^-_{te})_6$ model of $SF_6$ is reasonable, as judged by a single criterion: the compound's enthalpy of ionization.

### 1.2.5 Summary

"A theory is simple," Feynmann has said, "if you can describe it fully in several different ways without immediately knowing that you are describing the same thing" [59].

From the viewpoint of localized molecular orbitals, chemical bonding may be described in several, mutually complementary ways [54].

All bonding may be viewed as ionic: as the packing of large, negative "anions" (structurally transferable, localized, occupied electron-domains) about relatively small cations (e.g., $S^{6+}$, $F^{7+}$), with some anions, in order to complete the cations' valence-shell co-ordination polyhedra, having co-ordination numbers greater than one (two for conventional, two-center bonds; greater than two for multi-center bonds). Or, all bonding may be viewed as covalent; as the mutual sharing by electron-deficient atoms of localized electron-domains; i.e., as the simultaneous occupancy by some electron-domains of portions of the valence-shells of two (or more) atoms.

Localized electron-domains can, in turn, be described in several ways. They can be described in terms of atom-centered functions (atomic orbitals) belonging to one or the other or to both atoms of a two-center bond. The number of atomic orbitals used will depend on the accuracy with which the domains are to be described. Some expansions will converge to a desired accuracy much more rapidly than others. The distribution of charge in $SF_6$, for example, was described in Sect. 1.2.3 and 1.2.4, with the exception of the charge comprising the sulfur atom's kernel, solely in terms of fluorine-atom orbitals; however, the same charge distribution could be described, in a less rapidly converging expansion, even to the "fluorine" lone-pairs and inner-shell electrons, solely in terms of a set of sulfur orbitals. Generally, for large-core atoms, such as sulfur, expansions of bonding electron-domains in terms of conventional atomic orbitals are not expected to be rapidly convergent.

Localized domains can be described, also, often rather effectively, in terms of "bond orbitals" (e.g., "off-center" gaussian functions [48,49]), centered in bonding regions rather than on nuclei. They can be described [as in expressions (1.9)] in terms of symmetry-adapted, spectroscopic orbitals centered on the molecule as a whole. They can be described in terms of a set of continuum orbitals not centered on any center. Or, they can be described in terms of some combination of one, two, or many atom- and/or bond-centered functions, with, perhaps, some admixture of continuum orbitals.

Conversely, spectroscopic orbitals can be described in terms of linear combinations of localized orbitals [60] that, in turn, are built up from atom- and/or bond-centered functions.

The question as to which description of bonding is best may have no one, best answer. Best for what purposes? For computational purposes? For understanding visible and uv spectroscopy? Electron scattering? Molecular shapes? Chemical reactivity?

Pauling has said that, for understanding double bonds, "bent bonds are best". Recently, localized bonds have been used, also, in the interpretation of the Compton profiles of hydrocarbons [61]. On the other hand, for understanding interactions of molecules with electromagnetic radiation, in the visible and uv, delocalized, spectroscopic orbitals have been useful.

In pure science, what matters, it has been said, is essentially the *picture* that is eventually or at any given step obtained. Foretelling, predicting, observations are only means for ascertaining whether or not the picture that we have formed is useful [62]. From that point of view, the chief question for this chapter might be: Which description in its *simplest*, least elaborated form best captures for the chemist the essential features of electronic structure? Emphasis is on simplicity. For any description drawing upon a mathematically complete set of functions can in principle eventually yield, though perhaps with only many terms, if not understanding, at least a precise, arithmetical result. To *understand* bonding, a description is needed that, with minimum effort, creates instantly in the mind's eye a useful picture of the role in chemistry of coulombic forces and the space-filling property of electrons.

Space-filling — matter's most characteristic feature — is, according to current theory, a manifestation of, jointly, the wave-like character of electrons and spin-exclusion. It is a property of all electrons — not only of valence-shell electrons, which has been properly emphasized [52,53], but, also, of inner-shell electrons. They, too, occupy space, to the exclusion of other electrons of like spin. Witness the notably low ionization potential of lithium, compared to hydrogen and helium (Sect. 1.1.2).

The enormous difference between the chemistries of sulfur and oxygen arises, similarly, from differences in spin-exclusion between inner-shell and valence-shell electrons. Sulfur's kernel, with eight electrons, has a Pauling crystal radius of 0.29 Å; oxygen's, with only two electrons, has a crystal radius of only 0.09 Å. In those two numbers, the approximate, effective van der Waals-like radii of hard, spherical atomic kernels — hard to valence-shell electrons — can be seen, as will be sketched, nearly the entire difference between the chemistries of sulfur and oxygen.

## 1.3 LARGE-CORE MODEL

### 1.3.1 Historical Roots

Prior to the invention of the electron and an electronic interpretation of chemistry, chemists had been developing, slowly but surely, for over a century, two broad and still useful theories of structure, one for organic compounds, one, initially distinctly different, for inorganic compounds. These theories culminated, finally, in the Couper-Kekule-van't Hoff Doctrine of the Tetrahedrally Directed Tetravalence of Carbon [63] and Alfred Werner's Doctrine of Co-ordination [64].

In efforts to find a connection between the Doctrine of Valence and the Doctrine of Co-ordination, Werner suggested, with great insight, that —

> "The maximum co-ordination number must be considered as having reference to the space about the surface of the atom" [64].

So simple is Werner's suggestion, it is not, at first glance, illuminating. Yet, as can be seen today, and as Werner himself emphasized —

> "The maximum co-ordination number, when considered spacially, provides a means by which a very great number of phenomena may be grouped together, and is the first clear explanation to give any idea of the combining possibilities of the atom" [64].

Werner's ideas gained great clarity with (1) the introduction by G.N. Lewis of two seminal concepts: atomic kernels (or cores) and shared-electron pairs [15,16]; and (2) extensions of Lewis's atom-model to co-ordination compounds by, particularly, Huggins [65] and Sidgwick [66]. For, from the viewpoint of atomic cores and localized, non-interpenetrating valence-shell electron pairs, *all compounds are co-ordination compounds* [21]. All compounds are the result of the co-ordination of negatively charged electron domains by generally smaller, more highly, and always oppositely charged atomic kernels. In this view, *the bond diagrams of classical organic chemistry are co-ordination diagrams* − diagrams that show the co-ordination by small carbon cores of four valence-shell electron pairs [21]. In the language of Werner and Lewis, *combined carbon has an electron-pair co-ordination number of four.*

Applied at the level of atomic cores and localized electron domains, the doctrine of co-ordination summarizes in one simple, useful word, *co-ordination*, the major ideas of modern electron physics as they bear on structural chemistry.

"Co-ordination" of localized electron domains by atomic cores reflects the presence in molecules of coulombic forces. Simultaneously, the word "co-ordination" acknowledges the wave-like character of electrons and the Exclusion Principle, which, jointly, yield − so necessary for a visual, *geometrical* interpretation of chemical affinity − atomic cores and valence-shell domains of *finite size.*

The Werner-Lewis Doctrine of Co-ordination explains how a centrally symmetric potential function about an atomic core produces, in concert with the wave-like character of electrons and spin-exclusion, the two most salient features of chemical affinity: saturation and directional character. The following quotation from Werner [64] summarizes the important structural features of the modern doctrine of co-ordination. Where Werner used, from necessity, the word "atom", we have substituted the phrase "atomic core".

> "For the sake of simplicity we can suppose the atomic core to be a sphere. Such a supposition simplifies our mechanical picture of the structure of the molecule.
>
> "We will also make the following supposition on affinity. Affinity is an attractive force which acts from the centre of the atomic core, and [according to Coulomb's Law] is of equal value at all points on its surface. The amount of affinity saturated by the linking up of two atomic cores [through the mutual sharing of electrons] is distributed on a definite circular segment of the surface of the atomic core (binding zone [or valence shell]), and varies within wide limits with the nature [*i.e.*, the size] of these atomic cores."

To paraphrase Werner's first-quoted remark, *the maximum electron-pair co-ordination number of an atom must be considered as having reference to the size of the atom's kernel.* Kernel radii useful in a geometrical interpretation of chemistry are given in Fig. 1.2.

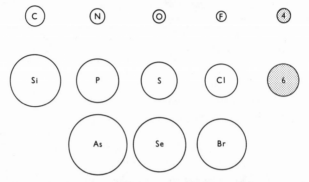

**Fig. 1.2.** Relative, van der Waals-like sizes of the atomic cores of the principal non-metals. Pauling core-radii, in hundredths of Ångstroms, are: Cl 5, N 11, O 9, F 7; Si 41, P 34, S 29, Cl 26; As 47, Se 42, Br 39. In *structural* (as opposed to molecular) formulas, the symbol "C", e.g., represents an *"ion"*, charge +4, radius 0.15 Å. Note the relatively large difference in sizes in going from the first to the second row. At the right are shown the approximate, *minimum* core sizes for tetrahedral and octahedral coordination (electron-pair core coordination numbers "4" and "6", respectively [22]). None of the small-core, octet-rule-obeying, first-row elements, C, N, O, and F, can support, without "rattling", an expanded octet (more than four valence-shell electron pairs). *All* other elements can [including the first-row elements Be and Li, core radii 0.31 and 0.60 Å respectively; cf. Fig. 1.3 (a) − (e)].

### 1.3.2 Quantitative Applications

#### 1.3.2.1 Spectra of Alkali-Atoms

The first − and still most important − quantitative application of Planck's doctrine of energy quantization to the problem of accounting for the *distinctive* properties of the elements was Bohr's discussion of the visible and uv spectra of hydrogen and the alkali-atoms [67]. Many rudimentary ideas regarding electronic structure stem from Bohr's pioneering papers.

The third element in the periodic table, lithium, presented Bohr with a severe problem [68]. From the viewpoint of wave mechanical principles *alone*, lithium's energy is least when its three electrons occupy a common orbital. The atomic volume curve, the chemical properties of lithium, and its ionization potential (Sect. 1.1.2) strongly suggest, however, that lithium holds one electron very loosely [68]. Bohr chose for its electronic structure a double-ring arrangement. His assignment of "ring" (or principal quantum) numbers for lithium was purely phenomenological, a direct translation into the terms of his model of lithium's observed properties[68].

"Today nobody really worries about the spectra of alkali-atoms," wrote Parsons and Weisskopf [69]. "Modern methods of solving many electron problems allow calculations of the spectrum and any other properties to [in principle] any desired accuracy. Still, there is a desire in many minds for a simple and clear insight into what are the essential reasons for the phenomena. In the case of the alkalis *the question is to formulate in a simple way the essential reason for the difference between the spectrum of the alkali-atoms and the hydrogen atom* [69] [emphasis added].

"The usual answer to the hydrogen – alkali-atom question," continue Parsons and Weisskopf, "refers to so-called *penetrating orbits*." One says that the valence electron finds itself in an $n$S orbit – the lowest that Pauli's statement admits, $n$ being taken as 2 for lithium – but that, compared to the corresponding Bohr energy values for hydrogen of $-13.6$ ev/$n^2$ ($-3.4$ ev for $n = 2$), the energy is depressed (to $-5.39$ ev for Li, Sect. 1.1.2), owing to the "fact" that the orbit dives into the internal region of the core where the effective strength of the Coulomb field is much stronger [70]. According to Slater's rules [70], for example, the effective atomic number, $Z_{eff}$, for lithium's third electron is not $3 - 2 = 1$ but $3 - 2(0.85) = 1.30$. The picture of penetrating orbits has weak points, however, for the valence electron is mostly outside the core [69].

Parsons and Weisskopf report a calculation based upon a picture of electronic structure that is almost the opposite of the conventional, penetrating-orbit model. They assume that, owing to the Exclusion Principle, *the valence-electron(s) cannot penetrate at all into the core* [69]. Such a model was first advanced in 1908 by Ramsay, who referred to an atom's outer electrons as a "rind" [71].

Parsons and Weisskopf calculate energy levels for a single electron – the alkali-atom's valence electron – in a non-local potential V(r), r being the distance of the valence electron from the atom's nucleus, that is Coulombic for r greater than a critical radius $r_c$ and infinitely repulsive for r less than $r_c$:

$$V(r) = \frac{-e^2}{r} \quad r > r_c$$

$$= +\infty \quad r < r_c \tag{1.10}$$

Results are reported in terms of calculated values for the parameter $\lambda$ in the Bohr-like expression $E_{Li} = -13.6$ ev/$\lambda^2$ (Table 1.3). The value of $r_c$, 0.43 Å, was determined by fitting the energy of the ground state.

The parameter $\lambda$ corresponds to the reciprocal of the parameter $Z_{eff}/n$ in the Bohr-Slater atom-model. For the ground state of lithium, $\lambda = [-13.6/(-5.39)]^{1/2} = 1.589$ (Table 1.3); from Slater's rules (see above), $(Z_{eff}/n)^{-1} = (1.30/2)^{-1} = 1.54$.

The simple, Ramsay-Parsons-Weisskopf Coulomb cutoff potential yields "surprisingly good agreement with the facts" (Table 1.3). Parsons and Weisskopf stress one "remarkable feature": the values of $r_c$, determined empirically from ionization energies, are proportional to the corresponding, structurally determined ionic radii of the alkalis. That $r_c$ is somewhat smaller than the conventional ionic radius of $Li^+$, and that it is a function of the angular momentum of the valence electron [69], can be understood, qualitatively, in terms of an angular-momentum dependent, partial penetration by a *single* valence electron of the partially anticoincident, split-shell, $1s1s'$, helium-like core of lithium.

**TABLE 1.3**

Calculated Electronic Energy Levels for an Impenetrable Core Model of Lithium [69]

$$E_{Li} = \frac{-13.6 \text{ ev}}{\lambda^2}$$

| Old Terms | New Terms | $\lambda$ (exp) | $\lambda$ (calc) |
|-----------|-----------|-----------------|------------------|
| 2S | 1S | 1.589 | 1.589 |
| 3S | 2S | 2.596 | 2.602 |
| 4S | 3S | 3.598 | 3.605 |
| 5S | 4S | 4.599 | 4.607 |
| 6S | 5S | 5.599 | 5.608 |
| 7S | 6S | 6.579 | 6.608 |

In summary, the calculations of Parsons and Weisskopf support the view that, like the familiar impenetrability of sodium ions to electrons shared, if very unequally, with kernels of adjacent chloride ions in sodium chloride — like, *i.e.*, the familiar, mutual impenetrability of $Na^+$ and $Cl^-$ ions in NaCl — atomic cores of atoms ($Li^+$, $Be^{2+}$, $B^{3+}$, $C^{4+}$, ... $S^{6+}$, $Cl^{7+}$ ...) are, as suggested by the diagram in Fig. 1.2, effectively impenetrable to *all* valence-shell electrons, shared (in covalent, ionic, or metallic bonds) or unshared.

The essential features of spin exclusion embodied in, *e.g.*, the useful *Valence-Shell*-Electron-Pair-Repulsion Model of electronic structure [53], can be extended to a *Valence-Shell-Inner-Shell*-Electron-Repulsion Model. About the kernel of a sulfur atom, *e.g.*, is a region, radius approximately 0.29 Å, that, owing to powerful spin-exclusion effects, cannot easily be penetrated by valence-shell electrons.

## 1.3.2.2 Interpretation of Interatomic Distances

In the "hard" domain approximation to an exclusive orbital representation of spin-exclusion and electronic structure, a system's energy is a minimum when valence-shell domains and adjacent, inner-shell domains are in mutual *contact*. In the hard domain approximation, therefore, bond angles and bond lengths are often related in a relatively simple, direct, geometrical manner to the shapes and sizes of a molecule's atomic cores and valence-shell orbitals.

The length d(A-B), for example, of a linear (unbent), conventional two-center, two-electron, single bond between atoms A and B, atomic radii $r_a$ and $r_b$ (generally highly transferable parameters; Fig. 1.2) is simply the sum of the two core radii plus twice the radius $R_{AB}$ of the intervening, approximately spherical, shared, valence-shell domain. Eq. (1.11) is

$$d(A\text{-}B) = r_a + r_b + 2R_{AB} \qquad (1.11)$$

applicable to *all* linear single bonds, whether ionic covalent, or metallic [55]. $R_{AB}$ is a transferable structural parameter in this sense: it may be expressed, empirically, as a constant, 0.60 Å, approximately equal to the Bohr radius, plus an increment that increases nearly linearly with the size of the *smaller* atomic core within whose immediate field (or "valence shell") the A-B bonding domain resides [55]. In Ångstroms, for $r_a \leq r_b$,

$$R_{AB} \cong 0.60 + 0.4\, r_a \qquad (1.12)$$

Usually the smaller atomic core, A in the discussion above, is the more highly charged of the two atomic cores of a two-center bond. Often, also, it is the core that belongs to the more electronegative partner of the bond. It is the core that largely determines the chemical and physical properties of the shared electron-pair.

In keeping with Lavoisier's remark that chemical nomenclature is an inventory of chemical knowledge [72], the smaller, electron-domain-governing atomic core usually belongs to the element whose name in the nomenclature for binary compounds carries the suffix *ide*. Thus, with the exception of hydrogen, a most noteworthy special instance, its atomic core having no inner-shell electrons, fluorine, of all the elements, has the smallest atomic core (Fig. 1.2). Correspondingly, without exception all binary compounds of fluorine are usually called fluor*ides*. Oxygen has the next smallest atomic core. Correspondingly, all binary compounds of oxygen are usually called ox*ides*, with the exception of $OF_2$.

We say "usually" because, in fact, $SF_6$, for example, like most compounds, can be named in several ways. Each name corresponds, by Lavoisier's Law, to a different way of partitioning the compound's electron density. Starting with the least familiar name first, the one that corresponds to a description of electronic structure in terms of mutually exclusive inner-shell and valence-shell electron domains, $SF_6$ can be formulated, following Ramsay [71], Lewis [15,16], and Fajans [73], as $S^{6+}(F^{7+})_6(E_2^{2-})_{24}$ and called *sulfur-fluorine electride*. $E_2^{2-}$ represents an "electride ion" [54,56] — a doubly-occupied, approximately spherical, localized molecular orbital. Correspondingly, the sulfur-fluorine internuclear distance $d$ is obtained by combining expressions (1.11) and (1.12). In Ångstroms,

$$d = 1.20 + 1.8\, r_a + r_b \qquad (1.13)$$

Taking for the S-F bond $r_a = 0.07$ and $r_b = 0.29$ (Fig. 1.2), one obtains from (1.13) for $d$ the value 1.61 Å; the experimental value is nearly the same, 1.57 Å. *The sulfur-fluorine internuclear distance in $SF_6$ is close to that expected for an electron-pair bond between $S^{6+}$ and $F^{7+}$.*

Or, following, *e.g.*, Kossel [74], Van Arkel [75], De Boer [76], and Ketelaar [77], $SF_6$ can be formulated as $S^{6+}(F^-)_6$ and called *sulfur hexafluoride*. $F^-$ represents a complex, tetrahedral anion $[F^{7+}(E_2^{2-})_4]^-$ — a cation $F^{7+}$ surrounded tetrahedrally by a co-ordination polyhedron of four electride ions. Correspondingly, the sulfur-fluorine internuclear distance is viewed as the sum of *ionic* radii: of the $S^{6+}$ cation [$r_b$ in eq. (1.13)] and the $F^-$ anion. The latter is, by the first model, the radius of $F^{7+}$ [$r_a$ in eq. (1.13)] plus twice the radius of an electron domain co-ordinated by $F^{7+}$ [*i.e.*, by eq. (1.12), twice $0.60 + 0.4\, r_a$]. Thus, the radius of $F^-$ is, by the first model, simply the first two terms on the right-hand side of expression (1.13). With $r_a = 0.07$, the radius of $F^-$ $(1.20 + 1.8\, r_a)$ is calculated to be 1.32 Å; the value assigned by Pauling is 1.36 Å [8]. *The sulfur-fluorine internuclear distance in $SF_6$ is close to that expected for an ionic bond between $S^{6+}$ and $F^-$.*

Finally, the compound in question can be formulated as a covalent compound "$SF_6$" and called, by analogy with, *e.g.*, the name tetrafluoro methane for $CF_4$, *hexafluoro "hypothetical $SH_6$"*. Correspondingly, the sulfur-fluorine internuclear distance is viewed as the sum of *atomic* radii of F and S. The former radius, in the previous notation, is $(r_a + R_{AB})$, the latter $(r_b + R_{AB})$; the sum of the two expressions is identically the right-hand-side of eq. (1.11) — which, with (1.12), yields (1.13). The numerical values for the atomic radii of F and S are, by the above expressions, 0.70 and 0.92 Å, respectively. Pauling lists 0.64 and 1.04 Å [8]. The former value has been corrected by Schomaker and Stevenson to 0.72 Å [78]. The latter value — Pauling's covalent single-bond radius for sulfur — differs from our value of 0.92 for sulfur in $SF_6$ by the relatively large amount 0.12 Å. That difference is precisely the Schomaker-Stevenson electronegativity correction [78] for sulfur-fluorine single bonds: namely, 0.08 times the absolute value of the difference in the electronegativity of fluorine, 4.0, and sulfur, 2.5. *The sulfur-fluorine internuclear distance in SF₆ is close to that expected for a covalent bond between sulfur and fluorine atoms.*

In summary, from the viewpoint of the electride-ion model of electronic structure, a bond can be described in several ways. Like the phrases "carbon tetrafluoride" and "tetrafluoro methane", the adjectives "ionic" and "covalent" correspond to different ways of describing the same thing.

*Broadly speaking*, however, a bond is usually considered largely "ionic" — the corresponding compound a "salt" — if the atomic cores associated with an electron domain differ widely in size (the larger — or largest — having a radius greater, approximately, than 0.5~0.7 Å), "covalent" if the atomic cores are similar in size and relatively small (smaller, approximately, than 0.5~0.7 Å), and "metallic" if all atomic cores are relatively large (larger, approximately, than 0.5~0.7 Å) [22,55].

By the above criteria, bonds between sulfur and non-metals (core-radii not larger than, approximately, 0.5~0.7 Å) are largely covalent.

The range of lengths of *linear, single bonds* will be, by eqs. (1.11)–(1.13), approximately 1.4~2.9 Å for covalent bonds; 1.9-4.0 Å usually, however, will be called covalent; longer than 4.0 Å will be metallic.

Lengths of linear single bonds between sulfur atoms and atoms of the relatively non-metallic elements of Groups IV-VII calculated by two methods, by eqs. (1.11)–(1.13) and from Pauling's single-bond covalent radii [8] corrected for the Schomaker-Stevenson electronegativity effect [78], are listed in Table 1.4.

The two sets of distances in Table 1.4 generally agree to within 0.05 Å. Increasing the numerical value of the empirical constant in eq. (1.12) from 0.60 to 0.625 Å cuts the difference between corresponding values in Table 1.4 to, on the average, 0.01 Å.

A sterically unhindered bond to sulfur that, after allowance for substituent-hybridization effects [79], is approximately 0.05-0.07 Å shorter (or longer) than the average of the values cited for that bond in Table 1.4 may generally be deemed to have significant multiple bond (or no-bond) character.

By the above criteria, the sulfur-nitrogen bond in sulfilimines, as discussed by Professor Oae in Chapter 8, corresponds, in length, chiefly, if not solely, to the single bond ylide structure [Sect. 8.3, structure **8.3**]. In contrast, the sulfur-oxygen bonds in sulfoxides exhibit large amounts of multiple bond character. Although probably close to a four-electron bond, the sulfur-oxygen bond in sulfoxides, as stressed by Professor Oae, is not a classical, olefinic-like double bond. The sulfoxide sulfur-oxygen linkage appears to correspond, approximately, to a superposition of two anticoincident spin-sets, one of which contributes to the bonding region one electron, in a single-bond orientation, the other three electrons, in a triple-bond orientation [39,44].

**TABLE 1.4**

Standard Lengths for Linear Single Bonds to Sulfur Atoms

| Bond | Calculated Value (Å) | |
|------|:---:|:---:|
| | From Eq. (1.13) | From Pauling's Covalent Radii[1] |
| S – C | 1.76 | 1.81 |
| – N | 1.69 | 1.74 |
| – O | 1.65 | 1.70 |
| – F | 1.62 | 1.64 |
| – Si | 2.13 | 2.17 |
| – P | 2.06 | 2.12 |
| – S | 2.01 | 2.08 |
| – Cl | 1.97 | 2.00 |
| – As | 2.19 | 2.23 |
| – Se | 2.14 | 2.20 |
| – Br | 2.11 | 2.17 |

[1] Corrected for the Schomaker-Stevenson electronegativity effect [78].

An excellent example of "no bond" resonance is cited by Dr. Ohno in Chapter 5. The sulfur-sulfur distances in structure **5.63a** are 2.35 Å [80], significantly longer than the standard sulfur-sulfur single bond distance (2.01 Å, Table 1.4), but significantly shorter than twice the van der Waals radius of sulfur (2 x 1.8 Å = 3.60 Å). Such interatomic distances, intermediate in length between standard single bond distances and conventional van der Waals contact distances, may be described in an electron-domain, LMO (*L*ocalized *M*olecular *O*rbital) representation of electronic structure as the partial formation about the primary electron-pair-co-ordination-shell of a large-core atom of a *secondary* electron-pair co-ordination polyhedron, yielding thereby relatively weak secondary chemical linkages that have been called *face-centered bonds* [44].

Face-centered bonds are analogous to hydrogen bonds [44]. In a hydrogen bond, the electron-pair accepter is a Brönsted acid, in a face-centered bond a Lewis acid. A hydrogen bond is the first step of a proton transfer — of a nucleophilic displacement at an exposed, acidic proton. A face-centered bond is the first step in the transfer of an atomic core with inner-shell electrons — of a nucleophilic displacement at a co-ordinatively unsaturated Lewis acid. An excellent example is cited in Chapter 6 by Professor Tagaki in Sect. 6.3.3 (see, also, Sect. 6.4.4). The electron-pair acceptor, Lewis-acid site is one of the co-ordinatively unsaturated atomic cores of a halogen molecule ($Cl_2$, $Br_2$, or $I_2$; never $F_2$). The electron-pair donor is the sulfur atom of a sulfide. The first step in the nucleophilic attack on the halogen molecule by sulfur is formation of a face-centered bond, structure **6.7**. The S...X distance is slightly shorter than a van der Waals contact distance; the halogen-halogen distance is slightly longer than a normal halogen-halogen bond. Further displacement of $X^-$ by the incoming sulfide yields structure **6.8**. Now the face-centered bond is between the halide ion $X^-$ and the halogen atom X, bound, almost covalently, to sulfur. In structure **6.8**, the sulfur-halogen distance is slightly longer than a normal sulfur-halogen bond; the halogen-halogen distance is slightly shorter than a van der Waals contact distance. From the viewpoint of the interior halogen atom, structures **6.7** and **6.8** are analogous to trihalide ions. The sulfide plays the role of a halide ion. It increases the electron-pair-co-ordination-number of the atomic core of the adjacent halogen atom from four to (approximately) five. Should, however, the sulfur center be a stronger Lewis acid than the halogen center, the partially, or nearly, displaced halide ion becomes co-ordinated to sulfur, yielding structure **6.6**. Or, the displaced halide ion may attack a carbon atom adjacent to the sulfonium sulfur (Ch.9, Sect. 9.2.5.1).

### 1.3.3 Chemical Implications

Most of the chemistry of sulfur can be rationalized in terms of its having in its lowest valence state, like oxygen, two *unshared valence-shell pairs*, but, relative to oxygen, a *relatively large atomic core*.

The presence of lone pairs on divalent sulfur accounts for the formation from sulfides of *sulfonium salts* (Ch. 9), *sulfoxides* (Ch. 8), and *sulfones* (Ch. 10). It accounts, also, for the *donor properties* of sulfur (Ch. 6, Sect. 6.3.2) toward Brönsted acids, in the formation of *H-bonds*; toward Lewis acids, in the formation of *face-centered bonds* (ref. cited in Sect. 1.3.2.3); and toward acidic carbon atoms, in *carbonium-ion stabilization, anchimeric effects* (Ch.6, Sect. 6.4.2, 6.4.3) and *enhancement of electrophilic aromatic substitution* (Ch.6, Sect. 6.4.5).

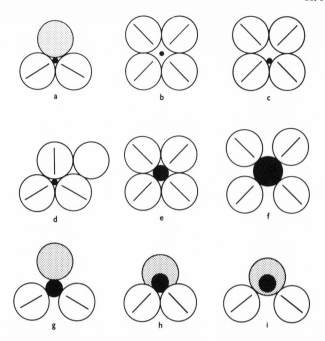

**Fig. 1.3.** Schematic representation in two dimensions of the Valence-Shell-Inner-Shell-Electron-Repulsion Model. Solid circles represent Pauli-impenetrable atomic cores. Other circles represent localized valence-shell electron-pairs, shared (bonding) with an electronegative element (e.g., fluorine) if occupied by a valence-stroke, unshared (a lone pair) if stippled.

In three dimensions the valence shells of the atomic cores would be completed (in stable compounds) with additional, localized orbitals: above *or* below the plane of the other three in (a) and (d) (yielding octet-rule, tetrahedral coordination); above *and* below the plane of the other four in (b), (e) and (f) (yielding expanded-octet, octahedral coordination).

(a) A coordinatively saturated small atomic core (e.g., in three dimensions, $O^{6+}$ in $OF_2$) with an angularly localized, Lewis-like, relatively basic lone pair (stippled), (b) A small atomic core (e.g., $O^{6+}$) with a hypothetical "expanded octet" (as in hypothetical $OF_6$). The "lattice energy" of (b) is slightly greater than that of (c), which, in turn, is much less stable than the non-rattling, non-hypervalent arrangement (d). (e) A coordinatively saturated medium-sized atomic core (e.g., $S^{6+}$ in extraordinarily unreactive $SF_6$). (f) A hypervalent but coordinatively unsaturated large-atomic core (e.g., $Te^{6+}$ in easily hydrolyzed $TeF_6$). (g) A hypothetical, angularly localized, Lewis-like lone pair attached to a coordinatively unsaturated, medium-sized atomic core. The energy of hypothetical structure (g) is greater than that of structure (h). (h) A radially contracted, angularly expanded, Gillespie-like, slightly basic lone pair. Note the increase in $s$-character of the lone pair orbital and, correspondingly, the decrease in $s$-character of the bonding orbitals in going from the small-core compound (a) [e.g., $H_2O$] to the larger-core compound (h) [e.g., $H_2S$]. (i) Incipient formation of a Sidgwick-like, inert pair (as, e.g., in $TeCl_6^{2-}$).

accounts for the formation of *relatively long bonds* to sulfur (longer than those to the corresponding oxygen compounds); the *large volumes* of sulfide groups (compared to oxygen groups; Ch. 6, Sect. 6.4.6) and thiocarbonyl groups (compared to carbonyl groups; Ch. 5); the *low ionization potential* of unshared electrons on sulfur (compared to oxygen; Ch. 5); and the *low electronegativity and electron-withdrawing inductive effect* of sulfur (compared to oxygen; Ch. 6, Sect. 6.4.6). The sulfur kernel's sheer size accounts, also, in part, owing to diminished nuclear-nuclear repulsions, for the *formation of sulfur-sulfur bonds* at the expense of oxygen-oxygen bonds (Ch. 7, Sect. 7.3.1).

Most significantly for chemistry, sulfur's kernel, unlike oxygen's, can accommodate in its valence-shell, without rattling [22], more than four localized electron pairs (Fig. 1.3). Sulfur can expand its octet (Sect. 1.2.5).

Classical octet, or incipient *octet expansion* about sulfur, however described, accounts for the formation of such *hypervalent compounds* as $SO_2$, $SO_3$, $SF_4$, $SF_6$, sulfoxides, and sulfones. It accounts, also, for the *activation of protons* on adjacent or conjugated centers (extensively examined by Professor Oae and coworkers; Ch. 6, 8, 9) and, generally, for the action of sulfur centers as *carbanion stabilizers* (Ch. 6, 8, 9, 10) in, *e.g.*, the *heightened acidity* of *para*-substituted phenols (Ch. 6, 10) and toluenes (Ch. 6) and in *enhanced rates* of, *e.g.*, decarboxylation, base-catalyzed elimination reactions, and additions to and isomerizations of double bonds (Ch. 6).

Because sulfur atoms in their lower valence states have both basic sites (unshared valence-shell electrons) and acidic sites (co-ordinatively unsaturated atomic cores), they have a high tendency to react with each other to form *sulfur-sulfur bonds* (Ch. 2).

Finally, and most important for understanding the chemistry of sulfur, although the electronegativity of sulfur is less than that of oxygen, a lone pair site on sulfur is (at first sight paradoxically) *less basic than oxygen,* owing to the sulfur atom's relatively large kernel, which allows *unusual angular dispersion of sulfur's unshared electrons* (Fig. 1.3).

The "diffused and polarizable" character of sulfur's unshared electrons (Ch. 6) accounts, not only for the relatively low basicity of the atom's unshared electrons but, also, for the *lessened participation of sulfur in reasonance stabilization of α-carbonium ions*, compared to oxygen (Ch. 6). It accounts, also, for sulfur atoms' *effectiveness in accommodating negative charge*, becoming, thereby, a relatively *good leaving group*, as manifested, *e.g.*, in the greater *stability of thiolate ions* compared to alcoholate ions (Ch. 5), the *readiness of thiocarbonyl compounds to form tetrahedral intermediates* (Ch. 5), and the relatively *large barrier to rotation about the carbon-nitrogen bond in sulfur analogues of amides* (Ch. 5).

Additionally, the diffused and polarizable character of sulfur's unshared electrons accounts for the generally *small bond angles* at sulfur, compared to oxygen [45] (cited in Ch. 6; see Fig. 1.3a,h); the *weak though short S-O bond in DSMO* (Ch. 8; competition for the potential energy space about sulfur's kernel by its unshared electrons and those on oxygen leads to a shallow minimum in the energy *vs.* bond-length curve); and it accounts for the occurrence of *O-atom transfer from sulfoxides to trivalent phosphorus* (Ch. 8; on phosphorus the lone-pair competition cited immediately above does not exist).

Finally, the diffused and polarizable character of sulfur's unshared electrons probably accounts in large measure for the *s-character rule* [74] as it applies to sulfur — namely, that in most of its compounds sulfur tends to concentrate its *s*-character in orbitals occupied by unshared electrons, the more so the more electronegative the attached substituents (Fig. 1.3h,i).

## 1.4 REFERENCES

1. K.A.R. Mitchell, *Chem. Rev.,* **69**, 157 (1969).
2. W.G. Salmond, *Quart. Rev.,* **22**, 253 (1968).
3. C.C. Price and S. Oae, "Sulfur Bonding" (Ronald Press, New York, N.Y., 1962).
4. G. Cilento, *Chem. Rev.,* **60**, 147 (1960).
5. C.K. Jørgensen, *Struct. and Bonding,* **6**, 94 (1969); E.A. Lucken, *ibid.,* p 1.
6. D.P. Craig, A. Maccoll, R.S. Nyholm, L.E. Orgel, and L.E. Sutton, *J. Chem. Soc.,* 332 (1954).
7. L. Pauling, *J. Amer. Chem. Soc.,* **53**, 1367 (1931).
8. L. Pauling, "Nature of the Chemical Bond" (Cornell University Press, Ithaca, N.Y., 1960).
9. C.E. Moore, "Atomic Energy Levels", Vol. I, NBS Circular 467 (U.S. Government Printing Office, Washington, D.C., 1949).
10. D.P. Craig and E.A. Magnusson, *J. Chem. Soc.,* 4895 (1956).
11. D.P. Craig and C. Zauli, *J. Chem. Phys.,* **37**, 601, 609 (1962).
12. R.E. Rundle, *Survey Prog. Chem.,* **1**, 81 (1963).
13. E.E. Havinga and E.H. Wiebenga, *Recueil,* **78**, 724 (1959).
14. G.C. Pimentel, *J. Chem. Phys.,* **19**, 446 (1951).
15. G.N. Lewis, *J. Amer. Chem. Soc.,* **38**, 762 (1916).
16. G.N. Lewis, "Valence" (Dover reprint, New York, N.Y., 1966).
17. M. Born, "The Constitution of Matter" (Methuen and Co., Ltd., London, 1923).
18. W.L. Bragg, *Phil. Mag.,* **40**, 169 (1920).
19. V.M. Goldschmidt, *Trans. Far. Soc.,* **25**, 253 (1929).
20. J. Lennard-Jones, *Advan. Sci.,* **11**, 136 (1954).
21. H.A. Bent, *J. Chem. Ed.,* **44**, 512 (1967).
22. H.A. Bent, *Fortsch. Chem. Forsch.,* **14**, 1 (1970).
23. F. Cajori, "A History of Physics" (Dover Publications Inc., New York, N.Y., 1962, p 376).
24. Cited by W.T. Scott, "Erwin Schrödinger" (University of Massachusetts Press, Amherst, Mass., 1967).
25. J.C. Phillips, *Rev. Mod. Phys.,* **42**, 317 (1970).

26. H.A. Bent, *J. Coll. Sci. Teach.*, **1**, No. 2, 20 (1971).
27. H. Margenau, *Phil. Sci.*, **11**, 187 (1944).
28. W. Pauli, *Z. f. Phys.*, **31**, 765 (1925).
29. H.A. Bent, *J. Chem. Ed.*, **40**, 446 (1963).
30. J.C. Slater, *Phys. Rev.*, **34**, 1293 (1929).
31. J. Lennard-Jones, *Proc. Roy. Soc. (London)*, **A198**, 1, 14 (1949).
32. J.A. Pople, *Quart. Rev. (London)*, **11**, 273 (1957).
33. C. Edmiston and K. Ruedenberg, *Rev. Mod. Phys.*, **35**, 457 (1963).
34. J.W. Linnett and A.J. Poe, *Trans. Far. Soc.*, **47**, 1033 (1951); C.E. Mellish and J.W. Linnett, *ibid.*, **50**, 657, 665 (1954). See, also, H.K. Zimmerman, Jr., and P. Van Rysselberghe, *J. Chem. Phys.*, **17**, 598 (1949).
35. J.M. Foster and S.F. Boys, *Rev. Mod. Phys.*, **32**, 300 (1966).
36. J.E. Lennard-Jones, *Rev. Mod. Phys.*, **20**, 1024 (1952).
37. K. Ruedenberg, *Mod. Quantum Chem.*, **1**, 85 (1965).
38. D. Peters, *J. Chem. Soc.*, 2003 (1963); *J. Chem. Phys.*, **43**, S115 (1965).
39. H.A. Bent, *Chemistry*, **40**, No. 1, 8 (1967).
40. J.W. Linnett, *J. Amer. Chem. Soc.*, **83**, 2643 (1961); "The Electronic Structure of Molecules. A New Approach" (John Wiley and Sons, Inc., New York, N.Y., 1964).
41. P.O. Lowdin, *Advan. Chem. Phys.*, **2**, 207 (1959).
42. E.A. Hylleraas, *Z. Physik.*, **54**, 347 (1929); C.C.J. Roothaan and A.W. Weiss, *Rev. Mod. Phys.*, **32**, 194 (1960); C.Eckart, *Phys. Rev.*, **36**, 878 (1930).
43. J.C. Slater, *Phys. Rev.*, **35**, 509 (1930).
44. H.A. Bent, *Chem. Rev.*, **68**, 587 (1968).
45. H.A. Bent, *J. Chem. Ed.*, **42**, 302 (1965).
46. H.A. Bent, *J. Chem. Ed.*, **40**, 523 (1963).
47. W.H. Adams, *J. Chem. Phys.*, **34**, 89 (1961); R.M. Pitzer, *ibid.*, **41**, 2216 (1964); M. Klessinger *ibid.*, **43**, S117 (1965); U. Kaldor, *ibid.*, **46**, 1981 (1967); V. Magnasco and A. Perico,*ibid.*, **47**, 971 and **48**, 800 (1968); R. Bonaccorsi, C. Petrongolo, E. Scrocco, and J. Tomasi,*ibid.*, **48**, 1500 (1968); J.H. Letcher and T.H. Dunning,*ibid.*, **48**, 4538 (1968); C. Trindle and O. Sinanoğlu,*ibid.*, **49**, 65 (1969).
48. A.A. Frost, *J. Chem. Phys.*, **47**, 3707, 3714 (1967); *J. Phys. Chem.*, **72**, 1289 (1968).
49. A.A. Frost, B.H. Prentice, III, and R.A. Rouse, *J. Amer. Chem. Soc.*, **89**, 3064 (1967); A.A. Frost and R.A. Rouse, *ibid.* **90**, 1965 (1968).
50. I. Langmuir, *J. Amer. Chem. Soc.*, **41**, 868 (1919).
51. N.V. Sidgwick and H.M. Powell, *Proc. Roy. Soc. (London)*, **176**, 153 (1940).
52. R.J. Gillespie and R.S. Nyholm, *Quart. Rev. (London)*, **11**, 339 (1957).
53. R.J. Gillespie, *J. Chem. Ed.*, **40**, 295 (1963); *Angew. Chem. Intern. Ed. Engl.*, **6**, 819 (1967).
54. H.A. Bent, *J. Chem. Ed.*, **42**, 348 (1965).
55. H.A. Bent, *J. Chem. Ed.*, **45**, 768 (1968).
56. N.V. Sidgwick, *Ann. Repts. Chem. Soc. (London)*, **30**, 116 (1933).
57. I am indebted to Mr. Robert Shen for this calculation.
58. Private communication.
59. R.P. Feynman, *Phys. Today*, **19**, 31 (1966).
60. H.B. Thompson, *Inorg. Chem.*, **7**, 604 (1968).
61. P. Eisenberger and W.C. Marra, *Phys. Rev. Letters*, **27**, 1413 (1971).
62. E. Schrödinger, "On the Peculiarity of the Scientific World-View" in "What is Life?" (Doubleday and Co., Inc., Garden City, N.Y., 1956).
63. O.T. Benfey, ed., "Classics in the Theory of Chemical Combination" (Dover Publications, Inc., New York, N.Y., 1963).
64. A. Werner, "New Ideas on Inorganic Chemistry", Engl. transl. by E.P. Hedley (Longmans, Green and Co., New York, N.Y., 1911).

65. M.L. Huggins, *Science,* **55,** 459 (1922).
66. N.V. Sidgwick, *J. Chem. Soc.,* **123,** 725 (1933).
67. N. Bohr, *Phil. Mag.,* **26,** 1 (1913); *ibid.,* **30,** 394 (1915).
68. J.L. Heilbron and T.S. Kuhn, "The Genesis of the Bohr Atom", in "Historical Studies in the Physical Sciences", R. McCormmach, Ed. (University of Pennsylvania Press, Philadelphia, Penna., 1969).
69. R.G. Parsons and V.F. Weisskopf, *Z. Physik*, **202,** 492 (1967).
70. J.C. Slater, *Phys. Rev.,* **36,** 57 (1930).
71. W. Ramsay, *J. Chem. Soc.,* **93,** 774 (1908).
72. A. Lavoisier, "Elements of Chemistry". R. Kerr, transl., (Dover Publications, Inc., New York, N.Y., 1965).
73. K. Fajans, *Chimia*, **13,** 349 (1959).
74. W. Kossel, *Ann. der Phys.,* **49,** 229 (1916).
75. A.E. Van Arkel, "Molecules and Crystals" (London, 1949).
76. A.E. Van Arkel and J.H. De Boer, "Die Chemische Bindung als Elektrostatische Erscheinung" (Leipzig, 1931).
77. J.A.A. Ketelaar, "Chemical Constitution" (Elsevier Publishing Co., New York, N.Y., 1958).
78. V. Schomaker and D.P. Stevenson, *J. Amer. Chem. Soc.* **63,** 37 (1941).
79. H.A. Bent, *Chem. Rev.,* **61,** 275 (1961).
80. S. Bezzi, M. Mammi, and C. Garbuglio, *Nature,* **182,** 247 (1958); S. Bezzi, C. Garbuglio, M Mammi, and G. Traverso, *Gazz, Chim. Ital.,* **88,** 1226 (1958); M. Mammi, R. Bardi, C. Garbuglio, and S. Bezzi, *Acta Cryst.,* **12,** 1048 (1960).

*Chapter 2*

# ELEMENTAL SULFUR AND
# ITS REACTIONS

Roland Mayer

*Department of Chemistry, Technical University of Dresden*
*Germany (GDR)*

## 2.1 INTRODUCTION

Although sulfur has been utilized since antiquity and later found widespread use in the preparation of gun powder, sulfuric acid, other sulfur-containing compounds and rubber (vulcanization), only in

33

recent years have we come to understand some of its properties and the special character of sulfur bonds. Especially during the last several years, the application of elemental sulfur has increased in many different fields, such as in agriculture and forestry, in construction and transportation, in industry of plastics and elastomers and last not least in synthetic chemistry [1]. This has resulted in a growing interest in the properties and reactions of this element [1-9,12a,18a].

Sulfur shows an extremely high tendency to react with itself and to form S—S bonds. Depending on external conditions cyclic and chain molecules of from two to more than $10^5$ sulfur atoms result. In the solid, liquid and gaseous states these species of elemental sulfur are in a complicated equilibrium [2-4]. Due to the S—S bond character, distance and bond angle, the formation of many different ring and chain molecules is possible [10]. The intermolecular interaction of sulfur molecules leads to a variety of similarly stable polymorphs. Difficulties arise, however, because sulfur reacts with many impurities and is photosensitive. Therefore, sulfur changes its physical and chemical properties variously in different environments.

First we will deal with some aspects of elemental sulfur and the different species it forms, then the problem of activating elemental sulfur will be discussed. Finally we will describe some special reactions of sulfur with C—H and C—Cl bonds.

## 2.2 DIFFERENT FORMS OF SULFUR

Sulfur has atomic number 16 and an atomic weight of $32.064 \pm 0.003$. The variability of $\pm 0.003$ is recommended on account of the variability of the isotopic composition in different sources.

Sulfur is readily available in elemental form. About 50% derives from elemental deposits, 30% from smelter gases, and 20% is recovered from pyrites. Normally this sulfur is 99.9% or higher in purity. During transport and storage, however, impurities are added (traces of metals, oxidation products, water, air inclusions, SH-compounds, etc.). For certain purposes sulfur of 99.9999% purity is available.

Today most of the Frasch sulfur produced is not shipped in the solid form but as liquid. Therefore, a complex transportation system for molten sulfur has been developed and the properties of this system have been very intensively studied.

### 2.2.1 Sulfur Molecule $S_8$

Commercial sulfur nearly exclusively consists of an eight-atomic cyclic molecule (cycloocta-S) crystallizing orthohombically ($S_\alpha$). This orthorhombic crown-shaped cyclooctasulfur is the only form thermodynamically stable at room temperature (up to 95°). All other sulfur molecules and modifications are converted to the $S_\alpha$ form with a half-life between $10^{-3}$ sec and about 1 year.

On heating, orthorhombic cycloocta-S is converted into monoclinic cycloocta-S. The monoclinic crystals ($S_\beta$) are another possibility for cyclooctasulfur molecules to build up crystals. Between 95° and the melting point the $S_\beta$ crystals are stable. Whenever sulfur crystallizes from solutions or melts, monoclinic sulfur crystals are formed within the temperature range of 95° to about 119°.

The $S_\alpha$ and $S_\beta$ molecules have been well studied. The average bond distance is $2.048 \pm 0.002$ Å, the bond angle $108 \pm 3°$ and the torsion angle 98°.

The "ideal" melting points of $S_\alpha$ and $S_\beta$ sulfurs are 112.8° and 119.3°, respectively. As $S_\beta$ is stable above 95°, a melting point of about 119° is detected at first. On heating for a longer time an equilibrium is established, leading to a product with a "mixed" m.p. of 114.6°. The kind and nature of the "impurities" responsible for this certain depression are not clear (see ref. 4). At elevated temperature cycloocta-S seems to be in an equilibrium with an open chain form (catenaocta-S), which is probably partially polymerized. But since this catenaocta-S has not been isolated actually, it is discussed only as a reactive intermediate:

$$\text{Cycloocta} - S \underset{\text{energy}}{\rightleftharpoons} \cdot \ddot{S} - \ddot{S_6} - \ddot{S} \cdot \text{ or } \ ^-\ddot{S} - \ddot{S_6} - \ddot{S}^+ \qquad (2.1)$$

Up to the melting point, elemental sulfur is predominantly in the form of the cycloocta-S. Such an eight-membered ring should be stable in different steric forms, but only the crown shape is known. It is noteworthy that the structure of certain thermodynamically unstable $S_8$-modifications is as yet unknown.

For details and further information on the structure of cyclo- and catenaoctasulfur ($S_8$) see refs. 3,4,10. For geometric structures of elemental sulfur see ref. 11.

## 2.2.2 Polymeric Sulfur $S_X$, and Behavior of Molten Sulfur and Sulfur Vapor

Sulfur forms at its melting point a moderately mobile yellow liquid. Although the yellow sulfur is the most common form, the color of sulfur in the solid state depends on the temperature. At $-200°$, for instance, sulfur is a snow-white solid. On further heating the color of the melt darkens, turns red, and the viscosity increases, passing a maximum at about 160 to 180°. At the boiling point (444°) sulfur is dark red-brown to nearly black and opaque.

Recent works [1,2], especially on spectroscopic investigations, indicate that the color in the solid state can fully be explained by different thermal excitations. However, in the liquid state the change of color must be due to the formation of new species.

At the melting point the melt predominantly consists of free $S_8$ units. At the maximum viscosity, caused by increasing polymerization, the viscous liquid contains units $> 100,000 \, S_8$ (mol.wt.$> 10^6$) [13].

Further heating leads to depolymerization of the polymer and at the boiling point (444°) finally a fairly fluid low-molecular ($< S_{12}$) liquid exists.

If sulfur is slowly cooled down from its boiling point, the reverse sequence of states appears.

All these facts indicate a complicated chemical equilibrium between different molecular species to exist in molten sulfur. The molecular weight is a function of temperature and time. Hitherto it has been noted that an equilibrium is established in the melt, especially between cycloocta-S and catenaocta-S as monomer and polycatena-S as polymer. However, in molten sulfur other rings like $S_8$ have also been detected, for instance, $S_{12}$ [4].

Of special interest is the finding[1,2] that the color of hot liquid sulfur ($>250°$) is due to the $S_3$ species. A portion of only about 1 mole % $S_3$ suffices to give rise to the color at the boiling point.

The presence of $S_3$ and other low-molecular species, rather than polymers, explains the high reactivity of liquid sulfur above 250°.

A knowledge of the chemical and physical behavior of sulfur molecules in the molten state (chain and cyclic molecules, oligomers, polymers) is important and interesting, $e.g.$, for the transportation of liquid sulfur.

If attempts were successful to produce a stabilized polymeric form of sulfur — because of its unusually good weathering and mechanical characteristics and its low costs — new applications would be possible.

In this field remarkable progress has recently been made leading to surprising results and absolutely new applications, such as thermoplastic product, rubber, glass, foamed sulfur and sulfur for construction materials.

In this connection the stabilization of polymeric sulfur elastomers ($e.g.$, stretched to long fibres) is of special interest. These $S_x$ species have been known for a long time. They are the result of molten sulfur being abruptly cooled. Thus, by kinetic control a thermodynamically unstable, polymeric sulfur is obtained which is insoluble in $CS_2$. Yield and molecular weight depend on starting and quenching temperatures.

At temperatures above $20°$ in the absence of stabilizers, plastic S-species are quickly converted into cycloocta-S $(S_\alpha)$ and elasticity diminishes.

To date it is not quite clear what molecular species actually exist in sulfur vapor and what equilibria between them are established. A surprising fact arose from mass-spectroscopic data: they indicate that at the boiling point the saturated sulfur vapor above the liquid phase consists of nearly 20% $S_8$, 40% $S_7$, 30% $S_6$, 6% $S_5$, and 4% $S_4, S_3, S_2$.

Sulfur vapor very probably contains cycles of $S_{>8}$. Traces of small rings $(S_5, S_6, S_7)$ have been detected at temperatures as low as $120°$. Above about $240°$ $S_2$ and above $350°$ $S_3$ and $S_4$ have also been found. Above $500°$ in sulfur vapor $S_2$ molecules predominate, and at high temperatures (e.g., $2000°$ and a pressure below $10^{-5}$ mm Hg) they are converted into atomic sulfur.

All these facts demonstrate that above the melting temperature sulfur is converted into energy-rich reactive sulfur species suitable for further reactions. For special information see refs. 3,4,103.

### 2.2.3 Other Sulfur Molecules

Summarizing the facts mentioned above we can state: in molten sulfur and sulfur vapor extremely different species of sulfur $(S_1$ to $S_{10^6})$ do exist. Until now, however, only some of them have been isolated in crystalline form and characterized. These are the cyclic structures of $S_6, S_7, S_8, S_9, S_{10}, S_{12}$, and $S_{18}$[4].

It should be added, however, that although formation of these sulfur species is favoured in certain kinetically controlled reactions, the only thermodynamically stable sulfur at room temperature is cycloocta-S in a certain crystal form $(S_\alpha)$. Now we are going to deal with simple syntheses of sulfur molecules other than $S_8$.

The attempts to synthesize $S_6$ and $S_{12}$ by a general method have been successful[14]. By kinetic control of the following reaction cyclohexasulfur $(S_6)$ and cyclododecasulfur $(S_{12})$ became available:

$$\underset{H}{\overset{H}{\underset{|}{\overset{|}{S_x}}}} + \underset{Cl}{\overset{Cl}{\underset{|}{\overset{|}{S_y}}}} \longrightarrow 2\,HCl + S_{x+y} \qquad (2.2)$$

$$x+y = 6 \text{ or } 12$$

*Cyclohexasulfur* $(S_6)$, already known[10,15] and synthesized by a classical method *via* $S_2O_3^{2-}$, crystallizes in orange yellow crystals that do not have a fixed melting point.

It is relatively unstable and is — especially in the presence of impurities — converted into rhombic cyclooctasulfur. With organic compounds it seems to react faster than $S_8$ but no exact relative rates are known [10,15]. For $S_6$, the bond distance is $2.057 \pm 0.02$ Å; the bond angle, $102 \pm 2°$; and the torsion angle, $74 \pm 3°$.

*Cyclododecasulfur* ($S_{12}$) is unexpectedly stable. It crystallizes in pale yellow lamellas (m.p. 146–148°). The bond distance is $2.050 \pm 0.005$ Å; the bond angle, $106.8 \pm 1°$; and the torsion angle, $107.5 \pm 1°$ [16]. $S_6$ and $S_{12}$ are also available by reacting bis(cyclopentadienyl)titanium(IV) pentasulfide with $SCl_2$ [17]:

$$(C_5H_5)_2TiS_5 \;+\; SCl_2 \longrightarrow (C_5H_5)_2TiCl_2 + S_6 + S_{12} \qquad (2.3)$$

$$+ \; S_2Cl_2 \longrightarrow S_7 \qquad\qquad\qquad\qquad\qquad (2.4)$$

$$+ \; S_4Cl_2 \longrightarrow S_9 \qquad\qquad\qquad\qquad\qquad (2.5)$$

$$+ \; SO_2Cl_2 \longrightarrow [S_5] \longrightarrow \tfrac{1}{2} S_{10} \qquad\qquad\quad (2.6)$$

This procedure is a general method, suitable also for preparations of $S_7$, $S_9$, and $S_{10}$.

$S_2Cl_2$ reacts analogously to form *cycloheptasulfur* ($S_7$) in relatively high yield of 20%.

$S_7$ crystallizes from toluene in long intensively yellow needles melting reversibly at 39°. Above this temperature polymerization and conversion into $S_8$ occur.

Under special conditions the pentasulfide in the presence of HCl reacts with $S_4Cl_2$ to form *cyclononasulfur* ($S_9$) [18]. $S_9$ — though somewhat more stable than $S_6$ — is also easily convertible into $S_8$. The intensively yellow needles show a melting behavior similar to $S_6$ and $S_{10}$.

$SO_2Cl_2$ reacts with the pentasulfide to form *cyclodecasulfur* ($S_{10}$). During the reaction $SO_2$ is split off[17]. The unstable $S_{10}$ shows neither a characteristic melting nor decomposition point. Above 60° polymerization occurs. After recrystallization from $CS_2$ cyclodecasulfur forms intensively yellow rhombic plates.

The existence of rings with an odd number of ring atoms ($S_7$ and $S_9$) is of special interest because of their deviating and distorted S–S bonds. Both species should be very reactive, but up till now there have been no data available.

A general comparison between the reactivity of $S_8$ and the other species is lacking, too.

### 2.2.4 Atomic Sulfur

Atomic sulfur $S_1$ is an extremely energy-rich form of sulfur because it forms relatively strong bonds with itself[15]. On bonding characteristics of the sulfur atom see ref. 19. The most stable form — the ground state — of the sulfur atom is the configuration of $s^2p^4$ with two unpaired electrons (= triplet sulfur)

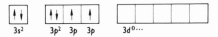

The lowest energy excited state lies about 26 kcal above the ground state. Formally this is the carbene-like singlet sulfur

Only these two states are of interest in our discussion because both are responsible for the specific reactions of atomic sulfur. The next excited state needs a very high energy of about 123 kcal.

$S_1$ exists at high temperatures or in extremely energy-rich systems.

Besides the thermal cleavage of elemental sulfur or sulfur compounds, atomic sulfur can also be obtained by photolysis of S-compounds[15,20]. The most important method is the photolysis of carbonyl sulfide COS, which yields two products, CO and $S_1$:

$$O=C=S \xrightarrow{h\gamma} CO + S_1^* \qquad (2.7)$$

The process to form S in the first excited singlet state (for spectroscopic states and term symbols see ref. 15) requires about 86 kcal/mole COS, corresponding to a wavelength of at least 332.5 nm. In this case, triplet sulfur is also formed, since this requires only about 60 kcal/mole COS.

In all cases, CO arises in the ground state since its first excited state is very high and lies about 186 kcal above the ground state.

The quantum yield of the CO formation is nearly two (1.8); that means for each quantum absorbed nearly two molecules of CO are formed. The second molecule of CO could result from the attack of S atom on COS:

$$COS + S \longrightarrow CO + S_2^* \qquad (2.8)$$

This idea is supported by the fact that the quantum yield is diminished to nearly unity (1) if captures for atomic sulfur (*e.g.*, olefins) are added. In analogy to carbenes, sulfur in the singlet state undergoes insertion reactions with hydrocarbons and also stereospecific additions to olefins.

On the other hand, under the conditions studied, sulfur in the triplet state does not react with saturated hydrocarbons at all but does with olefins to give only nonspecific addition products [15,20].

## 2.3 ACTIVATION OF ELEMENTAL SULFUR AND THE PROBLEM OF S–S BONDS

### 2.3.1 General Data

The pronounced tendency of sulfur to form bonds with itself and thus to show an unusually large variety of molecular species (allotropes) differs fundamentally from that of homologous oxygen. Oxygen forms only two molecular forms: ozone $(O_3)$ and the paramagnetic $O_2$. This remarkable difference between these two elements is due to their different positions in the periodic table.

While oxygen possesses only $sp$ hybrids, sulfur can also make use of free unoccupied $d$-orbitals. Sulfur in the ground state has the electron configuration shown on page 39.

In divalent sulfur and oxygen, two unpaired electrons of the $p$-level are necessary to form two $\sigma$ -bonds

$$\begin{array}{c} 3s^2 \\ \cdot\cdot \\ R - S - R \\ \cdot\cdot \\ 3p^2 \end{array}$$

### 2.1

Our meager knowledge of quantitative data in the field of orbital participation and hybridization in covalent bonds of sulfur, however, only leads to speculative formulations. Nevertheless, during the last several years qualitative working rules and formulations have been published and verified by means of numerous practical problems. These rules are derived mainly from experimental observations [5,6,18a,19,21,22,23]. As an especially successful and highly significant postulate – published [24,25] in the early 1960's – the concept has been proved that sulfur has a *strong* tendency to expand its electron shell by means of its $d$-orbitals. This possibility arises because of the existence of available vacant $3d$-orbitals which may either be occupied by its own $3s$ and $3p$ electrons or by external electrons.

In the bonds which elemental sulfur forms with itself this element is anxious to expand its valence shell and to try to form an electron decet.

Accordingly, the real S–S single bond should be very unstable and should not exist in elemental sulfur. Actually the apparent S–S single bonds seem to be a resonance hybrid approximately reflected by the formula

This means that for cycloocta-S the eight-membered sulfur ring is stable and of an energetically especially favored crown shape:

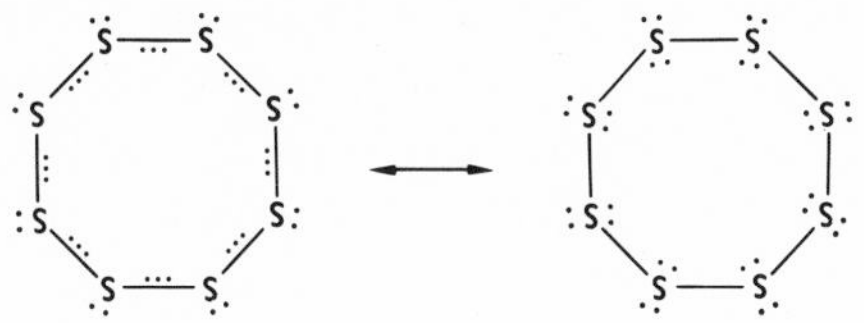

The eight S atoms are equal in their electronic state and each is formally surrounded by an electron decet according to the above-mentioned model. Thus, electrons are believed to be delocalized in the sulfur chain or ring. The 3s electron pair requires a high excitation energy ($3s \rightarrow 3d$: about 162 kcal/gram-atom) [26]. Therefore, obviously 3p electrons participate in the bonding ($3p \rightarrow 3d$: about 49 kcal/gram-atom). Hereby a neutral excited S atom ($s^2 p^3 d$) is formed. If two or more such S atoms in a molecule are bonded due to the overlapping of the 3d electrons, the energy necessary for excitation decreases.

The participation of lone electron-pairs in additional bonding accounts for the astonishing strength of the S—S bond (compared to the O—O bond), the shortened bond distance, the regulation of preferred angles and in this connection the hindered rotation.

This model of delocalized electrons explains also the color of sulfur and the special features which molecules with S—S bonds show in the uv spectrum [25].

Also a fact that will be discussed later indicates that sulfur in spite of its 16 lone electron pairs is a Lewis acid rather than a base.

These models and formulations lead to a clear understanding why no branched molecule of sulfur has been isolated. Such a branching would give rise to S—S single bonds much more unstable than O—O bonds. Then even under mild reaction conditions, cleavage or weakening of the S—S bond would take place.

For certain synthesis this consideration is of decisive importance: in principle a nucleophilic (+ X⁻) as well as an electrophilic (+ Y⁺) or a radical

(+ R·) attack leads in transition state to a branched structure and thus, to an incipient cleavage:

$$\begin{array}{c} S \\ | \quad \overset{\cdots}{S} \overset{.}{-} X \\ S_x \end{array} \longrightarrow {}^-\overset{\cdots}{S}{-}S_x{-}SX \qquad (2.9)$$

$$\begin{array}{c} S \\ | \quad \overset{\cdots}{S} \cdot \\ S_x \end{array} \xrightarrow{\ Y^+\ } \begin{array}{c} S \\ | \quad \overset{+}{\underset{\cdots}{S}}{-}Y \\ S_x \end{array} \longrightarrow {}^+\overset{\cdots}{S}{-}S_x{-}SY \qquad (2.10)$$

$$\begin{array}{c} S \\ | \quad \overset{\cdots}{S}{-}R \\ S_x \end{array} \longrightarrow \overset{\cdots}{S}{-}S_x{-}SR \qquad (2.11)$$

In principle the cleavage of an S—S bond occurs by homolytic or heterolytic scission:

$$\sim\sim S{-}S\sim\sim \quad \xrightarrow{\ R\cdot\ } \sim\sim SR \ + \ \cdot S \sim\sim \qquad (2.12)$$

$$\xrightarrow{\ X^-\ } \sim\sim SX \ + \ {}^-S \sim\sim \qquad (2.13)$$

$$\xrightarrow{\ Y^+\ } \sim\sim SY \ + \ {}^+S \sim\sim \qquad (2.14)$$

Since the nucleophilic attack on the sulfur molecule is favored, cleavage of the S—S bond predominantly occurs heterolytically. Thus, in practice only mercaptide sulfur is effective in subsequent reactions.

In addition to the above mentioned reaction paths other possibilities for the cleavage of an S—S bond or a C—S bond are under discussion [27]. Compared to the saturated carbon atom, the sulfur atom has a low electronegativity and a considerable tendency to become polarized or ionized. Hence, to a much greater extent than the corresponding oxygen compounds, sulfur structures having S–C or S–S bonds can cleave to form not only free radicals, but sulfur may also leave as an anion and even as a cation:

$$R{-}S{-}S{-}R \quad \longrightarrow \quad 2 \ R{-}S\cdot \qquad (2.15)$$

$$\longrightarrow \quad R{-}\overset{\cdots}{\underset{\cdots}{S}}{}^- \ + \ {}^+\overset{\cdots}{S}{-}R\cdot \qquad (2.16)$$

The possibility of forming a cation helps us to understand a number of fragmentation reactions of thiols, thioethers, thioketones, and polysulfides:

$$R{-}S{-}CR_3 \quad \longrightarrow \quad R{-}S^+ \ + \ :CR_3 \qquad (2.17)$$

If the residue R is H, the formation of a sulfur cation is conceivable in principle because of the special qualities of the sulfur. In practice this would be indentical to the desulfurization, and in this special case, the reduction of the thiol to the hydrocarbon:

$$
\begin{array}{ccccc}
HS-CR_3 & \longrightarrow & HS^+ & + & ^-:CR_3 \\
 & & \updownarrow & & \\
\tfrac{1}{8}S_8 & \longleftarrow & [S] & + & HCR_3
\end{array}
\qquad (2.18)
$$

In principle this reaction is reversible, but under the reaction conditions normally applied, sulfur becomes irreversibly stabilized to $S_8$.

In this reaction an essential difference between thiols and alcohols becomes apparent. The strongly electronegative oxygen does not leave the linking electron pair to the carbon atom; only sulfur is able to do this.

The sulfenyl cation searches for a nucleophilic reagent and like the S-anion it is a highly reactive, activated ionic intermediate. If no partner is available desulfurization results; that means stabilization and formation of elemental sulfur:

$$
R-S_x-S-S^+ \longrightarrow R-S_x-S^+ + \ddot{:}\dot{S} \; [\longrightarrow \tfrac{1}{8}S_8] \qquad (2.19)
$$

Formally the spontaneous desulfurization leads to the formation of $S_1$ at equilibrium. All attempts, however, to detect this species or another similar carbene-like singlet sulfur have failed.

### 2.3.2 Homolytic Scission of the S—S Bond

By energy (thermal, radiation, or chemical energy) elemental sulfur becomes activated in such a way that it not only splits the S—S bond and enters into reactions with itself, but also reacts with other chemical compounds or becomes attacked by these reagents.

About homolytic scission of the S—S bond see ref. 6; for thiyl radicals see also ref. 28.

On heating of elemental sulfur a homolytic scission of the S—S bond can be assumed to take place, predominantly leading to linear biradicals and cyclic sulfur molecules other than $S_8$:

$$
\text{Cyclo}-S_8 \rightleftharpoons \cdot S-S_6-S\cdot \rightleftharpoons \cdot S-S_6-S-S_7-S\cdot
$$
$$
\begin{array}{ccc}
 & \searrow \qquad \swarrow & \\
 & \cdot S-S_x-S\cdot \longrightarrow \text{cyclo}-S_x &
\end{array}
\qquad (2.20)
$$

S-radicals have been detected above $180°$, with a maximum concentration at about $300°$.

Induced by the initiator radical, thus, from elemental sulfur polymeric sulfur $S_X$ is formed, which is indicated by the increasing viscosity of the liquid sulfur and detected by direct isolation of $S_X$ by preparative methods.

The direct correlation of increased temperature and growing reactivity is indicative of the intermediate formation of radicals.

The strong tendency comprised in an S—S bond to act as an acceptor of electrons should be assumed on the basis of the above formulated biradicals

$$\cdot\ddot{S}{-}\cdot\ddot{S}_x{-}\ddot{S}\cdot \quad \underset{(\longrightarrow)}{\rightleftharpoons} \quad \ddot{S}{=}S_x{=}\ddot{S}^{\,+}$$

The result should be a stabilization or a transition into a dipolar hybrid molecule.

Another peculiarity is seen, because sulfur is easily polarized: under energetic conditions or in the presence of polar solvents the originally formed radical intermediates can also be transformed into ionic intermediate products:

$$\cdot\ddot{S}{-}S_x{-}S_3{-}\ddot{S}\cdot \quad \rightleftharpoons \quad \ddot{S}{=}S_x{=}S_3{=}S^{\,+}$$

$$\cdot\ddot{S}{-}S_x{-}\ddot{S}^{\,+} + \cdot\ddot{S}{-}\ddot{S}{-}\ddot{S}\cdot \quad \longrightarrow \quad \text{further cleavage} \qquad (2.21)$$

$$\ddot{S}{-}\ddot{S}^{\,+} + \cdot\ddot{S}\cdot$$

These multifarious possibilities are typical of elemental sulfur and explain its multiple reactivity. The dualistic character of sulfur also in homolytic or heterolytic scission should be taken into account for synthetic applications. In this connection the experimental experience that in principle a reaction initially radical in character can continue *via* ions, should be emphasized. The cleavage of sulfur-sulfur bonds by photolysis is also known [3]. Hereby the formation of high molecular chains of sulfur can be assumed because the viscosity (*e.g.*, in benzene) rapidly increases and the solubility decreases. This photosulfur is unstable and converted like other unstable modifications into cycloocta-S. Peculiarities of the reactivity of the photosulfur are not described [6]. Conversion and stabilization of thermodynamically unstable species of sulfur (*e.g.*, $S_6, S_{12}, S_x$) into the stable cycloocta-S also occur under conditions which favor homolytic scission of S—S bonds, thus, *e.g.*, at elevated temperatures in the presence of radicals or sources of radicals and by the action of light. Amines (Sect. 2.3.3) also catalyze this conversion.

Impurities, light and elevated temperatures favor the stabilization of unstable modifications of sulfur. About radical reactions in the sulfur series and radical additions of sulfur compounds to unsaturated hydrocarbons see refs. 6,15,29,30,31.

### 2.3.3 Heterolytic Scission of the S−S Bond

It is noteworthy that elemental sulfur as well as polysulfides and sulfanes are attacked and split easily by nucleophilic agents. In this regard common nucleophiles are: the cyanide ion ($^{\ominus}$CN), the hydroxyl and mercaptide ions ($^{\ominus}$OH, $^{\ominus}$SH, $S^{2\ominus}$), the sulfite ion ($^{\ominus}SO_3H$) as well as ammonia and amines (:$NR_3$), phosphines (:$PR_3$) and carbanions ($^{\ominus}CR_3$). This attack by Lewis bases or anions is the preferred reaction of sulfur, see refs. 6,15,24,32,33.

Accordingly, the S−S bond behaves as a Lewis acid and is extremely electrophilic.

The intermediate state during the addition of the nucleophilic partner $X^{\ominus}$ includes an expansion of the electron shell of the attacked S atom and therefore (possibly synchronous and at the rate determining step) cleavage of the ring occurs:

$$(2.22)$$

The result is an unsymmetrical octasulfane derivative with heterogeneous S atoms in the chain.

Further attack by an excess of $X^{\ominus}$ preferentially occurs then at the most electrophilic position of the intermediate, namely, in this example, due to the electron withdrawing action of the residue X at the opposite end of the chain. In contrast to the other positions which are provided with an electron decet, this particular S atom is only surrounded by an electron octet:

$$(2.23)$$

The decomposition takes place stepwise *via* heptasulfane, hexasulfane, *etc.*, up to the monosulfane derivative. It can be assumed that these subsequent reactions are not rate determining. This is the linearization reaction starting from the decet and measurable by kinetic methods.

In the course of nearly all these reactions elemental sulfur is decomposed to the monosulfane derivative:

$$S_8 + 8\,X^\ominus \rightleftharpoons 8\,{}^\ominus S - X \qquad (2.24)$$

Cyanide ($X = CN$) reacts with sulfur to form thiocyanate ions:

$$S_8 + 8\,{}^\ominus CN \rightleftharpoons 8\,{}^\ominus S - C \equiv N \qquad (2.25)$$

The rate constant at $40°$ is sufficiently high to titrate sulfur with cyanide.

The action of sulfite on sulfur yields monosulfane monosulfuric acid (thiosulfate):

$$S_8 + 8\,H_2SO_3 \rightleftharpoons 8\,HO_3S - SH \qquad (2.26)$$

As this reaction readily occurs even at $20°$ quantitatively, it is suited for determining elemental sulfur.

In a special way, however, nucleophilic partners can also activate elemental sulfur for certain subsequent reactions. For this purpose primary and secondary amines are especially suitable. These nucleophiles activate sulfur even at low temperatures in such a way that subsequent reactions often occur even at room temperature.

In principle the primary attack in these cases should follow a reaction path similar to that described for the other nucleophilic agents:

$$(2.27)$$

In this respect experimental experience shows that primary and secondary amines in conjunction with sulfur produce a highly reactive intermediate system consisting of sulfide ions as well as sulfenyl cations. About this problem see ref. 34a.

In this connection the possibility of concerted occurrence of radical and ionic intermediate states should be mentioned. In case of sulfur reacting with amines, not only ring cleavage followed by decomposition *via* ions, but also formation of linear polysulfur chains is possible. During the course of this conversion homolytic cleavage and decomposition may occur:

$$R_2N-S-S_X-H \rightarrow R_2N-S\cdot + \cdot S_X-H \qquad (2.28)$$

Actually the esr spectra of such a colored mixture shows signals caused by a certain amount of radical concentration. However, this finding is not definitively confirmed.

Elemental yellow cycloocta-S dissolves in primary and secondary amines with colors changing from deep yellow *via* orange-red to green. Conductivity measurements have shown that the solution contains different ions which slowly change their concentrations. These reactions seem to be reversible because on diluting these solutions with aqueous acids elemental sulfur regenerates in high yields.

Under comparable conditions tertiary amines do not react with elemental sulfur to form ions, nor do they noticeably activate sulfur. Evidently this is not only due to steric factors or the different nucleophilic power but is caused by the absence of H-atoms at the amino nitrogen leading to the final stabilization.

In additions to amines the $S_8$ ring can also be attacked and activated by other bases of corresponding nucleophilic strength, including bases formed during sulfurization, such as sulfide, polysulfide, and sulfite [6,24,35].

Specially thiophilic are carbanions potentially existing in metalorganic compounds [35]:

$$\overset{\delta^+}{Me} - \overset{\delta^-}{CR_3}$$

2.2

Carbanions should exist potentially in all Y–C bonds with residues Y more positive than the carbon atom:

$$Y\overset{\frown}{-}CR_3 \rightarrow Y^+ + {}^-:CR_3 \qquad (2.29)$$

A rough estimation by means of Pauling units shows that the donor Y should be more electropositive than 2.5 ($sp^3$–C:2.5; $sp^2$–C:2.8; $sp$–C: 3.3). This is the case with H(2.1), with all metals (*e.g.*, Li: 1.0; Mg: 1.2; Al: 1.5), with boron (2.0), silicon (1.8), phosphorus (2.1), and sulfur (2.5).

Therefore the following classes of substances are appropriate chemical species to attack and to react with sulfur:

| | |
|---|---|
| CH acids | $H—CR_3$ |
| metalorganic compounds | $Me–CR_3$ |
| boron organic compounds | $B—CR_3 \longrightarrow {}^{\ominus}CR_3$ |
| silicon organic compounds | $Si—CR_3$ |
| phosphorus organic compounds | $P—CR_3$ |
| sulfur organic compounds | $S—CR_3$ |

Little is known about these topics. Therefore, a wide field of synthetic possibilities and of interesting new synthetic routes is open.

## 2.4 REACTIONS OF SULFUR

The direct action of elemental sulfur on organic compounds has long been of scientific and technical interest. Besides oxygen, elemental sulfur alone or in the presence of basic additives is a frequently used agent for oxidations and dehydrogenations.

The sulfurization can also be carried out with a mixture of sulfur and $SO_2$ at about $200°$. Here, $SO_2$ oxidizes the $H_2S$ produced, and thus sulfur is recovered.

Even at low temperatures a reactive form of sulfur can be obtained by the reaction of hydrogen sulfide with oxygen or air:

$$2H_2S + O_2 \rightarrow 2H_2O + 2S* \qquad (2.30)$$

In the presence of hydrocarbons this mixture of $H_2S$ and $O_2$ reacts to form the dehydrogenated hydrocarbon and $H_2S$ is recovered:

$$R–CH_2–CH_3 \xrightarrow[-H_2S]{[S]} R–CH = CH_2 \qquad (2.31)$$

Oxides of alumina or cerite catalyze this special procedure.

Very often oxidation reactions with sulfur proceed more selectively than with oxygen. The resulting hydrogen sulfide — split off during dehydrogenation — is comparatively easily removed. By analytical determination of the $H_2S$ produced the progress of the reaction can be followed. In this way, for instance, alkanes can be dehydrogenated to olefins and the olefins — more reactive than alkanes — may again react with sulfur to form sulfur-free and sulfur-containing products, *e.g.*, sulfides, polysulfides, thiols, thiones, aromatic hydrocarbons, heterocycles, or undefined tars. Under special conditions the oxidation with sulfur leads to the formation of aldehydes, carboxylic acids, amides, or thioamides.

Vigorous conditions effect the complete oxidation of hydrocarbons and the formation of carbon disulfide and hydrogen sulfide. Subsequent reactions of the sulfurization may be dehydrogenation, cyclization, dehydrocyclization, aromatization, and the formation of new C–S or C=S bonds.

Especially during the last several years investigations in this field provided simple routes to synthesize new sulfur compounds; furthermore, efficient methods to prepare known classes of sulfur compounds, especially heterocycles, have been found. It has been shown that sulfur is a sufficiently strong oxidant to attack most of hydrocarbons as well as other carbon compounds[6]. About the action of elemental sulfur on organic compounds see refs. 15,34,37,38,39,40. In this connection see also Sect. 2.3 on activation of sulfur and the discussion given in ref. 36.

Sulfur not only can act as an oxidizing but also as a reducing agent. For instance, in the presence of air or oxygen molten sulfur undergoes rapid oxidation to sulfur dioxide. It has been established experimentally that a liquid phase reaction occurs between dissolved oxygen and the molten sulfur[41].

It is not known whether and to what extent radical reactions are actually involved. The overall result is the formation of unstable single bonds followed by splitting off $SO_2$:

$$\text{S}_x \xrightarrow{\ O_2\ } \ \text{S}_x SO_2 \longrightarrow SO_2 + \text{S}_x \qquad (2.32)$$

During the last several years the natural sulfur cycle has been cleared up[42].

This knowledge on natural sulfurization reactions and corresponding desulfurization is important for discussions of sulfur redox reactions carried out in the laboratory.

The main steps are the oxidation of elemental sulfur and organic sulfur compounds to form $SO_4{}^{2-}$ and the reduction of the sulfate ion or of elemental sulfur:

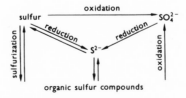

About oxidation of elemental sulfur and sulfur compounds by thiobacilli see ref. 43.

### 2.4.1 Thermal Reactions of Sulfur with C—H Bonds

Sulfur, like oxygen, reacts with alkanes only at elevated temperatures. The reaction starts at about 140 to 150° with formation of $H_2S$. On a preparative scale, however, it runs only above 200 to 300° and for lower alkanes even higher temperatures are necessary:

$$\text{alkane} \xrightarrow[-H_2S]{S_8, \ > 200^\circ} \text{dehydrated products} \qquad (2.33)$$

e.g.,

$$\text{cyclohexane} \xrightarrow{S_8, \ >200^\circ} \text{benzene} \qquad (2.34)$$

If methyl-substituted aromatic compounds are heated with sulfur, products with C=C bonds are formed. In this way, for instance,

$$R-CH_3 \longrightarrow 1/2 \ R-CH=CH-R \qquad (2.35)$$

starting from p-toluic acid, trans-4,4'-stilbene-dicarboxylic acid is easily prepared:

$$2 \ HOOC-C_6H_4-CH_3 \xrightarrow[-2H_2S]{S_8, \ 270^\circ} \begin{array}{c} HOOC-C_6H_4-CH \\ \parallel \\ HC-C_6H_4-COOH \end{array} \quad (2.36)$$

Methylene compounds react in an analogous manner (e.g., diphenylmethane → tetraphenylethylene).

The exact pathways of these reactions are yet unknown. The oxidation of the hydrocarbons is balanced by the reduction of sulfur to $H_2S$. This reaction can be utilized for the generation of $H_2S$ in the laboratory; advantageously a mixture consisting of solid paraffins and sulfur (mostly pressed to pilings) is heated.

The $H_2S$ thus formed is relatively pure, because under the reaction conditions the dehydration products of paraffins are not yet volatile.

All attempts to activate elemental sulfur in such a way that similar reactions readily proceed at low temperatures failed. Only olefins react below 200° in a relatively high yield. In all cases main reactions are dehydrations followed by dimerization, cyclization, and aromatization. Especially in the chemistry of natural products the dehydration with sulfur at about 220° was of interest in characterizing and isolating the skeleton of molecules. About this and the analogous dehydration with selenium ($>300°$) extensive data are available [6]. At higher temperatures, besides sulfur-free dehydrogenation products and $H_2S$, multi-sulfur-containing compounds arise (thiols, sulfides, polysulfides, tars, $CS_2$, thiophenes, dithiole-thiones).

Products and yields depend on temperature and holding period of the mixture in the hot region. With long periods thermodynamically stable S-compounds are mainly formed.

At high temperatures ($>500°$), analogous to the combustion ($\rightarrow CO_2 + H_2O$), sulfur oxidizes hydrocarbons to form $CS_2$ and $H_2$:

$$\text{Hydrocarbon} \xrightarrow{\;S_8,\;>500°\;} CS_2 + H_2S \qquad (2.37)$$

At very long holding periods and temperatures above $1000°$, free hydrogen has been detected (probably resulting from the cleavage of $H_2S$).

The end-products of the sulfurization depend not only on the temperature and the holding period, but also on the structure of the hydrocarbons.

When heated with sulfur many organic compounds react or cyclize, respectively, to form thiophenes or 1,2-dithiole-3-thiones ("trithiones"). Thiophenes are formed preferentially — mostly together with dithiole-thiones — if a system of 4 carbon atoms is applied:

$$(2.38)$$

On the other hand, dithiole-thiones are generated preferentially if there is a 3-carbon system present in the skeleton:

$$(2.39)$$

Generally, olefins are much more reactive than alkanes. Even around $130°$ olefins react with sulfur to form dithiole-thiones of dehydration products. The reaction of methylcyclohexene may be a good example [45]:

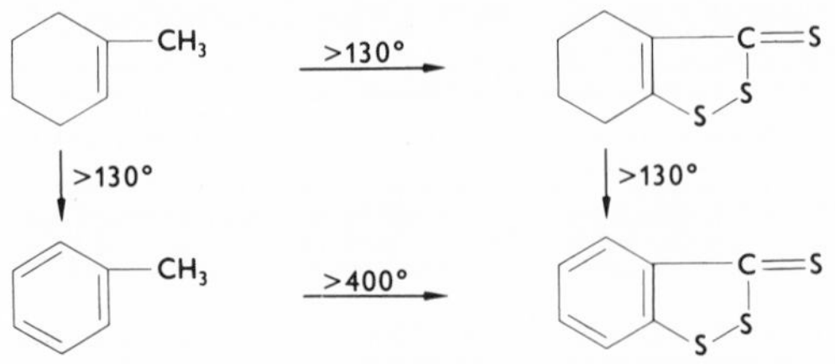

On heating hydrocarbons with sulfur, together with thiophenes, formation of dithiole-thiones is favored. But today other efficient methods for preparation of dithiole-thiones are available at room temperature or slightly elevated temperature. For details see ref. 44.

Besides thiones and dithioles, α-dithiopyrones are also formed as end-products in thermal sulfurization reactions. Thus, thiopyran-2-thione was first prepared by dehydrogenation of thiocyclohexane or thiocyclohexene with sulfur [46]:

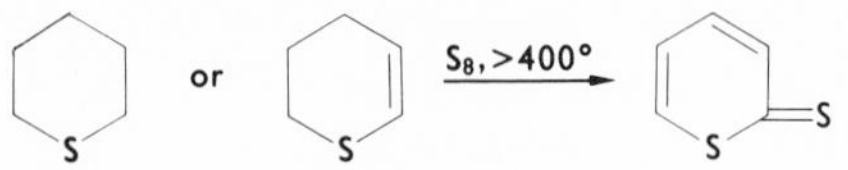

$$\text{(2.40)}$$

In spite of extensive efforts, the mechanisms and intermediate products of thiolation reactions are still highly speculative. In this connection see ref. 6 which has more than a thousand references and compare the above mentioned discussions on activating elemental sulfur. These reactions have been intensively studied because they are models for technical procedures, e.g., for vulcanization. The vulcanization is an extremely complex reaction in which polysulfide chains crosslink rubber and rubber-like molecules (see Chapter 3).

In principle there are only two possibilities for C—S bond formation by the action of sulfur on alkanes or C—H bonds and thus for the main steps in thiolation. On the one hand, it could be a direct insertion reaction of activated sulfur, formed by energy excitation.

This means a direct thiolation, such as

$$R-CH_3 \xrightarrow{S} R-CH_2-SH \tag{2.41}$$

This reaction should produce a larger variety of by-products by subsequent reactions. Furthermore, thiocarbonyl compounds are known to lose sulfur and to desulfurize and hence an insertion into S—H bonds must be considered:

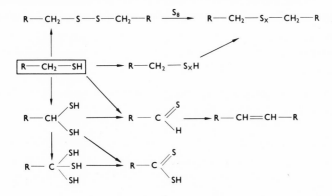

On the other hand the initial hydrogen abstraction and the resulting formation of a carbon radical, followed by stabilization (for instance by dimerization, formation of an olefin or subsequent reactions with the thiyl radical and so on) may have to be considered. However, intensive

$$R - CH_3 + S \rightarrow R - CH_2 \cdot + HS \cdot \qquad (2.42)$$

investigations will be necessary to establish the mechanism of these reactions and to make a choice among the controversial postulates. In this respect it is desirable to have a few fundamental new theories and considerations to provide new useful working hypotheses for current research.

Often the course of these complex reactions can also be influenced by the most different factors, especially by the analytical methods used for analyzing the products. An extreme example is the action of elemental sulfur on olefins and the mechanisms discussed on the basis of the isolated products.

Alkenes react with sulfur even below $160°$ to form different polysulfides. Up to $160°$ $H_2S$, R–S–H, monosulfides and thiiranes are generated only in small amounts. Virtually no new C–C bonds are formed.

At elevated ($>$ about $160°$) or high temperatures the formation of those products predominates which also result from reactions of sulfur with alkanes (thiophenes, dithiole-thiones, carbon disulfide, hydrogen sulfide, *etc.*)[6,37].

At first it was suggested that the reaction of sulfur with olefins at low temperatures occurs *via* allylic radicals. These carbon radicals were said to attack sulfur to produce $R-S_X\cdot$ which abstracts hydrogen from another olefin molecule. In an ionic reaction the resulting $R-S_xH$ should add to another alkene molecule to form products according to the Markovnikov rule:

$$
2 \quad
\begin{array}{c}
CH_2 \\
\parallel \\
C \\
\diagup \quad \diagdown \\
\qquad CH_2 \\
\qquad \vert
\end{array}
\longrightarrow
\begin{array}{c}
CH_2 \\
\parallel \\
C \qquad\qquad CH_3 \\
\diagup \quad \diagdown \qquad \vert \\
\qquad CH-S_x-C \\
\qquad \vert \qquad\qquad \diagdown CH_2 \\
\qquad\qquad\qquad\qquad \vert
\end{array}
\qquad (2.43)
$$

Careful analysis with modern methods, kinetic studies, the study of the influence of solvents, temperature, and light have shown a very complex picture of intermediates, pathways, and mixtures of end-products.

These studies led to the postulation of an ionic chain mechanism. The initiation involves heterolytic scission of the S–S bond to a sulfide anion and a sulfonium ion (see Sect. 2.3.3):

$$S_8 \rightarrow R-S^- + {}^+S-R \qquad (2.44)$$

The $R-S^+$ adds to a double bond to produce a carbonium ion. That means direct nucleophilic attack on olefins by sulfur:

$$
\begin{array}{c}
CH_2 \\
\parallel \\
CH \\
\diagup
\end{array}
+ \; {}^+S-R
\longrightarrow
\begin{array}{c}
CH_2 \quad + \\
\vert \quad\; \diagdown S-R \\
CH \diagup \\
\diagup
\end{array}
\text{ or }
\begin{array}{c}
CH_2-S-R \\
\vert \\
{}^+CH \\
\diagup
\end{array}
\qquad (2.45)
$$

The propagation involves the formation of new $R-S^+$ and further steps as for instance:

$$A + S_8 \longrightarrow R'-S^+ \qquad (2.46)$$

$$
A + H_2C'=C'H \longrightarrow
\begin{array}{c}
A-CH_2-\overset{+}{C}H \\
\vert \qquad\qquad \vert \\
B
\end{array}
\qquad (2.47)
$$

$$B + S_8 \longrightarrow R''-S^+ \qquad (2.48)$$

The termination could be:

$$
\left.
\begin{array}{l}
A \qquad + \; {}^-S-R \\
B \qquad + \; {}^-S-R \\
R'-S^+ + \; {}^-S-R \\
R''-S^+ + \; {}^-S-R
\end{array}
\right\} \longrightarrow \text{ polysulfides } etc.
\qquad (2.49)
$$

Of which kind and nature the sulfur intermediates actually undergo reaction at the respective temperatures is uncertain and mostly speculative, too.

In principle radical as well as ionic species could be the basic products. Most investigators assume diatomic sulfur $S_2$[46a] or atomic sulfur $S_1$ to be the reacting species at high temperatures. Although this is also speculative, we will mention some general aspects.

In fact, atomic sulfur produced by non-thermal methods reacts selectively with alkanes, alkenes and alkynes to form very specific sulfur products. Thus, a characteristic feature of the excited singlet sulfur is its ability to insert into C—H bonds.

This insertion reaction proceeds, if, *e.g.*, the photolysis of COS is carried out in the presence of alkanes, such as ethane, propane, isobutane, cyclopropane, cyclobutane, and cyclopentane[20].

In all these cases, in addition to the major products, CO and sulfur, the corresponding mercaptans are formed in approximately statistic distribution:

$$R_3 C\text{-}H \xrightarrow{[S]} R_3C\text{-}SH \qquad (2.50)$$

It is understandable that drastic conditions in the presence of an excess of sulfur lead to the subsequent production of thiols and finally to the above mentioned set of products. Experimental experience shows that not only new C—C bonds (cyclization, dehydrocyclization), but also new C—S, C—$S_X$—C, and C=S bonds are formed on a large scale.

In contrast to the reaction with alkanes the conversion of olefins with atomic sulfur is of preparative interest because of the formation of enethiols:

vinyl mercaptan        thioaldehyde        (2.51)

enethiol        thioketone        (2.52)

For the reactions, photolysis of, for instance, COS may be carried out in the presence of an appropriate olefin[20].

Monomeric aliphatic thiolaldehydes are as yet unknown today. In spite of many efforts this class of substances has not been successfully synthezied or characterized[47].

Therefore the observed amazingly high stability of vinyl mercaptans was of special interest. Vinyl mercaptans are the enethiol forms[47a] of the monomeric thioaldehydes. Aliphatic thioketones are generally unstable too, but can be isolated in the monomeric thione form as well as in the tautomeric enethiol form. Generally aliphatic unconjugated thiocarbonyl compounds are unstable in the monomeric form to such an extent because sulfur does not easily form a $p\pi - p\pi$ double bond with the carbon atom.

This is due to the fact that these two elements belong to different groups in the periodic table.

$$\begin{array}{c}\diagdown\\ \diagup C^{2p_\pi} = S^{3p_\pi}\diagdown\\ \big\uparrow\big\downarrow \qquad\qquad \text{chains, cycles}\\ \diagdown\\ \diagup C - SH \diagup\\ \text{enethiol} \end{array} \qquad (2.53)$$

Thus, the thiocarbonyl group tends to turn into a C–S single bond. This may occur by enethiolization or by dimerization, oligomerization, or polymerization. The subsequent products are either cyclic or chain (*cf.* Chapter 5).

Therefore, formation of simple thiones by heating methylene compounds with sulfur is not favored and hence this procedure is not significant [47]:

$$\begin{array}{c} R \\ \diagdown \\ \diagup CH_2 \longrightarrow \begin{array}{c} R\\ \diagdown\\ \diagup CH-SH\\ R \end{array} \longrightarrow \begin{array}{c} R\\ \diagdown\\ \diagup C{=}S\\ R \end{array} \qquad (2.54)\\ \\ \begin{array}{c} R \quad SH\\ \diagdown \diagup\\ C\\ \diagup \diagdown\\ R \quad SH \end{array} \end{array}$$

In general, products formed in the subsequent reactions predominate. Only thermodynamically stable thiones (certain aromatic types, *e.g.*, thiobenzophenones) are available in this way.

At present the thermal reaction of sulfur with acetylenes is not so important as are those with alkanes and alkenes (but see the base-catalyzed reaction: Sect.2.4.2.1). The thermal sulfurization of acetylene and substituted acetylenes leads to a reaction mixture which, however, is preparatively of no advantage.

Under extreme working conditions ($450^\circ$) the formation of $H_2S$, $CS_2$, thiophenes, 1,2-dithiole-3-thiones, and 1,3-dithiole-2-thiones has been observed.

The reaction of atomic sulfur with acetylenes mainly leads to a solid polymer besides $CS_2$, thiophene, and benzene.

### 2.4.2 Base-Catalyzed Reactions of Elemental Sulfur with C–H Bonds

Whereas only heat-resistant products are isolable after reacting sulfur on organic compounds at high temperatures, the base-catalyzed reaction of sulfur with suitable reactants – frequently taking place at room temperature — permits the isolation of thermally labile intermediates existing in the equilibrium. In these labile intermediates, $S_8$ or $S_X$ may be wholly or partly incorporated.

In all cases the first step in base-catalyzed sulfur reactions should be the cleavage of the $S_8$ ring, a reaction occurring at mild conditions. However, little is known about the operating mechanism of base-catalyzed thiolation reactions, particularly the action of sulfur on compounds containing nucleophilic C atoms. Thus, it is not even always clear whether the sulfur is attacked directly by the C-base and transferred to the substrate as activated sulfur.

### 2.4.2.1 Thiolation of Carbanions and Acetylenes

In activating elemental sulfur we have seen that carbanions are very suitable nucleophiles for attacking sulfur while at the same time sufficiently reactive to take up the sulfur at room temperature. Therefore if an auxiliary base is added for the abstraction of the proton, CH acids can readily form thiolation products:

$$R_3C-H \xrightarrow{-H^+} R_3C:^- \xrightarrow{S_8, H^+} R_3C-S_xH \qquad (2.55)$$
$$\underset{CH\,acid}{}$$

About the action of elemental sulfur on CH acids, see ref. 34. Organometallic compounds are also well known to be active chemical systems and potential carbanions [35].

Here we shall see only acetylenes as examples. It is well known that acetylenes and monosubstituted acetylenes in the presence of an auxiliary base react as CH acids with electrophiles. Acetylenes in the form of their salts, namely acetylides, are also readily thiolized. For references see citations in refs. 34,38.

$$R-C\equiv CH \xrightarrow{-H^+} R-C\equiv C:^- \xrightarrow{S_8} R-C\equiv C-S^- \qquad (2.56)$$

Because of its theoretically conceivable prototropic equilibrium the protonated thiolation product should be an extremely interesting intermediate.

$$R-C\equiv C-SH \rightleftharpoons R-CH=C=S \rightleftharpoons R-C=CH \underset{S}{\overset{}{\diagdown\diagup}} \qquad (2.57)$$
$$\underset{thioketene}{}$$

Polymers, observed in sulfur reactions with acetylenes could be derived from these intermediates. Hitherto the 3-membered ring is unknown.

In spite of many efforts, all attempts to prove the real presence of equilibrium between the mercaptoacetylene and the thioketene have failed [47a].

On acidifying, the thiolation product becomes stabilized especially by dimerization to dithiofulvenes: *e.g.*, in the presence of acids or other H-active compounds having a $pK_a$ value $<18$, sodium phenylethinyl thiolate dimerizes very rapidly to form 2-benzylidene-4-phenyl-1,3-dithiole:

$$2 \; C_6H_5 - C \equiv C - SH \longrightarrow C_6H_5 - C \cdots S \quad \text{...}$$

(2.58)

In the strong acid, an additional stabilization results from the formation of the 1,3-dithiolium salt.

Thus, dithiofulvenes and the corresponding dithiolium salts became easily accessible.

The reaction has general applicability. If selenium is used instead of sulfur the very stable selenium analogues are similarly obtained.

Some interesting results were obtained in the attempts to catalyze the reaction by means of amines at $20°$: addition occurs, followed by stabilization to form the thioamide:

$$R - C \equiv C - SH \rightleftharpoons R - CH = C = S \xrightarrow{HNR_2} R - CH_2 - C \overset{S}{\underset{NR_2}{\diagup}}$$  (2.59)

$$\xrightarrow{R - SH} R - CH_2 - C \overset{S}{\underset{SR}{\diagup}}$$  (2.60)

The similarity to the Willgerodt-Kindler reaction (see Sect. 2.4.2.2) is obvious, particularly since it is known that the reaction of phenylacetylene, sulfur and morpholine at about $150°$ leads to phenylthioacetomorpholide.

The analogous additions of mercaptans to alkynethiolates leads to dithioesters.

### 2.4.2.2 Willgerodt Reaction

A well-known base-catalyzed reaction of sulfur and methyl or methylene groups is the Willgerodt reaction. Willgerodt discovered[48] that ammonium polysulfide at $150°$ to $200°$ under pressure reacts with aryl alkyl ketones to form amides with equal numbers of carbon atoms:

$$C_6H_5 - CO - (CH_2)_n - CH_3 \xrightarrow{(NH_4)_2 \, S_x} C_6H_5 - CH_2 - (CH_2)_n - CSNR_2 \quad (2.61)$$

In this reaction a carbonyl group in the middle of the chain becomes reduced and the terminal methyl group becomes oxidized.

Kindler[49] improved this procedure by applying primary and secondary amines at about 130° or higher temperatures. This redox-amide reaction, intensively studied by several teams, can be discussed in two fundamentally different ways. On the one hand the primary step could be the formation of an enamine (or ketimine) resulting from the action of the amine on the ketone. On the other hand the primary reaction could be the activation of sulfur and its reaction with the ketone to form a mercaptoketone (*cf.* Sect. 5.2.3).

Therefore reactions of elemental sulfur with enamines, ketimines, and CH acids are of special interest. For a review see ref. 34.

### 2.4.2.2.1 Enamines

Enamines are sufficiently basic and nucleophilic to activate elementary sulfur without any additional catalyst and to react with sulfur to form a thiolation product:

$$S_8 \xrightarrow{\text{enamine}} \text{activated sulfur} \qquad (2.62)$$

$$\qquad (2.63)$$

Undoubtedly this intermediate is extremely reactive and should undergo stabilization. Besides, by dimerization it should become very stable by a special rearrangement, the formation of the thioamide:

$$\qquad (2.64)$$

This rearrangement involves a nucleophilic attack of an amine on a thiocarbonyl group or enethiol.

Actually even at room temperature elementary sulfur was found to react with suitable enamines (preferably in dimethyl formamide) and, thus, directly to form thioamides:

$$\qquad (2.65)$$

The similarity to the original Willgerodt-Kindler reaction (that is, the action of sulfur on the enamine components ketone and amine at about $130°$) is obvious, particularly since it was shown that those catalysts which favor enamine formation also give increased yields of thioamides. In this respect the high temperature required for the Willgerodt-Kindler reaction seems to be evidently necessary for the enamine formation. Using the enamine itself, thioamides are formed even at $20°$.

The Willgerodt reaction is especially interesting for aryl ketones since aryl ketones with longer aliphatic chains can also be converted into thioamides:

$$\text{Aryl} - CO - (CH_2)_n - CH_3 \xrightarrow[130°]{S_8, HNR_2} \text{Aryl} - (CH_2)_{n+1} - C \underset{NR_2}{\overset{S}{\diagup}} \qquad (2.66)$$

In this reaction the carbonyl group is formally reduced to a methylene group and the terminal methyl group is oxidized to a carboxylic derivative.

There have been many investigations of the mechanism and many groups have actually engaged in studying this conversion.

In this connection, an important observation appears to be that the reaction of enamines of the following type with sulfur at room temperature also gives thioamides:

$$\text{Aryl} - \underset{NR_2}{\overset{}{C}} = CH - (CH_2)_n - CH_3 \xrightarrow[20°]{S_8, DMF} \text{Aryl} - (CH_2)_{n+2} - C \underset{NR_2}{\overset{S}{\diagup}} \qquad (2.67)$$

However, enamines with longer chains give thioamides in much lower yields. An interesting observation is that enamines with a double bond in the middle of the molecule are isomerized with a catalytic amount of sulfur even at $20°$:

$$(2.68)$$

The double bond eludes conjugation. Using an equimolar amount of sulfur the equilibrium is irreversibly displaced to the right, as the enamine with the terminal double bond is thiolized and then irreversibly stabilized by thioamide formation.

## 2.4.2.2.2 Mercaptoketones

If the Willgerodt reaction starts by the activation of sulfur and the thiolization of the ketone is the initial main step, a mercaptoketone may be produced, and only indirectly detected:

$$R-CO-CH_3 \xrightarrow{S^*} R-CO-CH_2-SH \qquad (2.69)$$

With an excess of sulfur it is converted either into the disulfide (actually found in the reaction mixture) or into the thioaldehyde, formed in a direct way or *via* the geminate dithiol:

$$(2.70)$$

Sulfides of this type behave toward nucleophilic agents like any other S—S bond. Amines present, in the system, split off the bond to form the sulfenamide and the mercaptoketone:

$$\text{Disulfide} \xrightarrow{HNR_2} R-CO-CH_2-S-NR_2 + R-CO-CH_2-SH \quad (2.71)$$

The thioaldehyde — formed *via* either the mercaptoketone or the geminate dithiol — is unstable in the basic media. With excess amine, the Cannizzaro reaction leads to the oxothioamide and the mercaptoketone:

$$(2.72)$$

In the well-known subsequent reaction, the oxothioamide — detected as a further compound in the reaction mixture — is reduced by hydrogen sulfide to the hydroxythioamide or the proper end-product of the Willgerodt reaction, the thioamide.

In principle, the sulfenamide also could be dehydrated by elemental sulfur to form the thioamide.

About spontaneous base-catalyzed reduction of carbonyl groups by $H_2S$, see refs. 27,38.

2.4.2.2.3 Discussion of the Mechanism

Reviews about the Willgerodt-Kindler reaction [39,50,51,52] show that these complex and experimentally quite different reactions cannot be interpreted by just one mechanism alone (cf. refs. 34,53).

In view of the above mentioned experimental observations and the formation of enamines in the initial step of the reaction, we can interpret one of the main steps of the Willgerodt-Kindler reaction as follows:

$$
\begin{array}{l}
\text{ketone + sec. amine} \xrightleftharpoons{130°} R-\underset{\underset{NR_2}{|}}{C}=CH-CH_3 \\
\qquad\qquad\qquad\qquad\qquad\qquad\Big\updownarrow \text{ isomerization } S_8, 20° \\
R-CH_2-\underset{\underset{NR_2}{|}}{C}=CH-SH \xrightleftharpoons{S_8, 20°} R-CH_2-\underset{\underset{NR_2}{|}}{C}=CH_2 \\
\Big\downarrow \text{ irreversible stabilization} \\
R-(CH_2)_2-C\underset{NR_2}{\overset{S}{\diagdown}}
\end{array}
\qquad (2.73)
$$

The enamine is first formed from the ketone and the secondary amine. Then it is isomerized by the sulfur and subsequently thiolized. Finally the stable thioamide arises. Only the first step, the formation of the enamine, requires elevated temperatures. Isomerization, thiolation, and rearrangement occur at $20°$ as well as, of course, at higher temperatures. The migration of the former carbonyl group to the end of the chain is nothing but an isomerization of the enamine. After this isomerization, oxidation occurs; that is, the irreversible introduction of sulfur into the molecule takes place.

Based on the consideration of primarily occurring thiolization to the mercaptoketone, we feel quite safe in interpreting the main steps of this complex reaction to proceed via thiols and thioaldehydes or sulfenamides, respectively:

$$
\begin{array}{ccc}
\text{ketone + } S_8/\text{amine} & & \\
\Big\updownarrow & & \\
\text{mercaptoketone} & \rightleftharpoons & \text{disulfide} \\
\Big\updownarrow & & \Big\downarrow \\
\text{oxothioaldehyde} & \longrightarrow & \text{sulfenamide} \\
\Big\downarrow & & \\
R-CO-CSNR_2 & \xrightarrow{H_2S} & R-CH_2-CSNR_2
\end{array}
\qquad (2.74)
$$

Either the oxothioaldehyde undergoes disproportionation by the Cannizzaro reaction, thus yielding the oxothioamide, or the oxosulfenamide is converted to the oxothioamide by dehydration with sulfur. The subsequent reaction is the reduction by $H_2S$.

Naturally, the general mechanisms discussed above are only for borderline cases between numerous transitions and combinations. In principle all the mechanisms discussed in the literatures are based on these two extreme cases.

The close relationship between the reduction of the carbonyl group and the oxidation of the methyl group is demonstrated in the tautomeric equilibria:

$$R-\underset{\underset{O}{\|}}{C}-CH_2-SH \rightleftharpoons R-\underset{\underset{OH}{|}}{C}=CH-SH \rightleftharpoons R-\underset{\underset{OH}{|}}{CH}-C\underset{\diagdown H}{\overset{\diagup S}{}} \quad (2.75)$$

mercaptoketone                                                    hydroxythioaldehyde

$$R-\underset{\underset{O}{\|}}{C}-\underset{\diagdown SH}{\overset{\diagup SH}{CH}} \rightleftharpoons R-\underset{\underset{OH}{|}}{C}=\underset{\diagdown SH}{\overset{\diagup SH}{C}} \rightleftharpoons R-\underset{\underset{OH}{|}}{CH}-C\underset{\diagdown SH}{\overset{\diagup S}{}} \quad (2.76)$$

gem. dithiol                                                      hydroxydithioacid

## 2.4.2.3 Oxidations of Alkylated Aromatic and Heterocyclic Compounds and Olefins by Sulfur

The oxidizing power and ability of sulfur increases sharply with temperature. For oxidation reactions $320°$ is clearly better than $220°$, when oxidations are, as usual, carried out in the presence of basic additives or aqueous alkali hydroxide.

Under these conditions alkyl-substituted aromatics are oxidized by sulfur to afford oxidation products in high yields. This procedure is applicable on an industrial scale, e.g.,

$$C_6H_5-CH_3 \longrightarrow C_6H_5-COOH \quad (2.77)$$

$$H_3C-C_6H_4-CH_3 \longrightarrow HOOC-C_6H_4-COOH \quad (2.78)$$

These reactions have been carried out with different additives under different conditions.

Mostly the starting material is elemental sulfur or a compound splitting off sulfur above $300°$. Hydrogen sulfide formed during this reaction can act as a reducing agent. For instance, p-nitrotoluene reacts with alkaline polysulfides to form p-aminobenzaldehyde:

$$O_2N-C_6H_4-CH_3 \xrightarrow[-H_2O]{KS_xH} H_2N-C_6H_4-CHO \quad (2.79)$$

Generally, alkyl groups bonded to heteroaromatics are more reactive than alkyl groups of aromatic carbon rings. With these systems, oxidation reactions often occur without amines or bases. For details see the original literature cited in ref. 39.

Olefins which undergo thermal sulfurization in base-catalyzed reactions are very easily oxidized. Substituted olefins are also oxidized readily. Thus, cyclohexene carboxylic acid was oxidized by sulfur and caustic soda to pentane-1,5-dicarboxylic acid:

$$\text{(cyclohexene-COOH)} \xrightarrow[\text{NaOH}]{S_8, 350°} \text{HOOC'}-(CH_2)_5-C'OOH \qquad (2.80)$$

### 2.4.3 Reaction of Sulfur with S–H Bonds

Similar to the thiolation of a C–H compound it should also be possible to thiolate a SH group even at low temperature, but the presence of an auxiliary base may be necessary:

$$R_3C-SH \xrightarrow{S_8} R_3C-S_2H \text{ to } R_3C-S_xH \cdot \qquad (2.81)$$

Undoubtedly sulfurization of SH bonds commonly occurs more easily than thiolation reactions of C–H bonds; occasionally both reactions occur synchronously.

This has already been realized actually in the formation of trithiolanes, tetrathianes, pentathiepanes, and hexathiolanes. The McMillan-King reaction may also be regarded as involving thiolation of an intermediate (*cf.* ref. 34). Due to its decisive significance in preparation, transport, and treatment of elemental sulfur, the behavior of the sulfur–hydrogen sulfide-hydrogen polysulfide system is of interest [54]:

$$H_2S + S_x \rightleftharpoons H-S_{x+1}-H \qquad (2.82)$$

In natural sulfur deposits traces of hydrocarbons are responsible for the formation of $H_2S$.

The hydrogen polysulfides (sulfanes) form a homologous series $H_2S_x$ with unbranched sulfur chains.

At room temperature sulfanes are relatively stable liquids of high density. With increasing temperature the $H_2S_x$ molecules become unstable and decompose ($\rightarrow S_8 + H_2S$). A fast decomposition is observed near the boiling point of the hydrogen polysulfides.

Below $100°$ the equilibrium of the mixture of sulfur–$H_2S$ is determined by the solubility of sulfur in $H_2S$. In this unreacted solution sulfanes are not yet formed. But above $120°$ sulfur in considerable quantities reacts with hydrogen sulfide to form $H_2S_X$. In this temperature range the chemical equilibrium is largely determined by the processes taking place in the liquid sulfur phase.

At about $180°$ the equilibrium mixture in the molten sulfur phase consists of 23 $S_8$-units in the $H_2S_X$ chain compared to 100,000 $S_8$-units in the pure sulfur phase (that is, without $H_2S$) at the same temperature. The equilibrium mixture of sulfanes is essentially more reactive than pure sulfur melts, but in this field very few quantitative data have been published.

Increased pressure leads to an increased solubility of sulfur in hydrogen sulfide, and that means elevated reactivity.

### 2.4.4  Base-Catalyzed Oxidation of C—Cl Bonds by Sulfur

Under the conditions of the Willgerodt reaction various chlormethylated aromatic compounds and other halogen-hydrocarbons can be converted into carboxylic amides and carboxylic acids:

$$Cl-CH_2-\text{<benzene ring>}-SO_2NH_2 \xrightarrow[180°]{S_8, NH_3} H_2N-CO-\text{<benzene ring>}-SO_2NH_2 \quad (2.83)$$

Elemental sulfur is effective [40] also for the preparation of thioamides and dithiocarboxylic acids from organic halides.

Thiocarboxylic acids and their derivatives are of growing importance. These compounds are valuable in plant protection, in the field of pharmacological chemistry, and for the synthesis of numerous heterocycles.

*Thiocarboxylic amides* are formed by the action of elemental sulfur and $NH_3$, *prim.* or *sec.* amines with aromatic dichloromethyl compounds. Yields are nearly quantitative:

$$Aryl-CH\begin{smallmatrix}Cl\\Cl\end{smallmatrix} \xrightarrow[-HCl]{amine} \left[ Aryl-CH=NR \text{ or } Aryl-CH\begin{smallmatrix}NR_2\\NR_2\end{smallmatrix} \right] \xrightarrow{sulfur} Aryl-C\begin{smallmatrix}S\\NR_2\end{smallmatrix} \quad (2.84)$$

As intermediates in the reaction azomethines or geminate diamines, respectively, are conceivable. The choice of the most favorable temperature necessary for this reaction ($80°$ to $130°$) and the type of solvent depends mainly on the original amine. Similar to dichloro compounds the commonly more easily available monochloro derivatives also react with sulfur, *e.g.*,

$$C_6H_5-CH_2Cl \xrightarrow[-H_2S, \, -\, HCl]{S_8, HNR_2} C_6H_5-CSNR_2 \qquad (2.85)$$

The amine, which works as an auxiliary base in this reaction, captures the generated $H_2S$ and HCl.

The mechanism is believed to include the initial formation of a C–N bond followed by thiolation and stabilization:

$$R'-CH_2Cl \longrightarrow R'-CH_2NR_2 \rightarrow R'-CH(SH)-NR_2 \cdot R'-C(SH)_2-NR_2 \qquad (2.86)$$
$$\downarrow$$
thioamide

This base-catalyzed reaction is useful also for synthesizing certain aliphatic thiocarboxylic amides. Thus, methylene chloride yields thioformamides:

$$CH_2Cl_2 \xrightarrow[-2\, HCl]{S_8, \text{amine}} H-CSNR_2 \qquad (2.87)$$

In an analogous manner 1,1-dichloroethylene reacts with sulfur and amines to form thioacetamides:

$$CH_3-CHCl_2 \xrightarrow[-2\, HCl]{S_8, \text{amine}} CH_3-CSNR_2 \qquad (2.88)$$

A great variety of synthetic routes result from the conversion of α-halogenated acetic acids and their derivatives. These reactions yield thioamides of the oxalic acid, *e.g.*,

$$ROOC-CHCl_2 \xrightarrow{S_8, \text{amine}} ROOC-CSNR_2 \qquad (2.89)$$

Obviously, all these reactions are closely related to the Willgerodt-Kindler reaction.

*Dithiocarboxylic acids* are made available by reaction of sulfur with C–Cl bonds in the presence of nitrogen-free nucleophilic agents (*e.g.*, alcoholates). Thus, under relatively mild conditions benzyl chloride is oxidized by sulfur/alcoholate to dithiobenzoic acid

$$C_6H_5-CH_2Cl \xrightarrow[-NaCl, -2CH_3OH]{S_8, \ 2 \ NaOCH_3} C_6H_5-CSSNa \qquad (2.90)$$

To date only those dithiocarboxylic acids which are difficult to synthesize by other methods have been synthesized comparatively easily by this method. See ref. 40 for further references.

## 2.5 REFERENCES

1. For articles and reviews, see *The Sulfur Institute Journal*, The Sulfur Institute, Washington-London (on a quarterly basis).
2. G. Nickless (Editor), "Inorganic Sulfur Chemistry" (Elsevier Publishing Company, Amsterdam-London-New York, 1968).
3. B. Meyer, *Chem.Rev.*, **46**, 429 (1964). B. Meyer, "Elemental Sulfur" (Interscience Publishers, New York, 1965).
4. M. Schmidt, *Chemie in unserer Zeit*, **7**, 11 (1973).
5. C.C. Price and S. Oae, "Sulfur Bonding" (The Ronald Press Company, New York, 1962).
6. W.A. Pryor, "Mechanisms of Sulfur Reactions", (McGraw-Hill Book Company, Inc., New York-San Francisco-Toronto-London, 1962).
7. N. Kharasch (Editor), "Organic Sulfur Compounds", (Pergamon Press, Oxford-London-New York-Paris , Vol.1, 1961).
8. N. Kharasch and C.Y. Meyers (Editors), "The Chemistry of Organic Sulfur Compounds" (Pergamon Press, Oxford-London-Edinburgh-New York-Toronto-Sydney-Paris-Braunschweig, Vol.2, 1966; Vol. 3, in press).
9. M.J. Janssen (Editor), "Organosulfur Chemistry" (Interscience Publishers, John Wiley and Sons, New York-London-Sydney, 1967).
10. B. Meyer, "Elemental Sulfur", in reference 2, page 240.
11. J. Donohue, "The Structures of Elemental Sulfur", in reference 7.
12. B. Meyer and T.V. Oommen, "The Color of Liquid Sulfur", in reference 1, winter 1970/71, page 2. *cf.* B. Meyer and coworkers, "The Spectrum of Sulfur and Its Allotropes", in reference 12a, page 53.
12a R.F. Gould (Editor), "Advances in Chemistry, Sulfur Research Trends", series 110 (American Chemical Society, Washington, D.C., 1972).
13. A.V. Tobolsky and W. McKnight, "Polymer Sulfur and Sulfur Polymers" (Interscience Publishers, New York, 1966).
14. For principles see: M. Schmidt and E. Wilhelm, *Inorganic Nuclear Chem Letters*, **1**, 39 (1965). M. Schmidt and E. Wilhelm, *Angew. Chem.*, **78**, 1020 (1966).
15. R. E. Davis, "Mechanisms of Sulfur Reactions", in reference 2, page 85.
16. M. Schmidt, G. Kippschild and E. Wilhelm, *Chem.Ber.*, **101**, 381 (1968). A. Kutoglu and E. Hellner, *Angew.Chem.*, **78**, 1021 (1966); J. Buchler, *Angew.Chem.*, **78**, 1021 (1966).
17. M. Schmidt, H. Block, H.D. Block, H. Köpf and E. Wilhelm, *Angew.Chem.*, **80**, 660 (1968).
18. M. Schmidt and E. Wilhelm, *Chem. Commun.*, 1111 (1970).
18a A. Senning (Editor), "Sulfur in Organic and Inorganic Chemistry" (Marcel Dekker, Inc., New York, Vol.1,2, and 3, 1972).
19. A.B. Burg, "Bonding Characteristics of Sulfur Atom", in reference 7, page 30.

20. O.P. Strausz, "The Reactions of Atomic Sulfur", in reference 9, page 11. O.P. Strausz, "The Addition of Sulfuratomes to Olefins", in reference 12a, page 137. O.P. Strausz, "The Chemistry of Atomic Sulfur", in reference 18a, Vol.2.

21. W.J. Cruickshank and B.C. Webster, Orbitals in "Sulphur and its Compounds", in reference 2, page 7.

22. S. Oae, "3-$d$ Orbital Resonance Involving the Sulfur Atom in Organic Sulfides", in reference 8, Vol.3. J.L. Kice, "The Sulfur-Sulfur Bond", in reference 18a, Vol.1. J. Fabian, "The Quantum Chemistry of Sulfur Compounds", in reference 18a, Vol.3.

23. A. Mangini, "The Valence State of Sulfur in Organic Compounds", in reference 8, Vol.3, in press.

24. M. Schmidt, Österr. Chemiker-Ztg., 64, 226 (1963).

25. F. Feher and H. Münzer, Chem.Ber., 96, 1131 (1963).

26. cf. L. Pauling, "The Nature of Chemical Bond", 3rd ed., (Cornell Univ. Press, 1960).

27. R. Mayer, Quart. Reports on Sulfur Chemistry, 5, 125 (1970).

28. U. Schmidt, Angew.Chem., 76, 629 (1964); U. Schmidt, "Isolation and Characterization of Organic Sulfur Radicals and Radical Ions", in reference 9, page 75. cf. F. Feher and coworkers, Z. Naturforschung, 25 b, 1215 (1970).

29. A.A. Oswald and K. Griesbaum, "Radical Addition of Thiols to Diolefins and Acetylenes", in reference 8, Vol.2, page 233.

30. E. N. Prilezhayeva, "On the Thiylation of Multiple Bonds", in reference 9, page 57.

31. S.G. Cohen, "Action of Mercaptans and Disulfides in Free Radical Reactions", in reference 9, page 33.

32. O. Foss, "Ionic Scission of the Sulfur-Sulfur Bond", in reference 7, page 83.

33. O. Gawron, "On the Reaction of Cyanide with Cystine", in reference 8, Vol.2, page 351.

34. R. Mayer and K. Gewald, Angew.Chem., 79, 298 (1967), Angew.Chem. Internat. Edit., 6, 294 (1967).

34a R.E. Davis and H.F. Nakshbendi, J.Amer.Chem.Soc., 84, 2085 (1962). W.G. Hodgson, S.A. Buckler and G. Peters, J.Amer.Chem.Soc., 85, 543 (1963). F.P. Olsen and Y. Sasaki, J.Amer.Chem.Soc., 92, 3812 (1970).

35. H. Schumann and M. Schmidt, Angew.Chem., 77, 1049 (1965); H. Schumann, "Reactions of Elemental Sulfur with Inorganic, Organic, and Metal-Organic Compounds", in reference 18a, Vol.3.

36. E. Davis, "A Critique of the Reactions of Elemental Sulfur", in reference 8, Vol.3, in press.

37. L. Bateman and C.G. Moore, "Reaction of Sulfur with Olefins", in reference 7, page 210.

38. R. Mayer, "Synthesis and Properties of Thiocarbonyl Compounds", in reference 9, page 219.

39. R. Wegler, E. Kühle and W. Schäfer, "Reactionen des Schwefels mit araliphatischen sowie aliphatischen Verbindungen", in W. Foerst (Editor), "Neuere Methoden der präparativen organischem Chemie", Vol.III, page 1 (Verlag Chemie, Weinheim/Bergstraße, 1961).

40. F. Becke and H. Hagen, Chemiker-Ztg., 93, 474 (1969).

41. T.K. Wiewiorowski, reference 1, page 5 (summer 1968).

42. J.R. Postgate, "The Sulfur Cycle", in reference 2, page 259.

43. P.A. Trudinger, Rev.Pure and Appl.Chemistry, 17, 1 (1967).

44. N. Lozac'h and J. Vialle, "The Chemistry of the 1,2-Dithiole Ring", in reference 8, Vol.II, page 257; P.S. Landis, Chem.Rev., 65, 237 (1965).

45. R. Mayer, E. Hoffmann and J. Faust, J. Prakt. Chem., [4], 23, 77 (1964).

46. R. Mayer, W. Broy and R. Zahradnik, "Monocyclic Sulfur- Containing Pyrones", in Katritzky (Editor). "Adv. in Heterocyclic Chemistry", Vol.8., page 219 (Academic Press, Inc., New York, 1967).

46a B. Meyer, D. Jensen and I. Oommen, "Diatonic Species Containing Sulfur", in reference 18a, Vol. 2. *cf.* A.P. Ginsberg and W.E. Lindsell, *Chem. Comm.*, **232** (1971).

47. *cf.* R. Mayer, J. Morgenstern and J. Fabian, *Angew.Chem.*, **76**, 157 (1964).

47a R. Mayer, "Thione-Enethiol Tautomerism", in reference 18a; Vol.3.

48. C. Willgerodt, *Ber.Dtsch.Chem.Ges.*, **20**, 2467 (1887).

49. K. Kindler, *Ann.Chem.*, **431**, 187 (1923).

50. F. Asinger, W. Schäfer, K. Halcour, A. Saus and H. Trien, *Angew.Chem.*, **75**, 1050 (1963); F. Asinger, A. Saus and A. Mayer, *Mh.Chem.*, **98**, 825 (1967); F. Asinger, A. Saus and E. Michel, *ibid.*, **99**, 1436 (1968).

51. M. Carmack and M.A. Spielman, *Org.Reactions*, **3**, 83 (1946); F. Becke and H. Hagen, *Chemiker-Ztg.*, **93**, 474 (1969).

52. W. Walter and K.D. Bode, *Angew.Chem.*, **78**, 517 (1966).

53. K.F. Funk and R. Mayer, *J.Prakt.Chem.*, **21**, 65 (1963).

54. T.K. Wiewiorowski, *Endeavour* XXIX, 9 (1970).

*Chapter 3*

# VULCANIZATION OF RUBBER

## M. Porter

*The Natural Rubber Producers' Research Association*
*Welwyn Garden City, Hertfordshire, England*

## 3.1 INTRODUCTION

Although the improvement in the properties of natural rubber (NR) brought about by vulcanization with sulfur was discovered by Goodyear and by Hancock 130 years ago and the process has since become the basis of a world-wide commercial manufacturing industry, a scientific understanding of the changes taking place has begun to emerge only in the last 30 years. Many alternative reagents for effecting vulcanization have been discovered since 1840 but none of them has received widespread industrial acceptance and hence, for over a century, sulfur has remained the key reagent in the vulcanizing process — through which nearly all the rubber in use has to pass before its properties become commercially useful.

Goodyear's and Hancock's original treatments simply involved mixing the rubber with a few percent of its weight of sulfur and heating it at some elevated temperature for a period of hours. This conferred on the plastic, low-strength rubber a considerable degree of elasticity which made it useful for the manufacture of articles such as footwear, raincoats, hose, and solid tyres.

**TABLE 3.1**

Sulfur-Vulcanizable Rubbers and Related Model Olefins

A history of successive improvements and developments, nearly all empirical in nature, has brought the basic vulcanization process to its present high degree of versatility whereby NR and some seven different synthetic rubbers (Table 3.1) can be vulcanized in various forms at temperatures between 20 and 200°.

A rubber 'mix' may contain many components but only a few of them are included to effect vulcanization. These may be classed as: (i) the vulcanizing agent *per se* — usually sulfur but sometimes a 'sulfur donor' such as a tetra-alkylthiuram disulfide, **3.1**, or dithiobismorpholine, **3.2**; (ii) organic accelerators of vulcanization — these are generally derivatives of 2-mercaptobenzothiazole (MBT), **3.3**, or (the hypothetical) dialkyldithiocarbamic acids, **3.4**; (iii) activators of vulcanization — these include metal

$$\underset{\text{R}_2\text{N.C.S.S.C.NR}_2}{\overset{\text{S}\quad\ \text{S}}{\overset{||\quad\ ||}{\phantom{x}}}}$$

**3.1**

O⟨　⟩N.S.S.N⟨　⟩O

**3.2**

2-mercaptobenzothiazole structure C—SH

**3.3**

$$\underset{\text{R}_2\text{N.C.SH}}{\overset{\text{S}}{\overset{||}{\phantom{x}}}}$$

**3.4**

oxides (nearly always zinc oxide), the higher fatty acids, and nitrogen bases. A selection of practical formulations is given in Table 3.2. These formulations are examples of the rubber compounder's expertise, developed over many years and largely based on empiricism.

The characteristic changes in physical properties which the rubber undergoes during the vulcanization process are now known to be due to the covalent interlinking of the long flexible rubber chains. Just prior to vulcanization, after the preparation of the rubber 'mix', these chains have a number-average molecular weight of a few hundred thousand and form a mesh of randomly coiled molecules, entangled and intertwined but chemically discrete. Consequently, deformation of this mesh leads to slow disentanglement and partial alignment of the molecules which results in a fair degree of plastic flow but only very limited elastic recovery when the deforming force is removed. During vulcanization, the individual molecules become crosslinked to form a three-dimensional network in which some molecular alignment on deformation is still possible but in which the molecules are resistant to lateral movement and disentanglement. Removal of the deforming force therefore leads to rapid and nearly complete elastic recovery.

**TABLE 3.2**

Examples of Practical Formulations for Rubber Articles

| Formulation for: | Truck Tyre Carcass | Parts by wt. | Mechanical Rubber Goods | Parts by wt. | Cable Insulation | Parts by wt. | Fabric Proofing | Parts by wt. |
|---|---|---|---|---|---|---|---|---|
| Elastomer | NR | 100 | SBR | 100 | NR | 100 | NR | 100 |
| Vulcanizing agent | Sulfur | 2.5 | N,N'-Dithio-bismorpholine Tetramethyl-thiuram disulfide | 1.5 1.5 | Sulfur | 1.5 | Sulfur | 1.5 |
| Accelerators | N-oxydiethylene-benzothiazole-2-sulfenamide | 0.7 | Benzothiazol-2-yl disulfide | 1.5 | Benzothiazol-2-yl disulfide Bismuth dimethyl-dithiocarbamate | 0.5 0.5 | Amine-activated dithiocarbamate | 4.0 |
| Activators | Zinc oxide Stearic acid | 5 1.0 | Zinc oxide Stearic acid | 5 1.0 | Zinc oxide Stearic acid | 10 0.5 | Zinc oxide Stearic acid | 3.0 0.5 |
| Fillers | General purpose furnace carbon black | 45 | Hydrated silica Titanium dioxide | 60 20 | Fine thermal carbon black Calcium carbonate Kaolin (hard) | 25 75 25 | Fine thermal carbon black Calcium carbonate Kaolin (soft) | 10 75 50 |
| Processing aids | Process oil | 8 | Process oil | 5 | Paraffin oil/long-chain sulfonic acid mixture | 1.5 | Paraffin oil/long chain sulfonic acid mixture | 3 |
| Antioxidants Antiozonants | Acetone-diphenyl-amine reaction product | 2.0 | Hindered phenolic antioxidant | 1.0 | N,N'-di-2-naphthyl-p-phenylenediamine Polymerized 1,2-dihydro-2,2,4-trimethylquinoline Microcrystalline wax | 0.5 1.5 2.0 | N,N'-di-2-naphthyl-p-phenylenediamine | 0.5 |
| Perfume | | | | | | | Blend of essential oils | 0.1 |
| Vulcanization time & temperature | 30 min at 140° | | 11 min at 150° | | 20 sec at 200° | | 24–48 hr in warm room | |

The concentration of intermolecular crosslinks required to form such a network is quite small. To obtain the best physical properties from a rubber with a molecular weight of 250,000, about 25 crosslinks per chain are needed, *i.e.,* an average main-chain molecular weight between crosslinks ($M_c$) of $10^4$, or a chemical crosslink concentration ($n_{chem}$) of $5 \times 10^{-5}$ moles per gram. The main reason for this relatively low value is that the chain entanglements present in the unvulcanized rubber become trapped when the rubber is crosslinked and hence contribute to the elastic properties of the network thus greatly enhancing the effect of the actual crosslinks. However, this feature makes difficult the determination of $n_{chem}$, knowledge of which is essential for any structural analysis of the network in chemical terms. It would be inappropriate here to detail the methods devised for determining this quantity and the interested reader is referred to recent work on the subject [1-4] . Such methods depend, however, on relating the elastic properties of a series of vulcanizates from a particular rubber with the amounts of the reagent used to form the vulcanizates *in a case where the reagent is known to crosslink the rubber quantitatively.* This condition is a very stringent one and few, if any, quantitative crosslinking reagents are known. Most extant estimates of $n_{chem}$ should therefore be regarded as approximate only, although their relative accuracy is probably quite high.

## 3.2 STRUCTURAL FEATURES OF SULFUR VULCANIZATES

Structural analysis of sulfur vulcanizates is experimentally difficult. The low concentrations involved, the insolubility of polymer networks, and the difficulty of measuring chemical crosslinks all militate against the applicability of physical methods of analysis (such as infra-red, ultra-violet, and nuclear magnetic resonance spectroscopy) and of classical chemical

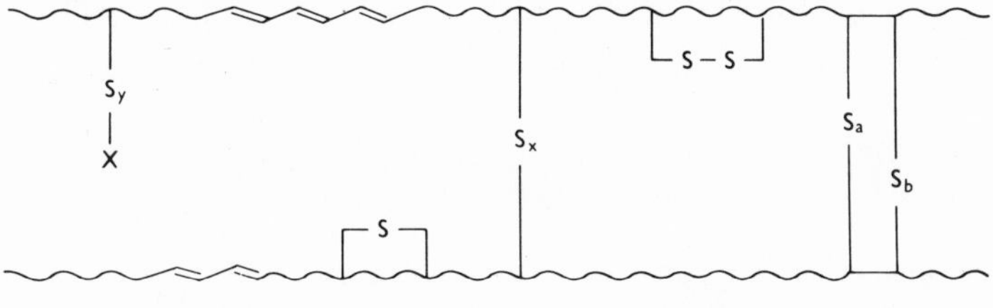

**Fig. 3.1.** Diagrammatic representation of the network structure of a sulfur vulcanizate ($x, y, a$ and $b = 1 - 6$; X = accelerator residue:

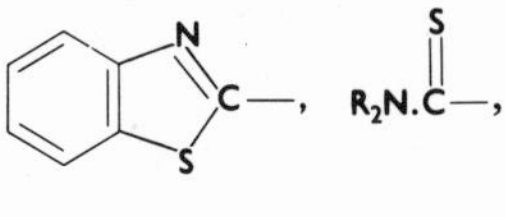

methods. It is therefore not surprising that much of our present knowledge has come from the use of model systems, that is to say, the reaction of sulfur and other vulcanizing ingredients with olefins structurally related to the rubbers but of low molecular weight. The structures of some of the olefins which have been commonly used and the rubbers for which they are models are shown in Table 3.1.

The combined results obtained from investigations on these model systems allow a vulcanized rubber network to be described in terms of the groups shown in Fig.3.1. The crosslinks may be simple mono-, di-, or polysulfides or they may consist of vicinally linked sulfides which act physically as a single crosslink. The rubber chains also become modified by cyclic mono- and disulfide groups, by accelerator-terminated pendent groups and by conjugated diene and conjugated triene groups (which represent dehydrogenation of the rubber). A further type of modification, not shown in Fig.3.1, involves the *cis,trans*-isomerization of olefinic double bonds. This isomerization has important consequences in the stereoregular rubbers, *cis*-BR, *cis*-IR and NR, in that it reduces the stereoregularity and hence the rubber's tendency to crystallize [5].

## 3.3 CHEMISTRY OF SULFUR-OLEFIN INTERACTION

The nature of the products isolated from reactions of sulfur with olefins depends markedly on the reaction temperature, the structure of the olefin, and the presence or absence of substances used as accelerators or activators of sulfur vulcanization. This section deals with the reaction between olefins and sulfur alone, usually at temperatures between 130 and 160°. Accelerated sulfuration is treated in Sect.3.4.

### 3.3.1 Reaction of Sulfur with Mono-Olefins

Interaction of sulfur with olefins below about 130° is extremely slow; above this temperature the primary products from mono-olefins are complex mixtures of organic polysulfides of three basic structural types, **3.5 – 3.7** [6-8]. Traces of episulfides [8] and cyclic disulfides [9] may also be formed, but thiols or hydrogen sulfide have not been observed when reaction temperatures have been maintained below about 150°.

$$\text{alkenyl}-S_x-\text{alkyl} \qquad\qquad \text{alkylene} \underset{S_z}{\overset{S_y}{\diagdown\diagup}} \text{alkylene}$$

$$\textbf{3.5} \qquad\qquad\qquad\qquad\qquad \textbf{3.6}$$

$$\text{alkenyl}-S_y-\text{alkylene}-S_z-\text{alkyl}$$

$$\textbf{3.7}$$

At a given temperature, the relative proportions of **3.5**, **3.6** and **3.7** formed depend on the reaction time and the degree of substitution at the olefinic double bond. At short reaction times, **3.5** is the principal type of product, in which the sulfur rank $x$, has an average value of about 4–5. At long times, or with less-substituted olefins, the proportions of **3.6** and **3.7** increase and the sulfur rank is reduced, monosulfides becoming apparent. Since the less-substituted olefins show reduced reactivity towards sulfur and their reaction rates are slow (Fig.3.2), this points to **3.6** and **3.7** being secondary products arising from further sulfuration of **3.5**. In several cases, indeed, the olefinic double bond in **3.5** ($x = 1$) has been shown to be several times more reactive towards sulfur than the corresponding double bond in the parent olefin[9,10].

The degree of substitution of the olefinic double bond also controls the detailed structures of the polysulfide mixtures. The latter have been deduced largely on the basis of the identities and yields of monothiols, dithiols, and hydrogen sulfide formed on reduction with lithium

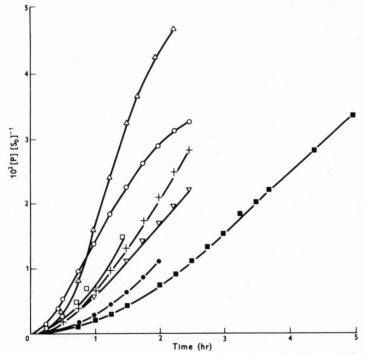

**Fig. 3.2.** Influence of olefin structure on rate of reaction with sulfur at 140°. P = polysulfide product; $S_0$ = initial sulfur; —■—■— = oct-1-ene; —|—|— = hept-2-ene; —▽—▽— = cyclohexene; —○—○— = methylcyclohexene; —△—△— = 2-methylpent-2-ene; —□—□— = 2,6-dimethyloct-2-ene; —●—●— = 2,6-dimethyl-octa-2,6-diene. (Courtesy, The Chemical Society[23].)

aluminium hydride [11] (Tables 3.3 and 3.4). In accordance with their formulation as mixtures of **3.5–3.7**, the yields of alkanethiols and alkenethiols are nearly equal in each case, the yields of dithiols reflecting the proportions of the more complex sulfides **3.6** and **3.7**. Irrespective of olefin structure, the alkyl-S- groups in the products are those derived by sulfur attachment exclusively or predominantly to the more substituted carbon of the original double bond. The structures of the alkenyl groups, however, vary with the degree of substitution in the olefin (Table 3.4). Thus, with the trialkylated olefin, 2-methylpent-2-ene, very little double-bond movement takes place whereas with oct-1-ene (the other extreme), double-bond movement is essentially complete. The disulfurated moieties of **3.6** and **3.7** are less well characterized but it seems likely that they are vicinal species derived by additive sulfuration of the double bond (Table 3.4).

**TABLE 3.3**

Compositions of Olefin Polysulfides and Yields of Reduction Products
(mole per mole of polysulfide)

| Polysulfide from | Formula | Mono-thiol | Di-thiol | $H_2S$ | Resi-due |
|---|---|---|---|---|---|
| Oct-1-ene | $(C_8H_{16.1}S_{2.33})_{2.12}$ | 0.62 | 0.60 | 2.3 | 0.32 |
| Cyclohexene | $(C_6H_{9.9}S_{2.87})_{2.1}$ | 0.90 | 0.53 | – | 0 |
| Hept-2-ene | $(C_7H_{14.0}S_{2.77})_{2.2}$ | – | – | – | – |
| 2,6-Dimethyloct-2-ene | $(C_{10}H_{20.7}S_{2.8})_{2.18}$ | 1.29 | 0.90 | 2.88 | 0 |
| 1-Methylcyclohexene | $(C_7H_{12.0}S_{2.68})_{2.17}$ | 1.14 | 0.42 | 2.90 | 0.12 |
| 2-Methylpent-2-ene | $(C_6H_{12.0}S_{2.18})_{2.06}$ | 1.12 | 0.31 | 2.30 | 0.05 |

In addition to polysulfides, sulfur-olefin reactions also produce small amounts of other olefins: position isomers, *cis,trans*-isomers, and conjugated dienes[9,10], *e.g.*,

$$\text{(3.1)}$$

$$\text{(3.2)}$$

**TABLE 3.4**

Structural Components of Polysulfide Mixtures

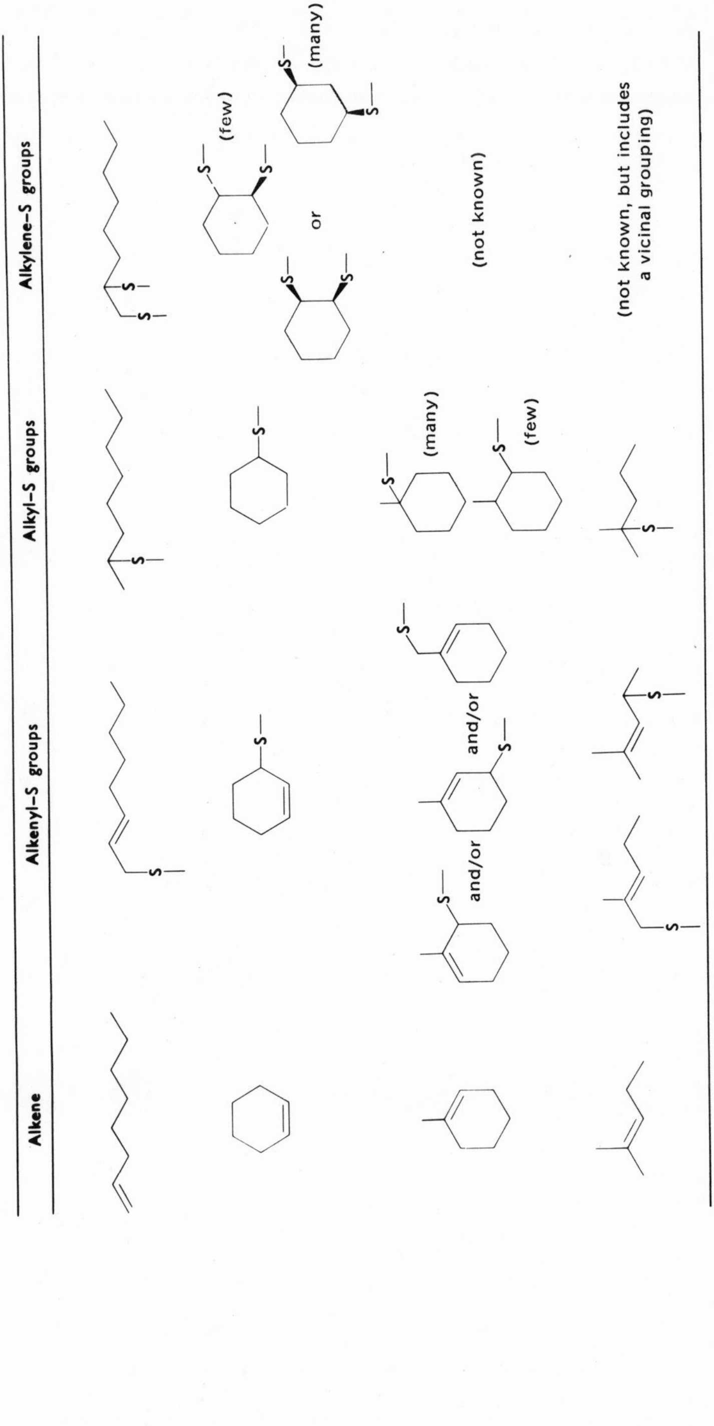

### 3.3.2 Reaction of Sulfur with 1,5-Dienes

Dehydrogenation products are more apparent in the sulfuration of 1,5-dienes; for example, *trans*-2, 6-dimethylocta-2, 6-diene, **3.8**, forms, *inter alia*, the two allo-ocimene isomers, **3.9** and **3.10**[12,13], and limonene, **3.11**, forms *p*-cymeme and *p*-isopropenyltoluene:[14]

$$(3.3)$$

$$(3.4)$$

**TABLE 3.5**

Cyclic Monosulfides formed by Sulfuration of 1,5-Dienes, **3.8** and **3.11**

Apart from these olefins, and crosslinked polysulfides analogous to
**3.5—3.7**, 1,5-dienes also form a new type of product — sulfur heterocycles.
The most abundant of these are saturated and unsaturated cyclic
monosulfides, as illustrated for the dienes **3.8** and **3.11** in Table 3.5[12,14].
In the case of the diene **3.8** at least, several other (probably unsaturated)
cyclic monosulfides are formed in trace quantities but have not been
identified as yet[13]. The diene **3.8** also forms small quantities of cyclic
disulfides but only one of these, **3.13**, has been characterized[15]. All
these cyclic mono- and disulfides (except **3.12**, see below) have one structural feature in common, *viz.*, one sulfur terminus occurs at the tertiary
carbon atom of one of the double bonds present in the diene. The other
sulfur terminus may be secondary or tertiary, saturated or unsaturated.

The conjugated hydrocarbons and the cyclic sulfides can be shown to
be largely, if not entirely, secondary products of sulfuration: not only
does their yield relative to polysulfides continue to increase after long
reaction times (Table 3.6) [12] but they can be shown to be formed from
the crosslinked polysulfides which accompany them by heating the latter
in an inert solvent at 140° [12,13]. Furthermore, by sulfurating the
1,5-diene, **3.8**, at low temperature in the presence of a very active
accelerator system, good yields of crosslinked polysulfides can be obtained
accompanied by only trace amounts of cyclic sulfides (see Sect. 3.6.1).

TABLE 3.6

Products from Reaction of Diene **3.8** (100 g) with Sulfur (10 g) at 140°

| Reaction time (hr) | Cyclic sulfides, CS (g) | Polysulfides, PS (g) | Triene, **3.10** (g) | Diene in CS / Diene in PS | Diene in **3.10** / Diene in PS |
|---|---|---|---|---|---|
| 5.0 | 5.0 | 12.0 | 0.48 | 0.56 | 0.07 |
| 50.0 | 20.6 | 24.4 | 1.6 | 0.92 | 0.09 |

Thermal treatment of these polysulfides produces the conjugated trienes,
**3.9** and **3.10**, and the sulfur heterocycles described above as well as an
additional type of sulfur heterocycle: the thiophene **3.14** [13].

**3.13**                                        **3.14**

While thiophenes and 1,2-dithiole-3-thiones have often been isolated
from high-temperature (180—220°) sulfurations [16,17], they have not

usually been observed at temperatures as low as 140°. Exceptions are the mass-spectrometric detection[18] of thiophene in the gas-phase sulfuration of 1,3-butadiene for 4 days at 130°, and the formation of 4,5-dimethyl-1,2-dithiole-3-thione on heating 2-methylbut-2-ene with sulfur at 142° for 24 hours[19]:

$$(3.5)$$

The fact that the thiophene, **3.14**, (and its structural analogue, **3.12**, Table 3.5) does not, of course, possess the saturated tertiary C–S grouping characteristic of the other heterocycles suggests that it arises by a different route. This has been confirmed by demonstrating its formation[15] when allo-ocimene, **3.10**, is heated with sulfur for 5 hours at 140°. Presumably the allo-ocimene formed by decomposition of the crosslinked polysulfides undergoes further reaction with sulfur in a polysulfide chain to form **3.14**. The latter may well be the precursor of the dithiole-thione, **3.15**, obtained[20] by heating allo-ocimene or linalool with sulfur at 210°.

**3.15**

### 3.3.3 Kinetics and Mechanism of Sulfur-Olefin Reactions

In view of the multiplicity of species formed on sulfuration of even simple mono-olefins, the kinetics of the process would be expected to be complex and this is the case. Although the rate of combination of sulfur with NR and other rubbers was originally thought[21] to follow simple zero or fractional order kinetics, it is now recognized[1,22] that sulfur combination has the kinetic features of an autocatalyzed reaction of some order greater than zero. This is consistent with the previously noted discrepancy[21] between the reaction order with respect to time and with respect to initial sulfur concentration. Autocatalysis is also a feature of the reaction of sulfur with low-molecular-weight olefins[23] (cf. the sigmoid shapes of the curves in Fig. 3.2) and can be shown to be due to the polysulfide products by adding them or synthetic polysulfides to the reactants, when the period of accelerating rate is eliminated and the overall reaction rate is increased[23]. The one rubber so far investigated which combines with sulfur with true first-order kinetics (both with respect to time and to initial sulfur concentration) is NBR (see Table 3.1) and in this case very few polysulfide crosslinks are formed (and hence the autocatalyst is absent)[24].

The Arrhenius activation energies for the combination of sulfur with rubbers (33–36 kcal/mole)[21,25] and with mono-olefins (37–38 kcal/mole)[23] are close to the energies required to break S–S bonds in organic polysulfides (37 kcal)[26] and in molecular sulfur (33 kcal)[27].

Attempts to establish the mechanistic nature of olefin sulfuration by investigation of the effects of additives on the kinetics have not led to a clear-cut result. The lack of response of the reaction[23] towards a range of free-radical initiators and inhibitors (or retarders) (see Table 3.7), coupled with the admittedly small positive effects of polar additives and solvents (Table 3.7) and the structures of the products detailed in Sect. 3.3.1 and 3.3.2, led workers[23,28] at the Natural Rubber Producers' Research Association to conclude in 1958 that a free-radical chain process such as had been postulated previously[7,29,30] was not operative; instead they put forward a polar chain mechanism (see below). This conclusion has been criticized[31,32] and the kinetic evidence must now be considered inconclusive.

TABLE 3.7

Effects of Additives and Solvents on the Accelerating Rate ($r_a$) and Maximum Rate ($r_m$) of Reaction of Cyclohexene with Sulfur[23]

| Additive (temp.) | Effect |
| --- | --- |
| 1,1'-Azobisisobutyronitrile (70°) | none |
| 1-Azobis-1-phenylpropane (110°) | none |
| Quinol (130°) | none |
| Benzoquinone (130°) | $r_a$ increased |
| Iodine (120°) | $r_a$ period eliminated |
| 2,2-Diphenyl-1-picrylhydrazyl (120°) | $r_a$ period eliminated, $r_m$ decreased |
| t-Butyl alcohol | none |
| Pyridine | none |
| Dioxane | none |
| Carboxylic acids | 20–50% overall increase in rate |
| Triethylamine (120–135°) | $r_a$ and $r_m$ increased |
| Propane-2-thiole (130°) | $r_a$ slightly increased |
| Polar solvents | small overall increase in rate |

Since 1958 very little additional evidence has been obtained on which to base a judgement. Addition of dicumyl peroxide to rubber-sulfur mixtures has been found to increase the rate of sulfur combination [33] but although polysulfide crosslinks were formed [34,35] the kinetics of their formation were different from those in the uninitiated reaction [35]. No structural investigation has been reported on the peroxide-initiated system so its relevance to the uncatalyzed reaction cannot be established. The mechanism may involve rapid scavenging of allylic carbon radicals by sulfur[36] followed by slow combination of the resulting persulfenyl radicals:

$$R'R''C{=}CHCH_2R''' \xrightarrow{RO\cdot} R'R''C{=}CH.CHR''' \xrightarrow{S_8} R'R''C{=}CHCHR'''S_x$$

$$\xrightarrow[\text{(slow)}]{\text{dimerization}} \text{dialkenyl polysulfide crosslink} \qquad (3.6)$$

Free-radical scavengers have been shown not to inhibit the unaccelerated sulfur vulcanization of SBR [37] and attempts [38-42] to demonstrate the presence of free radicals in vulcanizing rubbers by esr spectroscopy have also proved negative. Neither of these results should be regarded as conclusive, however, in view of the effectiveness of sulfur as a radical trap, in the first case, and in view of the possible short lifetime of any free-radical intermediates, in the second.

In summary then, all attempts to demonstrate the operation of a free-radical mechanism in olefin sulfuration have been unsuccessful to date. On the other hand, the effect on the reaction rate of changing the dielectric constant of the solvent is consistent only with such a reaction or with a polar reaction with very limited charge separation. However, the structure of the products — especially the Markovnikov orientation of the alkyl groups — restrict the number of possible mechanisms and only two have so far been proposed which are consistent with most of the features of the reaction:

$$TS_aS_bT \rightleftharpoons TS_a\cdot \ + TS_b\cdot \qquad (3.7)$$

$$TS_a\cdot \ + RH \rightarrow TS_aH + R\cdot \qquad (3.8)$$

$$R\cdot \ \ + S_8 \ \rightarrow RS_a\cdot \qquad (3.9)$$

$$RS_a\cdot \ + RH \rightarrow RS_aH + R\cdot \qquad (3.10)$$

$$RS_aH + RH \xrightarrow[\text{addition}]{\text{Markovnikov}} RS_aRH_2 \qquad (3.11)$$

$$(R = \text{alkenyl}; \ T = R \text{ or } RH_2; \ a, b \ = \ 2{-}8)$$

The first of these [7,29,30] is a two-part mechanism (eqs. 3.7 – 3.11) involving homolytic dissociation of S–S bonds and formation of an alkeneperthiol by a free-radical chain process, followed by polar addition of the perthiol to the olefin in a non-chain process. The process is formulated to yield the simple alkenyl alkyl polysulfide, **3.5**, but can be extended to cover **3.6** and **3.7** by postulating an alternative step whereby the persulfenyl radical, $RS_a \cdot$, adds to the olefin rather than abstracting hydrogen:

$$RS_a \cdot + RH \rightarrow RS_a RH \cdot \xrightarrow{\quad S_8, etc. \quad} 3.6 + 3.7 \qquad (3.12)$$

This mechanism cannot readily explain the formation of cyclic monosulfides, rather than cyclic polysulfides, from 1,5-dienes but these are, in any case, now best regarded as secondary products (see Sect. 3.3.2).

At the time when the polar mechanism of olefin sulfuration was proposed there was no unequivocal evidence for purely thermal homolysis of S–S bonds (eq.3.7) at temperatures of 140° and below [28]. This is not now the case. Tobolsky and coworkers [26,43,44] have shown, using the stable free radical **3.16** as scavenger, that dimethyl tetrasulfide undergoes thermal homolysis at temperatures as low as 52–80°. This reaction has an Arrhenius activation energy of 36.6 kcal/mole and produces two disulfenyl ($MeS_2 \cdot$) radicals by fission of the central S—S bond (eq. 3.13, R = Me).

$$(CH_3)_2 \, C - CH_2 - C \, CH_3$$
$$Ph \, N - O \cdot \overset{-}{O} - \overset{+}{N} \, Ph$$

**3.16**

Unsymmetrical fission to give one $MeS \cdot$ and one $MeS_3 \cdot$ radical would be energetically much more difficult [45]. In the absence of other reactants, the disulfenyl radicals initiate free-radical chains leading to disproportionation to sulfides of higher and lower sulfur rank (eqs.3.14, 3.15) and, in the case of unsymmetrical tetrasulfides, to interchange (eqs. 3.16, 3.17):

$$RS_4 R \quad \rightarrow \quad 2RSS \cdot \qquad (3.13)$$

$$RSS \cdot + RS_4 R \quad \rightarrow \quad RS_3 R + RSSS \cdot \qquad (3.14)$$

$$RSSS \cdot + RS_4 R \quad \rightarrow \quad RS_5 R + RSS \cdot \qquad (3.15)$$

$$RSS\cdot + RS_4R' \rightarrow RS_4R + R'SS\cdot \qquad (3.16)$$

$$RSSS\cdot + RS_4R' \rightarrow RS_4R + R'SSS\cdot \qquad (3.17)$$

Similar behavior is displayed by higher polysulfides, but trisulfides are less reactive because one of the pair of generated radicals is the high-energy monosulfenyl radical[43-45]:

$$RS_3R \rightarrow RS\cdot + RSS\cdot \qquad (3.18)$$

While the production of persulfenyl radicals from pre-formed polysulfides (eq.3.7) at $140°$ should therefore be rapid, the subsequent hydrogen abstraction from olefin (eq.3.8) should be energetically disfavored. Indeed, while monosulfenyl radicals are well established as abstractors of hydrogen atoms[46,47], no evidence has yet been obtained that persulfenyl radicals can perform the same function. Thus, $MeS_2\cdot$ radicals (generated by eq.3.13, R = Me) did not abstract hydrogen from triphenylmethane at 80 or $130°$ nor did they react with quinol[44].

The proposed polar addition of perthiol to olefin (eq.3.11) also has no direct experimental support. Various perthiols have been prepared[48] but their addition to olefins has not been reported. Eq. 3.11 is based on observations that elemental sulfur both inhibits anti-Markovnikov addition of thiols and hydrogen sulfide to olefins and catalyzes Markovnikov addition[49-51]. The mechanism of inhibition almost certainly involves the conversion of $RS\cdot$ radicals to the much more stable $RS_a\cdot$ radicals but the mechanism of catalysis is obscure. Elemental sulfur may not itself be the catalyst since organic polysulfides react directly with olefins to form products with Markovnikov orientation (see below)[30,52,9,53].

The second mechanism which explains most of the features of olefin sulfuration is the polar chain mechanism[23,28]. This is formulated in eqs.3.19 − 3.28 using the notation of eqs.3.7 − 3.11. Although the various entities are written formally as independent ions and are referred to as such, they were envisaged as being highly polarized molecules or, conceivably, as ion pairs in the relatively non-polar olefin solvent:

Initiation    $TS_aS_bT \rightarrow TS_a^+ + TS_b^-$ \qquad (3.19)

Propagation    $TS_a^+ + RH \rightarrow TS_aRH^+$ \qquad (3.20)

$$TS_aRH^+ + RH \xrightarrow[\text{H}^- \text{ transfer}]{\text{H}^+ \text{ transfer}} \begin{matrix} TS_aR + RH_2^+ \qquad (3.21) \\ TS_aRH_2 + R^+ \qquad (3.22) \end{matrix}$$

$$RH_2{}^+ + S_8 \rightarrow RH_2 S_a{}^+ \qquad\qquad (3.23)$$
$$(\equiv TS_a{}^+)$$
$$R^+ + S_8 \rightarrow RS_a{}^+ \qquad\qquad (3.24)$$

Termination $\quad RH_2{}^+$ $\left.\phantom{\begin{array}{c}a\\b\\c\\d\end{array}}\right\}$ $+ TS_b{}^- \rightarrow$ Products

$$\begin{array}{ll}
RH_2{}^+ & (3.25) \\
R^+ & (3.26) \\
TS_a{}^+ & (3.27) \\
TS_a RH^+ & (3.28)
\end{array}$$

In this mechanism, initiation involves heterolytic S–S bond fission in initially formed (or added) polysulfide to give 'incipient' or 'latent' persulfenium ions $(TS_a{}^+)$, which are postulated as the chain carriers, and similar persulfenyl anions $(TS_b{}^-)$, which are chain terminators. Reaction of the persulfenium ion at the olefinic double bond gives a cyclic persulfonium ion (eq.3.20) which, depending upon its structure and that of the olefin, can either lose a proton to the olefin (eq.3.21) or gain a hydride ion from the olefin (eq.3.22). In each case, polysulfide and an alkyl $(RH_2{}^+)$ or alkenyl $(R^+)$ carbonium ion are formed, the latter reacting with sulfur to continue the chain. Combination of eq.3.21 with 3.23 and eq. 3.22 with 3.24 shows that, in either case, the product is $RS_a RH_2$ —an alkenyl alkyl polysulfide, **3.5**. The formation of the more complex polysulfides, **3.6** and **3.7**, can be explained by alternative, but kinetically equivalent, reactions of the chain-carrying species, $TS_a RH^+$, with sulfur or a polysulfide chain rather than with olefin. Product **3.7** also arises from the termination reaction, eq.3.28.

The merit of this mechanism lies in its ability to explain consistently the detailed structures of the products from all types of mono-olefin (Table 3.4) and the 1,5-dienes, **3.8** and **3.11** (Table 3.5). At one extreme, the least-substituted type of olefin, exemplified by oct-l-ene, is postulated to undergo predominant proton transfer, eq.3.21 (Fig.3.3), giving rise to products **3.17** − **3.19** containing secondary octyl and oct-2-enyl groups as observed (Table 3.4).

At the other extreme, the trisubstituted olefin, 2-methylpent-2-ene, which reacts more rapidly with sulfur (Fig.3.2) is believed to suffer the hydride-ion transfer process, eq.3.22 (Fig.3.4), leading mainly to the single type of polysulfide, **3.20**, containing 1,1-dimethylbutyl, 1,3-dimethylbut-2-enyl, and 2-methylpent-2-enyl groups, as observed (Table 3.4).

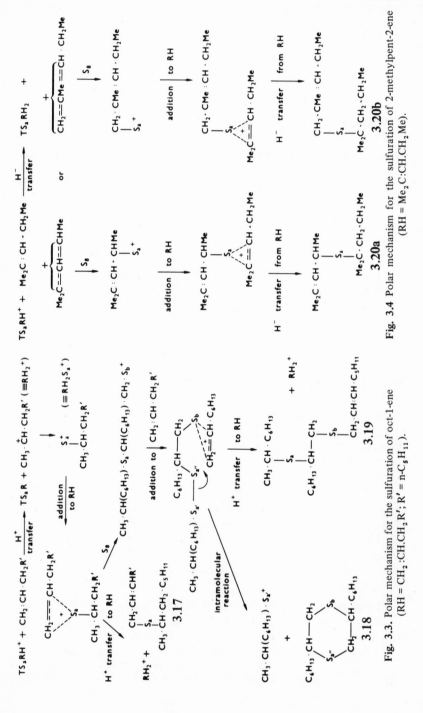

Fig. 3.4 Polar mechanism for the sulfuration of 2-methylpent-2-ene (RH = Me₂C:CH.CH₂Me).

Fig. 3.3. Polar mechanism for the sulfuration of oct-1-ene (RH = CH₂:CH.CH₂R'; R' = n-C₅H₁₁).

The formation of the conjugated trienes, **3.9** and **3.10**, from the 1,5-diene, **3.8**, has been explained [28,1] by a combination of proton and hydride-ion transfer. Although the trienes, like the cyclic sulfides, have since been established as being largely secondary products (Sect.3.3.2), this type of mechanism may still be operative. Thus, although heating the crosslinked polysulfides derived from **3.8** in an inert solvent affords both trienes and cyclic sulfides (the trienes in this case being formed by a process analogous to that of eq.3.29), their yields are quadrupled when heating is carried out in the diene, **3.8**, as solvent [13]. This points to the diene acting as a hydrogen source for the formation of the cyclic sulfides. An illustrative scheme for the hydrogen transfer process, based on the joint operation of eqs.3.21 and 3.22, is shown in Fig.3.5. In essence, this scheme involves a reduction in the sulfur rank of the crosslinked polysulfide with the concomitant conversion of two molecules of diene into one molecule of (hydrogen-poor) triene and one molecule of (hydrogen-rich) cyclic sulfide. The formation of the other cyclic sulfides from **3.8** and of the cyclic sulfides and conjugated hydrocarbons from **3.11** (eq.3.4 and Table 3.5) can be explained similarly [53a].

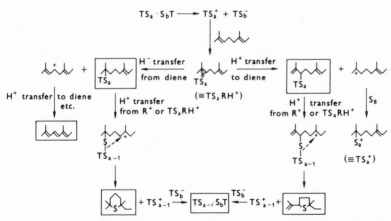

**Fig. 3.5.** Polar mechanism for triene and cyclic sulfide formation during sulfuration of 2,6-dimethylocta-2,6-diene (T = dimethyloctadienyl or dimethyloctenyl; final products are boxed).

## 3.4 CHEMISTRY OF ACCELERATED SULFURATION OF OLEFINS

Sulfuration of olefins in the presence of an accelerator-activator system, as described in Sect.3.1, proceeds much more rapidly than when sulfur alone is used and leads to simpler products. The greatest effects on product structure are observed with the more substituted olefins (such as

2-methylpent-2-ene and the diene, **3.8**) and when sulfur donors or high proportions of accelerators and activators relative to sulfur are used. Less-substituted olefins and lower concentrations of additives give products of intermediate overall structure.

### 3.4.1 Accelerated Sulfuration of Mono-Olefins

At temperatures of about 140°, in the presence of a complete sulfur-accelerator-activator system, 2-methylpent-2-ene forms 'crosslinked' sulfides and two additional types of product: nitrogen-containing derivatives and zinc sulfide. The organic sulfides are mixtures of simple dialkenyl mono-, di- and polysulfides; that is to say, the formation of saturated groupings and of complex sulfides like **3.6** and **3.7** is entirely suppressed. These features are evident from the absence of saturated monothiols and of dithiols, respectively, from the products of reduction with lithium aluminium hydride (see, for example, Table 3.8). The thiols observed are those derived by substitution at C−1 and C−4 of the 2-methylpent-2-ene skeleton, together with their isoallylic counterparts.

TABLE 3.8

Thiols Formed on Reduction of Organic Sulfides from MBT-Accelerated Sulfuration[a] of 2-Methylpent-2-ene[54]

| Thiol | Mole % |
|---|---|
| | 12.8 |
| | 25.9 |
| | 3.5 |
| | 52.9 |
| | 4.9 |

[a]2-Methylpent-2-ene (30 g), sulfur (1.5 g), MBT (1.5 g), zinc oxide (5 g), propionic acid (0.35 g), heated for 2 hr at 140°.

The same groupings are found in the monosulfide fraction of the sulfide products (Table 3.9).

TABLE 3.9

'Crosslinked' Monosulfides Formed on MBT-Accelerated
Sulfuration[a] of 2-Methylpent-2-ene[54]

| Monosulfide[b] | | Mole % |
|---|---|---|
| $B_1SB_1$ | (3 geometric isomers) | 27.1 |
| $B_1SA_1$ | (2 geometric isomers) | 10.4 |
| $A_1SA_1$ | | 12.5 |
| $B_1SB_2$ | (2 geometric isomers) | 39.6 |
| $A_1SB_2$ | | 10.4 |

[a] Conditions as in footnote $a$ to Table 3.8.
[b] The nomenclature, adopted for brevity, is as follows:

$A_1 =$    $A_2 =$    $B_1 =$    $B_2 =$

The composition and detailed structures of the sulfidic products vary with the precise composition of the sulfurating system (particularly the

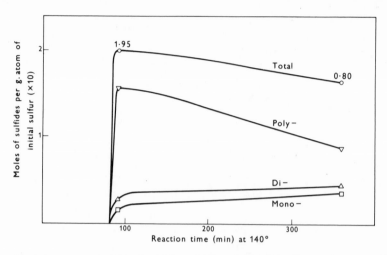

**Fig. 3.6.** Course of sulfuration of 2-methylpent-2-ene with a low accelerator : sulfur ratio [2-methylpent-2-ene, 50; sulfur, 2.5; N-cyclohexylbenzothiazole-2-sulfenamide (CBS), 0.6; zinc oxide, 5.0; propionic acid, 0.6 parts by wt. Figures on curves indicate ratios of sulfuration at C−1 *vs.* C−4 of 2-methylpent-2-ene] [55].

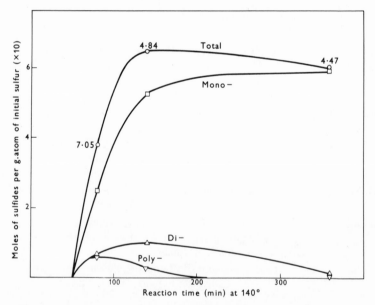

**Fig. 3.7.** Course of sulfuration of 2-methylpent-2-ene with a high accelerator : sulfur ratio [2-methylpent-2-ene, 50; sulfur, 0.4; CBS, 6.0; zinc oxide, 5.0; propionic acid, 0.6 parts by wt. Figures on curves indicate ratios of sulfuration at C−1 *vs.* C−4 of 2-methylpent-2-ene] [55].

nature of the accelerator, the molar ratio of accelerator: sulfur, and the nature and proportions of the activators), and reaction time and temperature. These factors alter the proportions of mono-, di- and polysulfides and the proportions of C−1 and C−4 substitution. Use of a low accelerator : sulfur ratio or a low proportion of activators leads to the formation of highly polysulfidic products with C−1 : C−4 substitution ratios close to unity or less (Fig.3.6). With a high accelerator : sulfur ratio, di- and polysulfides are formed initially but the products are mainly monosulfides over most of the reaction range; substitution at C−1 predominates greatly over that at C−4 (Fig. 3.7). It is apparent from Figs.3.6 and 3.7 that disulfides and, particularly, polysulfides are relatively long-lived intermediate, rather than final products. However, the fate of the polysulfides does not appear to be the same in the two cases. With a low accelerator concentration (Fig.3.6), mono- and disulfide concentrations rise only slowly between 100 and 360 minutes' reaction while the concentration of polysulfides is halved, causing an overall decrease in the yield of crosslinked sulfides. The fate of the polysulfides is to be found in the formation of 2-methylpenta-1,3-diene and 4-methylpenta-1,3-diene and in increased quantities of zinc sulfide (in excess of one mole per mole of crosslinked sulfides). This 'crosslink-destruction' reaction appears to follow the course of eq.3.29 and to be a purely thermal process since it

$$\text{(structure)} \xrightarrow[\text{elimination}]{\text{1,2- or 1,4-}} \left[ RS_x H \right] + \text{(diene)} + \text{(diene)}$$

$$\downarrow ZnO \qquad\qquad (3.29)$$

$$\tfrac{1}{2}(RS_{2x-1}R + H_2O + ZnS)$$

does not respond to the addition of base or of complexes such as 3.21 [56]. However, small quantities of various 2-methylpentenes are formed with the dienes; so eq.3.29 may well be an over-simplification of the process. This 'crosslink destruction' is responsible for the pronounced reduction in crosslink density or 'reversion' on overcure of polyisoprene vulcanizates when prepared with a conventional vulcanizing system (similar to that of Fig.3.6) and for the progressive reduction in maximum attainable degree of crosslinking as the temperature of vulcanization is increased.

During sulfuration with a high accelerator : sulfur ratio (Fig.3.7), the progressive disappearance of poly- and disulfides is accompanied by the formation of monosulfides and hence by little or no decrease in the yield of total sulfides. Experiments with synthetic examples (*cf.* Table 3.10) have clearly demonstrated that under 'sulfuration' conditions allylic di- and polysulfides are readily desulfurated to monosulfides and that the

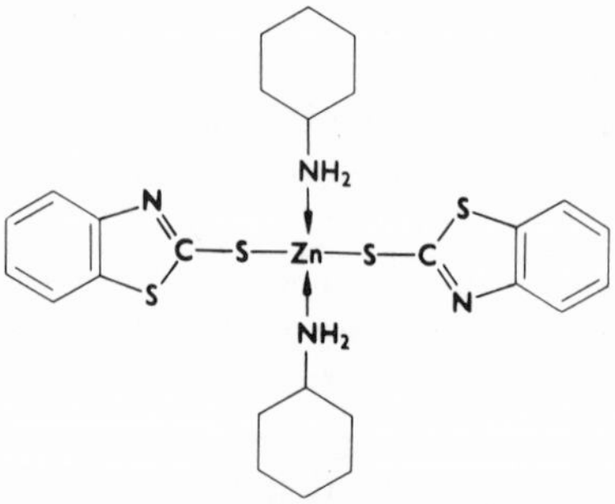

**3.22**

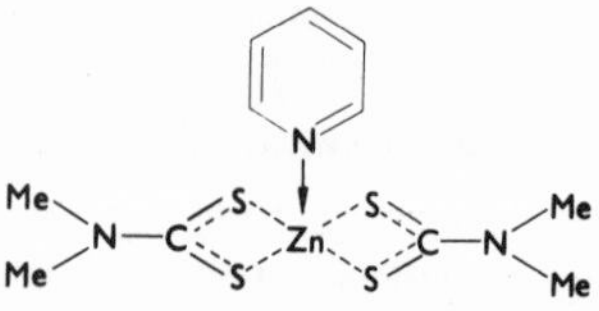

**3.21**

### TABLE 3.10

Product Yields and Compositions on Reaction of Methylpentenyl Trisulfides with ZMBT-Cyclohexylamine Complex **3.21** in the Presence of 2-Methylpent-2-ene and Zinc Oxide at 140° for 2 hr [56,57]

| Molar ratio, **3.21** : RSSSR | | 1:5 | | 2:1 | |
|---|---|---|---|---|---|
| Reactant RSSSR [a] | | $B_1SSSB_1$ | $A_1SSSA_1$ | $B_1SSSB_1$ | $A_1SSSA_1$ |
| *Yield (mole % of reactant RSSSR)* | | | | | |
| Sulfides | RSR | 22.8 | 12.2 | 148.2 | 95.3 |
| | $RS_2R$ | 20.0 | 25.8 | 22.5 | 50.2 |
| | $RS_3R$ | 58.2 | 65.3 | – | 3.0 |
| Total loss (−) or gain (+) of sulfides | | + 1.0 | + 3.3 | +70.5 | +48.5 |
| Methylpentadienes | | 6.5 | 16.1 | 3.5 | 11.2 |
| Zinc sulfide | | 20.5 | 11.6 | 95.9 | 76.6 |
| *Composition (mole % of reactant RSSSR)* | | | | | |
| | $A_2SB_2$ | – | – | 0.5 | – |
| | $A_1SB_2$ | 0.9 | 0.7 | 23.4 | 16.0 |
| | $A_1SA_2$ | 0.5 | 1.5 | 3.1 | 13.6 |
| | $A_1SA_1$ | 0.5 | 7.3 | 3.5 | 28.6 |
| | $B_1SB_2$ | 6.3 | 0.3 | 60.3 | 4.9 |
| | $A_2SB_1$ | 2.6 | 0.9 | 18.8 | 14.1 |
| | $A_1SB_1$ | 3.4 | 1.5 | 17.4 | 14.0 |
| | $B_1SB_1$ | 8.6 | – | 31.4 | 4.1 |
| | $C_1SSC_1$ | 0.9 | – | – | – |
| | $A_1SSC_1$ | – | 3.1 | – | – |
| | $A_1SSA_2$ | 1.4 | – | – | – |
| | $A_1SSA_1$ | – | 17.2 | 6.9 | 40.3 |
| | $B_1SSB_2$ | 2.5 | – | 0.3 | – |
| | $A_1SSB_1$ | 6.5 | 4.4 | 10.7 | 9.9 |
| | $B_1SSB_1$ | 8.7 | 1.1 | 4.5 | – |

[a]For key to structures (R = $A_1,A_2,B_1,B_2$), see footnote *b* to Table 3.9. $C_1 =$

extruded sulfur is used to resulfurate more olefin, forming additional crosslinked sulfides. This process is catalyzed by zinc benzothiazole-2-thiolate (ZMBT) or by zinc dialkyldithiocarbamates (ZDC), or their more soluble amine complexes, *e.g.*, **3.21**, **3.22**, and is accompanied by appreciable allylic rearrangement in the methylpentenyl moieties and by the formation of zinc sulfide (Table 3.10). Allylic rearrangement is evidently involved in the desulfuration–resulfuration pathway:

$$RS_3R' \xrightarrow{XSZnSX} RS_2R' \xrightarrow{XSZnSX} RSR + RSR' + XSS_xZnSX$$

olefin

additional crosslinks

(3.30)

$(R = A_1 \text{ or } B_1, R' = A_2 \text{ or } B_2, X SZ_nSX = \text{e.g., } \mathbf{3.21} \text{ or } \mathbf{3.22})$

and is believed to take place by a mechanism analogous to that operating in the desulfuration of allylic sulfides by triphenylphosphine[58-60] *viz.*, by complex formation between the zinc thiolate and the allylic sulfide followed by $S_N i$ or $S_N i'$ attack of thiolate sulfur on carbon:

(3.31)

During the course of olefin sulfuration, the intermediate di- and polysulfides therefore disappear by two competing processes: either they lose sulfur to zinc-accelerator-thiolate complexes, forming thermally stable monosulfides and additional sulfides (plus zinc sulfide), or they are thermally destroyed, forming conjugated dienes and zinc sulfide. Table 3.10 shows that the relative rates of these competing reaction paths are dependent, not only on the concentration of the catalyst complexes and on reaction temperature (as would be expected), but on the structures of the alkenyl di- and polysulfides themselves. Thus, $B_1SSSB_1$ which contains primary C–S bonds, is more rapidly desulfurated than $A_1SSSA_1$, which contains secondary C–S bonds. This is very evident with the high (2 : 1) ratio of **3.21** to trisulfide where the products from $B_1SSSB_1$ are nearly all monosulfides and a large proportion of these ($A_1SB_2$ and

$B_1SB_2$) contain the rearranged $B_2$ group. Fresh sulfuration of 2-methylpent-2-ene is indicated by the 70% increase in total sulfides and the appearance of a substantial proportion of $A_1$ and $A_2$ groups. These trends are very much less marked in the case of the trisulfide, $A_1SSSA_1$.

The figures for 2-methylpentadiene formation given in Table 3.10 also reflect the dependence of rates of desulfuration on both catalyst concentration and terminus structure. Thus, diene formation is less at the higher catalyst concentration and when the substrate is $B_1SSSB_1$ rather than $A_1SSSA_1$, for both of these features increase the rate of desulfuration of the polysulfides and thus reduce their propensity for thermal decomposition.

The nitrogen-containing derivatives formed with the 'crosslinked' sulfides have not generally been as well characterized as the latter, but there is little doubt that they consist mainly of compounds of the type, $RS_xX$, where R is an alkyl or, more usually, an alkenyl group derived from the olefin; X is an accelerator grouping, usually based on **3.3** or **3.4** (*cf.* Fig. 3.1); and $x$ has a value of $1-4$. This conclusion follows from the more complete structural analysis made on the corresponding products from 2-methylpent-2-ene, zinc oxide and the sulfur donor, tetramethylthiuram disulfide (Sect.3.5) and from studies with radioactively-labelled accelerators [61,62].

In the presence of less 'complete' sulfur-accelerator-activator systems, *i.e.*, where accelerators or activators are absent or their proportions are inadequate, characteristic differences in the structures of the 'crosslinked' sulfides may appear.

For example, any factor tending to limit the concentration of desulfurating agents will lead to relatively long-lived polysulfides which will tend to decompose faster than they are converted to sulfides of short chain length. Such factors may be low accelerator and low zinc concentration, but in the case of the accelerators, MBT **3.3** and MBTS (the corresponding disulfide), the low solubility of ZMBT, to which they are converted, is a limiting feature. In these cases, high concentrations of accelerator and zinc do not alone ensure rapid desulfuration because the bulk of the ZMBT formed is insoluble; the presence of sufficient primary or secondary amine or zinc carboxylate to render the ZMBT soluble by complex formation [1,63,64] is also necessary. The accelerator CBS (N-cyclohexylbenzothiazole-2-sulfenamide) is also converted to ZMBT, but in the form of its cyclohexylamine complex, **3.21**; additional activators are therefore not required.

More striking changes are apparent when either the accelerator or the zinc activators are omitted entirely (Table 3.11). The effect of the accelerator alone on the olefin-sulfur reaction is largely confined to

TABLE 3.11

Effects of Accelerators and of Zinc Activators on the Pattern of Sulfuration of 2-Methylpent-2-ene (2 hr at 140°) [65]

| Presence (+) or absence (−) of: | | | | |
|---|---|---|---|---|
| Accelerator | − | + | − | + |
| ZnO + Zn(OCOR)$_2$ | − | − | + | + |
| Rate of sulfuration | slow | fast | slow | fast |
| Efficiency of sulfuration (maximum yield of sulfides) | low | low | low | high |
| Yields of H$_2$S and thiols | none | v.small | none | none |
| Yield of ZnS (mole per mole of sulfides formed) | none | none | ca.1 | ca.1 |
| Yield of simple mono- and disulfides | v.low | v.low | v.low | high |
| Proportion of complex sulfides | ca.20 | ca.20 | 0 | 0 |
| Proportion of alkyl groups in simple sulfides (%) | 50 | ca.20 | ca.5 | 0 |
| Ratio of C−1 : C−4 substitution | 0.4 | 0.4 | 0.6 | >1 |

increasing the rate of sulfuration with very little change in efficiency or in product structure (there is some increase in the yield of dialkenyl sulfides at the expense of alkenyl alkyl sulfides). On the other hand, zinc activators by themselves hardly affect the rate of sulfur combination and do not increase efficiency, but they exert a marked effect on structure, making sulfuration almost entirely a substitutive process rather than a combined additive-substitutive process and thus nearly eliminating complex sulfides (such as 3.6 and 3.7) and alkyl-S- groups from the products. Only the combination of accelerator and zinc activator both increases rate and efficiency of sulfuration and makes the process entirely disubstitutive. Under these conditions, substitution at the α-methyl carbon atom (C−1) is enhanced vis-à-vis substitution at the α-methylenic carbon (C−4). The extent of this enhancement generally parallels the degree of improvement in efficiency which results when both accelerator and activator concentrations are raised relative to that of sulfur.

The above, quite detailed, picture of sulfuration refers to the trialkylated olefin, 2-methylpent-2-ene, which is a model for the polyisoprene rubbers. The sulfuration of symmetrical dialkylethylenes and terminal olefins — models for the bulk of the synthetic rubbers (Table 3.1) —has not been studied so comprehensively, but recent work[10,66] has allowed some of the principal features to be discerned. As a result, it is now clear: first, that these olefins generally give a product pattern intermediate between those observed for unaccelerated and fully accelerated sulfuration of the trialkylethylene, *i.e.*, they respond to the addition of accelerator but less so to the further addition of zinc activators (*cf.* Table 3.11); second, that no single mechanism of those proposed for sulfuration (Sect. 3.6) holds satisfactorily for all types of olefins. With respect to the first point, Skinner[10], using a fully-activated CBS-accelerated system, found that cyclohexene and hex-1-ene yielded only very small amounts of simple mono- and disulfides, the bulk of the sulfides being simple polysulfides or complex sulfides like **3.6** and **3.7**; with both olefins a substantial proportion of the sulfide termini were saturated and, in the case of hex-1-ene, none of the alkenyl-S- groups contained terminal unsaturation (*cf.* the unaccelerated sulfuration of oct-1-ene, Table 3.4). In contrast, *cis-* **3.23** and *trans*-hex-3-ene **3.24** gave product distributions which were almost identical to one another and similar to that obtained from 2-methylpent-2-ene except that small proportions of alkyl-S- groups and complex sulfides were present. However, the alkenyl-S- groups were comprised not only of hex-3-ene-2-thio groups **3.25** and **3.26**, but of substantial proportions of isoallylic hex-4-ene-3-thio groups **3.27** and **3.28**, even after short reaction times (Table 3.12). This contrasts with the sulfides from 2-methylpent-2-ene in which the content of the isoallylic

**3.23**                    **3.24**

groups $A_2$ and $B_2$ (Table 3.8) is usually very low. Furthermore, in the case of the *cis*-hexene, considerable *cis, trans*-isomerization of the double bond had taken place, a similar distribution of the four groupings **3.25 – 3.28** being observed from the two hex-3-enes (Table 3.12). Although some isomerization (both allylic and geometric) undoubtedly takes place during desulfuration of initially formed polysulfides (see eq.3.31), some also occurs at an earlier stage since sulfuration of the *cis*-hexene at room temperature (using a more active accelerator) under conditions in which desulfuration does not take place to a significant extent also results in the formation of the isomerized groups (Table 3.13)[10]. Examination of the recovered olefin from this experiment revealed that it contained less than 1% *trans*-hex-3-ene; hence isomerization occurs during the overall sulfuration process.

### 3.4.2 Accelerated Sulfuration of 1,5-Dienes

As with unaccelerated sulfuration, the products principally of interest here are the conjugated hydrocarbons and cyclic sulfides formed as secondary products by thermal decomposition of prior-formed crosslinked

TABLE 3.12

Composition and Structure of Sulfides Formed from *cis*- and *trans*-Hex-3-ene on
CBS-Accelerated Sulfuration for 100 min at $140°$ [10]

| Olefin | *cis*-hex-3-ene 3.23 | *trans*-hex-3-ene 3.24 |
|---|---|---|
| Composition of sulfides (mole %) | | |
| RSR | 3.9 | 8.6 |
| RSSR | 7.1 | 8.6 |
| $RS_xR$ | 89.0 | 82.8 |
| Structures of groups in di- and polysulfides [a] | | |
| 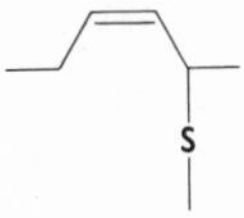 3.25 and/or 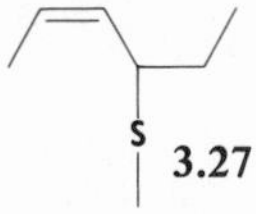 3.27 | 13.0 | 8.2 |
| 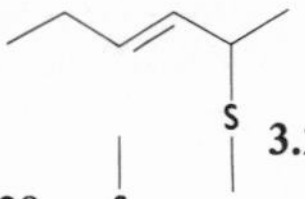 3.26 | 42.8 | 43.8 |
| 3.28 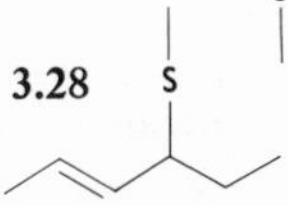 | 39.8 | 38.3 |
| 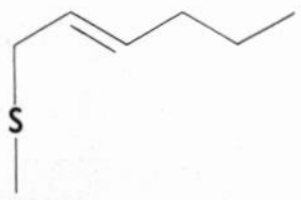 (allylic isomer) | 2.6 | 7.3 |
| hexanethio– | 1.8 | 2.4 |

[a] Area % by gas-liquid chromatography of thiols formed on reduction with lithium aluminium hydride.

TABLE 3.13

Structural Features of Di- and Polysulfides Obtained by Sulfuration of *cis*-Hex-3-ene
at Room Temperature [10]

| Grouping [a] | Proportion [b] |
|---|---|
| 3.25 and/or 3.27 | 7.5 |
| 3.26 | 45.1 |
| 3.28 | 46.3 |
| hexanethio- | 1.1 |

[a] See Table 3.12.
[b] See footnote to Table 3.12.

polysulfides. Inclusion of an accelerator and activators in a reaction mixture of *trans*-2,6-dimethylocta-2,6-diene (**3.8**) and sulfur does not lead to the formation of more than a small proportion of any new product. However, it drastically reduces the total amounts of conjugated hydrocarbons and cyclic sulfides formed, consistent with the theory that these compounds arise by decomposition of the polysulfides since, in the presence of the accelerator system, the latter are more rapidly desulfurated to thermally stable crosslinked monosulfides. This point is further. illustrated (Table 3.14) by the greater yields of hydrocarbons and sulfur heterocycles formed when **3.8** is treated with sulfuration systems which give a preponderance of substitution at C–4 in 2-methylpent-2-ene and thereby lead to less rapid desulfuration of polysulfides to monosulfides (p. 99).

TABLE 3.14

Dependence of Relative Molar Yields of Conjugated Hydrocarbons and Cyclic Sulfides, Formed by Sulfuration of **3.8**, on the Type of Crosslink Terminus

| Vulcan-izing System [a] | C–1:C–4 ratio in 2-Me-pent-2-ene | Relative Molar Yields from 3.8 | | | | |
|---|---|---|---|---|---|---|
| | | Cross-links | Conju-gated dienes | Conju-gated trienes | Cyclic mono-sulfides | Cyclic di-sulfides |
| 1 | 0.5 | 1.0 | 0.4 | 1.1 | 1.9 | 0.9 |
| 2 | 4.8 | 1.0 | 0.3 | 0.1 | 0.07 | 0.14 |

[a] 1 : Sulfur, 2; zinc dimethyldithiocarbamate, 2; zinc oxide, 3; **3.8**, 20 parts by wt. Heated 90 hr at 100°.

2 : Sulfur, 0.4; CBS, 6; zinc oxide, 5; propionic acid, 0.55; **3.8**, 20 parts by wt. Heated 2 hr at 140°.

Although new products are not formed in quantity, a range of some 10–15 hydrocarbons and up to 10 cyclic sulfides is generally obtained [13]. The complexity of the mixtures and low concentration of individual compounds has prevented identification in most instances but the following conclusions have been drawn [13] : (i) the pattern of hydrocarbons is similar to that obtained in unaccelerated sulfuration (the allo-ocimenes **3.9** and **3.10** are prominent); (ii) the principal cyclic sulfide formed in all cases is **3.29** (*cf.* Table 3.5); the two saturated cyclic sulfides **3.30** and **3.31** are absent; (iii) the new cyclic sulfides, formed in low concentrations, are probably di-unsaturated.

3.29          3.30          3.31          3.32

The incursion of features (ii) and (iii) on changing from unaccelerated to accelerated, activated sulfuration reflects the corresponding change in the structure of the crosslinked sulfides from alkenyl alkyl polysulfides to dialkenyl polysulfides (Sect.3.4.1). This, and the fact that all the cyclic monosulfides so far identified possess a common structural element, 3.32, suggests that the bulk of the cyclic monosulfides are formed by a process involving hydrogen transfer to an olefinic double bond in a pre-formed polysulfide followed by attack of the resulting carbon species (whether cation or radical) on the first sulfur atom in the polysulfide chain, e.g., eq.3.32 (cf. eq.3.33 for unaccelerated sulfuration; see also Fig.3.5).

$$(3.32)$$

$$(3.33)$$

## 3.5  THE REACTION OF OLEFINS WITH SULFUR DONORS

The tetra-alkylthiuram disulfides, 3.1, act as accelerators of sulfur vulcanization, but they are also used to vulcanize rubbers in the absence of elemental sulfur. In this case, they are effective only when used in conjunction with zinc oxide and, preferably, a carboxylic acid as well. Under such conditions, they yield thermally stable vulcanizate networks associated with a high proportion of monosulfide crosslinks (p.99) and this is borne out by the pattern of reaction products from 2-methylpent-2-ene [67,68] which is very similar to that of Fig. 3.7. The resemblance is closer than this, for the sulfides contain the same end-groups (footnote b, Table 3.9) as those from the accelerated sulfuration, saturated groups with complex sulfides being absent. Furthermore, the 1,5-diene, 3.8, gives small amounts of the same cyclic sulfides and conjugated hydrocarbons as are formed when an accelerated sulfur system is employed [13,67]. These olefins (and cyclohexene [67]) thus give an almost identical distribution of products irrespective of whether sulfuration is effected by a tetra-alkylthiuram disulfide-zinc oxide mixture or by an accelerator-activator-sulfur combination of high accelerator : sulfur ratio. The chemistry of these two vulcanizing systems is therefore concluded to be similar.

Of special interest are the nitrogenous compounds obtained by heating 2-methylpent-2-ene with tetramethylthiuram disulfide (TMTD) and zinc oxide (Table 3.15)[67,68]. In addition to simple decomposition products

TABLE 3.15

Relation Between Yields of Nitrogenous Compounds and of Crosslinked Sulfides During Reaction of TMTD and Zinc Oxide with 2-Methylpent-2-ene at 140°

| Compound[a] | Yield (Mole % of initial TMTD) | | | |
|---|---|---|---|---|
| | 1 hr. | 4 hr. | 10 hr. | b |
| RSS.CS.NMe$_2$ (3.33) | 18 | – | 0 | 0 |
| RS.CS.NMe$_2$ (3.34) | 12 | – | 12 | 13 |
| RS.CO.NMe$_2$ (3.35) | 1.3 | – | 2.8 | 0.8 |
| R.CS.NMe$_2$ (3.36) | 2.3 | – | 4.3 | 2.1 |
| Me$_2$N.CS.SS.CS.NMe$_2$ | 0 | 0 | 0 | – |
| Me$_2$N.CS.NMe$_2$ | 15.5 | 32.0 | 17.0 | – |
| CS$_2$ | 8.4 | 20.8 | 7.8 | – |
| Me$_2$N.CS.SZnS.CS.NMe$_2$ | 49.3 | 60.0 | 69.7 | 9.9 |
| RS$_x$R | 12.1 | 26.5 | 24.4 | 14 |

[a]R = (mainly) A$_1$ and B$_1$ (see footnote b, Table 3.9).
[b]Mixture of 3.33–3.36 isolated after 1 hr and reheated with zinc oxide and 2-methylpent-2-ene for 4 hr at 140°.

of TMTD (tetramethylthiourea and carbon disulfide) and zinc dimethyldithiocarbamate (ZDMC), three types of compound, 3.33 – 3.35, derived from the olefin have been identified and the presence of a fourth, 3.36, has been inferred. Comparison of yields after 1 and 10 hr suggests strongly that the dithiocarbamyl disulfides, 3.33, have reacted with zinc oxide to form crosslinked sulfides and ZDMC. Confirmation that 3.33 are the precursors of the crosslinked sulfides lies in the demonstration (column 5 of Table 3.15) that when the isolated mixture of 3.33 – 3.36 obtained after 1 hr is heated with zinc oxide and the olefin, a good yield of sulfides is obtained and only 3.33 disappears. ZDMC is also formed in quantity. Synthetic dithiocarbamyl disulfides 3.33 (R = A$_1$ or Et) behave similarly, whereas the monosulfide 3.34 (R = A$_1$) does not [67].

Clarification of the pathway by which these mixed disulfides form 'crosslinked' methylpentenyl sulfides was obtained by heating 3.33 (R = Et) with 2-methylpent-2-ene and zinc oxide at 140° when the following stoichiometry was observed[67]:

$$\text{EtSS.CS.NMe}_2 \xrightarrow[\text{ZnO}]{\text{RH}} \text{EtS}_{2.5}\text{Et} + \text{RS}_{1.4}\text{Et} + \text{ZDMC} \qquad (3.34)$$

1.78 mole          0.55 mole    0.45 mole    0.72 mole

Crosslinked sulfides are thus formed by two routes of approximately equal importance: one involving two molecules of the precursor, the other involving one molecule of the precursor and one molecule of olefin.

## 3.6 THE COURSE OF SULFUR VULCANIZATION

The results of the chemical studies with model olefins described in Sects.3.4 and 3.5 now allow a fairly clear picture to be drawn of the course (or pathway) of vulcanization by the technically important accelerated sulfur or sulfur donor systems. Thus the main sequence of reactions is that generalized in Fig.3.8 in which the term 'active sulfurating agent' denotes the species which interacts with the rubber to form

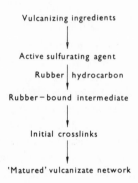

Vulcanizing ingredients

Active sulfurating agent

Rubber | hydrocarbon

Rubber–bound intermediate

Initial crosslinks

'Matured' vulcanizate network

**Fig. 3.8.** The sequence of reactions showing the course of sulfur vulcanization.

carbon-sulfur bonds; the product of this reaction, 'the rubber-bound intermediate', is identified with the polysulfide pendent groups of Fig.3.1 on the basis of the chemistry described in Sect.3.5. Each of the steps in the sequence appears to consist of at least two consecutive or competing reactions, some of which involve equilibria. The overall picture is therefore extremely complicated and many of the details, especially the mechanisms of individual reactions, remain unclarified or in dispute. This complexity is partly due to the number and variety of the vulcanizing ingredients which are used to obtain a crosslinked network but is largely the result of the generally high reactivity of C=C and S–S bonds, leading to reactions not involving the formation of crosslinks. Complexity is further increased by the fact that individual reactions assume more or less importance in rubbers of different structures. The parts of the sequence in most doubt are the nature of the active sulfurating agent and its reaction with the rubber.

### 3.6.1 Nature and Formation of the Active Sulfurating Agent

The presence of a polysulfide chain terminated by an accelerator fragment in the rubber-bound intermediate compound implies that this grouping already exists in the active sulfurating agent. There is controversy, however, as to whether the latter also contains zinc, it being regarded basically as either:

$$XS.S_y.SX$$

**3.37**

$$XS.S_a.Zn.S_b.SX$$

**3.38**

with ligand L on the Zn

[X = R$_2$N− (in **3.37** only) benzothiazolyl-C− or $R_2N.\overset{\displaystyle S}{\underset{\displaystyle \|}{C}}$− ;

L (ligand) = $\equiv$N: or O: of Zn(OCOR)$_2$]

In the case of trialkylethylene sulfuration, the overall effect of zinc, elaborated in Sect.3.4, is clear. It is to make sulfur crosslinking a disubstitutive process and to increase greatly the number of crosslinks obtained. However, even these effects could be ascribed in a rational way to the influence of zinc compounds on the processes taking place subsequent to formation of the rubber-bound intermediate (see p.95). Part of the difficulty of deciding whether zinc is implicated in initial crosslink formation is that crosslinking mechanisms not involving zinc do exist (*cf.* unaccelerated sulfuration) and in many cases can be invoked to interpret the experimental results satisfactorily. Evidence that zinc, when present, plays some part in the sulfuration step − at least in the case of trialkylethylenes − is as follows.

(1) Zinc accelerator thiolate complexes such as **3.21** and **3.22** are effective catalysts for removing sulfur from organic polysulfides and feeding it to olefins to form new polysulfides (see eq. 3.30); it therefore seems probable that they can perform the same function from elemental sulfur (or, at least, from polysulfide ions).

(2) Mixtures of complex **3.21** (or similar complexes) with sulfur and zinc oxide vulcanize *cis*-IR rapidly and effectively at 140°[64].

(3) In the presence of cyclohexylammonium benzothiazole-2-thiolate, sulfur, zinc oxide and propionic acid, 2-methylpent-2-ene and the diene,

**3.8**, are sulfurated at 20° – albeit at a very slow rate [13]. The products are the usual nitrogenous substances, zinc sulfide, and dialkenyl polysulfides. Under these conditions, desulfuration and thermal decomposition of polysulfides appear to be extremely slow reactions, as demonstrated by the very small concentrations of mono- and disulfides and of methylpentadienes formed from 2-methylpent-2-ene and by the similarly small concentrations of trienes and cyclic monosulfides formed from **3.8**. It is significant, then, that when zinc oxide and propionic acid are omitted from the reactants, the total amount of sulfuration (measured by the combined yields of organic polysulfides and nitrogenous compounds) taking place in 4 months falls nearly to zero (Table 3.16).

**TABLE 3.16**

Product Yields in the Accelerated Sulfuration of 2-Methylpent-2-ene at Room Temperature in the Absence and Presence of Zinc[13]

|  | Distribution of initial $S_8$ among products (%) | |
|---|---|---|
|  | Zinc oxide and propionic acid absent | Zinc oxide and propionic acid present |
| Crosslinked sulfides | 0.6 | 24.8 |
| Nitrogenous compounds | 4.7 | 2.5 |
| Zinc sulfide | – | 5.5 |
| (Recovered sulfur) | ca. 97 | 20.0 |

(4) TMTD reacts with cis-IR at 140° in the absence of zinc oxide, but no crosslinking takes place. Neither does crosslinking occur when the reaction product is subsequently heated with zinc oxide [67,69]. Hence the presence of zinc oxide is necessary not only for the conversion of the rubber-bound intermediate into crosslinks but also for the formation of the rubber-bound intermediate itself.

(5) Even in cases where zinc accelerators (ZDCs or ZMBT) are not added directly to the sulfuration mixture, their formation from accelerator and activators is rapid. For example, MBT is almost completely converted into ZMBT on heating in admixture with zinc oxide, sulfur and the saturated rubber, poly(isobutene), for one minute at 140° [70]. Evidence also exists for the formation of the zinc thiolate (or its complexes) during the induction period prior to onset of crosslinking during vulcanization of unsaturated rubbers accelerated by MBT, MBTS and thiazole-sulfenamides [70,71].

The above evidence points to **3.38** being the active sulfurating agent, rather than **3.37**. In the absence of zinc, however, when sulfuration is rapid but otherwise similar to unaccelerated sulfuration (Table 3.11), reaction of the accelerator with sulfur to form **3.37** is a likely initiating process and **3.37** may well be regarded as a special, and possibly more reactive, example of the $TS_a.S_bT$ of eqs.3.7 and 3.19. Polysulfides of the type **3.37** are formed during the vulcanization of NR, both without and with zinc[71-3], and they have been regarded as the active sulfurating agents in both cases[74]. On the basis of the above evidence, **3.38** is preferred when zinc is present.

The question of the direct involvement of zinc in the sulfuration step of mono- and dialkylethylenes is quite unresolved. These olefins show a relative lack of response to zinc activators, both in terms of the yields of crosslinked products formed (which is well illustrated for the dialkylethylene rubber, *cis*-BR, in Fig. 3.9[10]) and their structures (pp. 97 – 98). Even in the presence of zinc, olefins like cyclohexene and

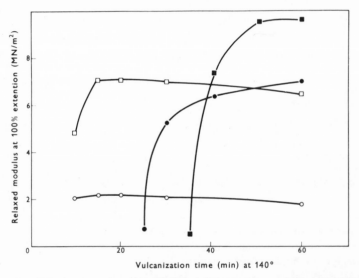

**Fig. 3.9.** Effect of zinc oxide on yield of crosslinks (as measured by relaxed modulus) in *cis*-BR and NR [Mix composition: *cis*-BR (●, ■) or NR (○, □), 100; sulfur, 1.8; CBS, 1.0; lauric acid, 1.4; zinc oxide, 0 (○, ●) or 3 (□, ■) parts by wt.] (After Skinner[10]).

hex-1-ene respond rather like 2-methylpent-2-ene does to a mixture of sulfur and accelerator (but without zinc). The reason for this may well be found in the recent observation[10,66] that the less-substituted olefins afford high yields of nitrogenous compounds of the type, RSX (where R = alkyl or alkenyl derived from the olefin and X = an accelerator fragment). Since these compounds do not react further to give crosslinked sulfides (and are otherwise stable), they represent loss of accelerator from the system. Consequently, the concentration of zinc-accelerator-thiolate desulfurating agents (such as **3.21** and **3.22**) is low and the reactions characteristic of unaccelerated sulfuration predominate.

In trialkylethylenes, the two principal routes by which **3.38** may be formed, depending upon the composition of the vulcanizing system, are illustrated in Fig.3.10.

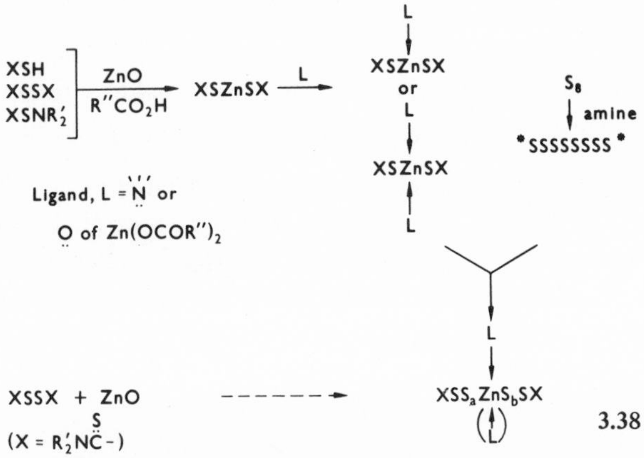

**Fig. 3.10.** Routes to the formation of the active sulfurating agent

$(X = \ \underset{S}{\overset{N}{\bigcirc}}C-$ , $R_2' N.CS-$, etc.).

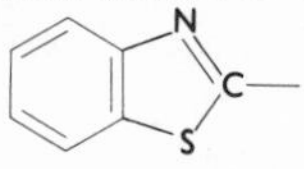

In one route, the accelerator (often a thiol, disulfide, or sulfenamide based on **3.3**) reacts with zinc activators to form a zinc thiolate (ZMBT or ZDC). As pointed out on p.96, the latter are only sparingly soluble in hydrocarbons (including rubbers), but they form readily soluble complexes with primary or secondary nitrogen bases or with zinc carboxylates. The nitrogen bases may be amines derived from sulfenamide accelerators or present in raw NR or they may be amines or guanidines added as so-called secondary accelerators. Many such complexes, *e.g.*, **3.21** and **3.22**, have been prepared[63,75-9]. The zinc carboxylate arises, of course, from the excess of zinc oxide and carboxylic acid present. In this case, the complexes have not been characterized, although there is good evidence for their existence [13,64,74,80]. They may well be non-stoichiometric.

Although the basic nitrogen complexes seem to be able to open the $S_8$ ring, the carboxylate complexes apparently cannot. Thus, the latter catalyze interchange reactions of disulfides but not the insertion of molecular sulfur into disulfides, whereas the former are good catalysts for sulfur insertion reactions [64]. Even so, it may well be that in practical vulcanization the $S_8$ ring is opened by more active species, such as the nitrogen bases themselves. Cleavage of cyclooctasulfur by primary and secondary amines, for example, occurs quite rapidly even at room temperature and forms polysulfide ions and, in some cases, radical-ions[81-4,13], *e.g.*, eqs. 3.35, 3.36, which should be much more reactive towards the zinc complexes than elemental sulfur is:

$$4\,R_2NH \;+\; zS_8 \;\rightleftharpoons\; (R_2NH_2)_2^+ \; S_x^= \;+\; R_2N.S_y.NR^2 \tag{3.35}$$

$$S_x^= \;\rightleftharpoons\; {}^-S_a\cdot \;+\; \cdot S_b^- \tag{3.36}$$

No examples of the postulated zinc perthiolate complexes **3.38** have yet been isolated, although analogs, *e.g.*, **3.39**, have recently been prepared and have been shown to possess rapidly exchangeable sulfur atoms[85-87]:

3.39

The perthiolate complexes **3.38** therefore probably exist in a series of equilibria (eq.3.37) which lie well on the side of the thiolate complex and free sulfur. The average values of *a* and *b* in **3.38** will therefore depend on

$$\tag{3.37}$$

(ligands have been omitted for simplicity)

the relative concentrations of *reactive* sulfur and *soluble* zinc complex. Consequently, provided sufficient nitrogen base is present to open the $S_8$ ring and to solubilize the ZMBT or ZDC present, a high ratio of accelerator : sulfur implies low values of *a* and *b* in **3.38** — and hence short sulfide crosslinks at all stages of the subsequent sulfuration. Conversely, a low ratio of accelerator : sulfur implies more polysulfide-sulfur in **3.38** and, initially at least, in the sulfuration products.

The second route for the formation of **3.38** is from a sulfur donor, *e.g.*, TMTD. TMTD reacts exothermically with zinc oxide at 136–147° to form zinc dimethyldithiocarbamate, dimethylthiocarbamate, **3.40** and sulfur [88], probably [1] *via* a sequence involving initial attack of oxygen at the 'hard' thiocarbonyl carbon of TMTD, displacing a perthioanion which subsequently attacks the 'soft' disulfide-sulfur atom of TMTD (eqs.3.38 and 3.39); further reaction of the thiuram trisulfide with zinc oxide then leads to the formation of higher thiuram polysulfides and,

$$\cdots Zn^{++}\cdots O = \overset{\overset{\displaystyle S}{\|}}{\underset{\underset{\displaystyle NMe_2}{|}}{C}} \overset{\frown}{S} - S - S - \overset{\overset{\displaystyle S}{\|}}{\underset{\underset{\displaystyle NMe_2}{|}}{C}} \longrightarrow \cdots Zn^{++}\cdots O^- - \overset{\overset{\displaystyle S}{\|}}{\underset{\underset{\displaystyle NMe_2}{|}}{C}} + {}^-ss - \overset{\overset{\displaystyle S}{\|}}{\underset{\underset{\displaystyle NMe_2}{|}}{C}} \qquad (3.38)$$

$$\overset{\overset{\displaystyle S}{\|}}{\underset{\underset{\displaystyle NMe_2}{|}}{C}} - S - S^- + Me_2N.\overset{\frown}{CS.S} - S.CS.NMe_2 \longrightarrow Me_2N.CS.SSS.CS.NMe_2 + {}^-S.CS.NMe_2 \quad (3.39)$$

$$\cdots Zn^{++}\cdots O^- - \overset{\overset{\displaystyle S}{\|}}{\underset{\underset{\displaystyle NMe_2}{|}}{C}} + {}^-S.CS.NMe_2 \longrightarrow Me_2N.CS.S.Zn.O.CS.NMe_2 \qquad (3.40)$$

eventually, of molecular sulfur. In a series of exchange reactions, **3.40** can then become disproportionate to zinc dimethylthiocarbamate and ZDMC, and the latter, by reaction with a thiuram polysulfide or $S_8$, can subsequently form a zinc dimethylperthiocarbamate **3.41** — a specific example of an active sulfurating agent **3.38** in which $a$ and $b$ are small:

$$2\, Me_2N.CS.S.Zn.O.CS.NMe_2 \rightleftharpoons (Me_2N.CS.O)_2 Zn + (Me_2N.CS.S)_2 Zn$$

$$(3.41)$$

$$(Me_2N.CS.S)_2 Zn + Me_2N.CS.S_x.CS.NMe_2 \rightleftharpoons Me_2N.CS.SS.CS.NMe_2\ +$$

$$Me_2N.CS.S.Zn.S_{x-1}.CS.NMe_2 \longrightarrow Me_2N.CS.S_a.Zn.S_b.CS.NMe_2$$

**3.41**

$$(3.42)$$

The other main type of sulfur donor, exemplified by dithiobis-morpholine **3.2**, is normally used in conjunction with an accelerator of the sulfenamide type since it is relatively slow-acting when used alone. It probably functions by decomposing to form morpholinium polysulfides (a process which is catalyzed both by free morpholine and by traces of hydrosulfide or sulfide ion [13] which then react with a ZMBT complex, formed from the accelerator in the usual way, to give **3.38**). Again, $a$ and $b$ will be small unless massive amounts of the sulfur donor are used.

### 3.6.2 Formation of Rubber-Bound Intermediates and their Conversion into Crosslinks

Good evidence was presented in Sect.3.5 that alkenyl dithiocarbamyl disulfides are the precursors to the crosslinked sulfides formed by reaction of 2-methylpent-2-ene with TMTD and zinc oxide. That the same process occurs in a polymeric environment is evident from Fig.3.11 which refers to the vulcanization of *cis*-IR by the same reagents. The effect of re-heating

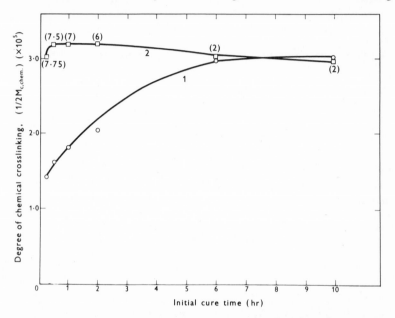

**Fig. 3.11.** Increase in degree of crosslinking induced by reheating extracted undercured *cis*-IR-TMTD-zinc oxide vulcanizates. Figures in parentheses on curve 2 are the heating periods (hr in vacuum at 140°) for the extracted vulcanizates of curve 1. (Courtesy Interscience Publishers).

undercured vulcanizates which have been freed from all non-network substances except zinc oxide and traces of zinc sulfide shows clearly that, even after a cure time as short as 20 minutes, the rubber network already has the potential for forming the maximum number of crosslinks. During the re-heating periods, sulfur and nitrogen are eliminated from the network in the form of one-molecule of ZDMC for each new crosslink produced.

The mechanism for the formation of the rubber-bound intermediate in this and analogous cases where ZMBT complexes are involved is still uncertain. Based on observations made in experiments with 2-methylpent-2-ene, *viz.*, (1) exclusive substitution of allylic hydrogen by sulfur, (2) absence of allylic rearrangement in the resulting alkenethio groups, and (3) widely varying ratios of α-methyl : α-methylene substitution occurring as the proportions of activators and of accelerators to sulfur are changed, Bateman, Moore, Porter and Saville [1] suggested a polar, largely concerted mechanism (eq.3.43) for the reaction between the active sulfurating agent **3.38** and olefin or olefinic rubber, R–H (where H is an α-methylene or α-methyl hydrogen atom). Nucleophilic attack of an incipient perthiolate ion on an allylic carbon atom forms the rubber-bound intermediate directly and displaces a hydrogen atom with its electron pair:

$$(3.43)$$

Receipt of the latter by the other sulfur chain of **3.38** is assisted by the simultaneous formation of zinc sulfide. This mechanism is able not only to provide a reasonable explanation [1] for the superiority of zinc over other metals in promoting vulcanization, but also to rationalize the effects of added ligands on the course of the reaction. Thus, the reaction rate will be controlled by the relative energetics of C–S bond formation and C–H bond fission; complexation of the zinc atom by electron-donating ligands such as amine-nitrogen or carboxylate-oxygen will increase the electron density on the $XS_x$ moieties of the sulfurating agent, which will in turn facilitate C–S bond formation by increasing the nucleophilicity of the attacking $XS.S_a$ group, but hinder C—H bond fission by reducing the electrophilicity of the receiving $XS_b$ group. The energetic contribution of bond formation will therefore increase relative to bond breakage. However, bond formation will also depend on steric accessibility at an allylic carbon atom, *i.e.*, it will be fastest at primary carbon, whereas bond fission will depend rather on no-bond resonance contributions of the type $R^+H^-$, *i.e.*, it will be fastest for a tertiary C–H bond. The enhancement of C–S bond formation relative to C–H bond fission will therefore increase substitution at primary α-carbon atoms, *i.e.*, at C–1 of 2-methylpent-2-ene (*vs.* C–4). This is the effect observed.

While this mechanism remains satisfactory for trialkylated olefins, some more recent results[10] with less-substituted olefins have called it into question. With these olefins, allylic rearrangement and *cis,trans*-isomerization of the double bond occur to a large extent during sulfuration even at room temperature (Table 3.13), suggesting that a free allylic ion or radical is involved. Such a feature would, of course, be inconsistent with the above mechanism. However, an alternative explanation exists for these isomerizations in that allylic di- and polysulfides have recently been found to undergo quite rapid thermal rearrangements (*e.g.*, eq.3.44)[89,90] which, with sulfides of suitable structure may lead to allylic rearrangement as well as to *cis,trans*-isomerization and racemization. The rate with which a dialkylethylenic sulfide undergoes this rearrangement at 20° is not known, but it could well be within the time scale of the room-temperature experiment.

$$\text{(structures)} \qquad (3.44)$$

It has already been established (see Sect.3.5) that the rubber-bound intermediate, $RS_a.SX$, is converted into crosslinks by two routes. The first of these is probably a simple disproportionation (eq.3.45), either catalyzed by a zinc-accelerator-thiolate complex[64, 77] or other nucleophile, or effected by a thermal free-radical chain reaction similar to the exchange reactions of dialkyl polysulfides (eqs.3.13-3.17). The second route, involving one molecule of $RS_a.SX$ and one of olefin, could well proceed *via* exchange of $RS_a.SX$ with **3.38** followed by a variant of eq.3.43:

$$2\,RS_a.SX \rightarrow RS_xR + XS.S_{2a-x}.SX \qquad (3.45)$$

$$RS_a.SX \; + \; XS.S_a.\overset{\overset{L}{\downarrow}}{Zn}.S_b.SX \longrightarrow RS_a.\overset{\overset{L}{\downarrow}}{Zn}.S_b.SX + XS.S_a.SX \qquad (3.46)$$

$$\qquad (3.47)$$

The $HS_bX$ species formed in eqs.3.43 and 3.47 will react with zinc oxide forming fresh sulfurating agent:

$$2\,XS_bH + ZnO \rightarrow XS_b.Zn.S_bX + H_2O \qquad (3.48)$$

### 3.6.3 Further Reactions of Polysulfide Crosslinks — Maturing of the Network

The crosslinks produced by these reactions are generally di- and polysulfides. The results described in Sect.3.4.1 and 3.4.2 show clearly, however, that dialkenyl di- and polysulfides so formed are subject to further reactions in the presence of accelerators, activators, and their transformation products. These reactions take place during the later part of the vulcanization cycle ('maturing of the network'), but they may also occur during subsequent service of the vulcanizate, if this is at an elevated temperature. They are therefore important in determining the rate at which physical properties change during usage and hence with the ultimate service life of the vulcanized rubber article.

The most important of the reactions are: (1) S—S bond interchange; (2) desulfuration, (3) decomposition.

*S—S Bond Interchange.* This well-established process may occur by a free-radical chain mechanism initiated by thermal homolysis of S—S bonds (*cf.* eqs.3.13—3.18) or by an anionic chain mechanism initiated by nucleophiles (*e.g.,* zinc-accelerator-thiolate complexes [64]). It may proceed quite rapidly even at moderate temperatures but is not detectable in an undeformed vulcanizate since it does not result in any change in the overall concentration of crosslinks or in their average sulfur rank. However, if interchange occurs while the vulcanizate is deformed or swollen, it may result in creep, stress-relaxation or incremental swelling or, after removal of the deforming force or on deswelling, permanent set.

*Desulfuration.* The progressive desulfuration of polysulfides to di- and monosulfides by zinc-accelerator-thiolate complexes (such as **3.21** and **3.22**) and the utilization of the extruded sulfur for new crosslinking have been described on pp. 93—95. Since this reaction produces changes in both total crosslink concentration and the relative proportions of mono-, di- and polysulfides, it causes alterations in the physical properties of the vulcanizate. The environmental factors favorable to desulfuration are: (1) moderate vulcanization or service temperatures, (2) high concentrations of soluble zinc-accelerator complexes relative to crosslinks, *i.e.,* high accelerator : sulfur ratios combined with adequate activator concentrations. With the common synthetic rubbers, removal of accelerator from the vulcanizing system by combination with the rubber in the form of inactive groups (p.106) denudes the system of desulfurating agent so that vulcanizates remain di- and polysulfides to a much greater extent than with NR (Fig.3.12) [10].

*Decomposition.* Thermal decomposition of di- and polysulfide crosslinks (pp.96 and 98—101) leads to changes in vulcanizate properties, not only by reason of the reduced concentration of crosslinks, but because the main polymer chains become modified by conjugated diene and triene groups and by cyclic sulfide groups. These partly destroy the stereoregularity of the rubber and the conjugated groups also increase the rubber's susceptibility to oxidation [91]. The only factor directly affecting the rate of crosslink decomposition is the vulcanization or service temperature but, in practice, desulfuration competes with decomposition and, since monosulfide crosslinks are stable thermally, rapid desulfuration leads to reduced degradation of the network. Low accelerator : sulfur

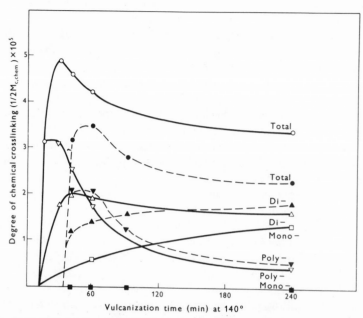

**Fig. 3.12.** Distribution of mono, di, and polysulfide crosslinks as a function of vulcanization time for *cis*-BR (filled symbols) and NR (open symbols). Mix composition: *cis*-BR or NR, 100; sulfur, 1.8; CBS, 1.0; lauric acid, 1.4; zinc oxide, 3.0 parts by wt. (After Skinner[10]).

ratios or inadequate concentrations of amine or zinc carboxylate activators therefore favor crosslink decomposition and main-chain modification. The influence of the degree of substitution at the olefinic double bond on the rate of decomposition of derived polysulfide crosslinks is not known since few studies with relevant low-molecular-weight olefins have been carried out. Cyclic monosulfides are formed from dialkylated olefins[66], but there is some evidence[10] that they are less prominent in the vulcanization of *cis*-BR compared to that of NR. In any case, with the less-substituted rubbers, decomposition of polysulfides is likely to lead to vicinal crosslinking based on the structures **3.6** and **3.7** (p.98), a feature unimportant in the accelerated vulcanization of NR. Again, removal of accelerator from the vulcanizing system is an important contributing factor.

### 3.6.4 Conclusion

The principal reactions taking place during the sulfur vulcanization of olefinic rubbers are presently best represented as shown in Fig.3.13, it being recognized that competition between the various pathways will occur and will be dependent on the vulcanizing conditions: structure of the rubber hydrocarbon, nature of non-rubber substances in the raw rubber, nature and relative concentrations of vulcanizing agent (sulfur or sulfur donor), accelerators and activators where present, and the vulcanizing temperature. In general, the reactions will each have proceeded to some arbitrary extent when the vulcanization process is terminated (after a time decided empirically by the achievement of some specified desirable combination of physical properties) and will therefore continue during the service life of the vulcanizate at rates determined by their individual activation energies and the environmental temperature. The complex situation this leads to can be readily appreciated by considering a

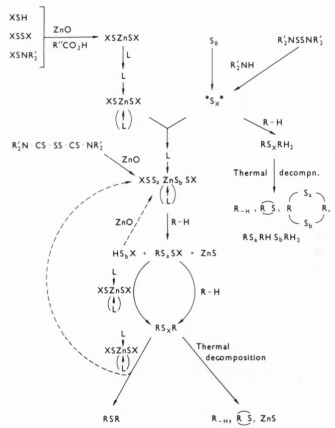

**Fig. 3.13.** Reaction pathways in the sulfur vulcanization of olefinic rubbers [R—H = rubber hydrocarbon in which H is an α-methylenic or α-methylic hydrogen atom; X = accelerator residue: $R_2'N.CS.—$, etc; L = ligand (basic nitrogen or zinc carboxylate)].

bulky rubber article such as a large truck tyre. Apart from the complex construction of the tyre (which often involves blends of different rubbers), it is clear, when the limiting rate of heat transfer in rubber is taken into account, that the vulcanization time-temperature profile must vary considerably throughout the tyre. When it is subsequently used on the road, the tyre is subjected continuously to a temperature-cycling process which may involve temperatures varying between ambient and at least $120°$ (if long distances are covered at high speeds). A description of the history of such a tyre in terms of the chemical processes taking place will indeed be complicated!

## 3.7 REFERENCES

1. L. Bateman, C.G. Moore, M. Porter and B. Saville, in "The Chemistry and Physics of Rubber-like Substances," Chapter 15, L. Bateman, Ed. (Maclaren, London, 1963).
2. B. Saville and A.A. Watson, *Rubber Chem. Technol.*, **40**, 100 (1967).
3. G. Kraus and G.A. Moczvgemba, *J. Polymer Sci.*, **2A**, 277 (1964).
4. F.P. Baldwin, P. Borzel and H.S. Makowski, *Rubber Chem. Technol.*, **42**, 1167 (1969).
5. L. Bateman, J.I. Cunneen, C.G. Moore, L. Mullins and A.G. Thomas, in "The Chemistry and Physics of Rubber-like Substances," Chapter 19, L. Bateman, Ed. (Maclaren, London, 1963).
6. R.T. Armstrong, J.R. Little and K.W. Doak, *Ind. Eng. Chem.*, **36**, 628 (1944).
7. E.H. Farmer and F.W. Shipley, *J. Chem. Soc.*, 1519 (1947).
8. L. Bateman, R.W. Glazebrook, C.G. Moore, M. Porter, G.W. Ross and R.W. Saville, *J. Chem. Soc.*, 2838 (1958).
9. C.G. Moore and M. Porter, *J. Chem. Soc.*, 6390 (1965).
10. T.D. Skinner, Ph.D.Thesis, Council for National Academic Awards, 1971, see *Rubber Chem. Technol.*, **45**, 182 (1972).
11. M. Porter, B. Saville and A.A. Watson, *J. Chem. Soc.*, 346 (1963).
12. L. Bateman, R.W. Glazebrook and C.G. Moore, *J. Chem. Soc.*, 2846 (1958).
13. M. Porter, unpublished work.
14. A.W. Weitkamp, *J. Amer. Chem. Soc.*, **81**, 3430 (1959).
15. G.M.C. Higgins, M. Porter and B.K. Tidd, unpublished work.
16. W.A. Pryor, "Mechanisms of Sulfur Reactions," Chapter 6 (McGraw-Hill, New York, N.Y., 1962).
17. N. Lozach and J. Vialle, in N. Kharasch and C.Y. Meyers, Eds., "Organic Sulfur Compounds," Vol. 2, Chapter 10 (Pergamon Press, New York, N.Y., 1966).
18. F.J. Linnig, E.J. Parks and L.A. Wall, *J. Res. Nat. Bur. Stand*, **65A**, 79 (1961).
19. M.L. Selker and A.R. Kemp, *Ind. Eng. Chem.*, **39**, 895 (1947).
20. N. Lozac'h and Y. Mollier, *Bull. Soc. Chim. France*, 1389 (1959).
21. Critical summaries of early work are to be found in:
    (a) A. Springer, *Monatsch.*, **78**, 200 (1948);
    (b) W. Scheele, *Rubber Chem. Technol.*, **34**, 313 (1961).
22. W. Scheele, *Rubber Chem. Technol.*, **43**, 588 (1970).
23. G.W. Ross, *J. Chem. Soc.*, 2856 (1958).
24. W. Scheele and R. Huischen, *Kaut. u. Gummi*, **21**, 423 (1968).
25. W. Scheele, H. Müller and W. Schulze, *Kaut. u. Gummi*, **14**, WT 364 (1961); *Rubber Chem. Technol.*, **33**, 910 (1964).

26. I. Kende, T.L. Pickering and A.V. Tobolsky, *J. Amer. Chem. Soc.*, **87**, 5582 (1965).
27. A.V. Tobolsky and A. Eisenberg, *J. Amer. Chem. Soc.*, **81**, 780 (1959).
28. L. Bateman, C.G. Moore and M. Porter, *J. Chem. Soc.*, 2866 (1958).
29. E.H. Farmer, *J. Soc. Chem. Ind.*, **66**, 86 (1947).
30. G.F. Bloomfield and R.F. Naylor, XIth. Int. Congr. Pure Appl. Chem., **2**, 7 (1947).
31. W.A. Pryor, "Mechanisms of Sulfur Reactions," Chapter 5 (McGraw-Hill, New York, N.Y., 1962).
32. R.E. Davis, in "Inorganic Sulfur Chemistry," Chapter 4, G. Nickless, Ed. (Elsevier, London, 1968).
33. G. Kaiser, K. Hummel and W. Scheele, *Kaut u. Gummi*, **19**, 347 (1966).
34. C.G. Moore and B.R. Trego, *J. Appl. Polymer Sci.*, **8**, 1957 (1964).
35. W. Scheele and E. Cep, *Kaut. u. Gummi*, **20**, 401 (1967).
36. For references, see W.A. Pryor, "Mechanisms of Sulfur Reactions," Chapters 2 and 3 (McGraw-Hill, New York, N.Y., 1962).
37. J.R. Shelton and E.T. McDonel, *Proc. Int. Rubb. Conf.* (Washington), 596 (1959).
38. G.A. Blokh, *Doklady Akad. Nauk*, **129**, 361 (1959); *Rubber Chem. Technol.*, **33**, 1005 (1960).
39. B.A. Dogadkin and V.A. Shershnev, *Polymer Sci. U.S.S.R.*, **1**, 21 (1960).
40. B.A. Dogadkin and V.A. Shershnev, *Kolloid Zhur.*, **21**, 244 (1959); *Rubber Chem. Technol.*, **33**, 398 (1960).
41. S.E. Bresler, B.A. Dogadkin, E.N. Kazbekov, E.M. Saminskii and V.A. Shershnev, *Vysokomol. Soed.*, **2**, 174 (1960); *Rubber Chem. Technol.*, **34**, 318 (1961).
42. L.S. Degtyanev and L.N. Ganyuk, *Polymer Sci. U.S.S.R.*, **6**, 31 (1964).
43. T.L. Pickering, K.J. Saunders, and A.V. Tobolsky, *J. Amer. Chem. Soc.*, **89**, 2364 (1967).
44. T.L. Pickering, K.J. Saunders and A.V. Tobolsky, in "The Chemistry of Sulfides," A.V. Tobolsky, Ed. (Interscience Publishers, New York, N.Y., 1968), p.61.
45. M. Porter and B.K. Tidd, Paper to IVth Organic Sulfur Symposium, Venice, 1970, to be published.
46. C. Walling, "Free Radicals in Solution " (John Wiley & Sons, New York, N.Y., 1957), pp.323–6.
47. W.A. Pryor, "Mechanisms of Sulfur Reactions " (McGraw-Hill, New York, N.Y., 1962), p.89 *et seq.*
48. J. Tsurugi, in "Mechanisms of Reactions of Sulfur Compounds," Vol. 2, N. Kharasch, B.S. Thyagarajan and A.I. Khodair, Eds. (Intra-Science Research Foundation, 1967), p.229.
49. S.O. Jones and E.E. Reid, *J. Amer. Chem. Soc.*, **60**, 2452 (1938).
50. R.F. Naylor, *J. Polymer Sci.*, **1**, 305 (1946).
51. R.F. Naylor, *J.Chem. Soc.*, 1532 (1947).
52. G.F. Bloomfield, *J. Chem. Soc.*, 1547 (1947).
53. S. Chubachi, P.K. Chatterjee and A.V. Tobolsky, *J. Org. Chem.*, **32**, 1511 (1967).
53a. C.G. Moore and M. Porter, *Tetrahredron*, **6**, 10 (1959).
54. M. Porter, Ph.D. Thesis, University of London, 1964.
55. A.A. Watson, unpublished work.
56. B.K. Tidd, unpublished work.
57. R. Cotton, unpublished results.
58. C.G. Moore and B.R. Trego, *Tetrahedron*, **18**, 205 (1962).
59. C.G. Moore and B.R. Trego, *Tetrahedron*, **19**, 1251 (1963).
60. D.N. Harpp, J.G. Gleason and J.P. Snyder, *J. Amer. Chem. Soc.*, **90**, 4181 (1968).
61. D.S. Campbell, *J. Appl. Polymer Sci.*, **14**, 1409 (1970).
62. C.R. Parks, D.K. Parker, D.A. Chapman and W.L. Cox, *Rubber Chem. Technol.*, **43**, 572 (1970).

63. B. Milligan, *J. Chem. Soc.*, **A**, 34 (1966).
64. B. Milligan, *Rubber Chem. Technol.*, **39**, 1115 (1966).
65. Based on the results of refs. 10 and 13.
66. E.C. Gregg and S.E. Katrenick, *Rubber Chem. Technol.*, **43**, 549 (1970).
67. A.A. Watson, Ph.D. Thesis, University of London (1965).
68. C.G. Moore, in "Proceedings of the Natural Rubber Producers' Research Association Jubilee Conference," Cambridge, 1964, L. Mullins, Ed. (Maclaren, London, 1965), p.168 *et seq.*
69. C.G.Moore and A.A. Watson, *J. Appl. Polymer Sci.*, **8**, 581 (1964).
70. I. Auerbach, *Ind. Eng. Chem.*, **45**, 1526 (1953).
71. R.H. Campbell and R.W. Wise, *Rubber Chem. Technol.*, **37**, 650 (1964).
72. R.H. Campbell and R.W. Wise, *Rubber Chem. Technol.*, **37**, 635 (1964).
73. E.N. Kavun, V.A. Shershnev and B.A. Dogadkin, *Soviet Rubber Technol.*, **28**, No. 6, 17 (1969).
74. A.Y. Coran, *Rubber Chem. Technol.*, **37**, 679 (1964); **38**, 1 (1965).
75. J.G. Lichty, U.S. Patent No. 2,129,621 (1938).
76. G. Spacu and C.G. Macarovici, *Bull. Sec. Sci. Acad. Roumaine*, **21**, 173 (1938); *Chem. Abst.*, **34**, 4007 (1940).
77. G.M.C. Higgins and B. Saville, *J. Chem. Soc.*, 2812 (1963).
78. E. Coates, B. Rigg, B. Saville and D. Skelton, *J. Chem. Soc.*, 5613 (1965).
79. S.K. Gupta and T.S. Srivastava, *J. Inorg. Nuclear Chem.*, **32**, 1611 (1970).
80. B.C. Barton and E.J. Hart, *Ind. Eng. Chem.*, **44**, 2444 (1952).
81. R.E. Davis and H.F. Nakshbendi, *J. Amer. Chem. Soc.*, **84**, 2085 (1962).
82. W.G. Hodgson, S.A. Buckler and G. Peters, *J. Amer. Chem. Soc.*, **85**, 543 (1963).
83. R. MacColl and S. Windwer, *J. Phys. Chem.*, **74**, 1261 (1970).
84. Y. Sasaki and F.P. Olsen, *Can. J. Chem.*, **49**, 283 (1971).
85. J.P. Fackler, D. Coucouvanis, J.A. Fetchin and W.C. Seidel, *J. Amer. Chem. Soc.*, **90**, 2784 (1968).
86. J.P. Fackler, J.A. Fetchin and J.A. Smith, *J. Amer. Chem. Soc.*, **92**, 2910 (1970).
87. J.P. Fackler and J.A. Fetchin, *J. Amer. Chem. Soc.*, **92**, 2912 (1970).
88. Y. Kawaoka, *J. Soc. Rubber Ind. Japan*, **16**, 397 (1943).
89. D. Barnard, T.H. Houseman, M. Porter and B.K. Tidd, *Chem. Commun.*, 371 (1969).
90. B.K. Tidd, *Int. J. Sulfur Chem.*, **C6**, 101 (1971).
91. D.S. Campbell, *J. Appl. Polymer Sci.*, **13**, 1013 (1969).

*Chapter 4*

# THIOLS

Atsuyoshi Ohno

*Institute for Chemical Research*
*Kyoto University, Uji, Kyoto 611, Japan*

and

Shigeru Oae

*Department of Chemistry*
*University of Tsukuba, Sakura-mura*
*Niihari-gun, Ibaraki-ken, Japan*

119

## 4.1 INTRODUCTION

Thiols, or mercaptans, are sulfur analogs of alcohols and may be characterized by their extraordinarily unpleasant odor. Indeed, for trained chemists, their stink is as sensitive an analytical tool as tlc. Probably because of this reason, not much research has been done on the chemistry of thiols. Nevertheless thiols play very important roles for organisms and it is no exaggeration to say that no organism can maintain its life without thiols.

In this chapter, physical properties of thiols are compared with those of their oxygen analogs at first, then their preparations and reactions are discussed. Some aspects of the biological functions of thiols will also be discussed.

## 4.2 PHYSICAL PROPERTIES

Bond dissociation energies of some thiols are listed in Table 4.1 as well as those of their oxygen analogs [1-3].

TABLE 4.1

Bond Dissociation Energies of Thiols and Alcohols

| Compound | Bond Dissociation Energy, kcal/mole | |
|---|---|---|
|  | X = S | X = O |
| HX-H | 89 | 116 |
| $CH_3X$-H | 89 | 100 |
| $C_2H_5X$-H | 87 | 63.5 |
| $CH_3$-XH | 67 | 90 |
| $C_2H_5$-XH | 63.5 | 90 |
| $PhCH_2$-XH | 53 | 73 |

TABLE 4.2

Bond Angles in Thiols and Alcohols

| Compound | Bond Angle, degree | |
|---|---|---|
| | X = S | X = O |
| /X\ H   H | 92.2 ± 0.10 | 104.45 ± 0.10 |
| /X\ CH₃   H | 100.3 ± 0.2 | 108   ± 2 |

Apparently, S—H (C—S) bond energy is less than that of O—H (C—O). The ease of free-radical hydrogen abstraction from S—H in comparison with that from O—H is accounted for by this fact. It is interesting to note that the bond dissociation energy of H—Br, which is also a good hydrogen donor, is 87.5 kcal/mole. As is seen in Table 4.2, bond angles around the sulfur atom in thiols are smaller than those around the oxygen atom in corresponding oxygen analogues [4,5]. This indicates that the bonding orbitals on a sulfur atom have greater p-character than those on an oxygen atom.

Hydrogen-bonds in thiols are much weaker than those in alcohols. This has been confirmed by nmr spectroscopy [6,7]. For example, the equilibrium constant for the equilibrium between monomeric and dimeric forms of thiophenol is only 0.0001 mole$^{-1}$ in carbon tetrachloride at room temperature. That thiols have less ability for hydrogen bonding than alcohols can also be seen from their boiling points, whcih are cited in Table 4.3.

Weaker bondings between sulfur and hydrogen in comparison with those between oxygen and hydrogen makes thiols more acidic than corresponding alcohols. The $pK_a$ values of thiols and alcohols are summarized in Table 4.4.[8-14]. More detailed results are reported in refs. 13 and 14. Interestingly, the difference in $pK_a$ of butanol and phenol is greater

TABLE 4.3

Boiling Points of Thiols and Alcohols

| Compound | Boiling Point, °C | |
|---|---|---|
| | X = S | X = O |
| $H_2X$ | −61 | 100 |
| $CH_3XH$ | 6 | 65 |
| $C_2H_5XH$ | 37 | 78 |
| $C_6H_5XH$ | 168 | 181.4 |

TABLE 4.4

Acidities of Thiols and Alcohols at 25°C

| Compound | pK$_a$ | |
| --- | --- | --- |
| | X = S | X = O |
| H$_2$X | 7.24[a], 14.92[b] | 14 |
| C$_2$H$_5$XH | 10.60 | 18 |
| n-C$_4$H$_9$XH | 11.51 | 16 |
| C$_6$H$_5$XH | 8.3 | 9.94 |

[a] First ionization constant.          [b] Second ionization constant.

($\Delta$pK$_a$ ≈ 6) than that of their corresponding sulfur analogues ($\Delta$pK$_a$ ≈ 4). The greater contribution of canonical forms **4.1b** and **4.1c** for the oxygen compound than its sulfur analogue accounts for this phenomenon (see Chapter 5).

**4.1a**              **4.1b**              **4.1c**

The above data invoke that alcoholate anion is a much stronger base than the corresponding sulfur analogue and the former has greater affinity toward a proton than the latter. However, the order of nucleophilicities, or affinities toward a carbon atom in nucleophilic substitution, of these anions is reversed. Namely, the nucleophilic constants, defined by Swain and Scott, are 4.20 and 5.1 for OH$^-$ and SH$^-$, respectively [15]. The greater nucleophilicity of thiolate anion than the oxygen analogue is accounted for by the greater polarizability of electrons on sulfur atom than that of electrons on oxygen atom. Note that electrons in an M shell are less strongly affected by positive charges on a nucleus than those in an L shell. Weaker solvation (hydrogen-bonding) to thiolate anion than to alcoholate anion is also responsible for the greater nucleophilicity of the former than the latter.

Molecular ionization potentials of thiols, E$_{RSH}$, are correlated with the inductive substituent constants, $\sigma_I$, by [16]:

$$E_{RSH} = 10.46 + 22.2\sigma_I \qquad (4.1)$$

The value of 22.2 obtained for thiols may be compared with that (37.5) observed in alcohols [17]. The smaller value for thiols than for alcohols indicates that the sulfur atom is less sensitive to inductive effect than is the oxygen atom, due to larger polarizability of the sulfur atom.

## 4.3 PREPARATION

Since many reviews are available [18-20], this topic is briefly surveyed in this text.

### 4.3.1 By Nucleophilic Substitution

Thiols are directly obtained by the reaction of organic halides with metallic hydrosulfides such as NaSH (eq. 4.2) [21]. Alkyl sulfates react similarly. Alkyl sulfides are often by-products of the reaction.

$$RX + NaSH \longrightarrow RSH + NaX \qquad (4.2)$$

The reaction of an alkyl halide with thiourea followed by alkaline hydrolysis of the intermediate thiouronium salt is one of the most common methods of preparation [22-25]:

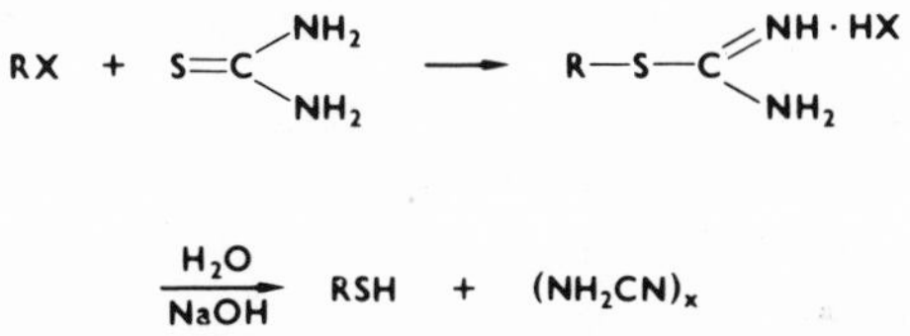

$$RX + S=C\begin{smallmatrix}NH_2\\NH_2\end{smallmatrix} \longrightarrow R-S-C\begin{smallmatrix}NH \cdot HX\\NH_2\end{smallmatrix}$$

$$\xrightarrow[NaOH]{H_2O} RSH + (NH_2CN)_x$$

(4.3)

Another important reaction concerns mainly the syntheses of aromatic thiols with the aid of xanthates [26,27]:

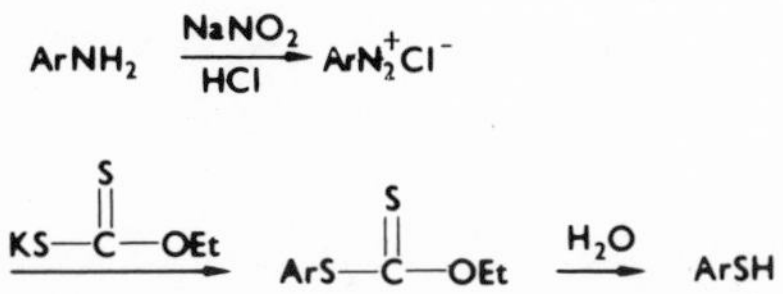

$$ArNH_2 \xrightarrow[HCl]{NaNO_2} ArN_2^+Cl^-$$

$$\xrightarrow{KS-\overset{S}{\overset{\|}{C}}-OEt} ArS-\overset{S}{\overset{\|}{C}}-OEt \xrightarrow{H_2O} ArSH \qquad (4.4)$$

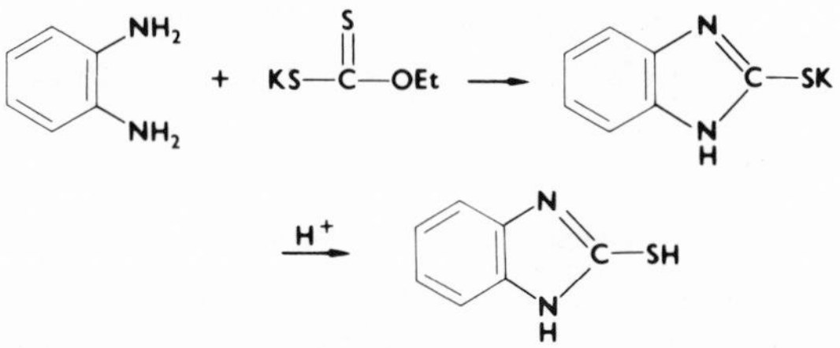

(4.5)

Ring-opening reactions of heterocyclic three-membered rings may be classified into these categories [18,28-30]:

$$H_2C\underset{O}{\overset{\diagdown\diagup}{\diagup\diagdown}}CH_2 \;+\; H_2S \;\longrightarrow\; HOCH_2CH_2SH \tag{4.6}$$

$$R-CH\underset{\underset{H}{N}}{\overset{\diagup\diagdown}{\diagdown\diagup}}CH_2 \;+\; H_2S \;\longrightarrow\; H_2NCH_2CH\overset{R}{\underset{SH}{\diagdown}} \tag{4.7}$$

$$H_2C\underset{S}{\overset{\diagdown\diagup}{\diagdown\diagup}}CH_2 \;+\; R_2NH \;\longrightarrow\; R_2NCH_2CH_2SH \tag{4.8}$$

### 4.3.2 By Addition to Olefins

Addition of hydrogen sulfide to an olefin affords small yield of thiol, which has the structure predicted by Markovnikov's rule [31,32]:

$$\underset{R'}{\overset{R}{\diagup}}C=C\underset{H}{\overset{R''}{\diagdown}} \;+\; H_2S \;\longrightarrow\; R'-\overset{R}{\underset{SH}{\underset{|}{\overset{|}{C}}}}-CH_2R'' \tag{4.9}$$

More sophisticated method in this category is radical addition of thioacetic acid followed by hydrolysis [33], although this reaction affords anti-Markovnikov adduct:

$$CH_3-\overset{O}{\overset{||}{C}}-SH \;+\; CH_2=C\underset{R'}{\overset{R}{\diagdown}} \;\longrightarrow\;$$

$$CH_3-\overset{O}{\overset{||}{C}}-SCH_2CH\underset{R'}{\overset{R}{\diagdown}} \;\overset{H_2O}{\longrightarrow}\; \underset{R'}{\overset{R}{\diagdown}}CHCH_2SH \tag{4.10}$$

### 4.3.3 By Reduction

Reduction of sulfonyl chloride is widely used for preparing aromatic thiols (eq. 4.11) [34,35]. Reducing agents may be Zn-HCl, LiAlH$_4$ and similar

$$ArSO_2Cl \;\overset{[H]}{\longrightarrow}\; ArSH \tag{4.11}$$

compounds. Disulfides are also potential precursors of thiols [36]:

$$RSSR \xrightarrow{2[H]} 2\,RSH \qquad (4.12)$$

### 4.3.4 From Organometallics

Thiols are obtained by the reaction of a Grignard reagent with elemental sulfur (eq. 4.13)[37]. Organolithium compounds work similarly[39-41].

$$RMgX \xrightarrow{S_8} RSMgX \xrightarrow{H_2O} RSH \qquad (4.13)$$

### 4.3.5 Miscellaneous Methods

Several other methods are possible for preparing thiols. Among them, noteworthy reactions are those in which a functional group separates from a sulfur atom. Thus, thiocyanates[42,43] and thiosulfates[44,45] yield thiols:

$$RSCN \xrightarrow{H_2SO_4} RS\overset{\overset{\displaystyle O}{\|}}{C}\text{-}NH_2 \longrightarrow RSH \qquad (4.14)$$

$$RSCN \xrightarrow{6\,[H]} RSH + CH_3NH_2 \qquad (4.15)$$

$$RSSO_3Na \xrightarrow{H_2O} RSH + NaHSO_4 \qquad (4.16)$$

Formations of thiols from O,O'-diarylthiocarbonates, which are prepared from phenols and thiophosgene, will be described in Chapter 5. Tautomerization of thione to enethiol will also be discussed in Chapter 5.

## 4.4 HYDROGEN TRANSFER

In Sect. 4.2, it is suggested that the sulfur-hydrogen bond is relatively weak and that the thiol may be a good hydrogen-transfer reagent. Yet, Walling proposed, in his review[46], that the reaction in eq. 4.17 would have negligible activation energy when polar and steric factors are favorable.

$$RS\cdot + H\text{-}R' \longrightarrow RS\text{-}H + \cdot R' \tag{4.17}$$

Since $C-H$ bond dissociation energy exceeds by only about 10 kcal/mole that of the $S-H$ bond, the hydrogen abstraction by $RS\cdot$ radicals from $R'-H$ may occur readily when the resulting $R'\cdot$ radicals are somewhat stabilized by resonance. One of the most elegant works supporting this postulation was done by Cohen and Wang[47]. When azobisdiphenylmethane, **4.2**, is decomposed in diphenyl-methane-[14]C, **4.3**, the resulting tetraphenylethane, **4.4**, has only 1.1% of the activity of the solvent. On the other hand, when the same reaction is carried out in the presence of thiophenol, the activity in **4.4** increases to 17%. The result is interpreted by the following scheme:

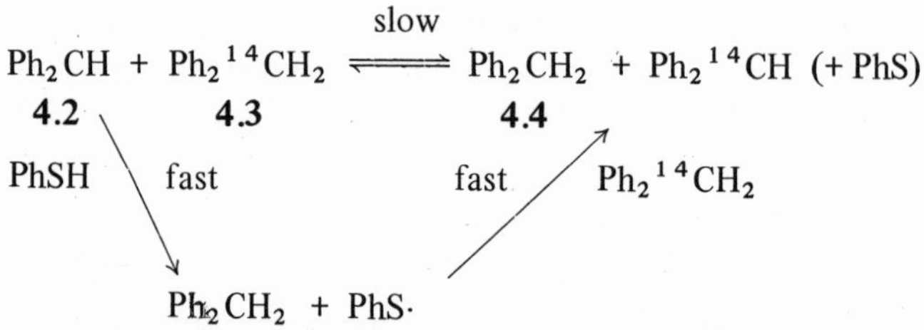

$$
\begin{array}{c}
\text{slow} \\
Ph_2CH + Ph_2{}^{14}CH_2 \rightleftharpoons Ph_2CH_2 + Ph_2{}^{14}CH\ (+PhS) \\
\textbf{4.2} \quad\quad \textbf{4.3} \quad\quad\quad\quad \textbf{4.4} \\
\end{array}
$$

PhSH ＼ fast      fast ／ $Ph_2{}^{14}CH_2$     (4.18)

$$Ph_2CH_2 + PhS\cdot$$

Another example can be seen in the thiol-catalyzed decarbonylation of aldehydes[48]. Here, only 0.5 mole % of benzyl mercaptan is effective in decomposing 80 – 90% of aldehydes, while only a trace of carbon monoxide is obtained without thiol. The following chain mechanism was suggested:

$$RS\cdot + R'CHO \longrightarrow RSH + R'\dot{C}O$$

$$R'\dot{C}O \longrightarrow R'\cdot + CO \tag{4.19}$$

$$R'\cdot + RSH \longrightarrow R'H + RS\cdot$$

The last step of the reaction in eq. 4.19 is the reverse of the reaction in eq. 4.17. Rate constants for reactions of this type have been studied by several groups and they are listed in Table 4.5[49-53]. Primary kinetic isotope effects of these reactions are used for the discussion on the nature of the hydrogen-transfer transition state[54,55].

TABLE 4.5

Rate Constants for Reactions of the Type: R· + R'SH → RH + ·SR', at 25°C

| Reaction | Rate Constant, liter mole$^{-1}$sec$^{-1}$ |
|---|---|
| $CH_3$· + $CH_3SH$ | $1.8 \times 10^4$ |
| $CH_3$· + $C_2H_5SH$ | $3.5 \times 10^4$ |
| $CH_3$· + iso-$C_3H_7SH$ | $4.1 \times 10^4$ |
| $CH_3$· + tert-$C_4H_9SH$ | $5.9 \times 10^4$ |
| $CH_3$· + $H_2S$ | $1.6 \times 10^6$ |
| $CF_3$· + $H_2S$ | $2 \times 10^5$ |
| H· + $CH_3SH$ | $4 \times 10^8$ |
| DPPH + tert-$C_4H_9SH$ | $4 \times 10^{-4}$ |

## 4.5 ADDITION TO OLEFINS

### 4.5.1 Nucleophilic Addition

Only electron-deficient olefins can react with thiols in basic conditions. For example, thiolate anions attach at $C\beta$-position of acrylonitrile or methyl acrylate [56,57]:

$$EtS^- + H_2C=CH\text{-}CN \longrightarrow EtSCH_2\text{-}\overset{-}{C}HCN$$

$$\xrightarrow{\ BH\ } EtSCH_2\text{-}CH_2CN + B^- \qquad (4.20)$$

The Michael addition of this type seems to be reversible and certain adduct sulfides eliminate thiol, giving an $\alpha,\beta$-unsaturated compound on heating with base [58]:

$$\xrightarrow[\substack{EtOH, \\ reflux}]{1\ N\text{—NaOH}} \underset{65\%}{H_2C=CH(CO_2Et)_2} \qquad (4.21)$$

Base-catalyzed addition of thiols to maleic anhydride was studied kinetically [59]. From the data shown in Table 4.6, it is proposed that

## TABLE 4.6

Acidities of Thiols and Third-Order Rate Constants for Triethylenediamine-Catalyzed
Addition of Thiols to Maleic Anhydride in Xylene at 27°C

| Thiol | $pK_a$ | log k, liter$^2$ mole$^{-2}$sec$^{-1}$ | $k_n{}^a$ |
|---|---|---|---|
| $n$-$C_4H_5SH$ | 12.6 | 2.11 | 1,000 |
| $C_2H_5$<br>$\quad$CHSH<br>$CH_3$ | 12.9 | 1.38 | 380 |
| $tert$-$C_4H_9SH$ | 13.1 | 1.13 | 340 |
| ⬡—SH | 12.7 | 1.47 | 300 |
| $PhCH_2SH$ | 11.8 | 2.98 | 1,200 |
| $CH_3$<br>$\quad$CHSH<br>Ph | 12.2 | 2.54 | 1,100 |
| (tetrahydrofuryl)CHSH | 11.3 | 3.55 | 1,400 |
| $Ph_3CSH$ | 11.1 | 1.73 | 14 |
| PhSH | 8.4 | 3.81 | 5 |
| (pyridyl)—SH | 10.6 | 3.21 | 13 |

$^a$ Relative thiolate anion nucleophilicities.

the dissociation equilibrium between thiol and base dominates the over-all rate of addition, although the addition of the complex to maleic anhydride is the slowest step:

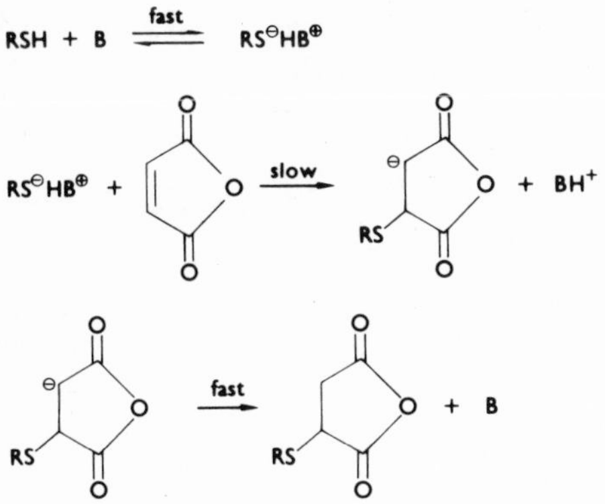

$$(4.22)$$

Correction of the third-order rate constants for the prior equilibrium allows relative rate constants to be determined for the addition of thiolate anions to maleic anhydride. These are also listed in Table 4.6. It should be noted that benzyl and 2-furyl mercaptans have relatively large nucleophilicities in spite of the relatively small $pK_a$ values. Lone pair electrons on the sulfur atom seem to be pushed out by neighbouring $\pi$-electrons, resulting in the increase of the nucleophilicity (*cf.* Sect. 4.2).

### 4.5.2 Electrophilic Addition

In contrast to nucleophilic additions, electrophilic addition of ordinary thiols and hydrogen sulfide take place easily with usual olefins in the presence of strong acid. Markovnikov adducts are obtained through the processes shown below:

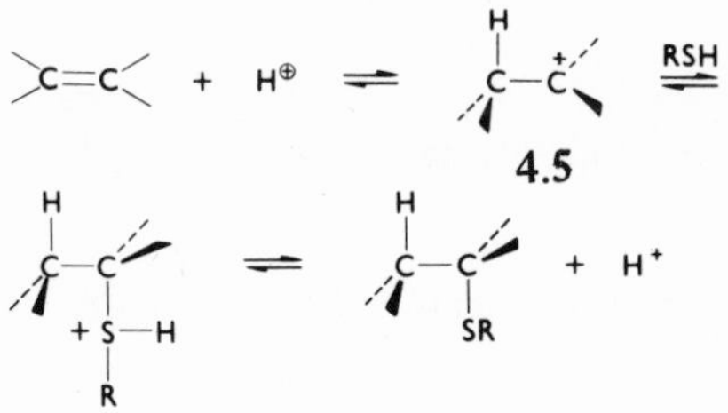

Since divalent sulfur compounds are stronger nucleophiles than alcohols or water, the addition of thiols to an intermediate carbonium ion,

**4.5**, takes place quite easily. Electrophilic addition of thiolacetic acid to a vinyl ether occurs without any strong acid and in the presence of small amounts of thionyl chloride and *p*-hydroquinone, it results in the formation of a thioketal, **4.6**, which cannot be obtained by either nucleophilic or radical additions. When the reaction is carried out under an atmosphere of air, the radical reaction yields only **4.7** [60].

$$
H_2C = CH - OR + AcSH \longrightarrow
\begin{cases}
\xrightarrow[\text{p-hydroquinone}]{SOCl_2} & CH_3CH \begin{matrix} OR \\ SAc \end{matrix} \quad \textbf{4.6} \\[2ex]
\xrightarrow{\text{air}} & ROCH_2CH_2SAc \\ & \textbf{4.7}
\end{cases}
$$

(4.24)

Acid-catalyzed additions of aliphatic thiols to conjugated dienes afford both 1,4- and 1,2-adducts [61].

Detailed studies have been done on the addition of sulfenyl chloride, a derivative of thiol, to olefins [62]. In this case thionium ion plays the role of an electrophile and the reaction proceeds as follows:

$$
>C=C< + RSCl \longrightarrow [C \overset{\underset{\displaystyle S}{|}{\overset{\displaystyle R}{|}}}{\triangle} C]^+ Cl^-
$$

$$
\longrightarrow Cl - \overset{|}{C} - \overset{|}{C} - SR
$$

(4.25)

The reaction rate increases in polar solvents and *trans*-addition is observed. Conjugated dienes also react with sulfenyl chloride *via* an episulfonium intermediate. The products, at low temperature, are 1,2-adducts, while at elevated temperatures they rearrange to 1,4-adducts:

$$
H_2C = CH - CH = CH_2 + CH_3SCl \longrightarrow \left[ H_2C \overset{\overset{\displaystyle CH_3}{\underset{\displaystyle |}{S}}}{\triangle} CH - CH = CH_2 \right]^+ Cl^-
$$

$$
\xrightarrow{\text{low temp.}} \underset{\underset{93\%}{\overset{|}{Cl}}}{CH_3SCH_2CHCH=CH_2} + \underset{\underset{7\%}{\overset{|}{SCH_3}}}{ClCH_2CHCH=CH_2}
$$

(4.26)

$$
\xrightarrow{\text{room temp.}} CH_3SCH_2CH_2CH_2CH_2Cl
$$

### 4.5.3 Radical Addition

Free radical addition reactions are the most extensively studied among the chemistry of thiols [63,64]. Since the addition is affected by peroxide [65] and affords anti-Markovnikov adducts, a chain mechanism has been proposed [66]:

$$\tag{4.27}$$

Fast reversible processes are proposed for the reaction based on the facts that the photo-induced *cis-trans* isomerizations of olefins are accelerated by thiols [67,68], that a G-value of more than $10^5-10^6$ is observed in γ-ray-induced *cis-trans* isomerizations of olefins in the presence of hydrogen sulfide [69], and that activation energies for the addition of thiols to olefins are small [70]. Hence, under certain conditions, such as with only small concentrations of a thiol or at high temperatures, geometrical isomerization may predominate over the addition. Thus, photo-induced isomerization of 2-butene in the presence of methyl mercaptan with partial pressure of less than 4 mm Hg obeys second-order rate law ($k_2 = 2 \times 10^7$ liter mole$^{-1}$ sec$^{-1}$), first-order in each reactant. The addition of a thiol to 1,3-butadiene proceeds one order of magnitude faster than the addition to 2-butene, and the activation energy of the reaction is estimated to be zero [71].

Walling and Helmreich found that elimination of MeS· radical from **4.8** and **4.9** are 80 and 20 times as fast as the hydrogen-transfer from methyl

$$\tag{4.28}$$

|  gauche  |  cis  |
|:---:|:---:|
| **4.8** | **4.9** |

mercaptan to these radicals, respectively [67]. Thus, radical-induced β-elimination of thiol from a sulfide takes place easily by abstracting a β-hydrogen atom. For example [72-75]:

$$(EtSCH_2\overset{\underset{\displaystyle CH_3}{|}}{\underset{\underset{\displaystyle CH_3}{|}}{C}}-N\!\!\neq_2 \longrightarrow EtSCH_2\overset{\underset{\displaystyle CH_3}{|}}{\underset{\underset{\displaystyle CH_3}{|}}{C}}\cdot \longrightarrow EtS\cdot + H_2C\!\!=\!\!C\!\!<^{CH_3}_{CH_3} \qquad (4.29)$$

$$\sim 100\%$$

$$R-CH_2-\overset{\underset{\displaystyle SR''}{|}}{\overset{\overset{\displaystyle R'}{|}}{C}}\!\!<^{SR''}_{SR''} + tert\text{-}BuO\cdot \longrightarrow R-CH\!\!=\!\!C\!\!<^{R'}_{SR''} \qquad (4.30)$$

$$+ \cdot SR + tert\text{-}BuOH$$

$$(CH_3)_3CSR + C_6H_5\cdot \longrightarrow \overset{CH_3}{\underset{CH_3}{>}}\!\!C\!\!=\!\!CH_2 + \cdot SR + C_6H_6$$

$$100\%$$

$$CH_3-\overset{\underset{\displaystyle}{\overset{\overset{\displaystyle OH}{|}}{C}}H}-CH_2SCH_3 + tert\text{-}BuO\cdot \longrightarrow \qquad (4.31)$$

$$tert\text{-}BuOH + CH_3-\overset{\overset{\displaystyle OH}{|}}{C}-CH_2SCH_3 \longrightarrow$$

$$CH_3S\cdot + CH_3-\overset{\overset{\displaystyle OH}{|}}{C}\!\!=\!\!CH_2 \longrightarrow CH_3-\overset{\overset{\displaystyle O}{\|}}{C}-CH_3 \qquad (4.32)$$

$$95\%$$

Since the radical addition of a thiol to an olefin is a two-step process, the rate of the reaction is a function of two rate constants. If $k_3$ is much larger than $k_2'$ (eq. 4.27), the observed rate constant for the formation of the adduct is equal to $k_1 k_3/(k_2 + k_3)$. The observed values are listed in Table 4.7[67]. The fact that styrene, methyl acrylate, and vinyl ether react faster than others seems to indicate the importance of the stability of the intermediate radical. However, since several alternative interpretations are possible, we cannot make any legitimate conclusion from the results obtained here. More sophisticated data are reported by Cadogan and Sadler [76] : the rates of additions of thioglycollic acid (and its ester) to substituted styrenes and stilbenes have linear free-energy relationships with $\rho = -0.4$ (using Brown-Okamoto's $\sigma^+$). Although small, this indicates that the transition state of the reaction has a somewhat polar character and the thiyl radical acts as though it's an electrophile. Indeed, photo-induced additions of thiols to diallyl maleate, **4.10**, and fumarate, **4.11**, afford products **4.12–4.15** (eqs. 4.33 and 4.34) [77]. No isomerization

## TABLE 4.7

Relative Rates for the Addition of $n$-$C_{12}H_{25}SH$ to Olefins at 60°C

| Olefin | Relative Rate |
|---|---|
| $PhCH=CH_2$ | 11.0 |
| $CH_3\!\!\diagdown$ $C=CH_2$ $\diagup$ $Ph$ | 5.5 |
| $n$-$C_4H_5OCH=CH_2$ | 3.9 |
| $CH_3O_2CCH=CH_2$ | 2.0 |
| $HOCH_2CH=CH_2$ | 1.2 |
| $CH_3\!\!\diagdown$ $C=CH_2$ $\diagup$ $C_2H_5$ | 1.2 |
| $n$-$C_6H_{13}CH=CH_2$ | (1.0) |
| $PhCH_2CH=CH_2$ | 0.99 |
| $CH_3CO_2CH=CH_2$ | 0.81 |
| $ClCH_2CH=CH_2$ | 0.71 |
| (cyclopentene) | 0.64 |
| $CH_3CO_2CH_2CH=CH_2$ | 0.59 |
| $NCCH_2CH=CH_2$ | 0.37 |
| (cyclohexene) | 0.25 |
| $cis$-$ClCH=CHCl$ | 0.20 |

$$\begin{matrix} HCCO_2CH_2CH{=}CH_2 \\ \| \\ HCCO_2CH_2CH{=}CH_2 \\ \mathbf{4.10} \end{matrix} \xrightarrow[RSH]{h\nu} \begin{matrix} HCCO_2(CH_2)_3SR \\ \| \\ HCCO_2(CH_2)_3SR \\ \mathbf{4.12} \end{matrix} + \begin{matrix} HCCO_2(CH_2)_3SR \\ \| \\ HCCO_2CH_2CH{=}CH_2 \\ \mathbf{4.13} \end{matrix} \quad (4.33)$$

$$\begin{matrix} H_2C{=}CHCH_2O_2CCH \\ \mathbf{4.11} \quad \| \\ HCCO_2CH_2CH{=}CH_2 \end{matrix} \xrightarrow[RSH]{h\nu}$$

$$\begin{matrix} RS(CH_2)_3O_2CCH \\ \| \\ HCCO_2(CH_2)_3SR \\ \mathbf{4.14} \end{matrix} + \begin{matrix} RS(CH_2)_3O_2CCH \\ \| \\ HCCO_2CH_2CH{=}CH_2 \\ \mathbf{4.15} \end{matrix} \quad (4.34)$$

R = Ph, Me, Et, Ac

between the maleate and fumarate can be observed. This suggests that no addition of a thiyl radical takes place at electron-deficient central double bonds.

A study on the direction of addition of thiyl radicals to substituted methyl β-mercaptoacrylates, **4.16**, provides insights into the electronic effect in the reaction [78]:

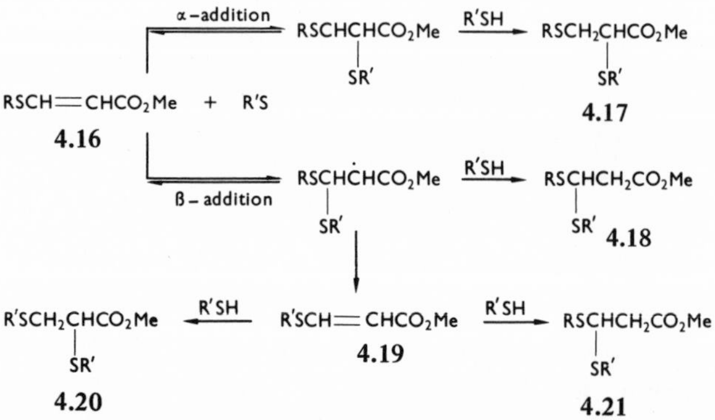

Addition may occur either at α- or β-position. α-Addition yields only **4.17**, while β-addition may afford a mixture of **4.18 − 4.21**. All possible products were isolated when R = Ac, and the ratio of α- to β-addition was 1. This ratio increases to 4.2 when R = Me and to 53 when R = Ph. The results are best interpreted by the varying abilities of the sulfur atoms of mercapto groups to stabilize an adjacent radical center. The electron-withdrawing carbonyl group decreases electron density on the sulfur atom, whereas the phenyl group promotes the electron-sharing participation of the sulfur as shown [79]:

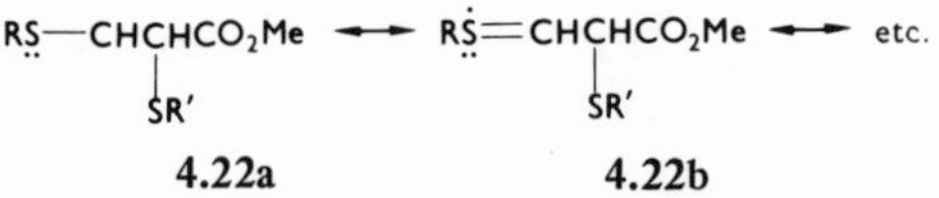

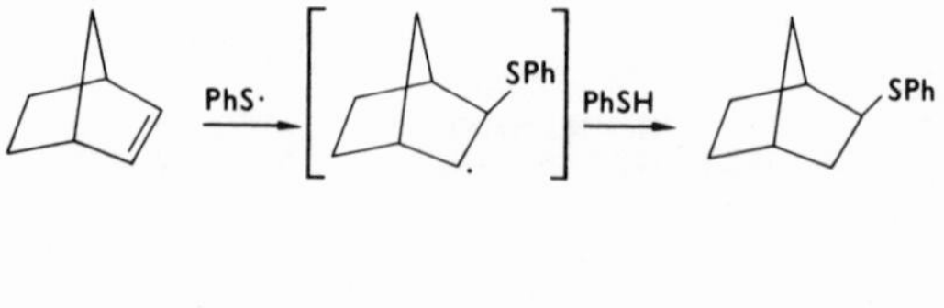

Stereochemistry of the addition has recently been subjected to extensive studies [80,81]. The addition of a thiol to norbornene results in the exclusive formation of the *exo*-adduct (eq. 4.36) [80]. The reaction is not accompanied by the Wagner-Meerwein rearrangement, which is commonly observed in reactions of carbonium ions. On the other hand, 1,2-alkyl shifts were observed in the reaction of polychlorinated norborna-2,5-dienes, where **4.23** affords **4.26** in addition to unrearranged products **4.24** and **4.25** [82]:

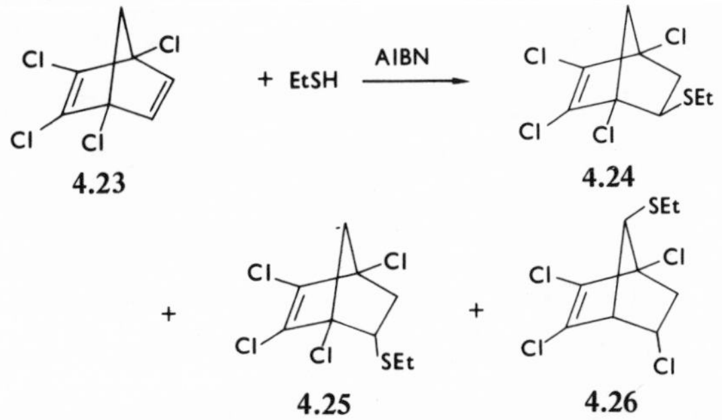

**4.23**     + EtSH     AIBN     **4.24**     (4.37)

+     **4.25**     +     **4.26**

Here arises a question whether a divalent sulfur atom can interact with a radical center at the $\beta$-position by electron-sharing conjugation. The answer is not unequivocal. Importance of the conjugation of this type seems to depend on the structure of the radical. It is interesting to see a few examples on this topic. The free-radical addition of thioacetic acid to either *cis-* or *trans*-2-chloro-2-butene yields 90% of *threo-2*-acetylthio-3-chlorobutane, **4.27**, and 10% of its *erythro*-isomer, **4.28**, without isomerization of the olefin at $-78°$ (eq. 4.38) [83]. These results

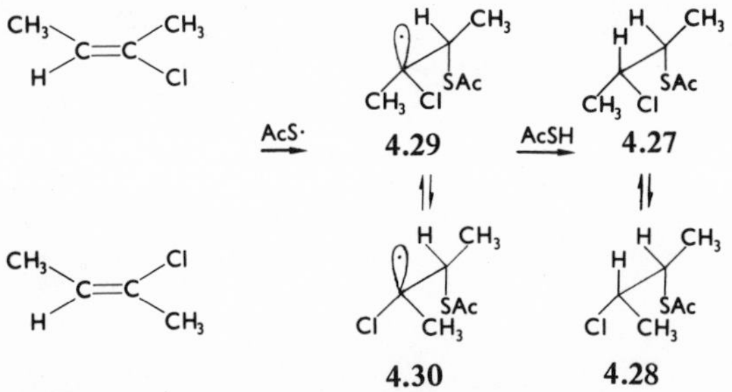

AcS·     **4.29**     AcSH     **4.27**

(4.38)

**4.30**     **4.28**

indicate that the rate of rotation around the carbon-carbon single bond in the intermediate radical is greater than the hydrogen transfer from a thiol. Neureiter and Bordwell suggested conformational preference for **4.29** over **4.30** and over-all *trans*-addition. The preference of **4.29** over **4.30** should be less when chlorine atom is replaced by bromine atom. This is indeed true and the *threo/erythro* ratio in the product decreases with 2-bromo-2-butene [64]. Both *cis-* and *trans*-2-butenes react with MeSD at $-70°$, giving the same mixture of *threo-* and *erythro*-adducts [84]. In the presence of DBr, however, *cis*- and *trans*-olefins afford exclusively *threo*- and *erythro*-adducts, respectively:

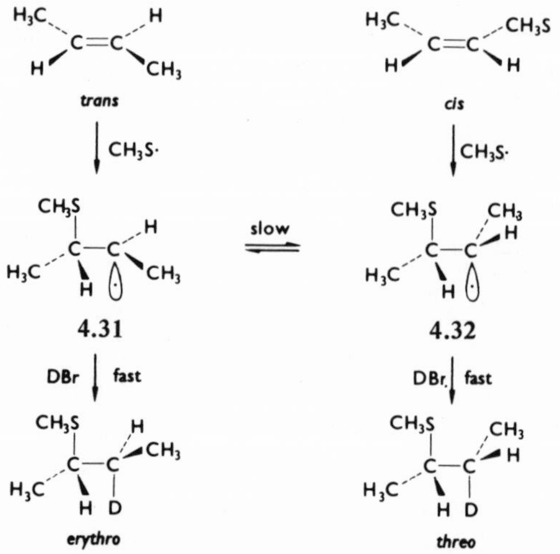

(4.39)

Thus, the equilibrium between intermediate radicals **4.31** and **4.32** takes place faster than the hydrogen transfer from methyl mercaptan to these radicals, while DBr transfers hydrogen much faster than the equilibration.

Methyl mercaptan was added to bicyclo[3.1.0]hexene-2, **4.33**, under photo-irradiation, giving products **4.34** − **4.38** (eq. 4.40) [85]. Product

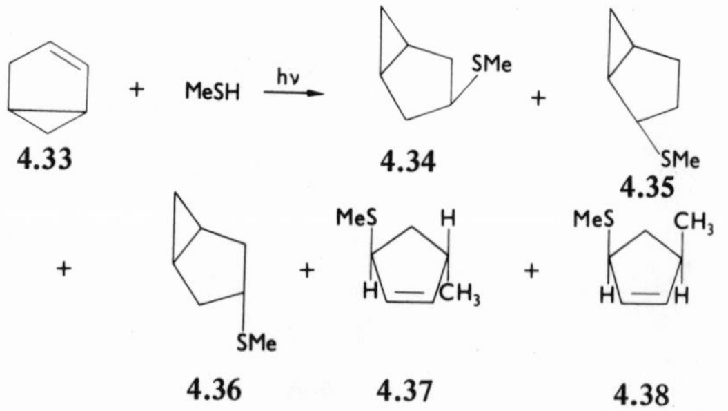

(4.40)

composition depends upon concentration of the thiol. This and other considerations on sterochemistry predicts that the equilibrium in eq. 4.41

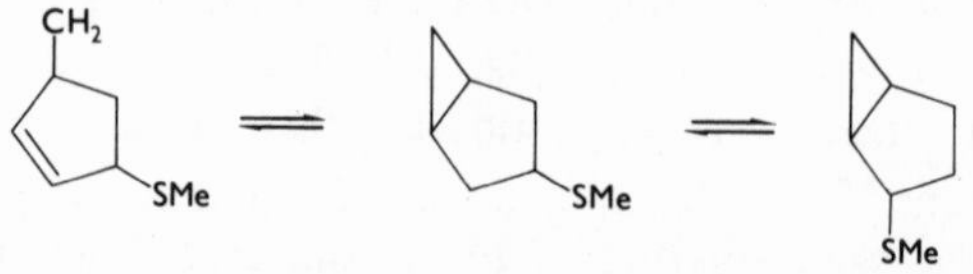

(4.41)

seems more probable for the intermediate radical than non-classical structures **4.39** and/or **4.40**.

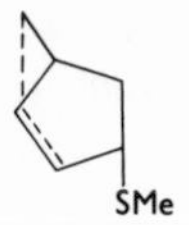

    4.39 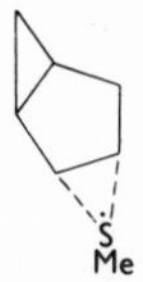                          4.40

Esr spectroscopy was employed to study the free-radical addition of thiols to olefins and no hyperfine splitting could be observed for elements of thiols, which implies that the intermediate radical has an open-structure [86].

Many studies have been done on the stereochemistry of the addition with conformationally fixed systems [87-90]. For example, addition of thiolacetic acid to 2-chloro-4-*tert*-butylcyclohexene yields all four possible isomers (eq. 4.42) [90]. Product composition depends both on thiol : olefin

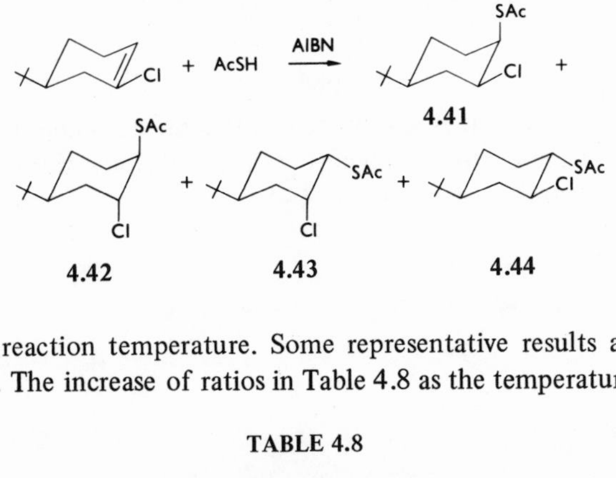

                                                            (4.42)

ratio and reaction temperature. Some representative results are listed in Table 4.8. The increase of ratios in Table 4.8 as the temperature decreases

**TABLE 4.8**

Free-Radical Addition of Thiolacetic Acid to 2-Chloro-4-*tert*-butylcyclohexene in Hexane

| AcSH : Olefin | Temp., °C | [4.41 + 4.42] / [4.43 + 4.44] | 4.41/4.42 | 4.43/4.44 |
|---|---|---|---|---|
| 1 : 10 | 63 | 1.2 | 6.3 | 1.9 |
| 1 : 1 | 63 | 1.4 | 9.3 | 2.2 |
| 10 : 1 | 63 | 2.5 | 10.0 | 2.2 |
| 1 : 1 | −78 | 4.1 | 49.2 | 1.8 |

indicates that the initial reversible step is important in addition of thiols and that *trans*-addition predominates over *cis*-addition. The fact that the ratio, **4.41/4.42**, is more sensitive to change in temperature than **4.43/4.44** may be accounted for by a sulfur-bridged radical intermediate, **4.45**, as a precursor of products **4.41** and **4.42** and an open-chain radical intermediate, **4.46**, as a precursor of **4.43** and **4.44**. Note that an equatorial-substituent can form a bridged structure with difficulty.

**4.45**                                    **4.46**

Several other reports support the existence of sulfur-bridged radicals [91-93]. However, as has been described above, no unequivocal conclusion can be drawn from these results. The most acceptable statement may be that a non-classical free-radical of high symmetry, such as **4.47**, has not been detected yet.

**4.47**

In this connection, it is worthy to mention the decomposition kinetics of sulfur-containing azo compounds [72,94] : enthalpies and entropies of activation for reactions in eqs. 4.43−4.45 suggest that no

$$\text{(EtSCH}_2\text{—}\underset{\underset{\text{Me}}{|}}{\overset{\overset{\text{Me}}{|}}{\text{C}}}\text{—N=})_2 \longrightarrow \text{EtSCH}_2\text{—}\underset{\underset{\text{Me}}{|}}{\overset{\overset{\text{Me}}{|}}{\text{C}}}\cdot \qquad (4.43)$$

$$\qquad (4.44)$$

$$\qquad (4.45)$$

participation of $\beta$-sulfur exists in the open-chain radical, while some ($\sim 2$ kcal/mole) stabilization by sulfur atoms at $\beta$- and even $\gamma$-positions is achieved in cyclic radicals, where sulfur atoms are more or less geometrically fixed.

The reaction of trans-$\Delta^2$-octalin provides an interesting example for the addition of thiols[95]. This olefin can afford only one axial and one equatorial adduct. Huyser and his co-workers found that the reaction of this olefin with methyl mercaptan or thiophenol results in the formation of about 8–12 parts of the axial adduct against 1 part of the equatorial isomer:

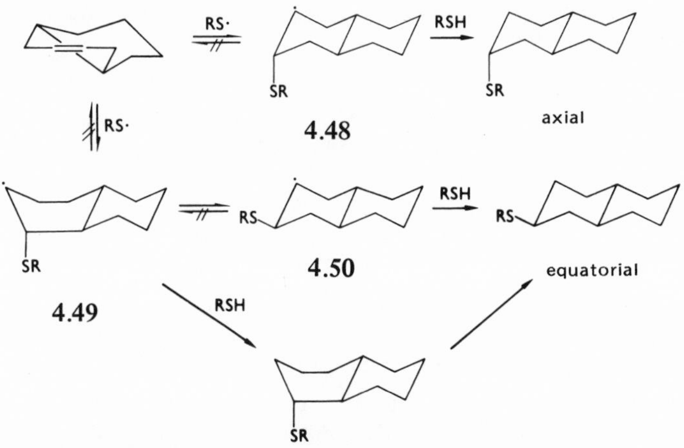

(4.46)

Product composition is independent of thiol/olefin ratio, indicating the absence of reverse reactions from **4.48** and **4.49**. The return from **4.48** to the starting olefin is prohibited because the former is sterically less hindered, while **4.49** does not regenerate the olefin because it isomerizes simultaneously to the conformationally more stable **4.50,** which is also more stable than the starting olefin.

Orientation and reactivity of diolefins toward free-radical addition of thiols have been extensively studied[96]. Benzenethiyl radical attacks exclusively the *endo*-double bond of 5-methylenenorbornene, affording **4.51–4.53** (eq. 4.47), while trichloromethyl radical attacks norbornene,

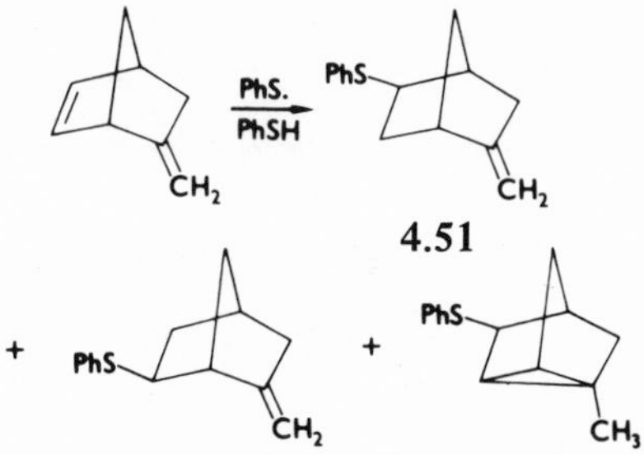

(4.47)

**4.54**, faster than methylenenorbornane, **4.55** [97] :

**4.54**                          **4.55**

Cristol and his co-workers explain this by strain relief and steric effect. On the other hand, Huyser and Kellogg prefer to interprete this by

$$\textbf{4.54} \quad + \text{ RS} \cdot \underset{k_{-1}}{\overset{k_1}{\rightleftharpoons}} \qquad \xrightarrow[\text{RSH}]{k_2} \qquad \qquad (4.48)$$

**4.56**

$$\textbf{4.55} \quad + \text{ RS} \cdot \underset{k'_{-1}}{\overset{k'_1}{\rightleftharpoons}} \qquad \xrightarrow[\text{RSH}]{k'_2} \qquad \qquad (4.49)$$

**4.57**

reversibility of the addition [98]. Thus, the conversion of **4.54** to **4.56** (eq. 4.48) is accompanied by a large relief of strain and $k_{-1}$ becomes small, while strain relief may be small in $k'_1$ step (eq. 4.49). This, as well as stability of **4.57** (*tertiary* radical) make $k'_{-1}$ large. Consequently, net reaction proceeds faster with **4.54** than with **4.55**.

A thiyl radical seems to attack a conjugated olefin so that the resonance-stabilized radical is produced:

$$\left[ \text{RSCH}_2\text{-}\overset{\cdot}{\text{C}}\text{H-}\overset{\overset{\displaystyle R}{|}}{\text{C}}\text{=CH}_2 \quad \longleftrightarrow \quad \text{RSCH}_2\text{-CH=}\overset{\overset{\displaystyle R}{|}}{\text{C}}\text{-}\overset{\cdot}{\text{C}}\text{H}_2 \right]$$

*minor*

$$\underset{\overset{|}{\text{CH}_2}=\overset{|}{\text{C}}\text{-CH-}\overset{\cdot}{\text{C}}\text{H}_2}{^{R\ SR}} \quad \xleftarrow{+\!\!\!\!+\text{RS}\cdot} \ + \ \overset{\overset{\displaystyle R}{|}}{\text{CH}_2\text{=C-CH=CH}_2} \ \xrightarrow{+\!\!\!\!+} \ \overset{\overset{\displaystyle R}{|}}{\overset{\cdot}{\text{C}}\text{H}_2\text{-C-CH=CH}_2}$$

$$\underset{\text{SR}}{\qquad\qquad\qquad\qquad\qquad\qquad\qquad\qquad} \qquad (4.50)$$

*major*

$$\left[ \text{RSCH}_2\text{-}\underset{\cdot\cdot}{\overset{\overset{\displaystyle R}{|}}{\text{C}}}\text{-CH=CH}_2 \quad \longleftrightarrow \quad \text{RSCH}_2\text{-}\overset{\overset{\displaystyle R}{|}}{\text{C}}\text{=CH-}\overset{\cdot}{\text{C}}\text{H}_2 \right]$$

For example, photo-induced reaction of 3,5-dimethylenecyclohexene with
$n$-butyl mercaptan proceeds as shown [99] :

$$(4.51)$$

The second step, hydrogen abstraction, controls the formation of
products. As has been discussed in Sect. 4.4, the activation energy for
hydrogen transfer is quite small in most cases. However, since allyl radical
is stabilized by resonance, activation energy for the second step of the
reaction in eq. 4.51 becomes relatively large and product composition is
thermodynamically controlled. For example, 1,3-butadiene, 2,3-
dimethyl-1,3-butadiene, and chloroprene afford 1,4-adducts (eq.
4.52) [96]. Since the initially formed radical, **4.58a**, has the radical center

$$(4.52)$$

$$(R = R' = H; \quad R = R' = Me; \quad R' = Cl, \quad R' = H)$$

**4.58a**                                **4.58b**

at a *tertiary* (or *secondary*) position, more reactive radical **4.58b**, which
has the radical center at primary position, abstracts a hydrogen from thiol
giving 1,4-adducts. Piperylene (1-methyl-1,3-butadiene) yields almost
equimolar amounts of 1,2- and 1,4-adducts [100]. Here, both canonical
forms, **4.59a** and **4.59b**, have the radical center at one of the secondary

positions. Furthermore, the addition of 2,5-dimethyl-2,4-hexadiene gives the 1,2-adduct (eq. 4.53). The reason is obvious.

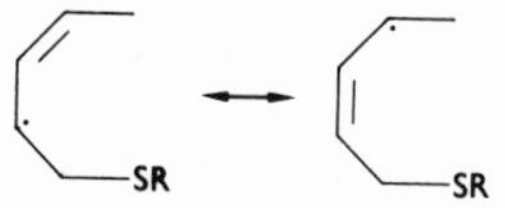

**4.59a**      **4.59b**

(4.53)

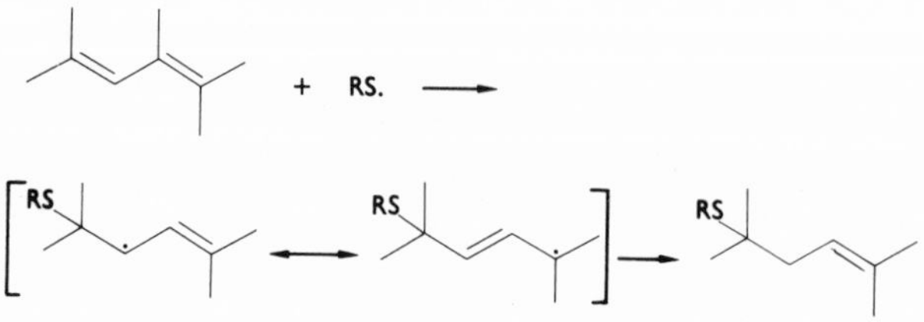

In this connection, it will be interesting to note the following reactions [102] :

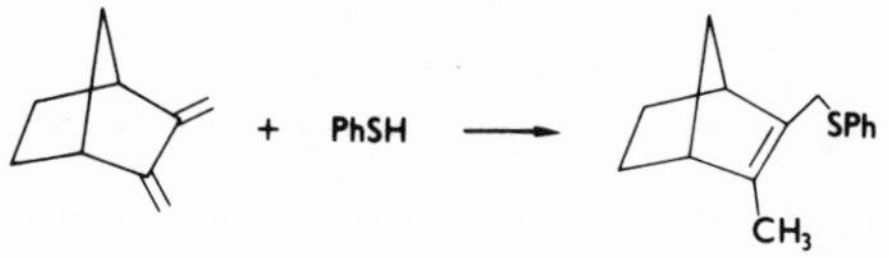

(4.54)

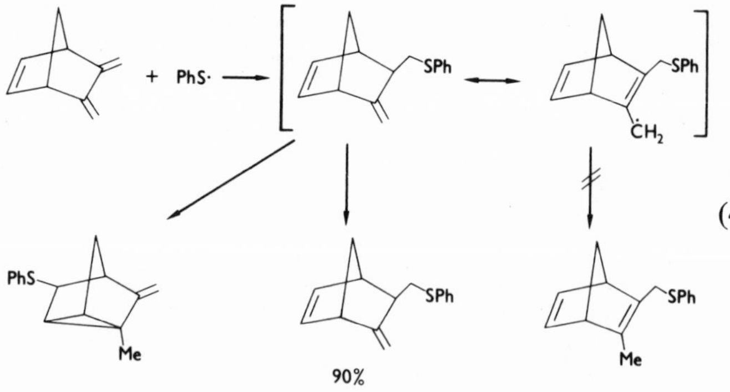

(4.55)

### 4.5.4 Oxidative Addition

The addition of thiols to olefins in the presence of excess oxygen provides a useful variant of thiol chemistry and is called "co-oxidation". The first example of co-oxidation was reported by Kharasch and his co-workers, who found that the reaction of n-propyl mercaptan with styrene in the presence of oxygen yielded **4.60**, a $\beta$-hydroxysulfoxide [103].

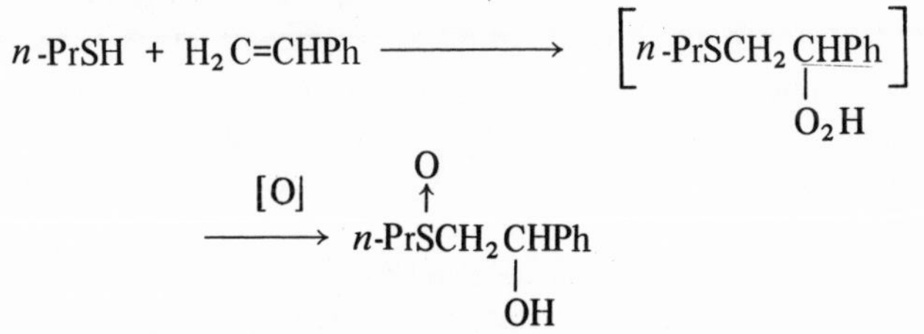

$$n\text{-PrSH} + H_2C=CHPh \longrightarrow \left[ n\text{-PrSCH}_2\underset{|}{\overset{}{CHPh}} \atop O_2H \right]$$

$$\xrightarrow{[O]} n\text{-PrSCH}_2\underset{|}{\overset{O\uparrow}{CHPh}} \atop OH \qquad (4.56)$$

**4.60**

Acrylonitrile and methyl acrylate react similarly [104]. The mechanism of the reaction has been extensively studied by Oswald and his co-workers [93,105-117], and the sequence shown is now accepted below:

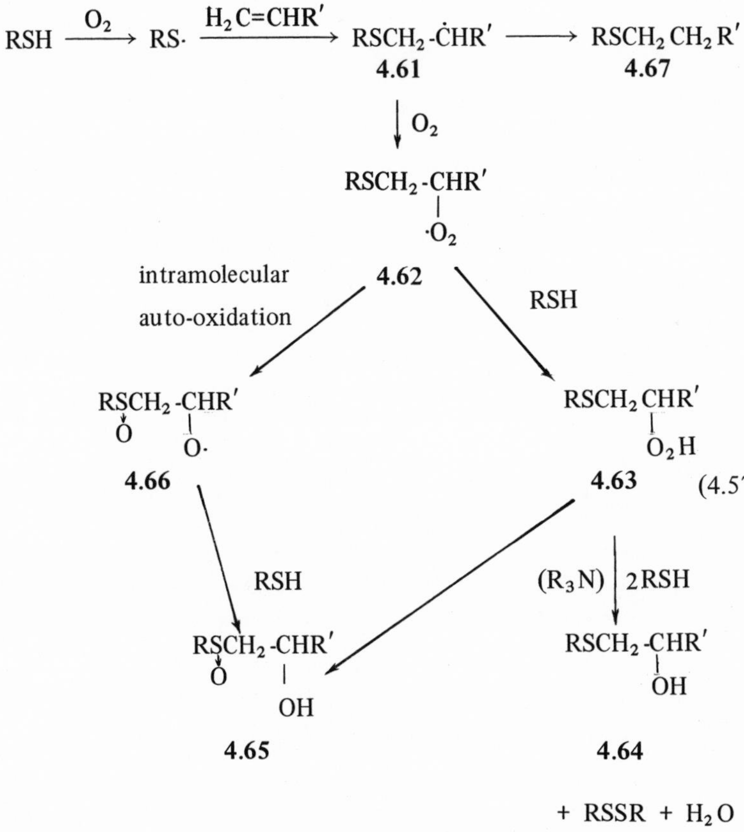

$$RSH \xrightarrow{O_2} RS\cdot \xrightarrow{H_2C=CHR'} \underset{\textbf{4.61}}{RSCH_2\text{-}\dot{C}HR'} \longrightarrow \underset{\textbf{4.67}}{RSCH_2CH_2R'}$$

$$\downarrow O_2$$

$$RSCH_2\text{-}\underset{|}{\overset{}{CHR'}} \atop \cdot O_2$$

**4.62**

intramolecular
auto-oxidation               RSH

$$\underset{\textbf{4.66}}{RSCH_2\text{-}\underset{\dot{O}\quad O\cdot}{\overset{}{CHR'}}} \qquad\qquad \underset{\textbf{4.63}}{RSCH_2\underset{O_2H}{\overset{}{CHR'}}} \quad (4.57)$$

RSH               $(R_3N)$ | 2 RSH

$$\underset{\textbf{4.65}}{RSCH_2\text{-}\underset{\dot{O}\quad OH}{\overset{}{CHR'}}} \qquad\qquad \underset{\textbf{4.64}}{RSCH_2\text{-}\underset{OH}{\overset{}{CHR'}}}$$

$$+ \; RSSR + H_2O$$

The hydrogen transfer from a thiol to **4.61** requires a small activation energy, while, since the reaction of this radical with oxygen is exothermic and very fast, the peroxy radical, **4.62**, is produced preferentially if excess oxygen is supplied to the system. The intermediate, **4.63**, was isolated and its conversion to **4.66** by intramolecular auto-oxidation was confirmed. The conversion of **4.63** to **4.64** was also demonstrated independently [108,109].

Since the over-all reaction is initiated by the addition of a thiyl radical to an olefin, the structures of thiols and olefins affect the reaction rate. When the intermediate radical, **4.61**, is resonance-stabilized, as occurs with styrene, indene, and conjugated dienes, the reaction proceeds smoothly even at low temperatures. On the other hand, $n$-octadecene-1 requires several days to complete the reaction. For this olefin, photo-irradiation promotes the speed of the reaction. Aromatic thiols, which are oxidized easily, are more reactive than aliphatic ones [117].

$$RSH \xrightarrow{\quad O_2 \quad} RS\cdot$$

$$\downarrow \quad \diagdown C=C\text{-}C=C\diagup$$

$$RS\text{-}C\text{-}C\text{-}C=C \longleftrightarrow RS\text{-}C\text{-}C=C\text{-}C\diagup \qquad (4.58)$$

**4.68**

$$O_2 \diagup \qquad \diagdown RSH$$

$$RS\text{-}C\text{-}C\text{-}C=C\diagup \quad + \quad RS\text{-}C\text{-}C=C\text{-}CO_2 \qquad RS\text{-}C\text{-}C=C\text{-}CH$$
$$\underset{O_2}{|} \qquad \qquad \qquad \qquad \qquad \qquad \qquad \qquad \qquad 1,4\text{-adduct}$$

$$\downarrow \qquad \qquad \qquad \downarrow$$

$$\downarrow \qquad \qquad \qquad \downarrow$$

$$RS\text{-}C\text{-}C\text{-}C=C\diagup \qquad RS\text{-}C\text{-}C=C\text{-}COH$$
$$\underset{OH}{|}$$

1, 2-adduct                1, 4-adduct

The direction of addition to conjugated dienes depends on the structure of the diene. As has been discussed in Sect. 4.5.3, the rate of hydrogen abstraction decreases in the order $1° > 2° > 3°$, whereas the reaction of a radical with oxygen is very fast and the activation energy is almost zero. Therefore, in co-oxidation, the more resonance-stabilized form of the intermediate radical, **4.68**, reacts with oxygen to give the final products. Thus, 2,3-dimethyl-1,3-butadiene affords the 1,2-adduct, while piperylene affords the equimolar mixture of 1,2- and 1,4-adducts [100a].

Stereochemistry of the reaction was studied [100a] and it was found that *cis-trans* isomerization of an olefin did not take place during the reaction. For example, *cis-* and *trans*-piperylenes afford *cis-* and *trans*-1,2-mercaptoalcohol, respectively. This is due to the extremely facile reaction of the resonance-stabilized intermediate radical with oxygen.

It is known that co-addition of carbon monoxide also takes place. For example, a mixture of propylene, ethyl mercaptan, and carbon monoxide affords 2-methyl-3-ethylthiopropanal, **4.69**, according to eq. 4.59 [119].

$$EtS\cdot + H_2C{=}CHCH_3 \longrightarrow EtSCH_2\dot{C}HCH_3$$

$$EtSCH_2\dot{C}HCH_3 + CO \longrightarrow EtSCH_2\underset{\underset{\textstyle CH_3}{|}}{CH}\dot{C}O$$

$$(4.59)$$

$$EtSCH_2\underset{\underset{\textstyle CH_3}{|}}{CH}\dot{C}O + EtSH \longrightarrow EtSCH_2\underset{\underset{\textstyle CH_3}{|}}{CH}CHO$$

**4.69**

Carbon monoxide is less reactive than oxygen so that high pressures are necessary for the reaction. The reaction in eq. 4.59 is reversible and when **4.69** is treated with di-*tert*-butylperoxide, it decomposes to propylene, ethyl mercaptan, and carbon monoxide.

## 4.6 ADDITION TO ACETYLENES AND ALLENES

### 4.6.1 Nucleophilic Addition

Nucleophilic addition of sodium *p*-tolyl mercaptide to phenylacetylene yields the *cis*-adduct, **4.70**, by *trans*-addition [119]:

$$PhC{\equiv}CH \ + \ ArSNa \ \longrightarrow \ \underset{H}{\overset{Ph}{>}}C{=}C\underset{H}{\overset{SAr}{<}}$$

(4.60)

**4.70**

$$Ar \ = \ p\text{-}CH_3\text{-}C_6H_4\text{-}$$

Ethoxyacetylene reacts similarly [120]. *Trans*-addition was also observed in the reaction of diacetylene with *n*-butyl mercaptan [121]. Thus, *trans*-addition to acetylenes seems to be established by many examples and interpreted as follows. The addition of a thiolate anion to an acetylene will form a carbanion such as shown by **4.71**. The intermediate carbanion,

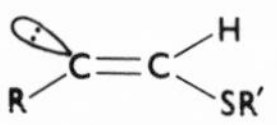

**4.71**

**4.71**, is isoelectronic to oximes and, as has been observed with the latter compounds, may have an anti-structure that reduces the repulsion force between lone pairs on carbon and sulfur atoms. In addition, since **4.71** is sterically stable, proton abstraction takes place from this structure, giving the *cis*-adduct.

Reactions of polyhalogenated ethylenes with thiolate anions are interesting from the view point of mechanism. Although reactions of this type proceed with net substitution on $sp^2$-carbon, the elimination-addition mechanism has been confirmed for 1,2-dichloroethylene [122,123]:

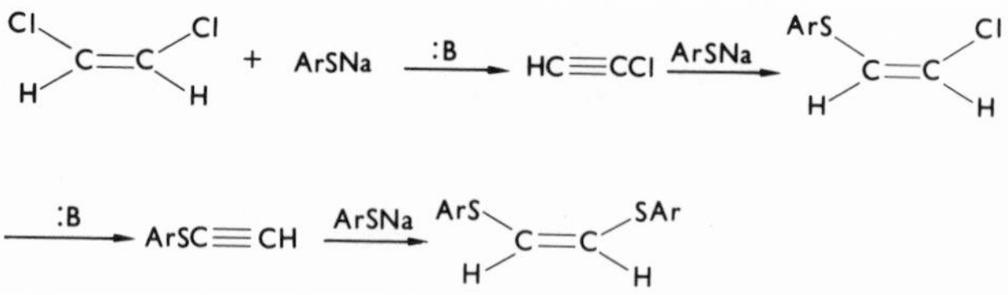

(4.61)

It should be noted that *trans*-addition is also observed here. On the other hand, addition-elimination mechanism is operating in reactions of 1,1,2-trichloroethylene and 1,1-dichloroethylene [122,123]:

$$Cl_2C=CHCl \xrightarrow{:B} ClC\equiv CCl \xrightarrow{ArS^-} \underset{Cl}{\overset{ArS}{\diagdown}}C=C\underset{H}{\overset{Cl}{\diagup}}$$

$$\xrightarrow{ArS^-} \left[ \underset{Cl}{\overset{ArS\overset{\ominus}{C}}{|}}{\overset{|}{-}}\underset{Cl}{\overset{CHSAr}{|}} \right] \xrightarrow{-Cl^-} \underset{Cl}{\overset{ArS}{\diagdown}}C=C\underset{SAr}{\overset{H}{\diagup}} \qquad (4.62)$$

$$\xrightarrow{:B} ArSC\equiv CSAr \xrightarrow{ArS^-} (ArS)_2C=CHSAr$$

$$H_2C=CCl_2 + ArS^- \xrightarrow{EtOH}$$

$$(4.63)$$

Arens and co-workers found that the addition of thiolate and other anions to mercaptoacetylenes proceeds smoothly (eq. 4.64)[125]. Electron-releasing conjugation of a sulfur atom is playing a role to stabilize an intermediate carbanion, **4.72**[126,127]:

$$HC\equiv CSR + R'S^- \longrightarrow [R'SCH=\bar{C}SR \longleftrightarrow R'SCH=C=\bar{S}R]$$

$$\textbf{4.72a} \qquad\qquad\qquad \textbf{4.72b}$$

$$(4.64)$$

$$\xrightarrow{R'SH} R'SCH=CHSR$$

Nucleophilic attack of thiols to allene occurs mainly at central carbon[128]. The fact that a considerable amount of methylacetylene was recovered from the reaction mixture and that methylacetylene yields the same product composition support the mechanism:

$$H_2C=C=CH_2 \rightleftharpoons CH_3\text{-}C\equiv CH \xrightarrow[\text{RSH}]{\text{RSNa}}$$

$$\underset{> 90\,\%}{\overset{\overset{\displaystyle CH_3}{|}}{RS\text{-}C=CH_2}} + RS\text{-}CH=CHCH_3 + \underset{\overset{\displaystyle |}{CH_3}}{\overset{\overset{\displaystyle CH_3}{|}}{RS\text{-}C\text{-}SR}} \qquad (4.65)$$

### 4.6.2 Electrophilic Addition

In contrast to olefins, acetylenes are barely susceptible to electrophilic attack and no example is reported for electrophilic attack of thiols to acetylenes. However, sulfenyl chlorides are known to add to acetylenes [129]. For example, arylsulfenyl chloride reacts with 1-butyne giving **4.73** (eq. 4.66). The reaction in chloroform proceeds three orders

$$ArSCl + CH_3CH_2C\equiv CH \longrightarrow \underset{}{CH_3CH_2C=CHSAr} \qquad (4.66)$$
$$\overset{\displaystyle Cl}{\overset{|}{\phantom{CH_3CH_2C}}}$$

**4.73**

of magnitude faster than when in carbon tetrachloride. 1-Butyne and 3-hexyne react 280 and 20,000 times faster than acetylene, respectively. In addition, values of activation energy (5–7 kcal/mole) and entropy of activation (-45~-47 eu) are typical to those of electrophilic bimolecular reactions.

### 4.6.3 Radical Addition

Free-radical addition of thiols to acetylenes easily occurs and proceeds with a chain mechanism [96,130]. For example, **4.74** reacts with ethyl mercaptan yielding **4.75** in the presence of AIBN or under photo-irradiation [131]:

$$RO_2C\text{-}C\equiv C\text{-}SEt + EtS\cdot \longrightarrow \underset{\overset{|}{SEt}}{RO_2C\text{-}C=\dot{C}\text{-}SEt}$$

$$\textbf{4.74}$$

$$(4.67)$$

$$\underset{\overset{|}{SEt}}{RO_2C\text{-}C=\dot{C}\text{-}SEt} + EtSH \longrightarrow \underset{\overset{|}{SEt}}{RO_2C\text{-}C=CHSEt} + EtS\cdot$$

$$\textbf{4.75}$$

Ethoxyacetylene behaves similarly [132]. However, the reaction with this compound does not stop at the formation of **4.76**, but proceeds further to yield **4.77**:

$$HC\equiv COEt + EtS\cdot \longrightarrow EtSCH=\overset{\cdot}{C}OEt \xrightarrow{\text{EtSH}}$$

$$EtSCH=CHOEt \xrightarrow{\text{EtS}\cdot} [EtS\overset{\cdot}{C}H\text{-}CHOEt \longleftrightarrow Et\overset{\cdot}{S}=CH\text{-}CHOEt]$$
$$\qquad\qquad\qquad\qquad\qquad\qquad\qquad | \qquad\qquad\qquad\qquad | $$
**4.76** $\qquad\qquad\qquad\qquad\qquad\qquad\qquad$ SEt $\qquad\qquad\qquad\qquad$ SEt

$$\longrightarrow EtSCH_2CH\big<{OEt \atop SEt} + EtS\cdot \qquad\qquad (4.68)$$
**4.77**

It is interesting to note that a thiyl radical attacks oxygen-linked carbon of **4.76**. Electron-sharing participation may be operating in the intermediate radical, as discussed in Sect. 4.5.3.

Stability of the intermediate vinyl radical, **4.78**, seems to control the

$$R\text{-}\overset{\cdot}{C}=CHSR'$$
**4.78**

reactivity and orientation of the addition and the following order is proposed for the stabilizing effect of a substituent [96,133]:

$$Ph > (RS >) CO_2H \approx CO_2Et > Me (> H)$$

Reactivities of triple and double bonds are compared (eq. 4.69) [134]. With thiophenol, the reaction affords **4.79** and **4.80** exclusively, but with aliphatic thiols, trace amounts of **4.81** and **4.82** are formed:

$$HC\equiv C\text{-}CH=CH_2 + RSH \xrightarrow{\text{DTBP}} HC\equiv C\text{-}CH_2CH_2SR$$
$$\qquad\qquad\qquad\qquad\qquad\qquad\qquad\qquad\qquad \textbf{4.79} \qquad\qquad (4.69)$$

$$+ H_2C=C=CHCH_2SR + H_2C=\underset{|}{C}\text{-}CH=CH_2 \quad RSCH=CH\text{-}CH=CH_2$$
$$\qquad\qquad\qquad\qquad\qquad\qquad\qquad SR$$

$$\textbf{4.80} \qquad\qquad\qquad\qquad \textbf{4.81} \qquad\qquad\qquad\qquad \textbf{4.82}$$

Since the addition to acetylenes produces vinylic radicals, stereochemistry of the reaction provides a new mechanistic problem. On the
basis of results obtained from other systems, it is known that *cis*- and
*trans*-vinylic radicals are in equilibrium at normal temperatures (eq.
4.70) [135-138]. Another possibility is a non-classical intermediate, **4.83**:

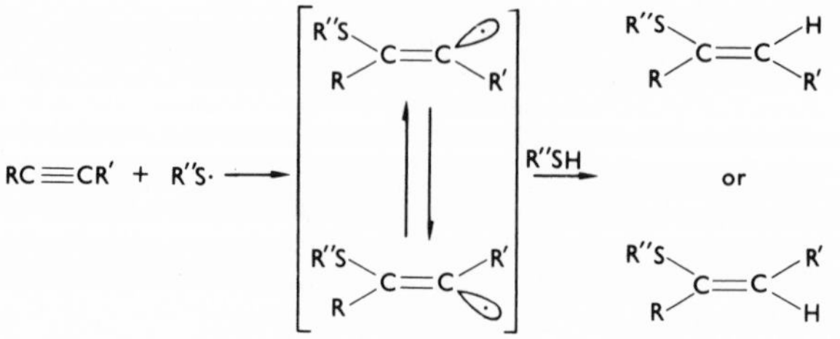

$$RC \equiv CR' + R''S \cdot \longrightarrow \qquad\qquad \xrightarrow{R''SH} \qquad or \qquad\qquad (4.70)$$

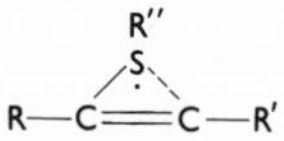

**4.83**

It is reported that the addition of ethyl mercaptan to ethoxyacetylene
affords a mixture of *cis*- and *trans*-adducts, while the addition of
ethylthioacetylene gives the *cis*-adduct exclusively [140]. More detailed
studies have been done on phenylacetylene [141,142]. The stereochemistry of products are affected by thiol : acetylene ratio, reaction time,
and temperature. When excess acetylene is used, more than 95% of
products have *cis*-configuration (at $0°$) regardless of the structure of thiol,
whereas increasing reaction time and/or thiol concentration increases the
proportion of the *trans*-isomer and finally product composition becomes
that of equilibrium (*cis* : *trans* = 20 : 80). 1-Hexyne reacts with thioacetic
acid affording 82% of the *cis*-adduct and 18% of the *trans*-isomer, provided
large excess of the former is present [142]. Apparently, the reaction is
controlled by kinetics instead of thermodynamics: hydrogen abstraction
by **4.84** is faster than that by **4.85** (eq. 4.71). Simamura and co-workers

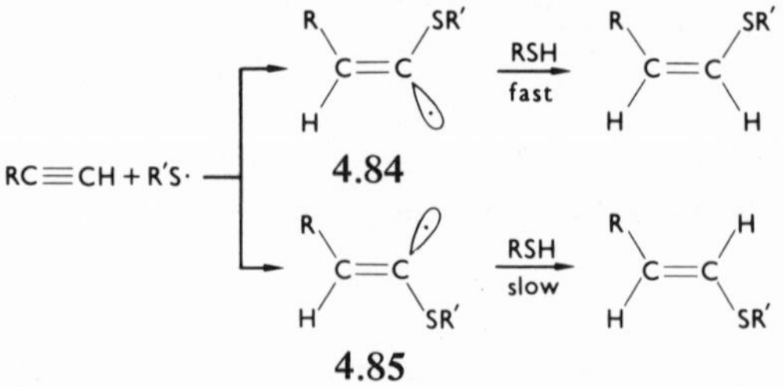

$$RC \equiv CH + R'S \cdot \longrightarrow \qquad\qquad \xrightarrow[fast]{RSH} \qquad\qquad (4.71)$$

**4.84**

**4.85**

found that **4.86** abstracts chlorine atoms faster than **4.87** does [143]:

$$
\begin{array}{cc}
\mathbf{4.86} & \mathbf{4.87}
\end{array}
$$

$$X + CH_3SH \xrightarrow{\ k_1\ } XH + CH_3S\cdot \qquad\qquad (4.72)$$

$$X \xrightarrow{\ k_2\ } A + CH_3S \qquad\qquad (4.73)$$

For reactions shown in eqs. 4.72 and 4.73, where A and X denote an unsaturated hydrocarbon and an intermediate radical formed from them by the reaction with methanethiyl radical, respectively, ratios of rate constants $k_1/k_2$ were calculated to be 291, 4.2, and 0.9 for acetylene, ethylene, and butene-2, respectively [144]. Although this seems to indicate very little occurrence of the reversible step for the addition to acetylenes, no definite evidence is available concerning this problem.

Numerous studies have been done on free-radical addition of thiols to allenes. Namely, hydrogen sulfide [145], thiolacetic acid [146,147], and various thiols [146-148] were studied. Although the earlier reports based on kinetics claimed that thiyl radicals attack the central carbon of allene, this conclusion is still open to question, because no product analysis was made [149,150]. Generally speaking, the addition takes place at terminal carbons. Thus, the addition of hydrogen sulfide to allene affords 3-propenyl mercaptan exclusively. However, normal thiols give about 10–20% of the product, in which a thiyl radical is bound to the central carbon [151]:

$$
\begin{array}{c}
RSCH_2\text{-}CH{=}CH_2 \xrightarrow{\ RSH\ } RSCH_2CH_2CH_2SR \\
\text{(a)} \quad \text{(b)} \qquad 2RSH \\
RSH + H_2C{=}C{=}CH_2 \ \rightleftharpoons\ CH_3\text{-}C{\equiv}CH \xrightarrow{\ } RSCH_2{=}CHSR \\
\text{(c)} \qquad\qquad\qquad RSH \qquad\ CH_3 \\
RS\text{-}C{=}CH_2 \\
CH_3
\end{array}
\qquad (4.74)
$$

Since control experiments excluded processes (a) and (b), there remains no doubt that the process (c) is actually operating.

The attack of a thiyl radical on the central carbon of allene seems to give an allyl radical, **4.88**, which is resonance-stabilized. However, since two double bonds in allene are orthogonal, the radical center in **4.88**

$$
\overset{\displaystyle \text{SR}}{\underset{\displaystyle }{\overset{|}{{\scriptstyle\diagup}\mathrm{C}\!=\!\mathrm{C}\text{-}\dot{\mathrm{C}}{\scriptstyle\diagdown}}}}
$$

**4.88**

cannot be stabilized by a double bond until the carbon-carbon single bond rotates 90°. The terminal-attack by a thiyl radical may be accounted for by its electrophilicity[96]. Electrophilic radicals such as trifluoromethyl[152] and dimethylamino[153] radicals are known to attack the terminal carbon. A molecular orbital calculation also supports the susceptibility of the terminal carbon toward the radical attack[154]. On the other hand, phosphinyl radicals give a very low yield of products resulting from the attack on the central carbon[155], while trimethyltin[156] and bromine[157] radicals afford both products.

With substituted allenes, orientation of the addition is controlled by the stability of the forming radical. This is illustrated by increase of the attack on the central carbon for a series of methyl-substituted allenes[147]:

$$
\begin{array}{ccc}
\mathrm{H_2C\!=\!C\!=\!CH_2} & \mathrm{H_2C\!=\!C\!=\!CH\text{-}CH_3} & \mathrm{H_2C\!=\!C\!=\!C}\overset{\diagup\mathrm{CH_3}}{\underset{\diagdown\mathrm{CH_3}}{}} \\
\uparrow & \uparrow & \uparrow \\
13\,\% & 48\,\% & 100\,\%
\end{array}
$$

It is noteworthy, however, that both radicals formed from 1,2-butadiene and 3-methyl-1,2-butadiene abstract a hydrogen at *primary* carbons. Rearrangements from secondary to primary and from tertiary to primary radicals seem to occur quite easily (eq. 4.75). Since hydrogen abstraction requires some small activation energy (*cf.* Sect. 4.4), a radical has enough life-time to rearrange to a more reactive one if the energy of rotation around the carbon-carbon single bond is reasonably smaller than that for hydrogen abstraction.

Activation energy for the attack of benzenethiyl radical on the central carbon of allene is only 1.1 kcal/mole more than that on the terminal carbon[57a]. However, since the former process yields an allylic radical after rotation around the carbon-carbon single bond, the reverse reaction takes place with difficulty. On the other hand, the latter process is similar to

$$(4.75)$$

the addition to olefins and an extensive reverse reaction is expected (*cf.* Sect. 4.5.3). Thus, increase of temperature and/or decrease of thiol concentration increases the proportion of the center-attack.

### 4.6.4 Oxidative Addition

$$RSH \xrightarrow{\text{O}_2} RS \cdot \xrightarrow{\text{HC} \equiv \text{CPh}} RSCH=\dot{C}Ph$$

$$(4.76)$$

Co-oxidation of phenylacetylene with a thiol and oxygen takes place giving a hemiketal, **4.89** (eq. 4.76) [114].

The process (b) seems most probable, but neither (a) nor (c) can be discarded. Interestingly, Oswald and his co-workers demonstrated that increased concentration of phenylacetylene results in increase of the yield of *cis*-l-mercapto-2-phenyl-ethylene [158] as has been seen in nucleophilic additions (*cf.* Sect. 4.6.1).

Acetylene also reacts with carbon monoxide in the presence of a thiol to give a low yield of unsaturated aldehyde [159]:

$$RS \cdot \; + \; HC \equiv CH \quad \longrightarrow \quad RSCH = \overset{\cdot}{C}H$$

$$RSCH = \overset{\cdot}{C}H \; + \; CO \quad \longrightarrow \quad RSCH = CH\overset{\cdot}{C}O \qquad (4.77)$$

$$RSCH = CH\overset{\cdot}{C}O \; + \; RSH \quad \longrightarrow \quad RSCH = CHCHO \; + \; RS \cdot$$

## 4.7 SUBSTITUTION ON AROMATIC SYSTEMS

### 4.7.1 Ionic Reaction

Since aromatic systems are barely susceptible to nucleophilic reactions, no example for this subject is available. No report on the electrophilic attack of a thiol itself on aromatic systems exists either. However, it is known that Friedel-Crafts type reaction proceeds with sulfenyl chlorides. The mechanism may be obvious [160-163]:

$$PhSCl \; + \; \langle C_6H_4 \rangle - OCH_3 \quad \xrightarrow{\text{Fe}} \quad PhS - \langle C_6H_4 \rangle - OCH_3 \qquad (4.78)$$

$$PhSCl \; + \; \langle furan \rangle \quad \xrightarrow{\text{Fe}} \quad PhS - \langle furan \rangle \qquad (4.79)$$

### 4.7.2 Radical Reaction

The difficulty of the free-radical addition of a thiol to aromatic systems is accounted for by unfavorable energetics involved in disrupting aromatic systems. This is not the only reason, however, since this cannot explain facile reactions with phenyl [164] and thienyl [165] radicals. Reversibility of the thiol addition may be responsible.

Aliphatic thiols add to thiophene in the presence of ferrous ion and hydrogen peroxide (eq. 4.80) [166]; *tert*-Butyl mercaptan affords only

$$RSH + \underset{S}{\text{[thiophene]}} \xrightarrow[\text{H}_2\text{O}_2]{\text{Fe}^{++}} RS-\underset{S}{\text{[thiophene]}}-SR \qquad (4.80)$$

the monoadduct and the addition to the 3-position of 2.5-dimethylthiophene is negligible. Since radicals are generated by the treatment with Fenton's reagent, it is not obvious whether they are free or complexed with transition metals.

Thiyl radicals generated by treating thiols with *tert*-butyl hydroperoxide and ferrous ion add to the 9-position of anthracene (eq. 4.81) [167], but yields of products are much higher when a mixture of thiol and anthracene is treated with oxygen [168,169].

(R = Ac PhCO CH₂CO₂H)

## 4.8 OXIDATION TO DISULFIDES

Oxidation of a thiol proceeds stepwise, giving a disulfide initially, then· thiolsulfinic acid, etc., and finally sulfonic acid. Under the title of oxidation, we will deal with only the formations of disulfides.

### 4.8.1 Oxidation by Air

When thiols are oxidized by air at low temperatures, they afford disulfides regardless of the presence or absence of catalysis (eq. 4.82). As with

$$2 \, RSH + \tfrac{1}{2}O_2 \longrightarrow RSSR + H_2O \qquad (4.82)$$

air-oxidation of carbanions, the formation of a thiolate anion is believed to be the initial step of the reaction. The second step is the formation of a thiyl radical by electron transfer between the thiolate anion and oxygen. Hence, the reaction is very much promoted under basic conditions:

$$RSH + R'O^- \rightleftharpoons RS^- + R'OH$$

$$RS^- + O_2 \rightleftharpoons RS\cdot + O_2^-\cdot$$

$$2\ RS\cdot \longrightarrow RSSR \tag{4.83}$$

$$RSH + O_2^- \longrightarrow RS\cdot + HO_2^-$$

$$RSH + HO_2^- \longrightarrow RS\cdot + HO^- + \cdot OH$$

$$RSH + \cdot OH \longrightarrow RS\cdot + H_2O$$

$$(R = H, alkyl)$$

The oxidation by hydrogen peroxide is also accelerated by base-catalysts[170,171]. N-(2,3-Dimethylthiopropyl)-O-ethyl-carbamate is oxidized by air in the presence of cupric ions. The reaction proceeds with the half-life of 6.5 min at pH 9.4 and increases to 26.5 min at pH 7, to 300 min at pH 5.4, and finally to infinite half-life at pH 4, which indicates the involvement of thiolate anions in the oxidation.

Solubility of thiols seem to affect the reaction rate: the longer the alkyl chain, the more difficult the oxidation. For example, the reactivity decreases in the order $n$-PrSH $>$ $n$-BuSH $>$ $n$-PenSH in the presence of 0 : 1–2.7 N NaOH at 29~35° [172].

Reaction rates are also affected by solvents as seen in Table 4.9, where

**TABLE 4.9**

Oxidation of $n$-Butyl Mercaptan in Various Solvents

| Solvent | $k \times 10^3$, min$^{-1}$ | Rel. Rate |
|---|---|---|
| Methanol | 3.22 | 1 |
| Tetrahydrofuran (THF) | 116 | 36 |
| Dioxane | 289 | 90 |
| Diglyme | 323 | 100 |
| Dimethylacetamide (DMA) | 936 | 291 |
| Dimethylformamide (DMF) | 1097 | 334 |

rate constants for oxidation of $n$-butyl mercaptan in the presence of sodium methylate in various solvents are listed [173].

In cation-solvating solvents such as THF and diglyme, methoxide ion becomes free and effective in comparison with that in methanol. Dipolar aprotic solvents such as DMA and DMF are more effective for complex formation than others.

There is no doubt, from the above discussion, that extent of dissociation — acidity — of thiols plays an important role in thiol reactivity. However, this is not the whole story, since the rate-determining step of the reaction is the process of electron transfer. For example, $p$-nitrothiophenol, which affords a resonance-stabilized anion, resists oxidation, while $p$-aminothiophenol reacts easily. The following order of reactivity seems acceptable: $ArSH > HO_2CCH_2SH > RCH_2SH > R_2CHSH > R_3CSH$ (*cf.* Sect. 4.6.4).

Amines are effective catalysts for oxidation and hydrogen-bonding complex, **4.90**, between an amine and a thiol is proposed as a reactive species (eq. 4.84) [117]. It is reported that $g$-aminoethyl mercaptans

$$\overset{S^+}{R_3N} \, \ldots \, \overset{S^-}{H\text{-}SR'} \; \rightleftharpoons \; R_3\overset{+}{N}H + {}^-SR' \qquad (4.84)$$

**4.90**

react with oxygen very rapidly (eq. 4.85) [174-176]. Cobalt (II) complexes of amines such as **4.91–4.93** and of pyrophosphate are

$$R_2N\overset{\displaystyle CH_2\text{-}CH_2}{\underset{\displaystyle H}{\overset{S^+/}{\diagdown}\quad\overset{\backslash S^-}{\diagup}S}} \; \rightleftharpoons \; R_2\overset{+/}{N}H \overset{\displaystyle CH_2\text{-}CH_2}{\overset{\backslash -}{\diagup}S} \qquad (4.85)$$

effective catalysts for the oxidation, because cobalt (II) has an unpaired electron in one of $3d$ orbitals [117,177].

Mechanisms of oxidations by peroxides have been the subject of discussions for many years. Some support ionic mechanisms and others prefer radical processes. No definitive evidence is available now. Whichever the mechanism of oxidation is, it is widely known that thiols are potential hydrogen-atom donors and very effective chain carriers. Thus, they are used as inhibitors of radical polymerizations, inhibitors of photoreduction of ketones [178], trapping agents for biradicals [179], and many other purposes.

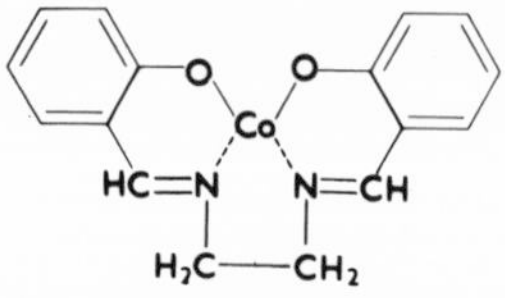

**4.91**

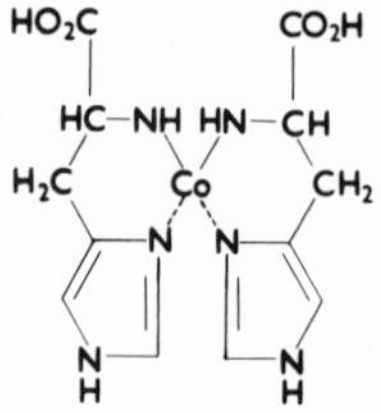

**4.92**

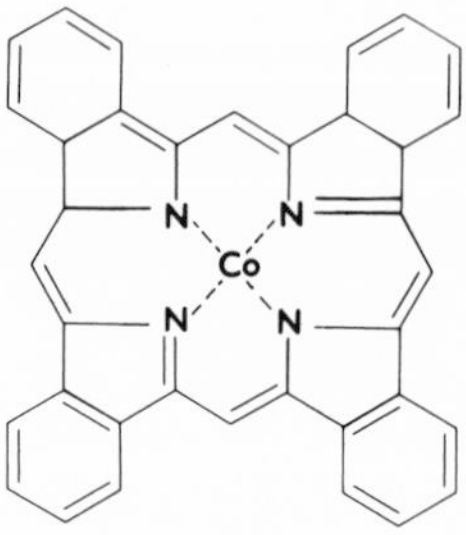

**4.93**

### 4.8.2 Oxidation by Metal Oxides

Oxidation of thiols to disulfides by $Fe_2O_3$, $MnO_2$, $Co_2O_3$, and CuO has been studied [180]. $MnO_2$ is the most effective of all. Reaction rate is not affected by structure and acidity of a thiol and is not catalyzed by amines. When an olefin is added to the system, addition of thiol to the olefin takes place by a chain mechanism. These results invoke, undoubtedly, that the reaction proceeds with rate-determining adsorption of thiols on the surface of metal oxides, followed by the formation of thiyl radicals. Ferric octanoate oxidizes thiols to disulfides with essentially the same mechanism [181]:

$$RSH + Fe(O_2CC_7H_{15})_3 \xrightarrow{\text{slow}} RS\cdot + Fe(O_2CC_7H_{15})_2$$
$$+ C_7H_{15}CO_2H$$

$$2\ RS\cdot \longrightarrow RSSR$$

(4.86)

Kinetics of the reaction follow second-order rate law, being first-order for the salt and the thiol each. Lead tetraacetate in glacial acetic acid is also an effective oxidant of thiols [182].

### 4.8.3 Oxidation by Organic Oxides

Thiols are oxidized to disulfide on heating with a sulfoxide [183,184]:

$$2\ RSH\ +\ R_2'SO\ \longrightarrow\ RSSR\ +\ R_2'S\ +\ H_2O \qquad (4.87)$$

Dimethyl sulfoxide (DMSO) is the most common reagent, simply because it is the most popular sulfoxide. Tertiary amines may improve the yield. Kinetics of the reaction is second-order, being first-order for each of reactants, and the rate increases with the increase of acidity of a thiol. In the reaction with tetramethylene sulfoxide, the reactivity of thiols decreases in the order: phenyl > 2-methylphenyl > benzyl > $n$-dodecyl. A mechanism in which an unstable intermediate is formed in the rate-determining step is proposed (eq. 4.88) [185]. The reaction is catalyzed

$$RSH\ +\ R_2'SO\ \underset{slow}{\rightleftharpoons}\ \left[\ R_2'S\diagdown^{\displaystyle OH}_{\displaystyle SR}\ \right]$$

$$(4.88)$$

$$\underset{fast}{\overset{RSH}{\rightleftharpoons}}\ RSSR\ +\ R_2'S\ +\ H_2O$$

not only by amines [185] but also by phosphoric acid, acetic acid [186] and perchloric acid [187]. For example, in the reaction of $n$-dodecyl mercaptan with tetramethylene sulfoxide, N,N-dimethylaniline and 2,4-lutidine accelerate the reaction only slightly, while the reaction becomes 84 and 269 times faster in the presence of $n$-dodecylamine and tri($n$-butyl)amine, respectively, than that without amines. Formation of a complex such as **4.94** is proposed. Complex formation is also proposed for acid catalysts, **4.95**.

**4.94**                    **4.95**

A. Ohno and S. Oae

The reaction of thiophenol with trimethylsulfoxonium iodide gives diphenyl disulfide (51%), methyl phenyl sulfide (45%), and dimethyl sulfide as well as iodine. The following mechanism is proposed [188]:

$$PhSH + Me_3\overset{+}{S}OI^- \xrightarrow[100°C]{DMF} [Me_3\overset{+}{\underset{O}{S}} \cdot \overline{S}Ph] + HI$$

$$\xrightarrow{\hspace{2cm}} Me_2SO + PhSMe \qquad\qquad (4.89)$$

$$Me_2SO + 2PhSH \xrightarrow{\hspace{1.5cm}} Me_2S + Ph_2S_2 + H_2O$$

$$Me_2SO + 2HI \xrightarrow{\hspace{1.5cm}} 2Me_2S + H_2O + I_2$$

Amine oxides also oxidize thiols to disulfide. Trimethylamine oxide, for example, reacts with cysteine and thioglycolic acid at 120~140° [189] Pyridine oxide reacts similarly [190].

## 4.9 DITHIOCARBOXYLIC AND DITHIOPHOSPHORIC ACIDS

As compared to ordinary thiols, both dithiocarboxylic and dithiophosphoric acids are fairly strong acids, and yet retain relatively high nucleophilic character. Because of the dual properties, they are not only far more reactive than normal thiols in addition reactions to unsaturated compounds and in reduction of sulfoxides, but also mild reducing agents cleaving such linkages as $\overset{+}{S}-\overline{C}$, S→N or $\overset{+}{N}-\overline{N}$. Therefore, the chemistry of these highly acidic thiols represented by dithiocarboxylic and dithiophosphoric acids [191] is treated separately in this section.

### 4.9.1 Preparation

#### 4.9.1.1 Dithiocarboxylic Acids

Treatment of Grignard reagents with carbon disulfide has been known for a long time as the most common procedure for preparing dithiocarboxylic acids [192] (eq. 4.90). Aromatic dithiocarboxylic acids can be obtained in

$$RMgX + CS_2 \xrightarrow{\text{ether}} RCS_2MgX \xrightarrow{H^+, H_2O} RCS_2H \quad (4.90)$$

good yields by this method; however, yields of aliphatic dithiocarboxylic acids are usually rather poor due mainly to the facile decomposition of the latter acids. In fact, alkane dithiocarboxylic acids are difficult to purify, unless they are distillable, and slowly decompose within a few days, even in a refrigerator.

Arene dithiocarboxylic acids can be prepared by the following simple procedure developed by Bost *et al.* [193], though yields are poor, *i.e.*, ~30% (eq. 4.91). Another simple and convenient method is the treatment

$$\text{ArCHO} + (\text{NH}_4)_2\text{S}_\text{X} \xrightarrow{\text{OH}^-} \text{ArCS}_2\text{H} \qquad (4.91)$$

of chloromethyl arenes with elemental sulfur in the presence of two equimolar amounts of alkali alcoholate (eq. 4.92). Though this procedure is limited to the synthesis of aromatic dithiocarboxylic acids, the work-up procedure is handy and the yields are quite good. Benzotrichloride also yields dithiocarboxylic acid (eq. 4.93) when treated with $K_2S–H_2S$ in alkaline media [195].

$$(4.92)$$

$$(4.93)$$

### 4.9.1.2 Dithiophosphoric Acids

Dithiophosphoric acids are commonly synthesized simply by treating an equimolar mixture of alcohols with phosphorus pentasulfide [196,197], while dithiophosphoric acid from alcohols of low molecular weights can be purified by distillation:

$$\text{EtOH} + \text{P}_2\text{S}_5 \xrightarrow{80-100°\text{C}} (\text{EtO})_2\text{PS}_2\text{H} \qquad (4.94)$$
$$(60\%)$$

In the case of sensitive alcohols, such as *t*-butanol and benzyl alcohol, mild conditions are necessary to ensure high yields as shown below (eqs. 4.95, 4.96). Dithiophosphoric acids are stable and can be stored.

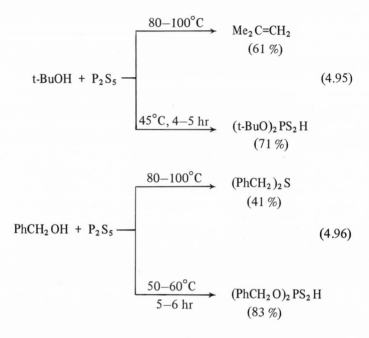

$$t\text{-BuOH} + P_2S_5 \quad \begin{cases} \xrightarrow{80-100°C} & Me_2C{=}CH_2 \\ & (61\%) \\ \\ \xrightarrow{45°C,\ 4-5\ hr} & (t\text{-BuO})_2PS_2H \\ & (71\%) \end{cases} \tag{4.95}$$

$$PhCH_2OH + P_2S_5 \quad \begin{cases} \xrightarrow{80-100°C} & (PhCH_2)_2S \\ & (41\%) \\ \\ \xrightarrow[5-6\ hr]{50-60°C} & (PhCH_2O)_2PS_2H \\ & (83\%) \end{cases} \tag{4.96}$$

### 4.9.2 Physical Properties

As is seen in Table 4.10, dithiocarboxylic acids are substantially more

**TABLE 4.10**

pK$_a$ Value of a Few Dithiocarboxylic Acids and Related Acids in Water at 25°C [191b]

| Compound | pK$_a$ | Compound | pK$_a$ | Compound | pK$_a$ |
|----------|--------|----------|--------|----------|--------|
| $CH_3CS_2H$ | 2.55 | $CH_3COSH$ | 3.33 | $CH_3COOH$ | 4.76 |
| $C_2H_5CS_2H$ | 2.52 | $PhCOSH$ | 2.48 | $PhCOOH$ | 4.21 |
| $H_2NCS_2H$ | 2.95 | $PhSH$ | 7.49 | $CH_3CS_2CH_2COOH$ | 2.90 |
|  |  | $C_2H_5OH$ | 18 |  |  |

acidic than corresponding thiols and carboxylic acids [191b]. The higher acidity of dithiocarboxylic acid is due partly to the strong electron-withdrawing inductive effect of dithiocarboxyl group, as witnessed by the small pK$_a$ value of $CH_3CS_2CH_2COOH$ (pK$_a$ = 2.90), and partly to the resonance stabilization of the resulting dithiocarboxylate anion, as seen in the comparison of the resonance energies of the following carboxylate anions: $RCOO^-$ (13.5 kcal/mole), $RCOS^-$ (5.5 kcal/mole) $RCSS^-$ (17.4 kcal/mole) [191b].

Table 4.11 summarizes $pK_a$ values of dithiophosphoric and dithiophosphinic acids [198], which are as acidic as dichloroacetic ($pK_a$ = 1.29) or nitroacetic acids ($pK_a$ = 1.68).

**TABLE 4.11**

$pK_a$ Values of Representative Dithiophosphoric and Dithiophosphinic Acids in 7% EtOH-Water at 25°C [198]

| Compound | $pK_a$ | Compound | $pK_a$ |
|---|---|---|---|
| $(MeO)_2PS_2H$ | 1.55 | $Et_2PS_2H$ | 1.71 |
| $(EtO)_2PS_2H$ | 1.62 | $Ph_2PS_2H$ | 1.75 |
| $(PhO)_2PS_2H$ | 1.81 | | |
| $(p\text{-}ClC_6H_4O)_2PS_2H$ | 1.79 | | |

There are four characteristic bands in ir spectra of dithiocarboxylic acids. For example, these four for dithioacetic acid are 2481 (−SH), 1216 (C=S), 860 (−SH) and 581 (C−S) $cm^{-1}$ [199]. The S−H stretching frequency of the very acidic trifluorodithioacetic acid is 2578 $cm^{-1}$, considerably shorter than that of dithioacetic acid [200], and is similar to that of ordinary thiols. Meanwhile, there is no indication of such hydrogen bonding as −S−H . . . S=C in the ir spectra [201] and dithiocarboxylic acids were found by Allen and Colalough not to be associated either in $CCl_4$ or neat [202]. Whereas, dithiophosphoric or dithiophosphinic acids appear to be associated by hydrogen bonding between −SH and P=S, since a broad band at 2420 $cm^{-1}$ for the associated acids in neat shifts to a sharp band at 2560 $cm^{-1}$ upon dilution with $CCl_4$ [202]. Apparently the magnitude of hydrogen bonding is in the following order: dithiophosphoric > dithiocarboxylic > monothio-carboxylic acids. In dilute aqueous solutions, dithiocarboxylic acid is considered to assume a symmetrical structure, shown by **4.96**, like a metalic salt, **497** [203]. The two bond distances of Hg−S linkages in mercuric salt of dithiocarbamate, **498**, were shown by X-ray diffraction to be identical [204].

**4.96**          **4.97**                **4.98**

There is a weak absorption at 469 nm in the uv spectrum of dithioacetic acid and this band is considered to be due to n-$\pi$* transition because of its blue-shift to 450 nm in polar media. There are two absorption bands of n-$\pi$* transition of C=S group, *i.e.*, 538 nm and 523 nm in the uv spectrum of dithiobenzoic acid, indicating that the two lone electron pairs on the sulfur are different in energy, perhaps due to possible internal hydrogen bonding with one of the two lone pairs [205]. Like thioketones [206] (see also Chapter 5), dithicarboxylic acids and their esters would be readily excited by visible light and hence display interesting photochemical behaviours, but practically no work has been done in this field.

### 4.9.3 Chemical Behaviors

Because of the dual properties, *i.e.*, highly acidic nature and nucleophilic character, both dithiocarboxylic and dithiophosphoric acids display rather unique chemical behaviors, sometimes common to both acids and other times different.

#### 4.9.3.1 Reactions with Amines

Reactions of dithiocarboxylic acids with amines or similar amino compounds have long been known to afford thioamides or their derivatives with evolution of $H_2S$, as shown in eqs. 4.97, 4.98, 4.99, 4.100. Other

$$RCS_2H \begin{cases} NH_3 \longrightarrow RCSNH_2 + H_2S & (4.97)\,[207] \\ NH_2NH_2 \longrightarrow RCSNHNH_2 + H_2S & (4.98)\,[208] \\ PhNHNH_2 \longrightarrow RCSNHNHPh + H_2S & (4.99)\,[209] \\ H_2NNHCONH_2 \rightarrow RCSNHNHCONH_2 + H_2S & (4.100)\,[210] \end{cases}$$

examples are also known in which thioamides are not formed (eqs. 4.101, 4.102). The latter example is the reaction at higher temperatures. A similar

$$(4.101)\,[211]$$

$$RCS_2H + H_2NNHCONH_2 \longrightarrow RCH=NNHCONH_2 + H_2S + S$$
$$(4.102)$$

reaction occurs with dithiobenzoic acid and $p$-methylphenylhydrazine (eq. 4.103) [212]. The reaction with hydroxylamine gives thioloxime as

$$PhCS_2H + p\text{-}MeC_6H_4NHNH_2 \xrightarrow[\text{ether}]{-15°C} [\text{salt}]$$
$$(4.103)$$

$$\xrightarrow{50°C} PhCH=NNCH_6H_4Me\text{-}p + H_2S + S$$

shown by eq. 4.104 [213]. In the presence of such a mild oxidizing agent as KI–I$_2$, even sodium salt of dithiocarboxylic acid reacts similarly, for example [214,215]:

$$RCS_2H + H_2NOH \longrightarrow RC(NOH)SH \qquad (4.104)$$

$$PhCS_2Na \; + \; NH_2\!\!-\!\!\langle\;H\;\rangle \xrightarrow{KI-I_2} Ph\!-\!\overset{\displaystyle S}{\overset{\|}{C}}\!-\!NH\!-\!\langle\;H\;\rangle \qquad (4.105)$$

There is an interesting reaction between $o$-phenylenediamine and $p$-toluenedithiocarboxylic acid to give benzoimidazole, mercaptobenzoimidazole and toluene with evolution of $H_2S$ [216]:

$$(4.106)$$

Two conceivable mechanisms have been suggested for the formation of thioamide. One involves the formation of the ammonium salt, **4.99**, subsequent proton-transfer from N to S, followed by the nucleophilic attack of amine on thiocarbonyl function to yield the thioamide with evolution of $H_2S$ (eq. 4.107) [209,211,212]. The other involves direct attack of amine on thiocarbonyl group to form an addition intermediate,

**4.100**, which upon extrusion of $H_2S$ gives the final product, as was suggested in the reaction with aniline (eq. 4.108) [217].

$$H_2NNHPh \xrightarrow{RCS_2H} \left[ RC\bar{S}_2 \ \overset{+}{N}H_3NHPh \right] \xrightarrow{-H_2S} \overset{S}{\underset{\parallel}{R}}CNHNH_2 \qquad (4.107)$$
$$\mathbf{4.99}$$

$$R-\overset{S}{\underset{\parallel}{C}}SH \ + \ \langle \ \rangle -NH_2 \longrightarrow \left[ \langle \ \rangle -\overset{H}{\underset{H}{N}}-\overset{R}{\underset{SH}{C}}-S^- \right] \qquad (4.108)$$
$$\mathbf{4.100}$$

Dithiophosphoric acid also react with hydrazines exothermically, but in a different fashion, giving N-N bond cleavage products (60% yield) (eq. 4.109) [218]. The reaction is presumed to proceed *via* the

$$R-\langle \ \rangle -NHNH_2 \xrightarrow{(EtO)_2PS_2H} R-\langle \ \rangle -NH_2 \ + \ \left[(EtO)_2PS\right]_2S_2 \ (4.109)$$

nucleophilic attack of dithiophosphate anion on terminal nitrogen of the protonated hydrazine and is considered to be a rare example of nucleophilic substitution on neutral nitrogen atom (eq. 4.110) [218]. Hydrazobenzenes also react similarly, affording, not only anilines but also

$$R-\langle \ \rangle -NHNH_2 \xrightarrow{(EtO)_2PS_2H} \left[ R-\langle \ \rangle -\overset{+}{N}H_2 \ NH_2 -\overset{S}{\underset{\parallel}{S}}P(OEt)_2 \right]$$

$$\longrightarrow R-\langle \ \rangle -NH_2 \ + \ NH_2\overset{}{\underset{\parallel}{S}}P(OEt)_2$$
$$\qquad\qquad\qquad\qquad\qquad\qquad S \qquad\qquad (4.110)$$

$$2NH_2\overset{S}{\underset{\parallel}{S}}P(OEt)_2 \longrightarrow \left[(EtO)_2PS\right]_2S_2 \ + \ \text{etc}$$

benzidine and azobenzene [218]. Another interesting reaction may be the following clear reduction of aziridine ring without changing the nitro group [218]:

$$\begin{array}{c} N\text{-}C_6H_4NO_2\text{-}o \qquad\qquad\qquad\qquad\qquad\qquad (4.111) \\ / \ \backslash \\ o\text{-}NO_2C_6H_4\text{-}N\text{---}N\text{-}C_6H_4NO_2\text{-}o \qquad \longrightarrow \ o\text{-}NO_2C_6H_4NH_2 \end{array}$$

$$+ \ [(EtO)_2PS]_2S_2$$

### 4.9.3.2 Reduction of Sulfoxide and Related Compounds

Thiols are known to reduce sulfoxides to corresponding sulfides [183,184]. However, it usually requires heating to complete the reduction with ordinary thiols. Even with acidic monothiolacetic acid, it takes about a week in order to complete the reduction of dimethyl sulfoxide at room temperature [219], whereas highly acidic dithiocarboxylic and dithiophosphoric acids reduce most sulfoxides exothermically just by mixing the two components at room temperature (eq. 4.112). Only in the case of diaryl sulfoxides may some heating be necessary. Similar reductions (eqs. 4.113, 4.114) take place smoothly with other trivalent sulfur compounds bearing semipolar linkages such as sulfilimines and sulfonium ylides [220]:

$$\underset{\underset{O}{\downarrow}}{R\text{-}S\text{-}R'} + 2\,R''CS_2H \longrightarrow R\text{-}S\text{-}R' + \underset{\underset{S}{\parallel}}{R''CS)_2} + H_2O \qquad (4.112)$$

$$\underset{\underset{NTs}{\downarrow}}{R\text{-}S\text{-}R'} + 2\,R''CS_2H \longrightarrow R\text{-}S\text{-}R' + \underset{\underset{S}{\parallel}}{R''CS)_2} + TsNH_2 \qquad (4.113)$$

$$\underset{\underset{C(CO_2CH_3)_2}{\overset{\mid}{\bar{C}}}}{R\text{-}\overset{+}{S}\text{-}R'} + R''CS_2H \longrightarrow R\text{-}S\text{-}R' + R''CS_2CH(CO_2CH_3)_2 \qquad (4.114)$$

The general scheme for all these reductions may be illustrated by:

$$(4.115)$$

Reductions of sulfoxides, sulfilimines and sulfonium ylides with dithiophosphoric acid occur still more readily in exactly the same fashions as in the case of dithiocarboxylic acids [221]. Pyridine N-oxide and N-iminopyridinium betaine are also reduced to pyridine upon gentle heating with dithiophosphoric acid [218,221].

Azobenzenes are also reduced by a large excess of O,O-diethyl dithiophosphoric acid to a mixture of benzidines and anilines (eq. 4.116) [218], and the suggested mechanism is shown in eq. 4.117 [218].

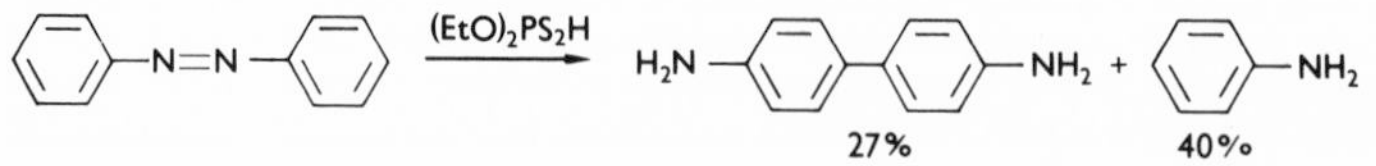

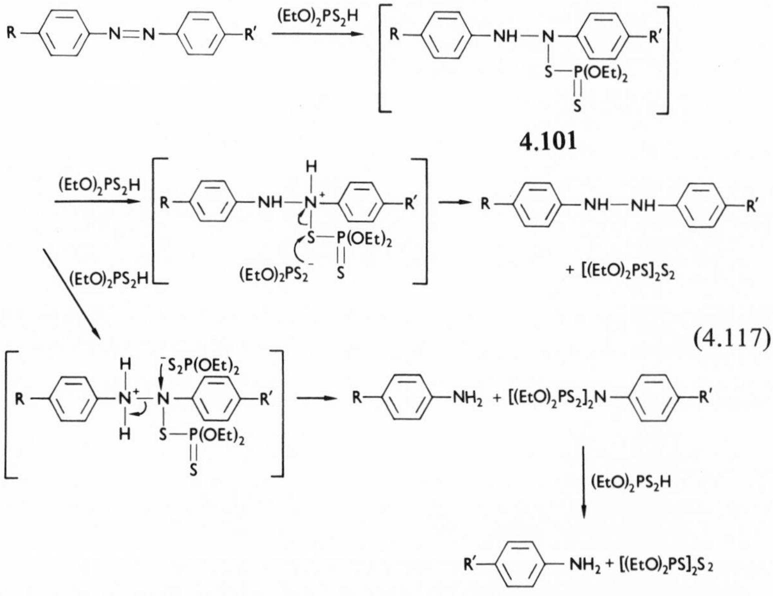

**4.101**

(4.117)

### 4.9.3.3  Addition Reactions

*With Olefinic Double Bond.* A remarkable feature of the addition of these dithioacids to olefins is that the addition takes place with both electrophilic and nucleophilic olefins without any catalyst, besides photochemical or homolytic chain addition reactions. With nucleophilic olefins such as styrene, *g*-methylstyrene, tolyl vinyl ether, vinyl acetate, acenaphthylene and indene, Markovnikov type addition takes place with dithiolacetic acid, while Michael type addition occurs with such olefins as acrylonitrile, acrylic acid, isobutenyl methyl ketone, methyl metaacrylate and vinylpyridine:

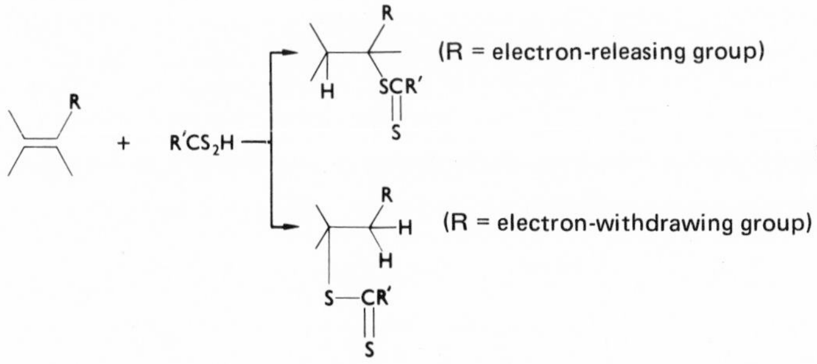

$$\text{(4.118)}$$

Although gentle heating may be necessary for some electrophilic olefins, additions are usually rapid and yields are nearly quantitative. A rough order of reactivities of a few representative olefins may be the following [220].

$$ROCH=CH_2 \approx RSCH=CH_2 > PhCH=CH_2 \approx ROCH_2CH=CH_2 > CH_2=CH\overset{O}{\overset{\|}{C}}R$$

Additions of dithiophosphoric acid to olefins were studied extensively and found to follow the same orientations as those with dithioacetic acid as shown below:

$$\text{(4.119)}$$

For the reaction scheme with $(RO)_2PS_2H$:

- $MeCOCH=CH_2 \rightarrow (RO)_2PS_2CH_2CH_2COMe$ [222]
- $R'_3SiCH=CH_2 \rightarrow R'_3SiCH_2CH_2S_2P(OR)_2$ [223]
- $R'_3SiCH_2CH=CH_2 \rightarrow R'_3SiCH_2\underset{\overset{|}{CH_3}}{CH}S_2P(OR)_2$ [223]
- $R'SCH=CH_2 \rightarrow (RO)_2PS_2\underset{\overset{|}{CH_3}}{CH}SR'$ [224]
- $CH_2=CHCN \rightarrow (RO)_2PS_2CH_2CH_2CN$ [225]

With an electrophilic olefin, $R{-}C_6H_4CH{=}C(CO_2Et)_2$, the following relative rates are obtained with R: *m*—$NO_2$ (1.47), *p*—$NO_2$ (1.09), *p*—Cl (1.16), H (1.00), *p*—Br (0.89), *p*—F (0.80) [206]. In view of the small effect of substituents the following mechanism may be considered for the Michael type addition. An interesting example may be the following

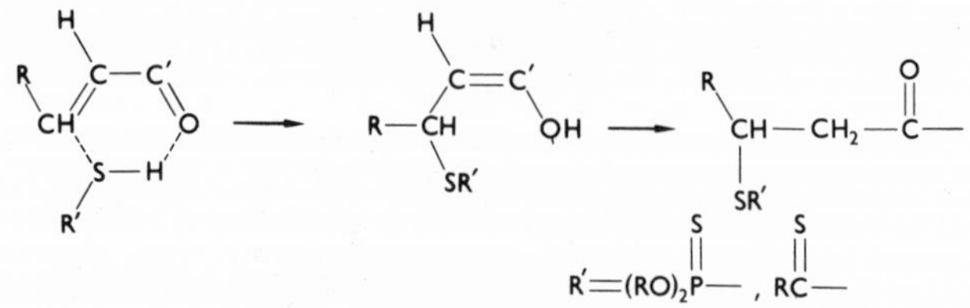


reaction which is presumed to proceed as shown below [220]:

$$\underset{\underset{O}{\downarrow}}{PhSCH{=}CH_2} + CH_3CS_2H \longrightarrow \left[\underset{\underset{OH}{|}}{Ph\overset{+}{S}CH{=}CH_2}\ {}^{-}S_2CCH_3 \longleftrightarrow \underset{\underset{OH}{|}}{PhS}{=}CH\text{-}CH_2S_2CCH_3\right] \qquad (4.120)$$

$$\xrightarrow{CH_3CS_2H} \left[\underset{\underset{{}^{+}OH_2}{|}}{Ph\text{-}S}{=}\overset{{}^{-}S_2CCH_3}{CH}\text{-}CH_2S_2CCH_3\right] \xrightarrow{-H_2O} \underset{\underset{S_2CCH_3}{|}}{PhSCH}CH_2S_2CCH_3$$

Photochemical additions of dithiphosphoric acid to olefins and acetylenes were studied in detail by Oswald *et al.* and found to take place readily to give anti-Markovnikov type addition products as shown below (eqs. 4.121, 4.122) [227]. A similar addition occurs with allene preferentially at terminal carbon (eq. 4.123).

$$RCH{=}CH_2 + (R'O)_2PS_2H \xrightarrow[\text{radical initiator}]{h\nu \text{ or other}} R\text{-}CH_2CH_2S_2P(OR')_2 \qquad (4.121)$$

$$CH_3C{\equiv}CH + (R'O)_2PS_2H \xrightarrow[\text{radical initiator}]{h\nu \text{ or other}} CH_3CH{=}CHS_2P(OR')_2 \qquad (4.122)$$

$$H_2C=C=CH_2 \; + \; (EtO)_2PS_2H \xrightarrow[48 \text{ hr}]{h\nu}$$

$$\longrightarrow (EtO)_2PS_2CH_2CH=CH_2 \xrightarrow{(EtO)_2PS_2H} [(EtO)_2PS_2CH_2]_2CH_2$$

$$(27\%) \hspace{5cm} (51\%)$$

$$\longrightarrow [(EtO)_2PS_2\underset{CH_3}{\overset{}{C}}=CH_2 \longrightarrow (EtO)_2PS_2\underset{CH_3}{\overset{}{C}}HCH_2S_2P(OEt)_2$$

$$(4.123)$$

$$(22\%)$$

Similar additions of dithiophosphoric acids were found to take place readily with conjugate olefins and are believed to be radical chain reactions [228].

*With Other Unsaturated Compounds.* O,O-Diethyl dithiophosphoric acid adds to nitriles, eventually affording corresponding thioamides (eq. 4.124) [229]. The addition apparently requires a minute amount of water,

$$R-C\equiv N \xrightarrow{(EtO)_2PS_2H} R-\underset{\overset{\|}{S}}{C}NH_2 \hspace{2cm} (4.124)$$

without which the reaction is very sluggish. The reason for it is not clear. The following reaction of phenyl isocyanate is a convenient way to prepare phenyl thioisocyanate [230]:

$$PhN=C=O \xrightarrow{(EtO)_2PS_2H} PhNH\underset{\overset{\|}{O}}{C}S_2P(OEt)_2 \xrightarrow{\Delta} PhN=C=S \hspace{1cm} (4.125)$$

By gently heating a mixture of a carbonyl compound with O,O-diethyl dithiophosphoric acid, the corresponding thiocarbonyl compound or its derivative is obtained [229]. For example, $p,p'$-dianisylketone gives the corresponding thioketone in 66% yield (eq. 4.126) [229]. When the

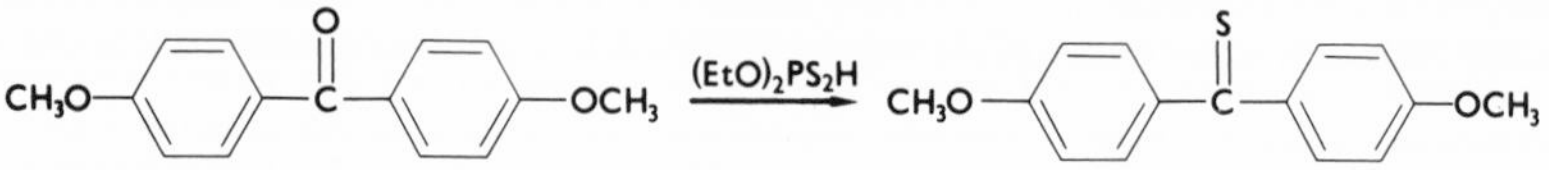

$$(4.126)$$

resulting thiocarbonyl compounds are unstable, stable dimers or thiols are the final products as seen in Table 4.12. This reaction is considered to proceed *via* the initial addition of the dithiophosphoric acid to carbonyl group as shown below (eq. 4.127). In fact, the adduct, **4.102**, was actually isolated in the case of benzaldehyde [229] .

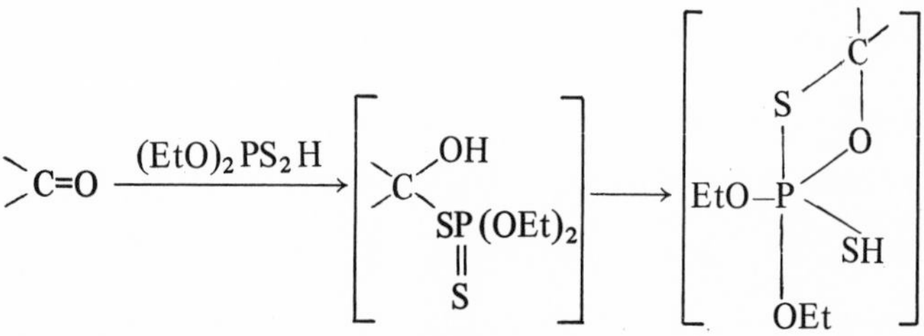

**4.102**

$$\longrightarrow \, \text{>C=S or dimer} + (\text{EtO})_2\text{PSOH}$$

Dithiophosphoric acid cleaves epoxy ring to afford corresponding addition product (eq. 4.128) [231]. Orientation of ring opening is controlled by substituent R.

Addition to dicyclohexylcarbodiimide (DCC) takes place readily to afford dicyclohexylthiourea nearly quantitatively [229].

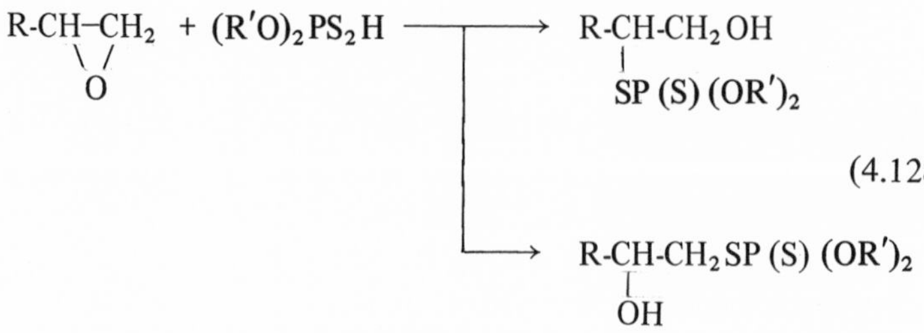

$$(4.128)$$

TABLE 4.12

Products and Yields in the Reaction of Carbonyl Compounds with $(EtO)_2PS_2H$

| Carbonyl Compound | Mole ratio $(C=O/PS_2H)$ | Temp. (°C) | Time (hr) | Product (%) |
|---|---|---|---|---|
| $(p\text{-MeOC}_6H_4)_2C=O$ | 1:6 | 80-85 | 50 | $(p\text{-MeOC}_6H_4)_2C=S$ (66 %) |
| | 1:3 | 80–85 | 50 | (60 %) |
| PhCHO | 1:1 | 80 | 24 | (quantitative) |
| | 1:2 | 80 | 48 | (quantitative) |

## 4.9.3.4 Other Reactions

Like other thiols, these dithioacids undergo facile oxidation with oxidizing agents and even with atmospheric oxygen to afford the corresponding disulfides. However, disulfides from dithiocarboxylic acids are usually unstable and often cannot be obtained easily [192].

Upon keeping an ether solution of dithiobenzoic acid at room temperature for a long period, several decomposition products are formed, among which biphenyl and phenyl derivatives of lenthionine are readily isolated (eq. 4.129) [232]. Dithioacetic acid kept standing for a prolonged

$$(4.129)$$

period affords tetramethylhexathioadamantane (eq. 4.130) [233]. In the presence of Lewis acids even monothiolacetic acid undergoes the

$$(4.130)$$

same reaction to give the same product [234]. Mechanism of the reaction is not known.

Upon heating dithiophosphoric acids, the Arbuzov reaction appears to take place in the manner shown by eq. 4.131. Since mercaptan can be obtained by the hydrolysis of the resulting rearranged products, this furnishes a convenient and easy way for preparing unaccessible *sec-* and *tert*-alkyl mercaptans.

$$(RO)_2PS_2H \longrightarrow (RO)(RS)\overset{S}{\overset{\|}{P}}OH \quad \text{or} \quad (RS)_2\overset{O}{\overset{\|}{P}}OH \qquad (4.131)$$

## 4.10 BIOLOGICAL FUNCTIONS

Although content of sulfur in living tissues is very small ($\sim$2%), nobody can deny the biological importance of sulfur compounds. Redox reactions between thiols and disulfides are essential to maintain the three-dimensional structure of proteins, which is closely related to enzymic actions. Quite a few subjects may be included under the heading of "biological functions of thiols", but, here, we will mainly deal with functions of coenzyme A and lipoic acid, two relatively well-understood coenzymes.

### 4.10.1 Coenzyme A and Acyl Carrier Protein

Coenzyme A (CoASH) has the structure shown in **4.103**. The active site of CoASH is the terminal mercapto group [235,236].

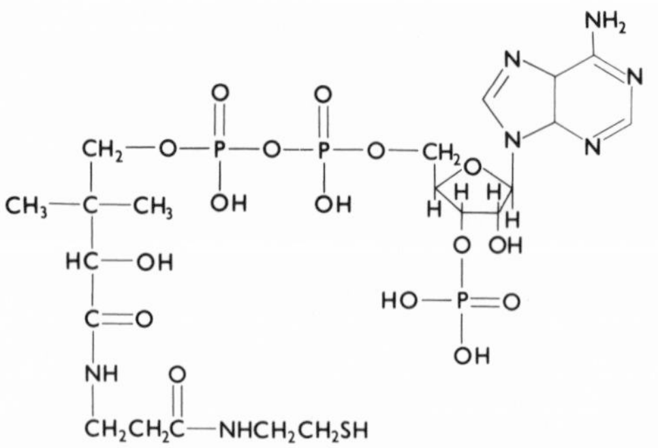

**4.103**

CoASH-catalyzed reactions may be classified into three groups. They are (i) acyl transfer, (ii) addition to double bond next to the acyl group, and (iii) condensation of $\alpha$-carbanion. These are nicely summarized in fatty acid biosynthesis[237]. Fatty acids are synthesized from acetyl coenzyme A (CoASAc) and malonyl coenzyme A, which is also produced from CoASAc by the reaction with carbon dioxide with CoASAc carboxylase and biotin, **4.104**, as catalysts:

(1)    biotin-Enzyme + **ATP** + **HCO$_3^-$**

$$\Longleftrightarrow CO_2 \sim \text{biotin-Enzyme} + ADP + Pi$$

(2)    $CO_2 \sim$biotin-Enzyme + $CH_3COSCoA$                    (4.132)

$$\Longleftrightarrow \text{biotin-Enzyme} + {}^-O_2CCH_2COSCoA$$

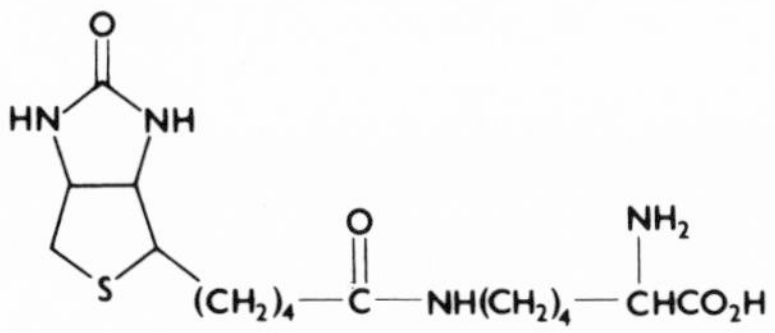

**4.104**

Fatty acid synthetase is an enzyme complex (multienzyme) composed of at least seven enzymes. There are two -SH groups in a fatty acid synthetase, one of which is called a "peripheral" -SH group and the other a "central" -SH group. Now it is known that the "central" -SH group operates as a supporter of an acyl group during the elongation of the chain (*vide infra*). Then, a protein that carries the "central" -SH group is called as acyl carrier protein (ACP). The structure of ACP from *Escherichia coli* has been studied and found that it has, interestingly, a constitution similar to that of CoASH as shown in **4.105**[238,239]. A schematic representation

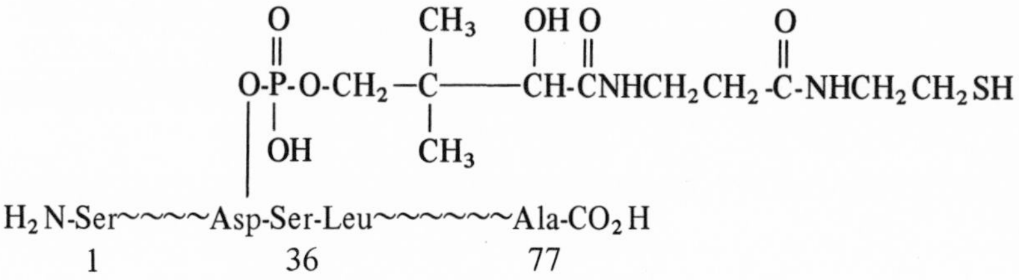

**4.105**

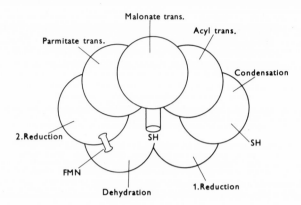

**Fig. 4.1.** Schematic representation of fatty acid synthetase from yeast.

of fatty acid synthetase is shown in Fig. 4.1 [240]. During a turn around the surface of enzymes, the malonyl moiety, which is bound to the "central" -SH group, accepts an acyl group from the "peripheral" -SH group, and, after several steps, the acyl group elongates its carbon atoms by two. In eq. 4.133 is shown a mechanism proposed by Lynen for yeast fatty acid synthetase [241].

Priming Reaction:

$$
\text{(1)} \quad CH_3COSCoA + \begin{matrix} HS^* \\ \diagdown \\ \diagup \\ HS \end{matrix}Enzyme \ \rightleftharpoons \ \begin{matrix} HS^* \\ \diagdown \\ \diagup \\ CH_3COS \end{matrix}Enzyme
$$

Chain Lengthening Reactions:

$$
\text{(2)} \quad \begin{matrix} CO_2H \\ | \\ CH_2COSCoA \end{matrix} + \begin{matrix} & HS^* \\ & \diagdown \\ & Enzyme \\ & \diagup \\ CH_3(CH_2CH_2)_nCOS \end{matrix}
$$

$$
\rightleftharpoons \ \begin{matrix} HO_2CCH_2COS^* \\ \diagdown \\ Enzyme \\ \diagup \\ CH_3(CH_2CH_2)_nCOS \end{matrix}
$$

(3)

$$\begin{array}{l} HO_2CCH_2COS^* \\ \quad\quad\quad\quad \backslash \\ \quad\quad\quad\quad\quad Enzyme \\ \quad\quad\quad\quad / \\ CH_3(CH_2CH_2)_nCOS \end{array} \rightleftharpoons \begin{array}{l} CH_3(CH_2CH_2)_nCOCH_2COS^* \\ \quad\quad\quad\quad\quad\quad\quad \backslash \\ \quad\quad\quad\quad\quad\quad\quad\quad Enzyme \\ \quad\quad\quad\quad\quad\quad\quad / \\ \quad\quad\quad\quad\quad\quad HS \end{array}$$

(4)

$$\begin{array}{l} CH_3(CH_2CH_2)_nCOCH_2COS^* \\ \quad\quad\quad\quad\quad\quad \backslash \\ \quad\quad\quad\quad\quad\quad\quad Enzyme + NADPH + H^+ \\ \quad\quad\quad\quad\quad\quad / \\ \quad\quad\quad\quad\quad HS \end{array}$$

$$\rightleftharpoons \begin{array}{l} CH_3(CH_2CH_2)_nCH(OH)CH_2COS^* \\ \quad\quad\quad\quad\quad\quad\quad \backslash \\ \quad\quad\quad\quad\quad\quad\quad\quad Enzyme + NADP^+ \\ \quad\quad\quad\quad\quad\quad\quad / \\ \quad\quad\quad\quad\quad\quad HS \end{array}$$

(4.133)

(5)

$$\begin{array}{l} CH_3(CH_2CH_2)_nCH(OH)CH_2COS^* \\ \quad\quad\quad\quad\quad\quad\quad \backslash \\ \quad\quad\quad\quad\quad\quad\quad\quad Enzyme \\ \quad\quad\quad\quad\quad\quad\quad / \\ \quad\quad\quad\quad\quad\quad HS \end{array}$$

$$\rightleftharpoons \begin{array}{l} CH_3(CH_2CH_2)_nCH{=}CHCOS^* \\ \quad\quad\quad\quad\quad\quad\quad \backslash \\ \quad\quad\quad\quad\quad\quad\quad\quad Enzyme + H_2O \\ \quad\quad\quad\quad\quad\quad\quad / \\ \quad\quad\quad\quad\quad\quad HS \end{array}$$

(6)

$$\begin{array}{l} CH_3(CH_2CH_2)_nCH{=}CHCOS^* \\ \quad\quad\quad\quad\quad\quad \backslash \\ \quad\quad\quad\quad\quad\quad\quad Enzyme + NADPH + H^+ \\ \quad\quad\quad\quad\quad\quad / \\ \quad\quad\quad\quad\quad HS \end{array}$$

$$\xrightarrow{(FMN)} \begin{array}{l} CH_3(CH_2CH_2)_{n+1}COS^* \\ \quad\quad\quad\quad\quad\quad \backslash \\ \quad\quad\quad\quad\quad\quad\quad Enzyme + NADP^+ \\ \quad\quad\quad\quad\quad\quad / \\ \quad\quad\quad\quad\quad HS \end{array}$$

(7)
$$CH_3(CH_2CH_2)_{n+1}COS^*$$
$$\diagdown$$
Enzyme
$$\diagup$$
HS

$$\rightleftharpoons$$

HS*
$$\diagdown$$
Enzyme
$$\diagup$$
$$CH_3(CH_2CH_2)_{n+1}COS$$

Terminal Reaction:

(8)
$$CH_3(CH_2CH_2)_{n+1}COS^*$$
$$\diagdown$$
Enzyme + HSCoA
$$\diagup$$
HS

$$\rightleftharpoons$$

HS*
$$\diagdown$$
Enzyme + $CH_3(CH_2CH_2)_{n+1}COSCoA$
$$\diagup$$
HS

The "central" -SH group is symbolized by an asterisk. Note that the process (2) in eq. 4.132 and those (1), (2), and (7) in eq. 4.133 are type (i) reactions mentioned above. Processes (6) and (3) are type (ii) and (iii) reactions, respectively.

### 4.10.2  Lipoic Acid and Glutathione

Lipoic acid, **4.106**, is one of coenzymes of an enzyme complex that catalyzes oxidative decarboxylation of α-ketocarboxylic acids [242]. For example, when pyruvic acid is decarboxylated, the acetyl group is transferred to thiamine pyrophosphate (TPP), **4.107**, initially; then it is

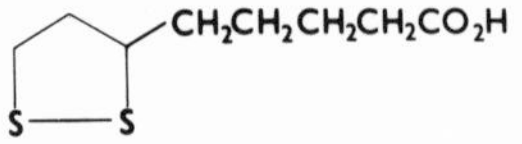

**4.106** 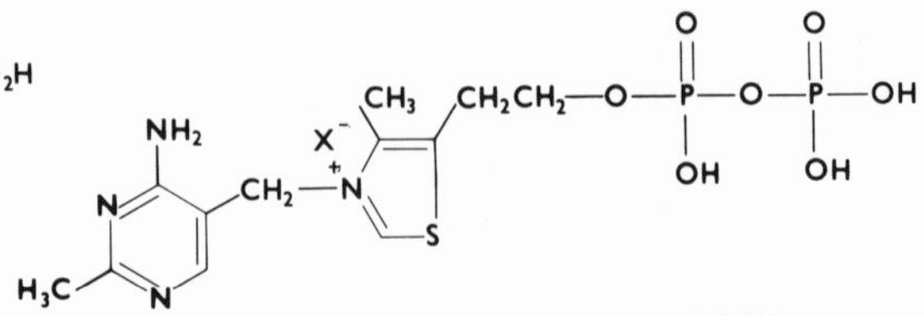                                                **4.107**

again transferred to lipoic acid (LipS$_2$). Acetyl lipoic acid (AcSLipSH) thus formed is reduced to dihydrolopoic acid (Lip(SH)$_2$) by CoASH. Finally, lipoic acid is regenerated from dihydrolipoic acid by oxidation with flavin adenine dinucleotide (FAD). The entire reaction is shown in eq. 4.134, in which TPP—Ald denotes TPP-bound activated aldehyde, **4.108** [243]:

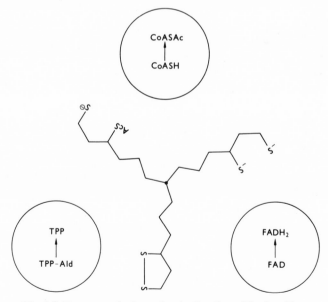

$$(4.134)$$

**4.108**

The function of lipoic acid may be best represented by a turn-around scheme (Fig. 4.2).

**Fig. 4.2.** Turn-around scheme for the function of lipoic acid.

The reactivity of lipoic acid was explained by ring strain [244]. However, its reliability is still a subject of controversy and some prefer the explanation based on entropy of ring-opening [242,245].

Glutathione (GSH), **4.109**, is also a coenzyme for several biochemical

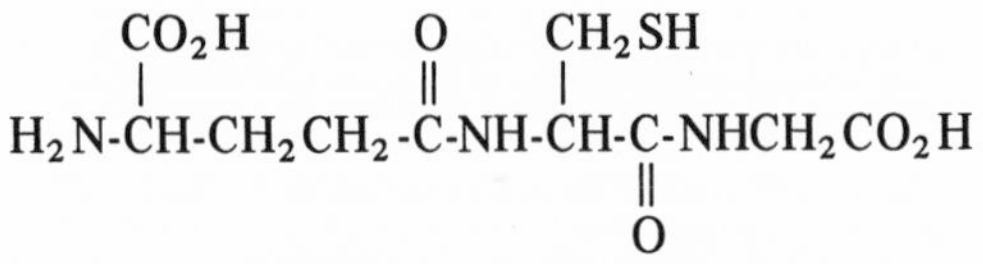

$$
\begin{array}{ccc}
\text{CO}_2\text{H} & \text{O} & \text{CH}_2\text{SH} \\
| & \| & | \\
\end{array}
$$

H$_2$N-CH-CH$_2$CH$_2$-C-NH-CH-C-NHCH$_2$CO$_2$H
$$\overset{\|}{\text{O}}$$

**4.109**

reactions, a typical example of which is the conversion of methyl glyoxal to lactic acid [246,247]:

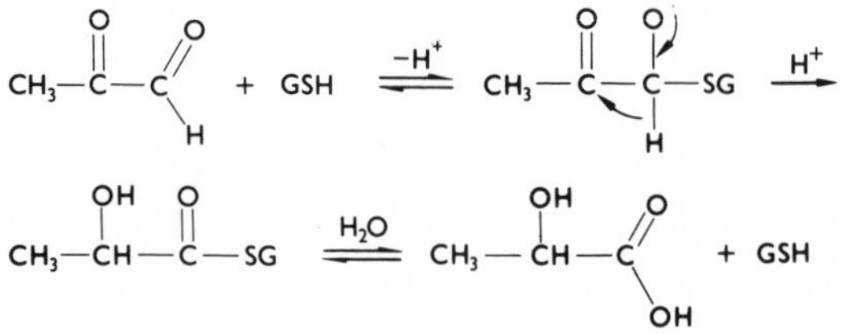

(4.135)

During the reaction, the α-hydrogen in lactic acid comes from glutathione, [248-250]. The mechanism shown in eq. 4.135 was supported by Franzen, who showed that phenyl glyoxal reacts with N,N-dimethyl-β-mercaptoethylamine at room temperature giving mandelic

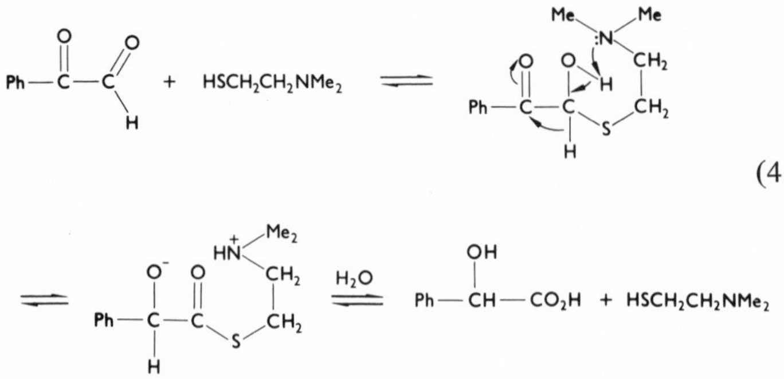

(4.136)

acid (eq. 4.136) [250]. The reaction requires not only a mercapto group, but also an amino group.

## 4.11  REFERENCES

1. T.L. Cottrell, "The Strength of Chemical Bonds", (Academic Press Inc., New York, N.Y., 1954).
2. J.L. Franklin and H.E. Lumpkin, *J. Amer. Chem. Soc.*, **74**, 1023 (1952).
3. (a) L.E. Sutton Ed., "Interatomic Distances", Special Publication No. 11, (The Chemical Society, London, 1958); (b) S.C. Abrahams, *Quart. Rev.*, **10**, 407 (1956).
4. L. Pauling, "The Nature of the Chemical Bond", (Cornell Univ. Press., Ithaca, N.Y., 1960), p. 110.
5. M. Kotake, Ed., "Constants of Organic Compounds", (Asakura Pub. Co., Tokyo, Japan, 1963), p. 511.
6. M.M. Rousselot, *Compt. Rend., Ser. C*, **262**, 26 (1966).
7. S.H. Marcus and S.I. Miller, *J. Amer. Chem. Soc.*, **88**, 3719 (1966).
8. W.H. Fletcher, *ibid.*, **68**, 2726 (1946).
9. J. Mauriu and P.A. Paris, *Compt. Rend.*, **232**, 2428 (1951).
10. H. Lumbroso, *J. Chim. Phys.*, **49**, 394 (1952).
11. L.P. Hammett, "Physical Organic Chemistry", (McGraw-Hill Book Co., New York, N.Y., 1940), p. 50.
12. G.R. Sprengling and C.W. Lewis, *J. Amer. Chem. Soc.*, **71**, 2624 (1949).
13. H.L. Loy and D.M. Himmelblau, *J. Phys. Chem.*, **65**, 264 (1961).
14. MM. Kreevoy, E.T. Harper, R.E. Duvall, H.S. Wilgus III, and L.T. Ditsch, *J. Amer. Chem. Soc.*, **82**, 4899 (1960).
15. C.G. Swain and C.B. Scott, *ibid.*, **75**, 141 (1953).
16. L.S. Levitt and B.W. Levitt, *J. Org. Chem.*, **37**, 332 (1972).
17. L.S. Levitt and B.W. Levitt, *Chem. and Ind. (London)* 990 (1970).
18. A. Schöberl and A. Wagner, in Houben-Weyl, "Methoden der Organischen Chemie", Vol. 9, (Georg Thieme Verlag, Stuttgart, Germany, 1955), pp. 7–42.
19. R.B. Wagner and H.D. Zook, "Synthetic Organic Chemistry", (John Wiley and Sons, Inc., New York, N.Y., 1953), pp. 778–786.
20. S.R. Sandler and W. Karo, "Organic Functional Group Preparations", (Academic Press, New York, N.Y., 1968), pp. 480–485.
21. L.M. Ellis, Jr. and E.E. Reid, *J. Amer. Chem. Soc.*, **54**, 1674 (1932).
22. J. Speziale, *Org. Synth.*, Coll. Vol. **4**, 401 (1963).
23. H. Lofod, *ibid.*, Coll. Vol. **4**, 491 (1963).
24. H.M. Foster and H.R. Snyder, *ibid.*, Coll. Vol. **4**, 638 (1963).
25. G.G. Urquhart, J.W. Gates, Jr., and R. Connor, *ibid.*, Coll. Vol. **3**, 363 (1955).
26. H.F. Wilson and D.S. Tarbell, *J. Amer. Chem. Soc.*, **72**, 5203 (1950).
27. J.A. van Allan and B.D. Deacon, *Org. Synth.*, Coll. Vol. **4**, 569 (1963).
28. H. Bestian, *Ann. Chem.*, **566**, 210 (1950).
29. H. Gilman and L.A. Woods, *J. Amer. Chem. Soc.*, **67**, 1844 (1945).
30. H.R. Snyder, J.M. Stewart and J.B. Ziegler, *ibid.*, **69**, 2672 (1947).
31. V.N. Ipatieff and B.S. Friedman, *ibid.*, **61**, 71 (1939).
32. S.O. Johns and E.E. Reid, *ibid.*, **60**, 2452 (1938).
33. F.D. Bordwell and W.A. Hewett, *ibid.*, **79**, 3493 (1957).
34. R. Adams and C.S. Marvel, *Org. Synth.*, Coll. Vol. **1**, 2nd Ed., 504 (1956).
35. O. Litvay, E. Riesz, and L. Landau, *Chem. Ber.*, **62**, 1863 (1929).
36. J. Strating and H.J. Backer, *Rec. trav. chim. Pays-Bas*, **69**, 644 (1950).
37. F. Taboury, *Compt. Rend.*, **138**, 982 (1904).
38. H. Wuyts, *Bull Soc. Chim. France*, [4], **5**, 405 (1909).
39. A. Mailhe and M. Murat, *ibid.*, [4], **7**, 288 (1910).
40. C.F. Koelsch and G. Ullyot, *J. Amer. Chem. Soc.*, **71**, 1478 (1949).
41. M. Seyhan, *Chem. Ber.*, **72**, 594 (1949).
42. A.W. Hofmann, *ibid.*, **1**, 177 (1868).
43. J. Strating and H.J. Backer, *Rec. trav. chim. Pays-Bas*, **69**, 638 (1950).

44. D.T. Gibson, *J. Chem. Soc.,* 12 (1930).
45. W. Alcaley, *Helv. Chim. Acta,* **30,** 578 (1947).
46. C. Walling, "Free Radicals in Solution", (John Wiley and Sons, Inc., New York, N.Y. 1957), p. 323.
47. S.G. Cohen and C.H. Wang, *J. Amer. Chem. Soc.,* **77,** 4435 (1955).
48. (a) E.F.P. Harris and W.A. Waters, *Nature,* **170,** 211 (1952); (b) K.E.J. Barrett and W.A. Waters, *Discussions Faraday Soc.,* **14,** 221 (1953).
49. T. Inabe and B. Darwent, *J. Phys. Chem.,* **64,** 1431 (1960).
50. K. E. Russell, *ibid.,* **58,** 437 (1954).
51. N. Imai and O. Toyama, *Bull. Chem. Soc. Jap.,* **33,** 652 (1960).
52. J.A. Kerr and A.F. Trotman-Dickenson, *J. Chem. Soc.,* 3322 (1957).
53. N.L. Arthur and P. Gray, *Trans. Faraday Soc.,* **65,** 434 (1969).
54. W.A. Pryor and K.G. Kneipp, *J. Amer. Chem. Soc.,* **93,** 5584 (1971).
55. E.S. Lewis and M.M. Butler, *J. Org. Chem.,* **36,** 2582 (1971).
56. C.D. Hurd and L.L. Gershheim, *J. Amer. Chem. Soc.,* **69,** 2328 (1947).
57. (a) E.A.I. Haiba, *J. Org. Chem.,* **31,** 776 (1966); (b) M.S. Kharasch and C.F. Fuchs, *ibid.,* **13,** 97 (1948).
58. C.F.H. Allen and W.J. Humphlett, *Can, J. Chem.,* **44,** 2315 (1966).
59. B. Dmuchovsky, F.B. Zienty, and W.A. Vredenburgh, *J. Org. Chem.,* **31,** 865 (1966).
60. (a) E.N. Prilezhaeva, N.P. Petukhova, and M.F. Shostakovskii, *Izvest. Akad. Nauk SSSR, Otdel. Khim. Nauk,* 728 (1962); *Chem. Abstr.,* **57,** 14930h (1962); (b) E.N. Prilezhaeva, N.P. Petukhova, and M.F. Shostokovskii, *Doklady Acad. Nauk SSSR,* **154,** 160 (1964); *Chem. Abstr.,* **60,** 9143g (1964).
61. B. Saville, *J. Chem. Soc.,* 5040 (1962).
62. N. Kharasch, "Sulfenium Ions and Sulfenyl Compounds", in "Organic Sulfur Compounds" Vol. 1, N. Kharasch, Ed., (Pergamon Press, New York, N.Y., 1961), pp. 375–396.
63. W.A. Pryor, "Mechanisms of Sulfur Reactions", (McGraw-Hill Book Co., New York, N.Y., 1962), pp. 75–93.
64. R.M. Kellogg, "Thiyl Radicals", in "Methods in Free Radical Chemistry", Vol. 2, E.S. Huyser, Ed., (Dekker, Inc., New York, N.Y., 1969), pp. 1–120.
65. F.R. Mayo and C. Walling, *Chem. Rev.,* **27,** 351 (1940).
66. M.S. Kharasch, A.T. Read, and F.R. Mayo, *Chem. and Ind. (London),* **57,** 792 (1938).
67. C. Walling and W. Helmreich, *J. Amer. Chem. Soc.,* **81,** 1144 (1965).
68. A. Ohno, T. Saito, A. Kudo, and G. Tsuchihashi, *Bull. Chem. Soc. Jap.,* **44,** 1901 (1971).
69. K. Sugimoto, W. Ando, and S. Oae, *ibid.,* **38,** 224 (1965).
70. S. Sivertz, W. Andrews, W. Elsdon, and K. Graham, *J. Polymer Sci.,* **19,** 587 (1956)
71. (a) D.M. Graham, R.L. Mieville, and C. Sivertz, *Can. J. Chem.,* **42,** 2239 (1964); (b) D.M. Graham, R.L. Mieville, R.H. Palleu, and C. Sivertz, *ibid.,* **42,** 2250 (1964); (c) D.M. Graham and J.F. Soltys, *ibid.,* **47,** 2719 (1969).
72. A. Ohno and Y. Ohnishi, *Int. J. Sulfur Chem.,* Part A, **1,** 203 (1971).
73. A.B. Terent'ev and R.G. Petrova, *Bull. Acad. Sci. USSR, Div. Chem. Soc.,* 1984 (1963).
74. J.A. Kampmeier, R.P. Geer, A.J. Meskin, and R.M. D'Silva, *J. Amer Chem. Soc.,* **88,** 1257 (1966).
75. E.S. Huyser and R.M. Kellogg, *J. Org. Chem.,* **31,** 3366 (1966).
76. J.I.G. Cadogan and I.H. Sadler, *J. Chem. Soc.,* (B), 1191 (1966).
77. A.A. Oswald and M. Naegele, *J. Org. Chem.,* **31,** 830 (1966).
78. W.H. Mueller, *ibid.,* **31,** 3075 (1966).
79. (a) A. Ohno, Y. Ohnishi, and N. Kito, *Int. J. Sulfur Chem.,* Part A, **1,** 151 (1971); (b) A. Ohno and Y. Ohnishi, *Tetrahedron Lett.,* 4405 (1969); (c) A. Ohno, N. Kito, and Y. Ohnishi, *Bull. Chem. Soc. Jap.,* **44,** 470 (1971).

80. (a) S.J. Cristol and J.A. Reeder, *J. Org. Chem.*, **26**, 2182 (1961); (b) S.J. Cristol and G.D. Brindell, *J. Amer. Chem. Soc.*, **76**, 5699 (1954); (c) G.D. Brindell and S.J. Cristol, "Additions of Thiols and Related Substances to Bridged Bicyclic Olefins", in "Organic Sulfur Compounds", Vol. 1. N. Kharasch, Ed., (Pergamon Press, New York, N.Y., 1961), pp. 134–145.
81. (a) H.L. Goering, P.L. Abell, and B.F. Aycock, *J. Amer. Chem. Soc.*, **74**, 3588 (1952); (b) H.L. Goering, D.I. Relyea, and W. Larsen, *ibid.*, **78**, 348 (1956).
82. D.I. Davies and P.J. Rowley, *J. Chem. Soc.*, (C), 2249 (1967).
83. N.P. Neutreiter and F.G. Bordwell, *J. Amer. Chem. Soc.*, **82**, 5354 (1960).
84. P.S. Skell and R.G. Allen, *ibid.*, **82**, 1511 (1960).
85. P.K. Freeman, M.F. Grostic, and F.A. Raymond, *J. Org. Chem.*, **36**, 905 (1971).
86. T. Kawamura, M. Ushio, T. Fugimoto, and T. Yonezawa, *J. Amer. Chem. Soc.*, **93**, 908 (1971).
87. P.D. Readio and P.S. Skell, *J. Org. Chem.*, **31**, 759 (1966).
88. E.S. Huyser and J.R. Jeffrey, *Tetrahedron*, **21**, 3083 (1965).
89. F.G. Bordwell, P.S. Landis, and G.S. Whitney, *J. Org. Chem.*, **30**, 3764 (1965).
90. N.A. LeBel and A. DeBoer, *J. Amer. Chem. Soc.*, **89**, 2784 (1967).
91. W.A. Pryor and K. Smith, *ibid.*, **92**, 2731 (1970).
92. A. Ohno, Y. Ohnishi, M. Fukuyama, and G. Tsuchihashi, *ibid.*, **90**, 7038 (1968).
93. J.F. Ford, R.C. Pitkethly, and V.O. Young, *Tetrahedron*, **4**, 325 (1958).
94. A. Ohno and Y. Ohnishi, *Tetrahedron Lett.*, 339 (1972).
95. E.S. Huyser, H. Berson, and H.J. Sinnige, *J. Org. Chem.*, **32**, 622 (1967).
96. A.A. Oswald and K. Griesbaum, "Radical Additions of Thiols to Diolefins and Acetylenes", in "Organic Sulfur Compounds", Vol. 2, N. Kharasch and C.Y. Meyers Eds., (Pergamon Press, New York, N.Y., 1966), pp. 233–256.
97. S.J. Cristol, T.W. Russell and D.I. Davies, *J. Org. Chem.*, **30**, 207 (1965).
98. E.S. Huyser and R.M. Kellogg, *ibid.*, **30**, 3003 (1965).
99. R.E. Benson and R.V. Linsey, Jr., *J. Amer. Chem. Soc.*, **81**, 4253 (1959).
100. (a) W. A. Thaler, A.A. Oswald and B.E. Hudson, Jr., *ibid.*, **87**, 311 (1965); (b) A.A. Oswald, K. Griesbaum, W.A. Thaler, and B.E. Hudson, Jr., *ibid.*, **84**, 3897 (1962).
101. p. 314 of Ref. 36.
102. J.A. Claisse and D.I. Davies, *J. Chem. Soc.*, 1045 (1966).
103. M.S. Kharasch, W. Nudenberg, and G.J. Mantell, *J. Org. Chem.*, **16**, 524 (1951).
104. H. Bredereck, A. Wagner, and A. Kottenhahn, *Chem. Ber.*, **93**, 2415 (1960).
105. A.A. Oswald, *J. Org. Chem.*, **24**, 443 (1959).
106. A.A. Oswald, *ibid.*, **25**, 467 (1960).
107. A.A. Oswald, *ibid.*, **26**, 842 (1961).
108. A.A. Oswald and F. Noel, *ibid.*, **26**, 3948 (1961).
109. A.A. Oswald, F. Noel, and A.J. Stephenson, *ibid.*, **26**, 3969 (1961).
110. A.A. Oswald, F. Noel, and G. Fisk, *ibid.*, **26**, 3974 (1961).
111. A.A. Oswald, B.E. Hudson, Jr., G. Rogers, and F. Noel, *ibid.*, **27**, 2439 (1962).
112. A.A. Oswald, K. Griesbaum, W.A. Thaler, and B.E. Hudson, Jr., *J. Amer. Chem. Soc.*, **84**, 3897 (1962).
113. A.A. Oswald, K. Griesbaum, and B.E. Hudson, Jr., *J. Org. Chem.*, **28**, 1262 (1963).
114. K. Griesbaum, A.A. Oswald, and B.E. Hudson, Jr., *J. Amer. Chem. Soc.*, **85**, 1969 (1963).
115. A.A. Oswald, K. Griesbaum, and B.E. Hudson, Jr., *J. Org. Chem.*, **28**, 2361 (1963).
116. A.A. Oswald, K. Griesbaum, and B.E. Hudson, Jr., *J. Org. Chem.*, **28**, 2355 (1963).
117. A.A. Oswald and T.J. Wallace, "Anionic Oxidation of Thiols and Co-oxidation of Thiols with Olefins", in "Organic Sulfur Compounds", Vol. 2, N. Kharasch and C.Y. Meyers, Eds., (Pergamon Press, New York, N.Y., 1966), pp. 205–232.

118. R.E. Foster, A.W. Larchar, R.D. Lipscomb, and B.C. McKusick, *J. Amer. Chem. Soc.*, 78, 5606 (1956).
119. W.E. Truce and J.A. Simms, *ibid.*, 78, 2756 (1956).
120. W.E. Truce and D.L. Goldhamer, *ibid.*, 82, 5798 (1960).
121. M.F. Shostakovskii, E.N. Prilezhaeva, L.V. Tsymbal, and L.G. Stolyarova, *Zhur. Obschei Khim.*, 30, 3143 (1960); *Chem Abstr.*, 55, 17474f (1961).
122. W.E. Truce, M.M. Boudakian, R.F. Heine, and R.J. MacManimie, *J. Amer. Chem. Soc.*, 78, 2743 (1956).
123. W.E. Truce and R. Kassinger, *ibid.*, 80, 1916, 6450 (1958).
124. W.E. Truce and M.M. Boudakian, *ibid.*, 78, 2748 (1956).
125. H.C. Volger and J.F. Arens, *Rec. trav. chim. Pays-Bas*, 77, 1170 (1958).
126. C.C. Price and S. Oae, "Sulfur Bonding", (Ronald Press, New York, N.Y., 1962).
127. H.J. Boonstra and J.F. Arens, *Rec. trav. chim. Pays-Bas*, 79, 867 (1960).
128. W.H. Mueller and K. Griesbaum, *J. Org. Chem.*, 32, 856 (1967).
129. A. Dondoni, G. Modena, and G. Scorrano, *Ric. Sci. Rend. Sez.*, A6 665 (1964); *Chem. Abstr.*, 65, 18447e (1966).
130. F.W. Stacey and J.F. Harris, Jr., *Organic Reactions*, 13, 150 (1963).
131. J. Bonnema and J.F. Arens, *Rec. trav. chim. Pays-Bas*, 79, 1137 (1960).
132. H.J. Alkema and J.F. Arens, *ibid.*, 79, 1257 (1960).
133. Y.-C. Liu, H.-K. Wang, and S.-C. Chu, *Hua Hseuh. Pao*, 30, 283 (1964); *Chem. Abstr.*, 61, 11865g (1964).
134. I.G. Sulimov and A.A. Petrov. Zhur. Organ. Khim., 2, 767 (1966); *Chem. Abstr.*, 65, 12099b (1966).
135. J.A. Kampmeier, *J. Amer. Chem. Soc.*, 88, 1959 (1966).
136. R.M. Fantazier and J.A. Kampmeier, *ibid.*, 88, 5219 (1966).
137. L.A. Singer and N.P. Kong, *Tetrahedron Lett.*, 2089 (1966).
138. L.A. Singer and NP. Kong, *J. Amer. Chem. Soc.*, 88, 5213 (1966).
139. J.F. Arens, A.C. Hermans, and J.H.S. Weiland, *Proc. Kon. Acad. Wetenschap*, B58, 78 (1955).
140. M.F. Shostakovskii, E.N. Prilizhaeva, and L.V. Tsymbal, *Trudy po Khim. i Him. Teknol.*, 4, 198 (1961); *Chem. Abstr.*, 56, 1331b (1962).
141. A.A. Oswald, K. Griesbaum, B.E. Hudson, Jr., and J.M. Bregmann, *J. Amer. Chem. Soc.*, 86, 2877 (1964).
142. J.A. Kampmeier and G.G. Cheu, *ibid.*, 87, 2608 (1965).
143. O. Simamura, K. Tokumaru, and H. Yui, *Tetrahedron Lett.*, 5141 (1966).
144. D.M. Graham and J.F. Soltys, *Can. J. Chem.*, 47, 2529 (1969).
145. K. Griesbaum, A.A. Oswald, E.R. Quiram, and P.E. Butler, *J. Org. Chem.*, 30, 261 (1965).
146. K. Griesbaum, A.A. Oswald, E.R. Quiram, and W. Naegele, *ibid.*, 28, 1952 (1963).
147. T.L. Jacobs and G.-E. Illingworth, Jr., *ibid.*, 28, 2692 (1963).
148. H.J. van der Pleog, J. Krotnerus, and A.F. Bickel, *Rec. trav. chim. Pays-Bas*, 81, 775 (1962).
149. A. Rajbenbach and M. Szwarc, *Proc. Roy. Soc. (London)*, 251A, 394 (1959).
150. A.P. Stefani, L. Herk, and M. Szwarc, *J. Amer. Chem. Soc.*, 83, 4732 (1961).
151. K. Griesbaum, A.A. Oswald, and E.R. Quiram, *J. Org. Chem.*, 28, 1952 (1963).
152. R.N. Haszeldine, K. Leedham, and B.R. Steele, *J. Chem. Soc.*, 2040 (1954).
153. R.S. Neale, *J. Amer. Chem. Soc.*, 86, 5340 (1964).
154. B. Pullmann, *J. Chim. Phys.*, 55, 790 (1958).
155. H. Goldwhite, *J. Chem. Soc.*, 3901 (1965).
156. H.G. Kuivila, W. Rahman, and R.H. Fish, *J. Amer. Chem. Soc.*, 87, 2835 (1965).
157. K. Griesbaum, A.A. Oswald, and D.N. Hall, *J. Org. Chem.*, 29, 2404 (1964).
158. A.A. Oswald, K. Griesbaum, B.E. Hudson, Jr., and J.M. Bregman, *Chem. and Eng. News*, Oct., 28, 42 (1963).

159. J.C. Sauer, *J. Amer. Chem. Soc.*, **79**, 5314 (1957).
160. T. Fujisawa, T. Kobori, N. Ohtsuka, and G. Tsuchihashi, *Tetrahedron Lett.*, 5071 (1968).
161. H. Brintzinger and M. Langheck, *Chem. Ber.*, **86**, 557 (1953).
162. B.S. Farah and E.E. Gilbert, *J. Org. Chem.*, **28**, 2807 (1963).
163. R.T. Wragg, *J. Chem. Soc.*, 5482 (1964).
164. W. Wolf and N. Kharasch, *J. Org. Chem.*, **30**, 2493 (1965).
165. R.M. Kellogg and H. Wynberg, *J. Amer. Chem. Soc.*, **89**, 3495 (1967).
166. Y.A. Gol'dfarb, G.P. Pokhil, and L.I. Belen'kii, *Doklady Akad. Nauk SSSR*, **167**, 822 (1966); *Chem. Abstr.*, **65**, 2196e (1966).
167. A.L.J. Beckwith and B.S. Low, *J. Chem. Soc.*, 2571 (1964).
168. A.L.J. Beckwith and B.S. Low, *ibid.*, 1304 (1961).
169. B.M. Mikailov and A.N. Blokhima, *Dokl. Akad. Nauk SSSR*, **80**, 373 (1951); *Chem. Abstr.*, **46**, 5025b (1952).
170. F. Frerichs and E. Wildt, *Ann. Chem.*, **366**, 105 (1908).
171. T. McAllan, T.W. Cullum, R.A. Dean and F.A. Fidler, *J. Amer. Chem. Soc.*, **73**, 3627 (1951).
172. J. Xan, E.A. Wilson, L.P. Roberts, and N.H. Horton, *ibid.*, **63**, 1139 (1941).
173. T.J. Wallace and A. Schiesheim, *J. Org. Chem.*, **27**, 1514 (1962).
174. S. Gabriel and J. Colman, *Chem. Ber.*, **45**, 1643 (1912).
175. J. Barnett, *J. Chem. Soc.*, 5 (1944).
176. H. Gilman, M.A. Plunkett, L. Tolman, L. Fullhart, and H.S. Broadbent, *J. Amer. Chem. Soc.*, **67**, 1845 (1945).
177. T.J. Wallace, A. Schreisheim, and H.B. Jonassen, *Chem. and Ind. (London)*, 743 (1963).
178. S.G. Cohen and S. Aktipis, *J. Amer. Chem. Soc.*, **88**, 3587 (1966).
179. P.J. Wagner and R.C. Zepp, *ibid.*, **94**, 285 (1972).
180. T.J. Wallace, *J. Org. Chem.*, **31**, 1217 (1966): *cf.* also G.N. Schrauzer and J.W. Sibert, *J. Amer. Chem. Soc.*, **92**, 3509 (1970).
181. T.J. Wallace, *J. Org. Chem.*, **31**, 3071 (1966).
182. L. Suchomelova and J. Zyka, *J. Electroanal. Chem.*, **5**, 57 (1963); *Chem. Abstr.*, **59**, 1482a (1963).
183. C.N. Yiannios and J.Y. Karabinos, *J. Org. Chem.*, **28**, 3246 (1963).
184. T.J. Wallace and H.A. Weiss, *Chem and Ind. (London)*, 1558 (1966).
185 (a) T.J. Wallace, *J. Amer. Chem. Soc.*, **86**, 2018 (1964); (b) T.J. Wallace and J.J. Mahon, *ibid.*, **86**, 4099 (1964).
186. T.J. Wallace and J.J. Mahon, *J. Org. Chem.*, **30**, 1502 (1965).
187. K. Balenovic and N. Bregant, *Chem. and Ind. (London)*, 37 (1964).
188. T.J. Wallace and J.J. Mahon, *ibid.*, 37 (1964).
189. S.D. Sokolov and N.M. Naideneva, *Zhur. Organ. Khim.*, **2**, 1123 (1966); *Chem. Abstr.*, **65**, 16849d (1966).
190. D.I. Relyea, P.O. Tawney, and A.R. Williams, *J. Org. Chem.*, **27**, 477 (1962).
191. (a) E.E. Reid, "Organic Chemistry of Bivalent Sulfur", (Chemical Publishing Co., N.Y., 1942) Vol. 4, Chapter 1; (b) M.J. Jansen, "The Chemistry of Carboxylic Acid and Esters", (Interscience-Publisher, 1969), Chapter 5.
192. H. Jouben and H. Pohl, *Ber.*, **39**, 3219 (1906), **40**, 1303 (1907).
193. R.W. Bost and O.L. Shealy, *J. Amer. Chem. Soc.*, **73**, 24 (1951).
194. F. Becke and H. Hagen, *Chem. Ztg. Chem. Appl.*, **93**, 474 (1969).
195. F. Kurzer and A. Lawson, *Org. Synth.*, **42**, 100 (1962).
196. M.I. Kabachnik and T.A. Kastryukova, *Izvest Akad. Nauk SSSR, Otdel. Khim. Nauk*, 121 (1953).
197. P. Hu and W. Chen, *Hua Hsueh Pao*, **24**, 112 (1958).
198. M.I. Kabachnik, S.T. Ioffe, and T.A. Mastryukova, *Zhur. Obshch. Khim.*, **25**, 684 (1955).
199. R. Mecke and H. Spiesecke, *Ber.*, **89**, 1110 (1956).
200. E. Lindener and H.G. Karmana, *Angew. Chem.*, **80**, 319 (1968).

201. L.J. Bellamy, "The Infra-red Spectra of Complex Molecules", (John Wiley & Sons, Inc., New York, N.Y., 1958), 2nd Ed., p. 351.
202. (a) G. Allen and R.O. Colalough, *J. Chem. Phys.,* **25,** 370 (1956); (b) G. Allen and R.O. Colalough, *J. Chem. Soc.,* 3912 (1957).
203. A. Hantzch and N. Bucerius, *Ber.,* **59,** 793 (1926).
204. E.A. Shuzam and V.M. Leviue, *Kristallografia,* **5,** 257 (1960).
205. K. Issleib and W. Grunder, *Z. Chem.,* **6,** 318 (1966).
206. See for example, A. Ohno, Y. Ohnishi, M. Fukuyama, and G. Tsuchihashi, *J. Amer. Chem. Soc.,* **90,** 7038 (1968).
207. A. Kaiser, *Acta. Chem. Scand.,* **4,** 1347 (1950), *ibid.,* **6,** 327 (1952).
208. K.A. Jensen and C.L. Jensen, *ibid.,* **6,** 957 (1962).
209. H. Wuyts, *Bull. Soc. Chim. Berg.,* **38,** 195 (1929).
210. L.A. Smith and J. Nichols, *J. Org. Chem.,* **6,** 489 (1941).
211. H. Wuyts, *Bull. Soc. Chim. Berg.,* **39,** 58 (1930).
212. H. Wuyts and M. Goldstein, *Bull. Soc. Chim. Berg.,* **40,** 497 (1931).
213. H. Wuyts and H. Koek, *ibid.,* **41,** 196 (1932).
214. G.E.P. Smith, Jr. U.S. 2647144; *Chem. Abstr.,* **48,** 7637 (1954).
215. G. Alliger and G.E.P. Smith, Jr., *J. Org. Chem.,* **14,** 962 (1949).
216. H. Wuyts and J. Van Vacrenlergh, *Bull. Soc. Chim. Berg.,* **48,** 329 (1939).
217. Y. Hirabayashi, M. Mizuta, M. Kojima, Y. Hirano and H. Ichihara, *Bull. Chem. Soc. Jap.,* **44,** 791 (1971).
218. S. Oae, N. Tsujimoto, and A. Nakanishi, *Bull. Chem. Soc. Jap.,* **46,** 535 (1973).
219. M. Mickolajczyk and M. Pare, *Angew. Chem. Int. Ed.,* **5,** 419 (1966).
220. S. Oae, T. Yagihara and T. Okabe, *Tetrahedron,* **28,** 32031 (1972).
221. A. Nakanishi and S. Oae, *Chem. and Ind. London,* 960 (1971).
222. E.I. Hoegberg, U.S. 2632020; *Chem. Abstr.,* **48,** 2759h (1954).
223. A.D. Petrov, V.F. Mironov, and V.G. Glukhoutsev, *Doklady Akad. Nauk SSSR,* **93,** 499 (1953).
224. T.A. Mastryukova, *Izvest. Akad. Nauk SSSR, Otdel. Khim. Nauk,* 443 (1956).
225. N.N. Mel'nikov, *Khim. i Primenenie Fosforogan soedinenii Akad. Nauk SSSR,* Trudy 1-oi Konferents, 1955, 50; *Chem. Abstr.,* **52,** 393g (1958).
226. A.N. Pudovik and R.A. Cherkasov, *Zh. Obshch. Khim.,* **38,** 2532 (1968).
227. A.A. Oswald, K. Griesbaum, D.N. Hall, and W. Naegele, *Can. J. Chem.,* **45,** 1173 (1967).
228. A.A. Oswald, K. Griesbaum, B.B. Hudson, Jr., *J. Org. Chem.,* **28,** 1262 (1963).
229. S. Oae, A. Nakanishi, and N. Tsujimoto, *Chem & Ind. London,* 575 (1972).
230. G.F. Ottman and W. Hecht, *Z. Electrochem.,* **24,** 65 (1918).
231. A.N. Pudovik, B.M. Faizulsin, and G.I. Zhuravlev, *Zh. Obshch. Khim.,* **36,** 718 (1966).
232. K. Morita and S. Kobayashi, *Tetrahedron Letters,* 573 (1966).
233. T. Okabe, T. Yagihara, and S. Oae, unpublished results.
234. (a) J. Bongartz, *Ber.,* **19,** 2182 (1886); (b) A. Fredga, *Arkiv. Kemi. Mineral. Geol.,* **25B,** 599 (1947).
235. T.C. Bruice, "The Chemistry and Biochemistry of the Acyl Thiols", in "Organic Sulfur Compounds", Vol. 1, Ed. N. Kharasch, (Pergamon Press, New York, N.Y., 1961), pp. 421–442.
236. T.C. Bruice and S. Benkovic, "Bioorganic Mechanisms", Vol. 1, (W.A. Benjamin, Inc., New York, N.Y., 1966), pp. 259–297.
237. S.J. Walkel, "Lipid Metabolism", (Academic Press, New York, N.Y., 1970), pp. 1–48.
238. E.L. Pugh and S.J. Wakil, *J. Biol. Chem.,* **240,** 4727 (1965).
239. P.W. Majerus, A.W. Alberts, and P.R. Vagelos, *ibid.,* **240,** 4723 (1965).
240. F. Lynen, *Biochem. J.,* **102,** 381 (1967).
241. F. Lynen, *Federation Proc.,* **20,** 941 (1961).
242. L.J. Reed, "Lipoic Acid", in "Organic Sulfur Compounds", Vol. 1, N. Kharasch, Ed., (Pergamon Press, New York, N.Y., 1961), pp. 443–452.

243. M. Koike and L.J. Reed, *J. Amer. Chem. Soc.,* **81,** 505 (1959).
244. J.A. Barltrop, P.M. Hayes and M. Clavin, *ibid.,* **76,** 4348 (1954).
245. A. Fava, A. Iliceto, and E. Camera, *ibid.,* **79,** 833 (1957).
246. E. Baker, *J. Biol. Chem.* **190,** 685 (1951).
247. K. Lohmann, *Biochem. Z.,* **254,** 332 (1932).
248. I.A. Rose, *Biochem, Biophys. Acta.,* **25,** 214 (1957).
249. E.E. Cliffe and S.G. Waley, *Biochem. J.,* **73,** 25 (1959).
250. V. Franzen, *Chem. Ber.,* **90,** 623 (1957).

Chapter 5

# THIONES

Atsuyoshi Ohno

*Institute for Chemical Research*
*Kyoto University, Uji, Kyoto 611, Japan*

## 5.1 INTRODUCTION

Bond energies of the carbon-sulfur double bond in carbon disulfide or in carbon oxysulfide and the carbon-oxygen double bond in carbon dioxide are 128 and 192 kcal/mole, respectively. Since bond energies of general carbon-sulfur and carbon-oxygen single bonds are 65 and 85.5 kcal/mole respectively, there exists more than 40 kcal/mole of difference between carbon-sulfur and carbon-oxygen $\pi$-bond energies [1]. The difference is undoubtedly due to the low efficiency of $2p$-$3p$ $\pi$-overlapping in comparison with that of $2p$-$2p$. Because of the inherent antibonding interactions, as shown in Fig.5.1, $\pi$-bonds from $2p$ and $3p$ orbitals are less

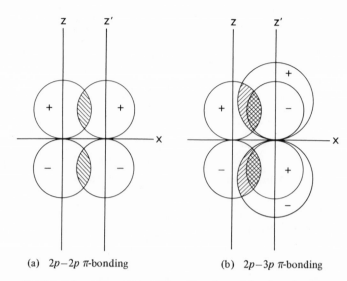

(a)  $2p-2p$ $\pi$-bonding                         (b)  $2p-3p$ $\pi$-bonding

**Fig. 5.1.** Schematic representation of $2p-2p$ and $2p-3p$ $\pi$-bondings.

favorable than from two $2p$ orbitals. The difference in bond energies of general carbon-carbon single and double bonds is 60–65 kcal/mole, which is comparable to that of carbon-sulfur bonding. This indicates that the resonance energy in a thiocarbonyl group is great enough to compensate for the unfavorable energies of $2p$-$3p$ $\pi$-overlapping. Note that no carbon-silicon double bond has been found yet in stable compounds.

It is reasonable to expect that because the electronegativity of oxygen (3.5) is greater than sulfur $(2.5)^2$, the canonical form **5.1b** will be more important when X is an oxygen than when it is a sulfur and the resonance energy will be greater for the carbonyl group than the thiocarbonyl group. However, all physical and chemical evidence indicates that **5.1b** and/or **5.1c** contributes more to the thiocarbonyl group than to the carbonyl group, as will be discussed below. Of course, the $\pi$-bond energy is greater for the carbonyl group than for the thiocarbonyl group. This is not, however, due to a greater contribution of **5.1b** or **5.1c** to the carbonyl group than the thiocarbonyl group, but due to the greater contribution of **5.1a** to the former than to the latter.

$$\overset{\backslash}{\underset{/}{C}}{=}X \longleftrightarrow \overset{\backslash}{\underset{/}{C}}{\overset{+}{-}}\bar{X} \longleftrightarrow \overset{\backslash}{\underset{/}{\dot{C}}}{-}\dot{X}$$

**5.1a**                    **5.1b**                    **5.1c**

It is interesting to see a few experimental data to support that the canonical form **5.1b** and/or **5.1c** is more important in the thiocarbonyl group than in the carbonyl group at their ground states. Dipole moments of several compounds are listed in Table 5.1. The substitution of N,N-

TABLE 5.1

Dipole Moments (in Debye) of Carbonyl and Thiocarbonyl Compounds in Benzene (25°)

| Compound | X = S | X = O |
|----------|-------|-------|
| $Ph_2C=X$ | 3.37 | 2.95 |
| $(p\text{-}Me_2NC_6H_4)_2C=X$ | 6.12 | 5.16 |
| $Cl_2C=X$ | 0.28 | 1.10 |

dimethylamino groups at $p$- and $p'$- positions of benzophenone and thio-benzophenone increases the dipole moments of these compounds by 2.21 and 2.15 D, respectively. The larger increment of the dipole moment for the oxygen compound than for the sulfur analogue shows that the resonance participation of the dimethylamino group is greater for the carbonyl group than for the thiocarbonyl group. The fact that an electron-withdrawing group makes the dipole moment of a thiocarbonyl compound less than that of a carbonyl compound while an electron-releasing group does the reverse, suggests that the carbonyl group has a greater double bond character than the thiocarbonyl group and that the oxygen atom participates in the conjugation more than the sulfur atom does. The same trend can be seen in the acidities of the following malonates:

|  5.2  |  5.3  |  5.4  |

The acidity of methylene protons in compounds **5.2–5.4** decreases in the order $5.2 > 5.3 > 5.4$[4,5]. Unfavorable $2p$-$3p$ $\pi$-overlapping will make the enethiol form, **5.5b**, more favorable than the thione form, **5.5a**, in the thione-enethiol tautomerism (eq. 5.1). Indeed, when cyclohexanone is treated with hydrogen sulfide in the presence of an acid catalyst below

192                                        Atsuyoshi Ohno

$$
\underset{\textbf{5.5a}}{RCH_2\text{-}\overset{\overset{\textstyle S}{\|}}{C}\text{-}R'} \;\; \rightleftharpoons \;\; \underset{\textbf{5.5b}}{RCH\text{=}\overset{\overset{\textstyle SH}{|}}{C}\text{-}R'} \qquad (5.1)
$$

$-40°$, thiocyclohexanone is obtained. However, this thioketone tautomerizes to 1-cyclohexenyl mercaptan simply on distillation [6]. Freshly prepared thiocyclopentanone has an absorption maximum at 235 nm with a small shoulder at 220 nm. As shown in Fig. 5.2, the intensity of the maximum

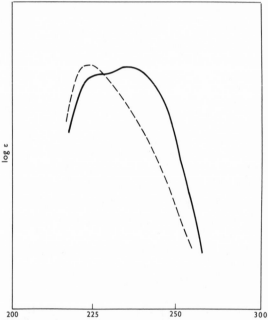

**Fig. 5.2.** Absorption spectra of thiocyclopentanone in cyclohexane. Fresh sample: ——— ; after 10 min: ----.

decreases gradually on standing at room temperature as the intensity of the shoulder increases and, after 10 min., the absorption at 235 nm disappears almost completely. The new absorption maximum at 220 nm is assigned to that of 1-cyclopentenyl mercaptan [7]. On the other hand, cyclohexanone and cyclopentanone contain their enol forms in only 1.2 and $8.8 \times 10^{-2}\%$, respectively, at room temperature [8]. Kunz and co-workers calculated the amounts of thione forms in **5.6**, **5.7**, and **5.8** by polarographic and potentiometric methods [9]. Results are summarized in Table 5.2.

TABLE 5.2

Contribution of Thione-Form in Several Thiocarbonyl Compounds

| Compound | R | R' | R'' | % Thione |
|---|---|---|---|---|
| 5.6 | $n$-Pr | H | Et | 33–34 |
| 5.6 | Ph | H | $CO_2$Et | 14–16 |
| 5.7 | $CO_2$Et | H | Et | 25–29 |
| 5.8 | – | – | – | 100 |

**5.6**          **5.7**          **5.8**

Systematic studies on thione-enethiol and keto-enol tautomerisms have been reported on six-membered heterocyclic systems [10]

$$(5.2)$$

(X=O, S)

**5.9a**          **5.9b**          **5.9c**

The $pK_T$ values, the logarithms of the equilibrium constants, of a few compounds are listed in Table 5.3. Surprisingly, the equilibrium constants for sulfur compounds are similar to those exhibited by oxygen analogues and the thione forms are usually slightly more favored than the corresponding keto forms. This is understandable when one takes into account

TABLE 5.3

The $pK_T$ Values of Six-Membered Heterocyclic Oxygen and Sulfur Compounds

| Position of XH group | Ring system | X = S | X = O |
|---|---|---|---|
| 2 | Pyridine | 4.7 | 2.9 |
| 3 | Pyridine | 2.2 | –0.08 |
| 4 | Pyridine | 4.5 | 3.3 |
| 1 | Isoquinoline | 5.8 | 4.3 |
| 3 | Isoquinoline | 3.0 | – |
| 2 | Quinoline | 5.1 | 3.5 |
| 3 | Quinoline | 1.5 | –1.1 |
| 4 | Quinoline | 5.0 | 4.4 |

the fact that the acidic properties of mercapto groups are considerably more pronounced than those of hydroxy groups. Thus, since stabilities of keto (thione) forms, **5.9b**, and zwitterion forms, **5.9c**, are reversed in oxygen and sulfur compounds, these two effects appear to cancel out mostly. Mukaiyama and co-workers applied this phenomenon beautifully to the peptide synthesis *via* redox condensation using 2,2′-dipyridyldisulfide as an oxidant [11]

$$
\underset{\substack{|\\R}}{XNHCHCOOH} + \underset{\substack{|\\R'}}{H_2NCHCOOY} + Ph_3P + \left( \underset{N}{\bigcirc}S\frac{}{}\right)_2
$$

(5.3)

$$
\longrightarrow \underset{\substack{|\ \ \ \ \ |\\R\ \ \ \ R'}}{XNHCHCONHCHCOOY} + Ph_3P=O + 2\ \underset{\substack{N\\H}}{\bigcirc}S
$$

Newman and Young have proved the importance of the structure **5.9c** for the sulfur compound by studying the energy barriers in rotation of C–N bonds of N,N-dimethylacetamide and its sulfur analogue [12]. The energy required for the rotation was 24.7 and 43.7 kcal/mole for the oxygen and the sulfur compounds, respectively. The greater barrier for the sulfur compound than the oxygen analogue indicates that the canonical form **5.10b** is more important when X = S than when X = O:

$$
\underset{\substack{\\N(CH_3)_2}}{CH_3\text{-}C}\overset{\substack{X\\ \|}}{} \longleftrightarrow \underset{\substack{\\\overset{+}{N}(CH_3)_2}}{CH_3\text{-}C}\overset{\substack{X^-\\ /}}{}
$$

(5.4)

       **5.10a**                                    **5.10b**

The predominancy of the thione form over the keto form has also been found in monothioacyl acylmethanes, **5.11**, through NMR studies [13]. Here, the enol form, **5.11b**, is the most abundant species at room temperature:

$$
\underset{\substack{/\\R}}{\overset{\substack{R\\ \backslash}}{}}C=S \ \ \ \underset{\substack{/\\R}}{\overset{\substack{\\ \backslash}}{}}C=O \ \ H_2C \rightleftharpoons HC \underset{\substack{/\\R}}{}C\text{-}O \ \ H \rightleftharpoons HC \ \ H \ \ C\text{-}S \ \ C=O
$$

(5.5)

       **5.11a**                 **5.11b**                 **5.11c**

(R = Me, Ph)

Ureas and thioureas, **5.12**, have been subjected to the electron impact:

$$\begin{array}{c} R_1 \quad X \\ \diagdown \quad \| \quad \diagup R_3 \\ N\text{-}C\text{-}N \\ \diagup \quad \quad \diagdown R_4 \\ R_2 \end{array}$$

$$(X = O, S)$$

**5.12**

The effect of methyl substitution on ionization potentials of these compounds are summarized in Table 5.4 [14]. As is seen in the table, the

TABLE 5.4

Ionization Potentials (in eV) of Substituted Ureas and Thioureas

| $R_1$ | $R_2$ | $R_3$ | $R_4$ | X = S | X = O |
|-------|-------|-------|-------|-------|-------|
| H | H | H | H | 8.50 | 10.27 |
| $CH_3$ | H | H | H | 8.29 | 9.73 |
| $CH_3$ | $CH_3$ | H | H | 8.34 | 9.10 |
| $CH_3$ | H | $CH_3$ | H | 8.17 | 9.42 |
| $CH_3$ | $CH_3$ | $CH_3$ | H | 7.93 | 8.94 |
| $CH_3$ | $CH_3$ | $CH_3$ | $CH_3$ | 7.95 | 8.74 |

increase in number of electron-releasing methyl substituents causes continuous decrease of the ionization potential in the urea series, while it remains almost constant in the thiourea series. This indicates that the ionization takes place from the nitrogen in urea, while it takes place from the sulfur in thiourea. Thus, electrons on the thiocarbonyl sulfur are more ionizable than those on the carbonyl oxygen.

Both uv [15] and nmr [16] spectroscopic studies indicate that thiocarbonyl group has greater volume than the carbonyl group.

The main characteristic features of thiocarbonyl and carbonyl groups may be summarized by the following:

1. $\pi$-Electrons in a thiocarbonyl group are more localized than those in a carbonyl group, so that the thiocarbonyl carbon bears more positive charge than the carbonyl carbon.
2. Unpaired electrons on the thiocarbonyl sulfur are more ionizable than those on the carbonyl oxygen.
3. The thiocarbonyl group has greater volume than the carbonyl group.

   In following sections, we shall see how these differences interplay in reactions of thiocarbonyl and carbonyl compounds.

## 5.2    PREPARATION [17]

### 5.2.1  Thioaldehydes

Thioaldehyde is so unstable that only two examples can be seen in literatures [18,19] and there is no universal method for their preparation.

### 5.2.2  Thioketones

Certain aromatic thioketones are prepared from their corresponding oxygen analogues by treatment with hydrogen sulfide and an acid catalyst (eq. 5.6) [20]. Aliphatic thioketones with no enolizable hydrogen, such as

$$\begin{array}{c}\diagdown\\ \diagup\end{array}C{=}O \ + \ H_2S \ \xrightarrow{\ H^+\ } \ \begin{array}{c}\diagdown\\ \diagup\end{array}C{=}S \qquad (5.6)$$

thiocamphor and thioadamantanone, are also prepared by this method [21,22]. However, when this reaction is applied to aliphatic ketones that have enolizable hydrogens, the reaction course is largely affected by temperature; below $-40°$ thioketones are obtained, while above $-25°$ *gem*-dithiols are formed [23].

More aliphatic thioketones are best synthesized from the corresponding ketals (eq. 5.7) [24].

$$\begin{array}{c}\diagdown \ \diagup OR\\ \quad C\\ \diagup \ \diagdown OR\end{array} \ + \ H_2S \ \xrightarrow{\ H^+\ } \ \begin{array}{c}\diagdown\\ \diagup\end{array}C{=}S \qquad (5.7)$$

*gem*-Dichlorides afford thioketones by reaction with certain thiols (eq. 5.8) [25,26]. The treatment of ketones with phosphorous

$$\begin{array}{c}\diagdown \ \diagup Cl\\ \quad C\\ \diagup \ \diagdown Cl\end{array} \ + \ \left\{\begin{array}{l}HS^-\\ HSCOR\\ KSCSOEt\end{array}\right\} \ \longrightarrow \ \begin{array}{c}\diagdown\\ \diagup\end{array}C{=}S \qquad (5.8)$$

pentasulfide is also effective for preparing heterocyclic thioketones [27,28], for example:

$$+ \ P_4S_{10} \ \longrightarrow \qquad\qquad\qquad (5.9)$$

*p*-Phenylthiobenzophenone is prepared from the thiocyanate, **5.13** (eq. 5.10) [29]. The kinetic isotope effect ($k_H/k_D$ = 3.1) indicates that the

$$Ph-\underset{H}{\overset{Ph}{\underset{|}{\overset{|}{C}}}}-SCN \ + \ Me_2CHO^-Na^+ \tag{5.10}$$

**5.13**

$$\xrightarrow[\textit{iso}-PrOH]{20°} \ Ph-\overset{S}{\overset{\|}{C}}-Ph \ + \ HCN$$

formation of the carbanion from **5.13** is the rate-determining step.

The Friedel-Crafts reaction with thiophosgene is a convenient method for the syntheses of aromatic thioketones (eq. 5.11) [30,31]. In several

$$ArH \ + \ CSCl_2 \ \xrightarrow{\ AlCl_3\ } \ Ar_2C=S \tag{5.11}$$

cases, elemental sulfur is also an effective reagent [32-34], for example:

$$\cdots + \ S_8 \ \xrightarrow{270°} \ \cdots =S \tag{5.12}$$

$$(CF_3)_2CF-Hg-CF(CF_3)_2 \qquad CF_3CF=CF_2$$
$$\underset{S_8}{\searrow} \qquad \underset{NaF/Pt}{\overset{S_8}{\swarrow}}$$
$$(CF_3)_2C=S \tag{5.13}$$

### 5.2.3 Thiocarboxylic Acids and Their Derivatives

The Grignard reaction with carbon disulfide is the most universal method to prepare dithiocarboxylic acids (eq. 5.14) [17]. O-Esters of thiocarboxylic

$$RMgX \ + \ CS_2 \ \longrightarrow \ RCS_2MgX \ \xrightarrow{H_2O} \ RCS_2H \tag{5.14}$$

acids are commonly prepared from either corresponding immonium salts [35,36] or trithio-orthoesters [37]:

$$
\underset{\substack{\| \\ \text{R-C-OR}'}}{\overset{\text{NH·HCl}}{}} + \quad H_2S \quad \longrightarrow \quad \underset{\substack{\| \\ \text{R-C-OR}'}}{\overset{S}{}} \qquad (5.15)
$$

$$
RC(SR')_3 + H_2S \xrightarrow{\text{Lewis acid}} \underset{\substack{\| \\ \text{R-C-OR}'}}{\overset{S}{}} \qquad (5.16)
$$

Triethyl orthoacetate was reacted with borosulfide yielding O-ethyl thioacetate in quantitative yield [38]:

$$
CH_3C(OEt)_3 + B_2S_3 \quad \longrightarrow \quad \underset{\substack{\| \\ CH_3\text{-C-OEt}}}{\overset{S}{}} + B(OEt)_3 \qquad (5.17)
$$

The B–O bond energy in $B_2O_3$ (302 kcal/mole) is far greater than B–S bond energy in $B_2S_3$ (57 kcal/mole), and hence the reaction proceeds smoothly. The strong affinity of silane to oxygen transforms dibenzoyl disulfide into thione esters [37], for example:

$$
\underset{\substack{\| \\ (\text{PHCS}}}{\overset{O}{}}{\longrightarrow}_2 + Et_3SiH \xrightarrow[110°]{ZnCl_2} \underset{\substack{\| \\ PhCOSiEt_3}}{\overset{S}{}} \qquad (5.18)
$$

Dithiocarboxylates are prepared by several methods [40-43] as shown below:

$$
ArCS_2H \xrightarrow{Me_2SO_4} ArCS_2Me \qquad (5.19)
$$

$$
\underset{\substack{\| \\ R\text{—C—SR}'}}{\overset{\text{NH·HCl}}{}} + H_2S \xrightarrow{\text{Pyridine}} \underset{\substack{\| \\ R\text{—C—SR}'}}{\overset{S}{}} \qquad (5.20)
$$

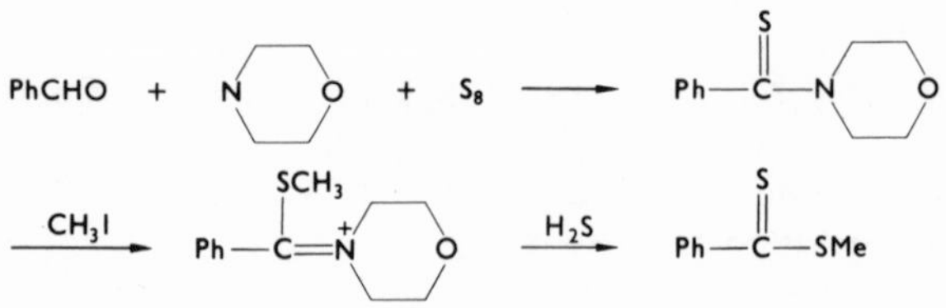

$$
PhCHO + \underset{\diagup\diagdown}{N \diagdown\diagup O} + S_8 \longrightarrow Ph\text{—C—N} \diagdown\diagup O
$$

$$
\xrightarrow{CH_3I} Ph\text{—C}\overset{SCH_3}{\underset{+}{=}}N \diagdown\diagup O \xrightarrow{H_2S} Ph\text{—C—SMe} \qquad (5.21)
$$

$$\text{BuC}\equiv\text{CSAc} + \text{EtS}^-\text{Na}^+ \xrightarrow[-30°]{\text{DMF}} \underset{\displaystyle \overset{\text{S}}{\|}}{\text{C}_5\text{H}_{11}\text{-C-SEt}} \qquad (5.22)$$

$$\text{RC}\equiv\text{CSLi} + \text{EtSH} \longrightarrow \underset{\displaystyle \overset{\text{S}}{\|}}{\text{RCH}_2\text{-C-SEt}} \qquad (5.23)$$

The methods for preparing thioamides were reviewed recently[44]. Among them, the reactions shown in eqs. 5.24 and 5.25 are the most com-

$$\text{RCN} + \text{H}_2\text{S (RSH)} \xrightarrow{\text{amine}} \underset{\displaystyle \overset{\text{S}}{\|}}{\text{R-C-NH}_2} \qquad (5.24)$$

$$\text{A(CH}_2)_n\text{CH}_3 + \text{R}_2\text{NH} + \text{S}_8 \longrightarrow \underset{\displaystyle \overset{\text{S}}{\|}}{\text{Ar(CH}_2)_{n+1}\text{C-NR}_2} \qquad (5.25)$$

mon. Gilbert and co-workers reported that the reaction in eq. 5.24 proceeds most satisfactorily with diethylamine as catalyst and dimethylformamide as solvent [45]. In addition, a catalytic amount of amine is effective for the reaction with aromatic nitriles, while aliphatic nitriles require equimolar amounts of amine and hydrogen sulfide. The reaction in eq. 5.25 is known as the Willgerodt-Kindler reaction [46] (c.f. Sect.2.4.2.2). Although the mechanism of this reaction is not well documented, recent results suggest the following [47]:

$$\text{RCOCH}_2\text{CH}_3 + \text{R}_2'\text{NH} \underset{\text{slow}}{\rightleftharpoons} \underset{\displaystyle \overset{\text{NR}_2'}{|}}{\text{R-C=CHCH}_3} \underset{\text{fast}}{\rightleftharpoons}$$

$$\underset{\displaystyle \overset{\text{NR}_2'}{\|}}{\text{RCH}_2\text{-C=CH}_2} \underset{\text{fast}}{\overset{\text{S}_8}{\rightleftharpoons}} \left[ \underset{\displaystyle \underset{\text{SH}}{\overset{\displaystyle \overset{\text{NR}_2'}{|}}{\text{RCH}_2\text{C=CH}}}}{} \right] \qquad (5.26)$$

$$\xrightarrow[\text{stabilization}]{\text{irreversible}} \underset{\displaystyle \overset{\text{S}}{\|}}{\text{RCH}_2\text{CH}_2\text{CNR}_2'}$$

### 5.2.4 Thiocarbonic Acids and Their Derivatives

Five different compounds are possible as sulfur derivatives of carbonic acid. Among them, only trithiocarbonic acid is known [48]. The most common method for the preparation of thioncarbonates is the reaction of carbon disulfide or thiophosgene with alcohols or phenols [17]:

$$CS_2 + RXM \longrightarrow RX\text{-}\overset{\overset{\displaystyle S}{\|}}{C}\text{-}SM \xrightarrow{H^+} RX\text{-}\overset{\overset{\displaystyle S}{\|}}{C}\text{-}SH \qquad (5.27)$$

$$(X = O, S; M = \text{alkali metal})$$

$$CSCl_2 + RXM \longrightarrow RX\text{-}\overset{\overset{\displaystyle S}{\|}}{C}\text{-}XR \qquad (5.28)$$

$$(X = O, S; M = \text{alkali metal})$$

Cyclic thioncarbonates are occasionally important intermediates in synthetic chemistry. They are prepared as follows [49]:

$$(5.29)$$

$$(5.30)$$

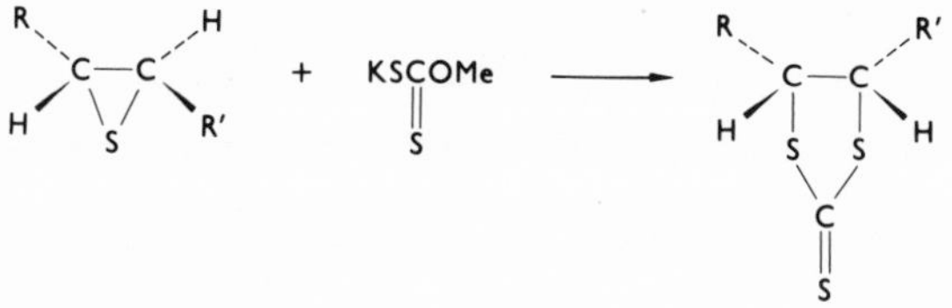

$$R \cdots \underset{H}{\overset{\cdots}{C}} - \underset{S}{\overset{H}{C}} \cdots R' \quad + \quad \text{KSCOMe} \quad \longrightarrow \quad \cdots$$

(5.31)

## 5.3 IONIC REACTIONS

### 5.3.1 Reactions with Nucleophiles

The general scheme for reactions of thiocarbonyl compounds with nucleophiles may be formulated [50]:

$$HB^- \quad + \quad \overset{\backslash}{\underset{/}{C}}{=}S \quad \longrightarrow \quad HB\text{-}\overset{|}{\underset{|}{C}}\text{-}S^-$$

**5.14**

(5.32)

$$\longrightarrow \quad B{=}\overset{/}{\underset{\backslash}{C}} \quad + \quad HS^-$$

In contrast to carbonyl compounds, tetrahedral intermediates, **5.14** [51], are stable and have often been isolated. For example, thioketones react with hydrogen sulfide giving *gem*-dithiols at $-40°$ (eq. 5.33). The reaction

$$\overset{\backslash}{\underset{/}{C}}{=}S \quad + \quad H_2S \quad \longrightarrow \quad \underset{SH}{\overset{SH}{C}}$$

(5.33)

with ethanol, however, affords ketals as final products (eq. 5.34), while the reaction with water yields ketones (eq. 5.35).

$$\overset{\backslash}{\underset{/}{C}}{=}S + EtOH \quad \longrightarrow \quad \left[ \underset{SH}{\overset{OEt}{C}} \right] \quad \longrightarrow \quad \underset{OEt}{\overset{OEt}{C}}$$

(5.34)

The importance of the tetrahedral intermediates has been demonstrated by kinetic studies on hydrolysis of O-ethyl thiobenzoate, **5.15** [52].

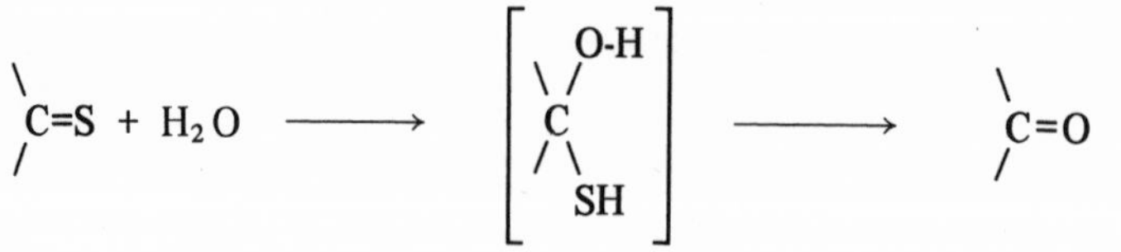

$$\begin{array}{c} \diagdown \\ \diagup \end{array} C{=}S \ + \ H_2O \ \longrightarrow \ \left[ \begin{array}{c} O\text{-}H \\ \diagdown \diagup \\ C \\ \diagup \diagdown \\ SH \end{array} \right] \ \longrightarrow \ \begin{array}{c} \diagdown \\ \diagup \end{array} C{=}O \qquad (5.35)$$

Interestingly, uncatalyzed hydrolysis is the predominant process in neutral solutions. Increase of rate constants both below pH 3 and above pH 10 indicates that substantial acid catalysis and direct hydroxide attack become important in these pH regions, respectively. Based on these results, Smith proposed the mechanism shown below:

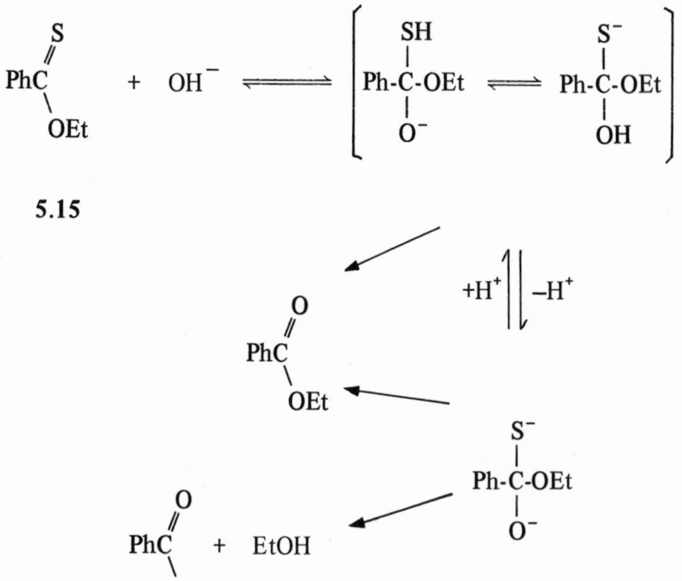

$$PhC\begin{array}{c}S\\ \| \\ \diagdown \\ OEt\end{array} \ + \ OH^- \ \rightleftharpoons \ \left[ \begin{array}{cc} SH & S^- \\ | & | \\ Ph\text{-}C\text{-}OEt & \rightleftharpoons \ Ph\text{-}C\text{-}OEt \\ | & | \\ O^- & OH \end{array} \right]$$

**5.15**

$$PhC\begin{array}{c}O\\ \| \\ \diagdown \\ OEt\end{array} \qquad\qquad +H^+ \Big\| -H^+ \qquad\qquad (5.36)$$

$$PhC\begin{array}{c}O\\ \| \\ \diagdown \\ S^-\end{array} \ + \ EtOH \qquad\qquad \begin{array}{c}S^-\\ | \\ Ph\text{-}C\text{-}OEt \\ | \\ O^-\end{array}$$

**5.16**

Since the formation of the dianion, **5.16**, is responsible for appearance of thiobenzoic acid, strong alkaline conditions are required to yield sufficient amount of thiobenzoic acid. It should be noted that **5.15** is more than a hundred times as reactive as its oxygen analogue in the neutral region. This is a general trend found in thiocarbonyl and carbonyl compounds. Thiocyclohexanone, for example, reacts with carbonyl reagents without any catalyst (eq. 5.37) [6], while reactions with carbonyl compounds

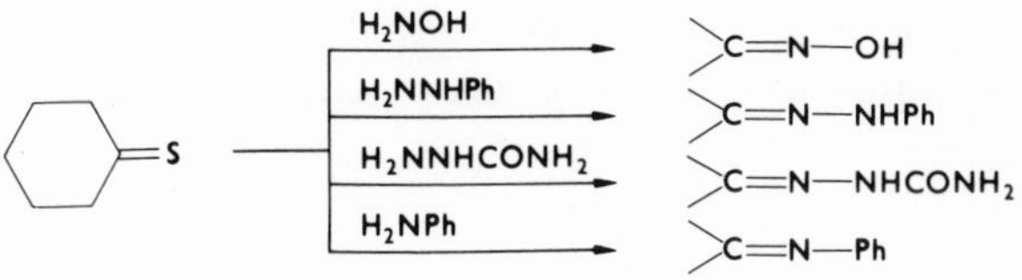

$$(5.37)$$

generally require catalysts. Monothiodibenzoylmethane (5.11, R = Ph) reacts with carbonyl reagents at the thiocarbonyl group [53]. This indicates that the thiocarbonyl group in 5.11b (and/or 5.11a) is more

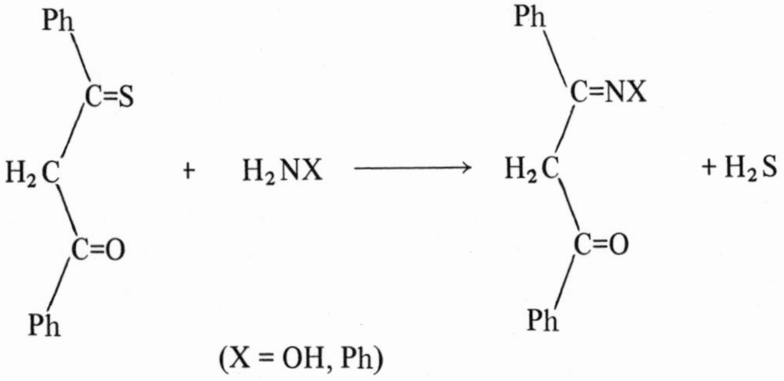

$$(5.38)$$

(X = OH, Ph)

reactive than the carbonyl group in 5.11c (and/or 5.11a). Powers and Westheimer studied the reactions of thiobenzophenone and benzophenone with phenylhydrazine quantitatively [54] and found that the reaction with thiobenzophenone is catalyzed by a base as well as by an acid, while the reaction with benzophenone proceeds only in acidic media. Rate constants, plotted against pH of the solutions, show that, at about pH 6, thiobenzophenone is about 2,000 times more reactive than benzophenone. Two possible mechanisms have been proposed for the base-catalyzed reaction of thiobenzophenone:

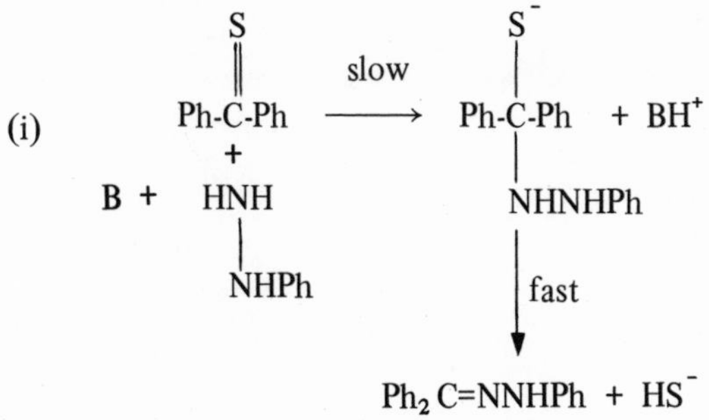

(i)

$$(5.39)$$

(ii)   $\underset{\text{Ph-C-Ph}}{\overset{\overset{\text{S}}{\|}}{}}$ + $H_2NNHPh$ $\underset{\text{fast}}{\rightleftharpoons}$ $Ph_2C\overset{\text{SH}}{\underset{\underset{\underset{\text{NHPh}}{|}}{\text{NH}}}{}}$

(5.40)

$Ph_2C\overset{\text{SH}}{\underset{\underset{\underset{\text{NHPh}}{|}}{\text{NH}}}{}}$ + B $\xrightarrow{\text{slow}}$ $Ph_2C=NNHPh$ +

$BH^+$ + $HS^-$

Although the choice between the two is very difficult, the greater stability of the thiolate ion than that of the alcoholate ion is responsible, in both schemes, for the more facile base-catalyzed formation of benzophenone phenylhydrazone. The spontaneous reaction of thiocarbonyl compounds with amines has been employed by Woodward and co-workers to a step of total synthesis of chlorophyll [18].

Thiobenzophenone is reduced to benzhydryl mercaptan by 1-benzyl-1, 4-dihydronicotinamide, 5.17 [55]. The rate of the reaction is unaffected by either typical free-radical chain breakers or change in pH, but increases in polar solvents, while an electron-releasing substituent in thiobenzophenone retards the reaction and a kinetic deuterium isotope effect of about 4 to 5 is observed. Based on these results, Abeles and co-workers proposed the mechanism which involves a direct hydride transfer. However, results, recently obtained by esr spectroscopy, revealed that the reaction proceeds through a three-step "electron-proton-electron transfer" mechanism as shown below [56]:

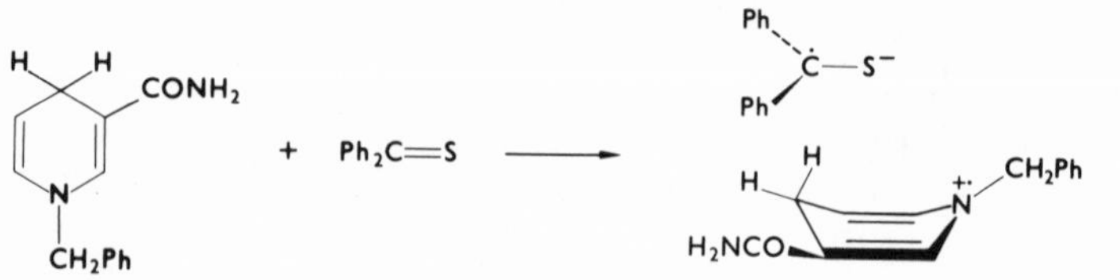

5.17

$$Ph_2CH—S·$$

$$Ph_2CH—S^-$$

$$(5.41)$$

Benzophenone does not react with **5.17** under the same condition. The facile reaction of thiobenzophenone can be accounted for by the ready availability of thiocarbonyl group toward direct reduction by electrons at low voltages [57,58].

The reaction of thiobenzophenone with Grignard reagents (eq. 5.42) [59] affords different type of products from those expected from the reaction of benzophenone. However, the reaction is analogous to the

$$2\,Ar_2C{=}S + 2\,RMgX \longrightarrow Ar_2C\underset{\diagdown\;S\;\diagup}{\text{—}}CAr_2 + R\text{-}R + MgX_2 + MgS$$

$$(5.42)$$

formation of benzopinacol in the reaction of benzophenone with a Grignard reagent; here, again, the pronounced ease of HS⁻ ion to leave plays an important role:

$$2\;\underset{\diagup}{\overset{\diagdown}{C}}{=}O + 2\,RMgX \longrightarrow 2\;\underset{\diagup}{\overset{\diagdown}{C}}\text{-OMgX} + 2\,R·$$

$$(5.43)$$

$$\longrightarrow \begin{array}{c} |\\ \text{-C-OMgX}\\ |\\ \text{-C-OMgX}\\ | \end{array} + R\text{-}R \xrightarrow{\;H_2O\;} \begin{array}{c} |\\ \text{-C-OH}\\ |\\ \text{-C-OH}\\ | \end{array}$$

The formation of 1,1-diphenylethylene from the reaction of thioacetophenone with phenylmagnesium bromide (eq. 5.44) can be interpreted similarly [60]:

$$\underset{Me}{\overset{Ph}{\diagdown}}C{=}S + PhMgBr$$

$$(5.44)$$

$$\longrightarrow \underset{Ph\;\;CH_2\text{-H}}{\overset{Ph(\;SMgBr}{\diagdown\;\diagup}}C \longrightarrow \underset{Ph}{\overset{Ph}{\diagdown}}C{=}CH_2$$

Thiobenzophenone reacts with phenyllithium giving, after treatment with water, benzhydryl phenyl sulfide in 30–40% yield. Similar reactions take place with phenyl dithiobenzoate and diphenyl trithiocarbonate [61]. The presence of the thioketyl radicals in the reaction of thiobenzophenone is shown by the observation of the corresponding signal in the esr spectrum of reacting solution. Beak and Worley propose two mechanisms, one of which involves a charge-transfer intermediate and the other which is based on the reverse polarization of the thiocarbonyl group:

$$2 \; \overset{\backslash}{\underset{/}{C}}{=}S \; + \; 2 \; RMgX \; \longrightarrow \; 2 \; \overset{\backslash}{\underset{/}{\dot{C}}}{-}SMgX \; + \; 2 \; R\cdot$$

$$\longrightarrow \quad \begin{matrix} \text{-C-SMgX} \\ | \\ \text{-C-SMgX} \\ | \end{matrix} \quad + \; R\text{-}R \; \longrightarrow \; \left\{ \begin{matrix} | \\ \text{-C-S-H} \\ | \\ \text{-C-S-H} \\ | \end{matrix} \right\} \qquad (5.45)$$

$$\longrightarrow \quad \begin{matrix} -C \\ | \quad \backslash \\ \quad \quad S \; + \; H_2S \\ | \quad / \\ -C \end{matrix}$$

$$\overset{\backslash}{\underset{/}{C}}{=}S \; + \; R^-M^+ \; \longrightarrow \; M^{+-}\overset{|}{\underset{|}{C}}\text{-S-R}$$

$$\uparrow$$

$$\overset{\backslash}{\underset{/}{C}}{=}S \; + \; R^-M^+ \; \longrightarrow \; \text{-}\overset{|}{\underset{|}{C}}\text{-S}^-M^+R\cdot \qquad (5.46)$$

$$\downarrow \overset{\backslash}{\underset{/}{C}}{=}S$$

$$R\cdot \; + \; M^{+-}\overset{|}{\underset{|}{C}}\text{-S-R} \; \longleftrightarrow \; \text{-}\overset{|}{\underset{|}{C}}\text{-S-R}$$

Hexafluorothioacetone is an interesting compound, since it is stable in the monomeric form and, of course, contains no enethiol form. This compound is highly susceptible to nucleophilic attack [62]. Here again arises the problem on the polarity of a thiocarbonyl group. The addition of bisulfite ion to hexafluoracetone yields a Bunte salt, **5.18**, instead of an α-mercaptosulfonic acid, **5.19**, as shown in eq. 5.47. This fact indicates that the canonical form **5.21** is the most important.

$$
(CF_3)_2 C=S + {}^-SO_3H
\begin{cases}
\longrightarrow \quad
\begin{array}{c}
CF_3 \\
| \\
H\text{-}C\text{-}S\text{-}SO_3^- \\
| \\
CF_3
\end{array} \\
\textbf{5.18} \\[2em]
\xcancel{\longrightarrow} \quad
\begin{array}{c}
CF_3 \\
| \\
HS\text{-}C\text{-}SO_3^- \\
| \\
CF_3
\end{array} \\
\textbf{5.19}
\end{cases}
\qquad (5.47)
$$

$$
\begin{array}{ccc}
\overset{\displaystyle S}{\underset{\displaystyle \|}{CF_3-C-CF_3}}
&
\overset{\displaystyle S^-}{\underset{\displaystyle |}{CF_3-\overset{+}{C}-CF_3}}
&
\overset{\displaystyle S^+}{\underset{\displaystyle |}{CF_3-\overset{-}{C}-CF_3}}
\\
\textbf{5.20} & & \textbf{5.21}
\end{array}
\qquad (5.48)
$$

This thioketone is so reactive that it can abstract an allylic proton even at −78° from the side of thiocarbonyl carbon (eq. 5.49) [63], which also

$$\text{(structure)} \longrightarrow \text{(structure)} \qquad (5.49)$$

supports the importance of the contribution of **5.21**. On the other hand, thiolate ion from mercaptoacetic acid attacks the thiocarbonyl carbon, indicating the predominancy of **5.20**. In addition, alkyl mercaptans as well as hydrogen chloride afford products expected from both **5.20** and **5.21**[62].

As previously discussed, $2p$-$3p$ $\pi$-bonding is relatively ineffective and it is unlikely that the polarizing ability of $\pi$-electrons in the thiocarbonyl group is greater than that of $\pi$-electrons in the carbonyl group. They would localize, if any, on an atom (or atoms) at the ground state of the compound[3]. Indeed, Minoura and Tsuboi have detected an esr signal from a THF solution of thiobenzophenone itself[64].

### 5.3.2 Reactions with Electrophiles
Almost all electrophiles attack the thiocarbonyl sulfur as schematically shown[50]:

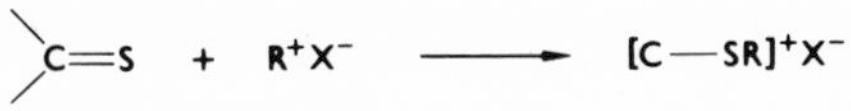

$$\text{C}{=}\text{S} \ + \ R^+X^- \ \longrightarrow \ [\text{C}{-}\text{SR}]^+X^- \qquad (5.50)$$

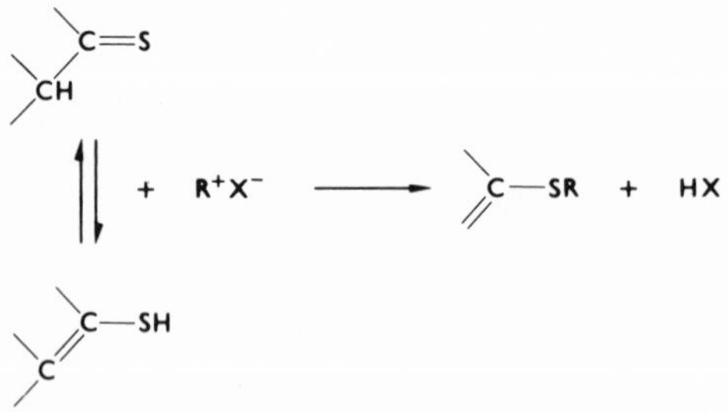

$$(5.51)$$

wherein RX may be alkyl halides, acyl halides, acid anhydrides, aldehydes, electron-poor olefins, oxygen, and so on. Most of the old examples were presented by Indian chemists[39,65,66].

More recently, the following reactions have been reported[6]:

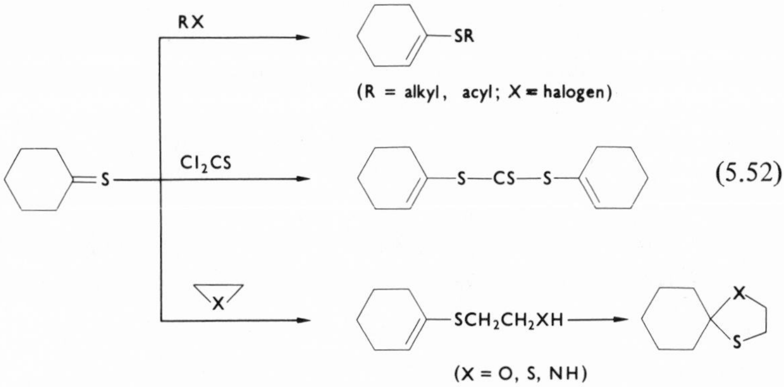

$$(5.52)$$

(R = alkyl, acyl; X = halogen)

(X = O, S, NH)

All these reactions undergo the enethiol form of the thioketone (eq. 5.53). Hence, electrophilic attack takes place at the α-carbon

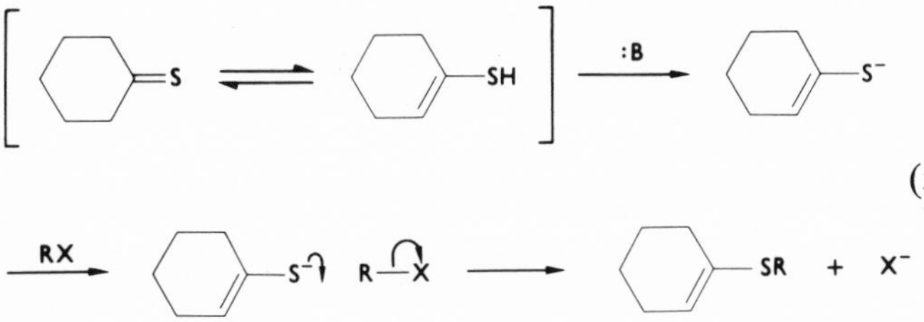

$$(5.53)$$

occasionally. For example, thiocamphor reacts with benzaldehyde yielding **5.22**, a simple aldol-condensation product, under basic conditions [65,67]:

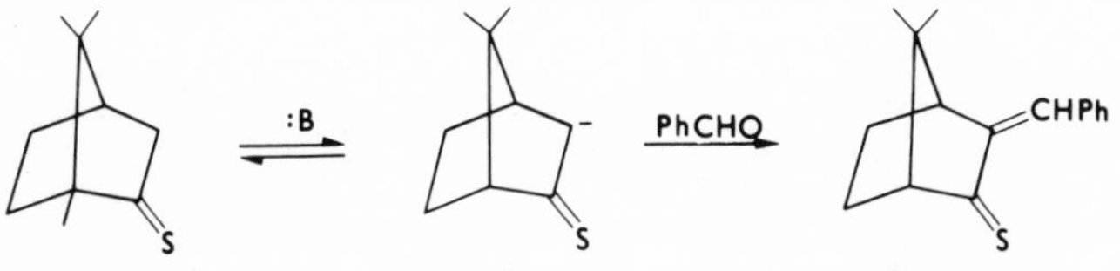

$$(5.54)$$

**5.22**

Isoamyl nitrite also affords α-isonitrosothiocamphor by the reaction with thiocamphor [68,69]. The reason for the preference of C-attack over S-attack with such an ambient ion is still ambiguous.

Thionesters that have an enolizable hydrogen react similarly, yielding S-alkylated enethiol products [70-73]. This is also true when an enolizable hydrogen is on a heteroatom (eq. 5.55). [74,75] This reaction is widely used for preparing a variety of alkyl mercaptans [76].

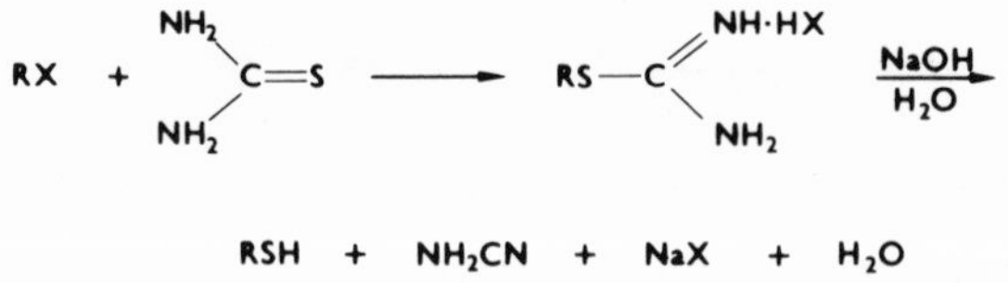

$$RX \quad + \quad \underset{NH_2}{\overset{NH_2}{C}}=S \quad \longrightarrow \quad RS-\underset{NH_2}{\overset{NH\cdot HX}{C}} \quad \xrightarrow[H_2O]{NaOH}$$

$$(5.55)$$

$$RSH \quad + \quad NH_2CN \quad + \quad NaX \quad + \quad H_2O$$

Thiones that have no enolizable hydrogen normally afford stable salts. Michler's thioketone, **5.23**, reacts with ethyl iodide affording a salt, **5.24**, which yields **5.25** by the subsequent reaction with potassium cyanide [77]:

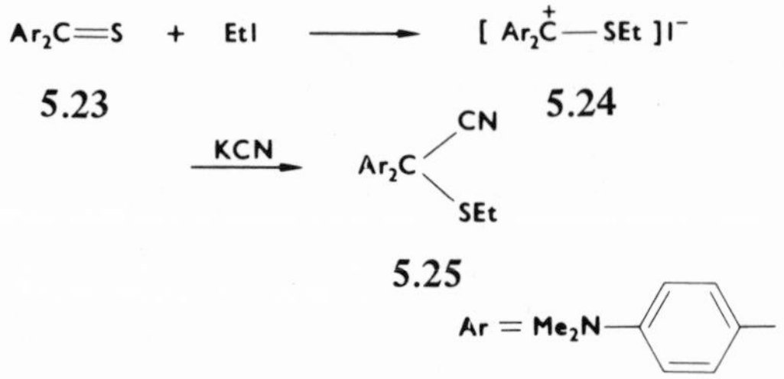

$$Ar_2C=S \quad + \quad EtI \quad \longrightarrow \quad [\ Ar_2\overset{+}{C}-SEt\ ]I^-$$

**5.23**                                                                    **5.24**

$$\xrightarrow{KCN} \quad Ar_2C\underset{SEt}{\overset{CN}{\big\langle}}$$

$$(5.56)$$

**5.25**

$$Ar = Me_2N-\!\!\!\langle\ \rangle\!\!\!-$$

Two possible mechanisms have been proposed for the reaction of **5.23** with *p*-nitrobenzyl chloride in alcoholic potassium hydroxide [78,79]:

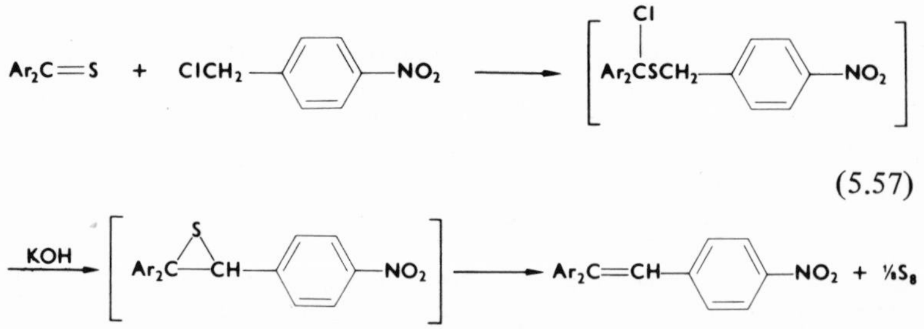

$$Ar_2C=S \quad + \quad CICH_2-\!\!\!\langle\ \rangle\!\!\!-NO_2 \quad \longrightarrow \quad \left[ Ar_2\overset{CI}{\underset{}{C}}SCH_2-\!\!\!\langle\ \rangle\!\!\!-NO_2 \right]$$

$$(5.57)$$

$$\xrightarrow{KOH} \left[ Ar_2C\overset{S}{\underset{}{\triangle}}CH-\!\!\!\langle\ \rangle\!\!\!-NO_2 \right] \longrightarrow Ar_2C=CH-\!\!\!\langle\ \rangle\!\!\!-NO_2 + \tfrac{1}{8}S_8$$

and

$$Ar_2C=S \quad + \quad CICH_2-\!\!\!\langle\ \rangle\!\!\!-NO_2 \quad \xrightarrow{KOH} \left[ Ar_2\overset{SH\ \ CI}{\underset{}{C}}-CH-\!\!\!\langle\ \rangle\!\!\!-NO_2 \right]$$

$$\longrightarrow Ar_2C=CH-\!\!\!\langle\ \rangle\!\!\!-NO_2 \quad + \quad HCI \quad + \quad \tfrac{1}{8}S_8$$

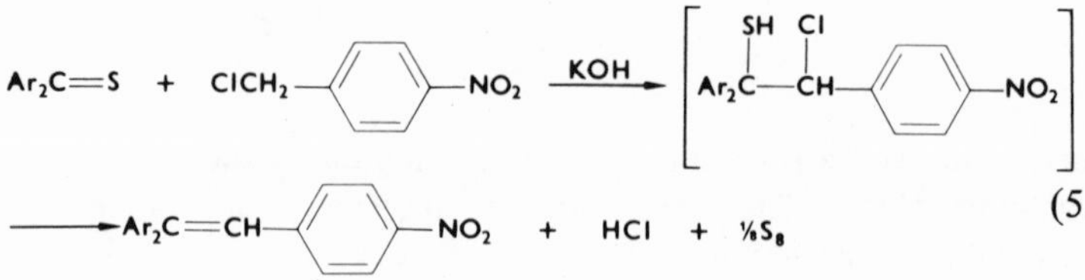

$$(5.58)$$

More recently, an unusual thioketone, **5.26**, was synthesized and reacted with methyl iodide and a proton, respectively [80]:

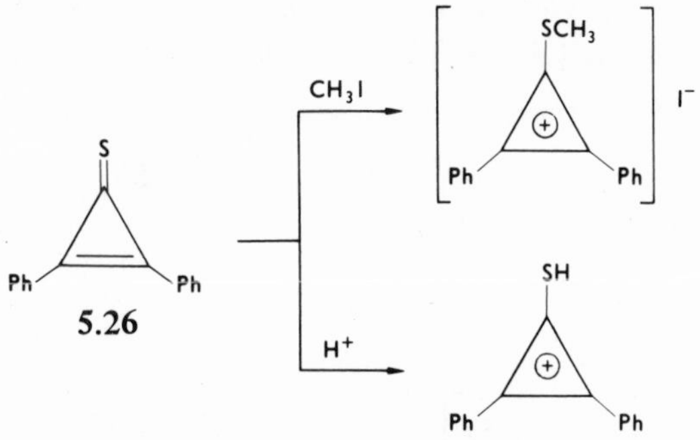

(5.59)

**5.26**

Products are relatively stable, because they have a $(4n + 2)\pi$ system with $n = 0$. Other salts with a $(4n + 2)\pi$ system, **5.27**, were isolated from the

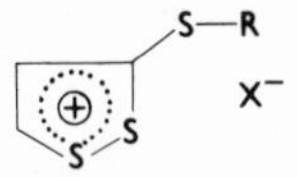

**5.27**

reaction mixtures of trithione with alkyl, benzyl, and phenacyl halides as well as with dialkyl sulfates[73].

Dittmer and Whitman tried to synthesize benzothietes, **5.28**, by a 1,2-cycloaddition of a thiocarbonyl group to benzyne (eq. 5.60)[81] (*cf.* Sect. 5.4.3). However, starting from benzenediazonium-2-carboxylate and

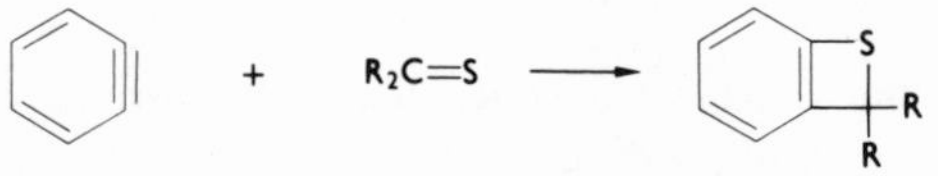

(5.60)

**5.28**

thiobenzophenone, they obtained 2,2-diphenyl-1,3-benzoxathian-4-one, **5.29**. Apparently, electrophilic attack of phenonium ion on thiocarbonyl sulfur prior to the formation of benzyne is responsible for this reaction:

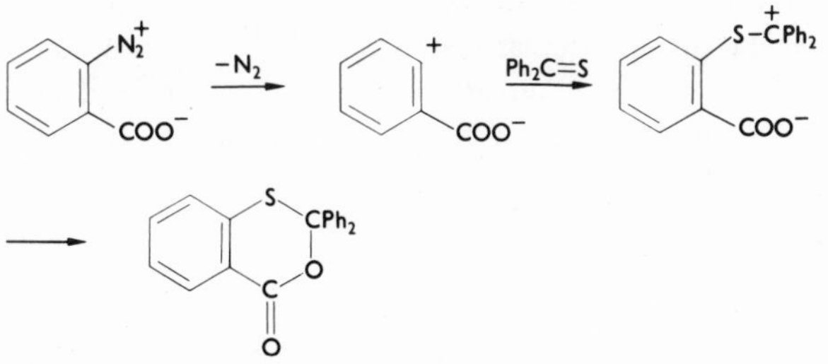

(5.61)

**5.29**

The thiocarbonyl group is so susceptible to the electrophilic attack that it reacts with acetylenes if the latter are reasonably electron-deficient [82,83] :

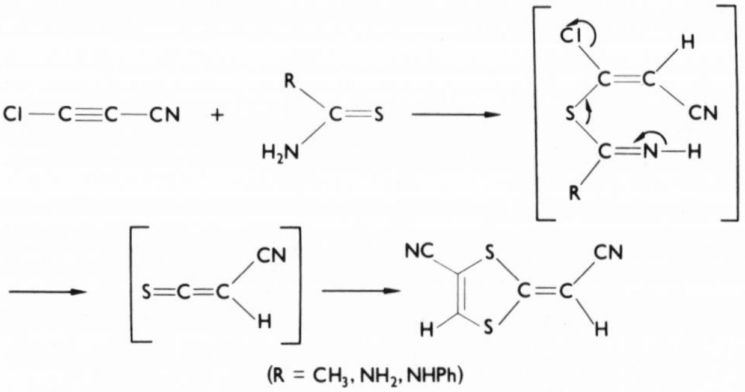

$$\text{(5.62)}$$

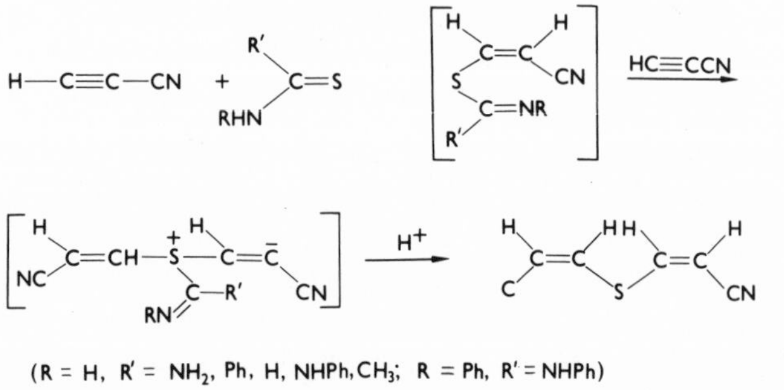

(R = CH₃, NH₂, NHPh)

$$\text{(5.63)}$$

(R = H, R' = NH₂, Ph, H, NHPh,CH₃;  R = Ph, R' = NHPh)

In the Schönberg rearrangement, diaryl thiocarbonates rearrange to diaryl thiolcarbonates on heating [84,85] :

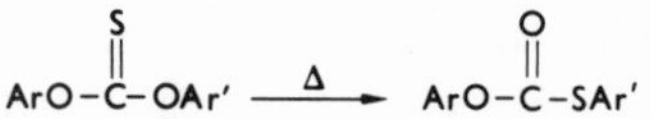

$$\underset{\text{ArO-C-OAr'}}{\overset{S}{\|}} \xrightarrow{\Delta} \underset{\text{ArO-C-SAr'}}{\overset{O}{\|}} \qquad\qquad (5.64)$$

Since the latter is easily hydrolyzed to thiophenols, this is a method of facile preparation of thiophenols from the corresponding phenols. Electron-withdrawing substituents facilitate the reaction [86]. Since this is an unimolecular intramolecular rearrangement [87], the reaction seems to be initiated by the nucleophilic attack of the thiocarbonyl group on the aromatic ring, forming a four-membered ring at the transition state as shown below [88] :

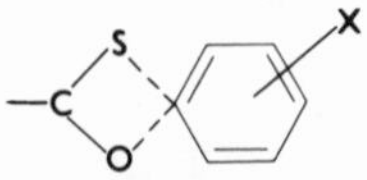

**5.30**

The reaction shown in eq. (5.65) proceeds intramolecularly with

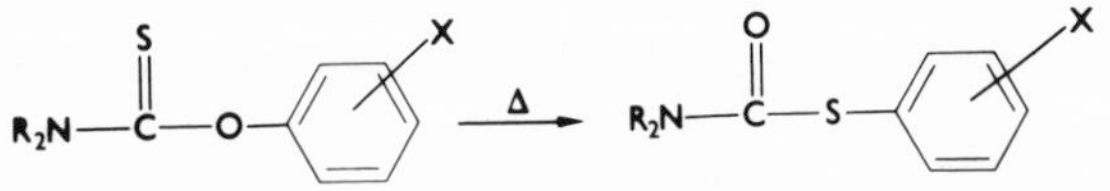

$$(5.65)$$

first-order kinetics. The rate of the reaction increases in the order of R = $iso$-$C_4H_9$ > $C_2H_5$ > $CH_3$, showing a good Hammett relationship with $\sigma^*$, while the Hammett plot with substituent X is nicely correlated with modified $\sigma^-$ [89]. Based on these and other results, the transition states of the Schönberg and similar rearrangements are supposed to have reactant-like structures [90]

A new method of converting aldoximes into nitriles involves the spontaneous rearrangement of **5.31** to the corresponding nitrile [91]. Similar

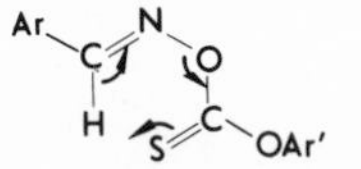

**5.31**

rearrangement occurs in the electron impact [92]. In this case, however, the most ionizable electron is that of lone pairs on the sulfur atom and the nucleophilic attack of the phenyl group on the sulfur is proposed as a driving force of the reaction:

$$(5.66)$$

As has been seen, most reactions in this category may be understood to be $S_N2$ reactions with *thiocarbonyl compounds as nucleophiles* instead of $S_E2$ reactions, wherein thiocarbonyl compounds are regarded as substrates.

In contrast to a carbonyl group, the thiocarbonyl group can be oxidized to corresponding mono- and di-oxides, sulfines and sulfenes [93]. The first S-oxide of thioketone, **5.32**, was prepared by Sheppard and Diekmann [94]:

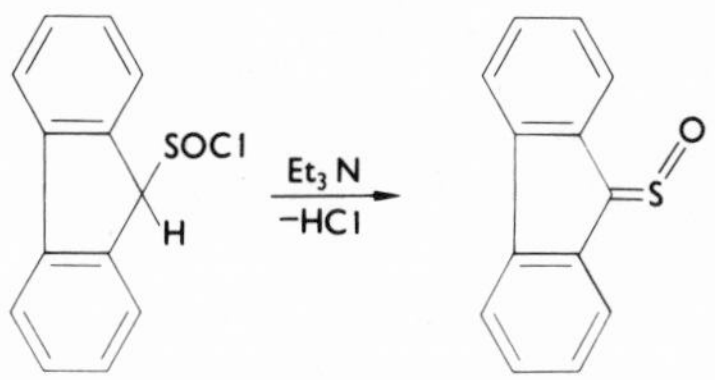

$$(5.67)$$

**5.32**

A similar reaction also affords thioaldehyde-S-oxide, **5.33** [95]:

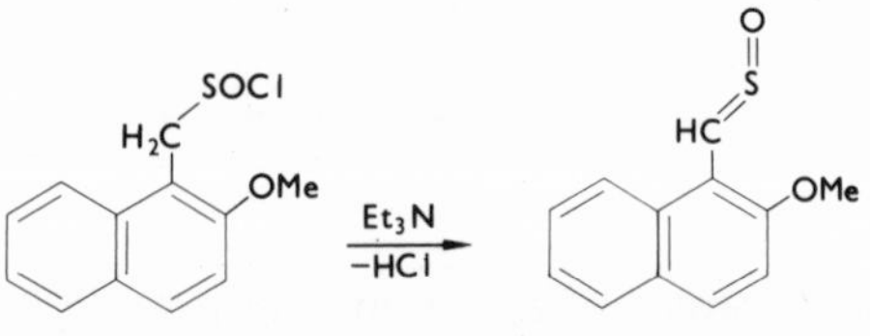

(5.68)

**5.33**

The direct oxidation of thioketones with monoperphthalic acid again results in the formation of sulfines [96,97]. *Cis*- and *trans*-isomers of methyl dithio-1-naphthoate-S-oxide, **5.34** and **5.35**, were synthesized by this method [98]. Methyl dithio-2-naphthoate reacts similarly:

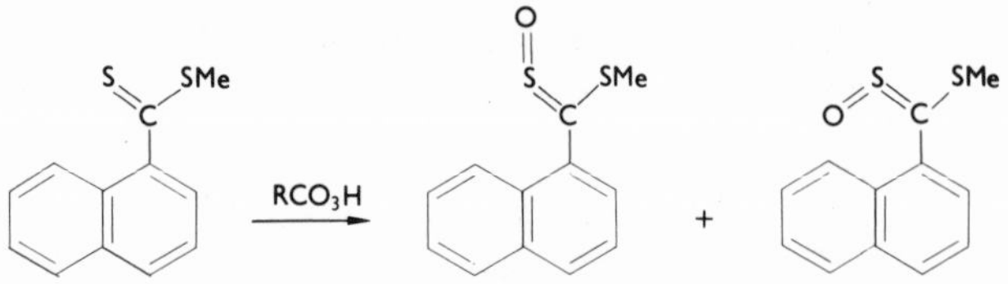

(5.69)

**5.34**              **5.35**

Sulfenes are unstable compounds and have been generated only as transients. However, recently, unsubstituted sulfene was trapped at $-196°$ and subjected to ir spectroscopy [99].

The complexity of reactions of thiocarbonyl compounds with organometallics has been mentioned in Sect. 5.3.1. Other reactions, not discussed here, in which thiocarbonyl compounds are attacked by electrophiles, at least formally, are formations of complexes with transition metals [60,100,101]. Readers are recommended to refer to the articles cited herein.

## 5.4 RADICAL AND RELATED REACTIONS

### 5.4.1 Reactions with Radicals

The importance of the canonical form **5.1c** may be most emphasized in radical reactions of thione. An old example is the coupling of **5.36** (eq. 5.70) [102]. Since the biradical form of **5.36** is expected to be stabilized by

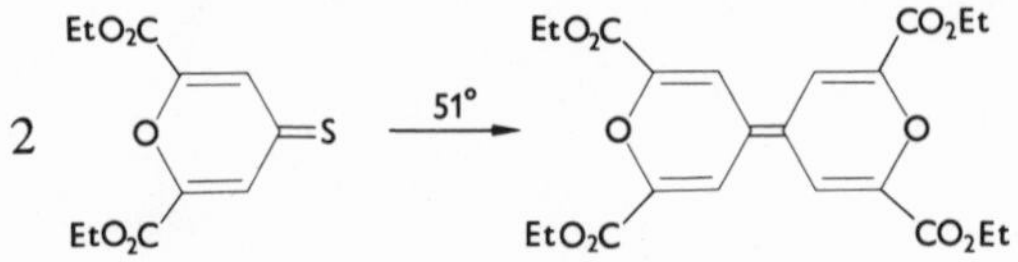

(5.70)

**5.36**

the contribution of many canonical forms (eq. 5.71), dithietane, which is

$$\text{(5.71)}$$

formed by coupling of two thiyl radicals, is suggested to be the intermediate (eq. 5.72). Recently, two research groups synthesized

$$+ \; \tfrac{1}{4}S_8 \qquad \text{(5.72)}$$

dithietane derivatives, **5.37**, independently [103,104]. Since these compounds are in resonance among the canonical forms **5.37**, **5.38**, and

$$(X = CF_3, CN) \qquad \text{(5.73)}$$

**5.37**          **5.38**          **5.39**

**5.39**, they react with certain olefins easily, affording cycloadducts:

$$\text{(5.74)}$$

Certain thiocarbonyl compounds are found to be powerful capturing agents for carbon radicals [105]. When azobisisobutyronitrile (AIBN) is decomposed in refluxing toluene containing an equimolar amount of thiobenzophenone, the yield of tetramethylsuccinonitrile, a dimerization

$$\text{(5.75)}$$

product of cyanoisopropyl radicals, decreases from 89% (without thiobenzophenone) to 21%. Cyanoisopropyl radicals are so efficiently captured [106] that the adduct, **5.40**, is produced in 78% yield. Similarly the addition of thiobenzophenone to a solution of azobis-1-phenylethane in toluene results in the marked decrease of the yield of 2,3-diphenylbutane from 90% (without thiobenzophenone) to 12%, and the adduct, **5.41**, is obtained in 80% yield. Thioacetophenone is also an efficient radical-capturing agent.

$$
\begin{array}{cc}
\text{Me}\ \ \text{Ph}\ \text{Me} & \text{Me}\ \ \text{Ph}\ \text{Me} \\
|\ \ \ \ |\ \ \ \ | & |\ \ \ \ |\ \ \ \ | \\
\text{NC-C-S-C--C-CN} & \text{H-C-S-C--C-H} \\
|\ \ \ \ |\ \ \ \ | & |\ \ \ \ |\ \ \ \ | \\
\text{Me}\ \ \text{Ph}\ \text{Me} & \text{Ph}\ \ \text{Ph}\ \text{Ph}
\end{array}
$$

**5.40**                                   **5.41**

It was found that a linear free-energy relationship holds between visible absorption maxima (in eV units) and radical-capturing powers (cp) of thiocarbonyl compounds. Here, cp is defined by:

$$cp = \frac{(D)_0 - (D)}{(D)} \qquad (5.76)$$

where $(D)_0$ and $(D)$ denote yields of dimers without and with thiocarbonyl compounds, respectively [107].

It is known that the visible absorptions of thiocarbonyl compounds are ascribed to $n \rightarrow \pi^*$ transitions of thiocarbonyl groups. Since the ionization potentials of the lone-pair electrons on the sulfur atom presumably do not differ greatly from each other among these compounds, the energy of the $n \rightarrow \pi^*$ transition mainly represents the relative energy level of the lowest vacant orbital of C–S $\pi$-bonding. Since the low level corresponds to the strong capturing power, it is apparent that carbon radicals behave as nucleophiles in this reaction.

The rate constant for the dimerization of cyanoisopropyl radicals was calculated to be $2.2 \times 10^9$ liter mole$^{-1}$ sec$^{-1}$ [108]. Then the rate of the reaction of thiobenzophenone with these radicals might be fast enough to compete with it.

### 5.4.2 Reactions with Carbenes

Reactions of thiocarbonyl compounds with diazo compounds have been extensively studied by Schönberg and co-workers [109]. There are two types of reactions depending on structures of thiocarbonyl and diazo compounds. They are:

$$\text{A:} \quad 2{>}C{=}S \;+\; N_2CHR \;\longrightarrow\; \underset{\underset{R \quad H}{S \quad S}}{\boxed{\phantom{x}}} \;+\; N_2 \qquad (5.77)$$

$$\text{B:} \quad {>}C{=}S \;+\; N_2C{<} \;\longrightarrow\; \triangle_S \;+\; N_2 \qquad (5.78)$$

Although the results were interpreted in terms of steric effect [109], an alternative mechanism is suggested [110]:

$${>}C{=}S \;+\; N_2C{<} \;\longrightarrow\; {>}\overset{\cdot}{C}{-}S{-}\overset{\cdot}{C}{<}$$

**5.42**

$$(5.79)$$

That is, when both radical centers in the biradical intermediate, **5.42**, are stable enough (for example, **5.43** + **5.46** → $Ph_2\overset{\cdot}{C}{-}S{-}\overset{\cdot}{C}Ph_2$), they have to combine with each other, giving a thiirane derivative. On the other hand, when one radical terminal can react with another molecule of thiocarbonyl compounds (*cf.* Sect. 5.4.3), the A-type process results, yielding a 1,3-dithiolane derivative (for example, **5.43** + **5.45** → $Ph_2\overset{\cdot}{C}{-}S{-}\overset{\cdot}{C}H_2$). Diaryl trithiocarbonates, **5.44**, are yellow compounds and their radical-capturing powers are as weak as that of O-ethyl thiobenzoate (*cf.* Sect. 5.4.1). Accordingly, they afford thiirane derivatives by reacting with **5.46** or **5.47**, which react with **5.43** to give a 1,3-dithiolane derivative.

$$\underset{\text{Ar-C-Ar}}{\overset{\overset{\displaystyle S}{\|}}{}} \qquad\qquad \underset{\text{ArS-C-SAr}}{\overset{\overset{\displaystyle S}{\|}}{}}$$

**5.43**                                **5.44**

$$CH_2N_2 \qquad\qquad PhCHN_2 \qquad\qquad MeCHN_2$$

**5.45**                       **5.46**                              **5.47**

Involvement of carbene in these reactions is, however, not clear.

Sometimes 1,3-dipolar adducts are isolated (eqs. 5.80 and 5.81)[111-113]. More recently, Diebert suggested a 1,3-dipolar adduct, **5.48**, as an intermediate of the reaction shown in eq. 5.82 [114]. This reaction exemplifies nicely the competitive nature of A- and B-type reactions.

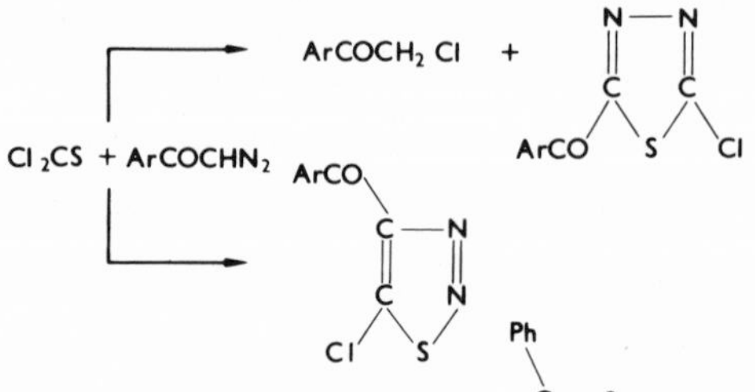

(5.80)

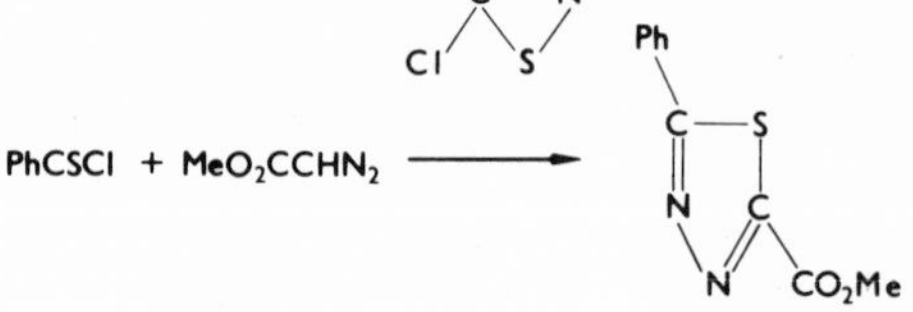

(5.81)

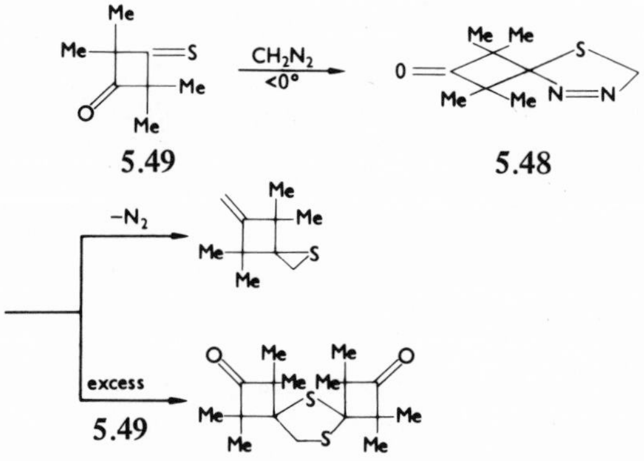

(5.82)

### 5.4.3 Photoreactions

Thiobenzophenone has three well-separated absorption bands. They are (in ethanol):

A: $\lambda_{max} = 235$ nm $\qquad$ $\epsilon = 9,000$

B: $\lambda_{max} = 316.5$ nm $\qquad$ $\epsilon = 15,800$

C: $\lambda_{max} = 599$ nm $\qquad$ $\epsilon = 181$

Absorptions A, B, and C correspond to $\pi \to \pi^*$, $\pi \to \pi^*$, and $n \to \pi^*$ transitions, respectively (Fig. 5.3) [115]. Since each absorption is

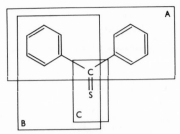

**Fig. 5.3.** A schematic representation of groups
contributing to transitions A, B, and C.

independent from the others, we can easily predict which transition is contributing to the reaction by applying light of an appropriate wavelength. This is an advantage of thioketones over ketones, in which transitions due to more than two origins are generally included in the photo-excitation.

When thiobenzophenone is irradiated with light of 5890 Å (sodium D-line), a singlet species of the n,π* state is produced, but it is rapidly deactivated to n,π* triplet state [116,117]. This species undergoes cycloadditions with certain olefins, affording 1,4-dithiane and/or thietane derivatives [110,118,119]. Some olefins, however, do not react with thiobenzophenone of n,π* triplet state, but react with its π,π* singlet state on irradiation with light of 3150Å, giving thietane derivatives [118b,120]. Yields of products are 40–100% and the reaction proceeds highly stereoselectively and regioselectively.

Olefins may be classified into three groups:

**Group I Olefins** are those which are substituted by electron-releasing groups. The olefin of this group reacts with thiobenzophenone of n,π* triplet state, resulting in the formation of a 1,4-dithiane derivative by the addition of two molecules of thiobenzophenone. However, when the concentration of thiobenzophenone is low enough, it affords a thietane derivative. Examples are cyclohexene, 2,3-dihydropyran, alkyl vinyl ethers, alkyl propenyl ethers, alkyl vinyl sulfides, mono-arylethylenes, and mono-alkylethylenes.

**Group II Olefins** are those which are substituted by electron-withdrawing groups. The olefin of this group reacts stereospecifically with thiobenzophenone of π,π* singlet state, giving thietane derivatives. Examples are acrylonitrile, methyl acrylate, vinyl acetate, dichloroethylene, dicyanoethylene, crotononitrile, and isocrotononitrile.

**Group III Olefins** are those which are substituted by electron-releasing groups. The olefin of this group reacts with thiobenzophenone of n,π* triplet state, resulting in the formation of thietane derivatives even when sufficient thiobenzophenone is supplied. Examples are 1,3-cyclooctadiene, propenylbenzene, $\alpha$-methylstyrene, and $\alpha$-phellandrene.

The following mechanisms are conceivable [118]:

$$(5.83)$$

**5.50**

Evidence obtained from kinetics [118b] and stereochemistry of olefins [121] supports that thiobenzophenone of n,$\pi^*$ triplet state behaves like thiyl radicals. The relationship between the reactivity of a radical center and radical-capturing power of thiobenzophenone discussed in Sect. 5.4.1 supports the proposed mechanism. Steric effect appears to play an important role in placing olefins into either group I or III categories. Propenylbenzene, $\alpha$-methylstyrene, $\alpha$-phellandrene, and 1,3-cyclooctadiene are group III olefins in contrast to styrene and cyclohexene (group I). Thus, the steric inhibition of substituent(s) against the *trans*-addition of a second molecule of thiobenzophenone to a composite biradical, **5.50**, controls the reaction path.

On the other hand, the reaction with group II olefins seems to involve nucleophilic attack of thiobenzophenone of $\pi,\pi^*$ singlet state, which behaves like thiolate anions, to an electron-deficient double bond. It is not clear whether the intermediate, **5.51**, is a charge-transfer complex as shown in eq. 5.84 or an incipient biradical, **5.50** [122,123]

Meanwhile, the reactions of acetone and of thiobenzophenone with acrylonitrile afford 2-cyano-4,4-dimethyloxetane [124] and 3-cyano-1,1-diphenylthietane [118b], respectively. From group I or III olefins, thietanes

$$Ph_2C=S \xrightarrow{3150\ \overset{\circ}{A}} \ ^1(Ph_2C=S)_{\pi,\pi^*}$$

$$^1(Ph_2C=S)_{\pi,\pi^*} + \ \underset{5.51}{\overset{\overset{C}{\parallel}}{\underset{C}{\longleftrightarrow}}} \rightleftharpoons \begin{matrix} C & S \\ | & | \\ C & CPh_2 \end{matrix} \tag{5.84}$$

$$\longrightarrow \begin{matrix} -C & S \\ | & | \\ -C & \underset{Ph}{C}-Ph \end{matrix}$$

are formed by anti-Markovnikov-type addition; styrene yields 1,1,3-triphenylthietane. These observations imply that the electron density of the $\pi^*$ orbital lobe on the carbon atom in $n,\pi^*$ states of thiocarbonyl and carbonyl groups is higher than that on a heteroatom, while, in the

$$\underset{R}{\overset{R}{>}}C\overset{\cdot\cdot}{\underset{Y}{=}}\overset{\cdot\cdot}{X} \longleftrightarrow \underset{R}{\overset{R}{>}}\overset{-}{\underset{Y}{C}}-\overset{+\ \cdot\cdot}{X} \tag{5.85}$$

$$n,\pi^* \text{ excited state}$$

$\pi,\pi^*$ states of these groups, the $\pi^*$ orbital lobe on the carbon atom is less electron-rich than the other [125]:

$$\underset{R}{\overset{R}{>}}\overset{\cdot}{C}-\overset{\cdot\cdot}{\underset{Y}{X}} \longrightarrow \underset{R}{\overset{R}{>}}\overset{+}{C}-\overset{-}{\underset{Y}{X}} \tag{5.86}$$

$$\pi,\pi^* \text{ excited state}$$

Thiobenzophenone of $n,\pi^*$ triplet state also reacts with acetylenic compounds yielding derivatives of isothiochromene [126]:

$$Ph_2C=S \xrightarrow{5890\ \overset{\circ}{A}} \ ^3(Ph_2C=S)_{n,\pi^*}$$

$$^3(Ph_2C=S)_{n,\pi^*} + R-C\equiv C-R' \longrightarrow$$

$$\tag{5.87}$$

The vinyl radical, which is a weak nucleophilic $\sigma$-radical, preferentially attacks aromatic rings over thiocarbonyl-sulfur.

Acetylenes studied are phenylacetylene, cyanoacetylene, propargyl alcohol, propiolic acid, and dimethyl acetylene dicarboxylate.

O-Alkyl thiobenzoates also afford cycloadducts with certain olefins on photo-irradiation [127]. However, when allylbenzene is employed as an

$$(5.88)$$

olefin, products of type **5.52** are obtained. Furthermore, the reaction with cyclooctadiene affords a mixture of thietane derivatives. The mechanism to explain these reactions universally is not yet known.

$$PhCH_2CH=CH_2 \quad + \quad \overset{\displaystyle S}{\underset{\displaystyle \|}{Ar\text{-}C\text{-}OR}}$$

$$\xrightarrow{3150\ \text{Å}} \quad PhCH_2CH_2\text{-}\overset{\displaystyle O}{\underset{\displaystyle \|}{C}}\text{-}Ar \quad + \quad (CH_2\text{=}S)_n \quad + \quad (CH_2)_n \quad (5.89)$$

**5.52**

$$(Ar = C_6H_5, p\text{-Cl-}C_6H_4 ; R = Me, Et)$$

Those carbonyl compounds which undergo cycloaddition with olefins are known to be photoreduced by various hydrogen donors such as alcohols, amines, and hydrocarbons [128]:

$$(5.90)$$

This is also true for thiocarbonyl compounds. Thiobenzophenone is reduced by ethanol on irradiation with light of 5890 Å, yielding products **5.53 – 5.57** [129]:

$$Ph_2C=S \ + \ EtOH \ \xrightarrow{\ 5890 \ Å\ }$$

$$Ph_2CHSSCHPh_2 \ + \ Ph_2CHSH \ + \ Ph_2CHSSC(Ph)_2SSCHPh_2$$

$$\textbf{5.53} \qquad\qquad\quad \textbf{5.54} \qquad\qquad\quad \textbf{5.55}$$

$$+ \ Ph_2CH_2 \ + \ CH_3CHO \qquad\qquad\qquad (5.91)$$

$$\textbf{5.56} \qquad\quad \textbf{5.57}$$

Since light of 5890 Å is effective for the reduction, it is apparent that thiobenzophenone of $n,\pi^*$ triplet state is involved. Although earlier works suggested the primary formation of diphenylmethanethiyl radical, **5.58** [130], recent studies including isotope technique, suggest the

$$Ph_2C=S \ + \ EtOH \ \xrightarrow{\ h\nu\ } \ Ph_2CHS\cdot \ + \ CH_3\dot{C}HOH \quad (5.92)$$

$$\textbf{5.58}$$

formation of diphenylsulfhydrylmethyl radical, **5.59**, as the first intermediate (eq. 5.93) [129,131,132]. Hydrogen abstraction by **5.59** affords **5.54**, which upon further free radical reactions gives **5.53**, **5.55**, and **5.56**.

$$Ph_2C=S \ + \ EtOH \ \xrightarrow{\ h\nu\ } \ Ph_2\dot{C}SH \ + \ CH_3\dot{C}HOH \quad (5.93)$$

$$\textbf{5.59}$$

The reduction of thiobenzophenone proceeds *via* essentially the same mechanism as discussed above in cyclohexane, diphenylmethane, benzhydrol, and benzhydrylmercaptan [129].

Contrary to carbonyl compounds, thiobenzophenone is reduced thermally by ethanol and other hydrogen donors [129]. Among these reactions, those with primary amines are quite fast and proceed at room temperature.

When O-ethyl thiopropionate is irradiated with light of 2537 Å, olefins **5.61** with *cis/trans* ratio of 2/3 are formed in 60–65% yield (eq.

5.94) [133] The mechanism involving the dimerized intermediate,

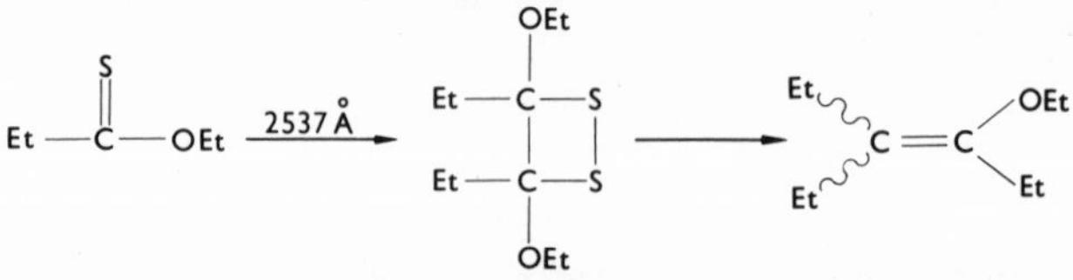

(5.94)

**5.60**                    **5.61**

**5.60**, seems to be more plausible than the carbene mechanism, because neither carbene nor atomic sulfur has been detected. Similar reactions have been found with other thiocarbonyl compounds [134]:

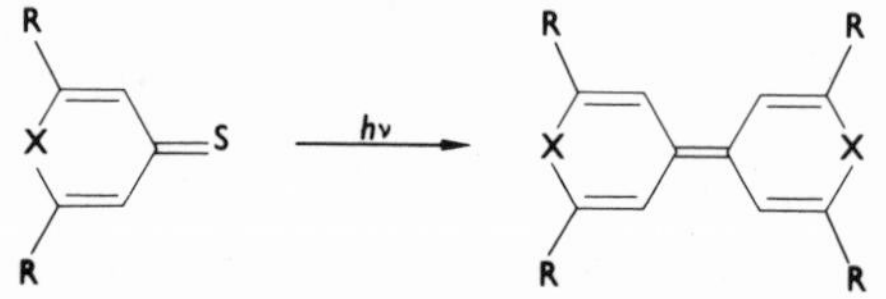

(5.95)

(X = O, S; R = Et, Ph)

### 5.4.4 Diels-Alder Reaction

Thiocarbonyl compounds react with dienes giving Diels-Alder adducts. The reactivity of these compounds is far greater than that of the corresponding oxygen analogues [135]. Although most reactions proceed thermally, they appear to be catalyzed by light. No detailed study on the mechanism has been made. However, the fact that unsymmetrically substituted dienes afford nearly equimolar amounts of two possible adduct isomers suggests a one-step radical-like mechanism. Typical examples are:

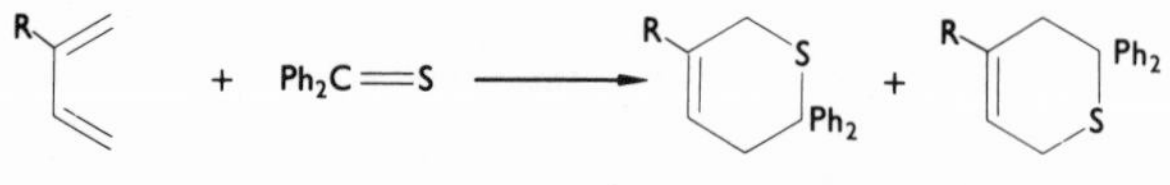

(5.96)

(R = Me, Cl)

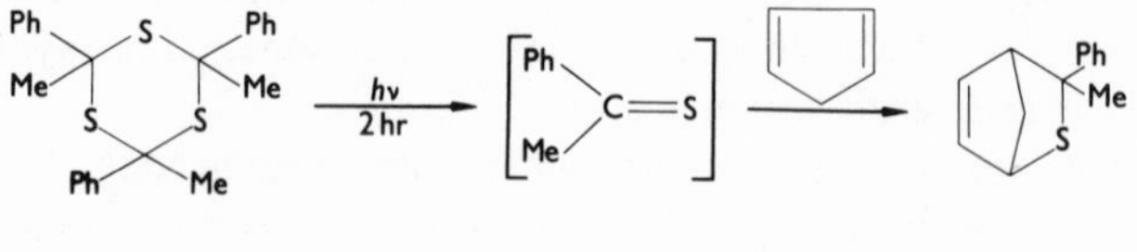

(5.97)

### 5.5 NO-BOND RESONANCE

The chemistry of thiones should not be concluded without touching no-bond resonance. The product of the reaction of 2,4,6-heptanetrione

with phosphorous pentasulfide was initially thought to have the
structure [137,138]:

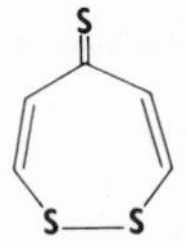

**5.62**

However, studies on X-ray diffraction [139], nmr [140], ir [141], and ESCA
[142] have revealed that the true structure of this compound is **5.63**.
Namely, two vinylic protons are magnetically equivalent and three sulfur

(5.98)

**5.63a**                          **5.63b**

atoms are colinear with equal distance. ESCA displays two sulfur-bands
with intensity ratio of 2:1. Since ylide and ylene structures are also
possible for the sulfur compounds, **5.63c** and **5.63d** can also be included in
the canonical structures [138,143]:

(5.99)

**5.63c**                          **5.63d**

When one of the sulfur atoms is displaced by an oxygen, such a
resonance can no longer be possible [144] and the carbonyl stretching
vibration appears at the normal position in the ir spectrum [141]:

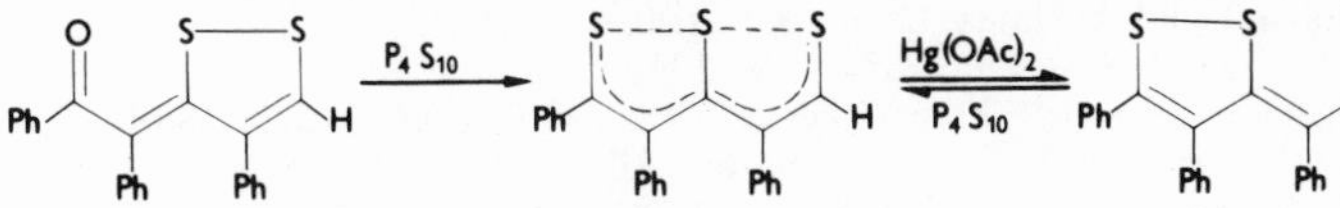

(5.100)

Interestingly, substitution of one of the sulfur atoms by a selenium atom
does not destroy the resonance [145]:

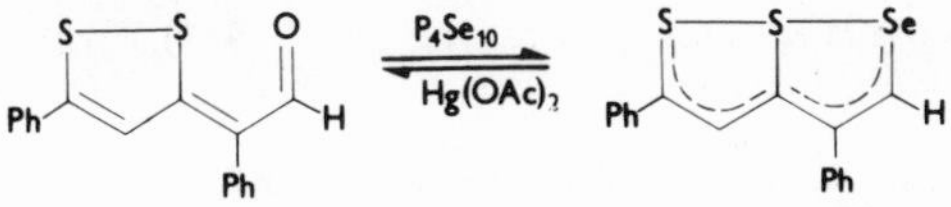

(5.101)

Participation of $3d$ orbitals is considered to be responsible for this
resonance.

## 5.6 REFERENCES

1. T.L. Cottrell, "The Strength of Chemical Bonds," 2nd Ed., (Butterworths Pub., London, 1959), pp. 275–276.
2. L. Pauling, "The Nature of the Chemical Bond," 3rd Ed., (Cornell Univ. Press, Ithaca, N.Y., 1960), p.90.
3. H. Lumbroso and C. Andrieu, *Bull. Soc. Chim. France,* 3201 (1966).
4. S. Scheithauer and R. Mayer, *Chem. Ber.,* **100**, 1413 (1967).
5. U. Schmidt, *ibid.,* **100**, 3825 (1967).
6. J. Morgenstern and R. Mayer, *J. prak. Chem.,* **34**, 116 (1966).
7. J. Fabian and R. Mayer, *Spectrochim. Acta,* **20**, 299 (1964).
8. A. Gero, *J. Org. Chem.,* **19**, 1960 (1954).
9. D. Kunz, S. Scheithauer, S. Bleisch, and R. Mayer, *J. prak. Chem.,* **312**, 426 (1970).
10. A.R. Katritzky and J.M. Lagowski, *Advan. Heterocyclic Chem.,* **1**, 339 (1963).
11. T. Mukaiyama, R. Matsueda, and M. Suzuki, *Tetrahedron Lett.,* 1901 (1970): *cf* also T. Mukaiyama and M. Hashimoto, *Bull. Chem. Soc. Jap.,* **44**, 196 (1971).
12. J. Newman, Jr. and L.B. Young, *J. Phys. Chem.,* **69**, 2570 (1965).
13. G. Klose, Ph. Thomas, E. Uhlemann, and J. Märki, *Tetrahedron,* **22**, 2695 (1966).
14. (a) M. Baldwin, A. Kirkien-Konasiewicz, A.G. London, A. Maccoll, and D. Smith, *Chem. Commun.,* 574 (1966); (b) M. Baldwin, A. Maccoll, A. Kirkien-Konasiewicz, and B. Saville, *Chem. and Ind., (London)* 286 (1966).
15. L. Kaper, J.U. Veenland, and Th. J. de Boer, *Spectrochim. Acta,* (A), **23**, 2605 (1967).
16. O. Korver, J.U. Veenland, and Th. J. de Boer, *Rec. trav. chim. Pays-Bas,* **84**, 310 (1965).
17. A. Schönberg and A. Wagner, in Houben-Weyl: "Methoden der Organischen Chemie," Vol. 9 (1954).
18. R.B. Woodward, *et al., J. Amer. Chem. Soc.,* **82**, 3800 (1960).
19. D.M. McKinnon and J.M. Buchshriber, *Can. J. Chem.,* **49**, 3299 (1971).
20. B.F. Gofton and E.A. Braude, *Org. Synth., Coll.,* **4**, 927 (1963).
21. P.C. Ray, *Nature,* **134**, 1010 (1934).
22. D.C. Sen, *J. Indian Chem. Soc.,* **12**, 647 (1935); *Chem. Zbl.,* **1**, 4916 (1936).
23. S. Bleisch and R. Mayer, *Chem. Ber.,* **100**, 93 (1967).
24. R. Mayer and H. Berthold, *ibid.,* **96**, 3096 (1963).
25. H. Staudinger and H. Freudenberger, *Org. Synth., Coll.,* **2**, 573 (1948).
26. A. Schönberg and E. Frese, *Chem. Ber.,* **101**, 694, 701 (1968).
27. L. Legrand, *Bull. Soc. Chim. France,* 1599 (1959).
28. C. Andrieu, Y. Mollier, and N. Lozac'h, *ibid.,* 2457 (1965).
29. A. Ceccon, U. Tonellato, and U. Miotti, *Chem. Commun.,* 586 (1966).
30. L. Gattermann, *Chem. Ber.,* **28**, 2869 (1895).
31. A. Senning, *Acta Chem. Scand.,* **17**, 2570 (1963).
32. A. Schönberg and W. Asher, *J. Chem. Soc.,* 272 (1942).
33. A. Mustafa and M.K. Hilmy, *Science,* **114**, 526 (1951).
34. W.J. Middleton, E.G. Howard, and W.H. Sharkey, *J. Org. Chem.,* **30**, 1375 (1965).
35. (a) M. Matsui, *Mem. Coll. Sci. Univ. Kyoto,* **B1**, 285 (1908); (b) M. Matsui, *ibid.,* **3**, 247 (1912); (c) Y. Sakurada, *ibid.,* **9**, 237 (1926); (d) Y. Sakurada, *ibid.,* **10**, 79 (1926).
36. U. Schmidt, E. Heymann, and K. Kabitzke, *Chem. Ber.,* **96**, 1478 (1963).
37. A. Ohno, T. Koizumi, and G. Tsuchihashi, *Tetrahedron Lett.,* 2083 (1968).
38. J.M. Lalancette and Y. Beauregard, *ibid.,* 5169 (1967).
39. (a) S.K. Mitra, *J. Indian Chem. Soc.,* **10**, 71 (1933); *Chem. Abstr.,* **27**, 3914 (1933); (b) S.K. Mitra, *ibid.,* **10**, 491 (1933); *Chem. Abstr.,* **28**, 1330 (1934).

40. R. Mayer, S. Scheithauer, and D. Kunz, *Chem. Ber.,* **99,** 1393 (1966).
41. H.J. Hedgley and H.G. Fletcher, Jr., *J. Org. Chem.,* **30,** 1282 (1965).
42. H.E. Wijers, C.H.D. van Ginkel, L. Brandsma, and J.F. Arens, *Rec. trav. chim. Pays-Bas,* **86,** 907 (1967).
43. P.J.W. Schuijl, L. Brandsma, and J.F. Arens, *ibid.,* **85,** 889 (1966).
44. W. Walter and K.D. Bode, *Angew. Chem. Int. Ed.,* **5,** 447 (1966).
45. E.E. Gilbert, E.J. Rumanowski, and P.E. Newallis, *J. Chem. Eng. Data,* **13,** 130 (1968).
46. For reviews, see (a) M. Carmack, *Org. Reactions,* **3,** 83 (1946); (b) J. March, "Advanced Organic Chemistry," (McGraw-Hill Book Co., New York, N.Y., 1968), pp. 910–911 and references cited therein.
47. R. Mayer, "Organosulfur Chemistry," M.J. Janssen, Ed., (Interscience Pub., New York, N.Y., 1967), p. 219.
48. H. Mills and P.L. Robinson, *J. Chem. Soc.,* 2330 (1928).
49. E.J. Corey and R.A.E. Winter, *J. Amer. Chem. Soc.,* **85,** 2677 (1963).
50. For a review, see R. Mayer, J. Morgenstern, and J. Fabian, *Angew. Chem. Int. Ed.,* **3,** 277 (1964).
51. M.L. Bender, *Chem. Rev.,* **60,** 53 (1960).
52. (a) S.G. Smith and M. O'Leary, *J. Org. Chem.,* **28,** 2825 (1963); (b) S.G. Smith and R.J. Feldt, *ibid.,* **33,** 1022 (1968).
53. S. Hauptmann, E. Uhlemann, and L. Widmann, *J. prak. Chem.,* **38,** 101 (1968).
54. J.C. Powers and F.H. Westheimer, *J. Amer. Chem. Soc.,* **82,** 5431 (1960).
55. R.H. Abeles, R.F. Hutton, and F.H. Westheimer, *ibid.,* **79,** 712 (1957).
56. A. Ohno and N. Kito, *Chem. Lett. (Tokyo),* in press: *cf.* also J.J. Steffens and D.M. Chipman, *J. Amer. Chem. Soc.,* **93,** 6694 (1971).
57. R.M. Elofson, F.F. Gadallah, and L.A. Gadallah, *Can. J. Chem.,* **47,** 3979 (1969).
58. G. Sartori and C. Furlani, *Ann. Chim. (Rome),* **44,** 95 (1954).
59. A. Schönberg, A. Rosenbach, and O. Schütz, *Ann. Chem.,* **454,** 37 (1927).
60. E. Campaigne, in "Chemistry of the Carbonyl Group," S. Patai, Ed., (Interscience Pub., New York, N.Y., 1966), pp. 917–959.
61. (a) P. Beak and J.W. Worley, *J. Amer. Chem. Soc.,* **94,** 597 (1972); (b) P. Beak and J.W. Worley, *ibid.,* **92,** 4142 (1970).
62. W.J. Middleton and W.H. Sharkey, *J. Org. Chem.,* **30,** 1384 (1965).
63. W.J. Middleton, E.G. Howard, and W.H. Sharkey, *J. Amer. Chem. Soc.,* **83,** 2589 (1961).
64. Y. Minoura and S. Tsuboi, *Kogyo Kagaku Zasshi,* **70,** 1955 (1967) (in Japanese).
65. (a) D.C. Sen, *Sci. and Cult.,* **2,** 582 (1936); *Chem. Zbl.,* **2,** 972 (1936); (b) D.C. Sen, *ibid.,* **4,** 134 (1938); *Chem. Abstr.,* **33,** 555 (1939); (c) D.C. Sen, *J. Indian Chem. Soc.,* **13,** 268 (1936); *Chem. Abstr.,* **30,** 6340 (1936); (d) D.C. Sen, *ibid.,* **13,** 523 (1936); *Chem. Abstr.,* **31,** 1384 (1937).
66. (a) S.K. Mitra, *ibid.,* **15,** 31 (1938); *Chem. Abstr.,* **32,** 4945 (1938); (b) S.K. Mitra, *ibid.,* **15,** 129 (1938); *Chem. Abstr.,* **32,** 7042 (1938); (c) K. Chandra, N.K. Chakrabarty, and S.K. Mitra, *ibid.,* **19,** 139 (1942); *Chem. Abstr.,* **38,** 4951 (1944).
67. D.C. Sen, *Sci. and Cult.,* **1,** 435 (1935); *Chem. Zbl.,* **1,** 4015 (1936).
68. D.C. Sen, *ibid.,* **1,** 158 (1935); *Chem. Zbl.,* **1,** 4916 (1936).
69. D.C. Sen, *J. Indian Chem. Soc.,* **12,** 751 (1935); *Chem. Abstr.,* **30,** 3806 (1936).
70. Y.G. Gololov and T.F. Dmitrieva, *Zh. Org. Khim.,* **2,** 2260 (1966); *J. Org. Chem. USSR,* **2,** 2219 (1966).
71. P.J.W. Schuijl and L. Brandsma, *Rec. trav. chim. Pays-Bas,* **87,** 929 (1968).
72. E.J. Smutny, *J. Amer. Chem. Soc.,* **91,** 208 (1969).
73. E.J. Smutny and W. Turner, *Tetrahedron,* **23,** 3785 (1967).
74. H.L. Wheeler and B. Barnes, *J. Amer. Chem. Soc.,* **22,** 141 (1899).

75. L. Field and J.D. Buckman, *J. Org. Chem.*, **32**, 3467 (1967).
76. G.G. Urquhart, J.W. Gates, Jr., and R. Conner, *Org. Synth., Coll.* 3, 363 (1955).
77. W. Madelung, *J. prak. Chem.*, **[2]**, **114**, 25 (1926).
78. G. Hahn, *Chem. Ber.*, **62**, 2485 (1929).
79. For a discussion, see E. Campaigne, *Chem. Rev.*, **39**, 1 (1946).
80. G. Laban, J. Fabian, and R. Mayer, *Z. Chem.*, **8**, 414 (1968).
81. D.C. Dittmer and E.S. Whitman. *J. Org. Chem.*, **34**, 2004 (1969).
82. T. Sasaki, K. Kanematsu, and K. Shoji, *Tetrahedron Lett.*, 2371 (1969).
83. Y. Kishida and A. Terada, *Chem. Pharm. Bull. (Tokyo)*, **16**, 1351 (1968).
84. A. Schönberg, L.v. Vargha, and W. Paul, *Ann. Chem.*, **483**, 107 (1930).
85. A. Schönberg and L.v. Vargha, *Chem. Ber.*, **63**, 178 (1930).
86. H.R. Al-Kazimi, D.S. Tarbell and D. Plant, *J. Amer. Chem. Soc.*, **77**, 2479 (1955).
87. D.H. Powers and D.S. Tarbell, *ibid.*, **78**, 70 (1956).
88. See also, H. Kwart and E.R. Evans, *J. Org. Chem.*, **31**, 410 (1966).
89. Y. Araki, *Bull. Chem. Soc. Jap.*, **43**, 252 (1970).
90. Y. Araki and A. Kaji, *ibid.*, **43**, 3214 (1970).
91. D.L.J. Clive, *Chem. Commun.*, 1014 (1970).
92. J.B. Thomson, P. Brown, and C. Djerassi, *J. Amer. Chem. Soc.*, **88**, 4049 (1966).
93. For a review, see G. Opitz, *Angew. Chem.*, **79**, 161 (1967).
94. W.A. Sheppard and J. Diekmann, *J. Amer. Chem. Soc.*, **86**, 1891 (1964).
95. J. Strating, L. Thijs, and B. Zwanenberg, *Rec. trav. chim. Pays-Bas*, **83**, 631 (1964).
96. J. Strating, L. Thijs, and B. Zwanenberg, *Tetrahedron Lett.*, 65 (1966).
97. B. Zwanenberg, L. Thijs, and J. Strating, *Rec. trav. chim. Pays-Bas*, **86**, 577 (1967).
98. B. Zwanenberg, L. Thijs, and J. Strating, *Tetrahedron Lett.*, 3453 (1967).
99. J.F. King, R.A. Marty, P. de Mayo, and D.L. Verdun, *J. Amer. Chem. Soc.*, **93**, 6304 (1971).
100. R.N. Hurd, *Mech. React. Sulfur Compds.*, **3**, 79 (1968).
101. H. Alper and A.S.K. Chan, *Chem. Commun.*, 1203 (1971).
102. F. Arndt and P. Nachtwey, *Chem. Ber.*, **56**, 2406 (1923).
103. C.G. Krespan, B.C. McKusick, and T.L. Cairns, *J. Amer. Chem. Soc.*, **82**, 1515 (1960).
104. H.E. Simmons, D.C. Bolmstrom, and R.D. Vest, *ibid.*, **84**, 4782 (1962).
105. G. Tsuchihashi, M. Yamauchi, and A. Ohno, *Bull. Chem. Soc. Jap.*, **43**, 996 (1970).
106. The cage effect of AIBN is known to be about 35%: S.F. Nelsen and P.D. Bartlett, *J. Amer. Chem. Soc.*, **88**, 137 (1966).
107. G. Tsuchihashi, M. Yamauchi, and A. Ohno, Abstracts, 4th Symposium on Organic Sulfur Chemistry, Gifu City, Japan, Feb. 1970, p.8 (in Japanese).
108. S. Weiner and G.S. Hammond, *J. Amer. Chem. Soc.*, **90**, 1659 (1968).
109. A. Schönberg, B. König, and E. Singer, *Chem. Ber.*, **100**, 767 (1967) and references cited therein.
110. A. Ohno, Y. Ohnishi, M. Fukuyama, and G. Tsuchihashi, *J. Amer. Chem. Soc.*, **90**, 7038 (1968).
111. H. Staudinger and J. Siegwart, *Helv. Chim. Acta*, **3**, 840 (1920).
112. T. Bacchetti, A. Alemagna, and B. Danieli, *Tetrahedron Lett.*, 3569 (1964).
113. W. Reid and B.M. Beck, *Ann. Chem.*, **673**, 124 (1964).
114. C.E. Diebert, *J. Org. Chem.*, **35**, 1501 (1970).
115. O. Korver, J.U. Veenland, and Th. J. de Boer, *Rec. trav. chim. Pays-Bas*, **81**, 1031 (1962).
116. G.N. Lewis and M. Kasha, *J. Amer. Chem. Soc.*, **67**, 994 (1945).
117. P. de Mayo, private communication (1971).

118. (a) A. Ohno, Y. Ohnishi, and G. Tsuchihashi, *Tetrahedron Lett.,* 283 (1969);
   (b) A. Ohno, Y. Ohnishi, and G. Tsuchihashi, *J. Amer. Chem. Soc.,* 91, 5038
   (1969): *cf.* also E.T. Kaiser and T.F. Wulfers, *ibid.,* 86, 1897 (1964).
119. G. Tsuchihashi, M. Yamauchi, and M. Fukuyama, *Tetrahedron Lett.,* 1971
   (1967).
120. A. Ohno, Y. Ohnishi, and G. Tsuchihashi, *Tetrahedron Lett.,* 161 (1969).
121. (a) A. Ohno, Y. Ohnishi, and G. Tsuchihashi, *ibid.,* 643 (1969); (b) A. Ohno,
   T. Saito, A. Kudo, and G. Tsuchihashi, *Bull. Chem. Soc. Jap.,* 44, 1901 (1971).
122. N.J. Turro and P.A. Wride, *J. Amer. Chem. Soc.,* 92, 320 (1970) and references
   cited therein.
123. J.C. Dalton, P.A. Wriede, and N.J. Turro, *ibid.,* 92, 1318 (1970).
124. J. A. Barltrop and H.A. Carless, *Tetrahedron Lett.,* 3901 (1968).
125. H.E. Zimmerman, *Advan. Photochem.,* 1, 183 (1963).
126. (a) A. Ohno, T. Koizumi, Y. Ohnishi, and G. Tsuchihashi, *Tetrahedron Lett.,*
   2025 (1970); (b) A. Ohno, T. Koizumi, and Y. Ohnishi, *Bull. Chem. Soc. Jap.,*
   44, 2511 (1971).
127. A. Ohno, Y. Akasaki, and T. Koizumi, Abstracts, 26th General Meeting of
   Chem. Soc. Jap., Hiratsuka City, Japan, April 1972, p. 1061 (in Japanese).
128. D.R. Arnold, R.H. Hinman, and A.H. Grick, *Tetrahedron Lett.,* 1425 (1964).
129. A. Ohno and N. Kito, *Int. J. Sulfur Chem.,* Part A, 1, 26 (1971).
130. G. Oster, L. Citarel, and M. Goodman, *J. Amer. Chem. Soc.,* 84, 703 (1962).
131. A. Ohno and N. Kito, *Chem. Commun.,* 1338 (1971).
132. T. Yonezawa, M. Matsumoto, Y. Matsumura, and H. Kato, *Bull. Chem. Soc.
   Japan,* 42, 2323 (1969).
133. (a) U. Schmidt and K.H. Kabitzke, *Angew. Chem.,* 76, 687 (1964); (b) U.
   Schmidt, K.H. Kabitzke, I. Boie, and C. Osterroht, *Chem. Ber.,* 98, 3819
   (1965).
134. N. Ishibe, M. Odani, and K. Teramura, *Chem. Commun.,* 371 (1970).
135. A. Ohno, Y. Ohnishi, and G. Tsuchihashi, *Tetrahedron,* 25, 871 (1969).
136. N. Sugiyama, M. Yoshida, H. Aoyama, and T. Nishio, *Chem. Commun.,* 1063
   (1971).
137. F. Arndt, P. Nachtwey, and J. Push, *Chem. Ber.,* 58, 1638 (1925).
138. For detailed discussions, see (a) N. Lozac'h and J. Vialle, in "The Chemistry of
   Organic Sulfur Compounds," Vol. 2, N. Kharasch and C.Y. Meyers, Eds.,
   (Pergamon Press, New York, N.Y., 1966), pp. 276–9; (b) E. Klingsberg, *Quart.
   Rev.,* 23, 537 (1969).
139. (a) S. Bezzi, C. Garbuglio, M. Mammi, and G. Traverso, *Gazz. Chim. Ital.,* 88,
   1226 (1958); (b) S. Bezzi, M. Mammi, and C. Garbuglio, *Nature,* 182, 247
   (1958).
140. (a) A.A. Bothner-By and G. Traverso, *Chem. Ber.,* 90, 453 (1957); (b) H.G.
   Hertz, G. Traverso, and W. Walter, *Ann. Chem.,* 625, 43 (1959).
141. G. Guillouzo, *Bull. Soc. Chim. France,* 1316 (1958).
142. D.T. Clark and D. Kilcast, *Chem. Commun.,* 638 (1971).
143. However, *cf.* also R. Gleiter and D. Schmidt, *ibid.,* 525 (1971), in which 5.63a
   and 5.63b are suggested to be in equilibrium.
144. E. Klingsberg, *J. Amer. Chem. Soc.,* 85, 3244 (1963).
145. J.H. van den Hends and E. Klingsberg, *ibid.,* 88, 5045 (1966).

*Chapter 6*

# SULFIDES

Waichiro Tagaki

*Department of Chemistry, Faculty of Engineering*
*Gumma University, Kiryu, Gumma-Ken, Japan*

## 6.1 INTRODUCTION

Sulfides treated in this chapter are thioethers having divalent sulfur atoms and the general formula RSR, where R may be an alkyl, vinyl, allyl, or aryl group. Mercaptals, $R_1 R_2 C(SR_3)_2$, and orthothioesters, $RC(SR')_3$, are also included. Heteroaromatic sulfides such as thiophenes are not treated, although they may also be considered to be sulfides in a broader sense.

Historically the chemistry of sulfides has been relatively well studied. It may be for this reason that the remarkable progress in the chemistry of organic sulfur compounds of other types has been achieved during the past ten years. Nevertheless the chemistry of sulfides is still important from both theoretical and practical viewpoints. Theoretically, only the divalent sulfur compounds can be compared with those of the first row elements, although they may seem to resemble more closely those of the third. Unfortunately little progress has been achieved during the past ten years in this field. Practically, organic sulfides can be readily prepared from inorganic sulfides which are the most abundant species of sulfur compounds on the earth, and the sulfides are the starting compounds for the preparations of other sulfur compounds, such as sulfoxides, sulfones, and sulfonium compounds, *etc.*

In this chapter is given a brief survey of well-known methods for preparation of sulfides, with greater emphases being placed on the recent advancements of physical organic chemistry.

## 6.2 PREPARATIONS

Most methods for preparation of sulfides seem to have been established before 1960. In the past ten years, fewer methods have been explored for sulfides than for other types of organic sulfur compounds. However, during this period, a variety of sulfides have been prepared, including those which may have been far too difficult to prepare without the recent progress in synthetic organic chemistry. The reader may consult the following books. The books in refs. 1–3 give laboratory directions for the preparation of a number of common organic sulfides. The books in refs. 4–5 compile extensive lists of compounds. In ref. 6 are classified synthetic reactions of sulfides with references to "Organic Synthesis". The books in ref. 7 are excellent reviews which may cover most of the literature up to 1968.

### 6.2.1 Reactions of Thiols

#### 6.2.1.1 Alkylation

The alkylation of thiol may be the most common laboratory method (eq.

6.1)[5]. R and R$'$ may be alkyl or aryl groups, and the alkylating agents,

$$RSH \xrightarrow{\text{NaOH}} RSNa \xrightarrow{\text{R}'\text{X}} RSR' \tag{6.1}$$

R,R$'$ = alkyl, aryl

R$'$X = halides, sulfates, esters of sulfonic acids,

$\alpha$-haloketones, $\alpha$-halosulfides, etc.

R$'$X, may be halides, sulfates, or esters of sulfonic acids. Common RSH and R$'$X are commercially available or easily prepared. The reactions are normally carried out under alkaline condition in polar solvents, and the overall yields usually amount to more than 70%. Aromatic halides are normally unreactive toward thiols in a protic solvent, but reactive enough in dimethylformamide at an elevated temperature [8].

$$C_6H_5Br + C_6H_5SK \xrightarrow[120-130°/5-10hr]{\text{DMF}} C_6H_5SC_6H_5$$
$$64.5\%$$

$t$-Alkyl mercaptans may be alkylated directly by $t$-alcohols in a strong sulfuric acid solution [5]. This method is suitable for the preparation of both symmetrical and unsymmetrical sulfides. Symmetrical sulfides may

$$R_3CSH + R_3'COH \xrightarrow{\text{H}_2\text{SO}_4} R_3CSCR_3'$$

also be prepared in 70–90% yield by refluxing an aqueous alcoholic solution of a halide and metallic sulfide [5]. Five and six membered cyclic sulfides may be prepared by this method using the corresponding dihalides, although the yields for other ring sulfides vary [9,10]:

$$2RX + Na_2S \longrightarrow RSR + 2NaX$$

$$XCH_2CH_2CH_2CH_2X + Na_2S \longrightarrow \left[\begin{array}{c} \\ \text{S} \end{array}\right] + 2NaX$$

Cyclization $via$ thiol esters gives much better yields:

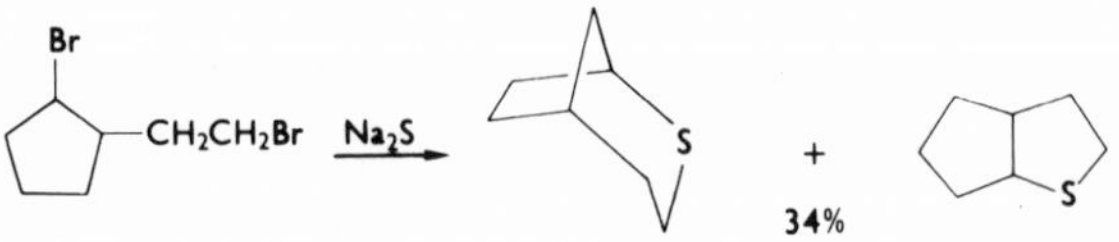

(Ref. 9a)

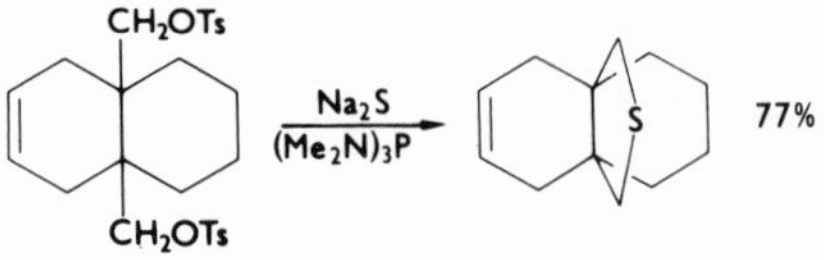

(Ref. 9b)

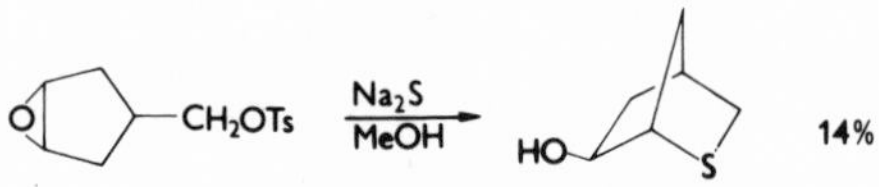

(Ref. 9c)

(Ref. 9d)

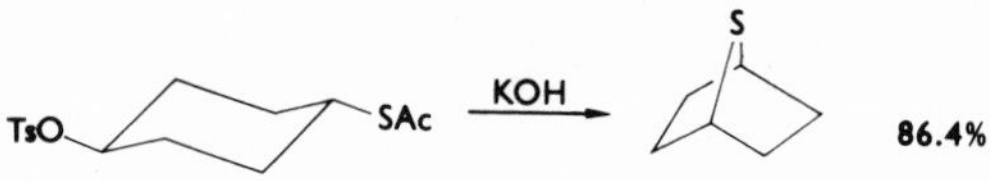

(Ref. 9d)

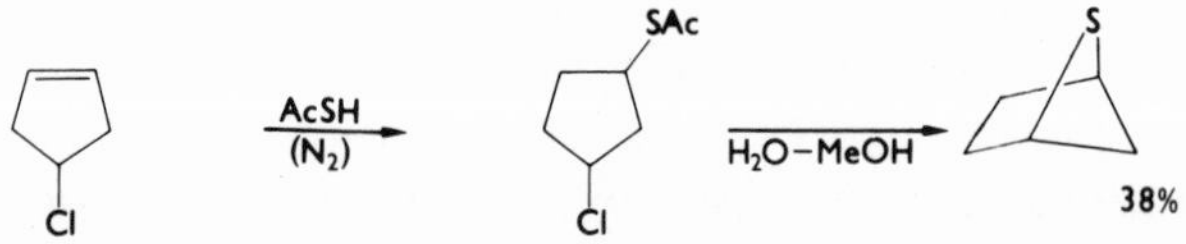

(Ref. 10)

Epoxide rings are readily cleaved by thiols in both acidic and basic conditions to give corresponding β-hydroxy sulfides in good yields (eq. 6.2) [5]. Thiourea or thiocyanates convert the oxides into episulfides at room temperature in aqueous solutions in good yields.

Aromatic diazonium salts may be convenient reagents for the arylation of thiols to give diazosulfides, which on heating may give aryl sulfides (eq. 6.3) [5], although colored materials are sometimes difficult to remove from the products.

$$\underset{O}{\overset{CH_2\text{-}CH_2}{\diagdown\diagup}} \quad \xrightarrow{\begin{array}{c} H_2S \\ \hline RSH \\ \hline KCNS \end{array}} \quad \begin{array}{l} HOCH_2CH_2SH \xrightarrow{\underset{O}{\overset{CH_2\text{-}CH_2}{\diagdown\diagup}}} S(CH_2CH_2OH)_2 \\ \\ RSCH_2CH_2OH \qquad\qquad\qquad\qquad (6.2) \\ \\ \underset{S}{\overset{CH_2\text{-}CH_2}{\diagdown\diagup}} \end{array}$$

$$ArN_2^+X^- \;+\; RSNa \;\longrightarrow\; ArN{=}NSR \;\xrightarrow[-N_2]{\Delta}\; ArSR \qquad (6.3)$$

Phenolic OH group may be replaced by mercaptans under strong acidic conditions (eq. 6.4)[11]. Severe conditions are necessary for simple phenols, but for phenols such as phloroglucinol and $\beta$-naphthol the conditions may be mild and yields are good.

R = CH$_3$, 99%; C$_6$H$_5$, 43%; CH$_2$COOH, 87%

Thiols are known to be readily acylated to give thiol esters. The esters may then be reduced to the sulfides with lithium aluminium hydride [12a]:

$$\underset{CH_3\overset{O}{\overset{\|}{C}}SC_6H_5}{} \;\xrightarrow[BF_3\cdot Et_2O]{LiAlH_4}\; CH_3CH_2SC_6H_5\,,\,80\% \qquad (6.5)$$

Thiolcarbonate esters are known to undergo base-catalyzed reaction to give sulfides [12b].

## 6.2.1.2 Addition

The addition reactions of thiols to multiple bonds are widely used for the preparation of various sulfides [2,5,13,14]. Reactions may occur by either ionic or free radical mechanisms depending on the types of substrates, thiols, and particularly on the catalysts (eq. 6.6). Ionic reactions are catalyzed by acids or bases and normally follow the Markovnikov rule. Free radical reactions are initiated in the presence of peroxides or azonitriles, or by irradiation, and give *anti*-Markovnikov addition products. The yields of sulfides are generally in the range of 60–90% [5].

$$
RCH=CH_2 + R'SH
\begin{cases}
\xrightarrow{H_2SO_4} R\text{-}CHCH_3 \;\; \overset{|}{SR'} \\
\\
\xrightarrow{peroxide} R\text{-}CH_2CH_2\text{-}SR'
\end{cases}
\tag{6.6}
$$

*Addition to Acetylenes.* Under basic conditions, thiols add to acetylenic triple bonds by nucleophilic mechanisms to give vinyl sulfides [14a]. The reactions normally involve stereospecific *trans* addition (eq. 6.7). Similar additions also occur on allenes [15]. The major products correspond to those derived from the addition of thiols to the central carbon.

$$
C_6H_5C\equiv CH + p\text{—}CH_3C_6H_4SH \xrightarrow{Base}
\underset{H}{\overset{C_6H_5}{>}}C=C\underset{H}{\overset{SC_6H_4CH_3(p)}{<}}
$$

$$
H_2C=C=CH_2 + RSH \xrightarrow{Base}
CH_3\underset{SR}{\overset{|}{-}}C=CH_2 + CH=CH-CH_3 + CH_3\underset{SR}{\overset{SR}{-}}C-CH_3
\tag{6.7}
$$

90%

Electrophilic addition of thiols to acetylenic triple bonds is usually difficult and of minor importance for preparative purposes. On the other hand, the free radical addition of thiols to acetylenes proceeds readily to give monoadducts, diadducts, or both, depending on the relative proportions of the starting materials [13,14b]. The *cis/trans* ratio of the monoadducts also depends on the ratio of [thiol]/[acetylene] and the reaction time. The *cis* isomer predominates when an excess of acetylene is used, while the *trans* isomer tends to predominate when the amount of thiol is increased [16]:

$$C_6H_5C\equiv CH \left\langle \begin{array}{c} \underset{RS\cdot}{\overset{C_6H_5}{\diagdown}}C=C\overset{SR}{\underset{H}{\diagup}} \xrightarrow{\quad RSH \quad} \underset{H}{\overset{C_6H_5}{\diagdown}}C=C\overset{SR}{\underset{H}{\diagup}} \quad cis \\ \\ \underset{C_6H_5}{\overset{\cdot}{C}}=C\overset{SR}{\underset{H}{\diagup}} \xrightarrow{\quad RSH \quad} \underset{C_6H_5}{\overset{H}{\diagdown}}C=C\overset{SR}{\underset{H}{\diagup}} \quad trans \end{array} \right. \qquad (6.8)$$

The complexities arise from such factors as the isomerization of the intermediate free radicals and the differences in the rates of their hydrogen atom abstraction. In the formation of the diadduct, the first molecule of thiol adds more readily than the second molecule. The first addition follows *anti*-Markovnikov's rule and the second addition generally yields the adduct with the thio group on the adjacent carbon atom. However, the additions to phenylacetylenes lead to products in which both RS groups are attached to the $\beta$-carbon atoms:

$$RS\text{-}CH=CHR' \xrightarrow{\quad RS\cdot \quad} RS\text{-}\overset{\cdot}{C}H\text{-}\underset{\underset{SR}{|}}{CHR'} \xrightarrow{\quad RSH \quad} RS\text{-}CH_2\text{-}\underset{\underset{SR}{|}}{CHR'}$$

$$CH\equiv CAr \xrightarrow{\quad RSH \quad} RS\text{-}CH=CHAr \xrightarrow{\quad RS\cdot \quad} (RS)_2 CH\text{-}\overset{\cdot}{C}HAr \xrightarrow{\quad RSH \quad}$$

$$(RS)_2 CH\text{-}CH_2 Ar$$

*Additions to Olefins.* Base-catalyzed nucleophilic additions of thiols occur with olefins which contain electron-withdrawing substituents such as methyl acrylate [17]. The orientations are analogous to those observed in Michael type addition reactions:

$$C_2H_5SH + CH_2=CH\text{-}CO_2CH_3 \xrightarrow{\quad Base \quad} C_2H_5\text{-}S\text{-}CH_2CH_2\text{-}CO_2CH_3$$
$$95\%$$

In acidic media, electrophilic addition of thiols to olefins occurs according to Markovnikov's rule. However, the method does not seem to be general for synthetic purpose.

Free radical addition of thiols to olefins occurs extremely readily and an enormous number of various sulfides have been prepared by this method (eq. 6.9) [13]. All types of thiols enter into the reaction and most olefins undergo thiol addition readily. The radicals generally attack at the position which gives the more stable radical forming *anti*-Markovnikov products. Internal olefins with no group present to

$$\begin{matrix} \backslash & / \\ C=C \\ / & \backslash \end{matrix} + RSH \quad \xrightarrow[\text{irradiation}]{\text{free radical initiator or}} \quad RS\text{-}\overset{|}{\underset{|}{C}}\text{-}\overset{|}{\underset{|}{C}}\text{-}H \qquad (6.9)$$

Olefins = linear, cyclic (regardless to the branching and the substituent)

RSH = alkyl, aryl; primary, secondary, tertiary, alkanethiols containing OH, F, Cl, $CO_2H$, COR, *etc.*

stabilize the radical usually give approximately a 1:1 mixture. The radical additions of thiols to non-cyclic olefins is not stereospecific. Apparently there is a rapid equilibrium between the intermediate radicals before the chain transfer step can occur. The addition to cyclopentenes or cyclohexenes is highly stereoselective. For example, the addition of thiophenols to 1-chlorocyclohexene gives 94–99% *cis* isomer, which indicates that the addition occurs largely in the *trans* sense [18]. Addition of *p*-thiocresol to norbornylene gives only *exo*-norbornyl *p*-tolyl sulfide, indicating no Wagner-Meerwein type of rearrangement [19]:

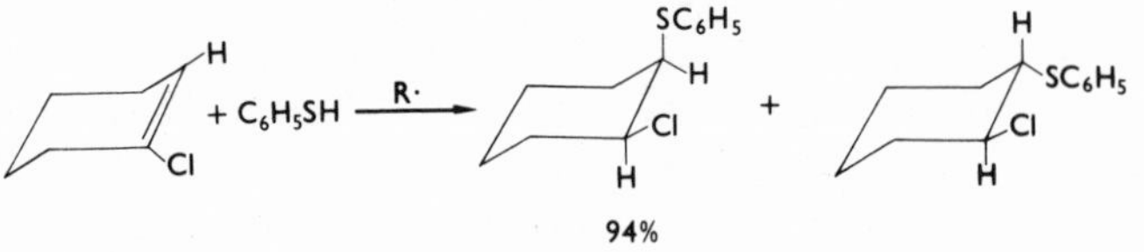

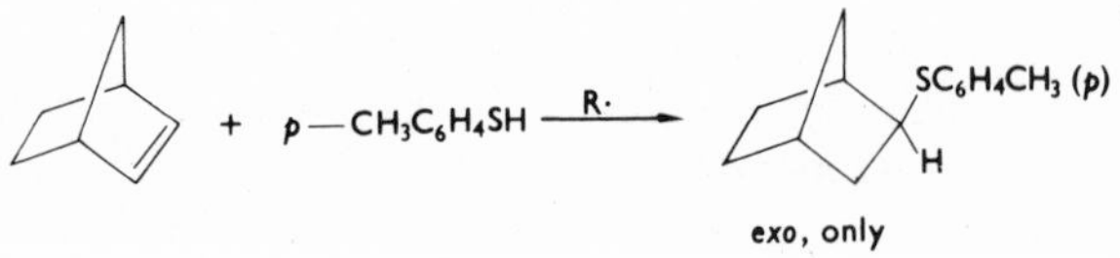

94%

exo, only

### 6.2.2 Addition of Sulfenyl Chlorides to Olefins

Sulfenyl chlorides add to olefins by electrophilic mechanism to give stereospecific *trans* addition products due to the formation of a cyclic sulfonium ion intermediate (eq. 6.10). Thus *trans*-2-butene reacts with 2,4-dinitrobenzene sulfenyl chloride to give the *erythro* adduct [20,21]. Similar *trans* additions have been observed in the reactions of acetylenes [22].

Sulfenyl chlorides also undergo free radical addition reactions. For example, trifluoromethane sulfenyl chloride adds to vinyl chloride to give two 1:1 adducts, in which the orientation of the addition with respect to

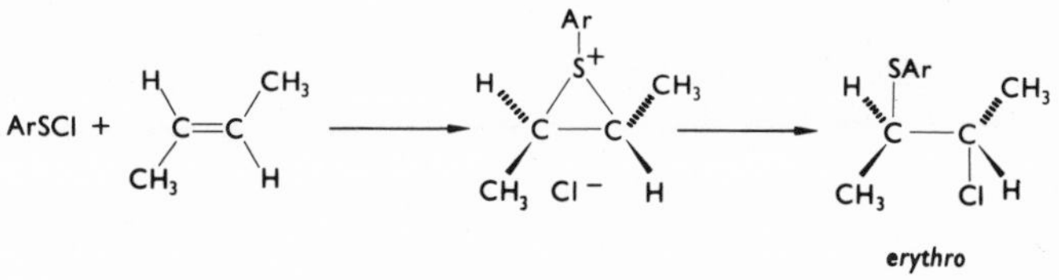

$$\text{ArSCl} + \quad \underset{CH_3}{\overset{H}{\diagdown}} C = C \underset{H}{\overset{CH_3}{\diagup}} \longrightarrow \longrightarrow$$

erythro

(6.10)

the $CF_3S$ group is opposite to that observed in an analogous thiol addition [23]. This is presumably due to the difference in chain-carrier species.

$$CF_3SCl \;+\; CH_2{=}CHCl \xrightarrow{\;R\cdot\;} \underset{\underset{SCF_3\; Cl}{|\quad|}}{CH_2{-}CHCl} + \underset{\underset{Cl\quad SCF_3}{|\quad|}}{CH_2{-}CHCl}$$

5%         95%

$$CF_3SH + CH_2{=}CHCl \xrightarrow{\;R\cdot\;} \underset{\underset{SCF_3}{|}}{CH_2{-}CH_2Cl} + \underset{\underset{SCF_3}{|}}{CH_3{-}CHCl}$$

100%         0%

Electrophilic additions of sulfur dichloride to olefins have opened new synthetic routes to a variety of cyclic sulfides (eq. 6.11) [24,26]. Transannular addition takes place. For example 1,4-cyclohexadiene, 1,3- and 1,5-cyclooctadienes, and norbornadiene give the corresponding bicyclic dichlorosulfides in good yields. Their reductions by metal hydrides give the unsubstituted bicyclic sulfides. The method has also been applied to linear olefins [25] and cyclic polyolefins [26]. Sulfur monochloride reacts similarly with monoolefins to give episulfides [27]. Any new approach to the preparation of episulfides [28] may be important in view of their high reactivities and their uses for the sulfur containing polymers. Sulfur dichlorides also add to acetylenes to give *trans* adducts in good yields [29].

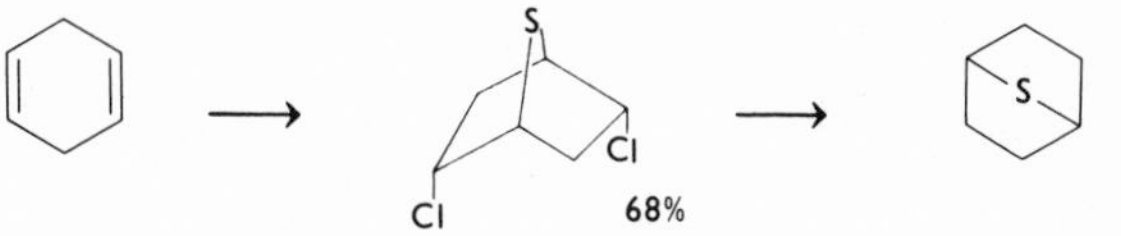

68%

(Ref. 24a, c)

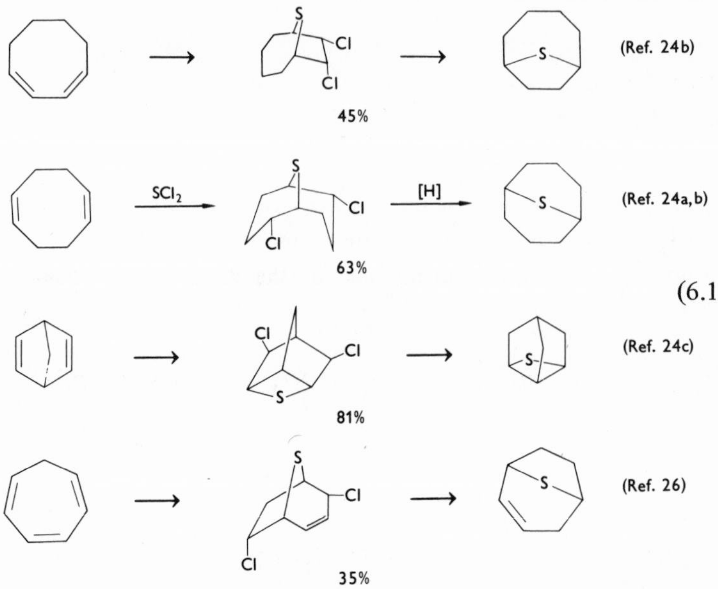

(Ref. 24b)

45%

(Ref. 24a,b)

63%

(6.11)

(Ref. 24c)

81%

(Ref. 26)

35%

### 6.2.3 Cycloaddition

Photochemical reactions of thioketones with olefins give certain cyclic sulfides in good yields. The reactions involve thiyl radicals, as we have seen in the previous chapter.

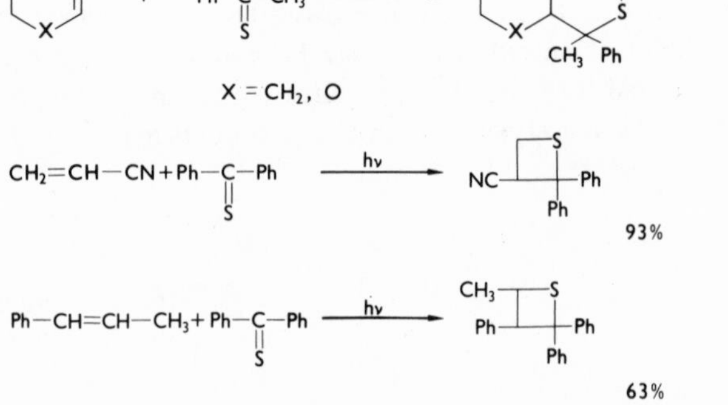

72−96%

X = CH$_2$, O

(6.12)

$CH_2$=CH—CN + Ph—C—Ph $\xrightarrow{h\nu}$

93%

Ph—CH=CH—CH$_3$ + Ph—C—Ph $\xrightarrow{h\nu}$

63%

Irradiation of thiobenzophenone with a high pressure mercury lamp in the presence of olefins gives 1,4-dithianes, while acetophenone gives 1,3-dithianes (eq. 6.12)[30]. With olefins which have an electron-poor double bond or bulky substituents, the monoadducts are the main products. Thioketones also undergo Diels-Alder type addition reactions with dienes. A few examples are shown below[30c,31-34]:

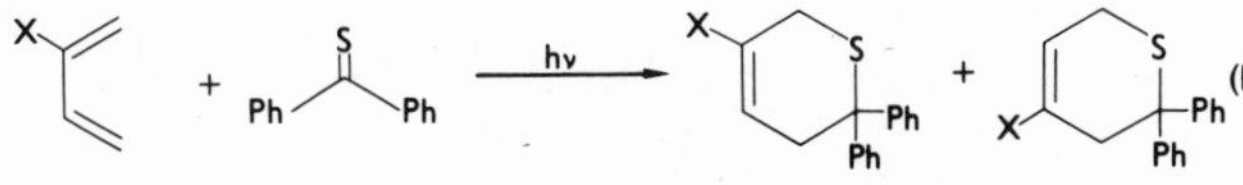


(Ref. 30c , 31)

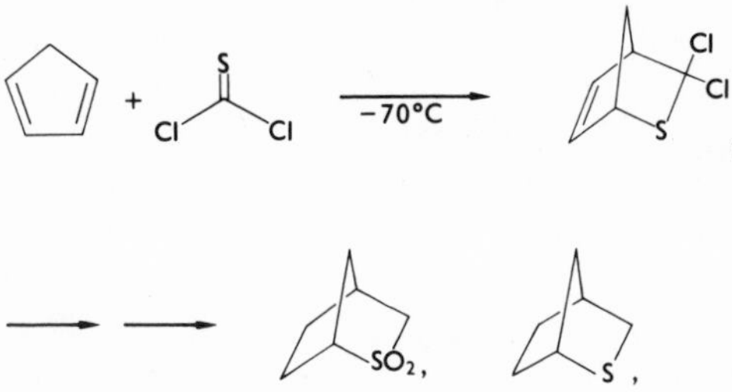


84% (Ref. 32)

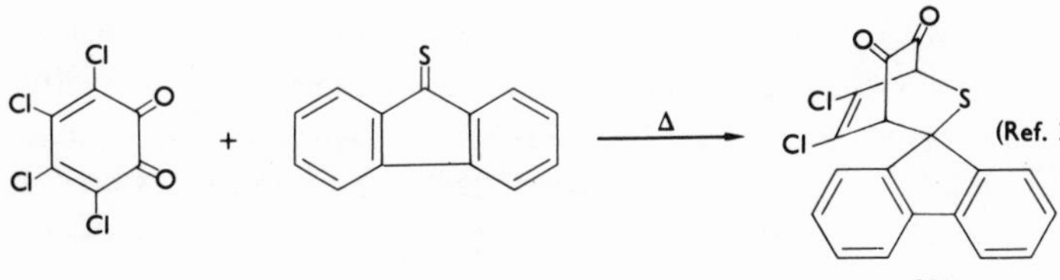


(Ref. 33)

82%

β,β′Diphenyldivinyl sulfide was found to be cyclized to the episulfide containing four-membered ring[35], although divinyl sulfide[35] and bis(1-propenyl) sulfide[36] did not give the cyclization product on irradiation:

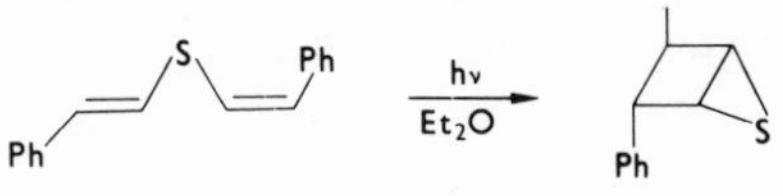


30−34%

### 6.2.4 Cleavage of Disulfides

Organic disulfides are not conventional starting materials for the preparation of sulfides. However, heterolytic or homolytic cleavage of the S—S bond by carbanions or carbon free radicals, or the abstraction of one of the sulfur atoms by certain reagents may be of use for the preparation of some sulfides (eq. 6.13).

Cleavage of disulfide bonds by carbanions is known to give sulfides in such reactions with phenyl lithium[37], lithium acetylides[38], and benzyne[39], *etc.* (eq. 6.13). Grignard reagents also cleave S—Cl bonds[40].

$$C_6H_5Li + C_6H_5SSC_6H_5 \xrightarrow[\text{liq. NH}_3]{} C_6H_5SC_6H_5$$

$$HC{\equiv}CLi + RSSR \longrightarrow HC{\equiv}CSR + RS^-$$

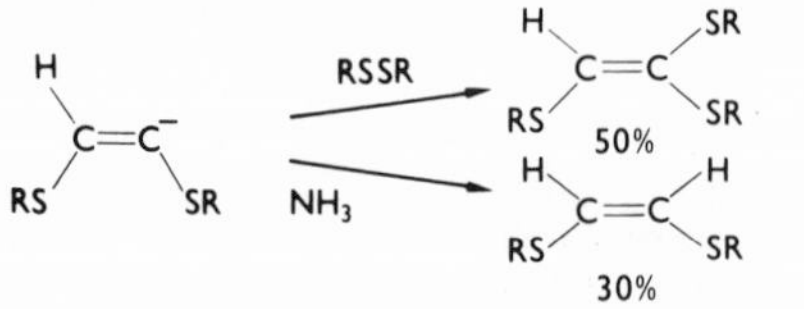

(6.13)

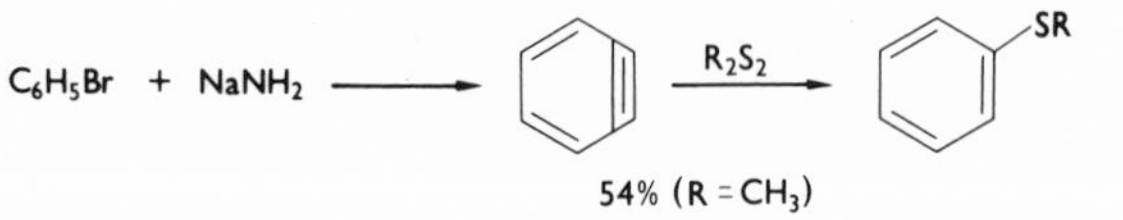

54% (R = CH$_3$)

$$C_6H_5MgBr + CF_3SCl \longrightarrow ArSCF_3 \sim 50\%$$

Cleavages of disulfide bonds by cyanide ion followed by the removal of thiocyanide ion give sulfides[41]. Similar examples are the desulfurization by triphenyl phosphine[42] and triethylphosphite[41]. Among these phosphorous compounds, tri(diethylamino) phosphine seems to be a particularly good reagent for this type of sulfur abstraction reaction[43].

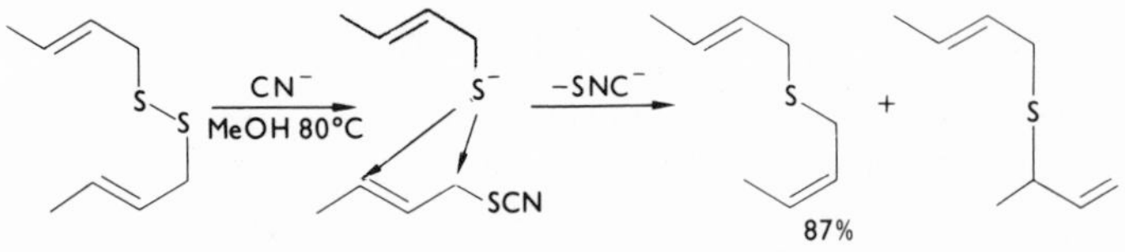

87%

$$RSSR + (C_6H_5)_3P \longrightarrow RSR + (C_6H_5)_3PS, \quad R = \text{allyl group}$$

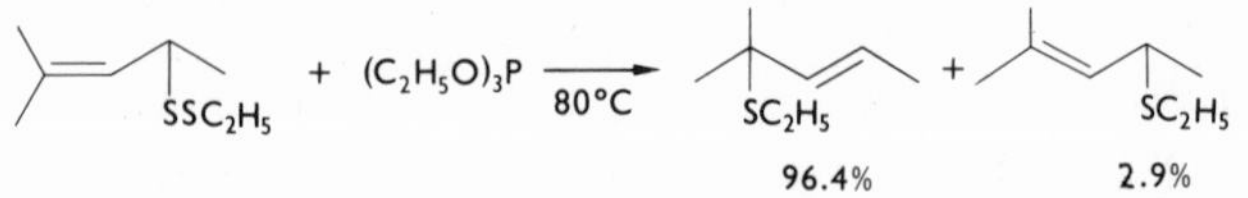

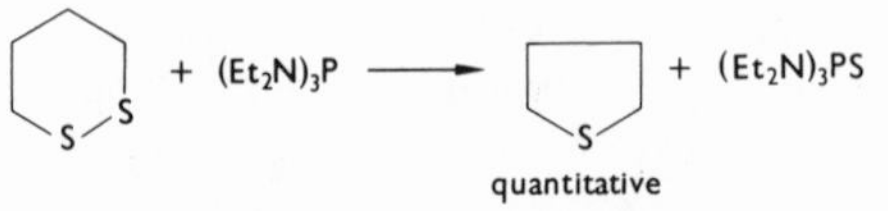

96.4%    2.9%

(6.14)

quantitative

This reagent transforms tetramethylene disulfide into tetrahydrothiophene in a quantitative yield (eq. 6.14)[43a]. This reagent may have a potential utility in the chemistry of peptides and proteins since it also abstracts sulfur atoms from peptide disulfide bonds[43c].

Trialkyl boranes are found to participate readily in free radical chain reactions with organic disulfides[44]. The reactions are initiated by oxygen or light and may provide a convenient synthetic route to a wide variety of alkyl and aryl sulfides (eq. 6.15). For example tri-$n$-butylborane reacts with diphenyl sulfide in $n$-hexane at room temperature to give $n$-butyl phenyl sulfide in 98.5% conversion (based on one butyl group).

$$n-Bu_3B + PhSSPh \xrightarrow[\text{in } n-\text{hexane}]{\text{24hr, room temp}} n-BuSPh \quad 98.5\%$$

$$+ CH_3SSCH_3 \xrightarrow{h\nu} RSCH_3 \qquad (6.15)$$

R = cyclohexyl, 94%; $n$-octyl, 82%

### 6.2.5 Aromatic Sulfides

Aromatic rings may undergo electrophilic or nucleophilic substitution reactions with various sulfur compounds in addition to those reactions shown in eqs. 6.2, 6.3, 6.4, 6.13. Commonly used reagents are sulfur chlorides, elemental sulfur and their analogues.

Diphenyl sulfides may be prepared by Friedel-Crafts type reactions of arenes with sulfur mono- and dichlorides, or with elemental sulfur (eq. 6.16)[45a]. Aluminum chloride is the usual catalyst, but iron powder

$$2C_6H_6 + S_2Cl_2 \xrightarrow{AlCl_3} C_6H_5SC_6H_5 + S + 2HCl$$

$$81\text{-}83\%$$

$$p\text{-CH}_3OC_6H_5 \xrightarrow{Fe} \begin{cases} \xrightarrow{S_2Cl_2} (p\text{-CH}_3OC_6H_4)_2\text{-}S_n \quad n = 1,2,3 \\ \\ \xrightarrow{C_6H_5SCl} p\text{-CH}_3OC_6H_4SC_6H_5 + C_6H_5SSC_6H_5 \end{cases} \qquad (6.16)$$

$$p\text{-ClC}_6H_4OCH_3(p)$$

may be a much better and selective catalyst than $AlCl_3$ [45b]. Phenols are particularly reactive and cleave S–S or S–Cl bonds in the reactions with polysulfides [46], sulfenyl halides [47], or disulfides [48] to give the corresponding *p*-hydroxyaryl sulfides:

$$C_6H_5OH \begin{array}{c} \xrightarrow{\quad S_8 \quad} (p\text{-}HOC_6H_4)_2\text{-}S \\[2ex] \xrightarrow{\quad X\text{-}C_6H_4SCl \quad} X\text{-}C_6H_4SC_6H_4OH(\text{-}p) \end{array} \qquad (6.17)$$

### 6.2.6 Thioacetals, Thioketals, and Orthothioesters

Acid catalyzed exchange reactions of acetals and ketals with thiol generally give the corresponding thioacetals and thioketals (eq. 6.18) [4b,5,49-51]. Aldehydes and ketones may be used [52], but the yields are not always high. The acids suitable for the catalyst are hydrogen chloride, *p*-toluene sulfonic acid, $BF_3$-etherate, zinc chloride or some other Lewis acids. A wide variety of carbonyl compounds and thiols may be used, and the reactions provide very useful means for a number of synthetic purposes, such as the protection of reactive C=O bonds, the conversion of C=O groups into methylene groups by desulfurization, or to utilize the products for another reaction (Sect. 6.5). The yields are generally satisfactory. A few examples are shown below:

$$R_2C(OR')_2 \xrightarrow[\text{acid}]{R''SH} R_2C(SR'')_2$$

$$(6.18)$$

$$R = H, \text{alkyl, aryl}; R' = CH_3, C_2H_5$$

$$C_2H_5CH(OC_2H_5)_2 + C_2H_5SH \xrightarrow[\text{benzene, reflux}]{p\text{-TsOH}} C_2H_5CH(SC_2H_5)_2$$

$$70\% \quad \text{(Ref. 50b)}$$

$$H_2C(OCH_3)_2 + HS(CH_2)_3SH \xrightarrow[\text{CHCl}_3, \text{reflux}]{BF_3(Et)_2O} \begin{array}{c} H \quad S \longrightarrow \\ \diagup\!\!\!\diagdown \\ H \quad S \longrightarrow \end{array}$$

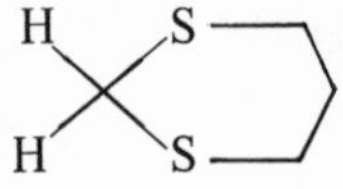

$$85\% \quad \text{(Ref. 51)}$$

*Gem*-dihalides [53] or α-halosulfides may also give either symmetrical or unsymmetrical thioacetals and thioketals according to the manner shown in eq. 6.1.

Orthothioesters may be prepared by the acid catalyzed exchange reactions of carboxylic acid esters or their orthoesters with thiols [5,54,55].

### 6.2.7 Other Methods

In addition to the above methods, various other means for preparing sulfides may be conceivable and some of them may be useful in certain special cases. Chlorination at the α-carbon of sulfides may provide starting materials for other sulfides (eq. 6.44). Methylation followed by substitution or condensation at the α-carbon of sulfides has opened a number of new synthetic routes to a variety of compounds (Sect. 6.5). Dimethyl sulfoxide may also be used since it is a commercial solvent and it undergoes a wide variety of reactions to give sulfides. The reduction of sulfoxides to the corresponding sulfides may be achieved by common reagents (Chapter 8) and also by trichlorosilane [56] and iron cobalt carbonyl [57].

## 6.3 PHYSICAL AND CHEMICAL PROPERTIES

### 6.3.1 Bond Dissociation Energies, Angles, Lengths, and Spectra

Bond dissociation energies are available for several sulfides [2,58], although the values may not be so reliable [58a] as those for the corresponding oxygen compounds (Table 6.1). The bond energies of S–H

TABLE 6.1

Bond Dissociation Energies of Some Divalent Sulfur and Oxygen
Compounds, kcal/mole

| Compound | Energy (Ref.) | Compound | Energy (Ref.) |
|---|---|---|---|
| S–H bonds: | | O–H bonds: | |
| HS–H | 90 (58a) | HO–H | 119 (58a) |
| $CH_3S–H$ | 88 (58a) | $CH_3O–H$ | 102 (58a) |
| | | | |
| C–S bonds: | | C–O bonds: | |
| $CH_3–SH$ | 74 (58b) | $CH_3–OH$ | 91 (58a) |
| $CH_3–SCH_3$ | 73 (58b) | $CH_3–OCH_3$ | 80 (58a) |
| $C_2H_5–SCH_3$ | 72 (58b) | $C_2H_5–OC_2H_5$ | 79 (58a) |
| $i\text{-}C_3H_7–SCH_3$ | 67 (58b) | $i\text{-}C_3H_7–OC_3H_7\text{-}i$ | 79 (58a) |
| $t\text{-}C_4H_9–SCH_3$ | 65 (58b) | | |
| $CH_3–SC_6H_5$ | 60 (58c) | | |
| $Allyl–SCH_3$ | 52 (58b) | $Allyl–OCH_3$ | 52 (58a) |
| $C_6H_5CH_2–SCH_3$ | 52 (58d) | | |

and C—S bonds are less than those of O—H and C—O bonds. The group on the heteroatom such as hydrogen or a simple alkyl group seem to have little effect on both C—S and C—O bond energies. However, aryl, benzyl, and allyl sulfides have much lower C—S bond energies which may be due to the stabilities of the resulting radicals (benzyl, allyl and thiyl).

TABLE 6.2

Bond Lengths and Angles of Sulfides and Ethers

| Compound | Length of C–S or C–O, Å | Angle of C–S–C or C–O–C | Ref. |
|---|---|---|---|
| $(CH_3)_2S$* | 1.82 | 105 | 60a |
| $(CF_3)_2S$ | 1.83 | 105.6 | 60b |
| $(p\text{-}CH_3C_6H_4)_2S$ | 1.75 | 109 | 60c |
| $(p\text{-}BrC_6H_4)_2S$ | 1.75 | 109 | 60d |
| $(C_6H_5)_2S$ | | 113 | 60e |
| $(CH_3)_2O$ | 1.41 | 111.7 | 60f |

* See also Table 6.3, Ref. 61a.

Bond lengths of C—S and C—O bonds for simple alkyl sulfides and ethers are generally close to 1.82 Å (C—S) and 1.41 Å (C—O) respectively (Table 6.2)[2,59,60]. The corresponding bond angles, $\angle$C—S—C and $\angle$C—O—C, are around 105° and 112° respectively. In general, the sulfur bonds have much longer bond lengths and somewhat smaller bond angles than the corresponding oxygen bonds. However, they show substantial variations, depending on substituents, electronic effects, or strains. For example, aryl sulfides have shorter bond lengths and larger bond angles than alkyl sulfides. The $p\pi$-$p\pi$ resonance involving sulfur unshared electrons and the $\pi$-electrons of the aromatic ring is apparently the major contributing factor. Substantial changes for both bond lengths and angles have also been observed by ring formation (Table 6.3)[61]. There appears to occur some compensation between the bond lengths and angles, presumably by ring strains, *i.e.*, a strain induced by ring contraction would be relieved to some extent by bond elongation. However, in the case of episulfide, the C—S bond length is normal in spite of the very small $\angle$C—S—C bond angle[61b], but here again there seems to exist some compensation by shortening of the C—C bond.

A large body of electronic spectra are available for various sulfides[7c,49a,62,65,70].

The spectra of simple dialkyl sulfides show strong bands in the region 200–215 nm ($\log_\epsilon$ ~3) with inflections near 230 nm ($\log_\epsilon$ ~2). The

## TABLE 6.3

Bond Lengths and Angles of Some Cyclic Sulfides

| Compound | Length, Å | | Angle | Ref. |
|---|---|---|---|---|
| | C–C | C–S | C–S–C | |
| | – | 1.802 | 98.9 | 61a |
| | 1.492 | 1.819 | 48.5 | 61b |
| | 1.55 | 1.85 | 78 | 61c |
| | 1.536 | 1.839 | 93.4 | 61d |
| | 1.553 | 1.865 | 69.7 | 61e |
| | 1.549 | 1.837 | 80.1 | 61e |

former bands are thought to be due to the transition of $\sigma - \sigma^*$ of C–S bond and the latter bands to the transition of $n_p-\sigma^*$ involving unshared electron of sulfur [63], although a question remains whether they are due to a single intravalence shell transition (N → V) or to several intervalence shell or Rydberg transitions (N → R) [63b].

The uv absorption bands of cyclic sulfides are shown in Table 6.4 [64]. Inspection of Tables 6.3 and 6.4 reveals that a greater red shift is obtained as the C–S–C bond angle decreases. The order of positions of the weak bands (230 nm) of cyclic sulfides (4 > 3 > 5 > 6) [65a] is also correlated with NMR $\delta$-values for the $\alpha$-hydrogen atom (4 > 3 > 5 > 6) [66], while the order of donor ability of sulfur atom in the complex-formation with Lewis acids is 4 > 5 > 6 > acyclic > 3 (Sect. 6.3.2). From these comparisons, the following types of resonance were suggested for the excited states of the three and four-membered sulfides [65a]:

$$\begin{array}{cc} CH_2 \quad CH_2^- & CH_2-CH_2^+ \\ \backslash\!\!\backslash \quad / & | \\ S & CH_2-S^- \\ + & \\ \mathbf{6.1} & \mathbf{6.2} \end{array}$$

**TABLE 6.4**

UV Spectra of Some Cyclic Sulfides (Ref. 64)

| Compound | $\lambda$max, nm ($\epsilon$) | Compound | $\lambda$max, nm ($\epsilon$) |
|---|---|---|---|
| $(C_2H_5)_2S$ | 229 (139) | | 257 (40) |
| | 229 (183) | | 270 (32) |
| | 235 (167) | | 278 (14) |
| | 239 (87) | | 230 (640) 291 (21) |
| | 239 (54) | | 232 (570) 294 (20) |
| | 242 (43) | | 240 (290) 295 (275) |
| | 247 (43) | | 303 (186) 246 (268) |
| | | | 233 (510) 285–290 (30–40) |

Allyl, benzyl, or $\beta$-keto sulfides have longer wavelength-absorptions than the simple sulfides where unshared electron pairs on the sulfur atom are not directly conjugated with double bond. Koch proposed a non-bonding hyper-conjugative resonance for the stability of the excited states of these sulfides [67]:

$$R\text{-}S\text{-}CH_2\text{-}CH{=}CH_2 \longleftrightarrow R\text{-}\overset{\pm}{S}CH_2{=}CH\text{-}\overset{\mp}{C}H_2 \longleftrightarrow R\text{-}\dot{S} \uparrow CH_2{=}CH\text{-}\dot{C}H_2 \downarrow \quad (6.19)$$

More interesting may be the red shift observed for mercaptals [49a,68,69], and they were explained by Fehnel and Carmack in terms of non-bonding hyperconjugation **(6.3)**[49a]. However since 2,2-dialkyl mercaptals show similar absorptions a direct non-bonding interaction between two sulfur atoms **(6.4)** is more likely to be involved [62,69]. In these types of interactions, the participation of $3d$ orbitals of sulfur are necessarily involved. However, Mangini and co-workers are strongly opposed to the views of the participation of $3d$ orbitals in divalent sulfides based on their extensive studies of uv spectra of aromatic sulfides [70].

**6.3**                                              **6.4**

Nmr $\delta$-values of $\alpha - H$ of sulfides are generally found at about $1-2$ ppm higher field than those of ether $\alpha - H$. They may be determined by the hybridization in C—S bonds or by a sterical arrangement of the unshared electron pairs of the sulfur atom with respect to the C—H bond in question [71].

### 6.3.2 Donor Properties

Basicity or donor strength of sulfur atom of sulfides is an important property and closely related to the reactivities of sulfides. This may be relatively unimportant for other types of sulfur compounds. The donor strength may be disclosed from the study of acid-base equilibria or formation of the charge-transfer complexes. Indirect information may also be obtained from the studies of electronic or nmr spectra.

#### 6.3.2.1 Hydrogen Bonding

Since a sulfide is a Lewis base, it may form hydrogen bond with an acid. However, since such bonding is weak for sulfides, the determination of the site of bonding may not be easy when a compound has more than one donor site.

Hydrogen bonding of sulfur donors with various phenols **(6.5)** has been studied by infrared spectroscopy as well as calorimetry [72]. For nitrogen and oxygen bases, a linear relationship was observed between the calorimetrically determined enthalpy change and the infrared frequency shift caused by hydrogen bonding, while the enthalpy change predicted from this relationship is larger for sulfides than the observed values [72c].

**6.5**

It was also found that the enthalpies of sulfides measured in $CCl_4$ are smaller than those measured in cyclohexane, which indicates a special interaction between sulfur donor and $CCl_4$. It was then proposed that since sulfur is a soft donor relative to oxygen and nitrogen, more electron density is transferred into the proton with a sulfur donor than with the corresponding oxygen and nitrogen donors for a given strength of interaction. In other words, proton becomes saturated with the sulfur donor even with a lower enthalpy change. Donor strength toward iodine and phenol is in reverse order for oxygen and sulfur donors, *i.e.*, the latter interacts more strongly with iodine than with phenol. Table 6.5 shows a few of these data [72c].

TABLE 6.5

Enthalpy Change for the Hydrogen Bonding between Phenol and Ethers, and Sulfides, 25°C (Ref. 72c)

| Donor | Solvent | $-\Delta H$ (kcal/mole) |
|---|---|---|
| $(C_2H_5)_2S$ | $C_6H_{12}$ | 4.6 |
| $(C_2H_5)_2S$ | $CCl_4$ | 3.6 |
| $(n\text{-}C_4H_9)_2O$ | $C_6H_{12}$ | 6.0 |
| $(n\text{-}C_4H_9)_2O$ | $CCl_4$ | 5.7 |
| | $C_6H_{12}$ | 4.9 |
|  | $CCl_4$ | 3.7 |
| | $C_6H_{12}$ | 6.0 |
|  | $CCl_4$ | 5.7 |

Two ir bands each were observed for the interactions of phenol with anisole and thioanisole [72]. A shorter wave length band was assigned as being due to the $\pi$-complex between the proton and the donor benzene ring and a longer one with much stronger intensity was assigned to the interaction between proton and the unshared electron pairs of sulfur. The donor capacity of cyclic donors was found to be cyclic imines $\gg$ cyclic ethers $>$ cyclic thioethers. In each group, the three-membered compounds are the weakest donors. Four-membered cyclic ethers and imines form the strongest hydrogen bonding, whereas the five-membered thioethers are the strongest donors.

## 6.3.2.2 Complexation with Metals

Sulfides are known to form sulfur-metal bonds of various strengths, depending on the types of metals. An enormous number of metal-sulfur chelate compounds are now known. The bonds with alkali metals are generally weak, while the bonds with heavy metals, such as mercury and copper, are strong and sometimes lead to C—S bond cleavage. The bonds of intermediate strengths are formed in the complexation with the halides of Group III metals.

Aliphatic sulfides form relatively strong complexes with trimethylaluminum or the halides of Group III metals (Al, Ga, B) [73,74]. Complexation becomes weak for aromatic sulfides because of the participation of unshared electron pairs of sulfur in $p\pi$-conjugation with aromatic rings. Such a $p\pi$-conjugation can be observed in the uv spectra of anisole and thioanisole in the region of 254—275 nm. By complexation with metal halides the bands shift and the intensities become reduced to those of the parent benzene ring [74]. Table 6.6 shows some data on

**TABLE 6.6**

Heats of Formation of the Complexes of Amines, Ethers, and Sulfides with Aluminum Bromide in Benzene, 25°C (Ref. 74)

| Types of Compound | $-\Delta H$ kcal/mole | Compound | $-\Delta H$ kcal/mole | $\Delta\Delta H$ |
|---|---|---|---|---|
| Aliphatic amines | 32.3[a] | N,N-Dimethyl aniline | 24.2 | 8.1 |
| Aliphatic ethers | 23.3[b] | Anisole | 15.5 | 7.8 |
| Aliphatic sulfides | 17.5[c] | Thioanisole | 13.1 | 4.4 |

Mean values for: [a]four amines; [b]four ethers; [c]nine sulfides.

the heat of formation with aluminum bromide. For both aliphatic and aromatic series, the order of donor strength is amines > ethers > sulfides. The differences in $\Delta H$ values between aliphatic and aromatic compounds are then considered to be a measure for $p\pi$-conjugation involving unshared electron pairs of heteroatom.

Pederson and co-workers have prepared a large number of macrocyclic ethers, amines and sulfides and examined the complex formation with alkali metal and some heavy metal ions [75]. In most cases, 1:1 complexes are formed and an alkali cation is held in the hole of the polyether ring by ion-dipole forces, while the covalent bonding plays a part in the $Ag^+$ complexes of polyethers containing N or S. As an example, the stability constants $(K_1)$ of $K^+$ and $Ag^+$ complexes with three polyethers are compared below (Table 6.7) [75b].

**TABLE 6.7**

Stability Constants of Metal Complex with Macrocyclic Ethers

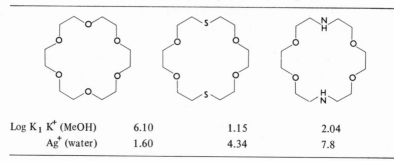

| | | | |
|---|---|---|---|
| Log $K_1$ $K^+$ (MeOH) | 6.10 | 1.15 | 2.04 |
| $Ag^+$ (water) | 1.60 | 4.34 | 7.8 |

## 6.3.2.3 Complexation with Halogens

Sulfides have long been known to form complexes with molecular halogens [76-78]. In some cases stable 1:1 complexes have been isolated.

The charge transfer (CT) spectra have been observed in the complexation of sulfides with iodine in nonpolar solvents [77,78]. The complex formation for sulfides [77] is stronger than for ethers and weaker than for amines (Table 6.8). The trend seems to be parallel with that seen in ionization potentials of sulfides and ethers [80] as observed generally for CT complexes [81].

**TABLE 6.8**

Iodine Complexes of Sulfides, Ethers, and Amines

| Donor | Sol-vent* | K(Temp., °C) 1/mole | $-\Delta H$ kcal/mole | Charge transfer band $\lambda$max (nm) $\epsilon$ | | Ref. |
|---|---|---|---|---|---|---|
| $(C_2H_5)_2O$ | C | 0.97 (20) | 4.3 | 250 | 5,700 | 79a |
| $(C_2H_5)_2S$ | H | 210 (20) | 7.82 | 302 | 29,800 | 77a |
| $(C_2H_5)_3N$ | H | 6,460 (20) | 12.0 | 278 | 25,600 | 79b |
| ⟨S⟩ | C | 70 (25) | 4.6 | | | 77b |
| (S) | C | 135 (25) | 7.1 | | | 77b |
| (S) | C | 182 (25) | 8.7 | | | 77b |

*C = carbon tetrachloride, H = $n$-heptane.

The crystalline metastable sulfide-halogen complexes have been isolated in several cases and their structures have been determined by X-ray crystallographic analysis [82-85]. Such complexes are usually labile, undergo thermal degradation or disproportionation, and are decomposed by moisture. Three types of structures (6.6, 6.7, and 6.8) have so far been proposed for such complexes.

Structure 6.6 is a tetracoordinate trigonal bipyramidal structure analogous to the structure of sulfur tetrafluoride [83]. Structure 6.7 is for a charge transfer type molecular complex, and structure 6.8 is for

6.6                6.7                6.8

a halosulfonium salt. It appears that preference of these structures for a particular complex depends both on the structure of sulfide and the relative strength of electronegativity between sulfur and halogen ligand [84,85]. The stronger the electronegativity of halogen, the more the complex tends to the trigonal bipyramidal structure, while less electronegative iodine favors formation of molecular complex [84]. For example, it has been shown by X-ray studies that the chlorine complex of bis($p$-chlorophenyl) sulfide (6.9) has a structure of type 6.6 [84], while the bromine complex of tetrahydrothiophene (6.10) has a structure of type 6.7 [85].

6.9                              6.10

## 6.4 REACTIONS

The reactions of sulfides may be divided into (1) those taking place directly on a sulfur atom and (2) those occurring at another part of the molecule. The former reactions include the oxidation to give sulfoxides and sulfones and the C—S bond formation to give sulfonium compounds and ylides, etc.; these are treated more in greater detail in Chapters 8,9 and 10. In this chapter, the latter class of reactions are discussed in terms of the effects of sulfide group as compared to those of the corresponding oxygen group. It may be true that the sulfur compounds should be compared with the corresponding selenium compounds rather than with the oxygen compounds because one cannot expect close analogy between a first-row and a second-row element [86]. The present choice is simply due to the abundance of data on oxygen compounds.

### 6.4.1 Effects of Sulfide Group

In considering the effects of sulfide group, one may take into account the facts that (1) the electronegativity of sulfur atom (2.44) is much less than that of oxygen (3.50), (2) the unshared electron pairs of sulfur are in $3s$, $3p$ or in their hybrid orbitals and hence less tightly bound as compared to the unshared electron pairs of oxygen, and (3) $d$-orbitals are available for sulfur. The longer bond length and smaller bond angle of sulfur bonds than the corresponding oxygen bonds may also be important factors.

The reactions involving carbonium ion may be facilitated by the presence of sulfide group due to ion stabilization. Two types of stabilization may be considered: (1) by $\pi$-bonding between sulfur and the $\alpha$-carbonium ion, and (2) by $\sigma$-bonding between sulfur and a carbonium ion generated at a position other than the $\alpha$-carbon. The former $\pi$-bonding is formulated as in the following "electron-releasing conjugation" involving unshared electrons, and this should be more effective for oxygen than for sulfur because of the more diffused nature of the sulfur orbital than the oxygen orbital (eq. 6.20). The latter $\sigma$-bonding (eq. 6.21) may reflect nucleophilicity of heteroatom, and sulfur is known to be a much stronger nucleophile than oxygen. A stable $\sigma$-bonding may result in the formation of sulfonium compounds and a reactive $\sigma$-bonding may result in the neighboring group participation of sulfide group.

$$R-\overset{..}{\underset{..}{S}}\overset{+}{\overset{|}{C}}-\ \longleftrightarrow\ R-\overset{+}{\underset{..}{S}}=C-\qquad\qquad R-\overset{..}{\underset{..}{O}}\overset{+}{\overset{|}{C}}-\ \longleftrightarrow\ R-\overset{+}{\underset{..}{O}}=C-\qquad\qquad(6.20)$$

electron − releasing conjugation

$$R\overset{..}{S}:\diagdown\ \overset{+}{\underset{\boxed{\phantom{xx}}}{}}\qquad\qquad\longrightarrow\qquad\qquad R\overset{+}{\underset{\boxed{\phantom{xx}}}{S}\overset{..}{}}\qquad\qquad(6.21)$$

$\sigma$−bonding

Meanwhile the reactions involving $\alpha$-carbanion may be facilitated by the presence of electron-withdrawing substituents. A strong electronegative group may stabilize the carbanion by an inductive effect. Electron-accepting conjugation involving $d$-orbitals may also stabilize the carbanion, while electron-releasing inductive effect and any repulsion between electron pairs and the carbanion may destabilize the carbanion. For a homolytic reaction, these effects may work in the positive or negative direction, depending on the polarity of the transition state of the reaction. $d$-orbital participation may be formulated by the following "electron-accepting" (eq. 6.22) and "electron-sharing" (eq. 6.23) conjugations for the stabilization of carbanion and free radical, respectively:

$$R\text{-}\ddot{S}\text{-}\overset{\frown}{C}\text{-} \quad \longleftrightarrow \quad R\text{-}\ddot{S}^{-}\text{=}C\text{-} \tag{6.22}$$

electron-accepting conjugation

$$R\text{-}\ddot{S}\text{-}\overset{\ast}{C}\text{-} \quad \longleftrightarrow \quad R\text{-}\dot{S}\text{=}C\text{-} \tag{6.23}$$

electron-sharing conjugation

Whatever the cause, in carbanion forming reactions sulfides are more reactive than the corresponding ethers, and in a number of cases sulfides are also more reactive than ethers in homolytic reactions.

### 6.4.2 Nucleophilic Substitution

$\alpha$-Chloroalkyl sulfides are known to be readily hydrolyzed, although the corresponding oxygen analogues are more reactive. Böhme found that $\alpha$-chloromethyl ether is hydrolyzed faster than the corresponding sulfide by a factor of $10^3$ in aqueous dioxane [87]. A reasonable mechanism proposed by Bordwell is of $S_N 1$ type solvolysis, and the carbonium ion is stabilized by a heteroatom through electron-releasing conjugation, which is more efficient for oxygen than for sulfur atom [88]:

$$R\text{-}S\text{-}CH_2 Cl \xrightarrow{\text{slow}} [R\text{-}\ddot{S}\text{-}\overset{\frown}{C}H_2^{+} \longleftrightarrow R\text{-}\overset{+}{\dot{S}}\text{=}CH_2 ]\, Cl^{-}$$

$$\text{fast} \bigg| H_2 O \tag{6.24}$$

$$\downarrow$$

$$RSH + CH_2 O + HCl$$

It was also suggested that the contribution of resonance types **6.11** and **6.12** is of minor importance based on the Hammett relationship in which no deviation of $p$-$NO_2$ and $p$-$CH_3O$ groups from the linearity was observed [78]. The former intermediate **(6.11)** has two positive charges on adjacent atoms and should be unstable, and the latter

**6.11**                              **6.12**

**(6.12)** involves an expansion of sulfur valence shell which is generally unimportant for divalent sulfide (however, see Sect. 6.4.7). It was also shown that the oxygen compounds are more reactive than the sulfur analogues in $S_N2$ type solvolysis [89]. Similar higher reactivity of oxygen compounds was also observed in the acid catalyzed hydrolysis of alkyl alkenyl ether and sulfide [90]. The mechanism may be formulated as follows:

$$R\text{-}C\equiv C\text{-}\ddot{X}\text{-}R' \xrightarrow[\text{slow}]{H^+} [R\text{-}CH=\overset{+}{C}\text{-}\ddot{X}\text{-}R' \longleftrightarrow R\text{-}CH=C=\overset{+}{\underset{..}{X}}\text{-}R']$$

$$\xrightarrow[\text{fast}]{H_2O} \quad R\text{-}CH_2\text{-}\overset{\overset{O}{\|}}{C}\text{-}\ddot{X}\text{-}R \qquad X = S, O$$

(6.25)

### 6.4.3 Neighboring Group Effects of Sulfide Group

The anchimeric effect by neighboring sulfur has long been known in the solvolysis of esters and/or halides [91,92]. A typical example is the very rapid hydrolysis of $\beta,\beta'$-bis-chlorethyl sulfide (mustard gas):

$$ClCH_2CH_2SCH_2CH_2Cl \longrightarrow \begin{matrix} CH_2 \\ | \\ CH_2 \end{matrix} \rangle \overset{+}{S}CH_2CH_2Cl \longrightarrow \begin{matrix} CH_2OH \\ | \\ CH_2 \\ | \\ SCH_2CH_2Cl \end{matrix} \quad (6.26)$$

Kinetic studies are consistent with a mechanism involving the rate-determining cyclization step to form the ethylenesulfonium ion [93,94] which is too reactive to permit isolation. As shown in Table 6.9 [95-98], those chlorides which on cyclization would give three-, five-, or six-membered sulfonium rings show definite rate enhancement, and the latter two intermediates have actually been isolated [96,97].

TABLE 6.9

Solvolysis of Chloroethers and Chlorosulfides* (Relative Rates: $n$-hexyl chloride = 1)

| R | R' | R–O–R' [95] Aqueous Dioxane, 100°C | R–S–R' [95] Aqueous Dioxane, 100°C | R–S–R' [96-98] 50% Acetone 80°C |
|---|---|---|---|---|
| $C_2H_5$ | $(CH_2)_2Cl$ | 0.21 | 3200 | – |
| $C_2H_5$ | $(CH_2)_3Cl$ | 0.91 | 1.2 | – |
| $C_2H_5$ | $(CH_2)_5Cl$ | – | – | 334 |
| $C_2H_5$ | $(CH_2)_6Cl$ | – | – | 3.5 |
| $C_6H_5$ | $(CH_2)_2Cl$ | 0.11 | 104 | – |
| $C_6H_5$ | $(CH_2)_3Cl$ | 0.43 | 0.55 | – |
| $C_6H_5$ | $(CH_2)_4Cl$ | 0.98 | 21 | 1900 |
| $C_6H_5$ | $(CH_2)_5Cl$ | – | – | 25 |

*Taken from Ref. 92a, p. 109.

Ethyl sulfides are more reactive than the phenyl sulfides, and the electron-releasing substituents on the phenyl ring increase the rates [99]. Thus the magnitude of the rate enhancement appears to depend on the nucleophilicity of sulfur atom as well as the ring size of the intermediate sulfonium ion. Lower reactivities of the ethers may reflect the smaller nucleophilicity of ether oxygen atom.

The effects of neighboring sulfur group have also been examined in more rigid systems. It has been shown that the transformation of substituted $syn$-benzaldoximes to the corresponding nitriles is accelerated by $ortho$ substituents [100]. The $ortho$-$para$ rate ratio $(k_o/k_p)$ lies in a range of 2–9 for X = F, Cl, Br and $CH_3O$; 119 for X = I; and 11,000 for X = $CH_3S$; it has been suggested that $o$-iodo and $o$-$CH_3S$ groups exert anchimeric assistance in the rate-determining fission of the N–O ester bond (6.13). A more dramatic $k_o/k_p$ ratio has also been observed in

$$(6.27)$$

6.13

the acetolysis of substituted benzyl chloride [101]. $o$-Dithiacyclopentyl group enhances the rate by a factor of $10^5$ as compared to that in the $para$ isomer. However, $k_o/k_p$ value is only 11 for $o$-$CH_3S$ group. In the former case, the carbonium ion is stabilized by the participation of two sulfur atoms involving stable five-membered ring (6.14), while in the latter case only one sulfur atom is involved with rather unstable four-membered ring (6.15). It becomes more apparent in rigid systems than in flexible systems that the neighboring group participation is quite sensitive to the proximity and the stereochemistry of the overlapping orbitals of unshared electrons of sulfur atom and the developing vacant $p$-orbital of carbonium ion center.

acetolysis rate, $k_O/k_P$     130,000           11             **6.14**           **6.15**

    The rates of solvolysis of *cis-* and *trans*-2-chlorocyclohexyl phenyl sulfide and *cis-* and *trans*-2-chlorocyclopentyl phenyl sulfide in 80% aqueous ethanol have been determined. In both cyclohexyl and cyclopentyl systems the *trans* isomer, in which the neighboring sulfide group can participate in the ionization step, is more reactive than the corresponding *cis* isomer by factors of $10^5$ to $10^6$ [102]. No such

anchimeric assistance was observed in the following bicyclic systems [103]. Presumably the rigidity of the system prevents coplanarity of *trans* carbon-chlorine and carbon-sulfur bonds. Perhaps the $C_7$-methylene group (6.16) and the benzene ring (6.17) prevent such *trans* participation. Transannular participation was also not observed in the

**6.16**                         **6.17**

following sulfur bridged bicyclic tosylate of seven-membered ring [104]. Flexible ring tautomerism may favor the conformation 6.18b for substitution rather than the conformation 6.18b where both the back side attack of nucleophile and the participation of sulfur bridge are hindered

**6.18a**                    **6.18b**

sterically while the participation should involve an unfavorable four-membered sulfonium ring (see also **6.15** and Table 6.9). Meanwhile in the acetolysis of compounds **6.19a–f**, a ring enlargement from five to six (**6.19d** and **19f**) results in a rate enhancement by a factor of $10^3$ [105]. Since the ionization of **6.19b** itself is estimated to be accelerated by a factor of $10^4$ as compared to that of an unassisted system, the rate of solvolysis of **6.19f** is estimated to be about $10^7$ faster than it would be if the transition state resembled a simple localized cation. A four-membered sulfonium ring **(6.20)** was then suggested to be involved. In other systems, there is no evidence to support such sulfur participation involving a four-membered ring. Presumably an enlargement of a bridging ring from 5 to 6 causes the sulfur and the carbonium ion center to be in closer proximity. Thus in **6.20**, all three conditions for an effective neighboring group participation, *i.e.*, favorable stereochemistry, proximity of the overlapping orbitals, and unstable nature of the intermediate, appear to be satisfied.

| k(rel.) | 1.0 | 4.9 | 8.4 | 32.2 |
|---------|-----|-----|-----|------|
|         | **6.19a** | **6.19b** | **6.19c** | **6.19d** |

|  | 1.8 | 5,564 |  |
|--|-----|-------|--|
|  | **6.19e** | **6.19f** | **6.20** |

The remote sulfur group participation has also been observed in the solovolysis of *endo*-4-thiatricyclo [5,2,1,0,$^{2,6}$] dec-8-yl *p*-nitrobenzoate **(6.21a)** [106]. About $10^3$ fold difference exists between the rates of the *endo* **(6.21a)** and the *exo* **(6.21c)** isomers as in the case of **6.19d** and **6.19f**. However, the only solvolysis product of **6.21a** is the *endo*-8-ol and none of the rearranged alcohol **(6.22)** is present. This may rule out the possibility of σ-bond formation since intervention of sulfonium ion (6.23) should lead to the formation of some 6.22.

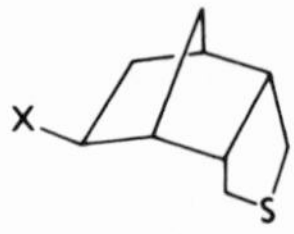

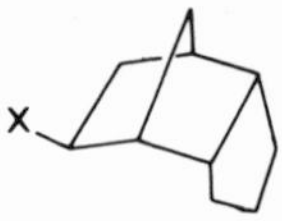

| k (rel.) | 752 | 1.0 |
|---|---|---|
| | 6.21a | 6.21b |

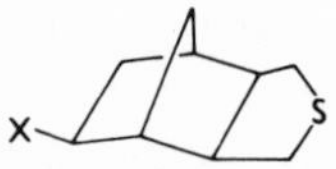

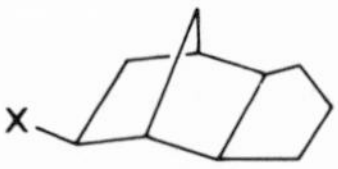

0.8
6.21c

0.7
6.21d

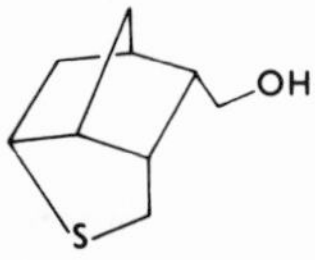

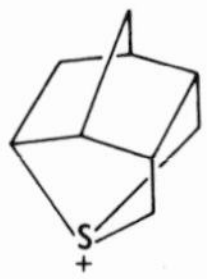

6.22

6.23

$X = p{-}O_2NC_6H_4CO_2$

### 6.4.4 Substitution with C—S Bond Cleavage

The C—S bond of sulfide is generally stable toward nucleophilic reagents. However when sulfur atom donates electrons to an electrophilic reagent the C—S bond of aliphatic sulfides becomes susceptible to the cleavage.

Most sulfides are resistant to the proton catalyzed C—S bond cleavage as expected from the rather poor donor ability of sulfur atom toward proton (Sect. 6.3.2), except for the thio-analogues of acetals, ketals and orthoesters.

The hydrolysis of thioacetals have been investigated as the model reactions for the hydrolysis of thioglycosides [107,108]. The rates of acid-catalyzed hydrolysis of oxygen compounds were found to be much greater than those of the corresponding sulfur analogues [108a,b]. For a

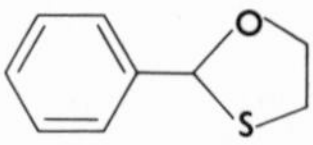

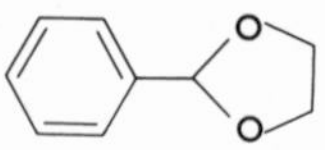

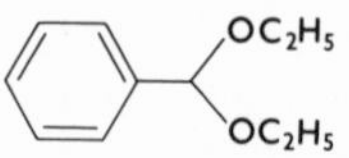

| k(rel.) | 1·0 | 380 | 20,316 |
|---|---|---|---|

mixed acetal, it was suggested that the C—S bond is first cleaved in the rate limiting step to form oxygen stabilized carbonium ion (eq. 6.28)[108c,d]. Consistent with this is the fact that hemithioacetals are generally formed

$$
\begin{array}{c}
\underset{\substack{| \\ S\text{-}C_6H_4\text{-}X}}{\overset{\substack{OCH_3 \\ |}}{C_6H_5\text{-}CH}}
\;\underset{}{\overset{H^+}{\rightleftharpoons}}\;
\underset{\substack{| \\ S^+C_6H_4\text{-}X \\ H}}{\overset{\substack{OCH_3 \\ |}}{C_6H_5\text{-}CH}}
\;\xrightarrow[\text{slow}]{-XC_6H_4SH}\;
C_6H_5CH^+\!-OCH_3
$$

$$
\Big\downarrow\; H_2O \quad \text{fast} \quad (6.28)
$$

$$
C_6H_5\text{-CHO}
$$
$$
+
$$
$$
CH_3OH
$$

much more readily and are much more stable than the corresponding hemiacetals[109]. An alternative rate limiting C—O bond cleavage may not be ruled out for other systems. When oxygen is substituted by nitrogen, the C—N bond breaking becomes preferential because of the predominant protonation on nitrogen as suggested by the following example[110]:

$$
ArSH + CH_2O + Ar\text{-}NH_2 \longrightarrow [Ar\text{-}S\text{-}CH_2\text{-}NH\text{-}Ar] \xrightarrow{\;H^+\;}
$$
$$
(6.29)
$$
$$
H_2N\!-\!\!\left\langle\!\!\bigcirc\!\!\right\rangle\!\!-\overset{+}{C}H_2SAr
$$

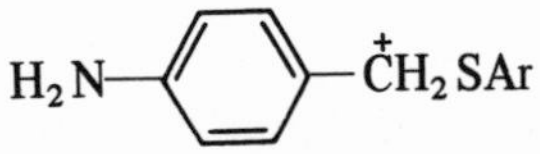

An acid catalyzed C—S bond cleavage may also take place when the resulting cation is stable[111]:

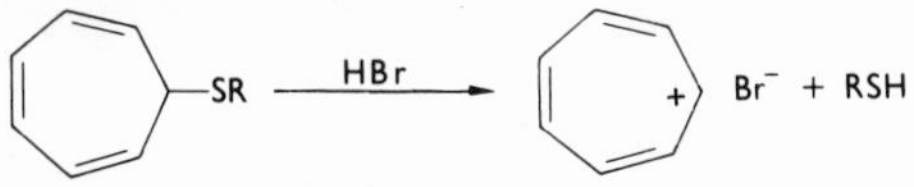

$$
(6.30)
$$

Chlorination of sulfides with various chlorinating agents is known to give relatively good yields of the corresponding α-mono-, α-di- or α-trichlorosulfides, depending on the reaction conditions (eq.

6.31)[87,112]. The reaction is considered to involve chlorosulfonium salts (**6.24**) as the intermediates which then undergo Pummerer type rearrangement to result in α-chlorination. Meanwhile analogous bromination of sulfides often results in C—S bond cleavage.

$$C_6H_5SCH_3 + SO_2Cl_2 \xrightarrow{-SO_2} [C_6H_5\overset{+}{\underset{}{S}}CH_3Cl^-] \quad \textbf{6.24} \qquad (6.31)$$

$$\xrightarrow{-HCl} C_6H_5SCH_2Cl \xrightarrow{SO_2Cl_2} C_6H_5SCHCl_2 \xrightarrow{SO_2Cl_2} C_6H_5SCCl_3$$

Brominolysis of episulfide[113-115] and of thiethane[116] leads to acyclic disulfides through sulfenyl bromide intermediates (eqs. 6.32, 6.33)[117]. Tetrahydrothiophene reacts with bromine in an

$$RCH\overset{S}{\overset{\diagup\diagdown}{-}}CH_2 + Br_2 \xrightarrow{CHCl_3} \left[ \underset{Br}{\overset{|}{RCHCH_2SBr}} \right] \quad RCH\overset{S}{\overset{\diagup\diagdown}{-}}CH_2 \longrightarrow \qquad (6.32)$$

$$(\underset{Br}{\overset{|}{RCHCH_2S}})_2 \quad 63-100\%$$

$$(6.33)$$

$$(BrCH_2CH_2CH_2S)_2$$

unexpected manner to form 3-bromo-3-butenyl-1-sulfenyl bromide (eq. 6.34)[118]. α-Bromination of methyl sulfide accompanies the C—S bond cleavage, where isolation of intermediate sulfonium salt may or may not be possible (eq. 6.35)[119]. The C—S bond cleavage occurs in the

$$(6.34)$$

reaction of aliphatic sulfides with NBS in carbon tetrachloride[120-122]. When the reaction is carried out in water certain sulfides give high yields of sulfoxides[123]. The reaction of dibenzyl sulfide with NBS may illustrate the characteristic features of the general reactions of sulfides with halogenating reagents (eq. 6.36)[122].

$$(CH_3)_2S + Br_2 \xrightarrow[0°]{CHCl_3} \underset{\text{isolated}}{(CH_3)_2\overset{+}{S}Br\overset{-}{Br}} \xrightarrow[\text{Reflux, 30hr}]{CHCl_3} \underset{16\%}{CH_3SCH_2Br}$$

$$+ CH_3Br$$
$$48\%$$

(6.35)

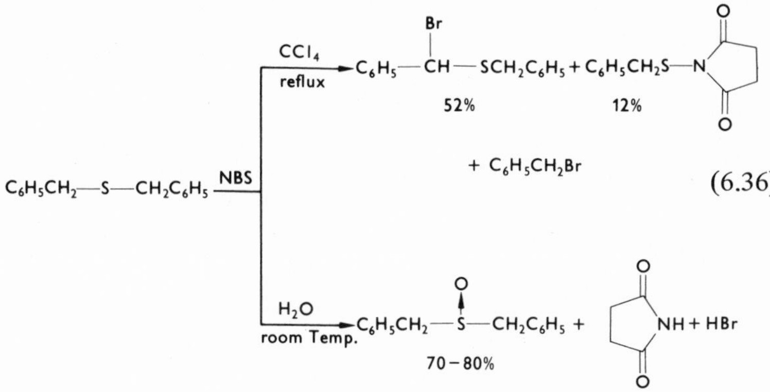

(6.36)

Aluminum chloride is known to cause C—S bond fission and the rearrangement of diaryl sulfides (eq. 6.37)[124-126]. Complete scrambling of [14]C labeled at the $C_1$—S position takes place[126].

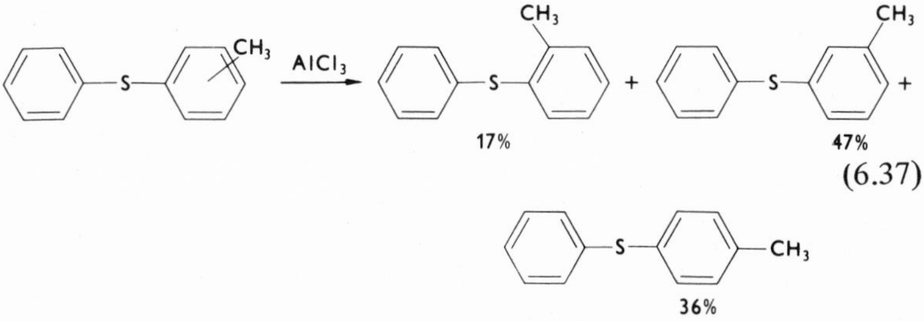

(6.37)

Heavy metal ions are also known to catalyze the cleavage of C—S bond in solvolysis, reduction, or desulfurization as will be described later (Sect. 6.5).

## 6.4.5 Electrophilic Aromatic Substitution

Bromination of methyl phenyl sulfide leads to the formation of sulfur-bromine complex which on further reaction results in the bromination at the *para* position of the benzene ring (eq. 6.38)[127]. Similar bromination at *para* position was also observed for other sulfides [128]. However the bromination of o- and p-methoxyphenyl

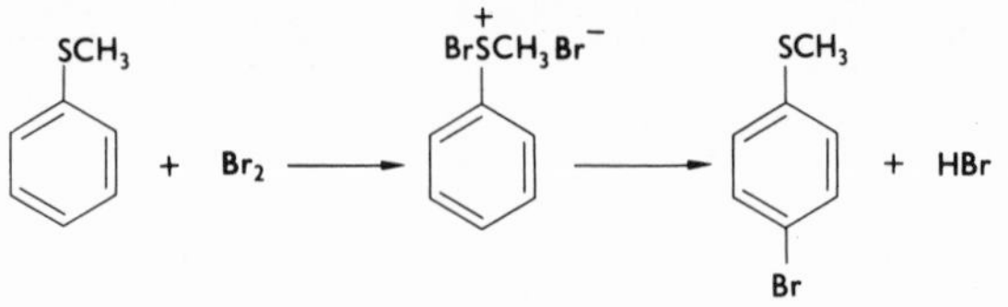

$$+ \quad (6.38)$$

methyl sulfide was shown to give *meta* bromination with respect to the mercapto group [129]. Even p-tolyl methyl sulfide undergoes *meta* bromination with respect to the mercapto group [130]. The latter observation appears to contradict the results observed in the solvolysis of α-halosulfides where sulfide group is far more effective for the resonance stabilization of α-carbonium ion than alkyl group. Presumably the formation of sulfur-bromine complex (6.25) [127] reduces the *ortho* and *para* directing power of neutral mercapto group. The sulfonation of o-methoxyphenyl methyl sulfide also was known to occur at the *para* position of methoxy group [131].

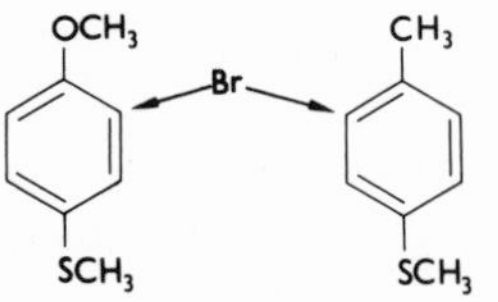

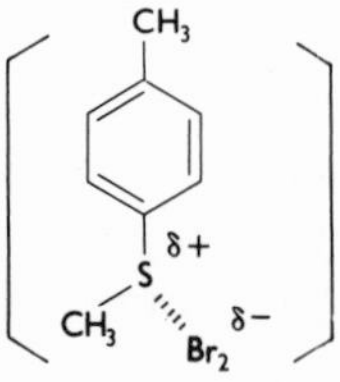

**6.25**

An acid-catalyzed H–D exchange reaction of thioanisole has been examined in an acetic acid–sulfuric acid mixture (Table 6.10) 9132. The relative rates for deuterobenzene, p-deuterothioanisole, p-deuteroanisole are 1,50, 2000, respectively. The corresponding *ortho* isomers were found to behave similarly [133].

**Table 6.10**

Relative Reactivities in Acid-Catalyzed H–D Exchange

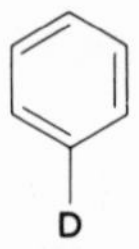

| k(rel.) in AcOH:$H_2SO_4$ = 50:1 (v/v) | | |
|---|---|---|
| 1 | 50 | 2,000 |

In these cases, the protonation of heteroatoms is not so important as to alter their normal orientation effects.

### 6.4.6 Structure and Reactivity

The electrophilic substituent constant, $\sigma_p^+$, values have been obtained for $CH_3O$ (−0.778), $CH_3S$ (−0.604) and $CH_3$ (−0.311) groups in the solvolysis of $t$-cumyl chloride (eq. 6.39)[134], *i.e.*, the electron releasing

$$(6.39)$$

conjugation of sulfide group with the developing carbonium ion through the benzene ring is less than that of the ether group. Similar trends were observed in the cyanhydrin formation of substituted benzaldehydes (eq. 6.40)[135] and also in the acid dissociation of substituted benzoic

$$(6.40)$$

$$Z = O, S, Se$$

acids[135,136]. Increasing order of the stability of cyanhydrin is O<S<Se as reflected in the same order of $K_p/K_m$ ratios. The major contributing factor may be the stabilization of aldehyde relative to the adduct which is in the order of O>S>Se. The acidities of $p$-methoxy and $p$-methylthiobenzoic acids are lower than those of the corresponding $m$-isomers. Here again the stabilization of the undissociated acid relative to the dissociated anion by the electron-releasing conjugation is important.

The effects of oxygen and sulfur groups in the acid dissociation of phenols and anilines are different from those observed in the above cases[136,137]. $p$-Methoxy group reduces the acidities of both phenol and aniline. Electron-releasing conjugation of methoxy group may hinder the cross conjugation of the dissociated anion with the benzene ring (Table 6.11). Meanwhile the *p*-methylthio group exerts acid strengthening effect.

**TABLE 6.11**

pK$_a$ Values (25°C) of *p*-Methoxy and *p*-Methylthio-Anilines and -Phenols

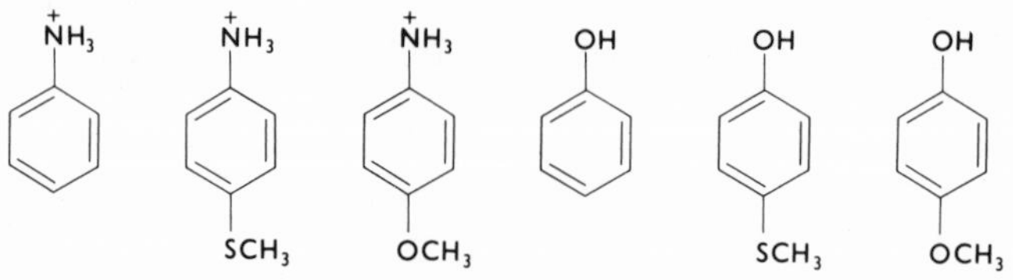

| $\overset{+}{N}H_3$ | $\overset{+}{N}H_3$ | $\overset{+}{N}H_3$ | OH | OH | OH |
|---|---|---|---|---|---|
| | SCH$_3$ | OCH$_3$ | | SCH$_3$ | OCH$_3$ |
| 4·58 | 4·40 | 5·29 | 9·95 | 9·53 | 10·20 |

This does not necessarily mean the absence of electron-releasing conjugation (acid weakening) of the sulfide group with the benzene ring. Such a conjugation may become apparent by looking at the steric hindrance. In the following example, Table 6.12, introduction of methyl

**TABLE 6.12**

pK$_a$ Values (25°C) of Hindered Phenols in 50% EtOH – H$_2$O

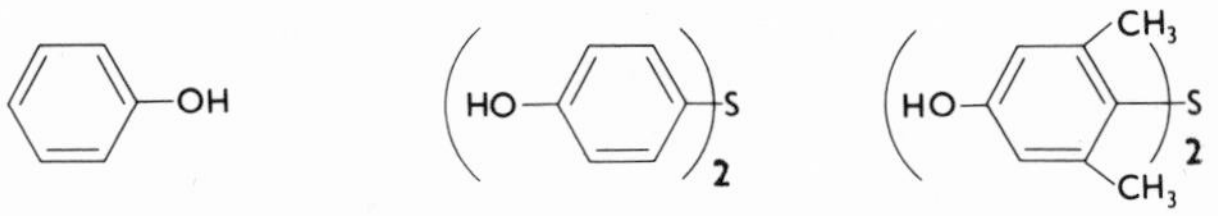

| ⬡—OH | (HO—⬡)$_2$S | (HO—⬡(CH$_3$)(CH$_3$))$_2$S |
|---|---|---|
| 11.22 | 10.28 | 9.43 |

group at the *o*-position of sulfur atom enhances the acidity of 4,4′-dihydroxydiphenyl sulfide in spite of the fact that the *m*-methyl group generally reduces the acidity of phenol[138]. A question may then arise for the origin of this acid strengthening effect of sulfide group; it is considered to be due to the fact that sulfide group can stabilize negative charge when connected with a conjugated system, as will be discussed later.

Electron-releasing conjugation of sulfide group with a conjugated system may affect the dipole moment of a molecule. For example, the dipole moment of nitrobenzene is substantially changed by the presence of *p*-methoxy or *p*-methylthio group[139].

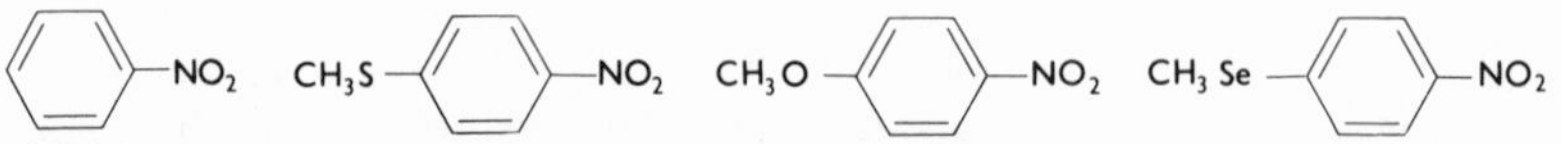

| ⬡—NO$_2$ | CH$_3$S—⬡—NO$_2$ | CH$_3$O—⬡—NO$_2$ | CH$_3$Se—⬡—NO$_2$ |
|---|---|---|---|
| **4.01D** | **4.36D** | **4.86D** | **4.38D** |

CH₃S —⟨benzene⟩          CH₃O —⟨benzene⟩          CH₃Se —⟨benzene⟩

**1.18D**                    **1.28D**                    **1.31D**

The polar substituent constant, $\sigma^*$, values of sulfide groups have been determined for the hydrolysis of $\alpha$-substituted ethyl acetates (Table 6.13) [140]. Among the $\sigma^*$ values for the mono- and disubstituted esters,

**TABLE 6.13**

$\sigma^*$ and Es Values for the Hydrolysis of $XCH_2CO_2C_2H_5$ and $XYCHCO_2C_2H_5$ in 50% EtOH-water (v/v), 25°C (Ref. 140)

| X | $\sigma^*$ | Es | X | Y | $\sigma^*$ | Es |
|---|---|---|---|---|---|---|
| EtO | 0.57 | −0.25 | EtO | EtO | 1.14 | −1.18 |
| EtS | 0.56 | −0.47 | EtS | EtS | 0.94 | −2.31 |
| PhO | 0.95 | −0.62 | PhO | PhO | 1.90 | |
| PhS | 0.77 | −0.79 | PhS | PhS | 1.54 | −3.29 |
| CH₃ | −0.10 | −0.07 | EtS | Me | 0.49 | −1.53 |
| Cl | 1.05 | −0.24 | PhS | Me | 0.67 | −1.76 |
| Ph | 0.23 | −0.45 | PhS | Ph | 1.00 | |
| | | | Cl | Cl | 2.10 | −1.54 |

* For the definition of $\sigma^*$ and Es values, see Ref. 141.

the following order of electron-withdrawing inductive effects can be clearly noticed: Cl>PhO>PhS>EtO>EtS>Ph>CH₃. Sulfide is a stronger electron-withdrawing group than phenyl group, and a phenyl ring enhances the electron-withdrawing effect of heteroatom attached directly to the ring. An interesting observation is made on the steric effects as measured by Es values. In any case the steric effect of a sulfide group is greater than that of the corresponding ether group.

### 6.4.7  Reactions Involving Carbanions and Related Species

The reactions so far described are such as to generate carbonium ions or related species as intermediates or at the transition state, where electron-releasing effect of sulfide group is the major contributing factor for facile reactions. In this section, the reactions involving carbanions or related species will be described, and here the electron-withdrawing effect of sulfide group is considered to be the major contributing factor.

6.4.7.1  Acidities of Phenols and Anilines

As mentioned already, p-mercapto group enhances the acidities of phenols and anilines, whereas the corresponding ether group reduces them (Table 6.11). In carboxylic acid series (eq. 6.40) (Table 6.13), the acid strengthening and/or weakening effects of sulfide and ether groups operate in the same direction.

Bordwell noticed an unexpectedly low value of $\Delta\sigma_p\text{-}\sigma_m$ for $CH_3S$ group ($-0.03$) in the acid dissociation of phenols as compared to the value for $CH_3O$ group ($-0.24$)[136b]. A difference is also apparent between $C_6H_5S$ ($+0.11$) and $C_6H_5O$ ($-0.13$) groups[138]. Meyers compared the effects of phenyl-mercapto group with the corresponding sulfinyl and sulfonyl groups on the acidity of phenols and found parallel effects between this divalent group and the positively charged sulfur groups[142]. Thus the following electron-accepting $2p\pi\text{-}3d\pi$ conjugation is also important in the divalent sulfide groups (eq. 6.41)[136b,138,142]. As for the positively charged sulfur groups such a conjugation (eq. 6.42) is believed to be well established[7c].

(6.41)                                        (6.42)

Electron-withdrawing effect of sulfide group appears to be enhanced when sulfur atom is attached to a strong electron-withdrawing R group in RS. Thus, in the acid dissociation[136b,143] and the dipole moments of anilines[144], $CF_3S$, $C_6H_5S$, NCS, and $CH_3COS$ groups are much stronger electron-withdrawing agents than $CH_3S$ group. An interesting observation may be that the $\Delta\sigma_p\text{-}\sigma_m$ value of $p$-nitrophenyldithio group is larger than that of $p$-nitrophenylmercapto group. Oae and co-workers explained it in terms of the following $3d$-orbital resonance effect of the dithio group[145]:

(6.43)

A question relating to the above observations is the possibility of "through conjugation" between the $\pi$-systems connected through sulfur atom in which expansion of sulfur valence shell is necessarily involved.

Based on the studies of dipole moments, Campbell, Baliah and their co-workers suggested the possibility of such a through conjugation when a powerful electron-releasing substituent is attached to the *para* position of a sulfide group as shown below (eqs. 44,45) [144]. However "through

$$(CH_3)_2\overset{..}{N}-\underset{}{\bigcirc}-\overset{..}{S}-C\equiv N \longleftrightarrow (CH_3)_2\overset{+}{N}=\underset{}{\bigcirc}=\overset{..}{S}=C=\overset{..}{N}^- \quad (6.44)$$

$$(CH_3)_2\overset{..}{N}-\underset{}{\bigcirc}-\overset{..}{S}-\overset{\overset{O}{\|}}{C}-CH_3 \longleftrightarrow (CH_3)_2\overset{+}{N}=\underset{}{\bigcirc}=\overset{..}{S}=\overset{\overset{O^-}{|}}{C}-CH_3 \,(6.45)$$

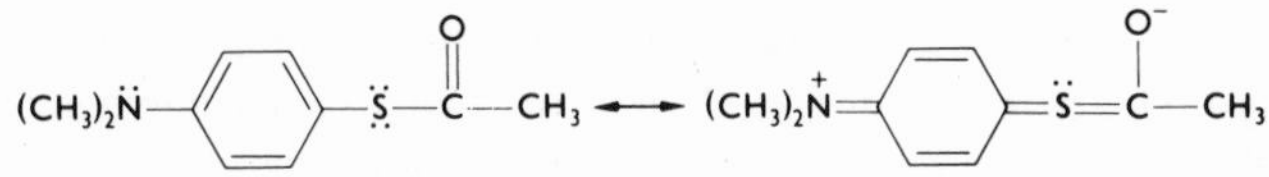

conjugation" has received considerable criticism. For example, Mangini and co-workers observed no indication of such a through conjugation in the electronic spectra of the following 2-amino-8-nitrobenzothiophene (eq. 6.46) [70d], whereas the through conjunction may be important for the resonance stabilization of thiophene ring [146].

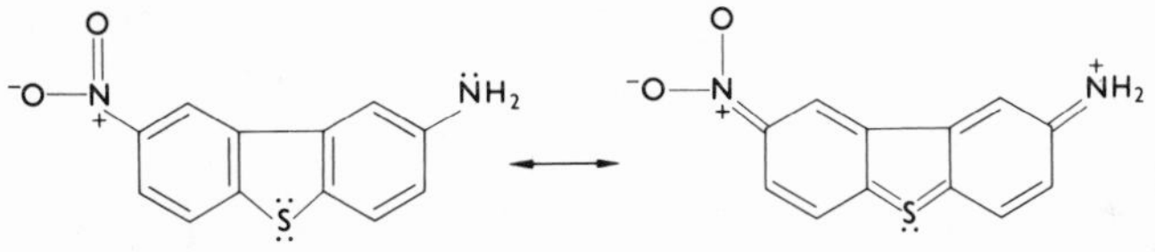

$$\phantom{xxxxxxxxxxxxxxxxxxxxxxxxxxxxxxxxxxxxxxxxxxxxxxxxxx} (6.46)$$

### 6.4.7.2 Hydrogen Exchange and Decarboxylation Reactions

Base-catalyzed hydrogen exchange reactions may be classified as carbanion reactions [147]. Shatenshtein and co-workers carried out extensive studies on the base-catalyzed hydrogen exchange reactions in liquid ammonia [148a]. They found that the rate for $C_6H_5SCD_3$ is much greater than those for $C_6H_5OCD_3$ and $C_6H_5N(CD_3)_2$ [148c]. They also examined the effects of $p$-substituent on the rates of H–D exchange of $CH_3$ group of toluene under similar conditions (Table 6.14) [148d]. The sulfur analogue is about $10^7$ times more reactive than the oxygen analogue. Similarly the phosphorus compound is about $10^4$ times more reactive than the nitrogen analogue. As compared to the unsubstituted toluene, sulfur group enhances the rate while oxygen or nitrogen group reduces the rate. Thus there is a clear distinction among the compounds containing the first and the second row elements in this type of carbanion forming reaction.

**TABLE 6.14**

Relative Rates of Strong-Base Catalyzed H-D Exchange

$$X-\boxed{\phantom{O}}-CH_3 \quad \xrightarrow[ND_3-diglyme]{KND_2} \quad X-\boxed{\phantom{O}}-CD_3$$

| X = | H | $CH_3$ | $CH_3O$ | $CH_3S$ | $(CH_3)_2N$ | $(CH_3)_2P$ |
|-----|---|--------|---------|---------|-------------|-------------|
| k(rel.) | 3 | 1 | 0.003 | 10,000 | 0.0001 | 2 |

Under less basic conditions the acidity of α-hydrogen of monosulfide is not enough to give measurable rate of hydrogen exchange reaction. Substitution by more than one sulfide group would enhance the acidity of the α-hydrogen atom. Thus Oae, Tagaki and Ohno have examined the H–D of H–T exchange reactions of the following mercaptals and orthothioesters in $t$-BuOK/$t$-BuOH system [50b,54,149a]. Their results are shown in Table 6.15. No measurable exchange was observed with the oxygen compounds except for the phenyl substituted acetal (**6.26f**). When the ethyl

$$C_2H_5-\overset{\overset{\displaystyle D}{|}\,\overset{\displaystyle SC_2H_5}{/}}{\underset{\displaystyle SC_2H_5}{C}} \qquad \textbf{6.26a}$$

$$C_6H_5-\overset{\overset{\displaystyle D}{|}\,\overset{\displaystyle SC_2H_5}{/}}{\underset{\displaystyle SC_2H_5}{C}} \qquad \textbf{6.26b}$$

$$D-\overset{\overset{\displaystyle SC_2H_5}{/}}{\underset{\displaystyle SC_2H_5}{C}}-SC_2H_5 \qquad \textbf{6.26c}$$

$$D-\overset{\overset{\displaystyle S-CH_2}{/}}{\underset{\displaystyle S-CH_2}{C}}\overset{\displaystyle \diagdown}{\underset{\displaystyle \diagup}{}}C-CH_3 \qquad \textbf{6.26d}$$

$$C_2H_5-\overset{\overset{\displaystyle D}{|}\,\overset{\displaystyle OC_2H_5}{/}}{\underset{\displaystyle OC_2H_5}{C}} \qquad \textbf{6.26e}$$

$$C_6H_5-\overset{\overset{\displaystyle D}{|}\,\overset{\displaystyle OC_2H_5}{/}}{\underset{\displaystyle OC_2H_5}{C}} \qquad \textbf{6.26f}$$

$$D-\overset{\overset{\displaystyle OC_2H_5}{/}}{\underset{\displaystyle OC_2H_5}{C}}-OC_2H_5 \qquad \textbf{6.26g}$$

$$D-\overset{\overset{\displaystyle O-CH_2}{/}}{\underset{\displaystyle O-CH_2}{C}}\overset{\displaystyle \diagdown}{\underset{\displaystyle \diagup}{}}C-CH_3 \qquad \textbf{6.26h}$$

**TABLE 6.15**

Kinetics for the Base-Catalyzed D–H and T–H Exchange Reactions*

| Compound | Base | Solvent | Temp. °C | $k(\sec^{-1})$ x $10^6$ | $k_{rel.}$ |
|----------|------|---------|----------|-------------------------|------------|
| 6.26a | $t$-BuOK | $t$-BuOH | 138 | 4.73 | 1.00 |
| 6.26b | EtONa | EtOH | 50 | 3.52 | $1.38 \times 10^5$ |
| 6.26f | $t$-BuOK | $t$-BuOH | 120 | 0.405 | 0.24 |
| 6.26c | $t$-BuOK | $t$-BuOH | 138 | 69,950 | $1.48 \times 10^4$ |
| 6.26d | EtONa | EtOH | 50 | 452 | $1.78 \times 10^7$ |
| 6.26g | $t$-BuOK | $t$-BuOH | 112 | No exchange | (in 20 hr) |
| 6.26h | $t$-BuOH | $t$-BuOH | 138 | No exchange | (in 5 hr) |

*References: 50b, 54, 149a.

group of **6.26a** is substituted by one more mercapto groups (**6.26c** and **6.26d**), the rate increases $10^4 - 10^7$ fold. Substitution by a phenyl group also increases the rate $10^5$ times (**6.26b**). Thus the effects of ethylmercapto and phenyl groups are nearly comparable in the rate enhancement. It is apparent that phenyl group stabilizes a carbanion by $2p$-$2p$ $\pi$-bonding. For the mercapto group $3d$- orbital overlapping effect seems to be the most important. Namely, the hydrogen exchange reaction of thioacetal may be formulated as follows where an intermediate carbanion is stabilized by the $3d$-orbital resonance involving two sulfur atoms (eq. 6.47). On the other hand, unexpectedly low reactivity of oxygen and nitrogen compounds

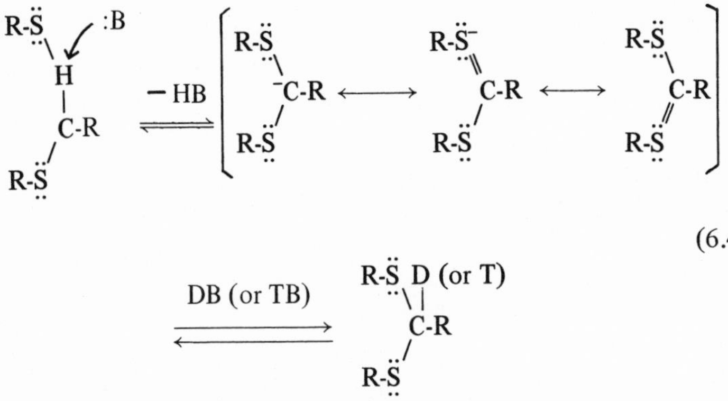

$$(6.47)$$

may suggest that (1) the electron-withdrawing inductive effect by a neutral heteroatom group is not very important for stabilizing a carbanion, and instead (2) the electron pair on nitrogen or oxygen atom destabilize a carbanion. It may also be noted that these *d*-orbital overlapping and electron-releasing conjugation effects operate regardless of whether a heteroatom group interacts with the carbanion directly or indirectly through a $\pi$-system.

An interesting observation is that the cage compound **(6.26d)** is much more reactive than its open chain analogue **(6.26c)**. One factor is the solvation for proton removal at the transition state which may be more favored for **6.26d** because of less steric crowding than in **6.26c**. There is another important difference between the structure of **6.26c** and **6.26d**. In **6.26d**, the $\alpha$—C—H bond or the resulting non-bonded lobe of the carbanion which would be of *sp³* configuration is coplanar with the adjacent three C—S—C planes. In **6.26c**, such complanarity can not be expected due to the considerable steric crowding around the three mercapto groups. A non-classical carbanion **(6.27)** that is stabilized not only by *2p-3d* overlapping, but also by *3p-3d* orbital sulfur-sulfur bonding due to the enforced proximity of the three sulfurs was proposed [54].

6.27                                6.28

For the corresponding sulfones, there is a classical work of Doering and Levy [150]. They found that the bicyclic sulfone is less acidic than the acyclic sulfone, although the acid strengthening effect of sulfonyl groups is remarkable for both cases. There is also a difference in the effect of ring structure of the acidities of cyclic mercaptals **(6.29)** [50b] and sulfones **(6.30)** [151], *i.e.*, the five-membered mercaptal is more reactive than other

6.29                                6.30

cyclic mercaptals in the T—H exchange reactions, while the five-membered sulfone is the weakest acid. Presumably, destabilization of carbanion by dipole-dipole repulsion between the non-bonding lobe of carbanion and the six S—O dipoles **(6.28)** is greater for the bicyclic trisulfone and the five-membered disulfone than the other sulfones. These comparisons may invoke questions about the preferred conformations of these sulfur stabilized carbanions. Further evidence suggests that a carbanion is stable when its non-bonding lobe is coplanar with the adjacent C—S—C planes, as in the cases of the bicyclic trisulfide (6.26d) and the five- and six-membered cyclic mercaptals (see next section). Breslow and co-workers also examined the H—D exchange reaction of ethyl (bis-mercapto)-acetates [152]. However, the effect of the mercapto group seems to be outweighed by the much stronger effect of the carboalkoxy group.

Base-catalyzed decarboxylation reaction is also considered to involve the formation of carbanion at the rate limiting step (eq. 6.48) [147,153]. Oae, Tagaki and co-workers have examined the effects of mercapto group on the decarboxylation of $\alpha$-mercapto acetic acids, and they obtained essentially the same results as those observed in the base-catalyzed hydrogen exchange reactions of mercaptals and orthothioformates [154].

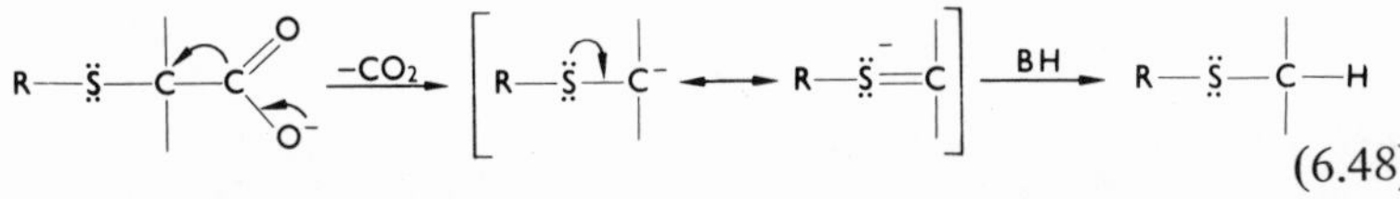

$$R-\overset{..}{S}-\overset{|}{\underset{|}{C}}-\overset{O}{\underset{O^-}{C}} \xrightarrow{-CO_2} \left[ R-\overset{..}{S}-\overset{|}{\underset{|}{C}}-C^- \longleftrightarrow R-\overset{-}{S}=\overset{|}{\underset{|}{C}} \right] \xrightarrow{BH} R-\overset{..}{S}-\overset{|}{\underset{|}{C}}-H$$

$$(6.48)$$

Their representative results are shown below. Here again the effect of mercapto group on the rate enhancement is much greater than that of the

| $(PhS)_2CHCO_2H$ | $(PhO)_2CHCO_2H$ | $(EtS)_2CHCO_2H$ | $Cl_2CHCO_2H$ |
|---|---|---|---|
| k (rel., 90°C) in DMSO: | | | |
| $1.90 \times 10^3$ | 0.73 | 1.00 | $1.2 \times 10^2$ |

| $PhSCH_2CO_2H$ | $(PhS)_2CHCO_2H$ |
|---|---|
| 1.1 | $1.9 \times 10^3$ |

| $(EtS)_2CHCO_2H$ | $(EtS)_3CHCO_2H$ |
|---|---|
| 1.00 | $3.8 \times 10^3$ |

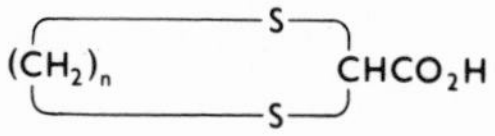

$$(CH_2)_n \overset{\displaystyle \diagup S \diagdown}{\underset{\diagdown S \diagup}{}} CHCO_2H$$

n = 2, 0.47; 3, 0.81; 4, 1.7

corresponding ether group. The addition of one more mercapto group enhances the rate $10^3$ fold. The electron-accepting conjugative effects of a few representative groups fall in the following order: PhS>Cl>EtS>PhO>EtO. Although a marked difference between the reactivities of cyclic and acyclic compounds has been observed in the H–D exchange reactions, there is no appreciable rate difference between the cyclic and the open chain α-mercapto carboxylic acids in this decarboxylation reaction. The lack of marked rate-difference may be due to the stabilization of the ground states of the cyclic acids, especially in the five-membered acid, due to a non-bonding interaction of carboxylate ion with mercapto groups as depicted [154b]:

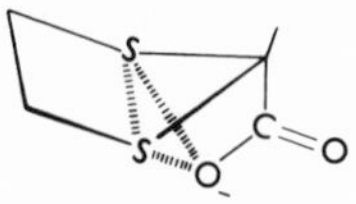

**6.31**

### 6.4.7.3 Structure and Conformation

Oae, Tagaki and Ohno found that the rate of hydrogen exchange of 2-ethyl-1,3-dithiane is some 200 times slower than that of the unsubstituted dithiane [50b]. They also observed a remarkable reactivity of the bridgehead hydrogen of a bicyclic orthothioester [54]. If a carbanion assumes $sp^3$ configuration, an axial conformation would be preferred for 2-ethyl-1,3-dithiane **(6.32)**, and a mixture of axial and equatorial conformations for the unsubstituted dithiane **(6.33a** and **6.33b)**, while only equatorial conformation is allowed for the bicyclic compound **(6.34)**. Then it was suggested that equatorial conformations are more stable than axial conformations. More convincing results have been

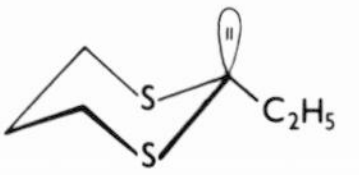

**6.32** 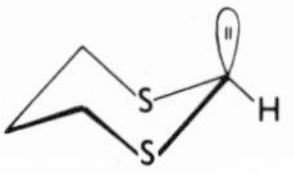            **6.33a** 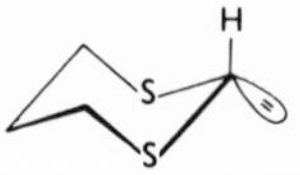            **6.33b** 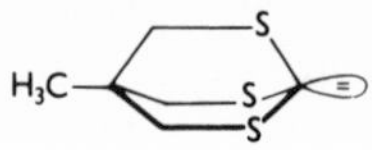            **6.34**

obtained by Hartmann and Eliel [155]. They studied the protonation and alkylation of conformationally fixed 2-lithio-1,3-dithianes and found that the reaction takes place only at the equatorial position. Thus the deuteration or dedeuteration of cis-4,6-dimethyl-1,3-dithiane occurs exclusively at the equatorial 2-position (more than 99%):

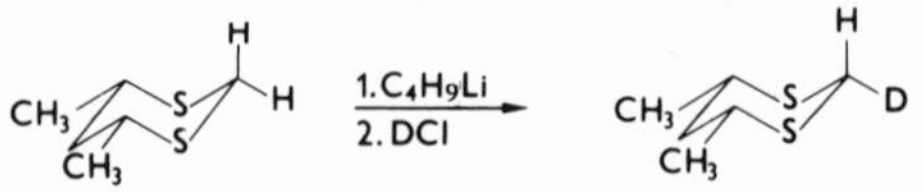

$$(6.49)$$

$$CH_3\text{-derivative} \xrightarrow[\text{fast}]{\substack{1.\,C_4H_9Li \\ 2.\,DCl}} CH_3\text{-derivative} \xrightarrow[\text{slow}]{\substack{1.\,C_4H_9Li \\ 2.\,DCl}} CH_3\text{-derivative} \qquad (6.50)$$

Furthermore lithiation of equatorial hydrogen of 2-methyl-*cis*-4,6-dimethyl-1,3-dithiane is much faster than the lithiation of the axial hydrogen (eq. 6.50). The latter reaction leads to a product of inverted configuration, and this suggests that the ionization of axial hydrogen occurs so as to form an inverted equatorial carbanion.

Another demonstration for the preferred formation of an equatorial carbanion is the reaction of tetrathiacyclooctane[156]. Treatment of tetrathiacyclooctane with butyllithium in the presence of tetramethylethylendiamine in THF was found to give a tetravalent carbanion which on deuteration or methylation gave an all *cis*-substituted (e,e,e,e) isomer as the major product (65%) (eq. 6.51). These results

$$\xrightarrow[\text{(CH}_3)_2N(CH_2)_2N(CH_3)_2]{C_4H_9Li} \qquad (6.51)$$

$$\xrightarrow{CH_3I}$$

suggest that the multivalent carbanion must be pyramidal and would eliminate a possibility that the equatorial alkylation occurs on a planar carbanion, since a planar structure having more than one carbanionic center at the same time is highly unlikely due to ring strain. A stable doubly lithiated carbanion was also prepared by the lithiation of tetrathiaadamantane by Bank and Coffen (eq. 6.52)[157]. Corey and Seebach prepared a stable carbanion from 1,3-dithiane and used it for organic syntheses (see Sect. 6.5).

$$\xrightarrow{C_4H_9Li} \xrightarrow{\substack{D_2O\ or \\ CH_3I}} \qquad (6.52)$$

R=D or CH₃

Thus, the structure and the conformation of a stable carbanion derived from a cyclic sulfide are now fairly well understood. However, a general consensus of opinion has not been reached about the role of sulfur. Oae, Tagaki and Ohno[50b] and Bank and Coffen[157] suggested that 3d-orbital conjugation is more effective when a carbanion takes an equatorial rather than an axial conformation with respect to the six-membered dithiane ring. Equatorial and axial conformations correspond to **6.35a** and **6.35b**, respectively, in Newman projection. In **6.35a** the non-bonding lobe of the carbanion bisects the two sulfur electron pairs. Meanwhile Wolfe and co-workers criticized the hypothesis of

**6.35a**                           **6.35b**

3d-orbital conjugation based on their molecular orbital calculations[158,159]. Their calculation on the energies of the rotational isomers of hypothetical sulfur stabilized carbanions (**6.36** and **6.37**) has led to the conclusion that the most stable conformation is such as to allow a maximum number of *gauche* interactions between electron pairs[158]. Thus among three possible conformations **6.38a** and **6.38c**

$$\bar{:}CH_2-SH \qquad\qquad \bar{:}CH_2-S-CH_2^-:$$

**6.36**                                **6.37**

**6.38a**            **6.38b**            **6.38c**

have an energy minimum while **6.38b** does not. Furthermore, **6.38c** is the most stable conformation and the energy difference between **6.38a** and **6.38c** is 6.1 kcal/mole and the rotational barrier from **6.38c** to **6.38a** is 18.8 kcal/mole. Essentially similar results were obtained in the calculation for **6.37** and it was pointed out that the conformation of sulfur stabilized carbanion need not be related to 3d-orbital conjugation. On the other hand, Koch and Moffitt, on the basis of molecular orbital calculations, concluded that the best overlap between a 2p-orbital of a carbon atom attached to a sulfone group and the 3d-orbitals of the sulfur atom occurs in the two conformations drawn as Case I and Case II (**6.39a**, **6.39b**)[160]. Available experimental data have well established that the Case II conformation is a stable form of α-sulfinyl and α-sulfonyl

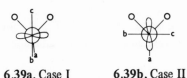

**6.39a,** Case I          **6.39b,** Case II

carbanions [147]. However, Wolfe and co-workers have also pointed out that the Case II conformation is not unique for sulfur compounds, since even hydrazines, hydroxylamines, and hydrogen peroxide, etc. have similar conformations with respect to electron pairs [159].

The works of Wolfe and co-workers may well explain why the equatorial carbanions (**6.35a** and **6.38c**) are preferentially formed while the axial carbanions (**6.35b** and **6.38b**) isomerize to the corresponding equatorial ones when forced to form [155]. However this may not answer why the sulfur carbanion is so readily formed as compared to the corresponding oxygen carbanion. Thus it is still tempting to accept 3d-orbital resonance as a probable explanation. However, it is also true that there are other possibilities. For example, the unshared electron pairs of oxygen are considered to be much more tightly bound than those of sulfur. If there is any repulsion between these unshared electron pairs and the carbanion, it should be greater for oxygen than for sulfur atom, since diffused and polarizable sulfur unshared electron pairs buffer such a repulsion. Furthermore the sulfur bond is much longer than the corresponding oxygen bond, and this would also contribute to the reduction of such a repulsion.

### 6.4.7.4 Elimination and Isomerization

Although base-catalyzed elimination reactions have been extensively studied and a substantial amount of information is available concerning the effects of polar substituents [16], very little is known on the effect of a heteroatom at the β-position. Hine and co-workers have found that the relative rates are 1:130:190 for fluoro, chloro and bromo compounds in the $E_2$ elimination reactions of β-haloethyl bromides [162]. When methyl β-hydroxyethyl sulfide is treated with a strong alkali, methyl vinyl sulfide is obtained, whereas under similar conditions methyl cellusolve does not react [163]. These observations suggest that an elimination reaction of a compound containing 2nd and 3rd row elements at the β-position proceed through a carbanion-like transition state which is stabilized by electron-accepting 3d-orbital conjugation.

Oae and Yano have made more detailed studies on the effects of β-mercapto group in the base catalyzed $E_2$ reactions of halosulfides [164,165]. They found that the rates of the $E_2$ reaction of p-substituted β-phenylmercaptoethylchlorides are $10^2$-$10^3$ times as great as those of the corresponding oxygen analogues in both $t$-BuOK/$t$-BuOH and EtONa/EtOH systems (eq. 6.53) [164]. The Hammett $\rho$ values for the sulfur compounds are 1.98 and 2.14 in $t$-butanol and ethanol respectively,

$$X-\langle\rangle-Z-CH_2CH_2Cl \xrightarrow[-HCl]{t-BuOK/t-BuOH \ (or \ EtONa/EtOH)}$$

$$X-\langle\rangle-Z-CH=CH_2 \qquad\qquad (6.53)$$

$$Z = S, O$$

while those for the oxygen analogues are 1.33 and 1.50 respectively. They discussed these higher rates and the greater Hammett $\rho$ values of the sulfur compounds, as compared with the oxygen counterparts, in terms of $3d$-orbital conjugative effect at the transition state. The substitution by one more mercapto group at the $\beta$-position greatly enhances the reactivity of the sulfur compounds [166], while the introduction of ethoxy or phenoxy group retards the reaction [164].

|                  | $PhCH_2CH_2Cl$ | $PhOCH_2CH_2Cl$ | $PhSCH_2CH_2Cl$ |
|------------------|:--------------:|:---------------:|:---------------:|
| $k_{E_2}$ (rel.) | 1              | 2               | 900             |

|                  | $(EtO)_2CHCH_2Cl$ | $(PhO)_2CHCH_2Cl$ | $(PhS)_2CHCH_2Cl$    |
|------------------|:-----------------:|:-----------------:|:-------------------:|
| $k_{E_2}$ (rel.) | 0.1               | 0.15              | too fast to be measured |

One may anticipate a non-bonding interaction of $\gamma$-mercapto group with a developing double bond (6.40) in elimination reaction in view of the suggestion that the red-shift of uv spectra of allyl and benzyl sulfides is due to such a non-bonding interaction (Sect. 6.3.1). Thus Oae and Yano

$$X-\langle\rangle-\overset{\delta-}{S}\cdots\overset{\overset{H}{\vdots}}{C}H\cdots\overset{\vdots}{C}H_2$$

**6.40**

have examined the rates of the base-catalyzed elimination of a series of γ-(p-substituted-phenylthia) propyl bromides and the corresponding oxygen analogues [165]. However, unlike the case of β-substituted compounds, the oxygen compounds are more reactive than the sulfur compounds. The ρ values are 0.37 for the sulfur and zero for the oxygen compounds. Although the ρ values appear to indicate a small non-bonding interaction as depicted in **6.40**, the interaction is anyway very small.

Evidence to support that γ-mercapto group does not participate in the formation of double bond has been obtained in the base-catalyzed elimination reaction of 3-chloro-1,1-bis-(ethylmercapto)propene (**6.41**). Rothstein found that **6.41** gave 1,1-bis(ethylmercapto)propene-1 (**6.42**) while the corresponding oxygen analogue (**6.43**) gave acrolein diethyl acetal

$$Cl\text{-}CH\text{-}CHT\text{-}CH(SEt)_2 \xrightarrow{t\text{-}BuO^-} CH_3\text{-}CH=C(SEt)_2$$

$$\textbf{6.41} \qquad\qquad\qquad\qquad \textbf{6.42}$$

$$k_H / k_T = 6.20$$

$$Cl\text{-}CH_2CHT\text{-}CH(OEt)_2 \xrightarrow{t\text{-}BuO^-} CH_2=CH\text{-}CH(OEt)_2$$

$$\textbf{6.43} \qquad\qquad\qquad\qquad \textbf{6.44}$$

$$k_H / k_T = 6.55$$

(**6.44**) [167]. The difference in the position of the double bond between the products of oxygen and sulfur compounds was explained later by Oae, Ohno and Tagaki as due to a facile isomerization of propenyl double bond formed in the rate limiting 1,2-elimination, based on their rate data including the observation of the nearly identical kinetic hydrogen isotope effects for the oxygen and sulfur compounds [168].

The effect of heteroatom group found in the olefin forming reaction may also be observed in the isomerization of olefinic double bond, since the isomerization depends on both the rate of proton removal and the equilibrium constants. The conjugative effect of unshared electrons of amines, ethers, and sulfides strongly favors the vinyl over the propenyl forms of these compounds, although in most cases equilibrium constants (K) are not available (eq. 6.54). On the other hand, it is known that there is a big difference between the rates of isomerization of olefins containing first and second row elements. Tarbell and co-workers found that allyl aryl [169a] and alkyl allyl [169b] sulfides are readily isomerized to the corresponding propenyl aryl and alkyl propenyl sulfides on treatment with

$$CH_2=CH\text{-}CH_2\text{-}\overset{..}{X}\text{-}R \overset{K}{\rightleftharpoons} CH_3\text{-}CH=CH\text{-}\overset{..}{X}\text{-}R \qquad (6.54)$$

$$CH_3\text{-}\overset{-}{C}H\text{-}CH=\overset{+}{X}\text{-}R$$

sodium ethoxide in ethanol, whereas the corresponding oxygen compounds are unchanged under the same reaction conditions. Meanwhile, allyl ethers have also been shown to undergo isomerization to the propenyl ethers but under more strongly basic conditions, where a high proportion of *cis* isomer was usually obtained. This suggests a concerted intramolecular 1,3-*cis* migration of $\alpha$-proton [147,170], whereas a

$$(6.55)$$

considerable amount of *trans* isomer was obtained in the isomerization of the former allyl sulfides [171]. These differences in the behavior of sulfides and ethers have been interpreted as being due to the ability of sulfur atom to stabilize the carbanion by $3d$-orbital conjugation (eq. 6.56) [169]. If $3d$-orbital conjugation is the principal factor determining the composition of an equilibrium mixture or the stereochemistry of sulfur containing olefins, the trends in sulfides should also be observed in

$$(6.56)$$

the corresponding sulfones and sulfoxides since $3d$-orbital conjugation is known to be stronger for the latter sulfur groups. However, the explanation based on $3d$-orbital participation apparently is not sufficient for the equilibria of the cyclic sulfones shown in eqs. 6.57 and 6.58 [172].

In order to avoid complexities due to ring effects, O'Connor and co-workers have made more detailed comparisons of the equilibria between open chain allyl and vinyl sulfides, sulfoxides, and sulfones (Table 6.16) [173]. The data indicate that the ability to stabilize the

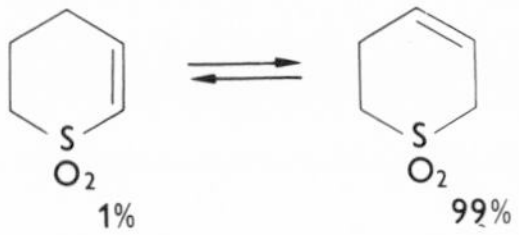

$$(6.57)$$

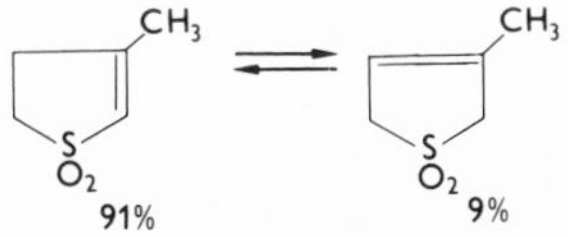

1%          99%

$$(6.58)$$

91%          9%

**TABLE 6.16**

Equilibria between Vinyl and Allyl Double Bonds (Ref. 173)*

$$RCH_2\text{-}CH\text{=}CH\text{-}X \underset{\text{Base}}{\overset{K}{\rightleftharpoons}} RCH\text{=}CH\text{-}CH_2\text{-}X$$

vinyl form          allyl form

| R | X | K |
|---|---|---|
| H | $SCH_3$ | $< 0.01$ |
| H | $SOCH_3$ | 0.25 |
| H | $SO_2CH_3$ | 0.80 |
| $C_3H_7$ | $SCH_3$ | 0.5 |
| $C_3H_7$ | $SOCH_3$ | 24 |
| $CH_3$ | $SOCH_3$ | 32 |
| $C_3H_7$ | $SO_2CH_3$ | $>99$ |

*Taken from Ref. 147, p. 203.

vinyl form of the olefin over the allyl form is $CH_3S>CH_3SO>CH_3SO_2$. The data also indicate a large effect of alkyl group on the K values favoring allyl form. The greater predominance of $\beta,\gamma$-isomers in the sulfoxides and sulfones was then claimed to be the convincing evidence that ground-state $3d$-orbital resonance contributes little or nothing to the stability of the $\alpha,\beta$-double bond. The major factor responsible for the above data was attributed to the destabilization of double bond by an inductive electron-withdrawing effect. The order of such effect is $CH_3SO_2>CH_3SO>CH_3S$, and $CH_3S$ and alkyl group stabilize vinylic double bond by electron-releasing inductive effect. Although these data and the interpretations seem to be reasonable, they do not necessarily rule out the possible $3d$-orbital effect since the factors stabilizing or destabilizing the $\alpha$-double bond also operate to destabilize or stabilize the $\alpha$-carbanion.

The base-catalyzed isomerization of acetylenic bond to allenic bond occurs readily when the allenic bond is in conjugation with a heteroatom carrying unshared electrons [174]:

$$CH_3CH_2OCH_2C{\equiv}CH \xrightarrow{\text{KOH}} CH_3CH_2OCH{=}C{=}CH_2 \qquad (6.59)$$
$$89\%$$

$$CH_3CH_2SCH_2C{\equiv}CH \xrightarrow{\text{NaNH}_2/\text{NH}_3} CH_3CH_2SCH{=}C{=}CH_2 \qquad (6.60)$$
$$70{-}85\%$$

A similar isomerization step may be involved in the following transannular cyclization reaction [175]:

$$(6.61)$$

The ready isomerization of double and triple bonds in allyl and propynyl sulfides appears to have a profound influence on the course of thio-Claisen rearrangement as compared to the usual oxy-Claisen rearrangement. The thermal oxy-Claisen rearrangement of allyl phenyl ethers is well known to give allyl phenols (eq. 6.62). Prop-2-ynyl phenyl ethers undergo similar rearrangement (eq. 6.63) in N,N-dimethylaniline [176].

$$(6.62)$$

$$(6.63)$$

Kwart [177] and Meyers [178], and their co-workers found that allyl phenyl sulfide undergoes thio-Claisen rearrangement when the reaction is carried out in quinoline to afford nearly an equal mixture of thiochromane and thiocoumarane (eq. 6.64). The isolation of the cyclic products suggests

$$(6.64)$$

that the reaction proceeds either through the rearranged thiophenol intermediates [178] or some other common intermediate [177b].

Kwart and George found that prop-2-ynyl phenyl sulfide (6.45a) also undergoes thio-Claisen rearrangement (eq. 6.65) [179a]. Thermolysis of 6.45a in quinoline at 200° yielded mainly two rearrangement products, 2-methyl-benzo [b] thiophene (6.45f) and phenyl allenyl sulfide (6.45d).

$$(6.65)$$

At higher temperatures or prolonged heating 6.45d is completely consumed and a second major component of the product, 2H-thiochromene (6.45c), is isolated. The multiplicity of the products is in contrast to the oxy-Claisen rearrangement of the corresponding ethers where only a single rearrangement product is found. The authors based their explanation of these data on the assumption of a prior thio-

$$\text{PhS-CH}_2\text{-C}\equiv\text{CH} \overset{\text{Base}}{\rightleftharpoons} \text{PhS-CH=C=CH}_2 \qquad (6.66)$$

propynylic rearrangement (eq. 6.66). Analogous thio-Claisen rearrangements were also observed in other heterocyclic compounds [179-181].

### 6.4.8 Free Radical Reactions

6.4.8.1 Hydrogen Abstraction from α-Carbon

Russell and co-workers found that hydrogen abstraction from methyl group by phenyl radical is facilitated by the presence of α-heteroatom and explained the rate-enhancement in terms of the stabilization of the incipient radical by electron-releasing conjugative effect of α-substituent as shown below (eq. 6.67)[182]. Oae and co-workers have made a more

$$
\underset{|}{\overset{\cdot\cdot}{-X}}\overset{\cdot}{C}\text{-} \longleftrightarrow \underset{|}{\overset{+}{-X}}\overset{\cdot\cdot}{\overset{-}{C}}\text{-}
$$

$$X = N, O, S$$

(6.67)

detailed study on the effects of heteroatom using the more polar butoxy radical rather than phenyl radical (Table 6.17)[183]. The rate enhancing

TABLE 6.17

Kinetics for the Hydrogen Abstraction Reactions (Ref.183)

$$Z\text{-}CH_3 + R\cdot \xrightarrow{k} Z\text{-}CH_2\cdot + RH$$

| ZCH$_3$ | k$_{rel}$ for R· | | Hammett |
|---|---|---|---|
| | C$_6$H$_5$· (60°C)[179] | t-BuO· (130°C)[180] | ρ t-BuO· |
| C$_6$H$_5$CH$_3$ | 1.00 | 1.00 | −0.55 (125°) |
| C$_6$H$_5$OCH$_3$ | 0.35 | 1.44 | −0.39 (130°) |
| C$_6$H$_5$SCH$_3$ | 1.43 | 2.12 | −0.15 (130°) |
| C$_6$H$_5$N(CH$_3$)$_2$ | 5.19 | 81.9 | −0.42 (130°) |
| C$_6$H$_5$SO$_2$CH$_3$ | − | 0.05 | −0.2 (130°) |

effect of the heteroatom group is in the order of $C_6H_5NCH_3$ $>C_6H_5O\sim C_6H_5S>C_6H_5SO_2$. The Hammett $\rho$ value is negative for each group of compounds. Thus these hydrogen abstraction reactions are electrophilic in nature, and any electron-releasing group, either at the α-position or on the benzene ring, enhances the reactivity. However it should be noticed that the reactivities of toluene and anisole are reversed toward phenyl and t-butoxy radicals, and for both radicals thioanisole is more reactive than anisole. One possible explanation is that an incipient free radical containing a mercapto group is additionally stabilized by an electron-sharing conjugation, as shown in eq. 6.68 and such a conjugation may become more important when the attacking radical becomes more neutral, as in the reaction of thioanisole with phenyl radical.

$$X-\langle\!\bigcirc\!\rangle-\overset{..}{\underset{..}{S}}-\dot{C}H_2 \longleftarrow X-\langle\!\bigcirc\!\rangle-\overset{..}{\underset{..}{S}}=CH_2 \qquad (6.68)$$

### 6.4.8.2 Addition

Price and co-workers found that vinyl sulfide and divinyl sulfide are much more reactive toward a growing polystyryl radical than the corresponding oxygen analogues [184]. They treated the data by an empirical Q-e scheme and proposed that a larger Q value is a measure of greater resonance stabilization of a growing radical. They also proposed that an electron-sharing conjugation involving $3d$-orbital is important for the stabilization of growing radical in the copolymerization of vinyl sulfides:

$$R\text{-}\overset{..}{\underset{..}{S}}\text{-}CH{=}CH_2 + \cdot R \longrightarrow R\text{-}\overset{..}{\underset{..}{S}}\text{-}CH\text{-}CH_2\text{-}R$$
$$\updownarrow \qquad \longrightarrow \text{Polymer} \qquad (6.69)$$
$$R\text{-}\overset{..}{\underset{..}{S}}{=}CH\text{-}CH_2\text{-}R$$

Tagaki, Oae and co-workers further studied the copolymerization of ketene diethylmercaptal and $p$-methylmercaptostyrene with styrene and observed much larger Q values than those for vinyl sulfides [185]. Some relevant Q–e values are shown in Table 6.18. Although $3d$-orbital

**TABLE 6.18**

Q–e Values for the Copolymerization Reactions of Substituted Ethylenes with Styrene (Ref. 185)

| Compound | Q | e |
|---|---|---|
| $CH_2{=}CH{-}OC_2H_5$ | 0.02 | −1.6 |
| $CH_2{=}CH{-}SCH_3$ | 0.34 | −1.5 |
| $(CH_2{=}CH)_2S$ | 0.68 | −1.1 |
| $CH_2{=}C(OC_2H_5)_2$ | (only homopolymerization of styrene) | |
| $CH_2{=}C(SC_2H_5)_2$ | 2.70 | −2.1 |
| $CH_2{=}CHSO_2CH_3$ | 0.11 | 1.2 |
| $CH_2{=}CHSi(CH_3)_3$ | 0.03 | −0.1 |
| $CH_2{=}CH\text{-}C_6H_4SCH_3\text{-}p$ | 3.29 | −1.65 |
| $CH_2{=}CH\text{-}C_6H_5$ | 1.0 | −0.8 |
| $CH_2{=}CH\text{-}C_6H_4OCH_3\text{-}p$ | 1.0 | −1.0 |
| $CH_2{=}CH\text{-}C_6H_4NO_2\text{-}p$ | 1.86 | 0.4 |
| $CH_2{=}CH\text{-}\langle\text{thienyl, S}\rangle$ | 3.0 | −0.80 |

resonance effect must be tested by further studies, it is clear that there is a sharp difference in Q values between sulfides and the oxygen analogues, like those observed in the carbanion forming reactions. Perhaps a growing radical in copolymerization reaction is more neutral than that in hydrogen abstraction reaction, since an odd electron is added to the system in the former reaction while an electron is removed from the system in the latter reaction. The lower Q values for vinyl sulfone and trimethylvinyl silane may be in accordance with the concept of electron-sharing conjugation because it requires the presence of an unshared electron pair in addition to $3d$-orbital.

The orientation effects of mercapto and alkoxy groups in a free radical addition reaction may be manifested in the following example (eq. 6.70). Arens and co-workers observed that a thiyl radical adds to the double bond of 1-ethylmercapto-2-ethoxy-ethylene in such a way to give 1,2-diethylmercapto-2-ethoxyethylene (6.46) [186].

$$C_2H_5S\text{-}CH\text{=}CH\text{-}O\text{-}C_2H_5$$

$$\downarrow \quad C_2H_5S\cdot$$

$$\left[ \begin{array}{c} C_2H_5\text{-}\ddot{S}\text{-}CH\text{-}CH\text{-}OC_2H_5 \\ | \\ SC_2H_5 \end{array} \longleftrightarrow \begin{array}{c} C_2H_5\text{-}\ddot{S}\text{=}CH\text{-}CH\text{-}OC_2H_5 \\ | \\ SC_2H_5 \end{array} \right]$$

$$\downarrow \quad C_2H_5SH \qquad\qquad (6.70)$$

$$C_2H_5S\text{-}CH_2\text{-}CH \overset{OC_2H_5}{\underset{SC_2H_5}{<}}$$

**6.46**

### 6.4.8.3 Decomposition

The comparison of the effects of mercapto and alkoxy groups on the stability of a free radical is a very difficult task in principle because the effects may be readily reversed, depending on whether the transition state is carbonium ion or carbanion like. The hydrogen abstraction and addition to double bond by a free radical may not generate a really neutral radical. Meanwhile it is believed that the unimolecular decomposition of a symmetric azo compound forms a neutral free radical [187]. From this viewpoint Ohno and co-workers examined the decomposition of heteroatom substituted azobis (2-propane) derivatives (eq. 6.71) [188]. They were able to follow the rate of decomposition in tetraline and obtained $\Delta H\ddagger$ and $\Delta S\ddagger$ values (Table 6.19). The $\Delta H\ddagger$ value for X=S is significantly smaller than those for X=O and $CH_2$ and the importance of

$$
\begin{array}{c}
\underset{\substack{| \\ CH_3}}{\overset{\substack{CH_3 \quad CH_3 \\ | \qquad |}}{Ph\text{-}X\text{-}C\text{-}N\text{=}N\text{-}C\text{-}X\text{-}Ph}} \xrightarrow[\text{in PhOPh}]{\Delta, -N_2} \left[ \underset{\substack{| \\ CH_3}}{\overset{\substack{CH_3 \\ |}}{Ph\text{-}X\text{-}C\cdot}} \right] \longrightarrow
\end{array}
$$

$$(6.71)$$

$$
\underset{\substack{| \quad | \\ CH_3 CH_3}}{\overset{\substack{CH_3 CH_3 \\ | \quad |}}{Ph\text{-}X\text{-}C - C - X\text{-}Ph}} + \underset{\substack{| \\ CH_3}}{\overset{\substack{CH_3 \\ |}}{Ph\text{-}X\text{-}C\text{-}H}} + \underset{\substack{| \\ CH_3}}{\overset{\substack{CH_3 \\ |}}{Ph\text{-}X\text{-}C\text{=}CH_2}} + \text{others}
$$

**TABLE 6.19**

$\Delta H\ddagger$ and $\Delta S\ddagger$ Values for the Decomposition of
Azobis (2-Propane) Derivatives (eq. 6.71)

| X | $\Delta H\ddagger$ (kcal/mole) | $\Delta S\ddagger$ (e.u.) |
|-----|-----|-----|
| S | 24.9 | −12.6 |
| O | 32.3 | − 2.5 |
| $CH_2$ | 35.6 | 4.1 |

electron-sharing conjugation involving $3d$-orbital was suggested to be responsible. The smaller $\Delta S\ddagger$ value for X=S was also claimed to support the freezing of the molecule at the transition state by conjugation. It is assumed in this case that fission of the two C—N bonds occurs simultaneously at the transition state, thus forming purely neutral free radicals.

Martin and co-workers found that the rate of homolysis of O—O bond of $t$-butyl benzoylperoxide is greatly increased by the presence of neighboring mercapto group (Table 6.20) [189]. $o$-Iodo and $o$-C=C bonds also cause some rate enhancement. In these three cases a larger $k_{rel}$ value

**TABLE 6.20**

Kinetics for the Homolytic Decomposition of $t$-Butyl Benzoyl Peroxides (Ref. 189d)

| | $o$-X-C$_6$H$_4$CO$_2$-O Bu-$t$ $\rightarrow$ $o$-X-C$_6$H$_4$CO$_2$· + $t$-BuO· | | |
|---|---|---|---|
| X | $k_{rel}$(40°C) | ΔH‡ kcal/mole | ΔS‡ e.u. |
| H | 1.0 | 34.1 | 10.0 |
| $t$-C$_4$H$_9$ | 2.4 | 34.4 | 12.5 |
| I | 80.5 | 28.0 | –0.8 |
| (C$_6$H$_5$)$_2$C=CH | 153 | 26.3 | –5.0 |
| C$_6$H$_5$S | 6.53 x 10$^4$ | 23.0 | –3.4 |
| C$_6$H$_5$SCH$_2$ | 5.6 | 32.2 | 7.2 |
| CH$_3$SO$_2$ | 0.4 | 38.0 | 19.5 |

is associated with a smaller ΔH‡ and a negative ΔS‡ values, as in the above unimolecular decomposition of azobis (2-propane) derivatives. They suggested an intramolecular anchimeric participation of sulfur atom involving an expansion of sulfur octet in the rate limiting C–O bond cleavage (eq. 6.72). Such a participation seems to be more favored in a

$$(6.72)$$

five-membered ring than a six-membered ring since the effect of phenylmercaptomethyl group is similar in magnitude to that of $t$-butyl group.

### 6.4.8.4 EPR Spectra

**Epr spectral studies may give more direct information on the structure of transient free radical intermediates.** Adams observed that the photolysis of $t$-butyl peroxide in the presence of diethyl and di-$n$-butyl sulfide gives only the α-radicals (eq. 6.73)[190]. The electron spin density on the α-carbon was found to be greater for the corresponding oxygen free radicals. However the interpretation of the spectra was not straightforward due to the complexity caused by the presence of β-substituent. Meanwhile Krusic

$$RCH_2SCH_2R' + (t\text{-}BuO)_2 \xrightarrow{\ h\nu\ } R\dot{C}HSCH_2R' \qquad (6.73)$$

and Kochi have succeeded in getting the epr spectrum of methylithiyl radical [191]. They interpreted the spectrum as showing a planar conformation (6.47) due to partial double bonding (6.48) between the trigonal carbon atom of the free radical center and the sulfur atom:

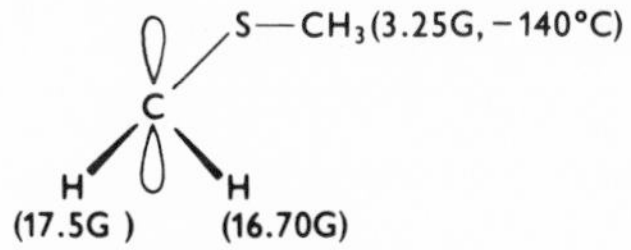

**6.47** 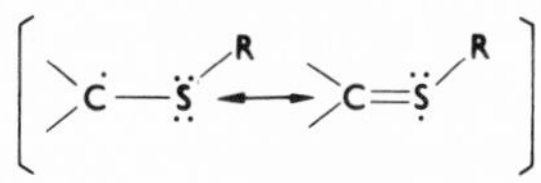                    **6.48**

Participation of symmetrical bridged radicals have been proposed by Skell in the addition of thiyl radical to olefin (6.49) [192]. However, no evidence to support the presence of such bridged radicals have been obtained from epr spectra [191,193], although Krusic and Kochi

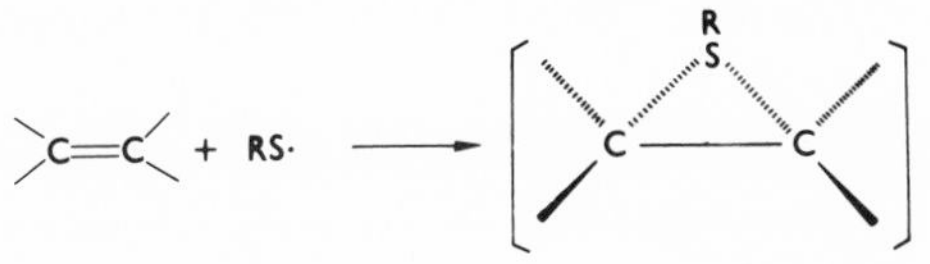

**6.49**

suggested an eclipsed interaction of heteroatom with the odd electron on the trigonal α-carbon atom [191]:

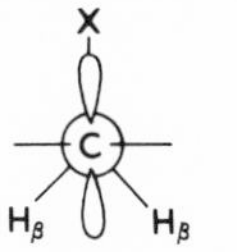

**6.50**

### 6.4.8.5 Cleavage of C—S Bond

The cleavage of C—S bond of sulfides is known to occur in thermal and photochemical reactions. In most cases, the mechanisms appear to involve free radical processes.

Bond dissociation energies of the C—S bond of sulfides can be obtained from the thermal decomposition reactions [58]. Irradiation of simple dialkyl sulfides yields products derived from alkyl and thiyl radicals [194]. A complex array of products are obtained from the photolysis of dibenzyl sulfides [195]. Diphenyl sulfide also undergoes photochemical C—S bond cleavage to give diphenyl disulfide along with the formation of benzene (eq. 6.74) [196].

Introduction of strain into alkyl sulfides enhances their photochemical reactivity. Thus, both thiacyclobutane and 6-thiabicyclo [3.1.1] heptane

$$C_6H_5\text{-}S\text{-}C_6H_5 \xrightarrow[h\nu]{\text{light petroleum}} C_6H_5\cdot + \cdot SC_6H_5$$

(6.74)

$$\xrightarrow{\hspace{2cm}} C_6H_6 + C_6H_5SSC_6H_5$$

undergo facile light-initiated radical polymerization [9b]. However, Corey and Block found that the reaction can be simplified when carried out in the presence of trivalent phosphorus compounds which convert the thiyl radicals into alkyl radicals [64a]. For example irradiation of 9-thiabicyclo [3.3.1] nonane (6.51a) in isooctyl phosphite gave in 47% total yield a mixture of hydrocarbons which are of cis-bicyclo [3.3.0] octane (6.51b, 46%), cyclooctene (46%) and cyclooctane (8%):

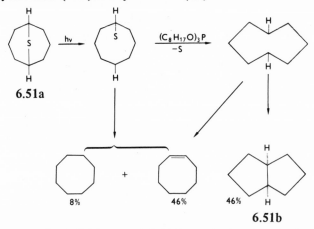

6.51a

6.51b

8%          46%          46%

Irradiation of 1,3-dithianes and dithiolanes also involves the free radical cleavage of C—S bond to give rearranged products [197]:

(6.75)

39%                    5%

Alkali metals are also known to cause C—S bond cleavage [198,199]. The reactions apparently involve electron transfer from the metal to the C—S bond as illustrated below [199b]:

$$PhSPh + K \xrightarrow{\quad e \quad} \left[ PhS^- + Ph^- + Ph\cdot \right] \xrightarrow{\quad HA \quad}$$

(6.76)

PhSH + PhH + Ph-Ph + other products (from the solvent)

## 6.5 SOME ASPECTS OF SYNTHESIS

The use of sulfides for organic synthesis has drawn increasing interests in recent years. The most useful methods appear to utilize the unique properties of the C—S bond: (1) to form sulfur stabilized carbanion; (2) to be replaced by another substituent by the C—S bond cleavage, reductively or oxidatively; (3) being stable under both alkaline and acidic conditions.

Dieckmann type condensation reaction is a classical example (eqs. 6.77, 6.78) [200]. The compound (6.52) gives exclusively the cyclic compound on treatment with sodium methoxide. The base preferentially attacks the methylene proton between S and carboalkoxy group [200]. This method allows one to prepare a number of cyclic sulfides [201].

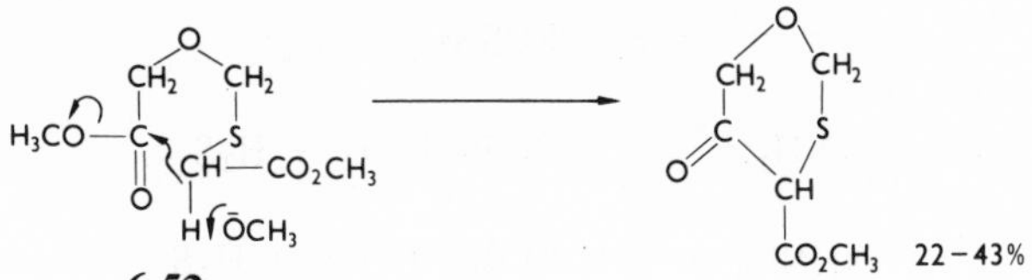

(6.77)

6.52

22–43%

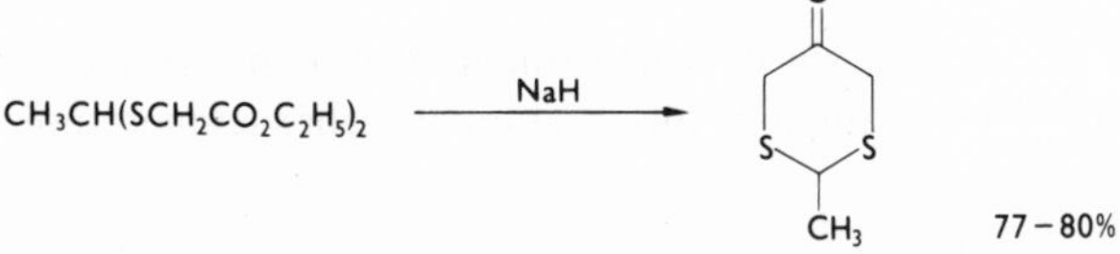

$$CH_3CH(SCH_2CO_2C_2H_5)_2 \xrightarrow{\quad NaH \quad}$$

(6.78)

77–80%

Arens and co-workers found that mercaptals can be alkylated at the α-carbon by treatment with alkali amide in liquid ammonia, followed by reaction with alkyl halides (eq. 6.79) [50a]. More active benzaldehyde diphenylmercaptal can be alkylated under milder conditions (eq. 6.80) [202]. Although pioneering, these methods are not satisfactory for synthetic purposes in view of limited applicability.

$$CH_2(SC_2H_5)_2 \xrightarrow[\text{liq. NH}_3]{\text{NaNH}_2} H\bar{C}(SC_2H_5)_2Na^+ \xrightarrow{\text{CH}_3\text{Br}} CH_3CH(SC_2H_5)_2 \qquad (6.79)$$

$$79\%$$

$$C_6H_5CH(SC_6H_5)_2 \xrightarrow[\text{DMF}]{\text{NaH}} C_6H_5\bar{C}(SC_6H_5)_2Na^+$$

$C_6H_5COCl$                $CH_3I$

$$\qquad (6.80)$$

$$\begin{array}{cc} \overset{\displaystyle COC_6H_5}{\underset{\displaystyle C_6H_5C(SC_6H_5)_2}{|}} & \overset{\displaystyle CH_3}{\underset{\displaystyle C_6H_5C(SC_6H_5)_2}{|}} \end{array}$$

Corey and Seebach extended the above reactions to a much more useful synthetic method of wide applicability [203]. They used the carbanion generated from dithiane by the action of alkyllithium in tetrahydrofuran at low temperatures. The carbanion reacts with $CO_2$, alkyl halides, esters, aldehydes and ketones, and imines to give the corresponding $\alpha$-substituted products, generally in high yields. The $\alpha$-substituted products can be hydrolyzed with a suitable catalyst to give the corresponding ketones as shown below:

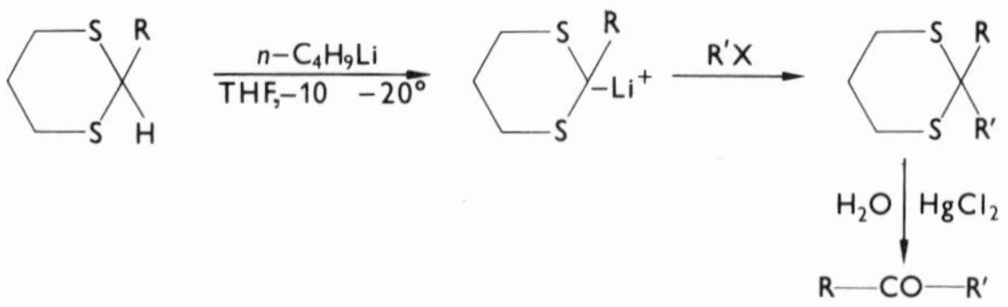

$$(6.81)$$

Types of compounds to be prepared: R–COR, RCOCO$_2$H, RCOCOR, RCOC(OH)R'R', RCOC(NH$_2$)R'R''; R = H, alkyl, allyl, benzyl aryl, or $CH_2CH(OC_2H_5)_2$ and $C_6H_5CH(O^-)$

Corey and co-workers have also found that 1,3-bis (methylthio) allyllithium (**6.53**) is a useful reagent for the synthesis of $\alpha,\beta$-unsaturated aldehydes (eqs. 6.82, 6.83) [204]. Another reagent, the lithium salt of allylthiomethylphosphonate can be used to introduce double chain

$$\text{6.53} + n\text{-}C_5H_{11}Br \xrightarrow[-75°,\ 2\,hr]{} n\text{-}C_5H_{11}\underset{\underset{SCH_3}{|}}{C}HCH=CHSCH_3 \qquad\qquad (6.82)$$

$$90\%$$

$$\xrightarrow[H_2O-acetonitrile]{HgCl_2} n\text{-}C_5H_{11}CH=CHCHO$$

$$trans \quad 84\%$$

$$\text{6.53} + CH_3CH_2CHO \longrightarrow C_2H_5\underset{\underset{OH}{|}}{C}H\text{-}\underset{\underset{SCH_3}{|}}{C}HCH=CHSCH_3$$

$$(6.83)$$

$$\xrightarrow[HgCl_2\ (or\ AgNO_3)]{H_2O} C_2H_5\underset{\underset{OH}{|}}{C}H\text{-}CH=CH\text{-}CHO$$

$$41\ \%\ (or\ 48\ \%)$$

branching at the carbonyl carbon [205]:

$$82\% \qquad\qquad 83\%$$

Mukaiyama and co-workers have shown that allyl 2-pyridyl sulfide can be used for carbon-carbon bond formation with a minor double bond migration (eq. 6.85)[206]. The sulfide group in the products can be removed (eq. 6.86) by the reduction with LiAlH$_4$ in the presence of heavy metal salts [207].

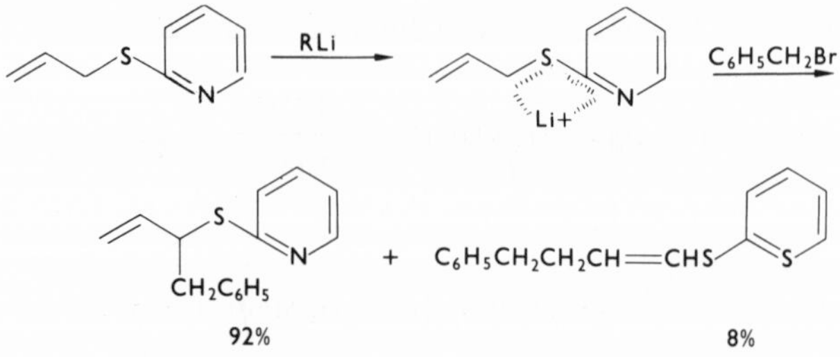

$$(6.85)$$

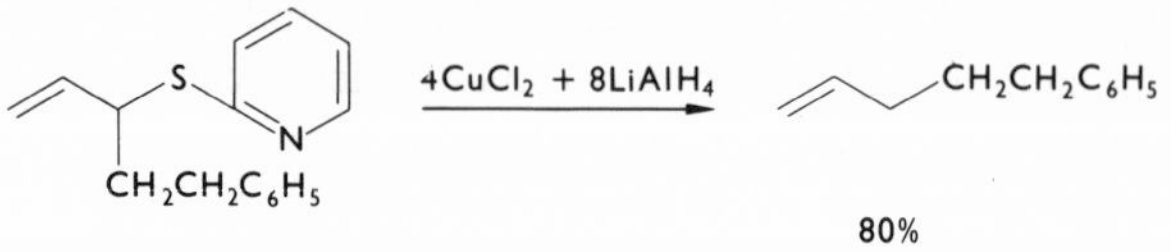

92%                                                        8%

$$(6.86)$$

80%

Trimethylene and ethylene dithiotosylate can be used for the protection of active methylene groups (eq. 6.87)[208]. The methylene group activated by carbonyl group can be made more active by the

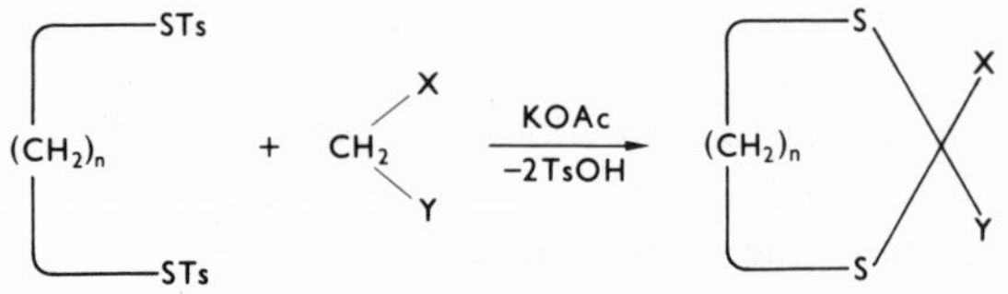

$$(6.87)$$

conversion into the enamines as shown below (eq. 6.88). The thioketal group may finally be hydrolyzed to the corresponding ketone [203,209] or

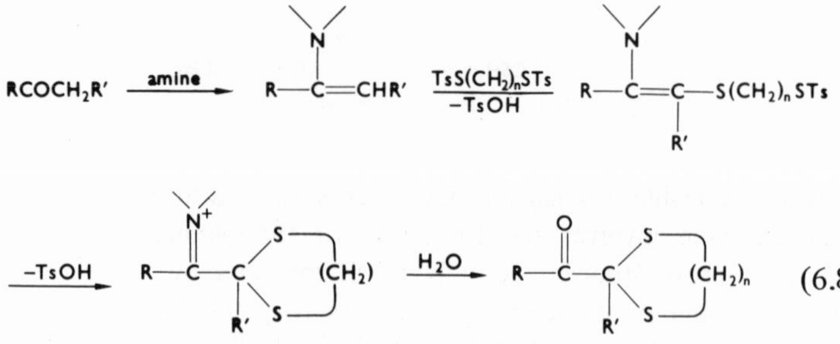

$$(6.88)$$

reduced to the starting methylene group by Raney nickel [209b,210], or by the method shown above in eq. 6.86.

## 6.6 REFERENCES

1. Houben-Weyl, Methoden der Organischen Chemie (Georg Thieme Verlag, Stuttgart) Vol. IX.
2. W.A. Pryor, Mechanism of Sulfur Reactions (McGraw-Hill, New York, 1962).
3. S.R. Sandler and W. Karo, Organic Functional Groups Preparation (Academic Press, New York, 1968) Chapt. 18, p. 478.
4. E.E. Reid, Organic Chemistry of Bivalent Sulfur (Chemical Publishing Co., Inc., New York, 1960) Vols. II and III.
5. R.B. Wagner and H.D. Zook, Synthetic Organic Chemistry (John Wiley, New York, 1953) Chapt. 32, p. 787.
6. J. Mark, Advanced Organic Chemistry (McGraw-Hill, New York, 1968).
7. a) N. Kharasch (Editor), Organic Sulfur Compounds (Pergamon Press, New York, 1961) Vol. I; b) N. Kharasch and C.Y. Meyers (Editors), Organic Sulphur Compounds (Pergamon Press, New York, 1966) Vol. 2; c) C.C. Price and S. Oae, Sulfur Bonding (Ronald Press, New York, 1962); d) S. Oae, Yuki Iokagobutsu no Kagaku, (Kagaku Dojin, Kyoto, Japan, 1968) Vol. 1.
8. J.R. Campbell, J. Org. Chem., 29, 1830 (1964).
9. a) S.F. Birch, R.A. Dean, N.J. Hunter and E.V. Whitehead, J. Org. Chem., 22, 1590 (1957); b) S.F. Birch, R.A. Dean and N.F. Hunter, ibid., 23, 1026 (1958); c) L.A. Paquette and J.C. Philips, Chem. Commun., 680 (1969); d) C.R. Johnson, J.E. Keiser and J.C. Sharp, J. Org. Chem., 34, 860 (1969); See also, T.W. Craig, G.R. Harvey and G.A. Berchtold, ibid., 32, 3743 (1967); e) E.J. Corey and E. Block, ibid., 31, 1663 (1966).
10. I. Tabushi, Y. Tamaru and Z. Yoshida, Tetrahedron Lett., 2931 (1970).
11. a) S. Oae and R. Kiritani, Bull. Chem. Soc. Japan, 38, 1381 (1965); b) F.M. Furman, J.H. Thelin, D.W. Hein and W.B. Hardy, J. Amer. Chem. Soc., 82, 1450 (1960).
12. a) E.L. Eliel and R.A. Faignault, J. Org. Chem., 29, 1630 (1964); K.A. Latif and P.K. Chakraberty, Tetrahedron Lett., 971 (1967); b) F.N. Jones, J. Org. Chem., 33, 4290 (1968).
13. a) F.W. Stacey and J.F. Harris, Jr., Organic Reactions (John Wiley, New York, 1963) Vol. 13, p. 150; b) G. Sosnovsky, Free Radical Reactions in Preparative Organic Chemistry (MacMillan Co., New York, 1964); c) B.A. Bohm and P.I. Abell, Chem. Rev., 62 599 (1962); d) E.N. Prilezhayeva, Organosulfur Chemistry (M.J. Janssen, Editor) (Interscience, New York, 1967) Chapt. 4.
14. a) W.E. Truce in Ref. 7a, Chapt. 12; b) G. Brindell and S.J. Cristol in Ref. 7a, Chapt. 13; c) A.A. Oswald and K. Griesbaum in Ref. 7b, Chapt. 9; d) A.A. Oswald and T.J. Wallace in Ref. 7b, Chapt. 8.
15. W.H. Mueller and K. Griesbaum, J. Org. Chem., 32, 856 (1967).
16. A.A. Oswald, K. Griesbaum, B.E. Hudson, Jr. and J.H. Bregman, J. Amer. Chem. Soc., 86, 2877 (1964).
17. C.D. Hurd and L.L. Gershheim, ibid., 69, 2328 (1947); M.S. Kharasch and C.F. Fuchs, J. Org. Chem., 13, 97 (1948).
18. H.L. Goering, D.I. Relyea and D.W. Larsen, J. Amer. Chem. Soc., 78, 348 (1956).
19. S.J. Cristol and G.D. Brindell, ibid., 76, 5699 (1954).
20. N. Kharasch in Ref. 7a, Chapt. 32.
21. A.J. Havlik and N. Kharasch, J. Amer. Chem. Soc., 78, 1207 (1956).

22. A. Dondoni, G. Modena and G. Seorrano, *Boll. Sci. Fac. Chim. Ind. Bologna*, 22, 26 (1964); K. Calo, G. Mallboni, G. Modena and G. Seorrano, *Tetrahedron Lett.*, 4399 (1965); L. Dinunno, G. Gelloni, G. Modena and G. Seorrano, *ibid.*, 4405 (1965); T.C. Fahey and D.J. Lee, *J. Amer. Chem. Soc.*, 88, 5555 (1966).
23. a) N. Kharasch, Z.S. Arian and A.J. Havlik, *Quarterly Reports on Sulfur Chemistry*, No. 2, 93 (1966); b) J.F. Harris, Jr., *J. Amer. Chem. Soc.*, 84, 3148 (1962).
24. a) Ref. 9e; b) E.D. Well, K.J. Smith and R.J. Gruber, *J. Org. Chem.*, 31, 1669 (1966); c) F. Lautenschlaeger, *ibid.*, 31, 1679 (1966); d) F. Lautenschlaeger, *Can. J. Chem.*, 34, 3991 (1966).
25. F. Lautenschlaeger, *J. Org. Chem.*, 33, 2620 (1968).
26. F. Lautenschlaeger, *ibid.*, 33, 2627 (1968).
27. F. Lautenschlaeger and N.V. Schwartz, *ibid.*, 34, 3991 (1969).
28. a) N. Latif, I. Fathy, N. Mishviky and B. Haggag, *Can. J. Chem.*, 44 629 (1966); b) D.D. Reynolds and D.L. Field, Heterocyclic Compounds with Three and Four-membered Rings (Interscience, New York, 1964); c) M. Sander, Chem. Rev., 66, 297 (1966); d) L. Goodman and E.J. Reist in Ref. 7b.
29. T.J. Barton and R.G. Zika, *J. Org. Chem.*, 35, 1729 (1970).
30. a) G. Tsuchihashi, M. Fukuyama and M. Yamauchi, *Tetrahedron Lett.*, 1971 (1967); b) A. Ohno, Y. Ohnishi and G. Tsuchihashi, *J. Amer. Chem. Soc.*, 91, 5038 (1969); c) A. Ohno, Y. Ohnishi and G. Tsuchihashi, *Tetrahedron*, 25, 871 (1969).
31. G. Laban and R. Mayer, *Z. Chem.*, 7, 227 (1967).
32. Ref. 9d.
33. J. Strating, L. Thijs and B. Zwanenburg, *Rec. trav. chim. Pays-Bas*, 86, 641 (1967).
34. For more examples, see: W.J. Middleton, E.G. Howard and W.H. Sharkey, *J. Amer. Chem. Soc.*, 83, 2589 (1961); G. Oster, L. Citrael and M. Goodman, *ibid.*, 84, 703 (1962).
35. E. Block and E.J. Corey, *J. Org. Chem.*, 34, 896 (1969).
36. R. Srinivasan and K.H. Carlough, *J. Amer. Chem. Soc.*, 89, 4932 (1967).
37. A. Schönberg, A. Stephenson, H. Kaltschmidt, E. Petersen and H. Schutten, *Chem. Ber.*, B66, 237 (1933).
38. J.R. Nooi and J.F. Arens, *Rec. trav. chim Pays-Bas*, 80, 245 (1961)
39. I. Tabushi, K. Okazaki and R. Oda, *Tetrahedron Lett.*, 3591, 3821 (1967).
40. W.A. Sheppard, *J. Org. Chem.*, 29, 895 (1964).
41. C.G. Moore and B.R. Trego, *J. Chem. Soc.*. 9205 (1962).
42. P.D. Bartlett and G. Meguerian, *J. Amer. Chem. Soc.*, 78, 3710 (1956); b) C. Walling and R. Rabinowitz, *ibid.*, 81, 1243 (1959); c) R.G. Harvey, H.I. Jacobson and E.V. Jensen, *ibid.*, 85, 1618, 1623 (1963).
43. a) D.N. Harp, J.G. Gleason and J.P. Snyder, *ibid.*, 90, 4181 (1968); b) D.N. Harp and J.G. Gleason, *ibid.*, 93, 2437 (1971); c) D.N. Harp and J.G. Gleason, *J. Org. Chem.*, 35, 3259 (1970); d) D.N. Harp and J.G. Gleason, *ibid.*, 36, 73 (1971).
44. H.C. Brown and M.M. Mildland, *J. Amer. Chem. Soc.*, 93, 3291 (1971).
45. a) W.W. Hartman, L.A. Smith and J.B. Dickey, Org. Syntheses, Coll. Vol.2, p. 242; b) T. Fujisawa, N. Ohtsuka, T. Kobori and G. Tsuchihashi, *Tetrahedron Lett.*, 4533, 5071 (1968); T. Fujisawa, T. Kobori and G. Tsuchihashi, *ibid.*, 4291 (1969); T. Fujisawa, N. Ohtsuka and G. Tsuchihashi, *Bull. Chem. Soc. Japan*, 43, 1189 (1970).
46. A.J. Neale, P.J.S. Bain and T.J. Rawlings, *Tetrahedron*, 25, 4583, 4593 (1969).
47. R.T. Wragg, *J. Chem. Soc.*, 5482 (1954).
48. B.S. Farah and E.E. Gilbert, *J. Org. Chem.*, 28, 2807 (1963).
49. a) E.A. Fehnel and M. Carmack, *J. Amer. Chem. Soc.*, 71, 84 (1949); b) R.M. Roberts and Ch.-Ch. Cheng, *J. Org. Chem.*, 23, 983 (1958).

50. a) A. Froling and J.F. Arens, *Rec. trav. chim. Pays-Bas,* **81,** 1009 (1962); b) S. Oae, W. Tagaki and A. Ohno, *Tetrahedron,* **20,** 427 (1964).
51. E.J. Corey and D. Seebach, *Angew. Chem.,* **77,** 1134 (1965).
52. T. Oda and K. Yamamoto, *J. Org. Chem.,* **26,** 4679 (1961).
53. G. Jeminet and A. Kergomard, *Compt. rend.,* **259,** 2248 (1964).
54. S. Oae, W. Tagaki and A. Ohno, *Tetrahedron,* **20,** 417 (1964).
55. A.L. Kramzfelder and R.R. Vogt, *J. Amer. Chem. Soc.,* **60,** 1714 (1938).
56. T.H. Chan, A. Melnyk and D.N. Harp, *Tetrahedron Lett.,* 201 (1969); T.H. Chan and A. Melnyk, *J. Amer. Chem. Soc.,* **92,** 3718 (1970).
57. H. Alper and E.C.H. Kenng, *Tetrahedron Lett.,* 53 (1970).
58. a) J.F. Kerr, *Chem. Rev.,* **66,** 465 (1966); b) J.L. Franklin and H.E. Lumpkin, *J. Amer. Chem. Soc.,* **74,** 1023 (1952); c) M.H. Back and A.H. Seon, *Can. J. Chem.,* **38,** 1076 (1963); d) E.H. Brayl, A.H. Seon and B. deB. Darwent, *J. Amer. Chem. Soc.,* **77,** 5282 (1957).
59. a) L.E. Sutton, Interatomic Distances (Special Publication, The Chemical Society of London, No. 11, 1958); b) S.C. Abraham, Quart. Rev. (London), 407 (1956).
60. a) L. Brockway and H.O. Jenkins, *J. Amer. Chem. Soc.,* **58,** 2036 (1936); b) H.J.N. Bowen, *Trans. Faraday Soc.,* **50,** 452 (1954); c) W.R. Blackmore and S.C. Abraham, *Acta Cryst.,* **8,** 329 (1955); d) J. Toussaint, *Bull. Soc. Chim. Belges,* **54,** 319 (1945); e) L.E. Sutton and G.C. Hampson, *Trans. Faraday Soc.,* **31,** 945 (1935); f) U. Blukis, P.H. Kasai and R.J. Myers, *J. Chem. Phys.,* **38,** 2753 (1963).
61. a) L. Pierce and M. Hayashi, *ibid.,* **35,** 479 (1961); b) G.L. Cunningham, Jr., A.W. Boyd, R.J. Myers, W.D. Gwinn and W.I. Levan, *ibid.,* **19,** 676 (1951); R.L. Shoemaker and W.H. Flyare, *J. Amer. Chem. Soc.,* **90,** 6263 (1968); c) E. Goldish, *J. Chem. Educ.,* **36,** 408 (1959); d) Z. Nahlovska, B. Mahlovska and M.M. Seip, *Acta Chem. Scand.,* **23,** 3534 (1969).
62. H.H. Jaffe and M. Orchin, Theory and Applications of Ultraviolet Spectroscopy (John Wiley, New York, 1962) Chapt. 17.
63. S.F. Mason, Physical Methods in Heterocyclic Chemistry (Academic Press, New York, 1963) Vol.2.
64. a) E.J. Corey and E. Block, *J. Org. Chem.,* **34,** 1233 (1969); b) E. Block, *Quart Reports on Sulfur Chemistry,* **4,** 239 (1969).
65. a) R.E. Davis, *J. Org. Chem.,* **23,** 216, 1380 (1958); b) J. Barrett and M.J. Hitch *Spectrochim. Acta,* Part A, **24,** 265 (1968).
66. H.S. Gutowsky, R.L. Rutledge, M. Tamres and S. Searles, *J. Amer. Chem. Soc.,* **76,** 4242 (1956).
67. H.P. Koch, *J. Chem. Soc.,* 387 (1949).
68. D. Welti and D. Whittaker, *ibid.,* 4372 (1962).
69. S. Oae, W. Tagaki and A. Ohno, *Tetrahedron,* **20,** 437 (1964).
70. a) A. Mangini and R. Passerini, *Experientia,* **12,** 49 (1956); b) A. Mangini and C. Zauli, *J. Chem. Soc.,* 2210 (1960); c) J. Degami, A. Mangini, A. Tronbetti and C. Zauli, *Spectrochim, Acta,* Part A, **23,** 1351 (1967); *Chem. Abstr.,* **67,** 37925g (1967), d) A. Mangini and R. Passerini, *Gazz. chim. ital.,* **84,** 606 (1954).
71. a) Y. Allingham, R.C. Cookson and T.A. Cragg, *Tetrahedron,* **24,** 1989 (1968); b) M. Brink and E. Larsson, *ibid.,* **26,** 5535 (1970); M. Brink, *ibid.,* **27,** 143 (1971).
72. a) T.D. Epley and R.S. Drago, *J. Amer. Chem. Soc.,* **89,** 5770 (1967); b) T.D. Epley and R.S. Drago, *ibid.,* **91,** 2883 (1969); c) W. Partenheimer, T.D. Epley and R.S. Drago, *ibid.,* **90,** 3886 (1968); G.C. Vogel and R.S. Drago, *ibid,* **92,** 5347 (1970).
73. C.H. Henrickson and D.R. Eyman, *Inorg. Chem.,* **6,** 1461 (1967).
74. I.P. Romm, E.M. Guryyanova and K.A. Kocheshkov, *Tetrahedron,* **25,** 2455 (1969).

75. a) C.J. Pederson, *J. Amer. Chem. Soc.*, **89**, 7017 (1967); C.J. Pederson, *ibid.*, **92**, 386, 391 (1970); C.J. Pederson, *J. Org. Chem.*, **36**, 254, 255 (1971); b) H.K. Frensdorff, *J. Amer. Chem. Soc.*, **93**, 600 (1971).
76. E. Fromm and G. Raiziss, *Ann. Chem.*, **74**, 90 (1910); H. Kwart, R.W. Body and D.M. Hoffman, *Chem. Commun.*, 765 (1967); C.R. Johnson and J.J. Rigan, *J. Amer. Chem. Soc.*, **91**, 5398 (1969); G. Edwin, Jr. and M.M. Chang, *Tetrahedron Lett.*, 875 (1971).
77. a) H. Tsubomura and R.P. Lang, *J. Amer. Chem. Soc.*, **83**, 2085 (1961); b) J.D. McCullough and I.C. Zimmermann, *J. Phys. Chem.*, **66**, 1198 (1962); see also, M. Good, A. Major, J. Nag-Chaudhuri and S.P. McGlynn, *ibid.*, **83**, 4329 (1961).
78. For more examples of $K_C$ values. see: N.W. Tideswell and J.D. McCullough, *J. Amer. Chem. Soc.*, **79**, 1031 (1957); J.D. McCullough and D. Mulvey, *ibid.*, **81**, 1291 (1959); D.W. Larsen and A.L. Allred, *ibid.*, **87**, 1216 (1965); J.M. Goodenow and M. Tamress, *J. Chem. Phys.*, **43**, 3393 (1965); M. Tamress and J.M. Goodenow, *J. Phys. Chem.*, **71**, 1982 (1967); I.G. Arzamanova and E.N. Gur'yanova, *Dokl. Akad. Nauk SSSR*, **166**, 1151 (1966); I.G. Arzamanova and E.N. Gur'yanova, *Zh. Obshch. Khim.*, **36**, 1157 (1966).
79. a) P.A.D. De Main, *J. Chem. Phys.*, **26**, 1192 (1957); J.S. Han, *ibid.*, **20**, 1170 (1952); K. Hartley and J.A. Skinner, *Trans. Faraday Soc.*, **46**, 621 (1958); b) S. Nagakura, *J. Amer. Chem. Soc.*, **80**, 520 (1958).
80. A. Walsh and W. Price, *Nature*, **148**, 372 (1941)
81. F.A. Matsen, Technique of Organic Chemistry (A. Weissberger, Editor) (Interscience, New York, 1956) Vol.9.
82. H.A. Bent, *Chem. Rev.*, **68**, 587 (1968).
83. E.L. Muetterties and R.A. Schunn, *Quart. Rev. (London)*, **20**, 245 (1966).
84. N.C. Baenziger, R.E. Buckles, R.J. Manner and T D. Simpson, *J. Amer. Chem. Soc.*, **91**, 5749 (1969).
85. G. Allegra, G.E. Wilson, Jr., E. Benedetti, C. Pedone and R. Albert, *ibid.*, **92**, 4002 (1970).
86. A. Senning (Editor), Sulfur in Organic and Inorganic Chemistry (Marcel Dekker, Inc., New York, 1971) Chapt. 1.
87. H. Böhme, H. Fischer and R. Frank, *Ann. Chem.*, **563**, 54 (1949); H. Böhme, *Chem. Ber.*, **74**, 248 (1941).
88. F.G. Bordwell, G.D. Cooper and H. Morita, *J. Amer. Chem. Soc.*, **79**, 376 (1957).
89. M. Murakami and S. Oae, *Proc. Japan Acad.*, **25**, 12 (1949).
90. W. Drenth and H. Hogeveen, *Rec. trav. chim. Pays-Bas*, **79**, 1002 (1960); T.L. Stamhuis, W. Drenth and H. Hogeveen, *ibid.*, **82**, 385, 394, 405, 410 (1963).
91. a) G.M. Bennett, *Trans. Faraday Soc.*, **37**, 795 (1941); b) A.C. Knipe and C.J.M. Stirling, *J. Chem. Soc.*, B, 1218 (1968).
92. a) A. Streitwieser, Jr., Solvolytic Displacement Reactions (McGraw-Hill, New York, 1962) p. 108–110; b) K.D. Gundermann, *Angew. Chem. Int. Ed. Engl.*, **2**, 674–683 (1963); c) B. Capon, *Quart. Rev. (London)*, **18**, 45 (1964).
93. P.D. Bartlett and C.G. Swain, *J. Amer. Chem. Soc.*, **71**, 1406 (1949).
94. A.G. Ogston, E.R. Holiday, J.S.L. Philpot and L.A. Stocken, *Trans Faraday Soc.*, **44**, 45 (1948).
95. H. Böhme and K. Sell, *Chem. Ber.*, **81**, 123 (1948).
96. G.M. Bennett, F. Heathcoat and A.N. Mosses, *J. Chem. Soc.*, 2567 (1929).
97. G.M. Bennett and E.G. Turner, *ibid.*, 813 (1938).
98. S.C.J. Olivier, *Rec. trav. chim. Pays-Bas*, **56**, 247 (1937).
99. G.M. Bennett and W. Berry, *J. Chem. Soc.*,, 1676 (1927).
100. R.I. Crawford and C. Woo, *Can. J. Chem.*, **43**, 3178 (1965).
101. M. Hojo, T. Ichi, Y. Tamaru and Z. Yoshida, *J. Amer. Chem. Soc.*, **91**, 5170 (1969).

102. H.L. Goering and K.L. Howe, *ibid.*, **79**, 6542 (1957).
103. S.J. Cristol and R.P. Argarbright, *ibid.*, **79**, 3441 (1959).
104. R.E. Ireland and H.A. Smith, *Chem. Ind. (London)*, 1252 (1959).
105. L.A. Paquette, G.V. Meeham and L.D. Wise, J. Amer. Chem. Soc., **91**, 3231 (1969).
106. R.F. Gratz and P. Wilder, Jr., *Chem. Commun.*, 1449 (1970).
107. a) C. Bamford, B. Capon and W.G. Overend, *J. Chem. Soc.*, 5138 (1962); R.L. Whistler and J. van Es, *J. Org. Chem.*, **28**, 2303 (1963); b) B. Capon, *Chem. Rev.*, **69**, 407 (1969).
108. a) T.H. Fife and L. Brod, *J. Org. Chem.*, **33**, 4136 (1968); b) N.C. de and L.R. Fedor, *J. Amer. Chem. Soc.*, **90**, 7266 (1968); c) T.H. Fife and L.K. Jao, *ibid.*, **91**, 4217 (1969); d) T.H. Fife and E. Anderson, *ibid.*, **92**, 5464 (1970).
109. E. Campaigne in Ref. 7a, p. 134.
110. G.F. Grillot and P.T.S. Lau, *J. Org. Chem.*, **30**, 28 (1965); P.T.S. Lau and G.F. Grillot, *ibid.*, **28**, 2763 (1963).
111. J. Deganin and R. Fochi, Boll. *Sci. Fac. chim. Ind. Bologna*, **22**, 7 (1964); *Chem. Abstr.*, **61**, 4261 (1964).
112. W.E. Truce, G.H. Birm and E.T. McBee, *J. Amer. Chem. Soc.*, **74**, 3594 (1952); F.G. Bordwell and B.M. Pitt, *ibid.*, **77**, 527 (1955); W.E. Lawson and T.P. Dawson, *ibid.*, **49**, 3119 (1927); H. Böhme and H.-J. Gran, *Ann. Chem.*, **577**, 68 (1952); H. Böhme and H.-J. Gran, ibid., **581**, 133 (1953); H. Richtzenhain and B. Alfredsson, *Chem. Ber.*, **86**, 142 (1953).
113. J.M. Stewart, *J. Org. Chem.*, **28**, 596 (1963).
114. J.M. Stewart, *ibid.*, **29**, 1655 (1964).
115. G.Y. Epshtein and I.A. Usov and S.Z. Ivin, *Zh. Obshch. Khim.*, **34**, 1948 (1964); *Chem. Abstr.*, **61**, 8178 (1964).
116. J.M. Stewart and C.H. Burnside, *J. Amer. Chem. Soc.*, **75**, 243 (1953).
117. P.S. Magee, in Ref. 86, Chapt. 9.
118. G.E. Wilson, *Quart. Report Sulfur Chem.*, **2**, 313 (1967).
119. F. Boberg, G. Winter and G.R. Schiltze, *Chem. Ber.*, **89**, 1160 (1956).
120. K.C. Schreiber and V.P. Fernandez, *J. Org. Chem.*, **26**, 2478 (1961).
121. W. Groebel, *Chem. Ber.*, **92**, 2887 (1959).
122. W. Tagaki, K. Kikukawa, K. Ando and S. Oae, *Chem. Ind. (London)*, 1624 (1964).
123. D.L. Tuleen and D.N. Buchanan, *J. Org. Chem.*, **32**, 495 (1967); D.L. Tuleen and V.C. Marcum, *ibid.*, **32**, 204 (1967); D.L. Tuleen, *ibid.*, **32**, 4006 (1967); D.L. Tuleen and T.P. Stephens, *Chem. Ind. (London)*, 1555 (1966).
124. C.H. Han and W.E. McEwen, *Tetrahedron Lett.*, 2629 (1970).
125. T. Fujisawa, N. Ohtsuka and G. Tsuchihashi, *Bull. Chem. Soc. Japan*, **43**, 1189 (1970).
126. S. Oae, M. Nakai and N. Furukawa, *Chem. Ind. (London)*, 1438 (1971).
127. E. Bourgeois and A. Abraham, *Rec. trav. chim. Pays-Bas*, **30**, 407 (1911).
128. T. Zincke and O. Krüger, *Chem. Ber.*, **45**, 3468 (1912); K. Fires and W. Vogt, *Ann. Chem.*, **381**, 337 (1911); J. Boeseken, *Rec. trav. chim. Pays-Bas*, **29**, 315 (1910).
129. T. van Hone, *Bull. acad. Belg.*, **37**, 98 (1928).
130. a) T. Zincke and W. Frohneberg, *Chem. Ber.*, **43**, 837 (1909); b) A. Arcoria and G. Scarlata, *Ann. chim. (Rome)*, **54**, 139 (1964); *Chem. Abstr.*, **61**, 11919h (1964).
131. R. Passerini and G. Purrello, *Gazz. chim. Ital.*, **90**, 1277 (1960); *Chem. Abstr.*, **56**, 5871 (1962).
132. S. Oae, A. Ohno and W. Tagaki, *Bull. Chem. Soc. Japan*, **35**, 681 (1962).
133. A.I. Shatenshtein, E.A. Ravinovich and V.A. Pavlov, *Reactionaya Sposobrot. Organ. Soedin Tartusk. Gos. Univ.*, **1**, 54(1964); *Chem. Abstr.*, **61**, 15949 (1964).
134. H.C. Brown and Y. Okamoto, *J. Amer. Chem. Soc.*, **80**, 4979 (1958).

135. J.W. Baker, G.F.C. Barrett and W.T. Tweed, *J. Chem. Soc.*, 2831 (1952).
136. a) F.G. Bordwell and G.C. Cooper, *J. Amer. Chem. Soc.*, **74**, 1058 (1952); b) F.G. Bordwell and P.J. Boutan, *ibid.*, **78**, 854 (1956).
137. N.F. Hall and M.R. Sprinkle, *ibid.*, **54**, 3469 (1932).
138. S. Oae, M. Yoshihara and W. Tagaki, *Bull. Chem. Soc. Japan*, **40**, 951 (1967).
139. L. Chierici, H. Lumbroso and R. Passerini, *Bull. Soc. chim. Fr.*, 643 (1955); For more data, see Ref. 7c and 7d.
140. I. Minamida, Y. Ikeda, K. Uneyama, W. Tagaki and S. Oae, *Tetrahedron*, **24**, 5293 (1968).
141. a) R.W. Taft, Jr., Steric Effects in Organic Chemistry (M.S. Newman, Editor) (John Wiley, New York, 1956) Chapt. 13; b) J.E. Leffler and E. Grunwald, Rates and Equilibria of Organic Reactions (John Wiley, New York 1963).
142. C.Y. Meyers, *Gazz. chim. ital.*, **93**, 120 (1963).
143. W.A. Sheppard, *J. Amer. Chem. Soc.*, **83**, 4860 (1961).
144. a) T.W. Campbell and M.T. Rogers, *ibid.*, **70**, 1029 (1948); b) V. Baliah and K. Gunapathy, *Trans. Faraday Soc.*, **59**, 1784 (1963); c) V. Baliah and M. Uma, *Tetrahedron*, **19**, 455 (1963).
145. S. Oae and M. Yoshihara, Bull. Chem. Soc. Japan, **41**, 2082 (1968).
146. For more discussions, see Ref. 7c, Chapt. 1.
147. D.J. Cram, Fundamentals of Carbanion Chemistry (Academic Press, New York, 1965).
148. a) A.I. Shatenshtein, *Advan. Phys. Org. Chem.*, **1**, 161 (1963); b) Ref. 130; c) A.I. Shatenshtein, E.A. Ravinovich and V.A. Pavlov, *Zh. Obshch, khim*, **34**, 3991 (1963); *Chem. Abstr.*, **62**, 8982 (1965); d) E.A. Yakovleva, E.V. Tsvetkov, D.I. Lobanov, M.I. Kabachnik and A.I. Shatenshtein, *Tetrahedron Lett.*, 4161 (1966).
149. a) Y. Yano and S. Oae, Mechanism of Reactions of Sulfur Compounds (N. Kharasch and B.S. Thyagarajan, Editors) (Intra-Sci. Res. Found., Santa Monica, Calif., 1969) Vol. 4, p. 167; L.R. Slaugh and E. Bergman, *J. Org. Chem.*, **26**, 3158 (1961).
150. W.E. Doering and L.K. Levy, *J. Amer. Chem. Soc.*, **77**, 509 (1955).
151. E.J. Corey, H. Koenig and T.H. Lowty, *Tetrahedron Lett.*, 515 (1962).
152. R. Breslow and E. Mohasci, *J. Amer. Chem. Soc.*, **85**, 431 (1963).
153. B.R. Brown, Quart. Rev. (London). 5, 131 (1951); b) J. Hine, Physical Organic Chemistry (McGraw-Hill, New York, 1962) 2nd Ed., Chapt. 13.
154. a) K. Uneyama, W. Tagaki, I. Minamida and S. Oae, *Tetrahedron*, **24**, 5271 (1968); b) S. Oae, W. Tagaki, K. Uneyama and I. Minamida, *ibid.*, **24**, 5283 (1968).
155. A.A. Hartmann and E.L. Eliel, *J. Amer. Chem. Soc.*, **93**, 2572 (1971).
156. R.T. Wragg, *Tetrahedron Lett.*, 4959 (1969).
157. K.C. Bank and D.L. Coffen, *Chem. Commun.*, 8 (1969).
158. S. Wolfe, A. Rauk, L.M. Tel and I.G. Cziazmadia, *ibid.*, 96 (1970).
159. A. Rauk, S. Wolfe and I.G. Czizmadia, *Can. J. Chem.*, **47**, 113 (1969).
160. H.P. Koch and W.E. Moffitt, *Trans. Faraday Soc.*, **47**, 7 (1951). For more discussions, see: Refs. 7c and 144; G. Cilento, *Chem. Rev.*, **60**, 147 (1960); G.E. Kimball, *J. Chem. Phys.*, **3**, 188 (1940); D.P. Craig, A. Maccoll, R.S. Nyholm, L.E. Orgel and L.E. Sutton, *J. Chem. Soc.*, 332 (1954).
161. D.V. Banthrope, Elimination Reactions (Elsevier, New York, 1963).
162. J. Hine and P.B. Langform, *J. Amer. Chem. Soc.*, 78, 5002 (1956).
163. C.C. Price and R.G. Gills, *ibid.*, 75, 4750 (1953).
164. S. Oae and Y. Yano, *Tetrahedron*, **24**, 5721 (1968).
165. Y. Yano and S. Oae, *ibid.*, **26**, 67 (1970).
166. See also, E. Rothstein, *J. Chem. Soc.*, 1522 (1940).
167. E. Rothstein, *ibid.*, 1558 (1940).
168. S. Oae, A. Ohno and W. Tagaki, *Tetrahedron*, **20**, 443 (1964).

169. a) D.S. Tarbell and M.A. McCall, *J. Amer. Chem. Soc.*, **74**, 48 (1952); b) D.S. Tarbell and W.E. Lovett, *ibid.*, **78**, 2259 (1956).
170. C.C. Price and W.H. Snyder, *ibid.*, **83**, 1773 (1961).
171. C.C. Price and W.H. Snyder, *J. Org. Chem.*, **27**, 4639 (1962).
172. a) E.A. Fehnel, *J. Amer. Chem. Soc.*, **74**, 1569 (1952); b) J. Boeseken and E. de R. van Zuydewin, *Proc. Acad. Sci. (Amsterdam)*, **40**, 23 (1937); *Chem. Abstr.*, **31**, 4953 (1937).
173. a) D.E. O'Connor and W.T. Lyness, *J. Amer. Chem. Soc.*, **85**, 3044 (1963); b) D.E. O'Connor and C.D. Broaddus, *ibid.*, **86**, 2267 (1964); c) D.E. O'Connor and W.I. Lyness, *ibid.*, **86**, 3840 (1964).
174. P.P. Montijn and L. Brandsma, *Rec. trav. chim. Pays-Bas*, **83**, 456 (1964); L. Brandsma, H.E. Wijers and J.F. Arens, *ibid.*, **82**, 1040 (1963).
175. G, Eglinton, I.A. Lardy, R.A. Raphael and G.A. Sin, *J. Chem. Soc.*, 1154 (1964).
176. I. Iwai and I. Ike. *Chem. and Pharm. Bull. (Japan)*, **10**, 926 (1962); I. Iwai and I. Ike, *ibid.*, **11**, 1042 (1963); B.S. Thyagarajan, K.K. Balasabramanian and R.B. Rao, *Tetrahedron*, **23**, 1893 (1967).
177. a) H. Kwart and C.M. Hackett, *J. Amer. Chem. Soc.*, **84**, 1756 (1962); b) H. Kwart and E.R. Evans, *J. Org. Chem.*, **31**, 413 (1966).
178. C.Y. Meyers, C. Rinaldi and L. Bonoli, *ibid.*, **28**, 2440 (1963).
179. a) H. Kwart and T.J. George, *Chem. Commun.*, 433 (1970); b) P.J.W. Schuijl, H.J.T. Bos and L. Brandsma, *Rec. trav. chim. Pays-Bas*, **88**, 597 (1969); L. Brandsma and D. Schuijl-Laros, *ibid.*, **89**, 110 (1970).
180. B.W. Bycroft and W. Landon, *Chem. Commun.*, 168 (1970).
181. Y. Makisumi and A. Murabayashi, *Tetrahedron Lett.*, 1971 (1969); Y. Makisumi and A. Murabayashi, *ibid.*, 2449, 2453 (1969).
182. R.F. Bridger and G.A. Russell, *J. Amer. Chem. Soc.*, **85**, 3754 (1963).
183. K. Uneyama, H. Namba and S. Oae, *Bull. Chem Soc. Japan*, **41**, 1928 (1968).
184. a) T. Alfrey and C.C. Price, *J. Polymer Sci.*, **2**, 101 (1947); C.C. Price, *ibid.*, **3**, 772 (1948); b) C.C. Price and J. Zomlefer, *J. Amer. Chem. Soc.*, **72**, 14 (1950); c) C.C. Price and T.C. Schwan, *J. Polymer Sci.*, **16**, 127 (1955); d) C.E. Scott and C.C. Price, *J. Amer. Chem. Soc.*, **81**, 2670, 2672 (1959).
185. W. Tagaki, T. Tada, T. Nomura and S. Oae, *Bull. Chem. Soc. Japan*, **41**, 1696 (1968).
186. H.J. Alkema and J.F. Arens, *Rec. trav. chim. Pays-Bas*, **79**, 1257 (1960).
187. W.A. Pryor, Free Radicals (McGraw-Hill, New York, 1966) p. 128.
188. A. Ohno and Y. Ohnishi, *Tetrahedron Lett.*, 4405 (1969).
189. a) W.G. Bentrude and J.C. Martin, *J. Amer. Chem. Soc.*, **84**, 1561 (1962); b) D.L. Tuleen, W.G. Bentrude and J.C. Martin, *ibid.*, **85**, 1938 (1963); c) J.C. Martin and J.W. Koenig, *ibid.*, **86**, 1771 (1964); d) T.H. Fischer and J.C. Martin, *ibid.*, **88**, 3382 (1966).
190. J.Q. Adams, *ibid.*, **92**, 4535 (1970).
191. P.J. Krusic and J.K. Kochi, *ibid.*, **93**, 846 (1971).
192. P.S. Skell and R.G. Allen, *ibid.*, **82**, 1511 (1960); P.S. Skell (Special Publication, No. 19, The Chemical Society, London, 1965) p. 137.
193. M. Ushio, J. Fukinoto and T. Yonezawa, *J. Amer. Chem. Soc.*, **93**, 908 (1971).
194. J.G. Calvert and J.N. Pitts, Jr., Photochemistry (John Wiley, New York, 1966) p. 488–492.
195. a) W.H. Laarhoven and T.J.H.M. Cuppen, *Tetrahedron Lett.*, 5003 (1966); W.H. Laarhoven, T.J.H.M. Cuppen and R.J.F. Nivard, *Rec. trav. chim. Pays-Bas*, **86**, 821 (1967).
196. W. Curruthers, *Nature*, **209**, 908 (1966).
197. J.D. Willett, J.R. Grunwald and G.A. Berchtold, *J. Org. Chem.*, **33**, 2297 (1968).
198. W.E. Truce and J.J. Breiter, *J. Amer. Chem. Soc.*, **84**, 1621 (1962).

199. a) R. Gerdil and E.A.C. Lucken, *J. Chem. Soc.*, 5444 (1963); b) R. Gerdil and E.A.C. Lucken, *ibid.*, 3916 (1964).
200. a) W.J. Brehm and T. Levenson, *J. Amer. Chem. Soc.*, **76**, 5389 (1954); b) A. Luttringhaus and H. Prinzbach, *Ann. Chem.*, **624**, 79 (1959).
201. For more examples, see: J.P. Shaefer, Organic Reactions (John Wiley, New York, 1967) Vol.15, p. 1–203.
202. W.E. Truce and F.E. Roberts, *J. Org. Chem.*, **28**, 961 (1963).
203. a) Ref. 51; b) E.J. Corey and D. Seeback, *J. Org. Chem.*, **31**, 4097 (1966); c) D. Seeback, B.W. Erickson and G. Singh, *ibid.*, **31**, 4303 (1966); d) D. Seebach, *Synthesis*, (1), 17 (1969).
204. a) E.J. Corey, B.W. Erickson and R. Noyori, *J. Amer. Chem. Soc.*, **93**, 1724 (1971); b) E.J. Corey and R. Noyori, *Tetrahedron Lett.*, 311 (1970).
205. E.J. Corey and J.I. Schulman, *J. Amer. Chem. Soc.*, **92**, 5522 (1970).
206. M. Mukaiyama, K. Narasaka, K. Maekawa and M. Furusato, Yukihanokiko Toronkai, 22nd Meeting, Nagoya, Japan, 1971.
207. T. Mukaiyama, K. Narasaka, K. Maekawa and M. Furusato, *Bull. Chem. Soc. Japan*, **44**, 2285 (1971).
208. R.B. Woodward, I.J. Pachter and M.L. Scheinbaum, *J. Org. Chem.*, **36**, 1137 (1971); See also, J.C.A. Chivers and S. Smiles, *J. Chem. Soc.*, 697 (1928); G.S. Brookes and S. Smiles, *ibid.*, 1723 (1926).
209. a) C. Djerassi, O. Halpern, G.R. Pettit and G.H. Thomas, *J. Org. Chem.*, **24**, 1 (1954); b) G.R. Pettit and E.E. van Tamelen, Organic Reactions (John Wiley, New York, 1962) Vol. **12**, p. 406.
210. D.J. Cram and H. Cordon, *J. Amer. Chem. Soc.*, **77**, 1810 (1955).

*Chapter 7*

# DISULFIDES AND POLYSULFIDES

Lamar Field

*Department of Chemistry, Vanderbilt University*
*Nashville, Tennessee 37203 USA*

## 7.1 INTRODUCTION

### 7.1.1 Scope and General Remarks

Disulfides and polysulfides have the structure $R^1SS_nR^2$, in which chains of sulfur atoms are terminated by two groups that may be the same, different, or connected in a ring. Chapter 7 considers only substances where the S-R bond involves a carbon linkage; a review is available of structures such as $ROS_nOR$ and $R_2NS_nNR_2$.[1] Disulfides (n = 1) and their oxidation products are by far the most important subclass, but trisulfides and higher sulfides have received much attention. This chapter discusses these bivalent sulfur compounds, with some attention to their oxidation products.

Disulfides might be thought to resemble peroxides, but actually the two have very little in common. Disulfides are reasonably reactive but do not share at all the highly unstable and often explosive nature of peroxides. A few bigots complain that volatile disulfides have disagreeable odors, but at least complaints are much less strident than against thiols. Repeated skin contact may cause a rash,[2] but ordinary precautions make such problems extremely rare.

The literature of di- and polysulfides is so vast that selection of material for an overview is difficult. Insight into what may be expected of this broad class must be the goal here, rather than comprehensiveness. Nevertheless, readers who wish to delve deeply into a topic can do so through the abundant reviews and leading references cited, combined with use of author and citation indexes. About a third of the references are to reviews and most others are to recent papers that cite earlier literature. This textbook-like device unfortunately often precludes properly crediting those who originally made the discoveries later elaborated upon by others;

the author complements his regret on this score with the hope that readers will work their way back to basic papers by further reading in areas that most interest them.

Comparatively little has been done so far with hydrodisulfides, hydropolysulfides, and their derivatives ($RSS_nH$ and $RSS_nX$). Since the chemistry is that of the -SS- as well as of the -SH or -SX moieties, the attention called for is provided at the end of this introduction.

### 7.1.2 Reviews

An early brief review remains useful.[2] A comprehensive treatment,[3] with experimental details, discusses hydrodisulfides and hydropolysulfides,[4] disulfides,[5] tri- and higher sulfides,[6] and various oxidation products.[3] Reid's encyclopedic coverage of di- and higher sulfides has about a thousand references.[7] Two reviews emphasize theoretical and mechanistic aspects.[8,9] The Kharasch-Meyers volumes of 1961 and 1966 list many reviews in appendices and include chapters on spectra,[10] lipoic acid,[10] fluoralkyl disulfides,[11] and many other topics cited later; a subsequent review provides continuity with these volumes.[12] Cyclic disulfides in general,[13] five- and six-membered multi-sulfur heterocycles,[14] and seven-membered sulfur-containing rings[15] have been reviewed. Other reviews cover metabolic[16a,c,f] and other biochemical aspects,[16f,g] the sulfur-sulfur bond,[16d] inorganic and organic polysulfides,[16e] a listing of reviews for 1961-1967,[16h] therapeutic uses,[17] and scission of C-S[18] or S-S[19] bonds. The last review, by Parker and Kharasch, will be referred to frequently. It was updated in 1966, with inclusion of a valuable listing of other reviews and key references.[20] Oae included a long chapter on di- and polysulfides in his treatise on organosulfur chemistry (1968); it is in Japanese, except for equations, tabular data, and references.[21] Fruitful sources of reviews, pertinent ones of which will be cited, are the periodicals *Quarterly Reports on Sulfur Chemistry* (1966-) and *Mechanisms of Reactions of Sulfur Compounds* (1966-). The first volumes of a highly promising biennial review cover disulfides and related substances (1970, 1973).[22]

Reviews are fruitful sources of research problems, it might be remarked for students, because restudy with modern tools is worthwhile of many observations that could not be understood earlier. As an example, several investigators studied the reaction of thiols with $SOCl_2$ about the turn of the century, with the result summarized in eq. 7.1.[6] This reaction looks like a curious but untidy one of little interest. However, dithiosulfites (7.1) actually are intermediates in the clean sequence shown by eq. 7.2.[23] The reaction proved to be quite general, with the stability of **7.1** being highly dependent on R. The decomposition of **7.1** is stoichiometric (R = $t$-butyl) and first order (R = Ph). It may occur as shown by eq. 7.3,[23] but further studies of the mechanism and of synthetic applications should be rewarding.

$$4RSH + SOCl_2 \longrightarrow RSSR + RSSSR + H_2O + 2HCl \qquad (7.1)$$

$$4RSH + 2SOCl_2 \xrightarrow{\text{Pyridine}} 2(RS)_2SO \xrightarrow{\Delta} RSSR + RSSSR + SO_2 \quad (7.2)$$

$$\mathbf{7.1}$$

$$\overset{O}{\underset{\displaystyle\text{RSSSR}}{\|}} \longrightarrow \overset{SO^-}{\underset{\displaystyle\text{RSSR}}{\underset{+}{\|}}} \xrightarrow{(RS)_2SO} RSSR + RSSSR + SO_2 \qquad (7.3)$$

### 7.1.3 Nomenclature

The practice of *Chemical Abstracts* is to name simple di- and higher sulfides in the same way as ethers. Thus $(PhS)_2$ is called phenyl disulfide, even though $(Cl_3CS)_2$ is called bis(trichloromethyl) disulfide.[24] Strong rebellion is afoot to use "di" even in simple instances, as with *di*phenyl disulfide, and we shall follow this less ambiguous practice. When a prefix is needed, for example, to show substitution on a grouping of higher precedence, it takes a form such as "alkyldithio," as with 2-(methyldithio)ethanol for $MeSS(CH_2)_2OH$.

Unsymmetrical compounds usually are named like ethers too, as with $t$-butyl 2-naphthyl disulfide for $t$-BuSS-2-$C_{10}H_7$, although the "dithio" approach is used for complex compounds.[24] Unsymmetrical disulfides often are called "mixed" disulfides, but the temptation to use this attractively short barbarism should be resisted; such usage is particularly misleading because unsymmetrical disulfides actually disproportionate to give "mixed" symmetrical disulfides (Sec. 7.5.6).

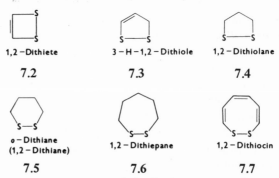

| 1,2–Dithiete | 3–H–1,2–Dithiole | 1,2–Dithiolane |
|:---:|:---:|:---:|
| **7.2** | **7.3** | **7.4** |

| o–Dithiane (1,2–Dithiane) | 1,2–Dithiepane | 1,2–Dithiocin |
|:---:|:---:|:---:|
| **7.5** | **7.6** | **7.7** |

**CHART 7.1   ILLUSTRATIONS OF HETEROCYCLIC NOMENCLATURE**

Systems that contain di- and higher sulfide moieties as part of a ring are named as heterocycles.[24] Chart 7.1 illustrates the handling of differing unsaturation in rings for which the usual characteristic endings signifying four- to eight-membered rings are -ete, -ole, -in, -epin and -ocin.[24] The naming of oxides is discussed in Sec. 7.5.2.

### 7.1.4 Hydrodisulfides and Hydropolysulfides, $RSS_nH$

In reviewing his undergirding research on the class $RSS_nH$, Böhme gives details for the typical syntheses illustrated by eqs. 7.4 and 7.5.[4] Hydrodisulfides and -trisulfides lose sulfur or $H_2S$ when heated, but those shown can be distilled under reduced pressure below 70°, preferably in quartz since glass accelerates decomposition.

$$Ac_2O + H_2S \xrightarrow[\text{(95\%)}]{\text{Pyridine}} Ac_2S \xrightarrow[\text{(50\%)}]{Cl_2, -20°} AcCl + AcSCl \qquad (7.4)$$

$$\downarrow \begin{array}{c} PhCH_2SH \\ (87\%) \end{array}$$

$$AcOEt + PhCH_2SSH \xleftarrow[\text{(88\%)}]{\text{Et OH-HCl}} AcSSCH_2Ph$$

$$(AcS)_2 \xrightarrow[\text{(90\%)}]{Cl_2, 0°} AcCl + AcSSCl \qquad (7.5)$$

$$\downarrow \begin{array}{c} EtSH \\ (91\%) \end{array}$$

$$EtSSSH \xleftarrow[\text{(57\%)}]{\text{MeOH-HCl}} AcSSSEt$$

Notice the two keys to entry into the hydrodisulfide or -trisulfide systems: chlorinolysis of the Ac-S bond to give AcCl and ClS-, and acid-catalyzed ester interchange to liberate the hydrosulfides.

Tsurugi recently reviewed continuing Japanese research on aralkyl hydrodisulfides.[25] Scheme 7.1 illustrates syntheses and reactions of this class.[25] Such compounds are stable for a few weeks.[26] Reaction (a), thermal decomposition, is stoichiometric and first order.[25] Homolytic dissociation of -SS- is the first and rate-determining step; the -SS- dissociation energy, thought to be only 20 kcal/mol, apparently is quite low (cf. Table 7.2). The tetrasulfide $RS_4R$ arises from coupling of $RSS\cdot$. Reaction (b) illustrates the frequent preference of nucleophiles like phosphines, phosphites and arsines for attacking at $S_\alpha$.[25] Attack at $S_\beta$ also can occur; it leads to hydrocarbons, disulfides and other products.

Predominance of attack at $S_\alpha$ *vs.* $S_\beta$ is governed by steric factors and by the
S-basicity (*cf.* Sec. 7.5.3.1) of the nucleophile. Cyanide and thiolate ions
attack at both $S_\alpha$ and $S_\beta$. Reaction (c) typifies reaction with a base.[25] The
course of reaction with amines depends on their basicity; they can act
either as nucleophiles to give RSH and several other products or as bases to
give $H_2S$, S, RSSR and polysulfides.[27]

**SCHEME 7.1**

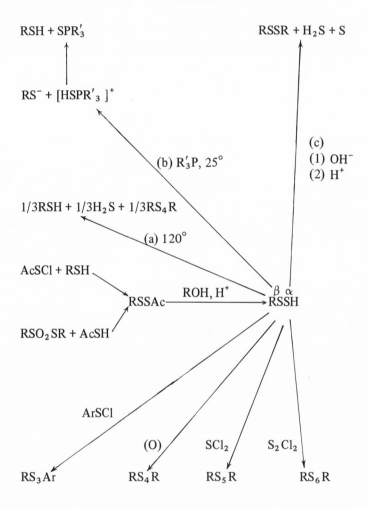

$R = PhCH_2$, $Ph_2CH$, $Ph_3C$[25]

Preparation and reactions of an alkyl hydrodisulfide are illustrated by eq. 7.6.[28] The analogue with OH instead of Cl was quite unstable; this and similar hydrodisulfides have a $\delta$ value of $\sim 3$ for RSSH and show ir absorption at $\sim 2500$ cm$^{-1}$.[26] It seems noteworthy that where $-CO_2H$ or -NHAc replaced Cl the analogues could not be obtained at all;[26] possible reasons for instability of such -OH, $-CO_2H$ and -NHAc compounds are mentioned in Secs. 7.6.1 and 7.6.2. Analogues of sulfenyl chlorides (*i.e.*, of RSCl) are known, as eq. 7.7 shows; they can be distilled at low pressure.[29] Reaction of $S_2Cl_2$ with suitable aromatic nuclei can lead to ArSSCl (for leading citations see ref. 30).

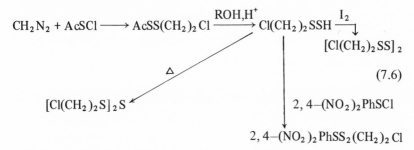

$$CH_2N_2 + AcSCl \longrightarrow AcSS(CH_2)_2Cl \xrightarrow{ROH,H^+} Cl(CH_2)_2SSH \xrightarrow{I_2}$$

$$[Cl(CH_2)_2SS]_2$$

$$(7.6)$$

$$[Cl(CH_2)_2S]_2S$$

$$2,4-(NO_2)_2PhSCl$$

$$2,4-(NO_2)_2PhSS_2(CH_2)_2Cl$$

$$\text{AlkSSAc} \xrightarrow{SO_2Cl_2} \text{AlkSSCl} \xrightarrow{I'} \text{AlkS}_4\text{Alk} \qquad (7.7)$$
$$\text{or}$$
$$\text{AlkS}_3\text{Alk} \qquad \text{RSH} \qquad \text{\backslash C=C\slash}$$

$$\text{AlkS}_3\text{R} \qquad \text{AlkSSC-CCl}$$

## 7.2 OCCURRENCE AND BIOLOGICAL SIGNIFICANCE

More than a few disulfides occur naturally, and study of their biogenesis should lead to fascinating results. Among acyclic ones, diallyl disulfide is the active constituent of garlic. An antibacterial disulfide monoxide (a thiolsulfinate) also present is allicin (7.8).[31] It smells like garlic and may arise from alliin (S-allyl-*L*-cysteine sulfoxide), a constituent that is cleaved

$$H_2C=CHCH_2\underset{\underset{O}{|}}{S}SCH_2CH=CH_2$$

7.8

by alliinase to products that have allicin-like activity.[31] Reid's review tells of several occurrences of odoriferous alkyl or alkenyl disulfides: dimethyl and diisopropyl (eucalyptus), dibutyl (skunks), allyl propyl (garlic) and allyl *sec*-butyl (asafetida).[7] One of the few genera of sea weed that possess odor contains two curious keto-disulfides, bis(3-oxoundecyl) disulfide and (-)-3-hexyl-4,5-dithiacycloheptanone; these lipids possibly are precursors of hydrocarbons present. [32] Trisulfides apparently occur as volatile constituents of onions. [33]

Several monocyclic disulfides occur naturally. Derivatives of 1,2-dithiolane (7.4) include the alkaloid brugine (7.9), [34] the neurotoxic compound nereistoxin from a marine worm (7.10), [35] the acid (7.11) from

7.4, $R^1 = R^2 = H$                   7.11, $R^1 = CO_2H; R^2 = H$
7.9, $R^1 = H; R^2 = CO_2$ Tropine       7.12, $R^1 = H; R^2 = (CH_2)_4CO_2H$
7.10, $R^1 = NMe_2; R^2 = H$

asparagus, [36] compounds from cabbage believed to be 1,2-dithiole-3-thiones (cf. Sec. 7.7), [37] and α-lipoic acid (7.12).

An extensive review states that "it would be hard to overestimate the importance of α-lipoic acid ... in biological systems today. It has been reviewed in every volume of the *Annual Reviews of Biochemistry* since 1952 ...". [38]    It outlines the isolation of the acid (only 30 mg of crystalline 7.12 from 10 tons of liver!), its crucial role in metabolism and growth, its possible involvement in photosynthesis, and its extensive chemistry. Much is known now about the biochemical functions of α-lipoic acid (cf. for example ref. 39). A major role is in the oxidative decarboxylation of α-keto acids. With pyruvic acid, the effect is as though decarboxylation to acetaldehyde (through action of thiamine) were followed by cleavage of 7.12 so that one sulfur atom of 7.12 acquired the Ac moiety of acetaldehyde and the other sulfur the H atom. The acetyl group is transferred later to the thiol group of Coenzyme A to give the thiol ester, "Acetyl CoA," a highly important metabolic intermediate.

The red dithiin (7.13) occurs in several species of plants; it gives the

7.13

corresponding thiophene readily. [40]    1,2,4,6-Tetrathiepane (7.14), the pentathiepane lenthionine (7.15), and the hexathiepane (7.16) have been

**7.14**             **7.15**             **7.16**

isolated from mushrooms.[15] Their identities were well established by nmr, mass spectra and synthesis. Lenthionine has been much studied by the Japanese because it is the characteristic odorous material of *Shiitake*, a mushroom prized in Asia for its edible qualities.

Heterocyclic disulfides with more than one ring occur naturally too. The intriguing alkaloid, cassipourine, has the structure **7.17**, based on X-ray evidence and reactions such as those shown.[41] A group of yellow

**7.17**

antibiotics isolated from a species of *Streptomyces* and based on pyrrolinonodithiole nuclei that has received much attention comprises **7.18-7.21**.[42] This group shows considerable activity against Gram-positive

**7.18, $R^1$ = $R^2$ = Me (Thiolutin)**          **7.20, $R^1$ = *i*-pr; $R^2$ = Me**
                                                    **(Isobutyropyrrothine)**

**7.19, $R^1$ = Et; $R^2$ = Me (Aureothricin)**   **7.21, $R^1$ = Me; $R^2$ = H**
                                                    **(Holomycin)**

and Gram-negative bacteria, fungi and protozoa,[42] but it is rather toxic (for example, the oral or subcutaneous dose of **7.18** that kills 50% of a group of mice is 25 mg/kg, in contrast to 1800 mg/kg for a much-used antibiotic, erythromycin).[43] Thiolutin (**7.18**) has the striking property of dissolving in aqueous $Na_2S$ solution and of reappearing upon acidification; a comparable reaction of the 1,2-dithiole-3-one (**7.22**) is believed to depend upon reversible formation of the hydrodisulfide (**7.23**)[42]:

                                                                           (7.8)

**7.22**                **7.23**

Another group of antibiotics and related metabolites, based this time on sulfur-bridged diketopiperazines, includes **7.24-7.27**.

**7.24 (Gliotoxin)[44]**     **7.25, n = 2 (Sporidesmin)[45]**     **7.27**

(Acetylaranotin,[47]
LL-S88α[48])

$$P_2S_5 + S \quad \Big\updownarrow \quad Ph_3P$$
(40%)                    (50%)

**7.26, n = 3 (Sporidesmin E)[45]**

Sporidesmin E (**7.26**) evidently can exist in different stable conformations of the trisulfide moiety, one of which reacts preferentially with triphenylphosphine.[46] It has high cytotoxic activity,[45] and **7.27** has antiviral activity.[47,48] The absolute configurations of **7.24**, **7.25** and **7.27** are identical, as are conformations of groups shared by **7.24** and aranotin.[49] X-ray study of **7.27** showed a short S-S nonbonded interaction distance of 3.27Å between sulfur atoms of adjacent molecules, indicating the interesting possibility of some $d$-orbital overlap between them.[48] Since structures like **7.24** and **7.27** inhibit multiplication of RNA viruses, the analogue **7.28** was synthesized as shown by eq. 7.9; it is highly effective and inhibits viral RNA synthesis by 50% at the low concentration of 0.0003 $\mu g/ml$.[50]    Activity of **7.28**, as well as of **7.24**, **7.25** and **7.27**, apparently depends upon the sulfur bridge.[50] Use of these configurationally understood mold metabolites of common chirality offers very intriguing chemical as well as biological possibilities. Thus conversion of the trisulfide (**7.26**) with triphenylphosphine to the disulfide (**7.25**), as shown, took place with retention of configuration at the asymmetric centers and therefore with preferential loss of the S-bonded rather than one of the C-bonded sulfur atoms.[46,51] By contrast, in conversion of the disulfide (**7.25**) to a monosulfide, circular dichroism strongly suggested inversion of the asymmetric carbon atoms.[46,51] Hitherto, the mechanisms of such "sulfur-stripping" reactions were unclear (*cf.* ref. 52, and also Sec. 7.5.3.2 which described further developments).

$$(7.9)$$

**7.28**

Sulfur-containing amino acids have been reviewed.[53] Those that contain the -SS- linkage are cystine (7.29) and the less important homocystine (7.30). Human hair contains about 18% cystine, and acid

$$\underset{\underset{NH_2}{|}}{HO_2CCH}(CH_2)_n SS(CH_2)_n \underset{\underset{NH_2}{|}}{CHCO_2H} \quad \underset{Oxid'n.}{\overset{Red'n.}{\rightleftharpoons}} \quad 2HS(CH_2)_n \underset{\underset{NH_2}{|}}{CHCO_2H}$$

7.29, n = 1 (cystine)         7.31, n = 1 (cysteine)
7.30, n = 2 (homocystine)     7.32, n = 2 (homocysteine)

hydrolysis affords a convenient means of getting it.[53] A recent report that cystine (or its reduction product, cysteine, 7.31) may be "essential" for the immature human, although neither is so for adults, also nicely outlines the roles of 7.29-7.32 in transsulfuration reactions of key importance to metabolism.[54]

Glutathione ($\gamma$-L-glutamyl-L-cysteinylglycine) occurs widely in plant and animal cells and is a very intriguing tripeptide.[55,56] In reactions that could turn out to be prototypes of "oxidative phosphorylation" in mitochondria (the powerhouses of cells), oxidation of glutathione by cytochrome c reportedly can be coupled to the formation of adenosine diphosphate from the monophosphate and inorganic phosphate (and of the triphosphate, ATP, from the diphosphate, ADP); glutathione disulfide is essential as a catalyst.[57] Processes of oxidative phosphorylation, which this one may simulate, are of transcendent biological importance because they somehow produce essential "high-energy" phosphate bonds by utilizing energy released upon oxidation of foods.

Occurrence of cystine in important small peptides is illustrated by 7.33-7.35 (in which amino acids are abbreviated as usual).

$H_2N$—Cys—Tyr—Ile                    $H_2N$—Cys—Tyr—Phe
       |        \                             |        \
       S         \                            S         \
       |          \                           |          \
       S           \                          S           \
       |            \                         |            \
      Cys—ArgNH$_2$ Glu                      Cys—ArgNH$_2$-Glu
       |                                      |
     Pro-Leu—GlyCONH$_2$                    Pro-Arg-GlyCONH$_2$

7.33 (Beef oxytocin)                  7.34 (Beef vasopressin)

$$(L)\text{Leu} \longrightarrow (D)\text{Leu}$$
$$(D)\text{Cys- SS - } (D)\,\text{Cys}$$
$$\diagdown \quad (L)\text{Val} \diagup$$

### 7.35 (Malformin A)

Oxytocin (7.33), the first polypeptide hormone to be synthesized, is the principal uterine-contracting and milk-ejecting hormone of the posterior pituitary gland; one microgram induces milk ejection in lactating females in 30 seconds, but activity is lost if the disulfide bond is reduced and the resulting dithiol benzylated.[58]  Vasopressins (e.g., 7.34) are a group of similar hormones that raise blood pressure and suppress urine flow.[59] Malformin A (7.35), a mold metabolite that causes malformation of plants, contains $D$-cystine instead of the $L$ form characteristic of proteins.[60]

On the larger scale of proteins, -SS- bonds of cystine residues frequently tie polypeptide chains together, or in single-chain proteins link two points in the amino-acid sequence.[61] In the permanent waving of hair, disulfide linkages are reduced (e.g., with ammonium thioglycolate) to permit curling and then are reoxidized to anchor the curl (e.g., with potassium bromate).[62]

Sulfur chemistry in relation to proteins has been reviewed.[63-65] One such review makes several chemically significant points.[64] Proteins for the most part can be divided into those with -SH groups (mostly cellular) and those with -SS- groups (mostly extracellular); since only a few contain both,[64] one wonders if there is not considerable significance still unsuspected in this curious fact. A main role of -SS- groups has long been thought to be in stabilizing three-dimensional structure,[64,65] but involvement of -SS- in biological activity is beginning to be recognized.[64] Thus the enzymes lipoic dehydrogenase and glutathione reductase participate in thiol-disulfide interactions (cf. Sec. 7.5.3.4; for discussion of these and other important "dithiol enzymes", see ref. 66); insulin (and vasopressin too) seems to react with tissue -SH.[64] Another view that emphasizes -SS- moieties of enzymes suggests that the unusual properties of -SH and of -SS- (to a lesser extent) promise to be important in enzymic catalysis.[65] For example, ribonuclease retains activity if only two -SS- bridges are reduced but not if all four are reduced, and -SS- linkages are essential for activity of several other enzymes as well.[65] Perhaps helping hands from organic chemists with these challenging areas of disulfide chemistry will enable reviewers of the future to write mechanisms for the actual interactions.

What is known of the nature of the -SS- bridges in these complex systems? The hormone insulin and the enzyme ribonuclease will serve as examples of at least ten important proteins that contain 3 to 25 -SS- bonds.[61] The -SS- bonds in these two are illustrated schematically in Fig.7.1 (*cf.* ref. 61). The sequence itself of amino acids predisposes a

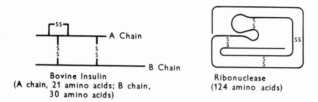

Bovine Insulin
(A chain, 21 amino acids; B chain, 30 amino acids)

Ribonuclease
(124 amino acids)

**Fig. 7.1.** Schematic representation of disulfide bonds in two proteins.

polypeptide to seek a particular thermodynamically stable conformation. For example, ribonuclease can be reduced and then can be unfolded in a solution of urea. When the reagents are removed, spontaneous refolding occurs and, after reoxidation, nearly all molecules are regenerated in the original conformation with the original enzymatic activity.[61] Had reformation of -SS- bonds been at random, only about 1% of the ribonuclease should have been reconstituted.[61] It is noteworthy that physiological properties of macrocyclic disulfides may depend not only on an intact -SS- bond, but even on its chirality (*cf.* Sec. 7.4.2).[67] Extensive reviews cover peptides that involve cystine,[68a] monocyclic disulfides (oxytocins, vasopressins)[68b] and polycyclic disulfides (insulin).[68c]

The -SS- linkage seems important to mitosis in cell division[69a] and to the clotting of blood plasma,[69b] in both perhaps through thiol-disulfide interchange reactions (see Sec. 7.5.3.4). It is important as well to the activity of antibodies,[64] to morphogenesis (the differentiation in an embryo that leads to various organs and parts),[70] and has even been suggested to be important in mammalian aging.[71]

Profound biological effects can occur even with disulfides not of natural origin. For example tetraethylthiuram disulfide, $[Et_2NC(S)S]_2$, is used under the name Antabuse (and over a dozen others) as an adjunct in treating alcoholism. The reason is that ingestion of alcohol by someone given the drug is followed by quite unpleasant effects, of which nausea is merely one.[72]

Hopefully, the biological relationships of -SS- bonds that have been sketched will suggest to organic chemists that their attention to such areas should generate valuable insights into intriguing and often critically important interdisciplinary problems. There is much room for research in these exciting areas that now are rife with dark shadows in chemical aspects.


## 7.3 PREPARATIONS

Chapter 4 dealt with reactions of thiols. Chapter 7 now considers such reactions, and others, for their preparative value. Methods useful for preparing unsymmetrical disulfides are signified in equations by use of $R^1SSR^2$. Those useful largely for symmetrical disulfides are indicated by use of RSSR, although they can be used for unsymmetrical disulfides when sufficient differences exist in components for practicable separation or when there is enough variation from statistical distribution in formation. Methods are compared in Sec. 7.3.5 , and various structural types are illustrated in Table 7.1.

### 7.3.1 By Oxidation

Mechanisms for the oxidation of thiols to disulfides by eq. 7.10a have been reviewed (see Chapter 4 and refs. 73, 74).

$$2RSH \xrightarrow{\text{(a)}} RSSR \xrightarrow{\text{(b)}} \text{Intermediates} \xrightarrow{\text{(c)}} 2RSO_3H$$
$$\text{(see Scheme 7.3)} \qquad (7.10)$$

Several reviews cover the older preparative literature and contain long lists of oxidants.[2,5,7,75] Among these oxidants are $O_2$, $H_2O_2$, $I_2$, $PbO_2$, $CuSO_4$, hypohalites, $Fe^{3+}$ salts, $K_3Fe(CN)_6$, NO, $SO_2Cl_2$, chromium and selenium compounds, peracids and their salts, sodium polysulfide, positive halogen compounds, $SO_2$, S and $PCl_5$. The first eight are usually the ones recommended for preparative use.[2,5] Less selective oxidants are $HNO_3$, $Br_2$, $Cl_2$, $H_2SO_4$ and $KMnO_4$. Actually, it would not be a surprise to learn that nearly everything with a reasonable thirst for electrons had been tried at one time or another!

Ease of oxidation usually decreases in the following order: ArSH $> 1°AlkSH > 2°AlkSH > 3°AlkSH >$ thiourea.[5] Even the highly hindered thiol 2,4,6-triadamantyl-benzenethiol can be oxidized to the disulfide, despite contrary indications of a model, although a radical (ArS·) does appear to persist for a long time.[76]

Oxidants that are too powerful or conditions that are unduly vigorous carry a desired disulfide through eqs. 7.10b,c to a sulfonic acid, the final product of oxidation. Oxidation of equimolar amounts of $R^1SH$ and $R^2SH$ ordinarily gives the statistically expected mixture of 1 mole of

$R^1 SSR^1$, 1 mole of $R^2 SSR^2$, and 2 moles of $R^1 SSR^2$, but if $R^1 SH$ and $R^2 SH$ differ enough (as with 1° and 3° thiols), $R^1 SSR^2$ is favored and can be practicably prepared; even with a 1°-1° mixture, efficient distillation can give a satisfactory separation. [77,78]

With $O_2$, pure thiols probably will be oxidized slowly if at all, [74,75] but even so catalytic impurities may make working with some thiols troublesome in actual practice; *t*-alkanethiols, however, are quite resistant to oxidation. Oxidation proceeds through $H_2O_2$ as an intermediate (eq. 7.11), is best done in aqueous or alcoholic ammonia, and is catalyzed by heavy-metal salts, especially $Fe^{3+}$ or $Cu^{2+}$.[5] Preparations based on $H_2O_2$ are similar, as eq. 7.11 implies; 30% $H_2O_2$ in alcohol or acetone often is used for water-insoluble thiols.[5] In the oxidation of· thiolate anions with $O_2$,[74] the rate-determining step is electron transfer to $O_2$, arenethiolates are oxidized more slowly than alkanethiolates because of greater anionic stability (in contrast to the sequence mentioned for thiols), and dipolar solvents that form complexes with the cation enhance the rate. Kinetic study of the oxidation of *o*-mercaptophenylacetic acid by excess $H_2O_2$ with $Fe^{2+}$ as catalyst showed a psuedo zero order and an inverse square-root hydrogen-ion dependence; it suggested formation of a complex, $(H_2O)(RSH)(OH^-)^{1/2} Fe^{2+}$, which then is oxidized. [79]

$$2RSH + O_2 \longrightarrow RSSR + H_2O_2$$

$$(7.11)$$

$$2RSH + H_2O_2 \longrightarrow RSSR + 2H_2O$$

Iodine oxidizes thiols stoichiometrically. Eq. 7.12 thus shows not only

$$2RSH + I_2 \longrightarrow RSSR + 2HI \qquad (7.12)$$

one of the best means for preparing disulfides,[75] but one of the most convenient means of determining thiols as well (or disulfides, after reduction with zinc and acetic acid).[80] Sodium or lead salts also can be oxidized,[5] including salts of vinyl thiols.[81] One usually uses a solution of $I_2$ in acetic acid, alcohol or aqueous potassium iodide. In anhydrous hydrocarbons, it is important to note that $I_2$ does *not* oxidize thiols to disulfides because HI, of course, is a powerful reducing agent. "Only when water or some other polar solvent is present can solvation of $H^+$ and $I^-$ ions drive the reaction to completion."[82] Heats of oxidation of thiols with $I_2$ in 90% ethanol range in kcal/mol from 9–10 for alkanethiols or that which forms 1,2-dithiane (7.5) to 12–14 for those that form the strained 1,2-dithiolane system (7.4).[83] However, pyridine or tertiary amines will accept HI as well, for example, in ether[84] or chloroform,[85] as eq. 7.13, illustrates for the preparation of an imidoyl disulfide.[84]

$$2\,MeC(S)NHPh + 2Bu_3N + I_2 \longrightarrow [MeC(=NPh)S]_2 + 2\,Bu_3NH^+I^-$$

$$(7.13)$$

Lead tetraacetate oxidizes most thiols cleanly to disulfides (eq. 7.14a).[86] In the presence of alcohols, reaction goes further and affords a one-step synthesis of arenesulfinate esters (eq. 7.14b);[87a] for alkanesulfinates,[87b] this method is less clean and that of eq. 7.15 is likely to be better.[88]

$$2RSH + Pb(OAc)_4 \xrightarrow{\text{(a)}} RSSR + Pb(OAc)_2 + 2AcOH \qquad (7.14)$$

$$\Big\downarrow \text{(b)}\ 3Pb(OAc)_4 + 4R'OH$$

$$2RS(O)OR' + 3Pb(OAc)_2 + 4AcOH + 2AcOR'$$

$$2RSH \xrightarrow{Cl_2} RSSR \xrightarrow{3Cl_2,\,2AcOH} 2RS(O)Cl \xrightarrow{R'OH} RS(O)OR'$$

$$(7.15)$$

Use of ferric chloride in ether is convenient and mild; for example, it works well with a hindered arenethiol[89] and is the reagent of choice for cyclizing 1,5-pentanedithiol.[85] Potassium ferricyanide, also mild, was used to synthesize an oxytocin.[90] Sulfur has no effect on thiols at $25°$, but with amines as catalysts it oxidizes $1°$ thiols smoothly.[91] If conditions are properly chosen, alkyl di-, tri-, and sometimes tetrasulfides can be prepared in good yields by oxidizing thiols using sulfur with amines; thiolate ion generated by the amine splits the $S_8$ ring, and the polysulfide formed ($RS_8SH$) then is attacked by thiolate ion to give final products that are determined by conditions and steric factors.[92] Recent promising oxidants include manganese dioxide[93] and manganic tris-(acetylacetonate).[94]

The oxidants mentioned so far are inorganic. Little attention was paid to organic ones until recently, except for quinoids.[95] Of several early ones, azodicarbonamide and nitrosobenzene were best.[95] An interesting mild one is 1,2-diiodoethane.[96,97] Dimethyl sulfoxide with $H_3PO_4$ catalysis oxidizes arenethiols rapidly to disulfides;[98a] uncatalyzed reaction takes much longer but succeeds with $o$-aminobenzenethiol and many other thiols (80-100% yield).[98b] Other methyl sulfoxides can be used, and the reaction leads to the usual relative ease of $RCH_2SH > R_2CHSH > R_3CSH$.[98c]

Oxidation of thiol derivatives sometimes furnishes a good preparation of disulfides. Bunte salts (7.36) are stable to air but can be oxidized without prior hydrolysis by $I_2$, $H_2O_2$ ($H_2SO_4$ catalysis), etc.[5,99] The sequence of eq. 7.16 thus often is attractive for converting halides to disulfides in essentially one step with good yields.[5]

$$RX + Na_2S_2O_3 \longrightarrow RSSO_3Na \xrightarrow{(O)} RSSR \qquad (7.16)$$
$$7.36$$

### 7.3.2 By Reduction

Reduction often provides convenient routes to disulfides, for example after chlorosulfonation (eq. 7.17). Reduction by HI of any arenesulfonyl chloride to its disulfide is probable, if the substituents resist HI[100] (alkanesulfonate salts can be reduced to disulfides with $PBr_3$-$PBr_5$ in 26-87% yield).[101] The basis with HI is that $I_2$ formed precludes reduction to thiols (recall eq. 7.12). A clever trick is to use $H_3PO_3$ with only a little KI; the $H_3PO_3$ reduces $I_2$ formed back to HI.[102] Reduction also occurs with HBr, which thus might be an alternative if R of eq. 7.17 contains groups incompatible with HI; to prevent bromination of the disulfide, bromine is scavenged by agents like $Na_2SO_3$ or aniline.[103] Benzenesulfonic anhydride has been reduced like the chloride, but since $Br_2$ was not scavenged the product was bis(p-bromophenyl) disulfide; methanesulfonic anhydride, in contrast, gave methanesulfonyl bromide (44% yield; however, $Me_2S_2$ may have reacted with $Br_2$ to give $MeSBr_3$, etc.)[104]

$$2RSO_2Cl \xrightarrow{HI, HBr, Mo(CO)_6, Cl_3SiH - Pr_3N, (RO)_3P, or R_3P} RSSR$$
$$(7.17)$$

Both arene- and alkanesulfonyl chlorides are reduced to disulfides nicely by hexacarbonylmolybdenum in dry tetramethylurea (55-80%).[105a] Carbonyl complexes of Ni, Fe or Cr reduce sulfenyl chlorides to disulfides below $0°$ in yields above 90%,[105b] but the synthetic value is dubious for most situations since sulfenyl chlorides not only are commonly made *from* disulfides but are readily and quantitatively reduced simply by using iodide ion[105c]:

$$2RSCl + 2KI \longrightarrow RSSR + I_2 + 2KCl \qquad (7.18)$$

An arenesulfonyl chloride (but not an ester) also is reduced nicely by trichlorosilane in benzene with amine catalysis (eq. 7.17), as are sulfinyl and sulfenyl chlorides or their esters (53-91%); the mechanisms are still unclear.[106] Phosphorus compounds are similarly flexible. Thus phosphite esters reduce arenesulfonyl chlorides, arenesulfenyl chlorides (ArSCl) or thiolsulfonates ($ArSO_2SAr$) to disulfides, along with other products.[107a] Phosphines ($R_3P$) reduce arenesulfonyl chlorides to thiols and sulfinic acids, as well as to disulfides, and also reduce $\alpha$-disulfones ($RSO_2SO_2R$), thiolsulfonates, or polysulfides to disulfides.[107a] Care is necessary because phosphites and phosphines can desulfurize disulfides to monosulfides.[107a]

An intriguing and promising reaction can be used to make unsymmetrical disulfides in 47-82% yield. In it, sulfenates are reduced as shown[107b]:

$$3PhSOMe + (RS)_3B \xrightarrow{\Delta} 3PhSSR + (MeO)_3B \qquad (7.19)$$

### 7.3.3 By Thioalkylation of Thiols

Foss, in 1947, first recognized that many diverse classes should show the "sulfenyl" behavior represented for RSZ in eq.7.20 that one expects for sulfenyl chlorides;[108] Kharasch has since emphasised the value of recognizing these similarities. [30,109]

$$R^1\underset{\phantom{x}}{S^-} + R^2S^{\delta+} \underset{\phantom{x}}{-Z^{\delta-}} \xleftarrow{\phantom{xxxxxxx}} R^1SSR^2 + Z^- \qquad (7.20)$$
$$7.37$$

Thioalkylation of a thiol by some kind of sulfenyl derivative usually gives the best synthesis for unsymmetrical disulfides (eq. 7.20). Among the groups Z that can be displaced are $R_2N^-$, halide ion, $SCN^-$, $RSO_2^-$ $ROC(O)S^-$ $SO_3^{2-}$, $RS^-$ and $CN^-$ (cf. Sec. 7.5.3.1). These displacements will be discussed in this order, since this seems likely to be a fairly reasonable approximation for relative ease of use and effectiveness in preparing an unsymmetrical disulfide from two different thiols.

#### 7.3.3.1 With Sulfenamides ($R^1SNR^2R^3$)

Eq. 7.21 illustrates the thioalkylation of a thiol by a sulfenamide.[110]

$$(7.21)$$

Several syntheses of unsymmetrical disulfides (7.37) have been achieved by ingenious uses of sulfenamides. The first, based on readily available diethyl azodicarboxylate, is shown by eq. 7.22. Mild neutral conditions lead to disulfides 7.37 in 75-90% yield.[111] Success is not unfailing however, because an aminothiol gave poor results (probably because it was so nucleophilic that it attached the first-stage sulfenamide as it was formed to give a symmetrical disulfide,[110] and since symmetrical impurities were serious in the preparation of acetyl phenyl disulfide.[112]

$$EtO_2CN=NCO_2Et + R^1SH \longrightarrow \underset{\underset{SR^1}{|}}{EtO_2CNHNCO_2Et} \longrightarrow R^1SSR^2 + (EtO_2CNH)_2$$
$$7.37$$

$$(7.22)$$

Two reports soon appeared simultaneously that made use of sulfenylphthalimides, obtained as shown by eq. 7.23a,[113] for thioalkylation by eq. 7.23b.[114] Alkyl aryl and dialkyl disulfides were prepared after several hours of reflux in yields of 71–98%, with only traces of symmetrical disulfides being formed. A thioacid also can be used as $R^2SH$,[114b] but unfortunately mixtures result if both $R^1$ and $R^2$ are aryl.[114a] Sulfenylsuccinimides or -maleimides also can be used.[114b] Precipitation of the imide drives the reaction to completion, and the disulfide 7.37 is readily recovered from the solution. Bisdisulfides can be prepared,[114b] and a trisulfide can be obtained by using a hydrodisulfide (98% yield).[114a] Advantages of the method are the easy preparation and unlimited shelf life of the precursor sulfenylphthalimides.[114a]

$$(7.23)$$

### 7.3.3.2 With Sulfenyl Halides (RSX)

The usual sequence used, shown by eq. 7.24, amounts really to a single step. It has excellent generality.[5] References 5, 7, 30 and 109 afford entry to the extensive literature, and ref. 115 reviews preparations of RSCl. Eq. 7.25 may offer an alternative that avoids the inconvenient and necessarily stoichiometric use of $Cl_2$ and deserves further exploration;[116] however, there seems to be a tendency for symmetrical disulfides to be formed.[116a, 122] Of seven approaches compared for synthesis of acetyl disulfides, that of Böhme and Clement was best (eq. 7.26; 77–97%);[112] base was not needed, and indeed the thiolate ion it would have engendered probably would have caused side reactions stemming from attack on the carbonyl and disulfide moieties.

$$R^1SH \xrightarrow{0.5\ Cl_2} 0.5\ R^1SSR^1 \xrightarrow{0.5\ Cl_2} R^1SCl \xrightarrow{R^2SH} R^1SSR^2 \qquad (7.24)$$

$$(7.25)$$

$$Ac_2S + Cl_2 \xrightarrow[\phantom{x}]{<10°,\ -AcCl} AcSCl \xrightarrow{RSH} AcSSR \qquad (7.26)$$

Sulfenyl chlorides, and RSSCl counterparts, also have been used with $H_2S$, etc., to give tri- and higher sulfides.[7] Reactions of thiols with $SCl_2$ or $S_2Cl_2$ have been used extensively too, presumably with sulfenyl-type halides as intermediates, but their attractiveness may be deceptive because $SCl_2$ and $S_2Cl_2$ as used are likely to be mixtures.[6,7]

### 7.3.3.3 With Sulfenyl Thiocyanates (RSSCN)

Thioalkylation of a thiol can be achieved by the sequence of eq. 7.27a. The power and flexibility of the sulfenyl thiocyanate route were enhanced greatly by the findings that the reaction succeeds even when the hydrogen of a thiol is replaced by a trityl, benzhydryl, α-pyranyl or isobutoxymethyl group; variations like that of eq. 7.27b thus became possible.[117] Advantages of such routes are high reactivity of thiols and certain thio ethers toward thiocyanogen and sulfenyl thiocyanates.[117] A disadvantage of thermal lability of the sulfenyl thiocyanates is modest because the thiocyanate rarely is isolated. Hiskey and his associates have used this method extensively for aliphatic disulfides and for sulfur-containing polypeptides (for the latest paper in the respective series see refs. 118 and 119). The method is powerful for stepwise formation of disulfide bridges in polypeptides because cystine peptides that contain S-trityl and S-benzhydryl protective groups can be selectively oxidized to disulfides either by thiocyanogen or by sulfenyl thiocyanates (which, since thiocyanogen is a psuedohalogen, evidently can react like sulfenyl chlorides; *cf.* eq. 7.77). Previously established -SS- bonds survive. An application is illustrated schematically by eq. 7.28, where dashes represent amino-acid units of a polypeptide and where amino and

$$Pb(SCN)_2 \xrightarrow[0°]{Br_2, EtOAc} (SCN)_2 \xrightarrow[\text{(or } R^1SR)]{R^1SH} R^1SSCN \xrightarrow{(a)R^2SH} R^1SSR^2 + HS\,CN$$

$$\text{(b)}\ R^2SCPh_3,\ 25° \downarrow$$

$$R^1SSR^2 + Ph_3CS\,CN$$

(7.27)

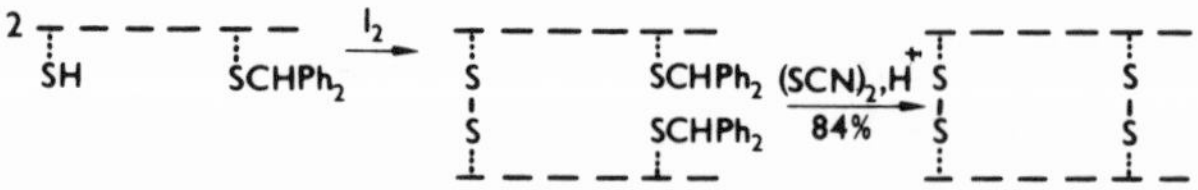

(7.28)

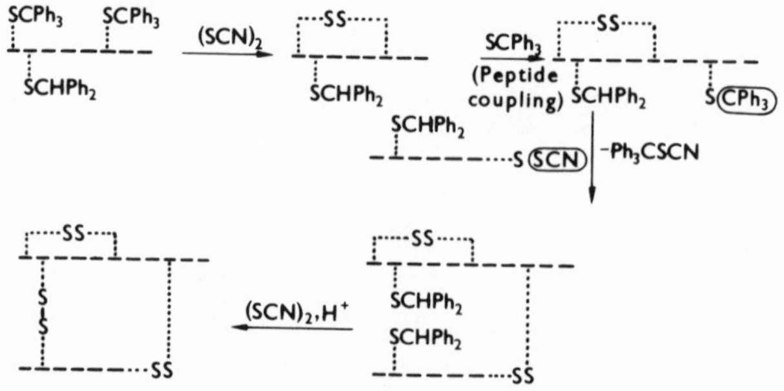

(7.29)

carboxyl groups are suitably protected.[120] Differing reactivities of trityl and benzhydryl groups permit clean stepwise introduction of at least three different -SS- bonds into polypeptides, as eq. 7.29 shows schematically.[121]

Like most methods, that of sulfenyl thiocyanates is not a panacea. For the preparation of acyl disulfides, acetylsulfenyl chloride worked better with a thiol (eq. 7.26) than did acetylsulfenyl thiocyanate (AcSSCN),[112] and for the preparation of $o$-carboxyphenyl disulfides, a thiolsulfonate was better than a sulfenyl thiocyanate.[122]

### 7.3.3.4 With Thiolsulfonates ($R^1SO_2SR^2$) or Thiolsulfinates ($R^1SOSR^2$)

Early literature of thiolsulfonates and thiolsulfinates has been reviewed.[123,124] Synthesis of disulfides by use of thiolsulfonates is illustrated by eq. 7.30. A chief advantage of this route, like that with sulfenamides, is that an agent desired for thioalkylating several thiols can be prepared once and kept indefinitely. A disadvantage, with a symmetrical thiolsulfonate, is that half is wasted as the displaced sulfinate (eq. 7.30b).

$$2R^1SH \xrightarrow{Cl_2} [R^1SSR^1] \quad \xrightarrow[H_2O;\ or\ H_2O_2]{(a)Cl_2\,(or\ SO_2Cl_2)\text{-HOAc-}} R^1SO_2SR^1$$

$$\xrightarrow{(b)R^2SNa} R^1SSR^2 + R^1SO_2Na \qquad (7.30)$$

A symmetrical aliphatic thiolsulfonate can be obtained smoothly by the Douglas-Farah reaction, in which one carefully chlorinates a disulfide at $-10°$ in acetic acid and then adds water (87% yield); unsymmetrical ones can be made by a modification (82-84%).[125] Aromatic thiolsulfonates can be prepared similarly from a disulfide or thiol (75-89%),[126] but $SO_2Cl_2$ seems to be a general and more convenient alternative to $Cl_2$ for these and other types as well (21-91%).[127] When $R^1$ of eq. 7.30 was the hindered 2,4,6-triisopropylphenyl group, however, only $H_2O_2$ oxidation succeeded and even then catalysis was necessary.[126] Fortunately, chlorinolysis and peroxide oxidation thus complement each other. As further examples of this complementarity for making thiolsulfonates, chlorinolysis succeeded when $R^1$ was $n$-decylaminoethyl (85%) but $H_2O_2$ failed,[128] and the reverse was true when $R^1$ was 2-acetamidoethyl.[129] The method with $H_2O_2$ often seems that of choice with aminoalkyl salts[128] and also was the better with a cyclic disulfide.[85]

Scheme 7.2 depicts preparations of unsymmetrical thiol-sulfonates. Some of these modifications of earlier procedures developed by Smiles, Lo, Kresze, Loudon, their co-workers, and others allow use of valuable thiol without the loss of half of it later (cf. Scheme 7.2e) that is inherent in eq. 7.30b. Selective oxidation of unsymmetrical disulfides also can give unsymmetrical thiolsulfonates practicably if the groups differ sufficiently (*cf.* ref. 133).

SCHEME 7.2

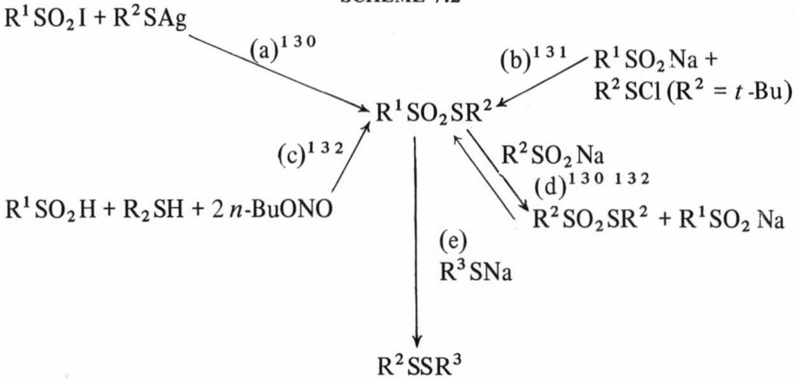
R¹SO₂I + R²SAg
(a)130
(b)131
R¹SO₂Na +
R²SCl (R² = t-Bu)
R¹SO₂SR²
(c)132
R²SO₂Na
(d)130 132
R¹SO₂H + R₂SH + 2 n-BuONO
R²SO₂SR² + R¹SO₂Na
(e)
R³SNa
R²SSR³

A thiolsulfonate can thioalkylate a thiolate salt rapidly and completely even at –86°.[131] Reaction is fairly rapid with a thiol; it may proceed only part way but can be pushed further by excess reagents.[131] With a *t*-butyl thiolsulfonate, in contrast, the same diminished reactivity is seen that occurs with other *t*-BuS- compounds, presumably because of a neopentyl-type steric effect (*cf.* Sec. 7.5.3.1); with a 2,4,6-triisopropylphenyl thiolsulfonate also, hindrance causes low reactivity.[131]

Eq. 7.31 shows a thioalkylation useful in working with thiols.[122] The reaction goes well with 1°, 2°, or 3°, alkyl, arene- or heterocyclic thiols, with mercapto amino acids, or with an enzyme. The products (**7.38**) are useful for purifying and characterizing thiols because they differ nicely in melting points and tlc values, can be titrated, and later can be either reduced to the original thiol or converted directly to the disulfide.[122] For the preparation of the products **7.38**, the route of eq. 7.31 was better than several others tried.[122]

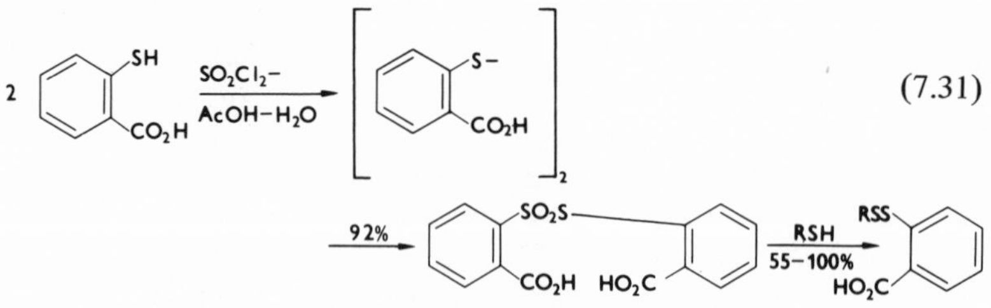
2
SH
CO₂H
SO₂Cl₂–
AcOH–H₂O
S–
CO₂H
2
92%
SO₂S
CO₂H HO₂C
RSH
55–100%
RSS
HO₂C

(7.31)

**7.38**

Cyclic thiolsulfonates also thioalkylate thiols[134]:

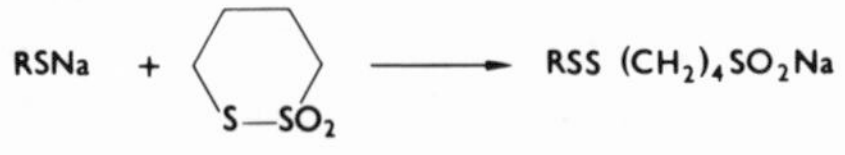
RSNa +
S—SO₂
RSS (CH₂)₄SO₂Na

(7.32)

The disulfides produced are terminated by sulfinate-salt moieties convertible to sulfones, sulfonate salts, or thiolsulfonates, so that disulfides containing these terminals become accessible; disproportionation of such products gives symmetrical disulfides hard to obtain by other means (cf. Sec. 7.5.6).[134,135.]

For synthesis of acyl-type alkyl disulfides, thioalkylation of thioacids with thiolsulfonates usually is less satisfactory than with the corresponding sulfenyl chloride (cf. eq. 7.24).[136] Reaction of AcSCl with a thiol, as mentioned, also affords a good route to acyl-type disulfides (eq. 7.26),[112] although sometimes when it fails a thiolsulfonate still can be used to thioalkylate the thioacid.[26] As an alternative to AcSCl, acetyl p-toluenethiolsulfonate ($AcSSO_2C_7H_7$) was synthesized, but it could not be purified and was unpromising.[112]

Symmetrical alkyl and aryl trisulfides can be prepared nicely from alkali sulfides using thiolsulfonates (80–95% yield),[135,137] but unsymmetrical ones could not be obtained by this route.[137]

Thiolsulfinates can be prepared from disulfides and used to convert thiols to unsymmetrical disulfides, all in one operation (eq. 7.33, 25–66% yield).[138] An interesting point is that *both* moieties of thiolsulfinates appear to react; for example, cyclic thiolsulfinates give bisdisulfides.[138] Hence there would be the prospect that half of a thiolsulfinate need not be wasted, as it is with a symmetrical thiolsulfonate (cf. eqs. 7.30b and 7.33). However, the particular reaction of eq. 7.34 could not be repeated.[139] Oxidation of disulfides to thiolsulfinates is facile if $R^1$ of eq. 7.33 is 1° alkyl, poor if it is 2°, and unsuccessful if it is 3°.[123,124]

Other aspects of the chemistry of thiolsulfonates and thiolsulfinates are discussed in Sec. 7.5.2.2.

$$2R^1SH \xrightarrow{H_2O_2} R^1SSR^1 \xrightarrow{RCO_3H} R^1S(O)SR^1 \xrightarrow{2R^2SH} 2R^1SSR^2 + H_2O$$

$$(7.33)$$

$$MeO_2CCH_2S(O)SCH_2CO_2Me + PhCH_2SH \xrightarrow{\quad\not\quad} MeO_2CCH_2SSCH_2Ph$$

$$(7.34)$$

### 7.3.3.5 With Sulfenyl Thiolcarbonates ($R^1SSCO_2R^2$)

In this elegant new route,[140] one first proceeds by the two steps of eq. 7.35 to a carboalkoxysulfenyl chloride (7.39)[141]:

$$Cl_3CSCl \xrightarrow{H_2O, H_2SO_4} ClSC(O)Cl \xrightarrow{ROH} ClSCO_2R \qquad (7.35)$$

$$\textbf{7.39}$$

$$7.39 \xrightarrow{\text{R}^1\text{SH, 0°}} \text{R}^1\text{SSCO}_2\text{R} \qquad (7.36)$$
$$\qquad\qquad\qquad \mathbf{7.40}$$

$$\underset{\mathbf{7.40}}{\overset{\displaystyle \underset{\text{O R}^1}{\overset{\curvearrowright \curvearrowright}{\text{ROCSS}}}\ \underset{\text{H}}{\text{SR}^2}}{}} \xrightarrow{25°} \text{R}^1\text{SSR}^2 + \text{COS} + \text{ROH} \qquad (7.37)$$
$$\qquad\qquad\qquad\qquad \mathbf{7.37}$$

The **7.39** next is converted to a sulfenyl thiolcarbonate (**7.40**, eq. 7.36), which then is used in a heterolytic thiol-mediated fragmentation to thioalkylate a second thiol (eq. 7.37; 93–99%). In the sense of Grob's conception of fragmentable systems,[142] R[1]S- is the electrofugal group, COS is the unsaturated fragment and RO- is the nucleofugal group (-Cl, -NR$_2$ and -SR are alternatives to -OR). A cyclic transition state (**7.41**) is favored presently; its synchronous fragmentation is thought to be rapid, with loss of COS serving as a driving force.[140]

**7.41**

Advantages of this method are the facile preparation, stability, high reactivity and compatibility with most functional groups of sulfenyl thiolcarbonates (**7.40**), as well as the absence of by-products.[140] Thus acetyl 2-aminoethyl disulfide hydrochloride was obtained in 99% yield,[140] although because of disproportionation it could not be obtained by the thiolsulfonate route[143] (it could be obtained in 85% yield, however, by the sulfenyl halide route of eq. 7.26[112,140]). The only disadvantage apparent so far to the sulfenyl thiolcarbonate route is the number of steps. However, once **7.39** or **7.40** are at hand, it would not be surprising if this method did not turn out to be the best of them all.

### 7.3.3.6 With Bunte Salts (RSSO$_3^-$)

A recent review lists several preparations of Bunte salts (RSSO$_3^-$, with a metal or ammonium counter ion), including reaction of thiosulfate ion with 1° or 2° alkyl halides, with activated multiple bonds, or with oxiranes and aziridines, as well as reaction of thiols with SO$_3$ or ClSO$_3$H (*cf.* also eq. 7.51).[99] Such methods might be useful in special situations to prepare

Bunte salts desired for thioalkylating thiols (eq. 7.38). Best results for eq. 7.38 ensue with short reaction periods in pH 8 buffer at ~0°, but even then yields may be only ~ 10–42% and disproportionation may be troublesome.[144] Formaldehyde or a formyl arenesulfonate help by removing sulfite ion that reverses eq. 7.38 and catalyzes disproportionation,[144,145] or by inhibiting thiol catalysis of disproportionation.[144] However, a Bunte salt seems likely to be the best choice for thioalkylation only when a special situation like those mentioned makes it more easily available than the corresponding thiol for use by another route, or when factors such as unsaturation preclude another approach (cf. ref. 145). Bunte salts also have been used to prepare tri- and tetra-, but not pentasulfides.[99]

$$R^1SSO_3^- + R^2S^- \rightleftharpoons R^1SSR^2 + SO_3^{2-} \qquad (7.38)$$

### 7.3.3.7 With Disulfides (RSSR)

Interchanges of thiols with disulfides are discussed in Sec. 7.5.3.4, but these are not devoid of preparative significance.[7] They are illustrated by eq. 7.39 and by eq. 7.40, which really is much the same thing. Simple unsymmetrical disulfides can be prepared by distilling mixtures of a thiol and disulfide or of two disulfides.[77,78] Such approaches can be useful when circumstances favor the formation or separation of the unsymmetrical product. For example, distillation of EtSH (the most volatile component) from a mixture of EtSSEt and t-BuSH leaves nearly pure EtSS-t-Bu in 95% yield.[77] After equilibration of EtSSEt and n-BuSS-n-Bu, these and EtSS-n-Bu were present in a 1:1:2 statistical mixture, from which fractional distillation separated EtSS-n-Bu.[77]

$$R^1SSR^1 + R^2SH \rightleftharpoons R^1SSR^2 + R^1SH \qquad (7.39)$$

$$R^1SSR^1 + R^2SSR^2 \xrightleftharpoons{RSH\ cat.} 2R^1SSR^2 \qquad (7.40)$$

### 7.3.3.8 With Thiocyanates (RSCN)

The method of eq. 7.41 apparently has been used mostly to make symmetrical disulfides.[5] It is rarely used now, although it proved reasonably effective for converting metal 1-alkenethiolates to unsymmetrical disulfides.[81] Disulfides result when a thiocyanate simply is heated with a base, perhaps because thiolate ion is generated and then is either thioalkylated by the thiocyanate or is oxidized.

$$R^1SCN + R^2SH \longrightarrow R^1SSR^2 + HCN \qquad (7.41)$$

### 7.3.4 By Other Methods

Arylation of $Na_2S_2$ with a suitably activated aryl halide can afford a good synthesis of a symmetrical disulfide (eq. 7.42).[146] Arenediazonium salts also arylate $Na_2S_2$, but one should beware of possible explosions, presumably caused by an accumulation of diazosulfides.[5,7] A review summarizes alkylation of $Na_2S_2$ with alkyl halides (yields are fair) and makes the important point that since "$Na_2S_2$" really is a "statistical compound," sulfides, trisulfides, and higher sulfides are formed too; fortunately boiling-point or melting-point differences often suffice for reasonable separation of simple disulfides (cf. also ref. 5).[7]

$$Na_2S + S \longrightarrow \text{``}Na_2S_2\text{''} \xrightarrow[66\%]{o\text{-}NO_2PhCl,\ EtOH} (o\text{-}NO_2\ PhS)_2 \qquad (7.42)$$

Reaction of alkylene dihalides with alkali polysulfides gives useful polymers that resist oils, solvents and oxygen.[147] Illustratively,[148] Thiokol A is made from ethylene chloride and sodium tetrasulfide, Thiokol ST from chloroethyl formal and sodium disulfide, and Thiokol FA as a copolymer from both chlorides. A little trihalide can be used to introduce cross links. High molecular-weight material can be reduced to liquids that can be cast and then reoxidized to elastomer; rocket motors are made by incorporating oxidizing agents.[148]

Sulfur monochloride ($S_2Cl_2$) with active methylene compounds, Grignard reagents, alkenes or arenes can give disulfides and chlorodithio compounds (RSSCl), but other products as well.[7] Its use has been reviewed.[149]

Disulfides form when a wide variety of compounds are heated with sulfur, along with higher sulfides.[7] Such reactions, however, are rarely clean or of much preparative interest.

### 7.3.5 Choice of Method

This section compares preparative methods discussed above and illustrates considerations that hopefully will help the reader select the best method for a particular di- or higher sulfide.

#### 7.3.5.1 Disulfides

If the requisite thiols are available, their use ordinarily furnishes the best path to disulfides of all types.

Uncomplicated symmetrical disulfides commonly are made by oxidation, e.g., of 1° and 2° alkanethiols, heterocyclic or arenethiols. Least trouble with overoxidation (eq. 7.10) is likely with $I_2$ (as its own indicator, in alcohol,[82] or benzene-water[75], or with amines as mentioned earlier), or $FeCl_3$ (ether, water, *etc.*), a ferricyanide (water), $Pb(OAc)_4$ (benzene, AcOH, or $CHCl_3$), one of the manganese compounds, or a selective organic oxidant. However, at least some 3° thiols give sulfenyl iodides with $I_2$[150] and rather low yields with $Pb(OAc)_4$;[86] cupric chloride,[151] hypoiodite ion,[5] or a ferricyanide[78] seem best. For allylic thiols, $I_2$ again was poor and a ferricyanide was used.[33] With other sensitive moieties, one must adapt to the situation; thus $Pb(OAc)_4$ gives its poorest yield with *o*-aminobenzenethiol (21%),[86] but dimethyl sulfoxide oxidizes it nicely (100%);[98b] $H_2O_2$ oxidizes *p*-aminobenzenethiol to the disulfide in fair yield (64%).[152]

For cyclic disulfides, the preferred method varies with ring size. A hydroperoxide or $FeCl_3$ often serves well, with high dilution being desirable for rings of more than six members.[153] Iodine, with $Et_3N$, also has been recommended.[154] When routes were compared for several α, ω-dithiols, 1, 2-dithiolane seemed best prepared using $H_2O_2$ at 75° (which minimizes polymerization), 1, 2-dithiane by using a hemitosylate of the dithiol (with base) or the lead dithiolate (with sulfur), and 1,2-dithiepane by using $FeCl_3$.[85]

Symmetrical acyl disulfides usually present no special problems. For example, benzoyl disulfide is readily obtained by oxidizing potassium thiobenzoate with $I_2$ (73%).[155]

If the requisite thiol is not available, symmetrical diaryl disulfides might be obtained by reducing the sulfonyl chloride with HI or other agents mentioned and dialkyl disulfides through reaction of the halide with sodium thiosulfate (or perhaps with the less useful $Na_2S_2$, although preparation and use of the thiol might be better).

Unsymmetrical disulfides usually are best obtained by thioalkylating one thiol with a sulfenyl-type derivative of another. Which of the sulfenyl derivatives will be best is difficult to predict and is likely to vary with circumstances. However, if one has both thiols, the order of discussion in Sec. 7.3.3 seems a reasonable one for trial. Thus the azodicarboxylate route requires essentially one step, the sulfenyl-halide route requires two, and so on. If only one thiol and an alkyl halide are available, on the other hand, the Bunte-salt approach deserves consideration. Table 7.1 illustrates the wide variety of types of unsymmetrical disulfides that have been prepared and shows the general approaches used (for other examples, see refs. 5 and 7).

## TABLE 7.1

Preparation of Typical Unsymmetrical Disulfides[a]

| $R^1SSR^2$ prepared | $R^1SH$ used | Route | Yield, % | Guiding reference |
|---|---|---|---|---|
| $Alk^1SSAlk^2$ | $Alk^1SH + Alk^2SH$ | 7.3.1 | ? | 78 |
| $HetSSC_6H_{11}$ | HetSH | 7.3.3.1 | 94 | 114b |
| HetSSAr | HetSH | 7.3.3.2 | 4-77 | 156 |
| $AlkSS(CH_2)_2NH_2^+\text{-}n\text{-}C_{10}H_{21}$ $Cl^-$ | AlkSH | 7.3.3.4[b] | 54-65 | 157 |
| $AlkSS(CH_2)_4SO_2^{-c}$ | $AlkSH^c$ | 7.3.3.4[b] | 86-97 | 134 |
| $NH_2CH_2CH_2SSSO_3H$ | $HSSO_3H$ | 7.3.3.5 | 93 | 140 |
| $PhCH_2SS\text{-}(CH_2CH\ NH_3^+Cl^-)\text{-}$ $CO_2H$ | $HO_2CCH(NH_3^+Cl^-)CH_2SH$ | 7.3.3.1 | 89 | 114a |
| $HO_2C(CH_2)_3SSCH_2CH(NH_2)\text{-}$ $CO_2H$ | Ditto | 7.3.3.4 | 66 | 138 |
| $HO_2CCH(NHCO_2CH_2Ph)\text{-}$ $CH_2SSCH_2CO_2Me$ | $HO_2CCH(NHCO_2CH_2Ph)\text{-}$ $CH_2SCPh_3^d$ | 7.3.3.3 | 56-60 | 158 |
| $MeO_2CCH_2SSAlk$ | AlkSH | 7.3.3.3 | 58-61 | 139 |
| $2\text{-}X\text{-}C_6H_{10}SS(CH_2)_2NH_3^{+e}$ | $2\text{-}X\text{-}C_6H_{10}SH^e$ | 7.3.3.4[b] | 20-57 | 159 |
| $t\text{-}BuSSAlk^c$ | $t\text{-}BuSH$ | 7.3.3.4[b] | 67 | 129 |
| $(CH_2)_{4-5}C(SSAlk)_2^c$ | $(CH_2)_{4-5}C(SH)_2$ | 7.3.3.4[b] | 87-99 | 160 |
| $AlkSS(CH_2)_nSSAlk^{c,135}$ | $HS(CH_2)_nSH$ | 7.3.3.4[b] | 76-93 | 160 |
| $R^1CH{=}CR^2SSR^3$ | $R^1CH = CR^2SH$ | 7.3.3.8 | 14-82 | 81 |
| | | 7.3.3.4[b] | 5-77 | 81,159 |
| $XPhCH_2SS(CH_2)_2NH_2^+\text{-}n\text{-}$ $C_{10}H_{21}Cl^{-f}$ | $X\text{-}PhCH_2SH^f$ | 7.3.3.4[b] | 15-66 | 161 |
| $o\text{-}XPhSSAlk^{c,g}$ | $o\text{-}XPhSH^g$ | 7.3.3.4[b] | 21-60[h] | 162 |
| $o(\text{or } m,p)\text{-}HO_2CPhSSAlk^c$ | $o(\text{or } m,p)\text{-}HO_2CPhSH$ | 7.3.3.4[b] | 34-77 | 163 |
| $p\text{-}XPhSSAlk^{c,i}$ | $p\text{-}XPhSH^i$ | 7.3.3.4[b] | 17-55[h] | 164 |
| $2,6\text{-}(MeO)_2PhSSAlk^c$ | $2,6\text{-}(MeO)_2PhSH$ | 7.3.3.4[b] | 81 | 162 |
| $1,2\text{-}Ar(SSAlk)_2^c$ | $1,2\text{-}Ar(SH)_2$ | 7.3.3.4[b] | 38-49 | 160 |
| $1,4\text{-}(CO_2^-)_2Ar\text{-}2,\text{-}5\text{-}(SSAlk)_2^{c,j}$ | $1,4\text{-}(CO_2H)_2\text{-}Ar\text{-}2,5(SH)_2^j$ | 7.3.3.4[b] | 66-85 | 165 |
| PhSSAr | PhSH | 7.3.3.2 | 83 | 5 |
| $R^1C(O)SSR^2$ [k] | $RCOSH^k$ | 7.3.3.2 | 55-100 | 136 |
| | | 7.3.3.4[b] | 52-100 | 143 |
| $AcSSCH_2X^l$ | $HSCH_2X^l$ | 7.3.3.2 | 77-97 | 112 |
| $RC(X)SS(CH_2)_4SO_2Na^m$ | $RC(X)SH^m$ | 7.3.3.4[b] | 66-100 | 166 |
| $R^1R^2NC(S)SSR^3$ [n] | $R^1R^2NC(S)SH^n$ | 7.3.3.4[b] | 58-88 | 167 |

[a]Alk = alkyl; Ar = aryl; Het = heterocyclic nucleus.

[b]The thiol was thioalkylated with the appropriate thiosulfonate.

[c]Alk = $(CH_2)_2NH_2^+R$ or $(CH_2)_2NHAc$.

[d]This thioether was allowed to react with $NCS\text{-}SCH_2CO_2Me$.

[e]X = $CO_2^-$ or Cl.

[f]X = H, $p\text{-}CH_3$, $p\text{-}MeO$, $p\text{-}Cl$, $p\text{-}NO_2$, $p\text{-}CN$, $m\text{-}NO_2$ or $m\text{-}CN$.

[g]X = $SO_3^-$, Cl, $NO_2$, Me, MeO or $CH_2OH$.

h Conversion % rather than % yield; low because base was omitted to obviate disproportionation.

$^i$X = H, $NO_2$, Cl, Me, or MeO.

jThe 2-mono disulfide also is included in the range of yields.

$^k$$R^1$ and $R^2$ = various combinations of Alk and Ar.

$^l$X = $CO_2H$, $(CH_2)_{1-3}CO_2H$, $CH_2NH_3^+Cl^-$ or CH = $CH_2$.

$^m$X = O or S.

$^n$$R^1$ and $R^2$ = combinations of Me and/or H to give di-, mono-, or unsubstituted N for $R^3$ = Ar, 1° alkyl or 3° alkyl.

### 7.3.5.2 Higher Sulfides

Mixtures that contain symmetrical higher sulfides are easy to obtain, but it may then be difficult or impossible to purify the components.[6,7] Symmetrical alkyl trisulfides can be obtained, however, by oxidizing thiols with sulfur *if* one attains the close control of variables that is essential; tetrasulfides thus prepared are likely to be impure, although disulfides can be obtained in high yield.[92] In a few instances where methods for symmetrical trisulfides were compared, thioalkylation of a metal sulfide with a thiolsulfonate gave better results than reaction of a Bunte salt or sulfenyl chloride with a sulfide, or of $SCl_2$ with a thiol, but unsymmetrical trisulfides could not be obtained this way.[137] The best routes to unsymmetrical compounds at present seem to be those based on RSSH or RSSCl (Sec. 7.1.4), but few have been made — in Reid's listing of 43 trisulfides, 25 tetrasulfides, 5 pentasulfides and one hexasulfide, only one, a trisulfide, is unsymmetrical.[7]

## 7.4 PHYSICAL PROPERTIES

### 7.4.1 Bonding and Stereochemistry

Three reviews have surveyed bonding and stereochemistry in di- and polysulfides,[8,67,168] so that attention here need be only to salient points from them (references not cited and further discussion can be found in these reviews).

Whether sulfur chains in form of $RS_nR$ are branched (**7.42**) or unbranched (**7.43**) was controversial for a long time. Chemical evidence could be interpreted either way. For example, the notable tendency of higher sulfides to lose sulfur to various nucleophiles (Nu⁻) and form lower sulfides might result either from preferential attack by Nu⁻ on the branched sulfur atom of **7.42** or on the internal sulfur atom of **7.43** (eq. 7.43). However, there is now quite an accumulation of strong physicochemical evidence that sulfur chains ordinarily are *un*branched. This evidence comes from dipole moments, ultraviolet, Raman and infrared spectra, electron and X-ray diffraction and magnetic susceptibility. Such work has been done with chains having n up to 8, but it is worth adding that some polythionates seem to have values of n as high as 50–100. Two exceptions are noteworthy: $S_2 F_2$ *can* exist in the branched form, FS(=S)F, evidently because $d$-orbital utilization by the sulfur atom is favored by strong electron withdrawal; also, the -OS(=OS)O- moiety evidently can occur in cyclic thionosulfites (for leading citations, see ref. 1).

$$\begin{array}{c} S \\ \uparrow \\ -SS- \end{array} \qquad -S\text{-}S\text{-}S- \; + \; Nu^- \; \longrightarrow \; -S\text{-}SNu \; n \; S^= \qquad (7.43)$$

$$-S\text{-}S\text{-} \; + \; SNu^-$$

**7.42**          **7.43**

There is an interesting indication, however, that branching may play a role *during* a reaction. It should first be said that branching does not occur when a disulfide (**7.37**) that contains only one [35]S-labeled moiety is prepared in various ways and then reduced; the label remains with the original moiety, showing that equilibria like that of eq. 7.44 ordinarily are not involved.[169] On the other hand, isomerization of one diastereoisomer of **7.44** leads to a 1:1 mixture of both diastereoisomers; the most attractive of several possible explanations invokes conversion of **7.44** to **7.45**, free rotation as indicated in **7.45**, and return to **7.44**.[170] No disulfide, tetrasulfide or allylic isomer (**7.46**) was formed; there appeared to be neither C-H or C-S fission involving charge separation or homolysis, nor S-S homolysis, and mixed trisulfides did not form when other trisulfides were added.[170]

$$R^1 SS^*R^2 \; \rightleftharpoons \; R^1 S^*R^2 \; \rightleftharpoons \; R^1 S^*SR^2 \qquad (7.44)$$
$$\text{\bf 7.37} \qquad\qquad \overset{\downarrow}{S}$$

**7.44**                    **7.45**                    **7.46**

Bond lengths require but brief comment, because for -SS bonds of organic di- and trisulfides lengths vary only slightly. The range is about 2.03–2.08 Å.[8,67,168]

Views on the nature of the bonds in acyclic di- and polysulfides have been discussed in a helpful review by Pryor;[8] only some major points of those developed there with references can be mentioned here. Pauling suggested that bivalent sulfur bonds are nearly pure $p$. This view is considered consistent with a bond angle for two atoms attached to sulfur of ~105°. Of the four electrons still left on a sulfur atom, two are assigned to the $s$ orbital and two to the third $p$ orbital. In an -SS- bond the pairs of the nonbonded $p$ electrons on adjacent atoms are considered to repel one another so that, in the minimization of repulsion, R groups of RSSR take a dihedral angle of ~90°. Fig. 7.2 depicts this situation (lobes represent the two electrons in each nonbonded $pz$ orbital; $s$ orbitals, spherically symmetrical about the nuclei, are not shown). The barrier to rotation around the S-S bond has been thought to vary from about 2.7 kcal/mol ($H_2S_2$) to 14.2 kcal ($S_2Cl_2$),[8] with values for $Me_2S_2$ and $Et_2S_2$ being 9.5 and 13.2 kcal, respectively.[168]

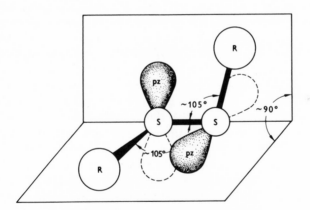

Fig. 7.2. Lowest-energy conformation of RSSR.

Recently, steric factors have begun to loom importantly. Barriers to rotation in disulfides of the structure $PhCH_2SSR$ were determined by nmr measurements of the coalescence temperatures of the -$CH_2$- protons, which would reflect steric effects involving these protons.[171] The lack of spectral changes down to −140° with some R groups, the variation from 8.0 to 9.3 kcal/mol in the barrier with several others, and the high value of 15.5 kcal reported earlier for the rotational barrier in bis(2,6-di-$t$-butylphenyl) disulfide, led to the conclusion that considerable *steric* hindrance to rotation can be present. Contrary to some of the previous estimates, the barrier to rotation in the *absence* of steric effects was thought to be less than 7 kcal; interestingly, the *cis* transition state for interconversion of enantiomers was concluded to be of lower energy than the *trans*.[171]

An alternative to the *p*-bond view, Pryor continued[8], is that of $sp^3$ hybridization, since bond angles often are not far from the tetrahedral value of $109°$ and since repulsion of nonbonding $sp^3$ orbitals might explain the dihedral angle. Still a third view invokes considerable *d*-orbital participation. It is worth adding that "no-bond resonance" has been suggested as an alternative to *d*-orbital involvement in polysulfides (7.47) (*cf.* ref. 8). While the question of bonding thus clearly cannot be thought of as settled, views fairly close to Pauling's perhaps are most in use at the moment, at least among those who are not quantum mechanicians.

$$R\ddot{S}—\ddot{S}—\ddot{S}—R \longleftrightarrow R\ddot{S}\bar{:} \quad \ddot{S}::\overset{+}{S}T$$

$$\text{7.47a} \qquad\qquad\qquad \text{7.47b}$$

As Fig. 7.2 shows, the groups R of acyclic disulfides are in planes at an angle of $\sim90°$, and the same kind of dihedral behavior holds for S-S bonds in tri- and higher sulfides. Stereochemical consequences have been summarized,[67] some of which should be mentioned here. Dihedral-angle variations from $\sim74°$ to $105°$ are seen for organic disulfides; the variations from $90°$ probably result from electronic or steric effects (as with R = *t*-butyl). Two enantiomeric forms of a disulfide are possible in principle, and nmr spectra reflecting these have been seen at low temperature. Indeed, an extremely large specific rotation of *L*-cystine relative to other amino acids may involve contributions to the optical rotation by relatively stable conformers. Trisulfides can exist in either a *cis* conformation (7.48a) or a *trans* conformation (7.48b); the *trans* form is dissymmetric and hence has two possible enantiomeric forms. X-ray crystallography shows that the several organic trisulfides studied are *trans*, but *cis* conformers exist among inorganic trisulfides. Tetrasulfides can exist as three pairs of enantiomeric conformers: *cis, cis* (which occurs in $S_8$); *cis, trans*; and *trans, trans* (which occurs in fibrous sulfur). There is X-ray evidence for inorganic *cis, cis* and *trans, trans* forms. Such behavior may well be seen even in ordinary work, since there is reason to believe that two different crystalline forms of dibenzhydryl tetrasulfide may be rotational isomers in the solid state.[172] Among cyclic disulfides, two enantiomeric conformers can exist in equilibrium, the rate of interconversion depending of course on the height of the rotational barrier; since the energy barrier to rotation usually is only $\sim$ 10–15 kcal,

|  |  |  |
|---|---|---|
| 7.48a | 7.48b | 7.49 |

however, it ordinarily has been too small to permit resolution. Nmr data at low temperature have provided good evidence for presence of such enantiomers. With **7.49**, on the other hand, nmr showed as AB quartet even at 130°. The △G‡ of inversion of **7.49** had the high value of 28.8 kcal/mol, and it therefore became possible for Lüttringhaus and his associates to effect the first resolution of optical isomers of such an enantiomeric disulfide. This reaction is important because most stereochemical consequences of non-coplanarity have been seen in the solid state or at low temperature.

1,2-Dithiolanes (**7.4**) are interesting cyclic disulfides. They are so nearly planar that the marked distortion from the preferred dihedral angle of 90° evidently figures importantly in the high degree of ring strain they manifest, variously estimated at 16-27 kcal/mol; for example, the 4-carboxylic acid has a dihedral angle of 26°.[67] It is not astonishing therefore, that pure 1,2-dithiolane polymerizes so readily that it can be kept well only in solution.[85,153] It is worth adding that nmr shows cyclic trisulfides to be more rigid than their disulfide counterparts.[67]

Typical bond energies are shown in Table 7.2. Values are not always consistent,[173] so ranges are given. Useful conclusions can be drawn about the energies of -SS- bonds in disulfides:[173] (1) With dialkyl disulfides, they are practically independent of chain length (2) Dibenzyl disulfide ranks with dialkyl disulfides. (3) Diphenyl disulfide has a bond energy about one-third those of dialkyl disulfides, in large part probably because a thiyl radical, $ArS\cdot$, can be stabilized by $p\pi$ interaction of the free electron with the ring. (4) *Para* substituents on diphenyl disulfide increase the bond strength. The order is $NO_2 > Ph > MeO > Me > H$.

Also probably playing a role in the lower bond strength of diaryl disulfides is conjugation of the sulfur atom with the $\pi$ system of the ring. Conjugation through $p\pi$ interaction of the unshared $3p$ electrons of the sulfur atom with the ring has been suggested as the cause of different increments of atomic refraction for sulfur in aryl and alkyl disulfides;[173] presumably related to conjugation also are the somewhat longer bonds and lower vibrational frequencies for aryl disulfides.[173] On the other hand, an S-C bond length of 1.767 Å in bis(*p*-nitrophenyl) disulfide, in comparison with lengths of 1.82 Å for a S-C single bond and of 1.62 Å for a double

**TABLE 7.2**

Bond Energies of Typical -SS- Linkages[a]

| Compound | Bond energy, kcal/mol | Compound | Bond energy, kcal/mol |
|---|---|---|---|
| $(HS)_2$ | 60-80 | $HS_5H$ | 62-63[b,c] |
| $(MeS)_2$ | 67-73 | $HS_6H$ | 62-63[b,c] |
| MeSSEt | 72 | Elemental S | 52-64 |
| $(EtS)_2$ | 64-70 | $(p\text{-}NO_2PhS)_2$ | 46-52 |
| $(PhCH_2S)_2$ | 62-68 | $(p\text{-}MeOPhS)_2$ | 29-35 |
| $(C_{18}H_{37}S)_2$ | 60-66 | $-S_n^-$ [d] | 33[b,c,d] |
| $HS_3H$ | 61-64[b,c] | $(p\text{-}MePhS)_2$ | 26-32 |
| $HS_4H$ | 62-63[b,c] | $(PhS)_2$ | 20-26 |

[a]Values from ref. 173 unless otherwise specified.
[b]Values from ref. 8.
[c]For the average S-S bond.
[d]For liquid sulfur at 300°, n = 50,000 (ref. 489 cited in ref. 8).

bond, led to the conculsion that $d$ orbitals of the sulfur atom participate in the $\pi$-electron system to the extent that the S-C bond contains about one-third double character; the S-S length of 2.019 implies $\pi$ character in the S-S bond also.[174] Another fact interpreted in favor of $d$-orbital involvement of sulfur is that the acidity of **7.50** exceeds that of **7.51** by ~0.4 $pK_a$ unit, suggesting that the $p$-nitrophenyldithio moiety may operate an electron-accepting conjugation through the benzene ring based on valence-shell expansion of the sulfur atoms *via* $3d$-orbital participation; stabilization of the ion **7.52a** by contributors like **7.25b** and **7.52c** was concluded.[175] Earlier literature contains further evidence for utilization of $3d$ orbitals in valence-shell expansion by the sulfur atoms of disulfides and polysulfides.[176]

Molecular-orbital treatment of $p\pi$ interactions and of hybridization possibilities in disulfides permitted interpretations of the torsional barriers, of the red shift in ultraviolet spectra as a function of decreasing dihedral angle (see Sec. 7.4.2) and of some of the chemical properties of -SS-bonds.[177]

$p$—O₂NPhSS—⟨benzene⟩—OH
($pK_a$, 9.91)

**7.50**

$p$—O₂NPhSS—⟨benzene⟩—OH
($pK_a$, 10.33)

**7.51**

7.52a            7.52b

7.52c

### 7.4.2 Ultraviolet and Visible-Region Spectra and Optical Rotation

A review on ultraviolet (uv) spectra offers an admonition worth repeating: that one should be wary of decomposition of compounds that contain -SS- linkages by uv radiation and, if spectra must be done in basic media, by attack of the base.[178] References on uv spectra not given below will be found in the review cited,[178] or in another.[9]

Uv spectra of most aliphatic disulfides show a rather broad maximum at $\sim$250 nm. This maximum is attributed to conjugation of unshared electrons of the sulfur atoms. Intense absorption also occurs at short wavelengths, perhaps associated with polarizations of the -SS- bonds. With the system of 7.53, lack of a bathochromic shift shows that the -SS- bond in effect is an insulator (though not completely so) and that the vinyl chromophores are not significantly conjugated through the -SS- bond.[179]

$$-\overset{|}{C}=\overset{|}{C}-S_n-\overset{|}{C}=\overset{|}{C}-$$

7.53, n = 2
7.54, n = 3–7

Fig. 7.3 shows an interesting feature of aliphatic spectra, that each replacement of a hydrogen atom of dimethyl disulfide results in a hypsochromic shift of 2.5 nm and in increased intensity.[180] These effects also are seen with unsymmetrical alkyl disulfides (see Table 7.3). They were ascribed to hyperconjugative interaction of the electrons on the $\alpha$-carbon atom with $3d$-orbitals of the sulfur atom to expand the sulfur octet; where a pair of electrons cannot be thus donated because of competitive resonance interactions, no peaks were seen, e.g., with diallyl or dibenzyl disulfide.[180]

Cyclic disulfides show intriguing differences in uv spectra from acyclic ones and among themselves as well. Table 7.3 illustrates that a seven-membered ring has a spectrum almost like that of a straight-chain disulfide. As one might guess from the previous discussion of bonding, a six-membered ring shows a considerable bathochromic shift and a five-membered ring still another. Such shifts were early attributed to an

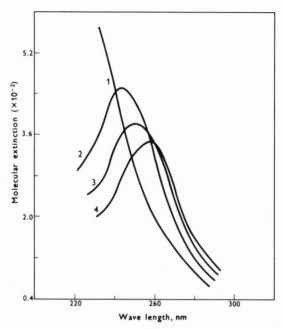

**Fig. 7.3.** Spectra of α-substituted disulfides in MeOH: **1**, di-*t*-butyl disulfide; **2**, diisopropyl disulfide; **3**, di-*n*-propyl disulfide; **4**, dimethyl disulfide. (Reprinted from ref. 180 by permission of the American Chemical Society, copyright owner.)

increasing degree of ring strain, which was suggested to be associated with the two non-bonding pairs of primarily *p* electrons on the sulfur atoms (recall Fig. 7.2).[181] In this still popular explanation, the idea is that the strain ensuing from repulsion of the unshared *p-p*-pairs, as they are more nearly constrained into coplanarity, should raise the energy of the ground state, accordingly decrease that for photoexcitation, and hence cause a shift toward lower energy frequencies (*cf*. ref. 9).

With phenyl disulfide, three bands occur, although they are not all readily seen (*cf*. Fig. 7.4). One interpretation is the following:[178] 240 nm ($\log_\epsilon$, 4.23), conjugation of Ph-S; 270 nm (less intense), the benzenoid chromophores; and 300–320 nm (inflection), excitation of electrons of the -SS- bond. Conversion of a thiol to the symmetrical diaryl disulfide gives much the kind of result seen with **7.53**, *i.e.*, as though two identical chromophores simply had been coupled by an insulating group — the molar intensity doubles but $\lambda_{max}$ does not change.[183] Even in unsymmetrical aryl disulfides, $Ar^1SSAr^2$, Fig. 7.4 makes it clear that the uv curves can be calculated by taking one-half the sum of those for $Ar^1SSAr^1$ and $Ar^2SSAr^2$ (except for slight differences attributed to

**TABLE 7.3**

Typical Ultraviolet Absorption Characteristics of Disulfides

| Compound | Remarks | $\lambda_{max}$, nm | $\epsilon_{max}$ | Solvent | Ref. |
|----------|---------|---------------------|------------------|---------|------|
| $(n\text{-PrS})_2$ | $1^\circ$ Alkyl | 250 | 480 | EtOH | 181 |
| 1,2-Dithiepane **7.6** | 7-Membered ring | 258 | 444 | EtOH | 85 |
| 1,2-Dithiane **7.5** | 6-Membered ring | ~293 | ~290 | EtOH | 181 |
| 1,2-Dithiolane **7.4** | 5-Membered ring | 330 | 147 | EtOH | 181 |
| MeSS-$n$-Bu | Unsym. alkyl | 252 | 356 | MeOH | 180 |
| MeSS-$sec$-Bu | Unsym. alkyl | 250 | 417 | MeOH | 180 |
| MeSS-$t$-Bu | Unsym. alkyl | 248 | 483 | MeOH | 180 |
| $(n\text{-BuS})_2$ + TCNE[a] | $K_c$ for complex, $0.09^b$ | $460^c$ | $5200^c$ | $CH_2Cl_2$ | 182 |
| $(t\text{-BuS})_2$ + TCNE[a] | $K_c$ for complex, $0.36^b$ | $530^c$ | $4600^c$ | $CH_2Cl_2$ | 182 |
| $(PhS)_2$ + TCNE[a] | $K_c$ for complex, $1.5^b$ | $510^c$ | $290^c$ | $CH_2Cl_2$ | 182 |
| $(PhCH_2S)_2$ + TCNE[a] | $K_c$ for complex, $2.6^b$ | $405^c$ | $310^c$ | $CH_2Cl_2$ | 182 |

[a]TCNE = Tetracyanoethylene.
[b]$K_c$ = formation constant at $25^\circ$, liters/mole.
[c]The $\lambda_{max}$ and $\epsilon_{max}$ quoted are for the charge-transfer band at $25^\circ$.

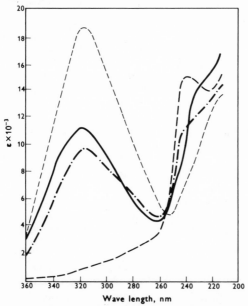

**Fig. 7.4.** Absorption spectra in 95% EtOH of bis(4-nitrophenyl) disulfide ———; diphenyl disulfide . . . .; 4-nitrophenyl phenyl disulfide (observed)_____; 4-nitrophenyl phenyl disulfide (calculated) – . – . (Reprinted from ref. 183 by permission of the American Chemical Society, copyright owner.)

inductive effects); transmission of electronic effects through the -SS- bond thus are not detectible in these systems.[183]

Molecular orbital treatment of saturated disulfides led to the conclusion that the first (*i.e.*, lowest frequency) uv absorption band is associated with an electronic transition from the antibonding $\pi$ orbital (formed by combination of $3p\pi$-atomic orbitals) to the antibonding $\sigma$ orbital.[177] Through this approach, the explanation was offered for the low excitation energy in cyclic disulfides that the energy of the antibonding $\pi$ orbital increases when the dihedral angle decreases, whereas that of the antibonding $\sigma$ orbital is independent of the angle.[177]

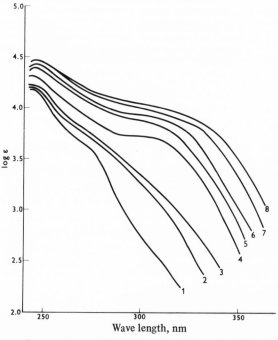

**Fig. 7.5.** Absorption spectra of bis(triphenylmethyl) polysulfides in $CHCl_3$: **1**, mono-; **2**, di-; **3**, tri-; **4**, tetra-; **5**, penta-; **6**, hexa-; **7**, hepta-; **8**, octasulfide. (Reprinted from ref. 184 by permission of the American Chemical Society, copyright owner.)

With polysulfides, as the number of sulfur atoms increases uv intensity also increases and a bathochromic shift occurs.[184] Fig. 7.5 illustrates both changes (the band at ~ 290–330 nm is assigned to -SS- linkages).[184] As a further example, while $Et_2S_2$ has a peak at 250 nm ($\epsilon$ 440), $EtS_4Et$ has one at 300 nm ($\log_\epsilon$ 3.39) and another at 209 nm.[178] Strong sulfur-sulfur conjugation has been inferred from such effects; non-coplanarity of polysulfide chains has been thought to indicate that the conjugation involves *d*-orbitals, which are considerably less subject than *p* orbitals to requirements of coplanarity.[9] With compounds of type **7.54**, it was concluded that sulfur chains do give rise to a conjugative effect, though one which (at most) is considerably less pronounced than with carbon-carbon double bonds[185] (the interpretation does not mention the seemingly relevant work reported in refs. 179 and 184).

Absorption progression toward the red in at least two series of polysulfides is sufficiently regular that the maximum ($\lambda_s$) can be correlated through eq. 7.45 with the number of sulfur atoms (n) and two constants ($k_s$ and $\lambda_s^\circ$); for a benzyl and a tolyl series, respectively, $k_s = 14$ and 18 nm, and $\lambda_s^\circ = 230$ and 259 nm.[9] Trisulfides and higher sulfides have spectra like that of $S_8$, and those for polysulfides of the same number of sulfur atoms are similar to each other.[9]

$$\lambda_s = k_s n + \lambda_s^\circ \tag{7.45}$$

In the visible region, most disulfides absorb negligibly and are colorless. A few disulfides show a low order of absorption that suffices to make them yellow, usually where strain would be expected (for examples, *cf.* refs. 76, 153 and 186). Additionally, among the compounds of Fig. 7.5 increasing numbers of sulfur atoms lead to color; thus 1 and 2 are colorless, 3-7 are pale yellow and 8 is yellow. On the other hand, many aryl disulfides (and some other types as well) *become* yellow or deepen in color when heated.[187] When such a change is reversible, it is called thermochromism. After a good deal of polemic as to the role played by dissociation into free radicals in thermochromism, it now seems clear that Beer's law is obeyed, indicating reversible dissociation into radicals does not occur, and that radicals are not responsible.[187] "Thermal broadening" evidently is the cause of thermochromism, *i.e.*, increased population of the higher vibrational levels of the ground state leads to broadening and flattening of the absorption curve so that just enough extension occurs in the tails of bands to produce a yellow color (*cf.* ref. 187).

Spectra afford good insight into the much investigated donor properties of sulfur compounds. Alkyl disulfides resemble sulfides in forming complexes with $I_2$, although sulfides are much the stronger donors.[182] Tetracyanoethylene, higher in electron affinity than $I_2$ but less effective in complexing disulfides because of less orbital overlap,[188] forms charge-transfer complexes with disulfides that are weak but show well separated absorption at 400-600 nm; Table 7.3 gives some examples.[182]

Let us turn now to spectra in relation to optical activity. With optically active disulfides, optical rotatory dispersion (ord) and circular dichroism (cd) frequently may be used interchangeably as tools. However, they also can be usefully complementary and can be better indicators of uv bands than uv spectra *per se*, for example when bands of optically active compounds are weak or hidden.[189]

Around 1962, great interest began to develop in relationships involving uv spectra, the chirality of disulfides, ord and (especially) cd. Advances to date, which have been summarized,[67,190] already give valuable insights. Gliotoxin, the first compound so studied, had a negative first band (*i.e.*, that of lowest frequency) in the cd spectrum and left-handed chirality at the -SS- linkage. Based on 1,2-dithianes (compounds of structure **7.5**) with known absolute configuration, Carmack and Neubert formulated the important rule that in simple 1,2-dithianes a positive cd band corresponding to the lowest frequency uv absorption band of the -SS- moiety ($\sim$ 280–290 nm for 1,2-dithianes) is associated with a right-handed screw sense (P) of the helix containing the atoms -CSSC- and a negative cd band with a left-handed screw sense (M). The rotary strength of 1,2-dithianes has the high values typical of "inherently dissymmetrical" chromophores. Others felt that since the optical activity results from inherent dissymmetry of the chromophore, which usually greatly outweighs effects of molecular environmental perturbations, the rule might be general for disulfides. Carmack and Neubert cautioned against premature use of the rule for ring sizes other than six or for acyclic disulfides, but later studies of five-membered rings and the eight-membered ring system of **7.49** have revealed no exceptions so far. In contrast, when dihedral angles are close to 90°, it now seems that the sign of the first uv band may be dominated by effects other than the screw sense of the -SS- moiety; consequently, in such compounds there may be no simple relationship between the screw sense and the sign of the first cd band. A quantum mechanical explanation has been offered for the related question of the lack of inherent optical activity of the first uv transition in acyclic disulfides but for strong inherent activity in cyclic disulfides.[190]

*L*-Cystine can show an optical activity ten-fold greater than other amino acids.[191] Certain deoxyuridilyl disulfides also show large optical rotations, which decrease with increasing temperature as one would expect if restricted rotation about the -SS- bond were involved (the optical rotations of the deoxyuridilyl methyl sulfides remain much the same).[192] Hence the explanation suggested in Sec. 7.4.1 for the high optical rotation of *L*-cystine, in terms of a preferred asymmetric conformation about the -SS- bond, seems more probable than earlier ones that invoked a ring system with stable hydrogen bonds or merely an unspecific proximity effect of -SS- on the asymmetric carbon.[192] *Cf.*, however, the recent statement that "The relatively high optical rotatory properties of *L*-cystine and derivatives in solution are due not to endocyclic interactions nor to biasing of screw sense in the disulfide bond but rather to unequal populations of three staggered rotamers" [J. P. Casey and R. B. Martin, *J. Amer. Chem. Soc.*, **94**, 6141 (1972)]. The reference includes considerable information on nmr and cd spectra. Cystine has a strong temperature coefficient of molecular optical rotation, indicating equilibrium of contributors.[191]

### 7.4.3 Infrared, Raman, Nmr and Mass Spectra

One authoritative source remarks that the S-S stretch of disulfides ($\sim 500$ cm$^{-1}$) does not lead to strong infrared bands, so that such linkages sometimes are difficult to detect.[193] The same is said for the C-S stretch (570-705 cm$^{-1}$). Fortunately, both bands are better in the Raman effect.[193] Another such source assigns 400-500 cm$^{-1}$ for the S-S stretch and 600-700 cm$^{-1}$ for the C-S stretch; it agrees that both absorptions are of very limited usefulness, pointing out that these bands usually are both extremely weak and variable and that indeed they are identifiable chiefly because the corresponding Raman lines are strong.[194] Values of S-S and C-S frequencies for a considerable number of aryl disulfides recently have been reported and helpfully discussed.[195]

Nmr work on di- and trisulfides has been reviewed.[196] In other studies, important analytical distinctions were found with various values of n in $RS_nR$, when R = Me, $t$-Bu or $PhCH_2$.[197] The chemical shift of the protons moves downfield in a theoretically predictable way as n increases. This technique was used to study equilibria between chains with n as great as 6 that resulted when disulfides were heated with sulfur (for other applications see Sec. 7.5.6 and ref. 92).[197b] This variation of $\delta$ for the terminals is illustrated in Table 7.4.

TABLE 7.4

Chemical Shifts of Protons in $MeS_nMe$[a,b]

| n | $\delta$ | n | $\delta$ |
|---|---|---|---|
| 1 | 2.16 | 6 | 2.66 |
| 2 | 2.38 | 7 | 2.69[c] |
| 3 | 2.52 | 8 | 2.71[c] |
| 4 | 2.62 | 9 | 2.72[c] |
| 5 | 2.63 | 10 | 2.73[c] |

[a]From ref. 197b.

[b]Ppm downfield from internal $Me_4Si$.

[c]Extrapolated value.

In relation to the discussion above of $L$-cystine, it is interesting that nmr spectra indicate that intramolecular interaction of the two moieties on the -SS- linkage stabilizes the actual configuration in acid solution, although no conclusion is possible yet as to which helical sense of the -SS- moiety may be favored.[191]

The value of nmr for deducing stereochemical features of cyclic di- and trisulfides was mentioned in Sec. 7.4.1, but a sampling of detail is appropriate here. [67] At $-50°$, 7.55 showed an AB quartet at 3.93 $\delta$ ($J_{AB}$, 14 Hz; $\Delta\gamma_{AB}$, 14 Hz) and a weaker one at $\delta$ 4.09 ($J_{AB}$, 14 Hz; $\Delta\gamma_{AB}$, 50 Hz). These values correspond to chair and boat forms of 7.55 respectively.

7.55

The relatively small amount of work on mass spectrometry thus far has been reviewed.[198-200] For alkyl disulfides, eq. 7.46 shows the main fragmentation; large or branched alkyl substituents lead to abundant hydrocarbon ions, with appropriate metastable peaks.[198] With unsymmetrical disulfides, the smaller group tends to remain as the RSSH fragment.[200] Dimethyl and diallyl disulfide undergo rearrangements, since olefin elimination is not feasible; the former also loses one or two methyl groups.[198] The cyclic disulfide 7.5 loses $HS_2 \cdot$, leaving a stable carbonium ion, but substitution can lead to loss of S or SH.[198] Acyl-type disulfides of structure $[R^1R^2NC(=S)S]_2$ usually show peaks for the molecular ion (as do most others mentioned above), with the main fission being of S-S and C-S; rearrangement products also result.[198] Among aryl compounds, methyl phenyl disulfide loses methyl and thiomethyl; apparently a tropylium species also results.[198] Diphenyl disulfide undergoes mainly fission between the two sulfur atoms, but loss of one or both sulfur atoms and of SH occurs also.[198]

$$RCH_2CH_2SSCH_2CH_2R \xrightarrow{-H_2C=CHR} RCH_2CH_2\overset{+\cdot}{S}SH \longrightarrow H\overset{+\cdot}{S}SH \quad (7.46)$$

### 7.4.4 Other Physical Properties

The *dipole moments* of several disulfides agree with dihedral angles of about 70° for 1° alkyl disulfides; with di-*t*-butyl disulfide, steric repulsions result in an increased angle of about 80° and, of course, in a smaller moment.[201] These moments and some other typical ones are given in Table 7.5. It is noteworthy that dielectric relaxation data confirm greatly hindered rotation about the S-S bond and rigidity of the screw conformation.[204] A discussion of *thermodynamic properties* contains a useful listing and considers estimation of such properties not yet reported.[205] *Molecular volume, molecular refraction,* and the logarithms of the *viscosity* within the series $RS_nR$ are linear functions of the value of n.[206] Reid's extensive survey includes physical properties for an abundance of specific di- and higher sulfides, among which are *melting and boiling points, densities, refractive indices* and *surface tensions* (*cf.* also ref. 200).[7] The -SS- moiety has about the same effect on boiling point as five methylene groups.[200]

**TABLE 7.5**

Typical Dipole Moments of $RS_nR$

| Compound | $\mu$(Debye) | Ref. | Compound | $\mu$(Debye) | Ref. |
|---|---|---|---|---|---|
| $(MeS)_2$ | 1.95 | 202 | $(p\text{-MeOPhS})_2$ | 3.11 | 203 |
| $(EtS)_2$ | 1.96 | 202 | $(p\text{-ClPhS})_2$ | 0.45 | 204 |
| $(n\text{-BuS})_2$ | 2.06 | 201 | $(\beta\text{-C}_{10}\text{H}_7\text{S})_2$ | 1.97 | 204 |
| $(i\text{-BuS})_2$ | 2.00 | 201 | $(PhCOS)_2$ | 1.1- | 203 |
| $(t\text{-BuS})_2$ | 1.86 | 201 | $(MeS)_2S$ | 1.66 | 202 |
| $(n\text{-C}_{18}\text{H}_{37}\text{S})_2$ | 2.07 | 203 | $(C_{16}H_{33}S)_2S^a$ | 1.63 | 202 |
| $(PhS)_2$ | 1.90 | 202 | $(C_{16}H_{33}SS)_2{}^a$ | 2.16 | 202 |

aPresumably the di-*n*-alkyl compound, but ref. 202 does not specify.

## 7.5 REACTIONS

Many of the reviews cited in Sec. 7.1.2 contain abundant information about reactions. Sec. 7.5 will deal almost exclusively with reactions of S-S bonds, since in general C-S bonds are relatively resistant to oxidation, reduction or hydrolysis.[207] An often-quoted review discusses C-S cleavages at length,[18] and some of the more important of them ought to be summarized here. Dialkyl disulfides at ~500° give thiols, $H_2S$, some of the sulfide, and occasionally thiophenes. Diaryl disulfides at ~300° yield the sulfide and perhaps the trisulfide. Reaction of diphenyl disulfide and diphenyl sulfone at 300° produces diphenyl sulfide and $SO_2$. Dibenzyl disulfide at 270° gives stilbene, $H_2S$ and S (which react further); other benzylic types also give aralkenes. A cyclic disulfide can lose sulfur to copper and give the sulfide. Acyl-type cleavages may occur readily; some of the most interesting are illustrated by eqs. 7.47-7.49. C-S cleavage involving bases is discussed in Sec. 7.5.3.7. With respect to -SS- cleavage, the value of the review by Parker and Kharasch (1959),[19] and its updating (1966),[20] deserves re-emphasis.

$$RC\overset{\displaystyle B}{\underset{\displaystyle X}{\overset{\|}{S}}}\overset{\displaystyle |}{\underset{\displaystyle Y}{\overset{\|}{S}}}CR + B^- \longrightarrow RC\overset{\displaystyle B}{\underset{\displaystyle X}{\overset{\|}{S}}}\overset{\displaystyle |}{\underset{\displaystyle Y}{\overset{\|}{S}}}CR \longrightarrow RC\overset{\|}{\underset{X}{S}}S^- + R\overset{\displaystyle B}{\underset{\displaystyle Y}{\overset{\displaystyle |}{C}}}$$

$$(X, Y = O, N \text{ or } S; BH = H_2O$$
$$NH_3 \text{ or amines; BH might}$$
$$\text{also attack and then lose } H^+)$$

$$RC\overset{\|}{\underset{X}{S}}^- + S$$

(7.47)

$$Me_2NC\overset{S}{\overset{\|}{}}S\overset{S}{\overset{\|}{C}}NMe_2 + KCN \longrightarrow Me_2NC\overset{S}{\overset{\|}{}}\overset{S}{\overset{\|}{C}}NMe_2 + KCNS$$

(7.48)

$$R^1N=C\underset{\underset{SR^2}{|}}{S}S\underset{SR^2}{\overset{|}{C}}=NR^1 \xrightarrow{25-130°} 2R^1NCS + (R^2S)_2$$

(7.49)

### 7.5.1 Reduction

Typical disulfides can be reduced easily and completely to thiols.[7] Since reoxidation of thiols usually is facile (Sec. 7.3.1), a disulfide-thiol combination can act as a redox buffer reminiscent of an acid-base buffer and can stabilize a system against other oxidants or reductants;[7] one wonders if biological consequences of this fact may not be more important than are now recognized.

Among older reducing agents for the -SS- bond are metal hydrosulfides or sulfides, thiols (see Sec. 7.5.3.4), phosphines, glucose-alkali, sodium arsenite, LiAlH$_4$ and Na(with possible complications).[7] Others are Zn-H$^+$, Sn-H$^+$ (easier removal of the metal ion, with H$_2$S), Na-Hg, Al, Fe and Na$_2$S$_2$O$_4$-alkali.[208] Hydroaromatics and amines can be dehydrogenated by disulfides, which are reduced thereby to thiols,[7] but such reactions are less attractive than formerly because of better dehydrogenative techniques.

Just as with oxidants, an array of reductants is desirable to meet special situations. For example, LiAlH$_4$ is a powerful reductant that probably will reduce nearly any -SS- bond, at least in a high-boiling solvent; since it reduces a nitro group as well, however, for a nitro disulfide one might choose Na$_2$S-NaOH or glucose-NaOH, which reduce -SS- but not -NO$_2$.[208] Again, Na-Hg with Na$_2$S will reduce a polysulfide when Zn may fail.[208]

Lately, other reductants have come into use. Hypophosphorous acid ($H_3PO_2$), in the presence of a small amount of a diselenide, is said to be a mild one.[209] Another is the elegant water-soluble "Cleland's reagent" (dithiothreitol, **7.56**).[210] It is widely used in biochemistry in preference to older alternatives like thioglycolic acid and mercaptoethanol, often simply to preclude oxidation of sensitive thiols. The *erythro* isomer also can be used. The equilibrium constant for eq. 7.50 with cystine is $1.3 \times 10^4$. Evidently oxidation of **7.56** ·proceeds so well because it is both intramolecular and sterically favored;[210] one wonders whether a favorable $\Delta S^{\ddagger}$ for the second phase of exchange, ring closure (*cf.* eq. 7.67), is not a key point. In at least one instance, however, reversibility led to reduction of a disulfide in only 35% yield with one molar proportion of **7.56**; two proportions led to a 66% yield, but use of still more was precluded by the expense of **7.56**.[122] $NaBH_4$ also shows promise as a mild reductant but has not yet been widely used (*cf.* refs. 122 and 200). $Na_2SO_3$ will dissolve keratins in the presence of cupric ammonium hydroxide, which oxidizes the thiolate ion expelled by attack of $SO_3^{2-}$ on the disulfide and thereby regenerates more disulfide (eq. 7.51); this reaction affords a means of tagging proteins that contain -SS- (or -SH moieties) with $^{35}S$, of showing that a chromatographic spot is a thiol or disulfide (since a new spot can be made to result) and of producing a Bunte salt for use in making another disulfide (*cf.* Sec. 7.3.3.6).[211] Although poisoning by bivalent sulfur compounds is a vexing problem with noble-metal catalysts, one can reduce a disulfide with $H_2$, slowly but in good yield, using 30% Pd on C at 50 psi.[212]

$$\underset{\textbf{7.56}}{\underset{\overset{|\;\;|}{HO\;\;H}}{HSCH_2\overset{\overset{H\;OH}{|\;\;|}}{C}CCH_2SH}} + RSSR \overset{\sim 25°}{\rightleftharpoons} 2RSH + \overset{\overline{\phantom{SCH_2CH(OH)CH(OH)CH_2}}}{SCH_2CH(OH)CH(OH)CH_2S} \tag{7.50}$$

$$RSSR + 2Na_2SO_3 + 2Cu^{2+} \longrightarrow 2RSSO_3Na + 2Cu^+ \tag{7.51}$$

Quite a few polarographic studies have been made of disulfides, and some attention has been given to higher sulfides (*cf.* Sec. 7.8)[12,200] Values of $E_{1/2}$ in one set of conditions range from $\sim-1.7$ to $-1.9$ V (*vs.* SCE) for 3° and 2° dialkyl, through $\sim-1.5$ to $-1.8$ V for 1° dialkyl, to $\sim-0.6$ V for diaryl disulfides.[200] A neat simple means of preparatively reducing aminoalkyl disulfides electrochemically in HCl has been developed.[213] Biochemical aspects of the polarography and electroreduction of disulfides have been reviewed.[64] In the polarographic reduction of disulfides at constant potential, presence of dimethyl sulfate permits the intriguing trapping as methyl sulfides of reduction products, many of which would otherwise be unstable; Table 7.6 shows some of the potentials and trapped products.[214]

A reaction that evidently involves reduction of disulfides with copper metal is mentioned later (cf. eq. 7.86).

**TABLE 7.6**
Polarographic Reduction of Disulfides[a]

| Starting Material | Product | Half-wave Potential (V vs SCE) |
|---|---|---|
| (PhCH$_2$S)$_2$ | PhCH$_2$SMe | −1.21 |
| PhSS-$n$-Bu | PhSMe + nBuSMe | −0.99 |
| (AcS)$_2$ | AcSMe | −0.85 |
| (cyclopentene with SSAc and CO$_2$Me) | (cyclopentene with SMe and CO$_2$Me) +AcSMe | −0.66 |
| (benzo ring fused to S–S–C=O) | (benzo ring with SMe and COSMe) | −0.64 |

[a]In the presence of (MeO)$_2$SO$_2$.

## 7.5.2 Oxidation

Oxidation of disulfides leads ultimately to sulfonic acids, but several of the intermediates (or their derivatives) shown in Scheme 7.3 frequently can be obtained. Oxygen-containing oxidants often are employed for such oxidations, but a halogen can be used with a disulfide (or thiol) for synthesis of several classes, as shown in Scheme 7.4.[125]

Good reviews of oxidation and of the many useful reactions of the products are available, and the reader is referred to these for references not cited below (especially ref. 207; see also Sec. 7.1.2).[7,20,207,215] Probably the most important reaction of the intermediates and their derivatives is with nucleophiles, such as thiols, amines, alcohols and carbanions.

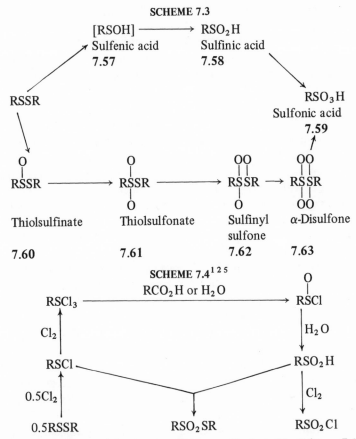

Names commonly used for intermediates are illustrated in Scheme 7.3. Some are lamentable, and it would simplify the lives of all concerned if there were a shift toward more systematic ones (e.g., for **7.61** with R = Me, 2,3-dithiabutane 2,2-dioxide instead of methyl methanethiolsulfonate; *cf.* ref. 207).

Names commonly used for intermediates are illustrated in Scheme 7.3. Some are lamentable, and it would simplify the lives of all concerned if there were a shift toward more systematic ones (e.g., for **7.61** with R = Me, 2,3-dithiabutane 2,2-dioxide instead of methyl methanethiolsulfonate; *cf.* ref. 207).

Let us look first at the top sequence and products of Scheme 7.3 (**7.57** and **7.58**), then at the bottom sequence and products (**7.60–7.63**). Cleavage obviously occurs in the top sequence. Actually in the bottom sequence, too, cleavage and recombination may well occur. At least in oxidation of a thiolsulfinate, $R^1 S(O)SR^2$, complete scrambling occurs – one gets *not only* both unsymmetrical thiolsulfonates ($R^1 SO_2 SR^2$, $R^1 SSO_2 R^2$), *but also* both symmetrical ones ($R^1 SO_2 SR^1$, $R^2 SO_2 SR^2$).[216] This important result implies that something more subtle than simple stepwise coordination of oxygen is likely to be involved in other reactions as well of Scheme 7.3. What will happen in any particular oxidation depends on the disulfide, oxidant, possible catalyst and solvent.[207] Hydrolytic conditions should favor the top sequence and anhydrous ones the bottom sequence, but interplay between the two sequences is likely; morever, even through C-S cleavage is rare, it cannot be ignored.[207]

### 7.5.2.1 With Cleavage

Sulfenic acids (7.57) generally are highly unstable—only a handful are known. The best entry into this series is through sulfenyl halides (eq. 7.24 and Scheme 7.4). Sulfinic acids (7.58) are obtainable as esters by oxidizing disulfides with $Pb(OAc)_4$ (eq. 7.14), or as chlorides or acids by oxidizing disulfides with $Cl_2$ (eq. 7.15 and Scheme 7.4); sulfinate salts, being ambident anions, are useful sources (through alkylation) of either sulfones or esters.[217] Free sulfinic acids decompose slowly, giving 7.59 and 7.61, and can be oxidized to $\alpha$-disulfones (7.63). Sulfonic acids, the ultimate products, can be obtained by oxidizing thiols, disulfides or intermediates with all sorts of agents—hot $HNO_3$ often is used, or perhaps $Cl_2$ with water if one prefers a sulfonyl chloride (Scheme 7.4). [7]

### 7.5.2.2 Without Apparent Cleavage

Allen was able to control oxidation of a symmetrical acyclic disulfide with 30% $H_2O_2$ to produce a monoxide (7.60) in 80% yield, a dioxide (7.61) in 77% yield or a tetroxide (7.63) in 16% yield.[218] Peracids also can be used.[207,218]

A fruitful field for research would be improvement of tetroxide syntheses and study of tetroxide reactions—relatively little has been done with this fascinating class. Curiously, for instance, $EtSO_2SO_2Et$ could not be obtained from the corresponding dioxide (7.61).[85] Why not? Monoxides, dioxides and tetroxides can be obtained from cyclic disulfides as well.[85] A variety of oxidants have been used to prepare oxides from disulfides, or from each other (reactions of $RSOCl$ or $RSO_2Cl$ also have been used).

Attention was given to thiolsulfinates and thiolsulfonates in Sec. 7.3.3.4 (see also ref. 218). Thiolsulfinates (7.60) are stable if the groups R are long-chain but not if they are short-chain or aryl;[218] the problem lies in disproportionation to disulfides and thiolsulfonates (7.61).[207] Thiolsulfonates usually are quite stable (although aminoalkyl thiolsulfonates as free bases are extremely unstable[128]). The older literature is replete with thiolsulfonates formulated as "$\alpha$-disulfoxides",

RS(O)S(O)R, but abundant evidence now proves that the products actually have the structure **7.61**. There seems to have been some evidence for transitory $\alpha$-disulfoxides, but no one has yet had one in a bottle and proved it; further study of whether they can exist and of their participation in Scheme 7.3 would be worthwhile, although probably quite challenging. Among his oxides, Allen did not find a sulfinyl sulfone **(7.62)**,[218] but these can be obtained as shown by eq. 7.52.[207] Sulfinyl sulfones once were thought to be sulfinic anhydrides, RS(O)OS(O)R. Much recent work by Kice has dealt with them as intermediates in the disproportionation of sulfinic acids. $\alpha$-Disulfones **(7.63)** are stable,[218] but they are readily cleaved by nucleophiles.

$$RSO_2Na + Cl(O)SR \longrightarrow RSO_2S(O)R + NaCl \qquad (7.52)$$
$$\textbf{7.62}$$

Oxidation by peracids has been suggested to be the equivalent of electrophilic attack by $OH^+$ on sulfur (eq. 7.53)[19]; in the absence of steric effects, oxidation occurs at the sulfur atom more distant from an electron-withdrawing group.[19,133] One wonders whether oxidations of

$$R^1SSR^2 + 2R^3CO_3H \longrightarrow R^1SO_2SR^2 + 2R^3CO_2H \qquad (7.53)$$

sulfides may not be a fair guide to those of disulfides vis-à-vis oxidants, mechanism, *etc*. Caution surely would be advisable in such thinking, however! For example, $N_2O_4$ nicely converts sulfides to sulfoxides, but there was no indication with it of stepwise oxidation of disulfides, as seen with $H_2O_2$ and peracids. Only sulfonic anhydrides, $(RSO_2)_2O$, were isolated (74–92% yield).[219] Since 1,2-dithiane **(7.5)** did give the 1,1-dioxide with $NO_2 (=N_2O_4)$,[85] further work is needed for understanding of how $NO_2$ functions.

Disulfides, for example $(PhCH_2S)_2$, are good oxidation inhibitors for commercial products like lubricating oils and polymers. By reducing the concentration of peroxides, they have been thought to break the oxidative chain reaction;[200] perhaps they also trap free radical chain carriers.

Polysulfides do interesting things. Two of these are shown in eqs. 7.54 and 7.55.[220] No hexoxides were found from trisulfides. Alkyl polysulfides with NOCl-$NO_2$ give alkanesulfonyl chlorides (67–73%).[220] An interesting research problem might lie in when $R(SO_2)_nR$ can exist and of how the properties might vary with n. Compounds of the structure $ArSO_2S_{1-3}SO_2Ar$ also can be made by condensation reactions of sulfinate or thiosulfate salts.[221]

$$AlkSSSAlk \xrightarrow{\quad H_2O_2,\ AcOH,\ 60^\circ \quad} (AlkSO_2)_2S \ (24\%) \qquad (7.54)$$

$$AlkS_{2-4}Alk + NO_2 \xrightarrow{\quad CCl_4 \quad} (AlkSO_2)_2O \ (\text{"high yield"}) \qquad (7.55)$$

### 7.5.3 With Nucleophiles and Bases

#### 7.5.3.1 Introduction

Kinetically first-order heterolytic scission of the -SS- bond in disulfides has been demonstrated only recently[222]:

$$(7.56)$$

On the other hand, there have been many studies of bimolecular heterolytic scission, both by nucleophiles and by electrophiles.[222] This section discusses scission by nucleophiles and Sec. 7.5.4 that by electrophiles. Eq. 7.57 typifies scission of the -SS- bond by nucleophiles (Nu$^-$). In all such cases of displacement from bivalent sulfur for which kinetic data are available, the reaction is of the $S_N2$ type.[19,223] Such reactions are fairly well understood, but it will soon be clear that there is ample room for further work, both on theory and application.

$$R^1SSR^2 + Nu \quad \rightleftharpoons \quad RS^1Nu + R^2S^- \qquad (7.57)$$

One immediately wonders about at least three aspects of eq. 7.57: (1) Among attacking nucleophiles, what are the relative potencies (i.e., the relative affinities for sulfur)? (2) Of the leaving groups, $R^2S^-$ vs. the other possibility, $R^1S^-$, which will be expelled? (3) What conclusions about the reversibility of eq. 7.57 are possible? Foss, in 1947, was the first to attack these questions effectively (cf. refs. 108 and 223). He gave answer to the first by relating anionoid reactivities to ease of oxidation. Thus a more easily oxidized "thio anion" in the following list displaces any thio anion after it from sulfur: $RC(S)S^-$, $RC(O)S^-$, $R_2NC(S)S^-$, $ROC(S)S^-$, $RS(C=NR)S^-$, $RS^- > S_2O_3^{2-} > ArSO_2S^- > EtSO_2S^- > SCN^-$. Foss also suggested two other related principles: (a) "Anthio anions" are those that can acquire sulfur to give "thio anions"; examples are $RSO_2^-$, $SO_3^{2-}$ and $CN^-$. Such anthio anions will displace the corresponding "thio anion", as eq. 7.58 illustrates. (b) One anthio anion can displace another anthio anion after it in the order: $CN^- > SO_3^{2-} > AlkSO_2^- > ArSO_2^-$.

$$RSSCN + CN^- \xrightarrow{83\%} RSCN + SCN^- \qquad (7.58)$$

As mentioned earlier (Sec. 7.3.3), Foss's recognition of the essential similarity of structures of the form RSZ also has been extremely helpful, so that one can consider the familiar thioalkylating capability of RSC1 (a sulfenyl chloride) to be shared in varying degree by RSSR (a sulfenyl thiolate), $RSSO_2R$ (a sulfenyl sulfinate), $RSSO_3^-$ (a sulfenyl sulfite) and so on.

Parker and Kharasch shed further light on these questions by studying the relative extents of cleavage of $R^1SSR^2$ by $Nu^-$ to give $R^2S^-$ (eq. 7.57), or $R^1S^-$.[224] They found that $Nu^-$ ultimately becomes attached preferentially to the sulfur more distant from the more electronegative group. The cleavage is thermodynamically controlled, with bond breaking being critical rather than bond making, so that the more stable (less basic) anion is displaced (almost exclusively in some instances studied). The same outcome has been found for several displacements discussed below (the possibility of a kinetically-controlled opposite prelude, however, is discussed in Sec. 7.5.3.5). The order of ease of departure as leaving groups is: $2,4\text{-}(NO_2)_2PhS^- > p\text{-}NO_2PhS^- > o\text{-}NO_2PhS^- > PhS^- > AlkS^-$. Increased electron-withdrawl increases the *rate* of cleavage, as well as the *extent*.

The position of equilibrium in eq. 7.57, they found, is determined by the relative "S-nucleophilicity" (or, better, sulfur basicity[225]) of $R^2S^-$ and $Nu^-$.[224] Here, the distinction should be emphasized that since displacements with disulfides involve equilibria and are controlled thermodynamically, the term "sulfur basicity" really is preferable to kinetically-suggestive ones like "sulfur nucleophilicity" (*cf.* ref. 20, pp. 320–321, and ref. 225). The terms "sulfur nucleophilicity" and "thiophilicity" have frequently been used as synonyms for "sulfur basicity" in discussions of equilibria like that of eq. 7.57, however, and the possibility that one may inadvertently infer a kinetic implication (involving rates) for an intended thermodynamic one (involving equilibria) should be borne in mind. Several synonyms thus have been used to describe relative effectiveness of $Nu^-$ in displacement equilibra: thiophilicity,[8] S-nucleophilicity,[19,224] base strength of anionoid reactivity,[108] and sulfur basicity.[225] For the sake of consistency, we shall use here the abbreviated form, "S-basicity", when an author has used any one of the synonyms. When rates seem the crux, we shall try to make that clear, although effects on rates of disulfides often seem to parallel those on equilibria.

Pryor has suggested that relative S-basicities toward a sulfur of an -SS-bond decrease in about the following order (see ref. 19 for further items and some differences in order):[8] $(EtO)_3P > R^-$, $HS^-$, $EtS^- > PhS^- > Ph_3P$, $CN^- > SO_3^{2-} > OH^- > 2,4\text{-}(NO_2)_2PhS^- > N_3^- > SCN^-$ (ineffective[224]), $I^-$, $PhNH_2$. Pryor also elaborated on two other noteworthy points: That the order of S-basicity is not that of "nucleophilicity" toward carbon, and that $t\text{-}BuSX$ shows the same lack of reactivity toward $S_N2$ displacements that characterizes neopentyl halides, $t\text{-}BuCH_2X$, for the same reason of steric resistance to backside attack (e.g., with $RSO_2S\text{-}t\text{-}Bu$, Sec. 7.3.3.4, and $t\text{-}BuSS\text{-}t\text{-}Bu$ in ref. 78). Later, Pryor reviewed the extensive effort on the general question of whether nucleophiles react with sulfur compounds *via* the usual direct $S_N2$ displacement or, with the same final outcome of inversion of the sulfur atom, by addition-elimination in which sulfur momentarily expands its octet; there is suggestive evidence for such metastable expanded intermediates, and an addition-elimination route with such an intermediate may account for the fact that nucleophilic attack on sulfur can be $10^9$ faster than on $sp^3$ carbon; there is some evidence for an expanded octet in the reaction of disulfides.[226]

A review by Davis of nucleophilic displacement at the -SS- bond describes correlation of activation energy with the inverse cube of the -SS-bond distance.[227] Discussed also is the "oxibase scale", through which rates may be correlated with the oxidation potential and basicity of a nucleophile X. The key equation for the scale, eq. 7.59, is based on an extension of Foss's ideas on the importance of redox potential by Edwards:

$$\log \frac{k_X}{k_{H_2O}} = \alpha\, E + \beta\, H \qquad (7.59)$$

In it, $k_X$ is the rate with the nucleophile X, and $k_{H_2O}$ is the rate with water. E and H are related to the $E_0$ and $pK_a$, respectively, of X. The constants $\alpha$ and $\beta$ are characteristic of the substrate and measure its sensitivity, respectively, to the oxidation potential and basicity of X; thus $\alpha$ probably measures ease of reduction of the substrate, and $\beta$ its acidity. No $\alpha$ or $\beta$ values are available for sulfur substrates, but E and H values for some nucleophiles are listed in Table 7.7 (in the order of Pryor's relative S-basicities, where available, for comparison).

**TABLE 7.7**
Values of E and H for Nucleophiles ($H_2O$, $25°$) [227]

| Nucleophile,X | E | H | Nucleophile,X | E | H |
|---|---|---|---|---|---|
| $HS^-$ | 2.40 | 8.88 | $SCN^-$ | 1.83 | 1 |
| $CN^-$ | 2.79 | 10.88 | $H_2O$ | $0.00^a$ | $0.00^a$ |
| $SO_3^{2-}$ | 2.57 | 9.00 | $I^-$ | 2.06 | $-9$ |
| $OH^-$ | 1.65 | 17.48 | $PhNH_2$ | 1.78 | 6.28 |
| $N_3^-$ | 1.58 | 6.46 | $ClO_4^-$ | $-0.73$ | $-9$ |

[a]Defined.

Davis also made the interesting point that 1,2-dithiolane (**7.4**) reacts $\sim 10^4$ faster than di-*n*-butyl disulfide with a thiolate ion. He estimated that the difference in activation energies accounted only for a factor of $\sim 10$

and attributed the remainder ($\sim 10^3$) to the more favorable $\Delta S^{\ddagger}$ for the ring opening of **7.4**.[227]

From the multitude of nucleophilic reactions reported, only a few suggestive types can be illustrated here, with leading references.

### 7.5.3.2 Compounds of Phosphorus

A review cites several reactions in which sulfur is removed.[107a] Eqs. 7.60–7.62 are illustrative. Triphenylphosphine will not reduce (EtS)$_2$. In the light of Pryor's S-basicity (thiophilicity) sequence, it is not astonishing that the phosphite of eq. 7.61 does so. A more recent review deals with the first phase of reaction of phosphines and its usefulness for mildly reducing disulfides to thiols when water is present (eq. 7.63); eq. 7.63 is useful for determining disulfides.[228]

$$2Ph_3P + RS_4R \xrightarrow{-2Ph_3PS} RSSR \xrightarrow{Ph_3P} Ph_3\overset{+}{P}SR \;\; \overset{-}{S}R \longrightarrow R_2S + Ph_3PS \tag{7.60}$$

$$(EtO)_3P + (EtS)_2 \xrightarrow{\Delta} (EtO)_3\overset{+}{P}SEt \;\; \overset{-}{S}Et \longrightarrow (EtO)_2P(O)SEt + Et_2S \tag{7.61}$$

$$(EtO)_3P + RSO_2SAlk \longrightarrow (EtO)_3\overset{+}{P}SAlk$$

$$\overset{-}{O}_2SR \longrightarrow (EtO)_2P(O)SAlk + RS(O)\text{-}OEt(+RSO_2\text{-}Et) \tag{7.62}$$

$$R_3P + R'SSR' + H_2O \longrightarrow 2R'SH + R_3PO \tag{7.63}$$

Aminophosphines seem very promising for desulfurizing di- and higher sulfides. They may circumvent the rearrangement or unreactivity often found with phosphites or other phosphines.[52] The reaction is stereospecific and second order. Careful study shows it can be represented by[229]:

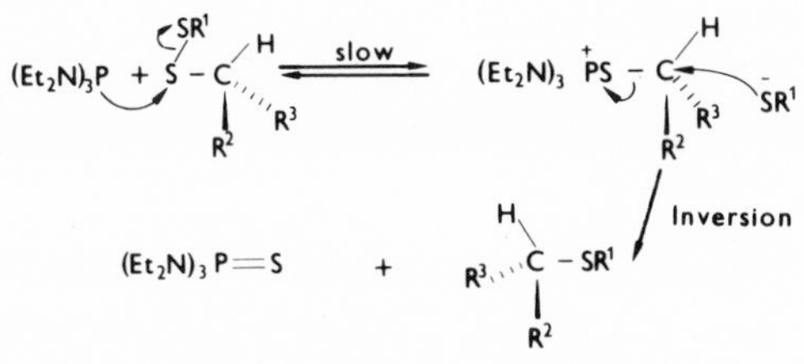

$$(7.64)$$

Aminophosphines also desulfurize thiolsulfonates and trisulfides.[229] Interestingly, as eq. 7.65 shows, different phosphines remove a sulfur atom from different positions of a [35]S-tagged trisulfide (*cf.* **7.26**).[230]

$$2PhCH_2SH + ClS^*Cl \longrightarrow PhCH_2—S—S^*—S—CH_2Ph$$

$$PhCH_2—S—S—CH_2Ph + Ph_3PS^* \quad PhCH_2\text{-}S\text{-}S^*\text{-}CH_2Ph + (Et_2N)_3PS \tag{7.65}$$
$$(90\% \text{ of } S^*) \qquad\qquad (96\% \text{ of } S^*)$$

(with $Ph_3P$ and $(Et_2N)_3P$ shown as reagents over the arrows)

A clever use of a phosphine for a promising peptide synthesis is typified by eq. 7.66. The reaction proceeds in 30 min at $25°$ to give a peptide of good optical purity in 91% yield.[231] Students can check their grasp of principles by comparing a mechanism with the published one.[231]

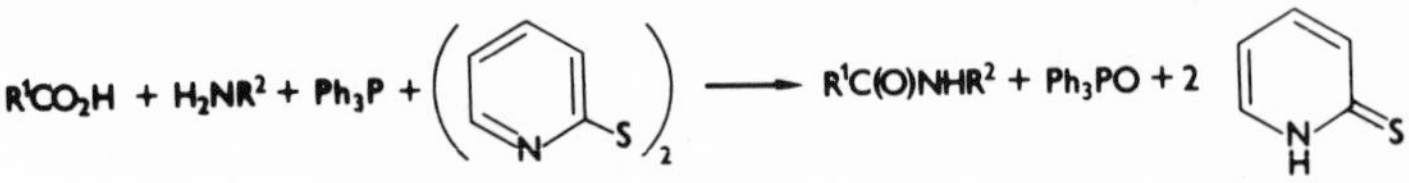

$$R'CO_2H + H_2NR^2 + Ph_3P + \left( \begin{array}{c} \end{array} \right)_2 \longrightarrow R'C(O)NHR^2 + Ph_3PO + 2 \tag{7.66}$$

Arsenic compounds such as arsines and arsenites also cleave -SS-bonds.[223]

### 7.5.3.3 Carbanions

**SCHEME 7.5**

$$RS^- + (R'SO_2)_2CHSR \xleftarrow{\;(R'SO_2)_2CH^-\;} RSSR \xrightarrow{\;R'MgX\;} RSR' + RSMgX$$

$$\downarrow PhLi$$

$$RSPh + RSLi$$

Scheme 7.5 shows some reactions typical of carbanions and related species.[7,19] Cleavage of unsymmetrical disulfides follows the usual pattern that the more stable (less basic) thiolate is displaced.[19] High reactivity of carbanions is evidenced by cleavage of disulfides with butyllithium in one minute or less at $15-70°$ (in determining them by assay of the thiol produced, the other product being RSBu).[232] A useful difference of a disulfide and sulfenyl chloride for thioalkylation of a carbanion is reflected by the reaction of 4-camphyllithium with diphenyl disulfide to give 4-camphyl phenyl sulfide (62% yield)—benzenesulfenyl chloride underwent $S_N2$ attack at chlorine and gave 4-chlorocamphane.[233] Also, a disulfide monosubstituted an enamine, giving a monosubstituted (by -SR) cyclohexanone after hydrolysis (33-79% yield)—sulfenyl chlorides mostly led to disubstitution.[234]

Thiolsulfonates thioalkylate active methylene groups as their carbanions, for example the carbanions of malonates, $\beta$-diketones and $\beta$-disulfones.[123,124] The reactions remind one of the thioalkylation of thiols (Sec. 7.3.3.4).

### 7.5.3.4 Thiols

Reduction of disulfides by thiols was mentioned in Sec. 7.5.1, and use of such interchanges to prepare disulfides in Sec. 7.3.3.7. Further discussion of such reactions here finds a good beginning in the earliest of three reviews.[19] Reactions of thiolates with disulfides were treated as simple redox processes for a long time, but the two-step displacement sequence of eq. 7.67 now is accepted for the usual overall reaction. In the overall reaction, one disulfide $(R^1SSR^1)$ is reduced to its thiolate, meanwhile oxidizing another thiolate to its disulfide $(R^2SSR^2)$. Confirmation for the sequence is afforded by numerous isolations of the intermediates, $R^1SSR^2$. The relative S-basicities of $R^1S^-$ and $R^2S^-$ determine the position of equilibrium, the more easily oxidized thiolate being the more basic and hence the more effective in displacing the other thiolate. Thus the reaction of eq. 7.68 goes essentially to completion, in contrast to that of eq. 7.69 where no observable reaction occurs $(NO_2 = -o$ or $-p-)$. Although thiol-disulfide interchange sometimes goes by free radical mechanisms (Sec. 7.5.5), nucleophilic displacement probably is far more common. No doubt, thiolate ion attacks the backside of a sulfur atom; hence, when $R^1$ of eq. 7.67 is $t$-butyl the usual neopentyl-like hindrance results in very slow reaction. The three sulfur atoms involved in the transition state are thought to lie on a straight line. Exchange of a thiol with a disulfide is second-order and depends on the concentration of thiolate ion, explaining why exchange is negligible in absolute EtOH, faster in aqueous EtOH, and very fast with thiolate ion itself. Interchanges of the foregoing kind are important in relaxation of polysulfide polymers under stress and (Sec. 7.2) in waving of hair, as well as in protein chemistry in general.[19]

$$R^1SSR^1 + {}^-SR^2 \;\rightleftharpoons\; R^1SSR^2 + R^1S^- \qquad (7.67a)$$

$$R^1SSR^2 + {}^-SR^2 \;\rightleftharpoons\; R^2SSR^2 + R^1S^- \qquad (7.67b)$$

$$PhS^- + PhSSPhNO_2 \longrightarrow (PhS)_2 + {}^-SPhNO_2 \qquad (7.68)$$

$$O_2NPhS^- + PhSSPhNO_2 \;\not\longrightarrow \qquad (7.69)$$

$$\text{2RS}^- + \left(\text{S}-\!\!\left\langle\!\!\!\begin{array}{c}\text{CO}_2\text{Na}\\\text{NO}_2\end{array}\!\!\!\right\rangle\!\right)_2 \quad\longrightarrow\quad \text{RSSR} + 2\ \ ^-\text{S}-\!\!\left\langle\!\!\!\begin{array}{c}\text{CO}_2\text{Na}\\\text{NO}_2\end{array}\!\!\!\right\rangle \qquad (7.70)$$

**7.64**

Widespread use of "Ellman's reagent" **(7.64)** for quantitatively determining thiol groups is an application of the general equation above, eq. 7.67, which features **7.64** as the disulfide (eq. 7.70; *cf.* ref. 235). The nitrothiophenoxide anion, expelled quantitatively, is determined spectrophotometrically. Too high a pH must be avoided because **7.64** also reacts very readily with OH⁻ (*cf.* Sec. 7.5.3.7); other nucleophiles, of course, can interfere too.[235] The same thing is true of two alternatives to **7.64**, 2- and 4-pyridyl disulfide.[235]

A later review stresses biological aspects of interchanges.[64] Thiols are the most specific reductants for -SS- moieties in proteins (*cf.* Sec. 7.2), although complete reduction of disulfides usually requires relatively high concentrations of thiols (Cleland's reagent, will be recalled as an exception; eq. 7.50). It is worth noting that vasopressin **(7.34)** and insulin (Fig. 7.1) appear to interact by exchange reactions with thiol groups of the tissues they affect, conclusions that proffer a challenge and inducement for further study of interchanges.

A third review emphasizes interchanges of aminothiols, which can protect animals against ionizing radiation; formation of unsymmetrical disulfides with proteins may play a key role in the protection.[236] This review suggests that the equilibrium constants for interchanges do not deviate more than a factor of ~2 from statistical prediction unless steric hindrance or ionization are unusual.

### 7.5.3.5 Cyanide Ion

The considerable volume of research on the reaction of CN⁻ with cystine and cystine peptides has been reviewed,[237] as have reactions with other types of disulfides.[19] Here again, both displacement and redox features are seen (eq. 7.71). With cystine itself, cyclization occurs, so that a thiazoline rather than a thiocyanate results.[237] Addition of the amino group to the cyanide ion may precede -SS- cleavage, but the matter still seems unsettled.[12] N,N-diacylcystine derivatives do give thiocyanates, but the equilibrium constant is only ~0.02 (pH 7, 35°).[237] Since HCN is a poor nucleophile, it is no surprise that rates increase with increasing pH.

$$R^1SSR_2 + CN^- \rightleftharpoons R^1SCN + R^2S^- \qquad (7.71)$$

For alkyl disulfides in general, eq. 7.71 lies well to the left unless the R group introduces special effects like the cyclization mentioned, or electron-withdrawal or ring strain.[19] Since CN$^-$ seems only slightly less nucleophilic than ArS$^-$, on the other hand, it will displace ArS$^-$ from an aryl disulfide, although it reportedly does not react with diphenyl disulfide itself; such a cleavage can be fairly to highly effective if the leaving group, ArS$^-$, bears -NO$_2$ or -CO$_2$H substituents, or if it is removed from the equilibrium (e.g., by formation of a sulfide with 2,4-dinitrochlorobenzene, which leads even with diphenyl disulfide to PhSCN in 25% yield).[19]

Attack of CN$^-$ on disulfides ordinarily displaces the more stable thiolate anion, as usual.[238] However a kinetically-controlled *opposite* result may be a *prelude* to this thermodynamically-controlled one. Thus in eq. 7.72 trapping showed the major product *first* displaced to be the less stable (more basic) *t*-BuS$^-$ anion.[238] Conceivably, steric resistance of the *t*-Bu group to attack of CN$^-$ on the adjoining sulfur atom could have led to this atypical result.[238] But the result may also imply, for the overall process of nucleophilic displacement, that a rapid, reversible, kinetically-controlled attack may occur *first,* in which CN$^-$ first attacks the *more* positive sulfur to displace the *less* stable (more basic) thiolate ion (*t*-BuS$^-$ in eq. 7.72). Subsequent attack of this thiolate upon the other product, **7.65**, then presumably could lead to the slower, more irreversible, thermodynamically-controlled step of eq. 7.73. Eq. 7.73 thus would produce the more stable (less basic) thiolate **7.66** that precedent leads one to expect.

$$MeO_2CCH_2SCN + t\text{-BuS}^- \qquad (7.72)$$
$$\textbf{7.65}$$

$$MeO_2CCH_2SS\text{-}t\text{-Bu} + CN^-$$

$$MeO_2CCH_2S^- + t\text{-BuSCN} \qquad (7.73)$$
$$\textbf{7.66}$$

Interactions of proteins with CN$^-$ have been reviewed.[64]

### 7.5.3.6 Sulfite, Thiosulfate and Sulfinate Ions

Several points, based largely on one review, will have to suffice as a summary[19]. Sulfite ion is less effective in displacement on sulfur than is sulfide ion, because of the effect of the oxygen atoms; however, this casts no great aspersion on it, because $S^{2-}$ readily cleaves diphenyl disulfide and must therefore be near the top of the S-basicity list (*cf.* ref. 19).

$$\text{RSSSSR} + \text{SO}_3{}^{2-} \underset{}{\overset{-\text{RS}^-}{\rightleftharpoons}} \text{RSSSSO}_3 \qquad (7.74)$$

$$\big\updownarrow \; {\scriptstyle +\text{SO}_3^{2-}} \; \Big\updownarrow \; {\scriptstyle -\text{S}_2\text{O}_3{}^{2-}}$$

$$\underset{-\text{S}_2\text{O}_3{}^{2-}}{\overset{+\text{RS}^-}{}} $$

$$\text{RSSR} \rightleftharpoons \text{RSSSO}_3$$

As an anthio anion displacing the corresponding thio anion, sulfite ion nevertheless desulfurizes polysulfides by displacing thiosulfate ion, so that eqs. 7.74 lie to the right. Sulfite ion does not cleave diphenyl disulfide under ordinary conditions and cleaves a mononitrophenyl disulfide only slowly. But it cleaves a 2,4-dinitrophenyl disulfide rapidly, displacing the less basic thiolate as usual. It also cleaves certain amino- or carboxyalkyl disulfides; these reactions are sensitive to pH because at low pH the less nucleophilic bisulfite ion forms, reminding one of the parallel effect with HCN. Interactions of sulfite ion with proteins have been reviewed;[64] a positive charge near the -SS- bond leads to a rapid reaction and a negative charge to a slow one,[64] as is seen also with simpler disulfides.[19]

Thiosulfate, $\text{SSO}_3{}^{2-}$, is not powerful and is displaced by $\text{CN}^-$, $\text{SO}_3{}^{2-}$, $\text{RS}^-$, etc.[19] It displaces $\text{SCN}^-$ and a few other anions from sulfur, even so. It will displace $\text{SO}_3{}^{2-}$, as well, if $\text{SO}_3{}^{2-}$ is removed from the equilibrium somehow. Sulfinate ions, $\text{RSO}_2^-$, are not very S-basic either, being displaced for example from $\text{RSO}_2\text{S}^-$ by $\text{SO}_3{}^{2-}$.

### 7.5.3.7 Hydroxyl and Alkoxyl Ions

Despite years of study, alkaline decomposition of organic disulfides still is not completely understood.[12] Simple dialkyl disulfides resist alkali quite well, and for rapid reaction activation is necessary by groups such as aryl or carbonyl, or by unsaturated groups at the $\alpha$-carbon atom; even then, reactions often are not clean.[19]

Decomposition of aliphatic disulfides in water can be initiated either by nucleophilic attack of $\text{OH}^-$ directly on one of the sulfur atoms or by proton abstraction; subsequent events frequently are quite complex and depend upon the structure of the disulfide.[239] Nucleophilic attack on the -SS- moiety usually occurs unless negative charge on the disulfide inhibits it; such a negative charge is strongly decelerating (as is steric hindrance), whereas a positive charge is strongly accelerating.[239,240] This effect of charge recalls that with $\text{SO}_3{}^{2-}$ and suggests that such effects on repulsion or attraction of $\text{Nu}^-$ are general.

Anaerobically, the major products of nucleophilic cleavage are about a 5:1 mixture of thiol and sulfonic acid.[239] If nucleophilic attack is inhibited, a disulfide may be stable, or it may lose an (activated) proton $\alpha$ or $\beta$ to sulfur and then undergo an elimination reaction.[239] Scheme 7.6 illustrates the three possibilities of nucleophilic attack (a), and $\alpha$- (b) or $\beta$ (c) elimination;[240] the alternative in (c) shown by the dotted lines also has been suggested.[19] In the reaction of rather complex disulfides containing keto and carboxylate functions, direct attack of OR⁻ on the S-S bond again was thought to be favored, although a product did undergo $\alpha$-elimination.[118]

### SCHEME 7.6

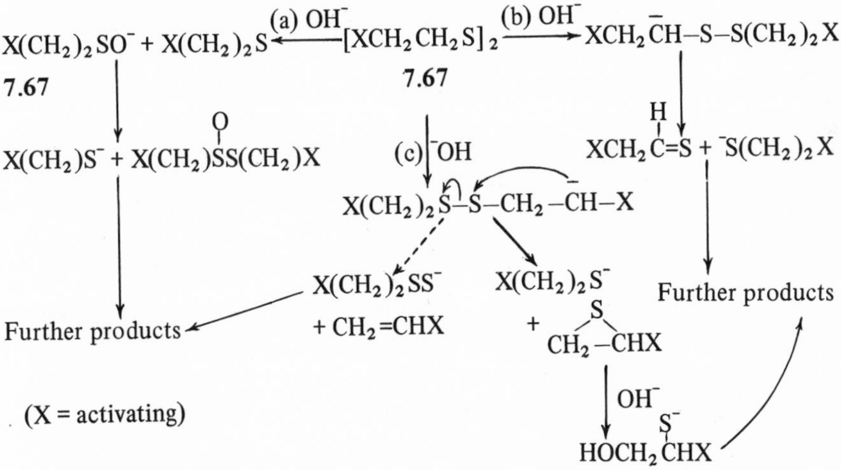

Diaryl disulfides are cleaved extensively by 0.1N alkali at ~35° according to eq. 7.75.[240] The usual principle applies that the more stable thiolate is preferentially displaced, but this time in the sense that those symmetrical disulfides that give thiols of lowest $pK_a$ are most *susceptible* to cleavage (this generalization applies also to those aliphatic disulfides that either decompose by nucleophilic attack or not at all).[240] A nice preparation of arenesulfinate salts was based on eq. 7.75, displaced thiolate being recycled to disulfide by the presence of $H_2O_2$.[240]

$$2(ArS)_2 + 4OH^- \longrightarrow 3ArS^- + ArSO_2^- + 2H_2O \qquad (7.75)$$

A cyclic disulfide, *meso*-1,2-dithiane-3,6-dicarboxylic acid, apparently decomposes by the interesting variant on $\alpha$-elimination shown by eq. 7.76;[241] the student may profit from formulating the reaction conformationally and predicting the behavior of the *rac*-isomer- (*cf.* ref. 241).

$$(7.76)$$

(trans − dicarboxylate)

    With acyl-type disulfides, RC(=X)SSR, the carbon atom seems likely to be more susceptible to attack than the S-S bond (cf. refs. 18, 19 and 112).

### 7.5.4 With Electrophiles

A review of Parker and Kharasch provides valuable background, stating that disulfides undergo acid hydrolysis and acid-catalyzed addition to alkenes and that they are electrophilically attacked by sulfenyl chlorides (eq. 7.77).[19] Similarly, they concluded that many reactions with electrophiles can be represented by eq. 7.78, where $X^+$ is an electrophile. Among these probably are equilibration of two symmetrical disulfides to an unsymmetrical one (and vice versa) in strong acids like conc. HCl or $H_2SO_4$, conversion of disulfides to sulfenyl halides with halogen, and conversion of a disulfide to a methyl sulfide by betaine (perhaps via $CH_3^+$). In the first of these, disulfide interchange, presumably the protonated disulfide is in equilibrium with $RS^+$ (the propagating entity) and RSH (cf. eq. 7.78); since this equilibrium is reversed by adding a thiol, presence of thiols inhibits such an acid-catalyzed interchange.[19,64]

$$\text{ArSCl} + \text{RSSR} \rightleftharpoons [\text{R(ArS)\overset{+}{S}SR} \quad \text{Cl}^-] \rightleftharpoons \text{ArSSR} + \text{RSCl} \quad (7.77)$$

$$\text{RSSR} + X^+ \rightleftharpoons \text{RSX} + \text{RS}^+ \qquad\qquad\qquad (7.78)$$

    An updating of the review mentioned reflects many advances in only seven years:[20] Disulfides react with $AgF_2$ to give structures of the type $RSF_3$. The latter are good synthetic tools for converting carbonyl to difluoromethyl groups and carboxyl to trifluoromethyl groups; they themselves can be converted to stable $RSF_5$ compounds (the group -$SF_5$ is a very intriguing new aromatic-substituent possibility). Disulfides with metal-halide catalysis will introduce RS groups into arenes; with iodine catalysis, they will add the two RS moieties across a carbon-carbon double bond or, in the case of β-aryl disulfides, they will cyclize to thiophenes; in such reactions, the catalyst of course is the electrophile, since it engenders an $RS^+$ species that in turn serves as the electrophile. Scarcely any attention has been given to possible reactions of carbenes with disulfides, despite the importance of both classes separately. Eq. 7.79 shows one such reaction. However, a synthetically useful reaction occurs with CO (which has some aspects of a carbene), with cobalt carbonyl or metal oxide catalysis (eq. 7.80). The insertion one might expect of a carbene does indeed occur with diazoalkane derivatives, with loss of nitrogen, but it may really involve action of the diazo compound as a nucleophile (eq. 7.81; see ref. 242).

$$(t\text{-BuS})_2 + Cl_2C: \longrightarrow (CH_3)_2 \; C\overset{+}{-}\!SS\text{-}t\text{-Bu} \longrightarrow (CH_3)_2C + SS\text{-}t\text{-Bu}$$

$$\begin{array}{cc} H_2C \; \big) \; C\,Cl_2 & H_2C \quad CHCl_2 \\ H^{\!\!\!\!\swarrow} & \end{array}$$

$$\tag{7.79}$$

$$RSSR + 2CO \xrightarrow{\text{Cat., } 250°, \; 1000 \text{ atm.}} RC(O)SR + COS \tag{7.80}$$

An interesting new reaction discovered by Helmkamp and his associates is shown by eq. 7.82; the sulfonium salt, **7.68**, undergoes the exchange of eq. 7.83 extremely rapidly ($k \geqq 10^5 M^{-1} \; sec^{-1}$), $\sim 10^9$ faster than for displacement by $Me_2S$ on any $sp^3$ carbon.[243] This exchange is faster by a factor of $\sim 10^5$ than that of the potent nucleophile $MeS^-$ with $(MeS)_2$, presumably because this leaving group is a much better one. The fact that salts like **7.68** undergo nucleophilic substitution "incredibly rapidly" should lead to interesting applications and, incidentally, "drive(s) home in a particularly forceful way just how fantastically rapid nucleophilic substitution at sulfenyl sulfur can easily be."[243]

$$Ar_2CN_2 + Ar'SSAr' \longrightarrow Ar_2C(SAr')_2 + N_2 \tag{7.81}$$

$$MeSSMe + Me_3O^+\bar{O}_3SPh\text{-}2,4,6\text{-}(NO_2)_3 \rightarrow Me_2\overset{+}{S}SMe^-O_3SPh\text{-}2,3,6\text{-}(NO_2)_3$$

$$\textbf{7.68} \tag{7.82}$$

$$Me_2S + Me\overset{+}{SS}Me_2 \rightleftarrows Me_2\overset{+}{S}SMe + Me_2S \tag{7.83}$$

**7.68**

A carefully studied reaction with the overall stoichiometry of eq. 7.84 seems to involve the sequence of eq. 7.85. The reactions of eq. 7.85 then are followed by further reactions in which a thiolsulfinate probably is an intermediate.[244]

$$4 \text{ArSO}_2\text{H} + (\text{ArS})_2 \xrightarrow{\text{H}^+} 3 \text{ Ar SO}_2 \text{SAr} + 2\text{H}_2\text{O} \tag{7.84}$$

$$\text{ArSO}_2\text{H} + \text{H}^+ \underset{}{\overset{-\text{H}_2\text{O}}{\rightleftharpoons}} \text{ArSO}^+ \underset{}{\overset{(\text{ArS})_2}{\rightleftharpoons}} \text{Ar S(O) } \overset{+}{\text{S}}(\text{Ar})\text{SAr} \tag{7.85}$$

Heavy-metal interations with disulfides have been of considerable biochemical interest but, because of the complexity, chiefly as side reactions one tries to avoid.[64] Formation of a complex, $[\text{RS(M)SR}]^+$, appears to occur first. With $\text{Hg}^{2+}$, $\text{Cu}^{2+}$ and $\text{Ag}^+$ in water, usually with participation of $\text{OH}^-$, thiolate salts and sulfinic acids finally result. Attractive syntheses are possible, however, using silver ion as an electrophile that will form a complex at one sulfur atom of RSSR while a nucleophile attacks at the other; thus silver ion leads with an amine to a sulfenamide $(\text{RSNR}_2')$ and with a sulfinate ion $(\text{R}'\text{SO}_2^-)$ to a thiolsulfonate $(\text{RSSO}_2\text{R}')$.[245a] This may be the best place also to call attention to the reaction of eq. 7.86; copper metal is thought to react with the disulfide to give a cuprous thiolate, which then reacts as a nucleophile.[245b] Disulfides also form complexes with salts of platinum, gold and iridium, as well as with diborane.[7]

$$\text{ArBr} + (\text{RS})_2 + \text{Cu} \xrightarrow{\text{AcNMe}_2, \Delta} \text{ArSR} \tag{7.86}$$

Since diphenyl disulfide can be brominated well, Friedel-Crafts acylation would seem plausible.[246] Use of acetyl chloride with $\text{AlCl}_3$ had failed earlier, however, and use of $\text{AlBr}_3$ led actually to phenyl thiolacetate (72%);[246] similar results occurred in other such reactions.[247] Attempts to effect a reaction analogous to the benzidine rearrangement by treating diphenyl disulfide with strong acids have led either to no reaction or to polymers; catalysis with a trialkyloxonium salt, the latest approach, apparently resulted in S-alkylation, but ultimately gave only various sulfides resembling those of semidine-rearrangement counterparts.[248]

An interesting reaction thought to involve electrophilic attack on unsymmetrical disulfides by pyridine N-oxide results in conversion to sulfonic acids and the symmetrical disulfides, concurrently with reduction of the N-oxide to pyridine; it gave a Hammett $\sigma^+$ correlation and is still under study.[249] Another reaction that probably involves electrophilic attack is the conversion of disulfides by ozone to thiolsulfonates (23–82% yield).[250] A number of reactions discussed in other sections probably involve electrophilic attack on disulfides too (e.g., eqs. 7.14b, 7.53, 7.96, and the reaction with MeI discussed in Sec. 7.6.3).

### 7.5.5 Homolytic Reactions

Among reactions of the -SS- bond, those of heterolysis with nucleophiles probably are the most notable (perhaps because sulfur then can so effectively use $3d$-orbitals in the transition state). Homolytic scissions nevertheless are important too; they occur in photolysis, in electrochemical and certain other reductions, and in such reactions as those with triphenylmethyl, sodium ketyls and metals.[223] The summary of some major features of homolysis that follows is based on Pryor's authoritative review, where not otherwise referenced.[8]

Mechanisms for homolysis are more controversial than for heterolysis, but two broad types certainly are recognizable: unimolecular scission (eq. 7.87) and bimolecular radical displacements (eq. 7.88). Unimolecular scission of some disulfides can be induced thermally in the neighborhood of $100-150°$ to give RS·, a thiyl radical (sometimes called a sulfenyl radical), as seen in eq. 7.87. Evidence for such dissociation is the initiation by some disulfides of vinyl polymerization with heat in the dark. Thus the common vulcanization accelerator tetramethylthiuram disulfide, $[Me_2NC(S)S]_2$, is nearly as efficient as a peroxide in initiating polymerization of methyl methacrylate $(70°)$. Benzothiazolyl disulfide initiates with the methacrylate also $(95°)$, although it fails with certain monomers where radicals of the growing polymer chains attack the -SS- bond in preference to monomer. As indicated, temperatures vary at which thermal homolysis of disulfides occurs—a further illustration is that MeSSEt will cause acrylonitrile to polymerize at $150°$ in the dark, though not at $100°$.

$$R^1SSR^2 \xrightarrow{\Delta \text{ or } h\nu} R^1S\cdot + R^2S\cdot \qquad (7.87)$$

$$R^1SSR^2 + R^3\cdot (\text{or } R^4S\cdot) \longrightarrow R^1SR^3 (\text{or } R^1SSR^4) + R^2S\cdot \qquad (7.88)$$

Photolytic conditions, on the other hand, can lead to initiation of vinyl polymerization by most disulfides at $\sim25°$. Indeed, disulfides can be even more effective initiators than benzoyl peroxide under such conditions. The quantum yield in photolysis of acyclic disulfides is relatively insensitive to structure or solvent. Its low value suggests that about 44 activated molecules are deactivated by collision for every homolysis. The cyclic 1,2-dithiolane 7.4, in marked contrast, has a quantum yield of $\sim1$.

Block has reviewed photochemical aspects of the chemistry of the -SS- bond, pointing out that more is known for disulfides and thiols than for any other classes of sulfur compounds.[251] Worth emphasis is the actual isolation of ArS· as colored solids at low temperature; it is noteworthy that electron donors or phenyl in the *para* position stabilize ArS· (presumably by making ArS· more like the stable species ArS⁻, in contrast to the effect of electron withdrawal which would make them more like the highly reactive species RS⁺; *cf.* ref. 164). Irradiation can lead to interchange of disulfides (eq. 7.89), which usually dominates, or to dismutation (eq. 7.90). The interchange involves participation of RS· in a chain process. Since interchange can be induced by heat or light, it can be useful in preparing unsymmetrical disulfides by equilibration, as will be recalled (Sec. 7.3.3.7), but it also can lead to very real difficulties in working with such compounds (*cf.* Sec. 7.5.6). Dismutation evidently follows C-S cleavage to R· and RSS·. It is worth adding that disulfide oxides have been little studied photochemically—as Block suggests, this neglect should be remedied. Another recent review considers the mobility of RS groups in systems of disulfides with thiols or trisulfides.[173] Exchange is considerably accelerated by uv radiation. Mobility depends much more on the properties of -SS- than of -SH bonds. Study of such rates of exchange, using thiols with isotopically labeled disulfides, was one of the means used to determine the bond energies of Table 7.2.[173]

$$R^1SSR^1 + R^2SSR^2 \underset{\longleftarrow}{\overset{h\nu}{\longrightarrow}} 2R^1SSR^2 \qquad (7.89)$$

$$RSSR \xrightarrow{h\nu} RSR + RS_3R + RS_4R \qquad (7.90)$$

Bimolecular radical displacements at -SS- bonds are very rapid, according to Pryor (e.g., $k = 2 \times 10^6$ $M^{-1}$ $sec^{-1}$).[8] These homolytic counterparts of familiar $S_N2$ heterolyses are analogously referred to as $S_H2$ displacements. Examples are the $S_H2$ reactions in which mixtures of disulfides are equilibrated (eq. 7.89) and in which disulfides effect chain transfer in polymerization. In the latter, the disulfide reacts with a polymer radical to terminate it, and the thiyl radical then displaced starts growth of a new chain. Evidence for this kind of reaction is that both dialkyl and diaryl disulfides incorporate only two sulfur atoms into each molecule of polymer. Pryor lists many "transfer constants" for disulfides, which are useful measures of upper limits for the rate of attack on the sulfur atoms. An important conclusion from such data is that $S_H2$ displacements differ markedly from $S_N2$ displacements in responding only slightly to backside hindrance, since an isopropyl group on sulfur slows attack by only 14-fold and a *t*-butyl group by only 66-fold.

Other reviews provide recent references on homolyses[20], including effects of ⁶⁰Co and X-rays and the reaction of N-bromoimides to give sulfenimides and bromine,[12] as well as on other advances.[252]

In yet another recent review, Pryor concludes that attack on sulfur usually occurs from the rear and, although the matter still seems likely to attract much debate, that many substitutions on sulfur by radicals (and by nucleophiles too, for that matter) proceed by addition-elimination rather than by direct one-step displacement.[226] A metastable intermediate is invoked, in which the octet of sulfur is expanded. Also considered is the relative extent of attack by radicals on the -SS- bond *vs.* attack on α-hydrogen atoms of dimethyl or diisopropyl disulfide, which show about the reactivity of benzylic hydrogen atoms—most attack occurs on the sulfur atom, even though the hydrogens are highly reactive.[226]

Other recent work ought to be mentioned. The rate of isotopic exchange between an arenethiol and the corresponding disulfide in nonpolar solvent free of a base is not appreciable. However, oxygen initiates a free-radical chain process that results in exchange; an alkanethiol-disulfide system contrasts in showing negligible exchange even in the presence of oxygen (because, for initiation, the dissociation energy for AlkS-H is ~14 kcal/mol greater than for ArS-H).[82] Facile participation of trialkylboranes in free radical chain reactions with disulfides, evidenced by initiation with oxygen or light and by inhibition with iodine, permits a new conversion of aryl or alkyl disulfides to sulfides in yields of 85-95% (eq. 7.91).[253] An older reaction that also converts diaryl disulfides to sulfides by a homolytic pathway is achieved by heating with degassed Raney nickel, but it has far less potential and, indeed, gives biphenyls at higher temperatures.[254] Thiol esters can be synthesized through a homolytic reaction of a disulfide with a phosphite under a high pressure of CO (eqs. 7.92; *cf.* eq. 7.80).[107a]

$$\text{PhSSPh} + n\text{-Bu}_3\text{B} \xrightarrow{\;25°,\ 1.5\ \text{hr (O}_2\ \text{cat.)}\;} n\text{-BuSPh} + n\text{-Bu}_2\text{BSPh} \qquad (7.91)$$

$$\text{RS·} + (\text{RO})_3\text{P} \longrightarrow (\text{RO})_3\text{PSR} \longrightarrow (\text{RO})_3\text{PS} + \text{R·}$$

$$\text{(RS)}_2 \qquad\qquad\qquad \downarrow \text{CO} \qquad (7.92)$$

$$\text{RS·} + \text{RC(O)SR} \longleftarrow\!\!\!\!\!\!\!\!\!\!\!\!\!\!\!\!\!\!\!\!\!\!\!\!\!\!\!\!\!\!\!\!\!\!\!\!\!\!\!\!\!\!\!\!\! \text{RĊO}$$

### 7.5.6 Disproportionation[7,19,164]

Disproportionation, a reaction in which like molecules are converted by simultaneous self-oxidation and reduction into two or more unlike ones, is the term used for conversion of an unsymmetrical disulfide $(R^1SSR^2)$ to two symmetrical ones $(R^1SSR^1$ and $R^2SSR^2)$. Disproportionation is a reasonable term for interconversions among higher sulfides as well, such as that of a tetrasulfide to a mixture of tri- and pentasulfide (the valence of sulfur shows formal changes appropriate to oxidation and reduction).

Disproportionation of an unsymmetrical disulfide is simply the reverse sense of eq. 7.89. Sometimes reactions like that of eq. 7.89 are referred to as "redistributions". As mentioned, disproportionation of disulfides can be induced heterolytically by electrophiles that engender $RS^+$ (cf. Sec. 7.5.4) or by nucleophiles that engender $RS^-$ (cf. eq. 7.57; especially by thiolate ions, as in eq. 7.67). Disproportionation also can be induced homolytically by heat or even ambient light (cf. eqs. 7.87, 7.88 and refs. 78, 164). For example, the reaction of 7.69 in hot water in the dark evidently is largely heterolytic (eq. 7.93).[164] It is accelerated by acid or a thiol, does not induce polymerization of acrylamide and shows a Hammett $\sigma$ correlation.

$$2\ XPhSS(CH_2)_2\ NH_3^+Cl^- \longrightarrow (XPhS)_2 + [-SCH_2CH_2\ NH_3^+Cl^-]_2 \quad (7.93)$$
$$\textbf{7.69}$$

The order of increasing resistance to disproportionation is $X = p\text{-}NO_2$ $\ll p\text{-}Cl < p\text{-}H < p\text{-}Me \lesssim p\text{-}MeO \lesssim 2,4,6\text{-}(i\text{-}Pr)_3$. The mechanism probably involves formation of the ions $RS^+$ (acid) or $RS^-$ (thiols), which then attack 7.69 in a chain-type process. It is noteworthy that a free base of 7.69 is extremely unstable (much less so than a salt 7.69, which in turn is less stable than a corresponding amide);[164] the reason appears to be that the free amino group can attack the -SS- bond intramolecularly to displace catalytically-active self-perpetuating thiolate ions (cf. ref. 161). Other groups in various aliphatic or aromatic systems also markedly accelerate disproportionation by what seems to be a similar neighboring-group induced expulsion of thiolate ion; examples are NHAlk, NHAc, $o\text{-}PhCO_2^-$,[112] and $(CH_2)_4SO_2^-$.[135,255] (see also ref. 138).

In contrast, photochemically induced disproportionation of 7.69 evidently goes mostly by a homolytic route. It results in virtual reversal of the order of substituent effects, and this time acrylamide is polymerized, indicating presence of free radicals.[164] In all probability, when one heats disulfides in ambient light heterolysis and homolysis proceed concurrently, in varied relationship. An illustration is that 7.69 with $X = p\text{-}NO_2$ and $p\text{-}Cl$ are the least stable thermally of several disulfides but nevertheless are the easiest to recrystallize because of greater resistance to ambient light.[164]

The effects of structure on the rapidity of disproportionation may be very great indeed.[112,129,136,138,143,157,160-164,167,255,256] Illustratively, the half life for the first-order disproportionation of 7.69 in the dark at $68°$ in water varies from 0.2 hr with $X = p\text{-}NO_2$ to 138 hr with $X = 2,4,6\text{-}(i\text{-}Pr)_3$.[164]

Disproportionation of disulfides ordinarily is an equilibrium reaction (for leading citations, see refs. 138 and 162). With eq. 7.89 as written, the statistically expected value for K is 4; K is temperature-independent, so that $\Delta H$ is zero.[257] The value of K is indeed close to 4 for groups like Me, Et and $i$-Pr. But it is ~24 for the mixture of $(EtS)_2$ and $(t\text{-}BuS)_2$, thus favoring EtSS-$t$-Bu. This result is attributed to the " 'release' of conformations when the $t$-butyl disulfide molecule breaks up".[257] Far more extreme deviations are readily possible. Thus eq. 7.93 is written as irreversible because virtual insolubility of the diaryl disulfide in water drives the reaction to the right—indeed, such a reaction cannot be significantly reversed.[162] Even when both symmetrical disulfides and the unsymmetrical one are soluble, however, essentially complete disproportionation of the latter $may$ be observed, although the reason for such instances of curious behavior is not clear.[161] Other systems are considered in Sect. 7.6.

Disproportionation reactions seem likely to play important roles in protein chemistry ($cf.$ Sec. 7.2 and ref. 64), and further studies along these lines should be valuable. Indeed, one wonders whether neighboring-group attack on biologically significant -SS- bonds, a kind of reaction mentioned above and later (Sec. 7.6), may not initiate disproportionation in a subtle but highly important way, as -SH apparently does (Sec. 7.2)—for example, perhaps subtle conformational changes of proteins uncover moieties like $-NH_2$, -NHAc or $CO_2^-$ which, by neighboring-group attack on -SS- bonds, then displace catalytically active $RS^-$ and thus lead to disproportionation and consequent reorganization of protein structure.

Reid states that "the most characteristic reaction of polysulfides is the taking up of sulfur by the lower and the giving up of sulfur by the higher".[7] He pointed out that owing to thermal instability only a few lower alkyl polysulfides can be distilled, even under high vacuum; illustratively, attempted distillation of diphenyl tetra- or pentasulfide at 1 mm gives only disulfide as distillate. (Agents such as amines or anions such as arsenite, $CH^-$, $S^{2-}$ and $SO_3^{2-}$ also desulfurize the higher sulfides to disulfides[7]).

Experiments with radioactive sulfur showed that in $RS_nR$ the C-S bond remains fixed but that S-S bonds are quite labile and take part in sulfur-exchange reactions.[7] Thus studies of equilibria between [35]S-labeled di- with trisulfides (eq. 7.94) and of tri- with tetrasulfides (eq. 7.95) revealed that the [35]S labels become distributed as shown.[173] Notice that the middle sulfur of the trisulfide remains unlabeled in eq. 7.94, but that internal sulfur atoms are exchanged in eq. 7.95.[173]

$$\overset{1*\ \ 1*}{RS-SR} + RS\text{-}S\text{-}SR \ \underset{\longleftarrow}{\overset{\longrightarrow}{\rule{1cm}{0pt}}}\ \overset{\tfrac{1}{2}*\ \ \tfrac{1}{2}*}{RS-SR} + \overset{\tfrac{1}{2}*\ \ \ \ \tfrac{1}{2}*}{RS-S-SR} \qquad (7.94)$$

1*                                      $\frac{1}{3}$*                    $\frac{1}{3}$* $\frac{1}{3}$*
$$RS\text{—}S\text{-}SR + RS\text{-}S\text{-}S\text{-}SR \;\rightleftharpoons\; RS\text{—}S\text{—}SR + RS\text{—}S\text{—}S\text{—}SR \qquad (7.95)$$

Recent research on the facile thermal decompositions of MeS$_4$Me and MeS$_3$Me helps one to understand eqs. 7.94 and 7.95.[258] In early stages at 80°, MeS$_4$Me gives dimethyl tri-, tetra-, penta- and hexasulfides. Cleavage occurs of MeS$_4$Me to MeSS·, but not to MeS· plus MeS$_3$· (since this latter unsymmetrical cleavage would produce MeS$_2$Me, none of which is found until late in the reaction; this finding, by the way, also argues against reactions such as MeS$_2$· + MeS$_4$Me →MeS· + MeS$_5$Me). Consistent with this result is the fact that the dissociation energy for the cleavage MeSS-SSMe is ~36 kcal/mol but for MeS-SMe it is ~70 kcal/mol. The reason for this difference in energies seems to be that unlike MeS·, MeSS, presumably is stabilized by resonance (if such stabilization is involved, however, it is envisioned as being limited to 2—3 sulfur atoms in a line). The radical MeSS·, stabilized by an estimated 17 kcal/mol, thus tends to form by cleavage or displacement, in preference to the unstabilized one, MeS·. Scheme 7.7 explains the decomposition of MeS$_4$Me.[258] This

**SCHEME 7.7**

$$MeS_4Me \;\rightleftharpoons\; 2\,MeS_2\cdot$$

$$MeS_2\cdot + MeS_4Me \;\rightleftharpoons\; MeS_3Me + MeS_3\cdot$$

$$MeS_3\cdot + MeS_4Me \;\rightleftharpoons\; MeS_5Me + MeS_2\cdot$$

$$2\,MeS_3\cdot \;\rightleftharpoons\; MeS_6Me$$

etc.

decomposition evidently is homolytic as shown, since it is suppressed by a free radical scavenger until the scavenger is consumed. Higher disulfides are believed to undergo similar reactions. On the other hand, decomposition of MeS$_3$Me is much slower than that of MeS$_4$Me, consistent with greater thermal stability of the tri- than that of the tetrasulfide linkage. The trisulfide initially gave MeS$_2$Me, MeS$_3$Me and MeS$_4$Me—this time, MeS$_2$Me *is* an early product, because cleavage to MeS· and MeS$_2$· *does* occur. However, the mechanism of trisulfide decomposition still is uncertain. No monosulfides, hydrocarbons or sulfur were found in decompositions of the foregoing kinds, despite the fact that sulfur is split out when distillation of polysulfides is attempted.[258] It is worth adding for students that analysis of these mixtures was based on the nmr differences of dialkyl polysulfides discussed in Sec. 7.4.3, because the point is so well illustrated that the right physical technique often affords precisely the sword needed to sever a chemical Gordian knot.

## 7.6 PRESENCE OF OTHER FUNCTIONAL GROUPS

Consideration of other functional groups vis-à-vis a disulfide moiety in the same molecule is desirable for two reasons: (1) there may be a need to modify such groups without damaging the disulfide moiety; (2) the group may interact with the -SS- moiety. Scarcely any attention has been given so far to exploration of physical and chemical intramolecular interactions, despite the interesting possibilities one can envision. The same is true to even greater extent of tri- and higher sulfides. Illustrations with several functional groups will suggest the dimensions of the problems and supplement scattered comments already made.

### 7.6.1 $CO_2H$, $CO_2R$, $NR_2$ and NHAc

It is appropriate to consider carboxyl and amine functions concurrently because of their mutual presence with the -SS- moiety in important amino acid systems. Although disulfide interchange is induced by a number of acidic and basic reagents, conditions have been developed under which the -SS- moiety survives reactions elsewhere in the molecule of esterification ($MeOH$-$H_2SO_4$), acid-catalyzed hydrolysis ($BF_3 \cdot Et_2O$-$AcOH$) and amide formation (carbodiimides, $Ac_2O$),[259] as well as of removal of protecting groups from carboxyl and amino functions.[260]

Marked acceleration of disproportionation of unsymmetrical disulfides by $CO_2^-$, $NH_2$, NHR and NHAc was discussed in Sec. 7.5.6 and effects on a hydrodisulfide in Sec. 7.1.4.

Schöberl and Gräfje made an extensive and useful study of unsymmetrical disulfides containing carboxy-, hydroxy-, amino-, and acylaminoalkyl groups, and cysteine or homocysteine units, among others.[138] Disproportionation was catalyzed by acid, light, and thiolate ions, and equilibrium constants could be determined by paper chromatography.[138] The free -$NH_2$, and evidently -$CO_2H$ sometimes, tend to cause disproportionation.[138]

### 7.6.2 OH and SH

Even though unsymmetrical alkoxycarbonylalkyl and carboxyalkyl disulfides could be kept for months without disproportionation, β-hydroxyethyl benzyl disulfide disproportionated in a few hours at ~25°.[139] The hydroxyl moiety evidently was implicated, perhaps through anchimeric assistance to thiolate displacement (cf. Sec. 7.5.6. and, for the effect of OH on a hydrodisulfide moiety, Sec. 7.1.4).

Intramolecular hydrogen bonding of OH to an SS moiety was inferred from infrared absorption spectra of various 4-hydroxy and 4,5-dihydroxy 1,2-dithianes.[261]

Since scarcely any mercapto disulfides are known,[262] this class offers interesting research prospects for brave souls not intimidated by probabilities of facile polymerization owing to thiol-disulfide interchange (Sec. 7.5.3.4)—actually, however, Sec. 7.5.3.4 lends a ray of hope that polymerization might not be a problem if bases and free radicals can be shunned.

### 7.6.3  S, S(O), $SR_3$, $SO_2^-$, $SO_2R$ and $SO_3^-$

Selective reactions of -S- in the presence of -SS- moieties are quite feasible. In keeping with conclusions that -S- is more readily oxidized than -SS-,[263] $NaIO_4$ oxidized an acyclic sulfide-disulfide to a sulfoxide-disulfide (71%),[264] as it did a cyclic counterpart (66%).[265] The acyclic sulfoxide could be selectively reduced too, but only in 6% yield ($Ph_3P$).[264] Intramolecular interactions of -SO- and -SS- in both the acyclic and cyclic compounds seem insignificant under ordinary conditions.[264,265] Similarly, no marked interactions were noticed in a cyclic system that contained -S- or $-S^+(Me)-$ together with -SS-.[265] It is noteworthy that a cyclic disulfide-sulfide and MeI gave a disulfide-sulfonium salt (72% yield), presumably reflecting much greater nucleophilic character of the -S- moiety, since it is well known that MeI cleaves simple dialkyl disulfides to form two moles of a dimethylalkylsulfonium iodide, plus $I_2$ (very slowly, unless catalyzed by $HgI_2$); students can test themselves both by developing a mechanism for the latter reaction and by locating a published one (see ref. 265).

The $-SO_2^-$ function strongly accelerates disproportionation of an unsymmetrical disulfide.[135,255] The unshared electrons of the sulfur can be inferred as displacing catalytically active thiolate ion as usual (cf. Sec. 7.5.6), since neither a sulfone moiety ($-SO_2R$) nor a sulfonate moiety ($-SO_3^-$) has much effect.[255] Structures of the type $RSO_2S_nSO_2R$ were mentioned earlier (Sec. 7.5.2.2).

### 7.6.4  C(O), C(S) and $C(=NR_2)$

Acyl disulfides, RC(O)SSR, are well known and are mentioned repeatedly above. Efforts to synthesize structures of the types Alk C(=NH)SSAr and RC(=S)SSR were discouraging, and the suggestion was made that they may be unstable.[143] Readers may recall, however, that disulfides of the form **7.70** have been made (cf. eq. 7.13). The stability of structures like **7.70** is enhanced by placing electron-withdrawing substituents on Ph and also by replacing $CH_3$ by Ar; the p-methoxy analogue of **7.70** begins to

$$[MeC(=NPh)S]_2 \qquad ArC(=NH_2Cl)SSCCl_3 \qquad ArC(=S)SSCCl_3$$
$$\textbf{7.70} \qquad\qquad\quad \textbf{7.71} \qquad\qquad\qquad \textbf{7.72}$$

decompose in ~2 hours.[84] An intriguing aside is that disulfides of structure 7.70 are formed by *Comamonas* sp. grown on thioacetanilide.[84] Reports of the structures 7.71 and 7.72 show that these variations as well can be put into bottles.[266] Structures of the form $R^1R^2NC(S)SSR^3$ can be obtained readily.[167]

### 7.6.5 Other Groups

With structures of the type $AcSS(CH_2)_2X$, there was little difference in ease of disproportionation for $X = Cl$, $=CH_2$ or $CH_3$.[112] Although an allyl group thus showed negligible neighboring-group effect on the -SS- moiety, interesting reactions might ensue with acid catalysis.

### 7.7    FURTHER REMARKS ON CYCLIC DISULFIDES AND POLYSULFIDES

Numerous cyclic disulfides and a few cyclic polysulfides are discussed in preceding sections. Here, only a few more need be added to illustrate special points or features of interest.

An older but still valuable guide to reactivity (and synthesis) of many ring sizes of disulfides classifies relative reactivities on a scale of 1 (least reactive) to 4.[153] Table 7.8 summarizes some of the results.[153] Substitution decreases reactivity and an endocyclic double bond increases it.

**TABLE 7.8**

Relative Reactivity of Cyclic Disulfides, $(CH_2)_nSS$

| Value of n | Reactivity based on: | | Value of n | Reactivity based on: | |
|---|---|---|---|---|---|
| | Cleavage by CN[-a] | Polym'n.[b] | | Cleavage by CN[-a] | Polym'n.[b] |
| 3 | 4 | 4 | 7 | 3 | 3 |
| 4 | 1 | 1 | 8 | 4 | 3-4 |
| 5 | 2 | 2 | 10 | 3 | 3 |
| 6 | 3 | 3 | 13 | 2 | 2 |

[a]On a scale of 1 (no reaction in time alloted) to 4 (rapid cleavage; like cystine).

[b]Based on estimated relative ease of polymerization: 1 = none; 2 = polym'n. only after adding catalyst; 3 = isolable monomer that polym. in a few hours; 4 = isolation difficult or impossible.

Chart 7.2 shows some interesting heterocycles considered in the leading references cited (the references overlap, and some not cited for a specific structure may be germane).

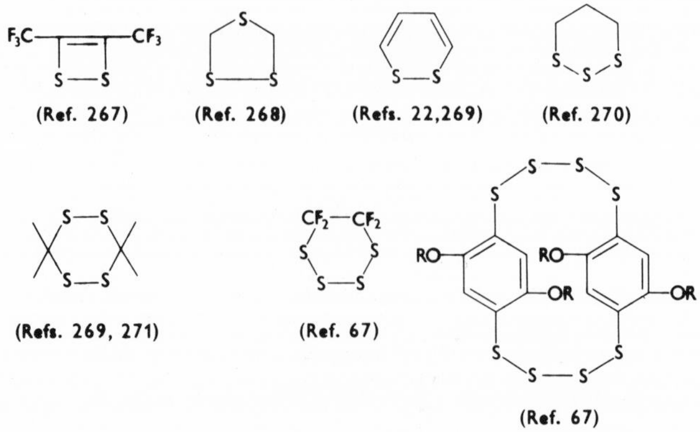

(Ref. 267)        (Ref. 268)        (Refs. 22,269)        (Ref. 270)

(Refs. 269, 271)        (Ref. 67)

(Ref. 67)

**CHART 7.2  ILLUSTRATIVE CYCLIC DI- AND HIGHER SULFIDES**

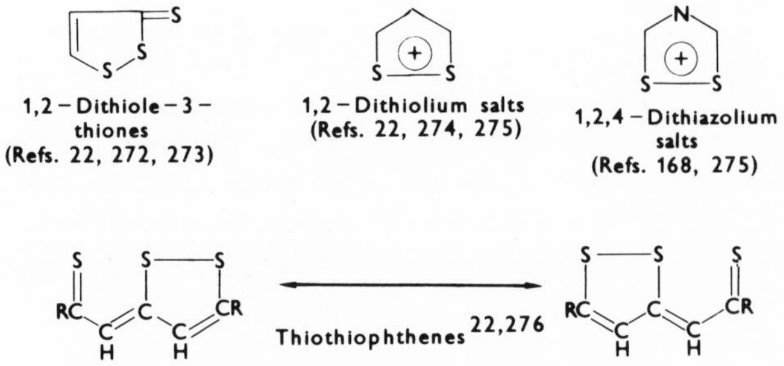

1,2 – Dithiole – 3 –
thiones
(Refs. 22, 272, 273)

1,2 – Dithiolium salts
(Refs. 22, 274, 275)

1,2,4 – Dithiazolium
salts
(Refs. 168, 275)

Thiothiophthenes [22,276]

**CHART 7.3  PSUEDOAROMATIC DISULFIDES**

Chart 7.3 shows some disulfides in the "pseudoaromatic" category, compounds that are planar and more stable than one would expect in the absence of delocalization of electrons. Structure and bonding at the sulfur atoms of thiothiophthene still are unsettled.[22]

## 7.8 ANALYTICAL ASPECTS

A comprehensive review by Karchmer on analysis of bivalent sulfur compounds gives much attention to di- and higher sulfides, in general as well as in analytical aspects;[200] another in prospect promises to be even more comprehensive and is to be eagerly anticipated.[277] One valuable feature is consideration of schemes for determining components in mixtures of disulfides, thiols, sulfides, $H_2S$, sulfur, *etc.* A few points of many are summarized here, together with some other references.

Let us first consider separation. In liquid chromatography, adsorptivity on activated carbon from isoctane increases in the order $Alk_2S < AlkSH < (AlkS)_2 < AlkS_nAlk < S$ (this order seems to imply that the greater the proportion of sulfur, the greater the adsorption). With silica gel or alumina, this order is almost reversed. The latter two seem less promising for separation of disulfides. Gas chromatography can effect good separations of disulfides and of trisulfides,[33] but it is not practicable for alkyl polysulfides higher than tetrasulfides.[278]

For qualitative analysis, Grote's test is a virtual classic for ordinary dialkyl disulfides (cleavage with cyanide and detection of the thiol by coloration of a nitroprusside reagent; *cf.* ref. 7). Grote's test fails with diaryl disulfides and certain other types, unfortunately.[279] However, diaryl disulfides make amends for this failing by showing useful differences in behavior toward concentrated $H_2SO_4$; one may see insolubility, a colorless solution, or widely different colors.[280] A promising means of characterizing even small amounts of unsymmetrical disulfides is to reduce with $LiAlH_4$, convert the two thiolate salts to 2,4-dinitrophenyl sulfides and separate the sulfides chromatographically.[281] The fascinating reaction of eq. 7.96 apparently permits one to form a sulfenamide derivative of at least 2,2'-bis (benzothiazolyl) disulfide and moreover, permits determination of the disulfide quantitatively by measurement of the $I_2$;[282] virtually no attention appears to have been given to study of the generality of this reaction, and it deserves further scrutiny.

$$RSSR + I_2 + 4 HNR_2' \longrightarrow 2RSNR_2' + 2R_2'NH_2I \quad (7.96)$$

For quantitative determination, Karchmer suggests that ir spectra are unpromising, but that uv spectra may be useful. Polarography has limited usefulness for water-insoluble disulfides. However, when one is working with only one disulfide, or a group of closely similar disulfides, polarography may afford a rapid, sensitive tool—with polysulfides, it may give helpful qualitative or semiquantitative results (*cf.* Sec. 7.5.1). The most common determination of disulfides, however, is to reduce them to thiols, for example with Zn-AcOH or with $NaBH_4$-$AlCl_3$ in non-aqueous media. The thiols then are determined by usual means, among which argentimetric potentiometry is popular. Mention was made earlier of reduction with BuLi (Sec. 7.5.3.3) or with $R_3P$ (eq. 7.63) for analysis, as well as of reduction followed by oxidation with $I_2$ (eq. 7.12). Cleavage with $SO_3^{2-}$, mentioned earlier for chromatographic-spot distinction (eq. 7.51), is useful for determining water-soluble disulfides—one reduces with $SO_3^{2-}$ and titrates with salts of mercury or copper.

Another review, biologically oriented, has much of general value about disulfides and in its analytical aspects emphasizes -SS- moieties in proteins.[64] Still another review summarizes recent developments with disulfides, including oxidative determination using bromine and a reductive one in which dithiothreitol engenders the thiol (eq. 7.50), which is determined with Ellman's reagent (eq. 7.70).[12]

## ACKNOWLEDGMENTS

For the support of research described that was done at Vanderbilt University, appreciation is expressed to the National Institute of Arthritis, Metabolism and Digestive Diseases (Public Health Service Research Grant No. AM11685), the U.S. Army Medical Research and Development Command, Department of the Army (Contracts No. DADA17-69-C-9128 and DA-49-193-MD-2030), and the U.S. Army Research Office, Durham, North Carolina. Thanks are due for helpful comments on the manuscript to Professors D.S. Tarbell and H.E. Smith of Vanderbilt University.

## 7.9 REFERENCES

1. Q.E. Thompson, *Quart. Rep. Sulfur Chem.*, 5, 245 (1970).
2. R. Connor in H. Gilman (Editor), "Organic Chemistry", Vol. I., 2nd Ed. John Wiley & Sons, New York, 1943, pp. 861–866.
3. E. Müller (Editor), "Methoden der Organischen Chemie" (Houben-Weyl), Vol. IX, 4th Ed. Georg Thieme Verlag, Stuttgart, 1955.
4. H. Böhme, *ibid.*, pp. 49–54.
5. A. Schöberl and A. Wagner, *ibid.*, pp. 55–82.
6. A. Schöberl and A. Wagner, *ibid.*, pp. 83–92.
7. E.E. Reid, "Organic Chemistry of Bivalent Sulfur", Vol. III, Chemical Publishing Co., New York, 1960, pp. 362–462.
8. W.A. Pryor, "Mechanisms of Sulfur Reactions", McGraw-Hill, New York, 1962, pp. 16–70.
9. C.C. Price and S. Oae, "Sulfur Bonding", Ronald Press Co., New York, 1962, pp. 38–47.
10. N. Kharasch (Editor), "Organic Sulfur Compounds", Vol. I, Pergamon Press, New York, 1961.
11. N. Kharasch and C.Y. Meyers (Editors), "The Chemistry of Organic Sulfur Compounds", Vol. 2, Pergamon Press, New York, 1966.
12. J.P. Danehy, *Mechanisms of Reactions of Sulfur Compounds*, 2, 69 (1968).
13. E.E. Reid in ref. 7, pp. 36–147.
14. D.S. Breslow and H. Skolnik, "Multi-sulfur and Sulfur and Oxygen Five- and Six-Membered Heterocycles", Interscience Publishers, New York, 1966: (a) Part One; (b) Part Two.
15. L. Field and D.L. Tuleen, in A. Rosowsky (Editor), "Seven-membered Heterocyclic Compounds containing Oxygen and Sulfur", Wiley-Interscience Publishers, New York, 1972, Chapter X.

16. (a) L. Young and G.A. Maw, "The Metabolism of Sulphur Compounds", Methuen, London, 1958 (unavailable to us); (b) A. Senning (Editor), "Sulfur in Organic and Inorganic Chemistry", Marcell Dekker, New York; c) G. A. Maw in ref. 16b, Vol. 2, 1972, pp. 113-142; (d) J.L. Kice in ref. 16b, Vol. 1, 1971, pp. 153-207; (e) T.L. Pickering and A.V. Tobolsky in ref. 16b, Vol. 3, 1972, pp. 19-38; (f) P.C. Jocelyn, "Biochemistry of the SH Group", Academic Press, New York, 1972; (g) M. Friedman "The Chemistry and Biochemistry of the Sulfhydryl Group in Amino Acids, Peptides and Proteins", Pergamon Press, New York, 1973: (h) J.L. Day, *Intra-Sci. Chem. Rep.*, 1, 297 (1967).
17. H. Herbst in E. Jucker (Editor), "Progress in Drug Research", Vol. 4 Birkhäuser Verlag, Basel and Stuttgart, 1962, pp. 9–219.
18. D.S. Tarbell and D.P. Harnish, *Chem. Rev.*, 49, 1 (1951).
19. A.J. Parker and N. Kharasch, *ibid.*, 59, 583 (1959).
20. N. Kharasch and A.J. Parker, *Quart. Rep. Sulfur Chem.*, 1, 285 (1966).
21. S. Oae, "Chemistry of Organosulfur Compounds", Kagakudojin, Kyoto, 1968, pp. 135–200.
22. D.H. Reid (Senior Reporter), "Organic Compounds of Sulphur, Selenium, and Tellurium", Vol. 1 The Chemical Society, London, 1970, pp. 102–108, 139–141, 144–145, 164, 177–180, 321–345, 364; vol. 2. 1973, pp. 90–99, 160–163, 176–179, 497–519, 551.
23. L. Field and W.B. Lacefield, *J. Org. Chem.*, 31, 3555 (1966).
24. "The Naming and Indexing of Chemical Compounds from Chemical Abstracts", *Chem. Abstr.*, 56, 22N–24N, 64N (1962).
25. J. Tsurugi, *Mechanisms of Reactions of Sulfur Compounds*, 2, 229 (1968).
26. J. Tsurugi, Y. Abe and S. Kawamura, *Bull. Chem. Soc. Japan*, 43, 1890 (1970).
27. J. Tsurugi, Y. Abe, T. Nakabayashi, S. Kawamura, T. Kitao and M. Niwa, *J. Org. Chem.*, 35, 3263 (1970).
28. H. Böhme and H. -D. Stachel, *Ann.*, 606, 75 (1957).
29. H. Böhme and G. v. Ham, *ibid.*, 617, 62 (1958).
30. N. Kharasch, Z.S. Ariyan and A.J. Havlik, *Quart. Rep. Sulfur Chem.*, 1, 93 (1966).
31. P.G. Stecher (Editor), "The Merck Index", 8th Ed. Merck and Co., Inc., Rahway, N.J., 1968, p. 33.
32. P. Roller, K. Au and R.E. Moore, *Chem. Commun.*, 503 (1971).
33. J.F. Carson and F.F. Wong, *J. Org. Chem.*, 24, 175 (1959).
34. J.W. Loder and G.B. Russell, *Austral. J. Chem.*, 22, 1271 (1969).
35. (a) H. Hagiwara, M. Numata, K. Konishi and Y. Oka, *Chem. Pharm. Bull.* (Tokyo), 13, 253 (1965); cited through *Chem. Abstr.*, 63, 4267 (1965); (b) T. Okaichi and Y. Hashimoto, *Nippon Suisan Gakkaishi*, 28, 930 (1962); cited through *Chem. Abstr.*, 60, 9666 (1964).
36. L. Schotte and H. Ström. *Acta Chem. Scand.*, 10, 687 (1956).
37. L. Jirousek and L. Stárka, *Collection Czechoslov. Chem. Communs.*, 24, 1982 (1959).
38. Ref. 14a, p. 328.
39. H.R. Mahler and E.H. Cordes, "Biological Chemistry", Harper and Row, New York, 1966, pp. 339, 439 and 533.
40. F. Bohlmann and K. -M. Kleine, *Chem. Ber.*, 98, 3081 (1965).
41. R.G. Cooks, F.L. Warren and D.H. Williams, *J. Chem. Soc.* (C), 286 (1967).
42. Ref. 14a, p. 417 ff.
43. Ref. 31, pp. 419, 1043.
44. Ref. 31, p. 491.
45. D. Brewer, R. Rahman, S. Safe and A. Taylor, *Chem. Commun.*, 1571 (1968).
46. S. Safe and A. Taylor, *J. Chem. Soc.* (C), 1189 (1971).
47. R. Nagarajan, N. Neuss and M.M. Marsh, *J. Amer. Chem. Soc.*, 90, 6518 (1968).
48. D.B. Cosulich, N.R. Nelson and J.H. Van Den Hende, *ibid.*, 90, 6519 (1968).
49. J.W. Moncrief, *ibid.*, 90, 6517 (1968).
50. P.W. Trown, *Biochem. Biophys. Res. Commun.*, 33, 402 (1968). [New syntheses of epidithiodiketopiperazines, including dehydrogliotoxin and sporidesmin A, have just been reported by Y. Kishi, T. Fukuyama, S. Nakatsuka and M. Havel, *J. Amer. Chem. Soc.*, 95, 6495 (1973)].

378                                                         Lamar Field

51.  S. Safe and A. Taylor, *Chem. Commun.*, 1466 (1969).
52.  D.N. Harpp, J.G. Gleason and J.P. Snyder, *J. Amer. Chem. Soc.*, **90,** 4181 (1968).
53.  A. Schöberl and A. Wagner in ref. 3, Vol. XI/2, Part III, pp. 429–457.
54.  J.A. Sturman, G. Gaull and N.C.R. Raiha, *Science*, **169,** 74 (1970).
55.  S. Colowick *et al.* (Editors), "Glutathione", Academic Press, New York, 1954.
56.  Ref. 31, p. 497.
57.  A.A. Painter and F.E. Hunter, Jr., *Science*, **170,** 552 (1970).
58.  V. Du Vigneaud, C. Ressler, J.M. Swan, C.W. Roberts, P.G. Katsoyannis and S. Gordon, *J. Amer. Chem. Soc.*, **75,** 4879 (1953).
59.  Ref. 39, pp. 27–28.
60.  K. Anzai and R.W. Curtis, *Phytochemistry*, **4,** 263 (1965).
61.  Ref. 39, pp. 74–83, 115–117.
62.  C.R. Noller, "Chemistry of Organic Compounds", W.B. Saunders Co., Philadelphia, 1965, p. 451.
63.  R. Benesch *et al.* (Editors), "Sulfur in Proteins. Proceedings of a Symposium held at Falmouth, Mass., 1958", Academic Press, New York, 1959.
64.  R. Cecil in H. Neurath (Editor), "The Proteins", Vol. I, 2nd ed., Academic Press, New York, 1963, pp. 379–476.
65.  P.D. Boyer in P.D. Boyer, H. Lardy and K. Myrbäck (Editors), "The Enzymes", Vol. 1, 2nd ed. Academic Press, New York, 1959, pp. 511–588.
66.  B.P. Gaber and A.L. Fluharty, *Quart. Rep. Sulfur Chem.*, **3,** 317 (1968).
67.  R. Rahman, S. Safe and A. Taylor, *Quart. Rev.* (London), **24,** 208 (1970).
68.  E. Schröder and K. Lübke, "The Peptides", Academic Press, New York, 1965: (a) Vol. I, pp. 235–239; (b) Vol. II, pp. 281–374; (c) Vol. II, pp. 374–396.
69.  (a) D. Mazia in ref. 63, pp. 367–389; (b) L. Lorand, A. Jacobsen and L.E. Fuchs, ref. 63, pp. 109–124.
70.  J. Brachet, *Nature*, **184,** 1074 (1959).
71.  C.I. Parhon and S. Oeriu, *Biokhimiya*, **25,** 61 (1960); cited through *Chem. Abstr.*, **54,** 21,387 (1960).
72.  Ref. 31, p. 393.
73.  D.S. Tarbell in ref. 10, p. 97.
74.  A.A. Oswald and T.J. Wallace in ref. 11, p. 205.
75.  E.E. Reid, "Organic Chemistry of Bivalent Sulfur", Vol. I, Chemical Publishing Co., New York, 1958, pp. 118–126.
76.  W. Rundel, *Chem. Ber.*, **102,** 1649 (1969).
77.  D.T. McAllan, T.V. Cullum, R.A. Dean and F.A. Fidler, *J. Amer. Chem. Soc.*, **73,** 3627 (1951).
78.  S.F. Birch, T.V. Cullum and R.A. Dean, *J. Inst. Petrol. London*, **39,** 206 (1953).
79.  I. Pascal and D.S. Tarbell, *J. Amer. Chem. Soc.*, **79,** 6015 (1957).
80.  D.P. Harnish and D.S. Tarbell, *Anal. Chem.*, **21,** 968 (1949).
81.  H.E. Wijers, H. Boelens, A. Van Der Gen and L. Brandsma, *Rec. trav. chim Pays-Bas*, **88,** 519 (1969).
82.  A. Fava, G. Reichenbach and U. Peron, *J. Amer. Chem. Soc.*, **89,** 6696 (1967).
83.  S. Sunner and I. Wadsö, *Amer. Chem. Soc., Div. Petrol. Chem., Preprints* **3,** No. 4, B, 59 (1958); cited through *Chem. Abstr.*, **54,** 23,700 (1960).
84.  J.R. Schaeffer, C.T. Goodhue, H.A. Risley and R.E. Stevens, *J. Org. Chem.*, **32,** 392 (1967).
85.  L. Field and R.B. Barbee, *ibid.*, **34,** 36 (1969).
86.  L. Field and J.E. Lawson, *J. Amer. Chem. Soc.*, **80,** 838 (1958).
87.  (a) L. Field, C.B. Hoelzel and J.M. Locke, *ibid.*, **84,** 847 (1962); (b) L. Field, J.M. Locke, C.B. Hoelzel and J.E. Lawson, *J. Org. Chem.*, **27,** 3313 (1962).
88.  (a) I.B. Douglass, *ibid.*, **30,** 633 (1965); (b) *Cf.* also I.B. Douglass and R.V. Norton, *ibid.*, **33,** 2104 (1968).
89.  C.S. Marvel, T.H. Shepherd, C. King, J. Economy and E.D. Vessel, *ibid.*, **21,** 1173 (1956).
90.  H. Schulz and V. Du Vigneaud, *J. Amer. Chem. Soc.*, **88,** 5015 (1966).
91.  F.H. McMillan and J.A. King, *ibid.*, **70,** 4143 (1948).
92.  B.D. Vineyard, *J. Org. Chem.*, **32,** 3833 (1967).
93.  E.P. Papadopoulos, A. Jarrar and C.H. Issidorides, *ibid.*, **31,** 615 (1966).
94.  T. Nakaya, H. Arabori and M. Imoto, *Bull. Chem. Soc. Japan*, **43,** 1888 (1970).

95. F.J. Smentowski, *J. Amer. Chem. Soc.*, **85**, 3036 (1963).
96. D. Yamashiro, D. Gilleson and V. Du Vigneaud, *ibid.*, **88**, 1310 (1966).
97. I. Photaki, *ibid.*, **88**, 2292 (1966).
98. (a) M.G. Burdon and J.G. Moffatt, *ibid.*, **88**, 5855 (1966); (b) C.N. Yiannios and J.V. Karabinos, *J. Org. Chem.*, **28**, 3246 (1963); (c) K. Balenović and N. Bregant, *Chem. Ind.* (London), 1577 (1964).
99. D.L. Klayman and R.J. Shine, *Quart. Rep. Sulfur Chem.*, **3**, 189 (1968).
100. W.A. Sheppard, *Org. Syntheses.* **40**, 80 (1960).
101. W.H. Hunter and B.E. Sorenson, *J. Amer. Chem. Soc.*, **54**, 3364 (1932).
102. J. Levy and J.H. Mayer, U.S. Patent 2, 986, 581 (1961); cited through *Chem. Abstr.*, **56**, 3416 (1962).
103. D. Klamann and G. Hofbauer, *Monatsh. Chem.*, **83**, 1489 (1952).
104. L. Field and P.H. Settlage, *J. Amer. Chem. Soc.*, **77**, 170 (1955).
105. (a) H. Alper, *Angew. Chem.*, Int. Ed. Engl., **8**, 677 (1969); (b) E. Lindner and G. Vitzthum, *ibid.*, **8**, 518 (1969); (c) N. Kharasch and M.M. Wald, *Anal. Chem.*, **27**, 996 (1955).
106. T.H. Chan, J.P. Montillier, W.F. Van Horn and D.N. Harpp, *J. Amer. Chem. Soc.*, **92**, 7224 (1970).
107. (a) J.I.G. Cadogan, *Quart. Rev.* (London), **16**, 223–233 (1962); (b) R.H. Cragg, J.P.N. Husband and A.F. Weston, *Chem. Commun.*, 1701 (1970).
108. O. Foss, *Acta Chem. Scand.*, **1**, 307 (1947).
109. N. Kharasch in ref. 10, p. 375.
110. N.E. Heimer and L. Field, *J. Org. Chem.*, **35**, 3012 (1970).
111. T. Mukaiyama and K. Takahashi, *Tetrahedron Lett.*, 5907 (1968).
112. L. Field, W.S. Hanley and I. McVeigh, *J. Org. Chem.*, **36**, 2735 (1971).
113. M. Behforouz and J.E. Kerwood, *ibid.*, **34**, 51 (1969).
114. (a) D.N. Harpp, D.K. Ash, T.G. Back, J.G. Gleason, B.A. Orwig, W.F. Vanhorn and J.P. Snyder, *Tetrahedron Lett.*, 3551 (1970); (b) K.S. Boustany and A.B. Sullivan, *ibid.*, 3547 (1970).
115. E. Kühle, *Synthesis*, 561 (1970).
116. (a) M.H. Benn, L.N. Owen and A.M. Creighton, *J. Chem. Soc.*, 2800 (1958); (b) J. Tulecki, J. Dabrowski and J. Kalinowska-Torz, *Diss. Pharm. Pharmacol.*, **18**, 473 (1966); cited through *Chem. Abstr.*, **67**, No. 63,971 (1967).
117. R.G. Hiskey and B.F. Ward, Jr., *J. Org. Chem.*, **35**, 1118 (1970).
118. R.G. Hiskey and A.J. Dennis, *ibid.*, **33**, 2734 (1968).
119. R.G. Hiskey, L.M. Beacham, III, V.G. Matl, J.N. Smith, E.B. Williams, Jr., A.M. Thomas and E.T. Wolters, *ibid.*, **36**, 488 (1971).
120. R.G. Hiskey, G.W. Davis, M.E. Safdy, T. Inui, R.A. Upham and W.C. Jones, Jr., *ibid.*, **35**, 4148 (1970).
121. R.G. Hiskey, A.M. Thomas, R.L. Smith and W.C. Jones, Jr., *J. Amer. Chem. Soa.*, **91**, 7525 (1969).
122. L. Field and P.M. Giles, Jr., *J. Org. Chem.*, **36**, 309 (1971).
123. A. Schöberl and A. Wagner in ref. 3, pp. 683–693.
124. E.E. Reid in ref. 75, pp. 330–335.
125. I.B. Douglass and B.S. Farah, *J. Org. Chem.*, **24**, 973 (1959).
126. L. Field and T.F. Parsons, *ibid.*, **30**, 657 (1965).
127. J.D. Buckman, M. Bellas, H.K. Kim and L. Field, *ibid.*, **32**, 1626 (1967).
128. L. Field, A. Ferretti, R.R. Crenshaw and T.C. Owen, *J. Med. Chem.*, **7**, 39 (1964).
129. L. Field, T.C. Owen, R.R. Crenshaw and A.W. Bryan, *J. Amer. Chem. Soc.*, **83**, 4414 (1961).
130. L. Field, T.F. Parsons and R.R. Crenshaw, *J. Org. Chem.*, **29**, 918 (1964).
131. T.F. Parsons, J.D. Buckman, D.E. Pearson and L. Field, *ibid.*, **30**, 1923 (1965).
132. L. Field and W.B. Lacefield, *ibid.*, **31**, 599 (1966).
133. L. Field, H. Härle, T.C. Owen and A. Ferretti, *ibid.*, **29**, 1632 (1964).
134. L. Field and R.B. Barbee, *ibid.*, **34**, 1792 (1969).
135. L. Field and Y.H. Khim, *J. Med. Chem.*, **15**, 312 (1972).
136. L. Field, W.S. Hanley, I. McVeigh and Z. Evans, *ibid.*, **14**, 202 (1971).
137. J.D. Buckman and L. Field, *J. Org. Chem.*, **32**, 454 (1967).
138. A. Schöberl and H. Gräfje, *Ann.*, **617**, 71 (1958).

139. R.G. Hiskey, F.I. Carroll, R.M. Babb, J.O. Bledsoe, R.T. Puckett and B.W. Roberts, *J. Org. Chem.*, **26**, 1152 (1961).
140. S.J. Brois, J.F. Pilot and H.W. Barnum, *J. Amer. Chem. Soc.*, **92**, 7629 (1970).
141. G. Zumach and E. Kühle, *Angew. Chem., Int. Ed. Engl.*, **9**, 54 (1970).
142. C.A. Grob, *ibid.*, **8**, 535 (1969).
143. L. Field and J.D. Buckman, *J. Org. Chem.*, **32**, 3467 (1967).
144. B. Milligan and J.M. Swan, *J. Chem. Soc.*, 6008 (1963).
145. A.A. Watson, *J. Chem. Soc.*, 2100 (1964).
146. M.T. Bogert and A. Stull, *Org. Syntheses*, Coll. Vol. I. 2nd Ed., 220 (1941).
147. E.M. Fettes in ref. 10, pp. 266–279.
148. Ref. 62, p. 790–791.
149. L.A. Wiles and Z.S. Ariyan, *Chem. Ind.* (London), 2102 (1962).
150. L. Field, J.L. Vanhorne and L.W. Cunningham, *J. Org. Chem.*, **35**, 3267 (1970).
151. W.A. Schulze, G.H. Short and W.W. Crouch, *Ind. Eng. Chem.*, **42**, 916 (1950).
152. C.C. Price and G. W. Stacy, *Org. Syntheses*, Coll. Vol. III, 86 (1955).
153. A. Schöberl and H. Gräfje, *Ann.*, **614**, 66 (1958).
154. D.N. Harpp and J.G. Gleason, *J. Org. Chem.*, **35**, 3259 (1970).
155. R.L. Frank and J.R. Blegen, *Org. Syntheses*, Coll. Vol. III, 116 (1955).
156. F. Runge, A. Jumar and F. Koehler, *J. Prakt. Chem. (4)*, **21**, 39 (1963).
157. L. Field, H.K. Kim and M. Bellas, *J. Med. Chem.*, **10**, 1166 (1967).
158. R.G. Hiskey, T. Mizoguchi and E.L. Smithwick, Jr., *J. Org. Chem.*, **32**, 97 (1967).
159. L. Field and P.M. Giles, Jr., *J. Med. Chem.*, **13**, 317 (1970).
160. L. Field, A. Ferretti and T.C. Owen, *J. Org. Chem.*, **29**, 2378 (1964).
161. M. Bellas, D.L. Tuleen and L. Field, *ibid.*, **32**, 2591 (1967).
162. L. Field and H.K. Kim, *J. Med. Chem.*, **9**, 397 (1966).
163. R.R. Crenshaw and L. Field, *J. Org. Chem.*, **30**, 175 (1965).
164. L. Field, T.F. Parsons and D.E. Pearson, *ibid.*, **31**, 3550 (1966).
165. L. Field and P.R. Engelhardt, *ibid.*, **35**, 3647 (1970).
166. L. Field, W.S. Hanley and I. McVeigh, *J. Med. Chem.*, **14**, 995 (1971).
167. L. Field and J.D. Buckman, *J. Org. Chem.*, **33**, 3865 (1968).
168. O. Foss in ref. 10, pp. 75–82.
169. T. Wieland and H. Schwahn, *Chem. Ber.*, **89**, 421 (1956).
170. D. Barnard, T.H. Houseman, M. Porter and B.K. Tidd, *Chem. Commun.*, 371 (1969).
171. R.R. Fraser and G. Boussard, "Abstracts, 161st National Meeting of the American Chemical Society, Los Angeles, Calif., March 29–April 2, 1971", No. Orgn 82; *cf.* R.R. Fraser, G. Boussard, J.K. Saunders, J.B. Lambert, and C.E. Mixan, *J. Amer. Chem. Soc.*, **93**, 3822 (1971).
172. J. Tsurugi and T. Nakabayashi, *J. Org. Chem.*, **25**, 1744 (1960).
173. E.N. Guryanova, *Quart. Rep. Sulfur Chem.*, **5**, 113 (1970).
174. J.S. Ricci and I. Bernal, *J. Amer. Chem. Soc.*, **91**, 4078 (1969).
175. S. Oae and M. Yoshihara, *Bull. Chem. Soc. Japan*, **41**, 2082 (1968).
176. G. Cilento, *Chem. Rev.*, **60**, 147 (1960).
177. G. Bergson, "Some New Aspects of Organic Disulfides, Diselenides and Related Compounds", Almqvist and Wiksell, Gamla Brogatan 26, Stockholm, 1962. This doctoral dissertation summarizes 18 papers that appeared during 1955–1962. See also D.B. Boyd, *J. Amer. Chem. Soc.*, **94**, 8799 (1972).
178. R.C. Passerini in ref. 10, pp. 66–67.
179. E. Campaigne and R.E. Cline, *J. Org. Chem.*, **21**, 32 (1956).
180. N.A. Rosenthal and G. Oster, *J. Amer. Chem. Soc.*, **83**, 4445 (1961).
181. J.A. Barltrop, P.M. Hayes and M. Calvin, *ibid.*, **76**, 4348 (1954).
182. W.M. Moreau and K. Weiss, *ibid.*, **88**, 204 (1966).
183. E. Campaigne, J. Tsurugi and W.W. Meyer, *J. Org. Chem.*, **26**, 2486 (1961).
184. T. Nakabayashi, J. Tsurugi and T. Yabuta, *ibid.*, **29**, 1236 (1964).
185. F. Fehér and E. Kiewert, *Z. Anorg. Allg. Chem.*, **377**, 152 (1970).
186. D.E. Pearson, D. Caine and L. Field, *J. Org. Chem.*, **25**, 867 (1960).
187. R.E. Davis and C. Perrin, *J. Amer. Chem. Soc.*, **82**, 1590 (1960).
188. W.M. Moreau and K. Weiss, *Mechanisms of Reactions of Sulfur Compounds*, **2**, 7 (1967).

189. C. Djerassi, H. Wolf and E. Bunnenberg, *J. Amer. Chem. Soc.*, **84**, 4552 (1962).
190. J. Linderberg and J. Michl, *ibid.*, **92**, 2619 (1970).
191. J.A. Glasel, *ibid.*, **87**, 5472 (1965).
192. C. Szantay, M.P. Kotick, E. Shefter and T.J. Bardos, *ibid.*, **89**, 713 (1967).
193. N.B. Colthup, L.H. Daly and S.E. Wiberley, "Introduction to Infrared and Raman Spectroscopy", Academic Press, New York, 1964, p. 306.
194. L.J. Bellamy, "The Infra-red Spectra of Complex Molecules", 2nd Ed. John Wiley and Sons, New York; Methuen and Co. Ltd., London, 1958, pp. 352−355.
195. T.A. Chibasova, A.Y. Zheltov, V.Y. Rodionov and B.I. Stepanov, *Zh. Org. Khim*, **7**, 143 (Consultants-Bureau translation, p. 142) (1971).
196. S.R. Heller, *Intra-Sci. Chem. Rep.*, **2**, 327−330 (1968). See also N.F. Chamberlain and J.J.R. Reed in ref.277, Part III.
197. (a) J.R. Van Wazer and D. Grant, *J. Amer. Chem. Soc.*, **86**, 1450 (1964); (b) D. Grant and J.R. Van Wazer, *ibid.*, **86**, 3012 (1964).
198. H. Budzikiewicz, C. Djerassi and D.H. Williams, "Mass Spectrometry of Organic Compounds", Holden-Day, Inc., San Francisco and London, 1967, pp. 292−296.
199. A.I.A. Khodair and A.A. Swelim, *Mechanism of Reactions of Sulfur Compounds*, **4**, 49 (1969).
200. J.H. Karchmer in I.M. Kolthoff and P.J. Elving (Editors), "Treatise on Analytical Chemistry", Part II, Vol. 13 Interscience Publishers, New York, 1966, pp. 337−517.
201. M.T. Rogers and T.W. Campbell, *J. Amer. Chem. Soc.*, **74**, 4742 (1952).
202. E.N. Guryanova and V.N. Vasileva, *Khim. Seraorgan. − Filial*, **4**, 24 (1961); cited through *Chem. Abstr.*, **57**, 5833 (1962).
203. V.N. Vasileva and E.N. Guryanova, *Zhur. Fiz. Khim.*, **33**, 1976 (1959); cited through *Chem. Abstr.*, **54**, 13,783 (1960).
204. M.J. Aroney, H. Chio, R.J.W. Le Févre and D.V. Radford, *Austral. J. Chem.*, **23**, 199 (1970).
205. J.P. McCullough, D.W. Scott and G. Waddington in ref. 10, pp. 20−29.
206. F. Fehér, G. Krause and K. Vogelbruch, *Chem. Ber.*, **90**, 1570 (1957).
207. W.E. Savige and J.A. Maclaren in ref. 11, pp. 367−402.
208. A. Schöberl and A. Wagner in ref. 3, pp. 23−28.
209. W.H.H. Günther, *J. Org. Chem.*, **31**, 1202 (1966).
210. W.W. Cleland, *Biochemistry*, **3**, 480 (1964).
211. J.M. Swan, *Nature*, **180**, 643 (1957).
212. T.P. Johnston and R.D. Elliott, *J. Org. Chem.*, **32**, 2344 (1967).
213. D.S. Tarbell, D.A. Buckley, P.P. Brownlee, R. Thomas and J.S. Todd, *ibid.*, **29**, 3314 (1964).
214. D. Kunz, H. Hartmann and R. Mayer, *Z. Chem.*, **9**, 60 (1969).
215. Ref. 5, and several other of the chapters in this volume.
216. D Barnard and E.J. Percy, *Chem. & Ind.* (London), 1332 (1960)
217. J.S. Meek and J.S. Fowler, *J. Org. Chem.*, **33**, 3422 (1968).
218. P. Allen, Jr. and J.W. Brook, *ibid.*, **27**, 1019 (1962).
219. N. Kunieda and S. Oae, *Bull. Chem. Soc. Japan*, **41**, 233 (1968).
220. F. Fehér, K.-H. Schäfer and W. Becher, *Z. Naturforsch.*, **17b**, 847 (1962).
221. N. Hofman-Bang, *Acta Chem. Scand.*, **12**, 861 (1958).
222. A.M. Kiwan and H.M.N.H. Irving, *Chem. Commun.*, 928 (1970).
223. O. Foss in ref. 10, pp. 83−96.
224. A.J. Parker and N. Kharasch, *J. Amer. Chem. Soc.*, **82**, 3071 (1960).
225. A.J. Parker, *Proc. Chem. Soc.*, London 371 (1961).
226. W.A. Pryor and K. Smith, *J. Amer. Chem. Soc.*, **92**, 2731 (1970).
227. R.E. Davis in A.F. Scott (Editor), "Survey of Progress in Chemistry", Vol. 2 Academic Press, New York, 1964, pp. 189−238.
228. M. Grayson and C.E. Farley, *Colloq. Int. Cent. Nat. Rech. Sci.*, No. 182, 275 (1970); *Chem. Abstr.*, 74 No. 13,196 (1971).
229. D.N. Harpp and J.G. Gleason, *J. Amer. Chem. Soc.*, **93**, 2437 (1971).
230. D.N. Harpp and D.K. Ash, *Chem. Commun.*, 811 (1970).
231. T. Mukaiyama, R. Matsueda and M. Suzuki, *Tetrahedron Lett.*, 1901 (1970).

382                                                                    Lamar Field

232.  S. Veibel and M. Wrónski, *Anal. Chem.*, **38**, 910 (1966).
233.  E.E. Van Tamelen and E.A. Grant, *J. Amer. Chem. Soc.*, **81**, 2160 (1959).
234.  M.E. Kuehne, *J. Org. Chem.*, **28**, 2124 (1963).
235.  J.P. Danehy, V.J. Elia and C.J. Lavelle, *ibid.*, **36**, 1003 (1971).
236.  G. Gorin, *Progr. Biochem. Pharmacol.*, **1**, 142 (1965).
237.  O. Gawron in ref. 11, p. 351–365.
238.  R.G. Hiskey and D.N. Harpp, *J. Amer. Chem. Soc.*, **86**, 2014 (1964).
239.  J.P. Danehy and W.E. Hunter, *J. Org. Chem.*, **32**, 2047 (1967).
240.  J.P. Danehy and K.N. Parameswaran, *ibid.*, **33**, 568 (1968).
241.  J.P. Danehy and V.J. Elia, *ibid.*, **36**, 1394 (1971).
242.  A. Schönberg, E. Frese, W. Knöfel and K. Praefcke, *Chem. Ber.*, **103**, 938 (1970).
243.  J.L. Kice and N.A. Favstritsky, *J. Amer. Chem. Soc.*, **91**, 1751 (1969).
244.  J.L. Kice and K.W. Bowers, *ibid.*, **84**, 2384 (1962).
245.  (a) M.D. Bentby, I.B. Douglass, J.A. Lacadie, D.C. Weaver, F.A. Davis and S.J. Eitelman, *Chem. Commun.*, 1625 (1971); (b) J.R. Campbell, *J. Org. Chem.*, **27**, 2207 (1962)
246.  A.H. Herz and D.S. Tarbell, *J. Amer. Chem. Soc.*, **75**, 4657 (1953).
247.  A.H. Weinstein, R.M. Pierson, B. Wargotz and T.F. Yen, *J. Org. Chem.*, **23**, 363 (1958).
248.  B. Miller and C.-H. Han, *ibid.*, **36**, 1513 (1971).
249.  K. Ikura and S. Oae, *Tetrahedron Lett.*, 3791 (1968).
250.  L. Horner, H. Schaefer and W. Ludwig, *Chem. Ber.*, **91**, 75 (1958).
251.  E. Block, *Quart. Rep. Sulfur Chem.*, **4**, 237 (1969).
252.  W.A. Pryor, J.P. Stanley and T.-H. Lin, *ibid.*, **5**, 305 (1970).
253.  H.C. Brown and M.M. Midland, *J. Amer. Chem. Soc.*, **93**, 3291 (1971).
254.  W.A. Bonner and R.A. Grimm in ref. 11, pp. 35–71.
255.  Y.H. Khim and L. Field, *J. Org. Chem.*, **37**, 2714 (1972).
256.  L. Field, P.M. Giles, Jr., and D.L. Tuleen, *ibid.*, **36**, 623 (1971).
257.  L. Haraldson, C.J. Olander, S. Sunner and E. Varde, *Acta Chem. Scand.*, **14**, 1509 (1960).
258.  T.L. Pickering, K.J. Saunders and A.V. Tobolsky, *J. Amer. Chem. Soc.*, **89**, 2364 (1967).
259.  R.G. Hiskey and M.A. Harpold, *J. Org. Chem.*, **33**, 559 (1968).
260.  R.G. Hiskey and E.L. Smithwick, Jr., *J. Amer. Chem. Soc.*, **89**, 437 (1967).
261.  A. Lüttringhaus, S. Kabuss, H. Prinzbach and F. Langenbucher, *Ann.* **653**, 195 (1962).
262.  L. Field and H.K. Kim, *J. Org. Chem.*, **31**, 597 (1966).
263.  B. Lindberg and G. Bergson, *Ark. Kemi*, **23**, 319 (1965).
264.  R.G. Hiskey and M.A. Harpold, *J. Org. Chem.*, **32**, 3191 (1967).
265.  L. Field and C.H. Foster, *J. Org. Chem.*, **35**, 749 (1970).
266.  F. Fischer and R. Gottfried, *Z. Chem.*, **6**, 146 (1966).
267.  G. Bergson, *Ark. Kemi*, **19**, 265 (1962).
268.  Ref. 14a, p. 68ff.
269.  W.G. Salmond, *Quart. Rev.* (London), **22**, 253 (1968).
270.  Ref. 14b, p. 689ff.
271.  Ref. 14b, p. 626ff.
272.  Ref. 14a, p. 347ff.
273.  N. Lozac'h and J. Vialle in ref. 11, pp. 257–285.
274.  Ref. 14a, p. 405ff.
275.  A. Hordvik, *Quart. Rep. Sulfur Chem.*, **5**, 21 (1970).
276.  Ref. 14a, p. 410ff.
277.  M.J. Cardone in J.H. Karchmer (Editor), "The Analytical Chemistry of Sulfur and its Compounds," in P.J. Elving and I.M. Kolthoff (Editors), "Chemical Analysis", Vol.29, Part II Wiley-Interscience, New York 1972, pp.89-431.
278.  See footnote 13 of ref. 92.
279.  W.D. Celmer and I.A. Solomons, *J. Amer. Chem. Soc.*, **77**, 2861 (1955).
280.  A. Fava, P.B. Sogo and M. Calvin, *ibid.*, **79**, 1078 (1957).
281.  J.F. Carson and F.F. Wong, *J. Org. Chem.*, **22**, 1725 (1957).
282.  L. Katz and W. Schroeder, *ibid.*, **19**, 109 (1954).

*Chapter 8*

# SULFOXIDES AND SULFILIMINES

Shigeru Oae

*Department of Chemistry, University of Tsukuba*
*Sakura-mura, Niihari-gun Ibaraki-ken, Japan*

## 8.1 INTRODUCTION

Sulfoxides are formed by partial oxidation of sulfides, which upon further oxidation are converted to sulfones, and hence may be considered as a very stable intermediate in the oxidation process. However, since the sulfur-oxygen linkage in sulfoxides is polarized and much weaker than that in sulfones, sulfoxides are relatively reactive and undergo many synthetically important reactions. Therefore, the chemistry involving this trivalent sulfur species has been growing rapidly in recent years.[1]

Another interesting trivalent sulfur species is sulfilimine which has a polar sulfur-nitrogen linkage like the polar sulfur-oxygen bond in sulfoxides. The chemistry of sulfilimines is relatively unexplored as compared to that of sulfoxides.[2] However, in view of the analogous properties, sulfilimines will soon be just as useful a chemical species as sulfoxides.

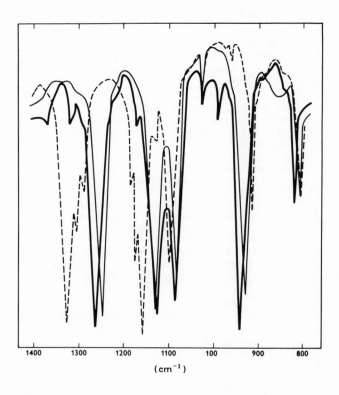

Fig. 8.1. IR spectra. ——— $p-CH_3C_6H_4SO_2N^-\!-S^+(CH_3)_2$ ; ———
$[p-CH_3C_6H_4SO_2NCl]^-\,Na^+$; - - - - - $p-CH_3C_6H_4SO_2NH_2$.

## 8.2 FORMATION OF SULFOXIDES AND SULFILIMINES

### 8.2.1 Sulfoxides

The standard method for the preparation of sulfoxides involves the oxidation of sulfides by various oxidizing reagents. The oxidation can be achieved not only by common oxidizing reagents such as peracids, nitric acid and periodate, but also by halogenating agents such as $t$-butyl hypochlorite and N-bromosuccinimide. Other methods for the synthesis of sulfoxides include the addition of sulfinyl carbanion to unsaturated groups, the reaction of aromatic sulfonyl chloride with alkylaluminium, the rearrangement of allyl or benzyl arylsulfinates, etc. Among these, the most significant is the synthesis of optically active sulfoxides by the reaction of menthyl sulfinates with Grignard reagents developed by K.K. Andersen. Another note-worthy development is the convenient incorporation of $^{18}O$ into the sulfoxides, since these two developments have provided a way to study the stereochemistry of substitution on the trivalent sulfur atom of sulfoxides.

### 8.2.1.1 Oxidation

The industrial process for making dimethyl sulfoxide is the direct air oxidation of dimethyl sulfide catalyzed by nitrogen dioxide. Both vapor and liquid phase oxidation procedures were devised.[3] The same procedure can be applied to the oxidation of dialkyl sulfides to the corresponding sulfoxides.[4] In the laboratory, sulfides are generally oxidized by peroxides, peracids and ozone.[5] Dialkyl, diaryl and alkyl aryl sulfides are oxidized to the corresponding sulfoxides with $H_2O_2$,[6] perbenzoic[7] and Caro's acids.[8] However, it is not always easy to stop the reaction at the stage of sulfoxide formation, especially in the case of diaryl sulfides. However, it is relatively easy to stop the oxidation at the sulfoxide stage for dialkyl sulfides, since the lone-pair electrons on the sulfur atom of diaryl sulfides are substantially delocalized over the entire molecule. Hence the first oxidation to the sulfoxide requires relatively high energy, which is substantially higher than that required for the oxidation of dialkyl sulfides and enough to result in further oxidation to the sulfone.[10] With the use of optically active peroxides, several partially optically active sulfoxides have been synthesized, as will be mentioned later (8.2.1.3). In order to avoid the concomitant formation of sulfones, several methods of oxidation have been devised.

Nitric acid was successfully used for oxidizing methanethiophenol to methanesulfinylphenol without any formation of the corresponding sulfonylphenol.[11] Similarly, many diaryl sulfides were also oxidized exclusively to the corresponding sulfoxides.[12] However, sometimes undesirable nitration takes place simultaneously.[13] Nitrogen tetroxide was found to be a very convenient selective reagent for oxidation of sulfides to sulfoxides without forming sulfones and other possible by-products.[14,15] A rough kinetic study on the oxidation of aryl sulfides indicates the importance of both steric and electronic effects.[16,17] While the steric effect can be seen in the remarkable retardation of the oxidation of dimesityl sulfide,[16] the importance of the electronic effect may be indicated by the large negative value of the Hammett $\rho = -2.71$, observed in the $N_2O_4$ oxidation of substituted thioanisoles.[17]

Although chromic acid is a powerful oxidizing agent frequently used to oxidize sulfides to sulfoxides, it gives undesired sulfones as by-products.[18] The use of pyridine as solvent is said to avoid sulfone formation.[19]

Oxidation of sulfides to sulfoxides by halogen in aqueous media has been documented[20]; however, undesired side reactions, such as cleavage of the C-S bond or formation of a C-halogen bond, often predominate.[21] In the case of aryl sulfides, halogenation at aryl ring is often observed.[22] The reaction of sulfides with N-bromosuccinimide in aqueous media also gives sulfoxides which are free form sulfones in the case of diaryl and benzyl aryl sulfides. However, C-S bond cleavage was the main reaction when the technique was applied to aryl alkyl and alkyl sulfides.[23] When the reaction was carried out under very mild conditions, even dialkyl sulfoxides were successfully prepared by either N-bromo- or N-chloro-succinimide.[24] Other N-chloro compounds such as chloramine-B[25] and 1-chlorobenzotriazole[26] have also been used for selective oxidation of both aryl and alkyl sulfides to the corresponding sulfoxides. When the bromine addition complex of diazabicyclo-[2.2.2]-octane, $N(CH_2CH_2)_3N \cdot 2Br_2$, was used in aqueous media, both dialkyl and diaryl sulfides were selectively oxidized to the corresponding sulfoxides.[27] Pyridine-bromine complex was also found to act similarly.[27] Bromine alone can also oxidize sulfides to sulfoxides if the reaction is carried out carefully in cold aqueous media.[28] Apparently the bromine oxidation is accelerated in the presence of a buffer such as sodium acetate and follows second-order kinetics, first order in bromine and first order in sulfide,[29] as in the case of iodine oxidation of sulfides.[30] The reaction is also quite sensitive to both steric and electronic effects. Thus, the rate of bromine oxidation of phenyl methyl sulfide is increased by an electron-donating substituent on the phenyl ring (Hammett $\rho = -3.2$) while isopropyl phenyl sulfide reacts almost 30 times more slowly than phenyl methyl sulfide.[29] A mechanism involving the fast initial formation of a bromosulfonium cation followed by its rate-determining hydrolysis was suggested[29]:

$$R_2S + X_2 \;\rightleftharpoons\; \begin{matrix} R_2{}^+SX + X^- \\ \text{or } R_2SX_2 \end{matrix} \;\xrightarrow[\substack{\text{Base or} \\ \text{nucleophilic buffer}}]{H_2O \text{ (slow)}}\; R_2SO + 2HX \qquad (8.1)$$

Carboxylate ions and phosphate dianion were suggested as being involved in the nucleophilic attack on the halosulfonium cation.[30,31] However, these buffer anions can act as general bases, and indeed such was found to be the case in the anchimerically assisted iodine oxidation of o-carboxythioanisole to the corresponding sulfoxide.[32] The formation of a stable bromosulfonium ion was recently made rather dubious by the successful characterization of the 1 : 1 bromine addition complex of tetramethylene sulfide (thiane) by X-ray diffraction,[33] which suggests the complex to be better illustrated as a charge-transfer type rather than a charge-separated salt.

Iodosobenzene is also known to be a good selective oxidizing agent for unsaturated sulfides.[34] Iodobenzene diacetate[35] and iodobenzene dichloride[36] are also used similarly, but undesirable C-S bond cleavage occasionally takes place.[36] In some cases, α-chlorosulfoxides are directly obtained.[37] Sodium periodate is another selective oxidizing agent often used.[38] t-Butyl hypochlorite was introduced for the selective oxidation of sulfide [39] and has often been used.[40] In the presence of water, however, the formation of sulfone is observed.[41]

### 8.2.1.2 Sulfinyl Carbanion Method

Addition of sulfinyl carbanion to unsaturated groups, such as carbonyl or olefin is a convenient way to introduce a sulfinyl group, especially at a terminal position. The following are a few examples[42-46]:

$$\underset{Ph}{\overset{Ph}{>}}C=O + CH_3\overset{O}{\underset{\uparrow}{S}}CH_2^- \longrightarrow \underset{Ph}{\overset{Ph}{|}}\underset{}{C}\text{--}CH_2\overset{OH}{\underset{}{|}}\overset{O}{\underset{\uparrow}{S}}\text{--}CH_3 \qquad (8.2)$$

$$PhCl + CH_3S(O)CH_2^- \longrightarrow PhCH_2S(O)CH_3 + Ph_2CHS(O)CH_3 \qquad (8.3)$$

$$R\overset{O}{\overset{\|}{C}}\text{--}OEt + CH_3S(O)CH_2^- \longrightarrow R\overset{O}{\overset{\|}{C}}\text{--}CH_2S(O)CH_3 \xrightarrow{Al\ (Hg)} R\overset{O}{\overset{\|}{C}}\text{--}CH_3 \qquad (8.4)$$

$$ArCH=CH_2 + CH_3S(O)CH_2^- \longrightarrow ArCH_2CH_2CH_2S(O)CH_3 \qquad (8.5)$$

$$PhC=CPh + CH_3S(O)CH_2^- \longrightarrow$$

$$\begin{array}{c} Ph \diagdown \qquad CH_2S(O)CH_3 \\ \qquad C=C \diagup \\ H \diagup \qquad \diagdown Ph \qquad + \end{array} \qquad (8.6)$$

89%

$$\begin{array}{c} Ph \diagdown \qquad Ph \diagup \\ \qquad C=C \\ H \diagup \qquad \diagdown CH_2S(O)CH_3 \end{array}$$

11%

Reaction with alkyl halides or tosylates also gives terminal sulfoxides[47]:

$$ROTs + CH_3SOCH_2^- \longrightarrow RCH_2SOCH_3 \qquad (8.7)$$

While the reaction of thionyl chloride with enol ethers gives bis($\beta$-chloro-$\beta$-alkoxyethyl) sulfoxides in high yields (eq. 8.8),[48] the reaction of Grignard reagents with thionyl chloride also gives the corresponding sulfoxides (eq. 8.9).[49,50,51] However, in most cases sulfides are also formed, apparently by further reaction of sulfoxides with thionyl chloride.[49]

$$2ROCH=CH_2 + SOCl_2 \xrightarrow[2\,hr]{0-5^{\circ}C} (RO-\underset{\underset{Cl}{|}}{C}HCH_2)_2SO$$

$$(R = Et, sec\text{-}Bu, n\text{-}Bu) \quad (8.8)$$

$$2RMgX + SOCl_2 \longrightarrow RS(O)R + RSR \qquad (8.9)$$

Highly substituted sulfoxides were prepared in high yields by treating sulfines with organolithium compounds, as shown below[51]:

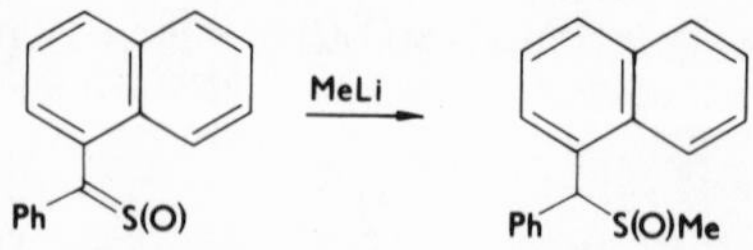

$$(8.10)$$

The reaction of trialkylaluminium or alkylaluminium chloride with aromatic sulfonyl chloride also gives aryl alkyl sulfoxide together with the corresponding sulfide.[52] N,N-Thionylimidazole is also used to prepare symmetrical diaryl sulfoxides by reaction with aryl Grignard reagents.[53]

The thermal rearrangement of sulfenates to sulfoxides is an interesting reaction, but occurs only in the case of benzylic,[54] allylic[55] or propargylic compounds (eq. 8.11).[56] These rearrangements proceed either *via* a homolytic path or Cope type sigmatropic path, as shown below:

$$CH_3C_6H_4\text{-S-O-CH}_2Ph \rightleftharpoons [CH_3C_6H_4SO\cdot + PhCH_2^\cdot] \rightarrow CH_3C_6H_4\overset{\overset{\displaystyle O}{\uparrow}}{S}CH_2Ph$$

$$\text{(8.11)}$$

$$CH_2\text{=CH-CH}_2OSMe \xrightarrow{\text{Cope rearr.}} CH_2\text{=CHCH}_2\text{-}\overset{\overset{\displaystyle O}{\uparrow}}{S}Me \qquad \text{(8.12)}$$

In another special method *p*-aminobenzenesulfinic acid is readily converted to *p,p'*-diaminodiphenyl sulfoxide upon heating[57]:

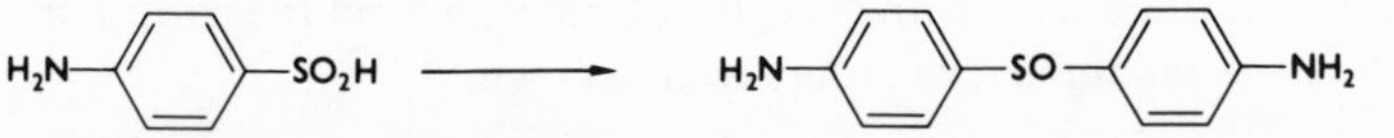

$$H_2N\text{—}\langle\bigcirc\rangle\text{—}SO_2H \longrightarrow H_2N\text{—}\langle\bigcirc\rangle\text{—}SO\text{—}\langle\bigcirc\rangle\text{—}NH_2 \qquad \text{(8.13)}$$

### 8.2.1.3 Stereoselective Methods

The most convenient method to synthesize optically active sulfoxides is the reaction of Grignard reagents with optically active menthyl arene- or alkanesulfinates. A typical example is shown below (eq. 8.14).[58] The reaction involves Walden inversion on the sulfur atom[59] without prior cleavage of O-alkyl linkage.[60]

$$(l)p\text{-CH}_3C_6H_4\overset{\overset{\displaystyle ^{18}O}{\uparrow}}{S}\text{-OC}_{10}H_{19} \xrightarrow{C_2H_5MgI} (d)p\text{-CH}_3C_6H_4\overset{\overset{\displaystyle ^{18}O}{\uparrow}}{S}\text{-C}_2H_5$$

$$\text{(8.14)}$$

Optically active sulfoxides were also obtained by treating diastereomerically enriched sulfinamides with organolithium reagents.[61] The characteristic of these reactions is to produce optically pure sulfoxides.

Asymmetric induction is usually observed when sulfoxides are oxidized by optically active oxidizing agents, and this method is often used for asymmetric synthesis of sulfoxides. Oxidizing agents used are monoperglutaric acid,[62] (+) and (−) α-cyclohexylperpropionic acid,[63] monopercamphoric acid,[64] (+) perhydratropic acid[64,65] and others.[66] Oxidation by biological fermentation was also found to be effective in asymmetric induction to give partially optically active sulfoxides. Thus, *Aspergillus niger* was used for the oxidation,[67] as shown in eq. 8.15. Other examples are also known.[68] Another microorganism used for selective oxidation is *Calonectria decora.*[69]

$$\text{Ph-SCH}_2\text{-Ph} \xrightarrow{\textit{Asper. niger}} \text{Ph-}\overset{\overset{\text{O}}{\uparrow}}{\text{S}}\text{-CH}_2\text{-Ph} \qquad (8.15)$$

$$[\alpha]_D^{21} -20.20$$

Asymmetric induction can be achieved by partial reduction of sulfoxides by optically active reducing agents[70] or by partial oxidation of sulfoxides to sulfones by asymmetric peracid such as percamphoric acid.[71]

## 8.2.2 Sulfilimines

### 8.2.2.1 N-Acylsulfilimines

The most common procedure for preparing sulfilimines is the reaction of sulfides with sodium N-$p$-toluenesulfonylchloramine (chloramine-T) discovered by Nicolet and Willard[72] and developed by Mann and Pope[73]; the reaction is often called the Mann-Pope reaction (eq. 8.16).

Generally, sulfides react with N-acylhaloamines to form the corresponding sulfilimines, and numerous examples are known.[74-87]

$$\underset{R'}{\overset{R}{\diagdown}}S + \overset{Cl}{\underset{Na^+}{\diagdown}}NSO_2\text{-Ar} \longrightarrow \underset{R'}{\overset{R}{\diagdown}}\overset{+}{S}\text{-N}^-\text{-SO}_2\text{Ar} + NaCl \qquad (8.16)$$

Sulfilimines are obtained by the reaction of dialkyl sulfides with cyanogen azide (eq. 8.17),[88] while dialkyl sulfides react with N-chlorobenzimidates to afford the corresponding sulfilimines.[89]

$$R_2S + NCN_3 \longrightarrow R_2S \to NCN \quad R = C_{1-12} \text{ } n\text{-alkyl group} \qquad (8.17)$$

Sulfilimines can be obtained by the reaction of sulfoxides with acid amides (eq. 8.18),[90] sulfinylsulfonamides (eq. 8.19),[84,91,92,93] sulfonylisocyanates (eq. 8.20)[84,94,95,96] or sulfonylazides (eq. 8.21)[84]

$$\underset{R'}{\overset{R}{\diagdown}}S \to O + H_2NSO_2Ar \longrightarrow \underset{R'}{\overset{R}{\diagdown}}S \to NSO_2Ar + H_2O \qquad (8.18)$$

$$\underset{R'}{\overset{R}{\diagdown}}S \to O + R''SO_2NSO \longrightarrow \underset{R'}{\overset{R}{\diagdown}}S \to NSO_2R'' + SO_2 \qquad (8.19)$$

$$\underset{R'}{\overset{R}{\diagdown}}S \to O + R''SO_2NCO \longrightarrow \underset{R'}{\overset{R}{\diagdown}}S \to NSO_2R'' + CO_2 \qquad (8.20)$$

$$\begin{array}{c} R \\ \backslash \\ S \rightarrow O \\ / \\ R' \end{array} + R''SO_2N_3 \rightarrow \begin{array}{c} R \\ \backslash \\ S \rightarrow NSO_2R'' \\ / \\ R' \end{array} + N_2 \qquad (8.21)$$

The mechanism of the reaction of aryl methyl sulfides with chloramine-T (Mann-Pope reaction) has been investigated in some detail.[85,87] The reaction is affected very much by changes in pH and facilitated by the substitution of an electron-releasing group on the phenyl ring. The following mechanism (eq. 8.22) has been suggested. The method is convenient for the synthesis of diaryl sulfilimines bearing strong electron-withdrawing substituents.

$$\begin{array}{c} Cl \\ \backslash \\ \phantom{.}^{-}N\text{-}Ts \\ Na^+ \end{array} + H_2O \underset{}{\overset{fast}{\rightleftharpoons}} \begin{array}{c} Cl \\ \backslash \\ N\text{-}Ts \\ H \end{array} + NaOH$$

$$(8.22)$$

$$\begin{array}{c} R \\ \backslash \\ S \\ / \\ R' \end{array} + \begin{array}{c} Cl \\ \backslash \\ N\text{-}Ts \\ H \end{array} \xrightarrow{slow} \begin{array}{c} R \\ \backslash \\ \overset{+}{S}\text{-}Cl \\ / \\ R' \end{array} + HN^{-}\!\!-Ts \xrightarrow{fast} \begin{array}{c} R \\ \backslash \\ S \rightarrow N\text{-}Ts \\ / \\ R' \end{array} + HCl$$

Optically active and inverted sulfilimines can be obtained by the reaction of optically active sulfoxides, either with sulfonamides in the presence of phosphorus pentoxide (eq. 8.23) or with sulfinylsulfonamides (eq. 8.24)[93]:

$$p\text{-}CH_3C_6H_4\overset{\overset{O}{\uparrow}}{S}\text{-}CH_3 + H_2N\text{-}Ts \xrightarrow[\substack{P_2O_5,\ Et_3N \\ 66\%}]{0°C,\ 1\ hr} p\text{-}CH_3C_6H_4\underset{CH_3}{\overset{\diagdown}{\underset{\diagup}{S}}} \rightarrow NTs$$

$[\alpha]_{546}^{25}$ 180.5                                    (8.23)

(optical purity 70%)

$$p\text{-}CH_3O_6H_4\overset{\overset{O}{\uparrow}}{S}CH_3 + TsNSO \xrightarrow[95\%]{0°C,\ 3.5\ hr} p\text{-}CH_3C_6H_4\underset{CH_3}{\overset{\diagdown}{\underset{\diagup}{S}}} \rightarrow NTs$$

$[\alpha]_{546}^{25}$ 180.5          (optical purity 98%)                (8.24)

$[\alpha]_{546}^{25}$ $-320$

On the basis of kinetic and sterochemical observations, a mechanism involving a nucleophilic substitution at the sulfur atom along two radial-radial bond axes was suggested[93]:

$$
\text{:-}\overset{+}{\underset{C_6H_4CH_3}{\overset{CH_3}{S}}}\hspace{-0.2cm}{\overset{O}{\diagdown}} \;+\; 2TsNSO \;\xrightarrow{\hspace{1cm}}\; \underset{C_6H_4CH_3\text{-}p}{\overset{\overset{\displaystyle Ts}{\underset{\displaystyle |}{CH_3}}}{\overset{N----S}{\underset{O-S}{\overset{\diagup}{S}}}\hspace{-0.2cm}{\overset{O}{\diagdown O}}}}\;\xrightarrow[\;-TsNSO\;]{-SO_2}\; \text{:-}\underset{C_6H_4CH_3\text{-}p}{\overset{CH_3}{S}}\text{----NTs} \hspace{1cm} (8.25)
$$

(+)                                                                          (−)

### 8.2.2.2 Unsubstituted Sulfilimines

The reaction of sulfides with hydroxylamine sulfate in the presence of bases gives the corresponding sulfilimines[98-101]:

$$
\overset{R}{\underset{R'}{S}}\;+\; H_2NOSO_3H \;+\; NaOCH_3 \;\longrightarrow\; \left[\overset{R}{\underset{R'}{S\text{-}NH_2}}\right]^{+}_{2} SO_4{}^{2-}
$$

$$
\xrightarrow[\text{ammonia}]{\text{liquid}}\; \overset{R}{\underset{R'}{S\rightarrow NH}} \hspace{2cm} (8.26)
$$

Only dialkyl sulfilimines, such as the dimethyl or diethyl derivatives, have been prepared so far by this reaction. Di-(p-methoxyphenyl) sulfide was successfully converted to the corresponding sulfilimine by treating its halogen addition complex with liquid ammonia[102]:

$$
[(p\text{-}CH_3OC_6H_4)_2S^+Cl]\; Cl^- \;+\; NH_3 \rightarrow\; [(p\text{-}CH_3OC_6H_4)_2\overset{+}{S}\text{-}NH_2]\, Cl^-
$$

$$
\xrightarrow{NaNH_2}\; (p\text{-}CH_3OC_6H_4)_2S\rightarrow NH \hspace{2cm} (8.27)
$$

The similar reaction of sulfides with chloramine is, however, known to give sulfondiimines instead of sulfilimines:

$$
\overset{R}{\underset{R'}{S}}\;+\; NH_2Cl \rightarrow\; \left[\overset{R}{\underset{R'}{S\text{-}NH_2}}\right]^{+}\; Cl^- \;\longrightarrow\; \left[\underset{R'\;\;NH_2}{\overset{R\;\;NH}{S}}\right]^{+}\; Cl^- \hspace{1cm} (8.28)
$$

The best method of preparing free sulfilimines is the hydrolysis of N-sulfonylsulfilimines with concentrated sulfuric acid and subsequent neutralization with ammonia. By this procedure, most free sulfilimines can be synthesized.[102a]

## 8.3 NATURE OF SULFOXIDE AND SULFILIMINE BONDS

The nature of the sulfur-oxygen linkage in sulfoxides has been a matter of controversy. Double-bond character (8.2) was suggested by early workers[103] for the sulfur-oxygen linkage on the basis of observations such as the shorter bond length (1.45 Å average), the smaller bond moment of the sulfur-oxygen linkage and others. Later, the bond moment was recalculated to be $\mu_{SO}$ = 3.0 average; however, this rather small value led Cumper and Walker[104] to favor the double bonded structure (8.2) for the sulfur-oxygen linkage in sulfoxides.

8.1                                    8.2

However, the rather small bond force constants ($7.00 \times 10^{-5}$ dyne/cm average) calculated for the frequencies of infrared absorption spectra (1055 $cm^{-1}$ average),[105] the strong hydrogen bonding property of sulfoxides,[105,106,107] and the smaller value of the calculated bond order[108] seem to suggest that the sulfur-oxygen linkage in sulfoxides is best described as a semipolar single bond (8.1), though that in sulfones is strong (force constant = $9.6 \times 10^{-5}$ dyne/cm average) and would be an essentially covalent double bond. The bond force constant of the sulfur-oxygen linkage in sulfoxides is of similar magnitude to that of a semipolar nitrogen-oxygen linkage in heteroaromatic N-oxides, ($6.8 \times 10^{-5}$ dyne/cm).[109] Such observations as bond refraction[110] and parachor[111] also suggest the semipolar nature of the sulfur-oxygen linkage in sulfoxides. A recent work on dipole moments of substituted diphenyl sulfoxides[112] indicates the bond moment of the sulfur-oxygen linkage to be 4.76 D and definitely suggests the semipolar character of the linkage.

Although it appears that the S-O bond in sulfoxides is better represented by the semipolar structure, (8.1), this does not mean that "back-bonding" involving $p$ orbitals on oxygen and $d$ orbitals of the sulfur is not important, but is convenient for interpreting physico-chemical phenomena involving the sulfur-oxygen bond. The "back-bonding" is, in fact, believed to be quite important not only on theoretical grounds,[108,113] but also for explaining the marked stereochemical stability of optically active sulfoxides.[114] This "back-bonding" is commonly expressed by the double bond formula (8.2), but the double bond in this case is not of a classical $p$-$p$ type and does not require any coplanarity for those groups attached to the S-O bond.

The nature of the sulfur-nitrogen linkage in sulfilimines is also a matter of controversy. N-$p$-Toluenesulfonylsulfilimine may be expressed as either the ylide structure (**8.3**), the ylene structure (**8.4**), or by a resonance hybrid of both. The ylide structure was favored on the basis of ir spectroscopic data, which indicate that an average $SO_2$ band in sulfilimines appears at a wave number considerably higher (40–50 cm$^{-1}$) than that in sulfonates.[115]

However, the similarity of the ir spectra of S,S-dimethyl-N-$p$-toulenesulfonylsulfilimine and chloramine-T shown in Fig. 8.1, appears to suggest that the sulfilimine and the chloramine-T anion have similar electronic environments around the $SO_2$ group, and seems to favor the ylide structure.[116] The ir absorption band for stretching vibration of the semipolar S-N bond lies between 930–950 cm$^{-1}$ for most sulfilimines. A recent X-ray crystallagraphic study of S,S-dimethyl-N-methanesulfonyl-sulfilimine revealed that the bond lengths of the two N-S linkages are 1.63Å and 1.58Å respectively,[117] which are normal values for N-S dis-

$$(CH_3)_2 S \qquad SO_2 CH_3$$
$$1.63\text{Å} \diagdown \diagup 1.58\text{Å}$$
$$N$$

tance of sulfonamide such as N-phenylmethanesulfonamide[118] or sulfamide.[119] These data also support the consideration that the S-N bond is of semipolar nature. The small force constant for the S-N bond (ca. 4.4 x 10$^5$ dyne/cm) also suggests the bond to be rather weak and hence semipolar. The markedly higher reactivity of sulfilimines than that of corresponding sulfoxides is undoubtedly due to the much weaker S-N linkage in sulfilimines than the S-O linkage in sulfoxides. Unfortunately no data on the dipole moment is available to estimate the extent of permanent polarization of the S-N linkage. However, in view of the relatively large bond length and other chemical behavior, the S-N bond is considered more polarized than the S-O bond in sulfoxides. It is worth noting that the difference[112,113] between the calculated value of an S-N single bond(1.74Å)[120] and that of the observed (1.63Å) is much smaller than the corresponding difference of S-O bond length in sulfoxide (1.70 calc. − 1.45 obs. = 0.25Å).

## 8.4 STEREOCHEMISTRY OF SULFOXIDES AND SULFILIMINES

### 8.4.1 Sulfoxides

It has long been known that optical isomerism can exist in sulfoxides and sulfinates. For example, optically active enantiomers of ethyl and *n*-butyl *p*-toluenesulfinate,[121] *m*-carboxyphenyl methyl, and *p*-aminophenyl *p*-tolyl sulfoxides[122] were successfully resolved several decades ago. Since

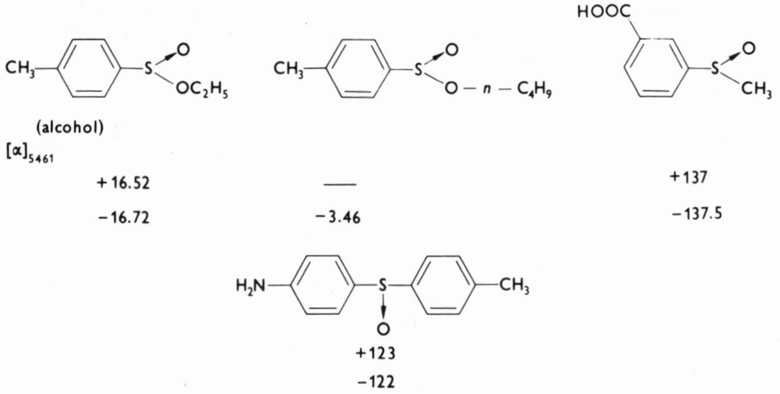
HOOC
CH3
O
OC2H5
CH3
O
O − n − C4H9
O
CH3
(alcohol)
[α]5461
+16.52
−16.72
—
−3.46
+137
−137.5
H2N
CH3
O
+123
−122

then, a large number of sulfoxides bearing either an amino or a carboxyl group have been resolved. For amino-substituted sulfoxides, camphorsulfonic acid has often been used,[123] while carboxyl-bearing sulfoxides have mostly been resolved with brucine.[124,125,126] Sulfoxides which have no other functional group can also be resolved by forming optically active platinum complexes.[127] Thus, ethyl *p*-tolyl sulfoxide was resolved with optically active α-methylbenzylamine-platinum complex.[126]

The most convenient method for obtaining optically active sulfoxides today is the Andersen procedure of treating an optically active menthyl alkane- or arenesulfinate with a Grignard reagent.[128a] Many optically active sulfoxides with high optical purities have been prepared by this method.[128b]

In the case of cyclic sulfoxides, geometrical isomerism can exist. For example, both *cis* and *trans* isomers of 3-substituted thietane 1-oxides were separated,[129] while predominant formation of the *trans* isomer was found to take place when 3-carboxythietane was oxidized to the sulfoxide

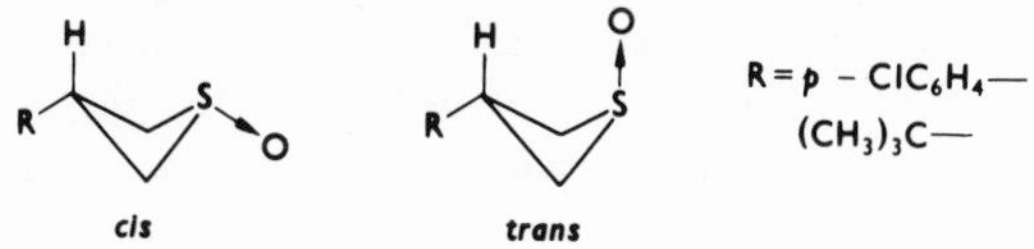
H
R
S
O
cis
H
O
R
S
trans
R = p − ClC6H4—
(CH3)3C—

by hydrogen peroxide.[130] A mixture of both *cis* and *trans* isomers of 4-carboxythiane 1-oxide is also known.[130]

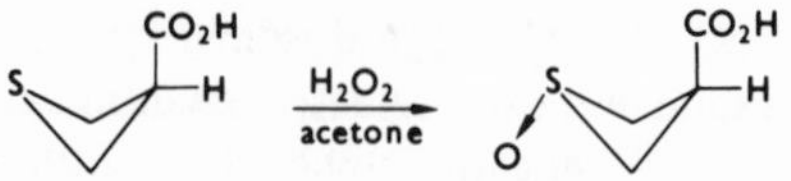

$$
\begin{array}{cc}
\underset{\text{acetone}}{\xrightarrow{\text{H}_2\text{O}_2}}
\end{array}
\tag{8.29}
$$

In the case of 4-substituted thiane 1-oxides the, *cis* isomers are usually more stable than the corresponding *trans* isomers. For example, when the R group is *t*-butyl, the *cis* isomer is stabilized by 1.3 kcal/mol more than the *trans* isomer.[131]

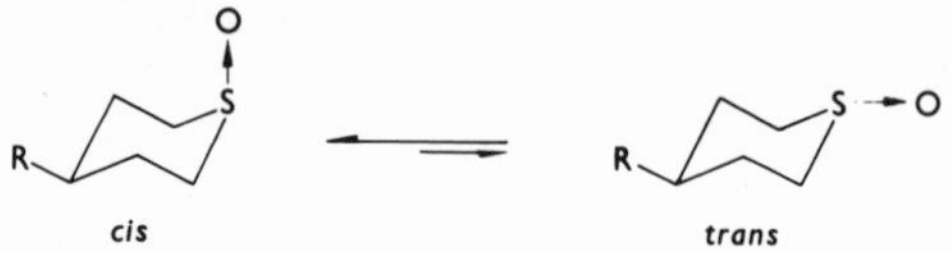

$$
\tag{8.30}
$$

cis                              trans

In the case of thioxanthene 10-oxide and related compounds, however, the sulfinyl oxygen atom appears to prefer the pseudo equatorial position.[132] Similar geometrical isomerism exists in the cases of five-membered cyclic sulfoxides[133] and a 6-membered nitrogen containing ring system.[134]

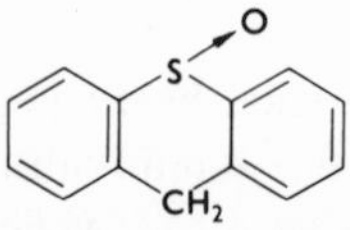

All these stereochemical observations indicate clearly that sulfoxides are of stable pyramidal configuration which is, in fact, illustrated by the structure of the most simple sulfoxide, DMSO, as shown below.[135]

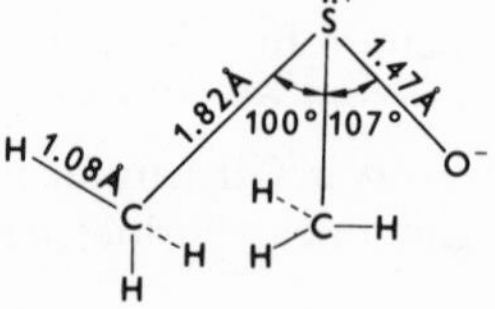

Sulfoxides are sterically much more stable than other isoelectromeric compounds such as amines, phosphines, or carbanions and do not undergo thermal racemization readily except for benzylic and allylic sulfoxides. Ordinary sulfoxides generally undergo racemization substantially slower than the analogous sulfonium salts. For example, dialkyl sulfoxides start to racemize above 190°C,[128,137] while methylethyl-*t*-butylsulfonium perchlorate undergoes facile thermal racemization at 50°C.[136] Unlike the sulfonium ion, the sulfoxide has a semipolar sulfur-oxygen linkage which may be used for dimerization or association. This together with the back donation of an electron pair on the oxygen into a $d$ orbital of the sulfur may contribute to the stability of the pyramidal structure of the sulfoxide.

Thermal racemizations of $p$-tolyl alkyl, aryl and allyl sulfoxides were studied carefully by Mislow *et al.*, and the relative rates of racemization at 210°C are shown in Table 8.1.[137]

**TABLE 8.1**

Relative Rates of Racemization of $p$-Tol-$\overset{*}{\underset{\downarrow}{S}}$-R

O

| R | $C_6H_5$ | $m$-or $p$-subst. $C_6H_5$ | $o$-subst. $C_6H_5$ | $\alpha$-$C_{10}H_7$ | Menthyl | $CH_3$ | $C_6H_5CH_2$- or $CH_2=CH\text{-}CH_2$- |
|---|---|---|---|---|---|---|---|
| rel. rate $(k_R/k_{C_6H_5})$ | 1 | 0.5 – 2.0 | >1.0 | ~2 | 60 | 0.06 | 2000 |

It appears that the racemization becomes more facile with the increase of the size of the R group in accordance with the I-strain concept, suggesting the occurrence of pyramidal inversion without cleavage of bonds.

However, the markedly increased rates of racemization of $p$-tolyl benzyl sulfoxide is undoubtedly due to the facile homolytic cleavage shown below[135,138]:

$$(R) \text{ or } (S)\text{-PhCH}_2\text{-}\overset{\overset{O}{\uparrow}}{S}\text{-}C_6H_4CH_3\text{-}p \rightleftharpoons (PhCH_2^{\cdot} + p\text{-}CH_3C_6H_4SO\cdot)$$

$$\rightleftharpoons (S) \text{ and } (R)\text{-PhCH}_2\underset{\downarrow}{\overset{}{S}}C_6H_4CH_3\text{-}p \quad (8.31)$$
$$O$$

The unusually facile racemization of allylic sulfoxides was shown to proceed via the sequence shown in eq. 8.32, which involves the formation of the sulfenates.[135,139] The rearrangement appears to proceed via an intramolecular cyclic $\alpha,\gamma$-shift of the allylic group between the sulfoxide oxygen and sulfur terminals, with partial retention of configuration (37%) when $(S)$ $\alpha$-methyl allyl derivative is used.[139] *Cis-trans*-isomerization of 4-substituted thiane 1-oxides is also known to take place at around 190°C.[136]

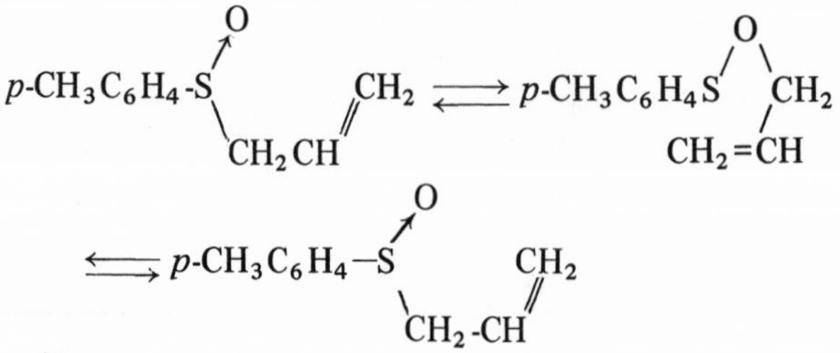

$$p\text{-}CH_3C_6H_4\text{-}S \overset{O}{\underset{CH_2CH}{\diagdown}} CH_2 \rightleftarrows p\text{-}CH_3C_6H_4\overset{O}{\underset{CH_2=CH}{S}} CH_2$$

$$\rightleftarrows p\text{-}CH_3C_6H_4\text{-}S \overset{O}{\underset{CH_2\text{-}CH}{\diagdown}} CH_2 \qquad\qquad (8.32)$$

Photochemical racemization is known to take place with $p$-tolyl aryl and alkyl sulfoxides under the irradiation of light of wave length longer than 285 nm; however, (+) butyl methyl sulfoxide does not racemize under the same condition.[140] From a study of both intra- and inter-molecular photosensitized racemization, it was determined that the excited singlet mainly undergoes rapid inversion.[141] Perhaps the initial rapid electronic excitation of the lone-pair electrons on the sulfur atom to an $n\text{-}\pi^*$-like state is followed by transformation of the molecule around the sulfur atom to a coplanar form that allows resonance interaction with the benzene ring. Although this scheme does not fit aliphatic sulfoxides which do not undergo facile racemization, photo-induced isomerization does take place with the highly strained sulfoxide shown below[142]:

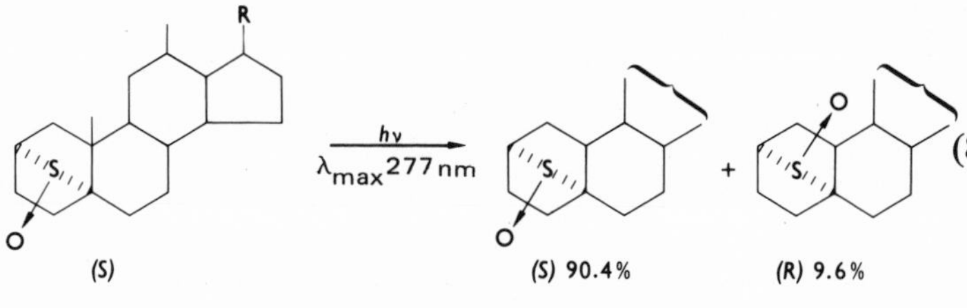

$$(8.33)$$

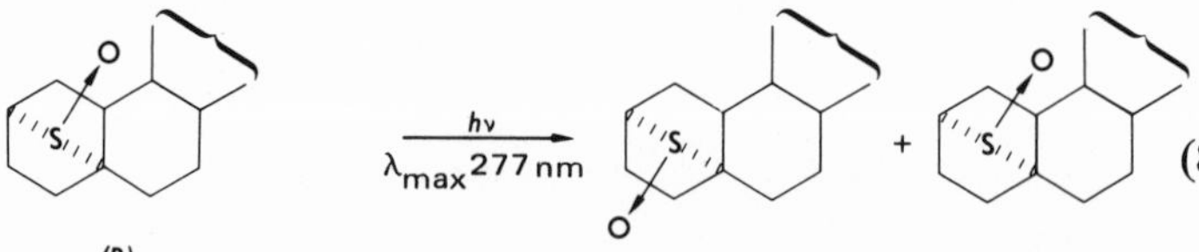

$$(8.34)$$

Sulfoxides undergo stereomutation with many chemical agents. (-)Methionine-$d$-sulfoxide was shown to be racemized completely upon treatment with concentrated hydrochloric acid at 40°C for an hour.[143] This racemization, however, is not caused by pyramidal inversion, but by the nucleophilic replacement on the sulfur atom by chloride ion which accompanies the concurrent oxygen exchange.[144] Hydrobromic acid is also similarly effective for the stereomutation which accompanies oxygen exchange.[145] Other weak acids such as acetic and polychloroacetic acids are also found to be effective for racemization when the solution is heated.[146] When d-p-tolyl p-aminophenyl sulfoxide was dissolved in a concentrated ($>90$ %) sulfuric acid, it racemized in a few minutes even at $0°C$.[147] In this case too, the racemization is caused by the oxygen-exchange reaction.[148,149] Similar racemization reactions have been found to occur with phosphoric acid[150,151] and nitrogen tetroxide[152,156]; however, the racemizations are actually caused by the oxygen–exchange reactions.[151,153] Another concurrent oxygen–exchange and racemization reaction is that with acetic anhydride.[154] All these concurrent oxygen-exchange and racemization reactions involve either prior heterolysis (A-1 type mechanism) or the simultaneous formation and cleavage ($S_N2$ type mechanism) of S-O bond in sulfoxides, as we shall see later in detail. Hydrogen flouride,[155] trifluoroacetic acid,[156] and acetic anhydride with Lewis acids[157] are also known to racemize optically active sulfoxides, presumably involving similar concurrent oxygen exchange.

Another interesting observation is that racemization is often assisted markedly by a neighboring carboxyl group.[158] Thus, the halide ion catalyzed epimerizations of optically active 3-($\omega$-toluenesulfinyl)-butyric acid[159] and 2-methanesulfinylbenzoic acid[160] in acid media are markedly faster (usually more than $10^4$ fold) than those of the corresponding parent skeletal compounds without the neighboring carboxyl group.

The optical rotation of sulfoxide is known to be influenced very little by the polarity of the solvent[161] and is usually high due to the inherently dissymmetric nature of the sulfoxide chromophore.[162] Since the sign of the Cotton effect of a dissymmetric chromophore directly depends on the chirality of the chromophore, the determination of optical rotatory dispersion[162] or circular dichroism[163] has been successfully used to assign absolute configuration or configurational correlation to many sulfoxides.[162]

The inherently dissymmetric nature of the sulfinyl function in unsymmetrical sulfoxides causes magnetic nonequivalence of the $\alpha$-methylene protons of aliphatic groups in organic sulfoxides.[163,164] The nmr proton chemical nonequivalence shift $\delta$ AB in, for example, phenyl ethyl or isopropyl sulfoxides, is known to be affected by the polarity of the solvent, especially by the protonating solvents; it is also affected markedly by the electronic properties of the substituent adjacent to the coupling protons. For example, in the case of benzyl p-substituted phenyl sulfoxides, the proton *gauche* to the lone pair on the sulfur ($H_A$) is more sensitive to the polarity of substituents than the other, while the proton *gauche* to the oxygen atom ($H_B$) is more influenced by the addition of trifluoroacetic acid.[163,165] This nonequivalence is also reflected in the difference in the rates of base-catalyzed hydrogen isotopic exchange of the two methylene hydrogen atoms of benzyl methyl sulfoxide[166] and of benzenesulfinylacetic acid[167] and also leads to the stereoselective deuteration of the methylene hydrogen ($H_B$) of benzyl p-chlorophenyl sulfoxide,[165] as will be mentioned later in detail.

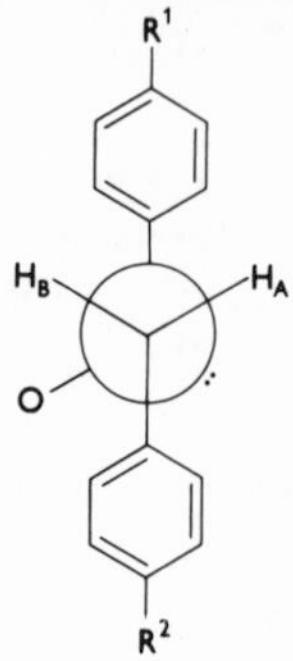

The nonequivalent nature of methylene protons $\alpha$ to the sulfoxide function has been utilized for configurational assignment of the diastereomeric isomers of methyl benzyl p-chlorophenyl sulfoxide by nmr[168] and also for the direct determination of optical purities and absolute configurations of sulfoxides by the measurement of the nmr chemical shift with the use of an optically active hydrogen bonding solvent.[169] Other examples are also known.[170]

The valence bond angles around the sulfur atom of a sulfoxide usually do not change much when sulfoxide is oxidized further to the corresponding sulfone. Since many sulfoxides are isomorphous to the corresponding sulfones, one often finds that melting points of the mixtures of these sulfoxides and sulfones do not fall, which puzzles unfamiliar organic chemists.[171] Among many examples,[171,172] dibenzothiophene monooxide and dioxide and the *anti-cis* and *trans* disulfoxides of thianthrene are but two representative pairs to show typical isomorphism.

### 8.4.2  Sulfilimines

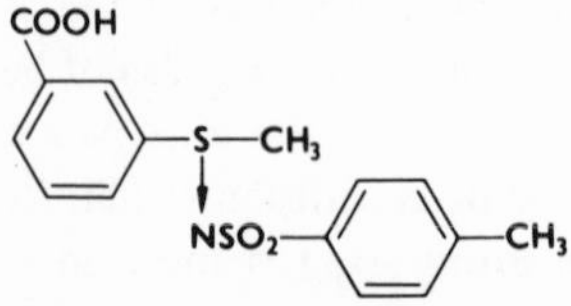

The first example of an optically active sulfilimine is *m*-carboxyphenyl methyl-N-p-tosyl sulfilimine, shown below.[173] As in the case of the analogous sulfoxides, the sulfur atom in this compound is asymmetric and two optical isomers were isolated. Symmetric cyclic sulfilimines also have geometrical isomers like the analogous sulfoxides. Following are two known examples[174,175]:

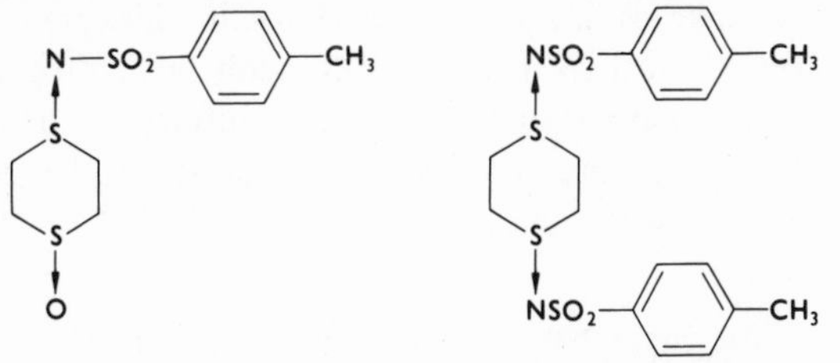

α-form; m.p. 230–240°          α-form; m.p. 250–255°
β-form; m.p. 176–177°          β-form; m.p. 200–210°

A convenient way of preparing optically active sulfilimines is the conversion of optically active sulfoxides to the corresponding sulfilimines. Thus, optically active (+) (R) p-methylphenyl methyl sulfoxide was converted by two reactions to the (−) (S) sulfilimine of inverted configuration (eq. 8.35)[176], while the reaction of N-phthaloylmethionine sulfoxide with N-sulfinyl-p-toluenesulfonamide gave the corresponding sulfilimine with retained configuration.[177] The reaction undoubtedly involves a nucleophilic displacement on the sulfur atom and leads to either a clean inversion or a retention, according to the reaction condition and the nature of surrounding substituents around the central sulfur atom. This procedure is at present the best method to synthesize optically active sulfilimines from optically active sulfoxides.

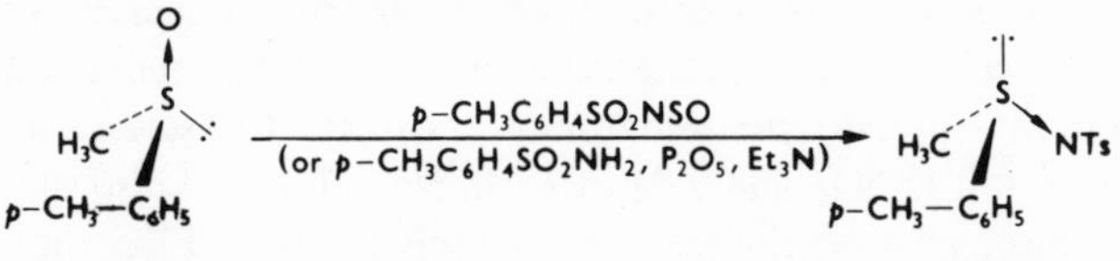

$$(8.35)$$

Although no chemical racemization has been investigated with optically active sulfilimines, stereomutation by pyramidal inversion has been found to occur readily with aryl alkyl and aryl benzyl N-tosyl sulfilimines.[178]

The rate of stereomutation by pyramidal inversion of a common N-tosylsulfilimine is $10^7$ times higher than that of a similar sulfoxide. This is believed to be caused mainly by the difference in electronegativities of oxygen and nitrogen, for $d\pi$-$p\pi$ interaction stabilizing the pyramidal structure should be larger with the more electronegative oxygen than with nitrogen.

## 8.5  CHEMICAL REACTIVITIES OF SULFOXIDES

### 8.5.1  Cleavage of S-O Bond

#### 8.5.1.1  Reduction

Unlike the S-O linkages in sulfones, those in sulfoxides are relatively weak and easily cleaved. Just by heating with alcoholic hydrogen chloride at 100°C, sulfoxides may be reduced to the sulfides[179] or may undergo fission.[180] Sulfoxides are also readily reduced to the corresponding sulfides by treating with zinc and acetic acid[179b,181] or with hydrogen iodide[182,183]:

$$R_2SO + 2HI \longrightarrow R_2S + I_2 + H_2O \qquad (8.36)$$

The latter reaction is of third-order kinetics (first-order in sulfoxide and second-order in HI) and acid catalyzed.[184] The strong-acid-catalyzed reduction of dimethyl sulfoxide was suggested to involve the double-protonated species shown below,[185] while in a high acid region, a mono-

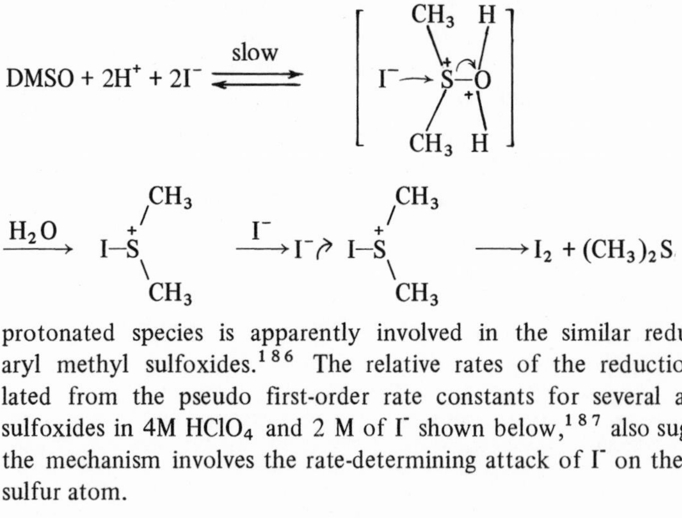

$$DMSO + 2H^+ + 2I^- \underset{slow}{\rightleftharpoons} \left[ I^- \rightarrow \overset{CH_3 \quad H}{\underset{CH_3 \quad H}{\overset{\backslash \quad /}{\underset{/ \quad \backslash}{S^+\!-\!O}}}} \right]$$

$$\qquad (8.37)$$

$$\xrightarrow{H_2O} I\!-\!\overset{+}{\underset{CH_3}{\overset{CH_3}{S}}} \xrightarrow{I^-} I^-\ ?\ I\!-\!\overset{+}{\underset{CH_3}{\overset{CH_3}{S}}} \longrightarrow I_2 + (CH_3)_2S$$

protonated species is apparently involved in the similar reduction of aryl methyl sulfoxides.[186] The relative rates of the reduction, calculated from the pseudo first-order rate constants for several aryl alkyl sulfoxides in 4M $HClO_4$ and 2 M of $I^-$ shown below,[187] also suggest that the mechanism involves the rate-determining attack of $I^-$ on the trivalent sulfur atom.

Relative rates of HI reduction:

|            | MeSOPh | EtSOPh | $i$-PrSOPh | $t$-BuSOPh |
|------------|--------|--------|-----------|-----------|
| rel. rates | 1      | 0.64   | 0.021     | 0.00055   |

|            | MeSO-$C_6H_4$OMe-$(p)$ | MeSO-Tol-$(p)$ | MeSO-Tol-$(m)$ |
|------------|------------------------|----------------|----------------|
| rel. rates | 1.2                    | 1.4            | 0.59           |

|            | MeSO-$C_6H_4$Cl-$(m)$ | MeSO-$C_6H_4$-$NO_2$-$(m)$ | MeSO-$C_6H_4$-$NO_2$-$(p)$ |
|------------|-----------------------|----------------------------|----------------------------|
| rel. rates | 0.43                  | 0.6                        | 0.23                       |

Since the reduction is so clean and stoichiometric, the reduction with hydroiodic acid either in acetic acid[188] or with acetic anhydride[189] has been suggested as a convenient method for the determination of sulfoxides.

As in the case of iodine oxidation of sulfoxides, the reduction of sulfoxide function with iodide ion is assisted by proper neighboring groups. A typical scheme is shown by[190]

$$I_2 + I^- \rightleftharpoons I_3^- \tag{8.38}$$

($S$)-Methyl-L-cysteine sulfoxide undergoes disproportionation upon treatment with hydrochloric acid.[191] A similar disproportionative decompositon is known to take place with dimethyl sulfoxide in ethylene glycol or acetamine.[192]

A catalytic reduction of dimethyl sulfoxide by molecular hydrogen with rhodium (III) complexes is known.[193] Iron carbonyl, $Fe(CO)_5$, is also known to reduce both dialkyl and aryl sulfoxides to the corresponding sulfides in good yields,[194] with initial formation of complexes of one of the two types:

$$R_2SO[Fe(CO)_3] \quad \text{or} \quad [(R_2SO)_6Fe][Fe(CO)_3]$$

Aromatic sulfoxides are reduced by trichlorosilane in high yields to the corresponding sulfides, while aliphatic sulfoxides give mercaptals in good yields together with small amounts of sulfides[195]:

$$(R = H, CH_3, Cl)$$

$$R-CH_2-S-CH_2-R \xrightarrow{HSiCl_3} RCH \begin{matrix} \diagup SCH_2R \\ \diagdown SCH_2R \end{matrix} + RCH_2-S-CH_2R$$

$$(R = C_6H_5, C_2H_5 \text{ or } C_3H_7)$$

$$\tag{8.39}$$

Reduction of sulfoxides to sulfides by $LiAlH_4$ or $NaBH_4 \cdot CoCl_3$ is also known.[195b]

Sulfoxides are readily reduced to the corresponding sulfides by treatment with thiols,[196] and sulfoxides from low-boiling sulfides are usually used for the oxidation of thiols. Thus, not only DMSO, but also diisopropyl sulfoxide and tetramethylene sulfoxide are often used.[197]

The use of a tertiary amine, such as tributylamine, increases the oxidation yield. The reaction is not only useful for the "sweetening" of fuels[198] in petroleum refining, but also for the oxidation of biologically active compounds such as a thiamine-like molecule.[199]

A detailed study on the oxidation of thiols to disulfides by sulfoxides has been made.[200] The reaction follows second-order kinetics, first-order with respect to both thiol and sulfoxide, with the rate markedly dependent on the acidity of the thiols as in the following order, ArSH > ArCH$_2$SH > RSH, and was suggested to proceed via the following route:

$$R_2SO + R'SH \;\xrightleftharpoons{\text{slow}}\; \left[ R_2S\underset{SR'}{\overset{OH}{\diagup\!\!\!\!\diagdown}} \right]$$

$$\left[ R_2S\underset{SR'}{\overset{OH}{\diagup\!\!\!\!\diagdown}} \right] + R'SH \rightleftharpoons R'\text{-}S\text{-}S\text{-}R' + R_2S + H_2O \qquad (8.40)$$

Diphenyl sulfoxide is reduced to the sulfide nearly quantitatively by heating with sulfur at 280°C. Diphenyl sulfone also reacts similarly above 300°C to give the sulfide and sulfur dioxide. However, the latter reaction was found mainly to involve the cleavage of phenyl-sulfur linkages according to the $^{35}$S-tracer experiment (eqs. 8.41, 8.42).[201] Such a difference is undoubtedly caused by the difference of bond strengths of the S-O linkages in the sulfoxide and the sulfone.

$$\qquad\qquad\qquad\qquad\qquad\qquad\qquad\qquad\qquad\qquad (8.41)$$

$$\qquad\qquad\qquad\qquad\qquad\qquad\qquad\qquad\qquad\qquad (8.42)$$

Di-$n$-butyl sulfoxide is reduced similarly at an even lower temperature, 200°C, to the sulfide[202]:

$$n\text{-Bu-}\overset{*}{\underset{\displaystyle O}{S}}\text{-Bu-}n + 1/16\, S_8 \xrightarrow{200°C} n\text{-Bu-}\overset{*}{S}\text{-Bu-}n + H_2S + H_2O \text{ etc.} \quad (8.43)$$

In the case of diphenyl sulfoxide, the reaction proceeds at a lower temperature, below 200°C, but the benzylic C-H linkages appear to be the weakest and undergo initial cleavage and stilbene, benzaldehyde, benzylmercaptan, toluene and hydrogen sulfide are formed.[203]

Sulfoxides are reduced by sulfides, which are oxidized to sulfoxides. In other words, oxygen-transfer reactions occur readily between sulfoxides and sulfides. For example, when a mixture of thiacyclopentane-1-oxide and dibenzyl sulfide is refluxed in acetic acid and thiacyclopentane is distilled off, dibenzyl sulfoxide is obtained in a good yield (eq. 8.44).[204] When dimethyl sulfoxide is used, most alkyl sulfides are readily oxidized to the sulfoxides upon removing the dimethyl sulfide formed at around 160–175°C.[205] The reaction is presumed to be acid catalyzed.

$$(8.44)$$

Thianthrene monoxide is readily reduced to thianthrene upon treatment with arenesulfenyl chloride or sulfur dichloride (eq. 8.45). The sulfinyl chloride formed undergoes further decomposition to give the thiolsulfonate and other sulfur derivatives. Apparently the S-O linkage of this sulfoxide is rather weak and readily reduced. Even on heating with acetic anhydride, thianthrene monoxide loses its oxygen. Similarly, diaryl disulfides also reduce the sulfoxide[206]:

$$(8.45)$$

$$(8.46)$$

The oxygen-transfer reaction between sulfoxides and thionyl chloride is also known[207]:

$$\overset{O}{\underset{\uparrow}{R\text{-}S\text{-}R}} + SOCl_2 \longrightarrow R\text{-}S\text{-}R + SO_2Cl_2 \qquad (8.47)$$

Trivalent phosphorus compounds also reduce the sulfoxide linkage. Thus, dimethyl sulfoxide and diphenyl sulfoxide are reduced to the sulfides by triphenyl phosphine.[208,209] A careful study[210] indicates that the reaction is acid catalyzed and favored by electron-withdrawing substituents in the sulfoxides, suggesting that the reaction proceeds by the nucleophilic attack of the phosphorus center on an acid-complexed sulfoxide (eq. 8.48). Aromatic sulfoxides are reduced to the sulfides on treatment with phosphorus trichloride without being chlorinated.[211]

$$(8.48)$$

### 8.5.1.2 Pummerer-Type Reactions

Sulfoxides bearing a methyl or methylene group directly attached to the sulfur atom readily react with acylating agents, resulting in the formation of an α-acyloxy sulfides. The reaction was discovered by Pummerer[212] and hence bears his name.[213,214] When acetic anhydride is used, the reaction is formulated as shown in eq. 8.49.

α-Acetoxymethyl sulfide is readily hydrolyzed upon dissolving in aqueous media to form the corresponding thiol, formaldehyde and acetic acid (eq. 8.50). The whole reaction is so facile and proceeds so smoothly in mild conditions that it has been suggested as a possible pathway for enzymic demethylation of methionine,[215,216] like the similar demethylation of tertiary amine oxides.[217]

$$R\text{-}S\text{-}CH_3 + (CH_3CO)_2O \longrightarrow R\text{-}S\text{-}CH_2O\overset{\overset{\displaystyle O}{\|}}{C}\text{-}CH_3 + CH_3COOH \qquad (8.49)$$
$$\downarrow O$$

$$R\text{-}S\text{-}CH_2O\overset{\overset{\displaystyle O}{\|}}{C}\text{-}CH_3 \xrightarrow{\cdot H_2O} R\text{-}SH + CH_2O + CH_3COOH \qquad (8.50)$$

In an earlier study using $^{18}$O-labeled acetic anhydride, it was suggested[217] that the Pummerer reaction of dimethyl sulfoxide with acetic anhydride proceeds through an intermolecular nucleophilic attack of the acetoxy group on the methylene carbon of the ylide-ylene intermediate, according to the mechanisim suggested by Bordwell and Pitt.[218] A kinetic study with p-substituted phenyl methyl sulfoxides was carried out and revealed[219] that, although the rate is accelerated by electron-releasing substituents with the Hammett $\rho = -1.5$, the rate-determining step is undoubtedly the proton removal on the basis of the kinetic isotope effect, e.g.., $k_H/k_D = 2.85$.

Though a noticeable acetoxy interchange does take place, the rate of the Pummerer reaction is 5 to 7 times larger than that of the exchange under the normal reaction conditions. On the basis of these observations, the following mechanism has been suggested:

$$(8.51)$$

The migration of acyloxy group to the site of the more acidic proton, as found by Johnson,[220] also support this mechanism.

Cyclic sulfoxides, upon Pummerer rearrangement, often give the corresponding cyclic olefinic sulfides. A typical example is the following reaction[221] in which α-acyloxy-thiacyclohexane is presumed to be formed in the initial step:

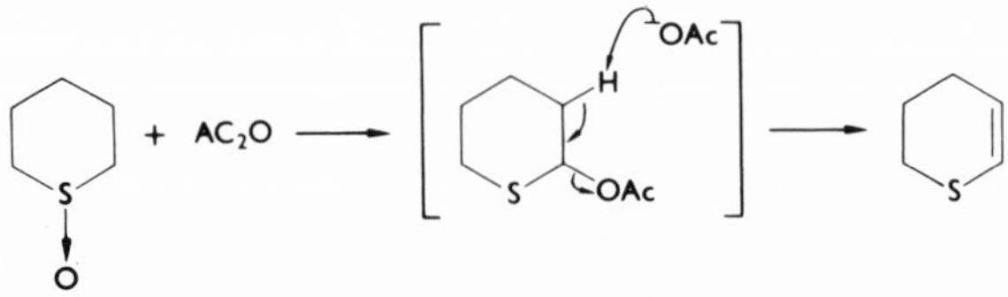

$$(8.52)$$

A few more examples are shown below[222,223]:

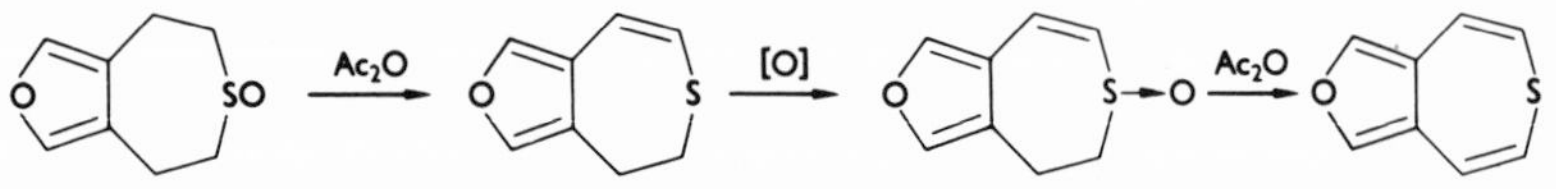

$$(8.53)$$

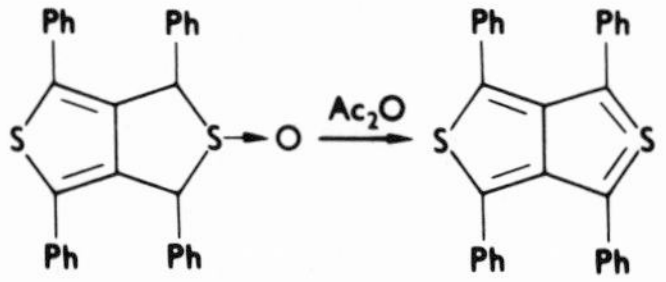

$$(8.54)$$

Sulfoxides which have acidic methylene groups undergo the Pummerer reaction just by heating with acid. Thus, benzenesulfinylacetic acid reacts with boiling acetic acid giving bis-(benzenethia)-acetic acid and glyoxalic acid.[224] Similarly, α-(α-toluenesulfinyl)-p-nitroltoluene gives α,α-bis(α-toluenethia)-p-nitroluene and p-nitrobenzaldehyde.[225] Even dimethyl sulfoxide is decomposed upon heating with sulfuric acid at 190°C to give dimethyl disulfide, dimethyl sulfide and formaldehyde.[226] An interesting example is the direct oxidative decomposition of substituted benzenthiaacetic acid by treatment with hydrogen peroxide in acidic media[227]:

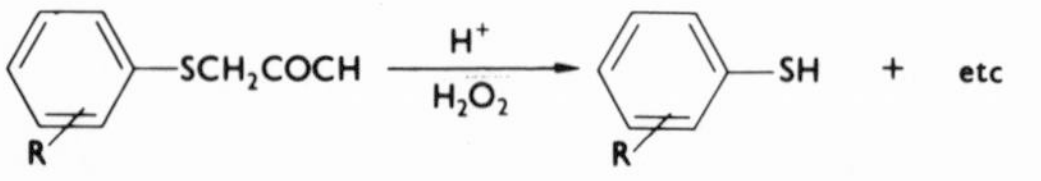

$$(8.55)$$

The reaction undoubtedly involves the oxidation to sulfoxide and subsequent Pummerer rearrangement.

Another interesting modification is the following reaction of β-ketosulfoxides with iodine in refluxing methanol[228]:

$$
\underset{\text{R-C-CH}_2\text{-SOCH}_3}{\overset{\text{O}}{\overset{\|}{}}} \xrightarrow{\text{CH}_3\text{OH, I}_2} \left[ \underset{\text{RCOCH=SCH}_3}{\overset{\text{OH}}{\overset{|}{}}} \right] \xrightarrow{\text{CH}_3\text{OH}} \underset{\text{R-C-CH-SCH}_3}{\overset{\text{O OCH}_3}{\overset{\|\ |}{}}}
$$

$$
\xrightleftharpoons{\text{CH}_3\text{OH}} \overset{\text{O}}{\overset{\|}{\text{RCCH}}}(\text{OCH}_3)_2 + \text{CH}_3\text{SH}
$$

$$
\text{CH}_3\text{SH} \underset{}{\overset{\text{I}^2}{\rightleftharpoons}} \text{CH}_3\text{SSCH}_3 + 2\text{HI} \tag{8.56}
$$

Examples of ordinary Pummerer rearrangements are numerous.[229]

Cyclic sulfoxides which do not possess an α-hydrogen undergo an internal oxidative-reductive reaction in which a carbon atom β to the sulfur atom becomes oxidized. Thus, treatment of 2,2-dimethyl thiochromane 1-oxide with freshly distilled refluxing acetic anhydride gives the acetate in 80% yield[230]:

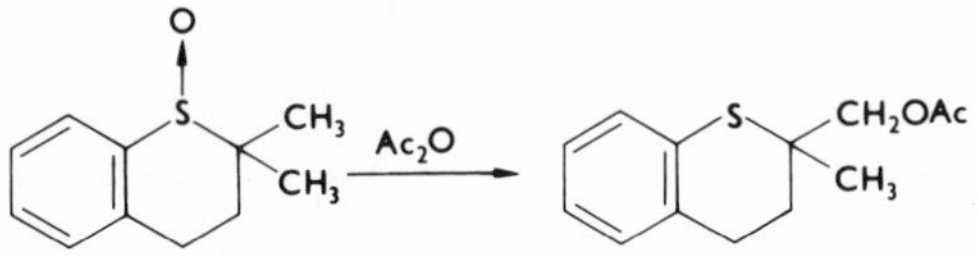

$$\tag{8.57}$$

Another similar reaction is shown below[231]:

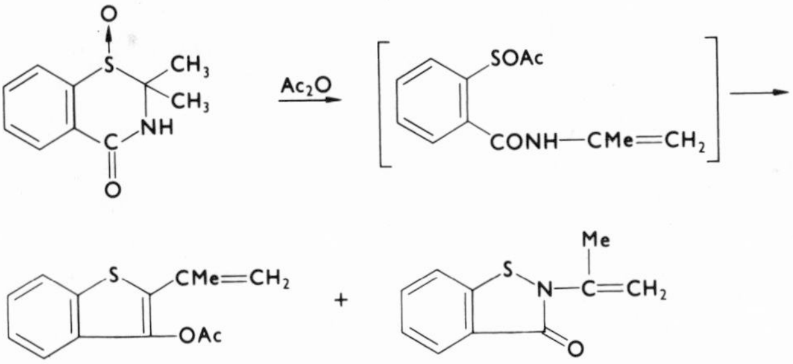

$$\tag{8.58}$$

An application of this reaction is found in the valuable transformation of penicillin S-oxide to cephalexin[232]:

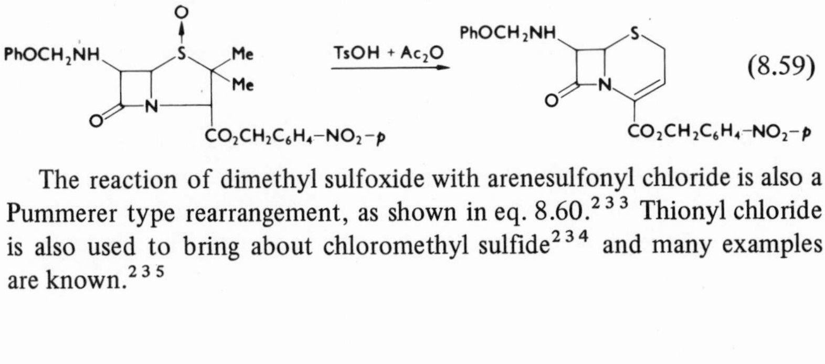

$$\tag{8.59}$$

The reaction of dimethyl sulfoxide with arenesulfonyl chloride is also a Pummerer type rearrangement, as shown in eq. 8.60.[233] Thionyl chloride is also used to bring about chloromethyl sulfide[234] and many examples are known.[235]

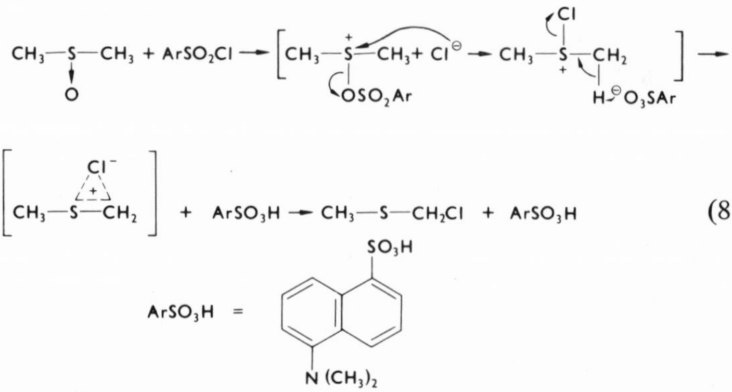

$$CH_3\text{—}S\text{—}CH_3 + ArSO_2Cl \longrightarrow \left[ CH_3\text{—}\overset{+}{S}\text{—}CH_3 + Cl^{\ominus} \longrightarrow CH_3\text{—}\overset{+}{S}\text{—}CH_2 \right] \longrightarrow$$

$$\left[ CH_3\text{—}\overset{+}{S}\text{—}CH_2 \right] + ArSO_3H \longrightarrow CH_3\text{—}S\text{—}CH_2Cl + ArSO_3H \qquad (8.60)$$

$$ArSO_3H \ = \ \text{(naphthalene ring with } SO_3H \text{ and } N(CH_3)_2 \text{ substituents)}$$

Benzoyl and acetyl chlorides are also used for this Pummerer type rearrangement of dimethyl sulfoxide to form α-chloromethyl methyl sulfide.[236]

In any of these Pummerer type reactions, the initial step is acylation of a sulfinyl oxygen to form an acyloxysulfonium salt. This is also the case with dicyclohexylcarbodiimide (DCC). Such an intermediate is generally very reactive and susceptible to nucleophilic attack on the sulfur atom cleaving the S-O linkage. There are a few interesting reactions which proceed *via* such an acyloxysulfonium intermediate.[237] An interesting example is the following reaction:

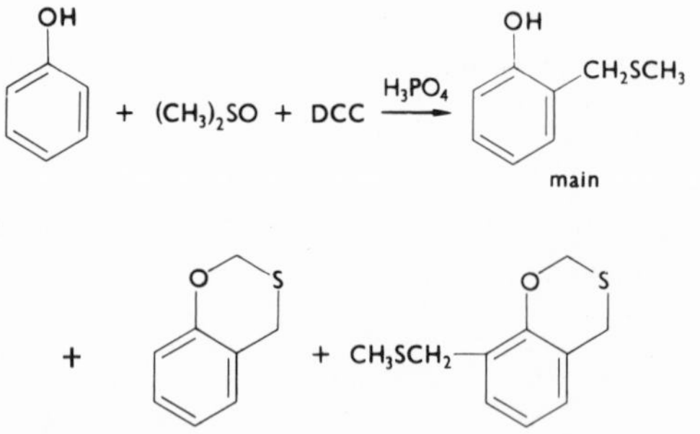

$$\text{(phenol, OH)} + (CH_3)_2SO + DCC \xrightarrow{H_3PO_4} \text{(ortho } CH_2SCH_3 \text{ phenol, OH)} \quad \text{main}$$

$$+ \ \text{(benzothioxane)} \ + \ CH_3SCH_2\text{—}\text{(benzothioxane)} \qquad (8.61)$$

The initial step apparently involves the formation of the DCC-DMSO adduct (A) upon protonation, which in the subsequent step undergoes nucleophilic attack by phenol on the sulfur atom, followed by Pummerer-Sommelet type rearrangement to the *ortho* position. The formation of benzothioxane is also suggested to proceed as shown below:

$$(8.62)$$

In the presence of a strong acid such as hydrochloric acid, phenol reacts with dimethyl sulfoxide to give p-dimethylsulfoniophenol, which upon pyrolysis gives p-methylthiophenol[238]:

$$(8.63)$$

A similar reaction takes place with a more complicated molecule intramolecularly[239]:

$$(8.64)$$

The reaction of sulfoxides with a compound possessing an active methylene group gives the corresponding ylides. A typical example is the reaction shown in eq. 8.65.[240] Other examples are also known.[241,242,243]

$$(8.65)$$

Another interesting modification may be the reaction between oximes and DCC–DMSO complex.[244] While benzophenone oxime reacts readily giving thiomethoxymethylnitrone and the isomeric O-alkyl oxime (eq. 8.66), those of *syn-* and *anti-p*-bromobenzaldoximes give the corresponding nitrites.

$$(8.66)$$

17-Oximinosteroids give the lactam and the unsaturated nitrile:

$$(8.67)$$

Grignard reagents react with sulfoxides in the following fashion[245]:

$$p\text{-}CH_3C_6H_4SOR + C_6H_5MgX \xrightarrow[18\ hr]{25°C} C_7H_7SR,\ C_7H_7SH,\ C_7H_7SC_6H_5,\ C_7H_7\overset{+}{\underset{R}{S}}C_6H_5$$

$$p\text{-}CH_3C_6H_4SOCH_3 + C_6H_5MgX \longrightarrow p\text{-}CH_3C_6H_4CH_2C_6H_5\ (50\%)$$

$$p\text{-}CH_3C_6H_4SOCH_2C_6H_5 + C_6H_5MgX \longrightarrow p\text{-}CH_3C_6H_4SCH(C_6H_5)_2\ (90\%)$$

$$p\text{-}CH_3C_6H_4SCH_2CH(C_6H_5)_2 + C_6H_5MgX \longrightarrow p\text{-}CH_3C_6H_4SCH_2C(C_6H_5)_3\ (50\%)$$

From the reaction of Grignard reagents (RMgX) with dimethyl sulfoxide, compounds possessing a general formula of $RCH_2SMe$ are obtained.[246] The reaction is believed to involve the initial formation of sulfonium methylide, as shown in the following scheme[247]:

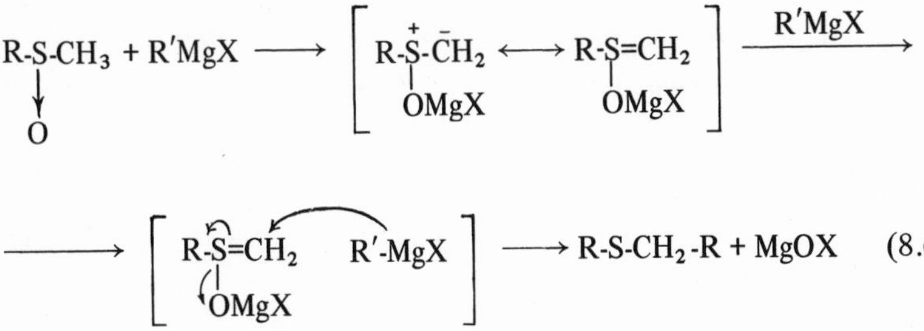

$$R\text{-}\underset{\underset{O}{\downarrow}}{S}\text{-}CH_3 + R'MgX \longrightarrow \left[ R\text{-}\overset{+}{\underset{OMgX}{S}}\text{-}\overset{-}{C}H_2 \longleftrightarrow R\text{-}\underset{OMgX}{S}=CH_2 \right] \xrightarrow{R'MgX}$$

$$\longrightarrow \left[ R\text{-}\underset{OMgX}{S}=CH_2 \quad R'\text{-}MgX \right] \longrightarrow R\text{-}S\text{-}CH_2\text{-}R + MgOX \qquad (8.68)$$

### 8.5.1.3 Oxygen Exchange and Related Reactions[248]

Diphenyl sulfoxide ionizes in concentrated sulfuric acids in the following manner, giving a deep blue green solution.[249-251]

$$(C_6H_5)_2S \rightarrow O + H_2SO_4 \longrightarrow (C_6H_5)_2\overset{+}{S}\text{-}OH + HSO_4^- \qquad (8.69)$$

When diphenyl sulfoxide was dissolved in a large excess of concentrated [18]O-enriched sulfuric acid. Other sulfoxides such as dimethyl, phenyl resulting blue-green solution was diluted with a large excess of ordinary water, [18]O-labeled diphenyl sulfoxide recovered quantitatively was found to incorporate roughly the same concentration of [18]O as that of the [18]O-enriched sulfuric acid. Other sulfoxides such as dimethyl, phenyl methyl and phenyl benzyl sulfoxides also undergo a similar oxygen exchange. An optically active $(-)p$-aminophenyl $p$-tolyl sulfoxide was found to be racemized concurrently.[252]

A careful kinetic study with an optically active and [18]O-labeled p-tolyl phenyl sulfoxides and other substituted diphenyl sulfoxides revealed that the rate of oxygen exchange in 95.5% sulfuric acid is identical to that of racemization ($k_{ex}/k_{rac} = 0.97$) and correlates nicely with the Hammett acidity function, $H_0$, with a slope of nearly one. In contrast, the effect of a polar substituent is very small, but that of steric inhibition by o-substituents is substantial. Based on these observations the following mechanism has been suggested for the reaction[253]:

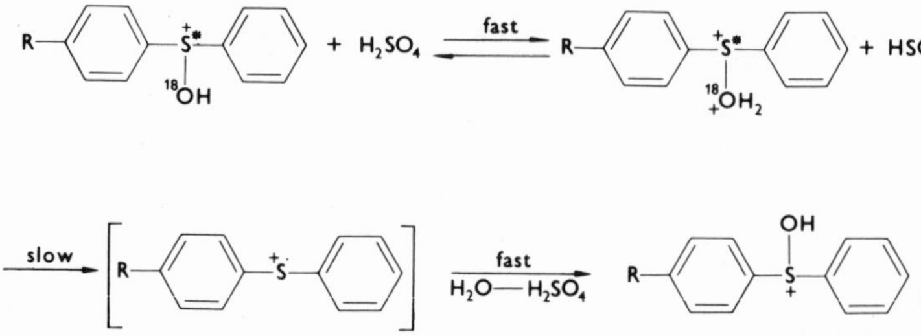

(8.70)

Though the formation of any radical cation intermediate from diphenyl sulfoxide was not observed with esr spectroscopy, noticeable esr signals were observed with 96% sulfuric acid solutions of substituted diphenyl sulfoxides bearing electron-releasing groups at *para* positions.[253,254,255]

In case of thianthrene 5-oxide, the oxygen exchange proceeds *via* the formation of the di-cation[256]:

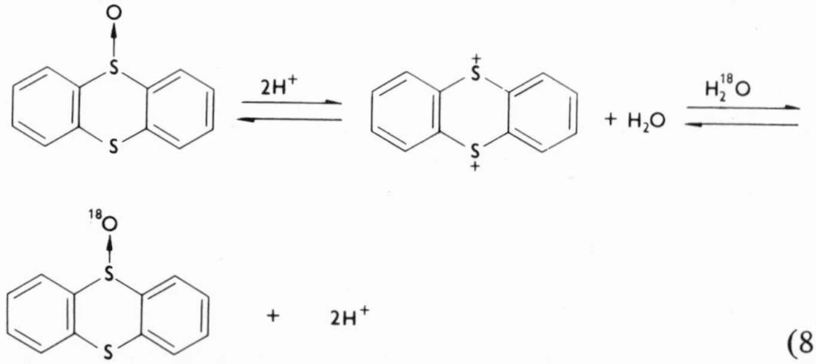

(8.71)

In a less concentrated sulfuric acid, for example in 70% sulfuric acid, the rate of oxygen exchange becomes half that of racemization, while the effect of a polar substituent is also small and the Bunnett w value is 1.28.

These data suggest that the reaction proceeds through a different mechanistic route, an $S_N2$-type path as shown below (eq. 8.72).[257] In sulfuric acids ranging between 95% and 70% of acid, the mechanism changes gradually from A-1 to A-2 type, and this is illustrated by the gradual change of $k_{ex}/k_{rac}$ and also the Hammett acidity correlations.

$$\underset{Ar\text{-}\overset{\displaystyle\uparrow}{\underset{}{S}}\text{-}Ar'}{\overset{O}{}} + H^+ \underset{\xleftarrow{\hspace{1em}}}{\overset{fast}{\xrightarrow{\hspace{1em}}}} \left[ \underset{Ar\text{-}\overset{\displaystyle}{\underset{+}{S}}\text{-}Ar'}{\overset{OH}{}} \right] + H_2O + H_2SO_4 \overset{slow}{\underset{\xleftarrow{\hspace{1em}}}{\xrightarrow{\hspace{1em}}}}$$

$$\left[ \begin{array}{c} HOH \\ | \\ Ar\text{-}S\text{-}Ar' \\ | \\ HOH \end{array} \right]^{2+} \overset{slow}{\underset{\xleftarrow{\hspace{1em}}}{\xrightarrow{\hspace{1em}}}} \left[ \begin{array}{c} Ar\text{-}\overset{+}{S}\text{-}Ar' \\ | \\ OH \end{array} \right] + H_2O + H_2SO_4 \qquad (8.72)$$

$$\left[ \underset{Ar\text{-}\underset{+}{S}\text{-}Ar'}{\overset{OH}{}} \right] + H_2O \overset{fast}{\underset{\xleftarrow{\hspace{1em}}}{\xrightarrow{\hspace{1em}}}} \underset{Ar\text{-}\overset{\displaystyle\uparrow}{\underset{}{S}}\text{-}Ar'}{\overset{O}{}} + H_3O^+$$

Similar concurrent oxygen-exchange and racemization reactions take place with various acids, such as $p$-toluenesulfonic, phosphoric,[258] acetic, monochloro-, dichloro- and trichloroacetic acids.[259] In all these cases, the effect of polar substituents is small while $k_{ex}$ equals $k_{rac}$ and the reactions are acid-catalyzed. The reactions are suggested to proceed *via* homolytic S-O bond cleavage at the rate-determining step.[258,259]

Aqueous hydrochloric[260] and hydrobromic acids[261] similarly effect the concurrent oxygen-exchange and racemization reactions in which the rate of exchange is also identical to that of racemization.[262] In this case, however, the reaction is acid catalyzed and retarded markedly by bulky substituents around the sulfur atom.[260,262,263]

Based on these observations, the following mechanism which involves the formation of a halogen-sulfide complex is suggested:

$$\underset{\underset{O}{\overset{\displaystyle\downarrow}{}}}{\overset{}{R\text{-}S\text{-}R'}} \xrightarrow[2X^-]{2H^+} \underset{\underset{R'}{|}}{\overset{+}{R}\text{-}\overset{+}{S}\text{-}\overset{}{O}H_2} + X^- \longrightarrow \left[ X^- \searrow \overset{R}{\underset{R'}{\overset{|}{S}}} \text{-} \overset{+}{O}H_2 \right]$$

$$(8.73)$$

$$\longrightarrow H_2O + \left[ X\text{-}\overset{R}{\underset{R'}{\overset{/}{S}}}{}^+ \right] X^- \longrightarrow \underset{X_2}{\overset{R}{\underset{}{}}}\overset{}{S}\text{-}R' \xrightarrow{H_2O} \underset{\underset{O}{\overset{\displaystyle\downarrow}{}}}{\overset{}{R\text{-}S\text{-}R'}} + 2HX$$

The identical rate for both the oxygen-exchange and racemization reactions may be due to the following rapid $S_N2$-type interconversion in polar media:

$$X-\overset{\overset{\displaystyle R}{+/}}{\underset{\underset{\displaystyle R'}{\backslash}}{S}} + X^- \rightleftharpoons X^- + \overset{\overset{\displaystyle R}{\backslash+}}{\underset{\underset{\displaystyle R'}{/}}{S}}-X \qquad (8.74)$$

The concurrent oxygen-exchange and racemization reactions of diaryl sulfoxides in dinitrogen tetroxide are somewhat different in that the polar effect of substituents on the rates is large and correlates with $\sigma^+$ values with $\rho = -1.30$, though the rate of oxygen exchange is identical to that of racemization. The reaction is presumed to involve the rate-determining heterolysis of the S-O linkage.[264]

Sulfoxides undergo concurrent oxygen exchange and racemization in acetic anhydride, and the rate of racemization is twice that of oxygen exchange, as in the case of dilute sulfuric acid.[265] The reaction is clearly and $S_N2$ process as shown in eq. 8.75. The addition of either a Brönsted or a Lewis acid markedly accelerates the reaction.[266,267]

$$R\text{-}S\text{-}R' + Ac_2O \rightleftharpoons R\text{-}\overset{+}{S}\text{-}R' + OA\bar{c} \rightleftharpoons \left[ R\text{-}\overset{+}{S}\text{-}R' \atop {}^{18}\overset{|}{O}Ac \right] \rightleftharpoons$$
$$\overset{\displaystyle\downarrow}{{}^{18}O} \qquad\qquad {}^{18}OAc \qquad\qquad \left[ {}^{18}\overset{|}{O}Ac \right]$$

$$(8.75)$$

$$\overset{\overset{\displaystyle OAc}{|}}{R\text{-}\underset{+}{S}\text{-}R'} + {}^{18}OA\bar{c} \rightleftharpoons R\text{-}\overset{\overset{\displaystyle O}{\uparrow}}{S}\text{-}R' + Ac_2{}^{18}O$$

Dimethyl sulfoxide undergoes oxygen exchange readily with potassium t-butoxide and even with sodium hydroxide under the drastic condition of heating at 150°C. Other sulfoxides such as di-n-butyl, p-tolyl methyl were also found to undergo oxygen-exchange reactions, through much more slowly.[268] The reaction is believed to proceed through an $S_N2$-type process.

In the case of alkoxysulfonium salt, the alkoxy exchange is so facile that it results in the inversion of the configuration around the sulfur atom[269]:

$$RO-\overset{\overset{\displaystyle CH_3}{+/}}{\underset{\underset{\displaystyle CH_3}{\backslash}}{S}} + RO^- \longrightarrow RO^- + \overset{\overset{\displaystyle CH_3}{\backslash+}}{\underset{\underset{\displaystyle CH_3}{/}}{S}}-OR \qquad (8.76)$$

Stereochemistry of all these oxygen-exchange reactions is not different in any way from that of nucleophilic substitution reactions on carbon atom. However, the nucleophilic substitution on a second-row element, in this case the sulfur atom in trivalent sulfur compounds, may proceed *via* the typical bipyramidal activated complex formation either as an intermediate or as an incipient transition complex at the transition state. Then other stereochemical consequences could arise, depending upon the steric arrangement of both incoming and outgoing groups.

When both the entering and leaving groups assume axial positions, the process is similar or identical to an $S_N2$ reaction on carbon atom, and the net result is the inversion of configuration.

A similar net inversion will also result when both incoming and outgoing groups assume radial position, as shown below. A probable example is provided by Cram and his coworkers,[270] who showed that the formation of *p*-tolyl methyl N-*p*-tosylsulfilimine from the corresponding sulfoxide resulted in net inversion. The reaction is first order with respect to the sulfoxide and second order with respect to the sulfur diimide reagent.

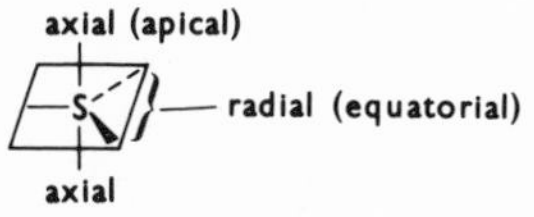

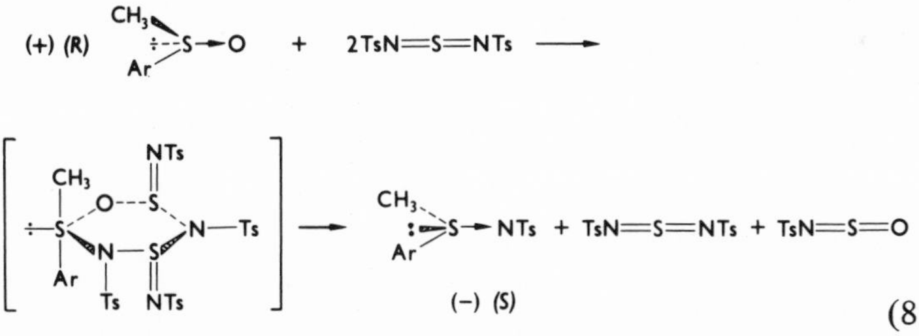

$$(8.77)$$

No oxygen-exchange reaction of this type has been found yet. However, careful examination of kinetic data on other possible oxygen-exchange reactions of sulfoxides with carbonyl or nitroso-type compounds may open the way for finding one in the future.

If both incoming and outgoing groups can assume axial and radial positions, namely the perpendicular arrangement, the reaction will lead to the retention of configuration. The first example of this type may be the oxygen-exchange reaction between *p*-tolyl methyl sulfoxide and dimethyl sulfoxide[271]:

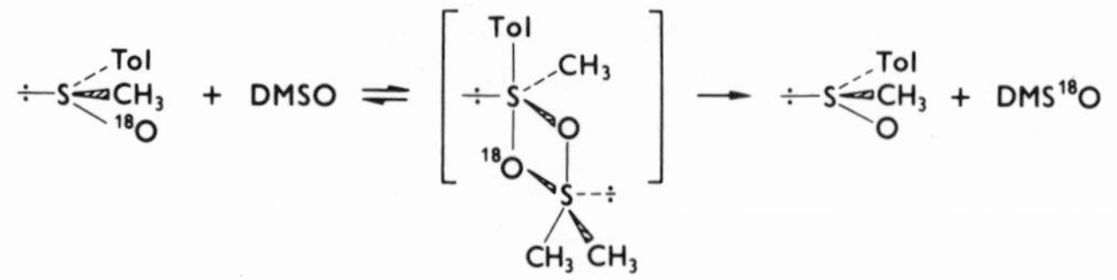

$$\text{—S}\overset{\text{Tol}}{\underset{^{18}\text{O}}{\text{CH}_3}} + \text{DMSO} \rightleftharpoons \left[ \begin{array}{c} \text{Tol} \\ \text{—S} \overset{\text{CH}_3}{\underset{^{18}\text{O}}{\diagdown}}\text{O} \\ \diagup \diagdown \\ \text{CH}_3 \ \text{CH}_3 \end{array} \right] \longrightarrow \text{—S}\overset{\text{Tol}}{\underset{\text{O}}{\text{CH}_3}} + \text{DMS}^{18}\text{O} \qquad (8.78)$$

Another example of this type is the formation of the sulfilimine in the reaction between N-phthaloylmethionine sulfoxide and N-sulfinyl p-toluenesulfonamide.[272] The reduction of dimethyl sulfoxide shown below could also be a reaction of this type, though further examination is necessary to confirm it[273]:

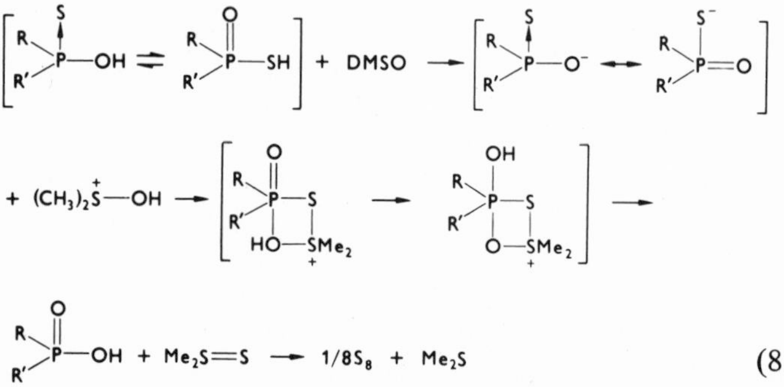

$$\left[ \overset{R}{\underset{R'}{>}}\overset{S}{\underset{}{P}}\text{—OH} \rightleftharpoons \overset{R}{\underset{R'}{>}}\overset{O}{\underset{}{P}}\text{—SH} \right] + \text{DMSO} \longrightarrow \left[ \overset{R}{\underset{R'}{>}}\overset{S}{\underset{}{P}}\text{—O}^- \longleftrightarrow \overset{R}{\underset{R'}{>}}P{=}O \overset{S^-}{\underset{}{}} \right]$$

$$+ (\text{CH}_3)_2\overset{+}{\text{S}}\text{—OH} \longrightarrow \left[ \overset{R}{\underset{R'}{>}}\overset{O}{\underset{}{P}}\text{—S} \\ \underset{\text{HO—}\overset{+}{\text{SMe}}_2}{} \right] \longrightarrow \left[ \overset{R}{\underset{R'}{>}}\overset{OH}{\underset{}{P}}\text{—S} \\ \underset{\text{O—}\overset{+}{\text{SMe}}_2}{} \right] \longrightarrow$$

$$\overset{R}{\underset{R'}{>}}\overset{O}{\underset{}{P}}\text{—OH} + \text{Me}_2\text{S}{=}\text{S} \longrightarrow 1/8\text{S}_8 + \text{Me}_2\text{S} \qquad (8.79)$$

As in the case of solvolytic reactions of carbon compounds, the effect of neighboring groups is also observed in the acid-catalyzed oxygen exchange reaction. Thus, the rate of oxygen exchange of o-carboxyphenyl phenyl sulfoxide is ca. $10^4$ times greater than that of racemization and also than that of oxygen exchange of unsubstituted diphenyl sulfoxide. The reaction is presumed to involve the formation of a five-membered cyclic intermediate, as shown below[274]:

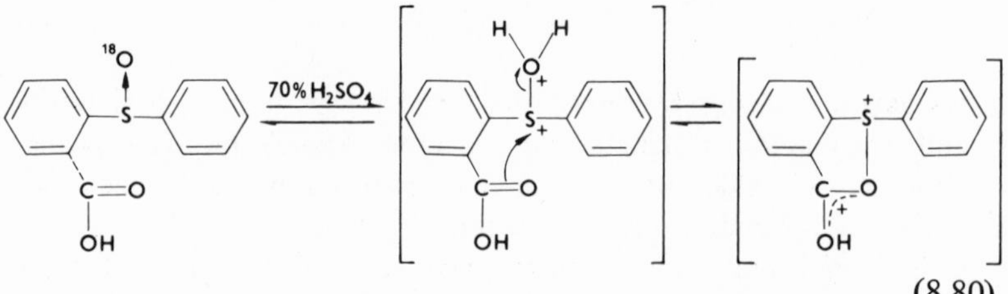

$$(8.80)$$

o-Alkylthia group also displays a substantial neighboring group participation in the heterolysis of the S-O bond. However, in this case, the reaction is a net oxygen transfer, as shown below[275]:

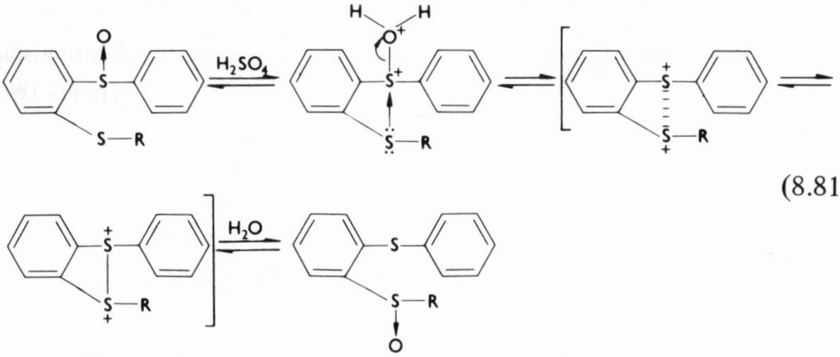

(8.81)

## 8.5.1.4 Kornblum Reaction and Its Modifications

When alkyl halides are heated in dimethyl sulfoxide, alkylation takes place either on the sulfur atom or on the oxygen atom, depending upon the nature of alkyl group. Only with methyl halides is the exclusive formation of the S-alkylation product observed,[276] while with others the reaction appears to proceed *via* an unstable alkoxysulfonium salt, which upon further reaction affords aldehyde and dimethyl sulfide[277,278]:

$$RCH_2X + (CH_3)_2SO \xrightarrow{\quad 100-160^\circ C \quad} \left[RCH_2\overset{+}{O}S(CH_3)_2\right]X$$

$$\xrightarrow{\hspace{3cm}} RCHO + (CH_3)_2S + HX \qquad (8.82)$$

Thus, this reaction can be used as a convenient procedure for the preparation of aldehydes. Since Kornblum showed that treatment of *p*-bromophenacyl bromide with DMSO gives phenyl glyoxal in a good yield (eq. 8.83),[279] this reaction has been used extensively[280] and called the Kornblum reaction. The reaction can be applied not only to alkyl halides, but also to alkyl tosylates and other similar compounds.[281-285] Since the reaction is carried out without any strong acid or base, unlike other methods for aldehyde synthesis, this procedure has found convenient use for preparing many delicate aldehydes which are sensitive to both acid and base.[288]

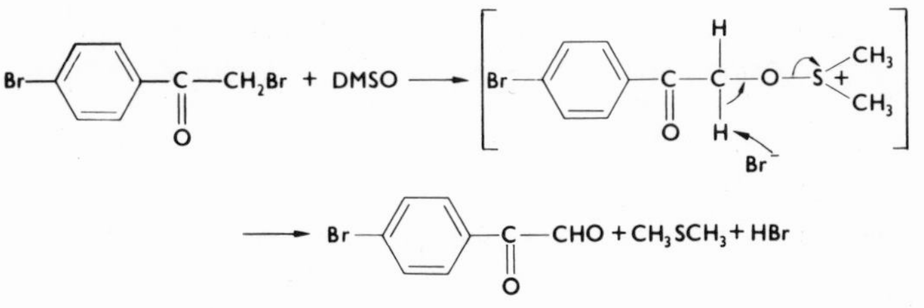

(8.83)

The reaction has been shown to proceed through the initial formation of the S-alkoxysulfonium salt. For the subsequent reaction, there are two conceivable routes, i.e., (a) and (b):

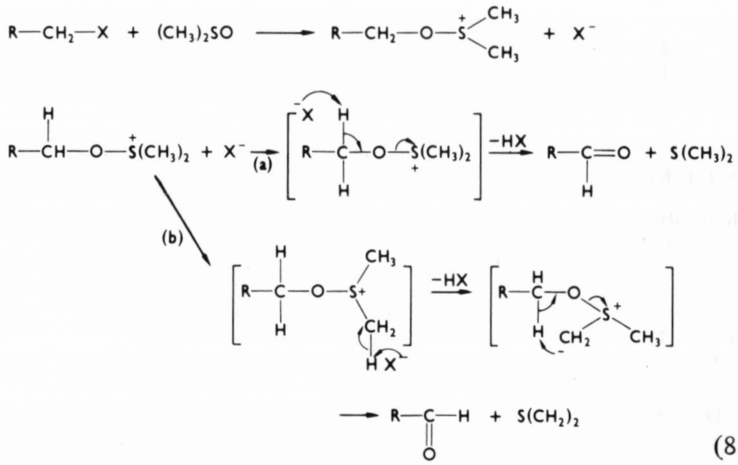

(8.84)

The use of $(CD_3)_2SO$ for this reaction revealed that the reaction of primary halides such as the isobutyl derivative apparently proceeds *via* route (b) and results in the formation of $CD_3SCHD_2$ (eq. 8.85), while that of the *p*-bromophenacyl compound proceeds *via* route (a) and hexadeuterated dimethyl sulfide is recovered intact, as shown below (eq. 8.86).[287] Oxygen-18 tracer experiments also confirmed that the carbonyl oxygen of the aldehyde formed in this reaction comes from the dimethyl sulfoxide.[288]

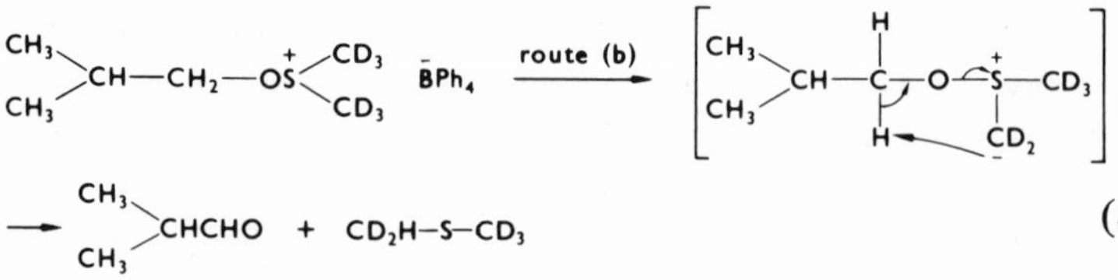

(8.85)

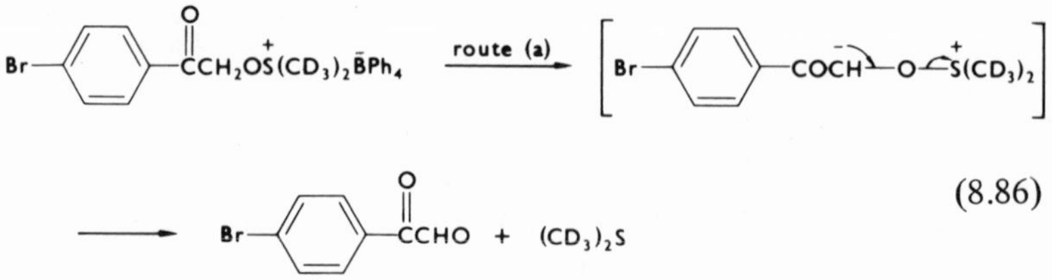

(8.86)

In the following reaction and even in the modified Kornblum reactions of cholestanone-3α-3H(T)-3β-ol with either acetic anhydride or dicyclohexylcarbodiimide, the incorporation of tritium into dimethyl sulfide from the starting steroid compound is about 30–40%, suggesting that the reaction *via* route (a) is favored over the (b) route[289]:

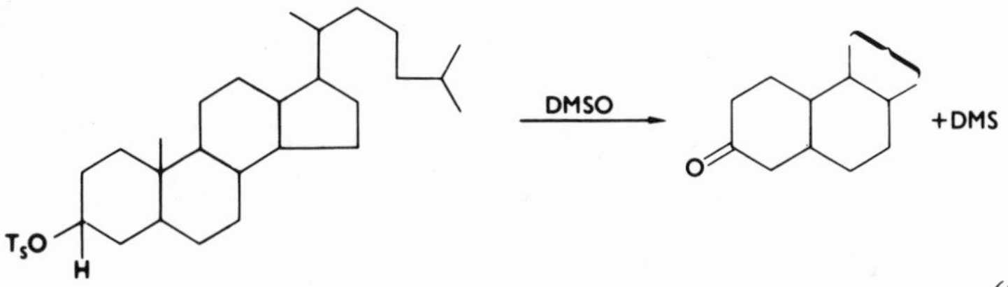
DMSO
+DMS
$T_sO$
H

(8.87)

Among a number of elegant modifications in which the oxidation by DMSO is accomplished at room temperature, the earliest one was developed by Barton, Barner and Wightman.[290] Hydroxy compounds are at first converted to their chloroformate esters, which are treated with DMSO, followed by triethylamine at room temperature to give the carbonyl compounds in up to 80% yield:

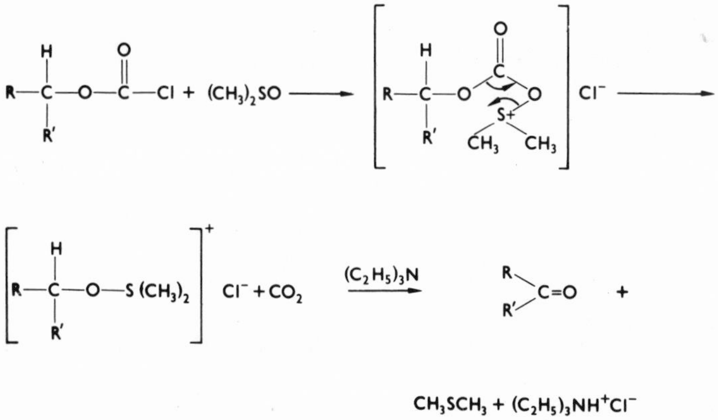
$(CH_3)_2SO$
$Cl^-$
$Cl^- + CO_2$
$(C_2H_5)_3N$
C=O
$CH_3SCH_3 + (C_2H_5)_3NH^+Cl^-$

(8.88)

Pfitzer and Moffatt developed the following modification in which dicyclohexylcarbodiimide and anhydrous phosphorous acid are used[291]:

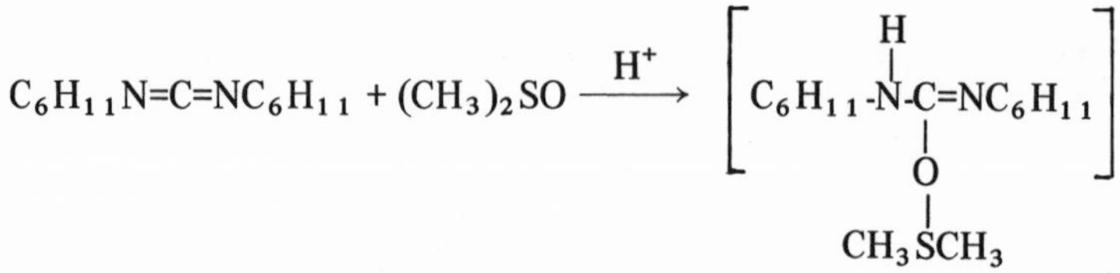

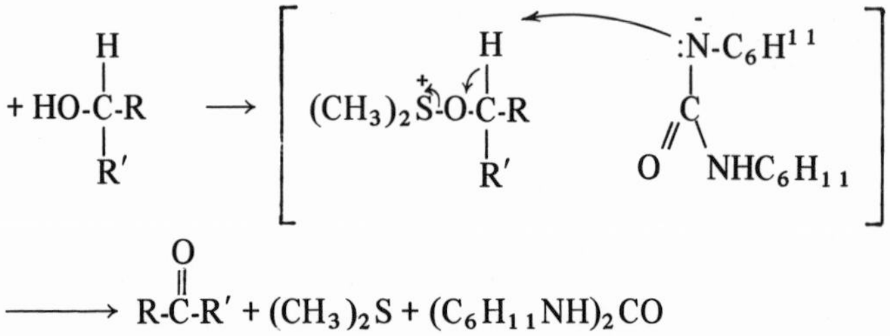

$$C_6H_{11}N=C=NC_6H_{11} + (CH_3)_2SO \xrightarrow{H^+} \left[ \begin{array}{c} H \\ | \\ C_6H_{11}\text{-}N\text{-}C=NC_6H_{11} \\ | \\ O \\ | \\ CH_3\overset{+}{S}CH_3 \end{array} \right]$$

$$+ \begin{array}{c} H \\ | \\ HO\text{-}C\text{-}R \\ | \\ R' \end{array} \longrightarrow \left[ \begin{array}{cc} H & :\text{N-}C_6H^{11} \\ | & \\ (CH_3)_2\overset{+}{S}\text{-}O\text{-}C\text{-}R & C \\ | & // \quad \backslash \\ R' & O \quad NHC_6H_{11} \end{array} \right]$$

$$\xrightarrow{\phantom{xxxx}} \begin{array}{c} O \\ \| \\ R\text{-}C\text{-}R' \end{array} + (CH_3)_2S + (C_6H_{11}NH)_2CO \qquad (8.89)$$

Another convenient procedure, developed by Albright and Goldman, is to use acetic anhydride, as shown below[292]:

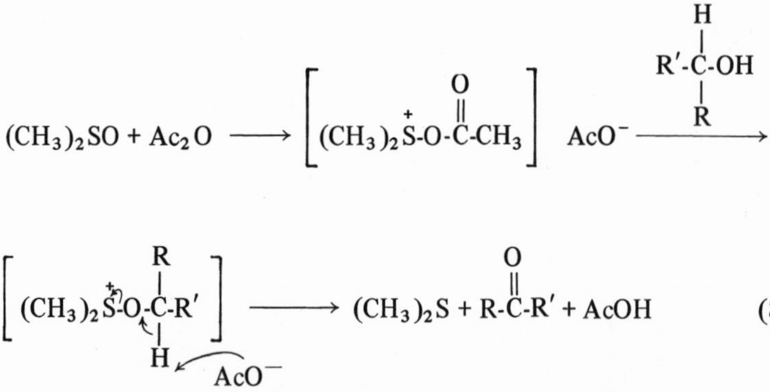

$$(CH_3)_2SO + Ac_2O \longrightarrow \left[ \begin{array}{c} O \\ + \quad \| \\ (CH_3)_2S\text{-}O\text{-}C\text{-}CH_3 \end{array} \right] AcO^- \xrightarrow{\begin{array}{c} H \\ | \\ R'\text{-}C\text{-}OH \\ | \\ R \end{array}}$$

$$\left[ \begin{array}{c} R \\ | \\ (CH_3)_2\overset{+}{S}\text{-}O\text{-}C\text{-}R' \\ | \\ H \quad \text{\ } AcO^- \end{array} \right] \longrightarrow (CH_3)_2S + \begin{array}{c} O \\ \| \\ R\text{-}C\text{-}R' \end{array} + AcOH \qquad (8.90)$$

The use of phosphorous pentoxide together with DMSO was found by Onodera *et al.* to effect the oxidation of various hydroxy groups of carbohydrates and other hydroxy compounds to the corresponding aldehydes, ketones and carboxylic acids.[293]

When epoxides are dissolved in anhydrous DMSO and reacted with catalytic amounts of boron trifluoride, the corresponding keto-alcohols are obtained in fairly good yields.[294] The reaction is presumed to proceed in the following fashion:

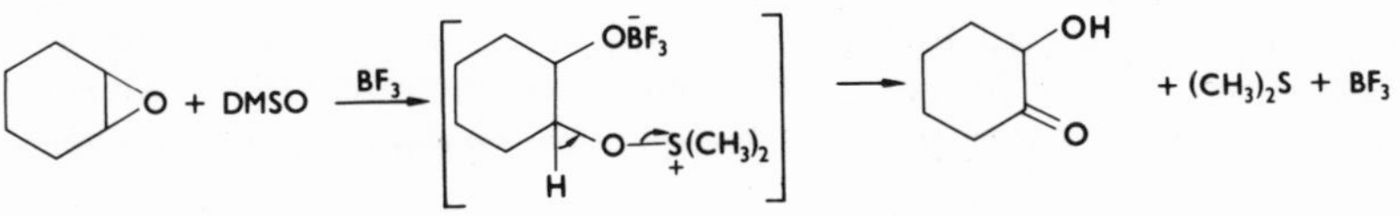

$$(8.91)$$

The base-catalyzed oxidation of 2,5-dimethylquinyl acetate by DMSO is an interesting example[295] and is presumed to proceed in the following manner:

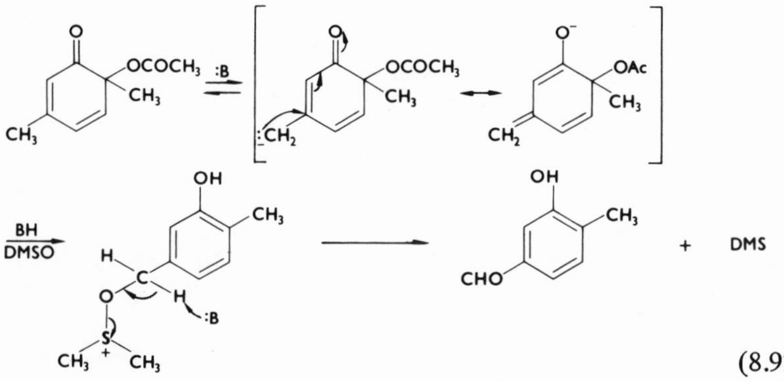

$$(8.92)$$

Following is another similar reaction in which an $\alpha,\beta$-unsaturated steroid ketone is converted to the corresponding $\gamma$-keto-alcohol[296]:

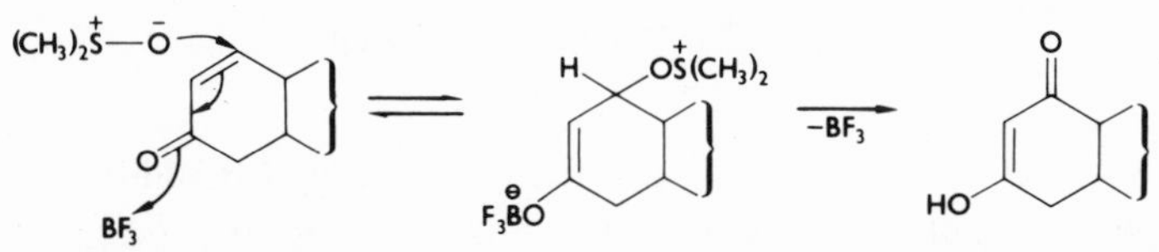

$$(8.93)$$

Benzyl alcohols were shown by Traynelis and his coworker to be selectively oxidized to the corresponding aldehydes by refluxing the alcohols in dimethyl sulfoxide.[297] One possible explanation is that the reaction involves the nucleophilic attack of benzyl alcohol on the sulfoxide sulfur atom, as shown below (eq. 8.94). The rate-enhancement by electron-releasing substituents at *para* position may be rationalized on the basis of the above mechanism.

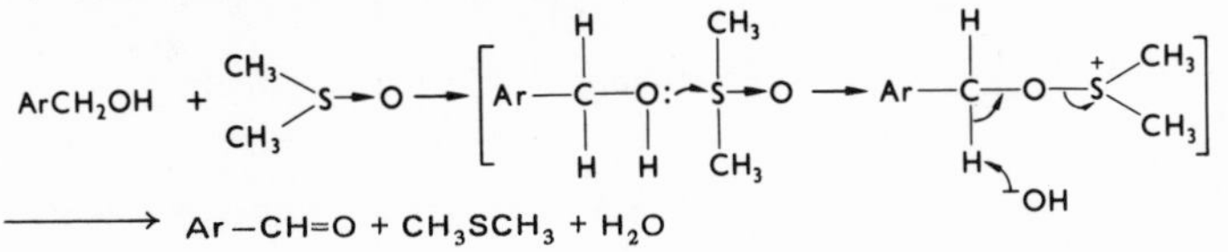

$$(8.94)$$

However, it is hard to explain why the reaction is facilitated by the passage of a stream of air. A similar direct oxidation of highly cleavable alcohols, is known,[298] while the addition of a bivalent metal salt such as mercuric acetate facilities the oxidation of benzyl alcohol, cyclohexanol and *sec*-butyl alcohol to benzaldehyde, cyclohexanone and methyl ethyl ketone, respectively.[299]

Tertiary alcohols are known to be converted to the corresponding olefins in good yields upon refluxing with DMSO.[301] The reaction would proceed similarly, as shown below:

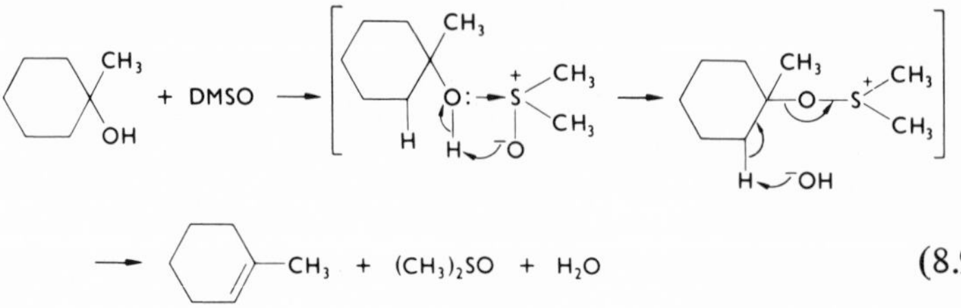

$$(8.95)$$

The aziridine ring is also oxidized by treatment with DMSO as illustrated by the following example[302]:

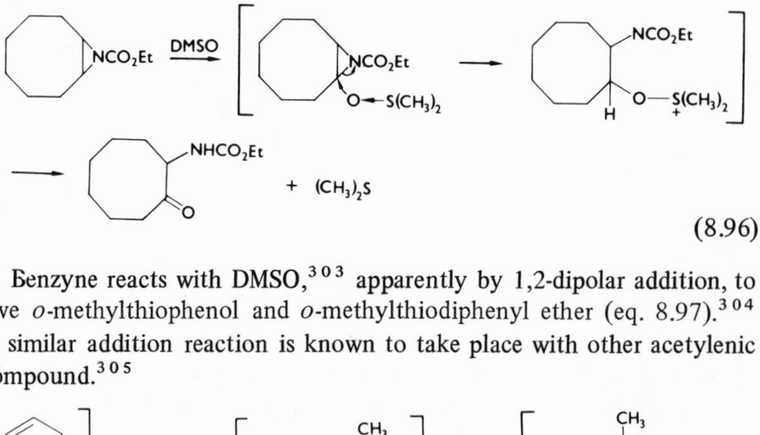

$$(8.96)$$

Benzyne reacts with DMSO,[303] apparently by 1,2-dipolar addition, to give *o*-methylthiophenol and *o*-methylthiodiphenyl ether (eq. 8.97).[304] A similar addition reaction is known to take place with other acetylenic compound.[305]

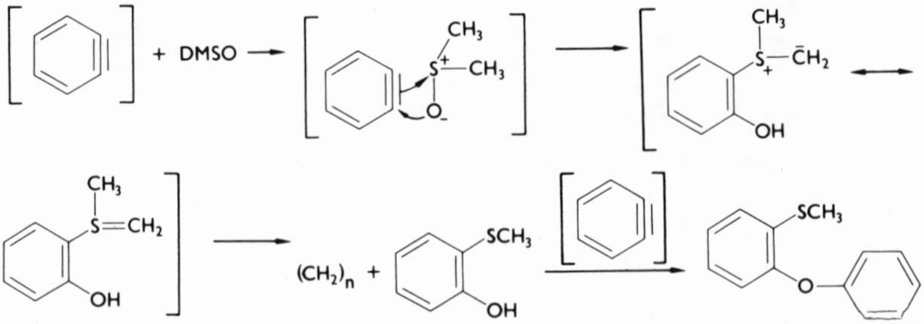

$$(8.97)$$

## 8.5.2 Basicity of S-O Bond

### 8.5.2.1 $pK_a$ Values

Because of the semipolar nature of the S-O linkage in sulfoxides, the terminal oxygen atom should be basic like that in tertiary amine oxides or the carbonyl oxygen. In fact, the oxygen atom in sulfoxides is the host site for strong hydrogen-bonding,[306] while the affinity of sulfoxides toward Lewis acids, such as $BF_3$, is larger than the corresponding sulfides or sulfones [307]

Sulfoxides have been regarded as weak bases and the basicities of dialkyl sulfoxides, for example, diethyl sulfoxide, fall in the same order of magnitude as those of acetanilide and acetamide.[308] However, there have been many controversies in the assigning of actual base strengths of common sulfoxides.

Values of $pK_a$ of common sulfoxides have been estimated either by potentiometric titration with perchloric acid in glacial acetic acid,[309,310] or by plotting the chemical shifts of methyl groups of aryl methyl or alkyl methyl sulfoxides against the $H_0$ scale,[311] or by plotting the optical densities against $H_0$ scale[312,313] or by other methods,[314a] and a few sets of different $pK_a$ values have been obtained. However, the sulfoxides are not Hammett bases and do not follow the $H_0$ acidity function. Thus, thermodynamic $pK_a$ values have recently been calculated[314b] by the following linear free energy relationship, proposed by Bunnett and

$$\log\left([BH^+]/[B] + H_0\right) = \phi\left(H_0 + \log[H^+]\right) + pK_{BH^+}$$

Olsen.[315] A new set of $pK_a$ values of several commonly occurring sulfoxides are shown in Table 8.2.

The effect of polar substituents on the basicity of the sulfinyl function is rather small and the Hammett $\rho$-value obtained from the plot of $\sigma$ values for substituted phenyl methyl sulfoxides is only $-0.85$. This means that the sulfinyl group is much less polarizable than the carbonyl group in substituted acetophenones where $pK_a$ values are correlated by $\sigma^+$ values with $\rho = -2.17$,[316] and hence are of a more semipolar nature than of $\pi$-bonding type.

Molecular association between sulfoxides and phenols by hydrogen-bonding is well known.[317,318,319] A detailed thermochemical study indicates that the heat of association between DMSO and phenol is 7 kcal/mol and twice as much as that between phenol and acetone and even slightly greater than that with DMF.[320]

Sulfoxides form 1 to 1 addition complexes with iodine,[321,322] and their Lewis basicities far exceed those of acetone and DMF, but not that of diethyl sulfide. Numerous other reagents such as $SO_3$,[323]

**TABLE 8.2**
pKa Values of Several Sulfoxides, R-SO-R'[314]

| R | R' | $pK_a$ | R | R' | $pK_a$ |
|---|---|---|---|---|---|
| CH₃O–⟨C₆H₄⟩ | CH₃ | −2.05 | O₂N–⟨C₆H₄⟩ | CH₃ | −2.96 |
| CH₃–⟨C₆H₄⟩ | CH₃ | −2.22 | ⟨C₆H₅⟩ | ⟨C₆H₅⟩ | −2.54 |
| ⟨C₆H₄(o-CH₃)⟩ | CH₃ | −2.26 | CH₃–⟨C₆H₄⟩ | ⟨C₆H₅⟩ | −2.39 |
| ⟨C₆H₄(m-CH₃)⟩ | CH₃ | −2.27 | CH₃ | CH₃ | −1.80 |
| Cl–⟨C₆H₄⟩ | CH₃ | −2.45 | | | |
| ⟨C₆H₄(o-Cl)⟩ | CH₃ | −2.61 | CH₃ | t–Bu | −1.62 |

tetracyanoethylene,[324] cobalt (II), nickel (II)[325] SnCl₃, FeCl₃, MgCl₂,[326] TeX₄[327] and alkali metal salts[328] form complexes with sulfoxides, while sulfoxides form hydrogen bonding with such weak acids as chloroform, phenylacetylene or N-methyl acetamide.[329]

Lewis base strengths of several common organic oxides are shown to fall in the following sequence[330]: $Ph_2SeO > Me_3PO > Me_2SO > (MeO)_3PO > (MeO)_2SeO > Me_2CO > AcOEt > (MeO)_2SO > Me_2SO_2 > Me(MeO)SO_2$. The unusually high Lewis basicity of dimethyl sulfoxide[331] is reflected in its unusual solvent properties.[332]

While DMSO associates strongly with hard metallic ions or acids, hard anions, such as hydroxide, alkoxides, fluoride or chloride, are poorly solvated in DMSO, which is highly polar with a dielectric constant of 48.9 (at 20°C).[333] This results in enhanced reactivity in a number of reactions involving hard nucleophiles, because the energy level of the initial state of the reactants is not much reduced by solvation in DMSO. For example, in such reactions as base-catalyzed E2 reactions,[335] carbanionic reactions,[336] nucleophilic substitutions[337] or rearrangements[338] and the Wolf-Kischner reaction,[336,339] the rates are markedly enhanced, sometimes over a million fold in DMSO and other aprotic solvents such as DMF or HMPA, and DMSO is now being widely used.[340] Care must be taken when sodium hydride is added to dimethyl sulfoxide for the preparation of dimsyl sodium, for the mixture might occasionally explode,[341] due perhaps to the spontaneous reduction of DMSO by hydride. For drying DMSO, a molecular sieve is most conveniently used.

## 8.5.2.2 Pyrolysis

Pyrolysis of alkyl or aralkyl sulfoxides is another interesting reaction caused by the basic nature of the sulfoxide oxygen atom. Kingsbury and Cram showed that, when the following compounds were heated at 80°C, predominant *cis*-elimination took place,[341] similar to the well-known Ei-type Cope elimination reaction of *t*-amine oxides (eqs. 8.98, 8.99). However these stereospecificities are substantially lost when the pyrolysis is carried out at 130°C due to the facile radical cleavage of C-S bond.[343] The corresponding sulfones do not undergo pyrolysis under the same condition. Even sulfoxides which do not possess α-benzylic or allylic group do not pyrolyze readily as compared to the corresponding *t*-amine oxides or sulfilimines.[344] Meanwhile, the predominant formation of 2-butene from *n*-propyl 2-butyl sulfoxide[345] indicate the importance of both C–α-S bond cleavage and the C–β-H attack by the sulfinyl oxygen, and suggests the reaction to be nearly an ideal Ei reaction.

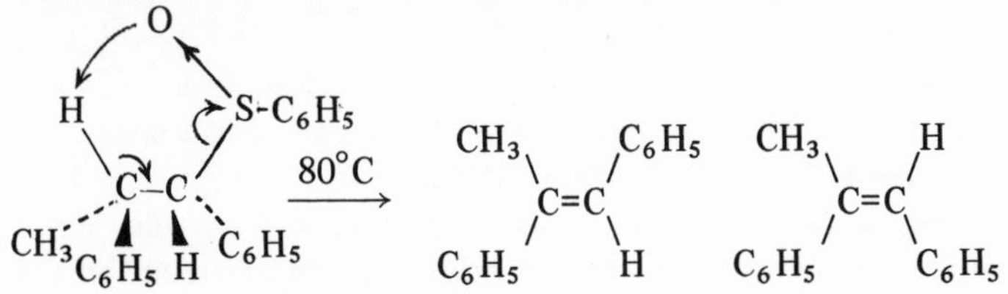

$$\text{(8.98)}$$

*erythro*      90%      10%

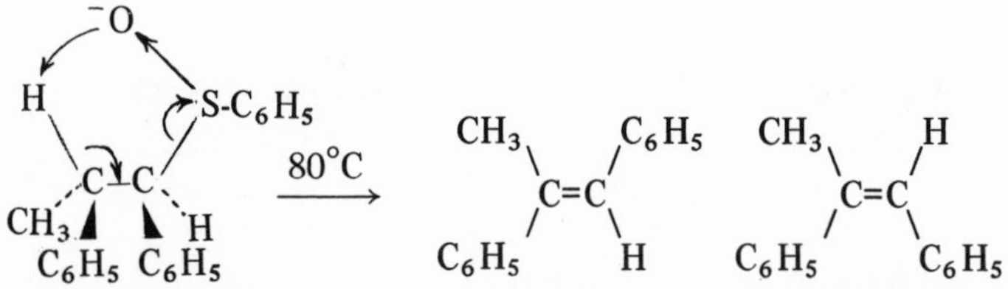

$$\text{(8.99)}$$

*threo*      10%      90%

The effect of polar substituents were examined in the pyrolysis of aryl *n*-propyl sulfoxides and a small Hammett $\rho$ value, $\rho = 0.51$, was obtained,[346] as is expected from the nearly ideal Ei-type reaction. Steric effect, however, seems to be quite important in determining the direction of pyrolytic elimination, especially with cyclic systems. The following are interesting examples illustrating this point[347]:

$$(8.100)$$

$$(8.101)$$

The rate of pyrolysis of cycloalkyl phenyl sulfoxide varies with the ring size of cycloalkyl group. Thus, the rates at $129.6°C$ fall in the following order: cycloheptyl > cyclopentyl > cyclohexyl.[348] As in the pyrolysis of sulfilimines, the distance between the sulfinyl oxygen atom and the $\beta$-hydrogen in the stable conformation of each sulfoxide appears to determine the orientation of the pyrolytic elimination.

Useful applications of this reaction combined with condensation reaction with dimethyl sulfoxide or with other reagents are found in the following examples[349-351]:

$$CH_3(CH_2)_nCH_2Br \text{ (or OTs)} + Na\overset{+}{C}\overset{-}{H}_2SOCH_3 \longrightarrow$$

$$CH_3(CH_2)_nCH_2CH_2SOCH_3 \xrightarrow{\Delta} CH_3(CH_2)_nCH=CH_2 \quad [349](8.102)$$

$$(C_6H_5)_2C=CH_2 + Na\overset{+}{C}\overset{-}{H}_2SOCH_3 \xrightarrow{DMSO} (C_6H_5)_2CHCH_2CH_2SOCH_3$$

$$\xrightarrow{\Delta} (C_6H_5)_2CHCH=CH_2 \quad (8.103)$$

$$RCH_2CHO \xrightarrow[-H_2O]{R'SH,H^+} RCH_2CH(SR')_2 \xrightarrow{[O]} RCH_2CH(SR')SOR'$$

$$\xrightarrow{\Delta} \begin{matrix} R \\ \diagdown \\ \diagup \\ R \end{matrix} C=C \begin{matrix} H \\ \diagup \\ \diagdown \\ SR' \end{matrix} \quad (8.104)$$

In some cases, not only olefins but also aldehydes were obtained by the pyrolysis of alkyl sulfoxides.[352,353] For example, thermal decomposition of di-$n$-butyl sulfoxide gives not only 1-butene, but also $n$-propionaldehyde, di-$n$-butyl sulfide, disulfide and water. Although the mechanism of aldehyde formation has not been clarified, the following scheme of proton removal and subsequent Pummerer type rearrangement can give the products:

$$CH_3CH_2CH_2CH_2\text{-}\underset{\underset{O}{\downarrow}}{S}\text{-}CH_2CH_2CH_2CH_3 \xrightarrow{\Delta} [CH_3CH_2CH_2\overset{-}{C}H\text{-}\overset{+}{\underset{\underset{OH}{|}}{S}}\text{-}C_4H_9\text{-}n$$

$$\longleftrightarrow CH_3CH_2CH_2CH=\underset{\underset{OH}{|}}{S}\text{-}C_4H_9\text{-}n \longleftrightarrow CH_3CH_2CH_2CH\overset{+}{=}\underset{\phantom{O}\overset{-}{OH}}{S}\text{-}C_4H_9\text{-}n] \longrightarrow$$

$$\longrightarrow CH_3CH_2CH_2\underset{\underset{OH}{|}}{C}H\text{-}SC_4H_9\text{-}n \xrightarrow{H_2O} CH_3CH_2CH_2CHO + HSC_4H_9\text{-}n$$

$$(8.105)$$

Formation of unstable sulfenic acid has been postulated for the pyrolysis of alkyl sulfoxides and, in fact, was detected by nmr and ir spectroscopic analysis. Sulfenic acid was successfully trapped by allowing it to react with olefinic or acetylenic compounds as shown below[354]:

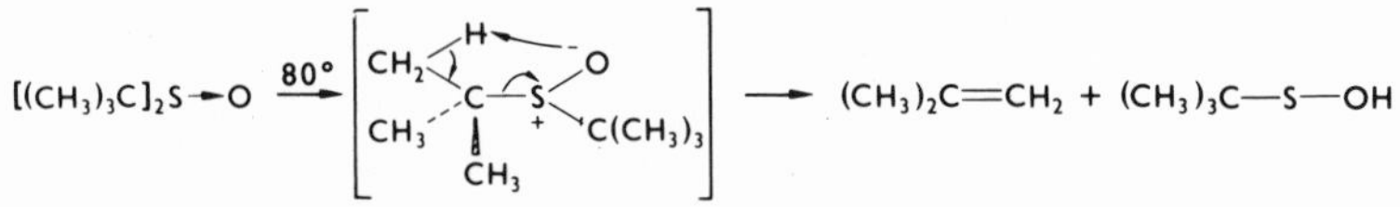

$$(CH_3)_3C\text{-}S\text{-}OH + CH_2=CHCOOEt \longrightarrow (CH_3)_3C\overset{O}{\overset{\uparrow}{\text{-}S}}\text{-}CH_2CH_2COOEt$$

$$(CH_3)_3C\text{-}S\text{-}OH + CH\equiv CCO_2Me \longrightarrow (CH_3)_3C\overset{O}{\overset{\uparrow}{\text{-}S}}\text{-}CH=CHCO_2Me$$

$$(8.106)$$

Episulfoxides can be obtained readily now by the oxidation of episulfides.[355-358] The cyclic sulfoxides undergo facile pyrolysis upon heating and no clean stereospecificity was found, for example, in the pyrolysis of 2-butene episulfoxides, as shown below; the pyrolysis is believed to be an E1-type reaction[356]:

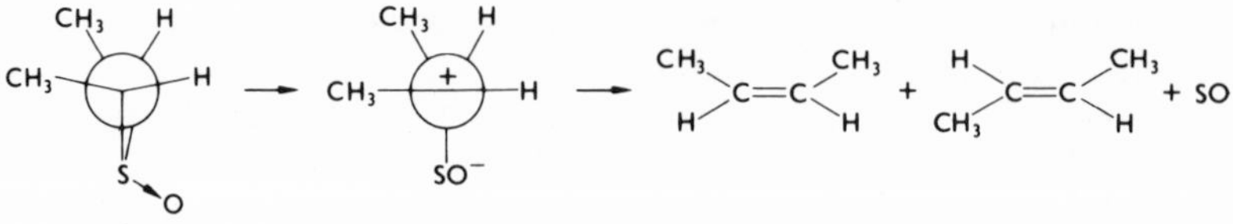

*cis* isomer                          89%           11%    (8.107)

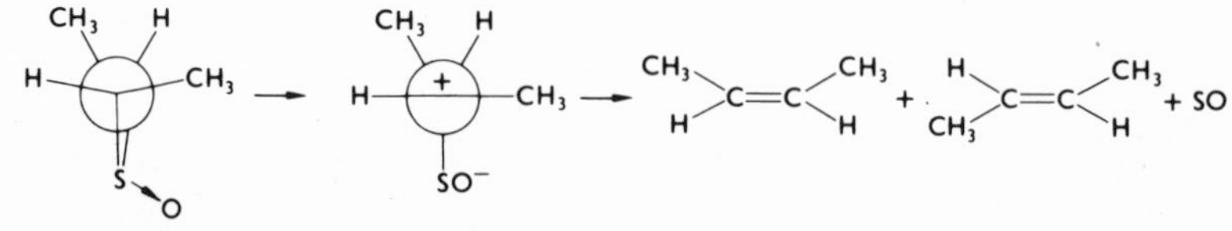

*trans* isomer                        42%           58%    (8.108)

An interesting variation of this reaction may be found in the deuterium incorporation reaction of penicillin sulfoxide (eq. 8.109).[358] Apparently, there is an equilibrium between the sulfoxide and the sulfenic acid which undergoes hydrogen isotopic exchange.

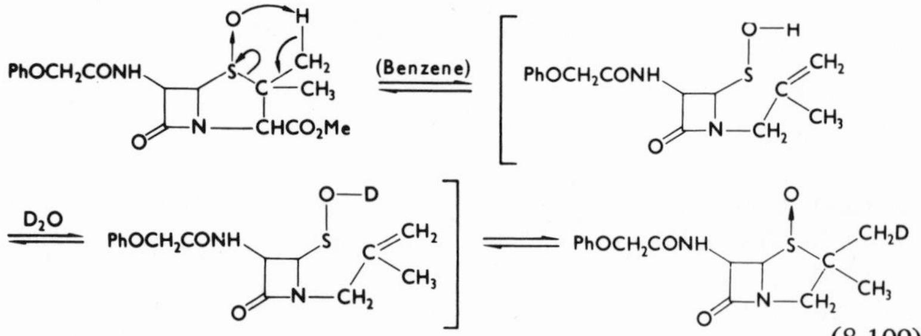

(8.109)

### 8.5.2.3 Neighboring Group Effect

The nucleophilic nature of the sulfinyl group can be seen in the following few examples of neighboring group participation. Treatment of the following cyclic ketosulfoxide with perchloric acid gives the O-alkyl product (A) and not the S-alkyl product (B),[359,360] while the bridging oxygen atom is shown to originate from the sulfinyl oxygen on the basis of an [18]O-tracer experiment[359]:

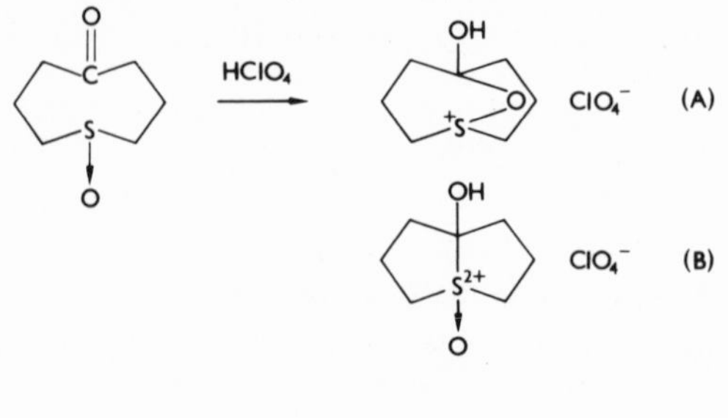

(8.110)

The rate of solvolysis of *trans*-4-chlorothiane-1-oxide in 50% aqueous ethanol is 650 times greater than that of the *cis*-isomer, which undergoes normal unassisted $S_N1$ type solvolysis[361] :

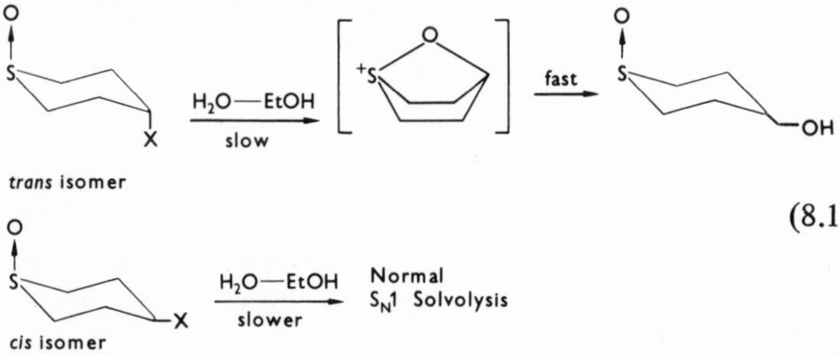

*trans* isomer

(8.111)

*cis* isomer

The facile iodolactonization of the following *syn*-ester also confirms the neighboring group participation of the sulfinvl oxygen[362] :

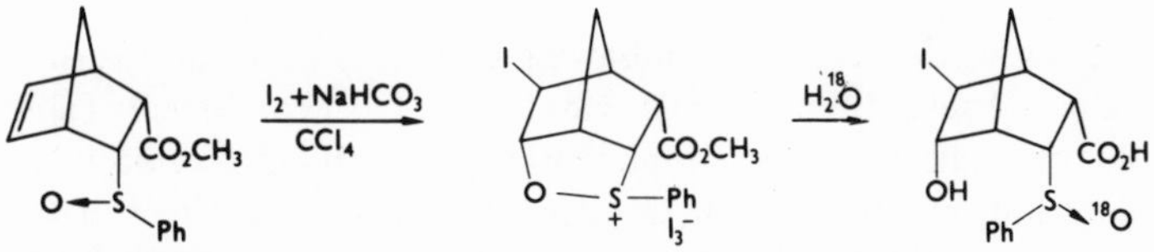

(8.112)

The rates of solvolysis of $PhSO(CH_2)_n$ $CMe_2Cl$ (n = 1, 2, 3, 4) and $PhSOCH_2$-$CH_2CRR'Cl$ (R,R' = H,H; Me,H; Me,Me; *t*-Bu,H) measured at 35°C in 80% DMF or 80% sulfolane were found to fall in the following order: $\delta > \gamma > \beta > \epsilon \leq$ *t*-Bu,[363] similar to the $S_N1$ reactivity sequence of $PhCO(CH)_n$ - Br series,[364] indicating the favorable neighboring group participation by $\delta$- and $\gamma$-sulfinyl groups.

## 8.5.3 Electron-Withdrawing Effect of SO Group

The polar effect of sulfinyl and sulfonyl groups has been of interest to organic chemists for some time. Shriner *et al.* showed that bis(phenylsulfinyl)methane (A) is not soluble in strong alkaline solution, but bis(phenylsulfonyl)methane (B) is soluble.[365]

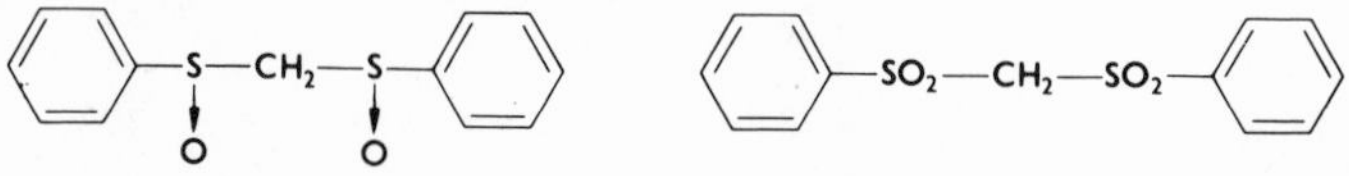

A                                B

The $pK_a$ values of the following three compounds[336] also indicate the order of the magnitudes of the polar effects of sulfenyl, sulfinyl and sulfonyl functions.

|     | C | D | E |
|-----|---|------|------|
| pK$_a$ |   | 3.12 | 2.72 |

However, the facile proton exchange between dimethyl sulfoxide and dimsyl sodium or lithium,[367] the complete deuteration of DMSO in alkaline $D_2O$,[368] and the apparently easier proton exchange of sulfoxide compared to sulfone[369] make it appear as though the sulfinyl group is unique in stabilizing the $\alpha$-carbanion.

### 8.5.3.1 pK$_a$ Values of Sulfinyl Benzoic Acids and Phenols

For comparison of the electron-withdrawing effects of polar substituents, the pK$_a$ values of substituted benzoic acids are the most convenient. Table 8.3 lists the pK$_a$ values of several representative substituted benzoic acids, and the data indicate clearly that the electron-withdrawing effect of the sulfinyl group is substantial, but not as large as those of the corresponding sulfonyl and nitro groups. Table 8.4 lists the pK$_a$ values of substituted phenols. The higher acidity of $p$-sulfinyl or $p$-sulfonylphenols as compared to the corresponding $m$-isomer is due to the electron-pair-accepting conjugation involving a $3d$ orbital of the sulfur atom:

TABLE 8.3

pK$_a$ Values of a Few Substituted Benzoic Acids in 50% Ethanol at 25°C

| Substituent | pK$_a$ | $\sigma$ | Reference |
|-------------|--------|----------|-----------|
| H | 5.73 | 0.00 | 370 |
| $p$-NO$_2$ | 4.53 | 0.82 | 371 |
| $m$-NO$_2$ | 4.66 | 0.72 | 371 |
| $p$-(CH$_3$)$_3$N$^+$ | 4.42 | 0.88 | 372 |
| $m$-(CH$_3$)$_3$N$^+$ | 4.22 | 1.02 | 372 |
| $p$-CH$_3$SO$_2$ | 4.68 | 0.75 | 370,373 |
| $m$-CH$_3$SO$_2$ | 4.78 | 0.65 | 370,373 |
| $p$-CH$_3$SO | 5.01 | 0.51 | 373,374 |
| $m$-CH$_3$SO | 4.90 | 0.48 | 373,374 |
| $p$-C$_6$H$_5$SO | 4.97 | 0.47 | 375 |
| $p$-C$_6$H$_5$SO$_2$ | 4.63 | 0.70 | 375 |

Such an electron-pair-accepting conjugation effect of the sulfonyl group is also greater than that of the sulfinyl group, as can be seen from the data of $\sigma_p$-$\sigma_m$ and also $\sigma_{OH}$-$\sigma_{COOH}$.

TABLE 8.4

$pK_a$ Values of $m$-, $p$-Methylsulfinyl- and Methylsulfonylphenols in $25°C$ Water[370,374]

| Substituent | $pK_a$ | $\sigma$ | $\sigma_p$-$\sigma_m$ | $\sigma_{OH}$-$\sigma_{COOH}$ |
|---|---|---|---|---|
| $m$-CH$_3$SO | 8.75 | +0.53 | | 0.05 |
| $p$-CH$_3$SO | 8.28 | +0.73 | +0.20 | 0.22 |
| $m$-CH$_3$SO$_2$ | 8.40 | +0.70 | | 0.05 |
| $p$-CH$_3$SO$_2$ | 7.83 | +0.98 | +0.28 | 0.23 |

The operation of the $3d$-orbital resonance effect may be confirmed by the lack of stereo sensitivity in the acidity constants of hindered sulfinyl phenols shown in Table 8.5, in which the calculated values of the Hammett

TABLE 8.5

$pK_{a_1}$ Values of A Few Hindered Bis(4-Hydroxyphenyl)sulfoxides in 50% Aqueous Ethanol at $25°C$[377]

| Phenol | $pKa_1$ | $\sigma$ (obs.) | $\sigma$ (calc.) |
|---|---|---|---|
| 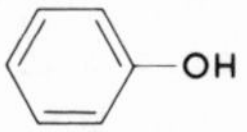 ⟨benzene⟩—OH | 11.22 | 0 | 0 |
| 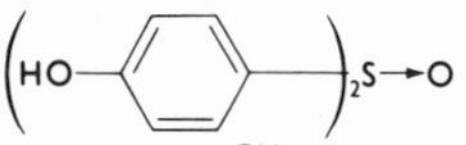 (HO—⟨benzene⟩—)$_2$S→O (F), CH$_3$ | 9.09 | 0.78 | − |
| 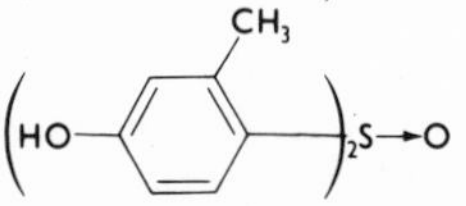 (HO—⟨benzene⟩—)$_2$S→O (G), CH$_3$ | 9.28 | 0.71 | 0.71* |
| 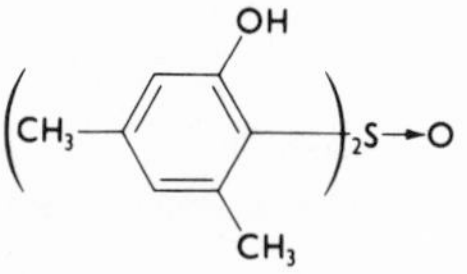 (CH$_3$—⟨benzene⟩—)$_2$S→O (H), OH, CH$_3$ | 9.14 | − | − |
| 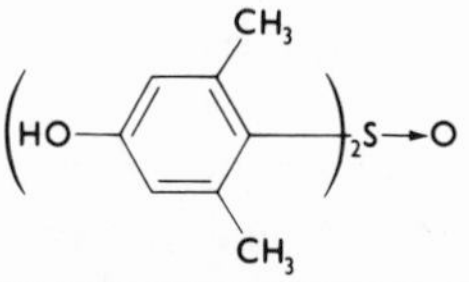 (HO—⟨benzene⟩—)$_2$S→O (I), CH$_3$, CH$_3$ | 9.57 | 0.61 | 0.64** |

*The observed (obs.) $\sigma$ for (F) + $\sigma$ $m$-CH$_3$ (−0.03).
**The observed $\sigma$ for (F) + 2$\sigma$ $m$-CH$_3$(−0.03).

σ constants are shown to be nearly identical to the observed values.[376,377] This is quite different from that of hindered phenol, e.g., 3,5-dimethyl-4-nitrophenol whose $pK_a$ (obs.) is quite small as compared to the calculated value, as shown below. Sterically small p-cyano- and stereo-insensitive p-sulfonyl-substituted phenols give nearly identical values for both calculated and observed σ.[378]

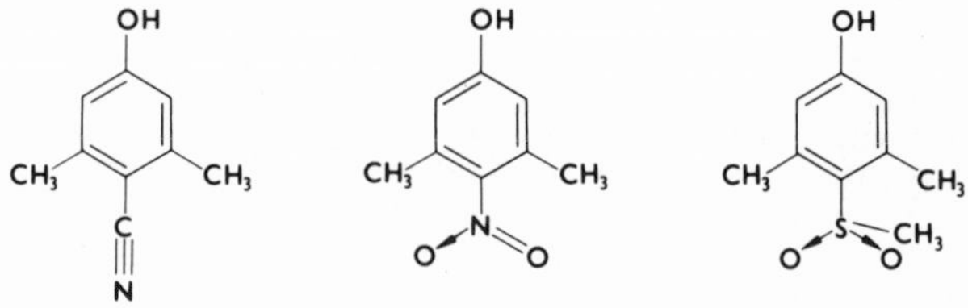

|            |  8.21  |  8 25  |  8 13  |
|------------|--------|--------|--------|
| $pK_a$     |  8.21  |  8 25  |  8 13  |
| σ (obs.)   |  0 79  |  0.77  |  0.83  |
| σ (calc.)  |  0.77  |  1.14  |  0.83  |

The electron-withdrawing effects of sulfinyl and sulfonyl groups appear to be enhanced in the excited state. Thus, the $pK_a$ value of photoexcited p-methanesulfinylphenol is substantially less than that of the m-isomer, and is affected very little by the substitution of two methyl groups at the 2 and 6 positions, ortho to the sulfinyl group.[379] The situation is somewhat different for the sulfenyl group in which the conjugative interaction is of p,π type and which is affected by the steric hindrance of the two ortho methyl groups. The nmr data also support this.[380]

### 8.5.3.2 Activating Effect of Sulfinyl Group on Aromatic Nucleophilic Substitution

Earlier, it was shown that p-iodophenyl phenyl sulfoxide is hydrolyzed by alkali under conditions where the meta-isomer is unaffected.[381] More quantitative data on the alkaline ethanolysis of m- and p-benzenesulfinyl- and sulfonylchlorobenzenes are listed in Table 8.6.[382,383] The data reveal that the activating effect of the sulfonyl group is substantially higher than that of the sulfinyl group.

However, the markedly greater relative reactivities of the p-isomers of both sulfoxides and sulfones support the view that the following 3d-orbital resonance is in operation to stabilize the incipient addition complex in the hydrolysis, as shown in eq. 8.113.

**TABLE 8.6**

Relative Rates of Hydrolysis of Chlorophenyl Phenyl Sulfoxides and Sulfones with
Hydroxide Ion in Aqueous DMSO, [DMSO] (2) : $H_2O$ (1), V/V Ratio at 158°C

| Compound | Relative Rate | Relative Rate within the Isomers |
|---|---|---|
| 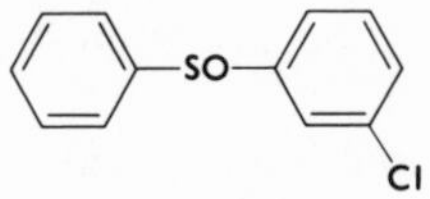 | 1 | 1 |
| 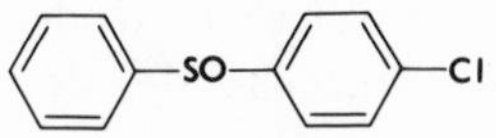 | 81 | 81 |
| 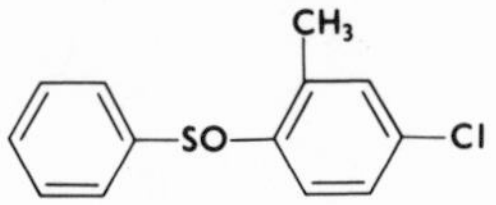 | 59 | 67 |
| 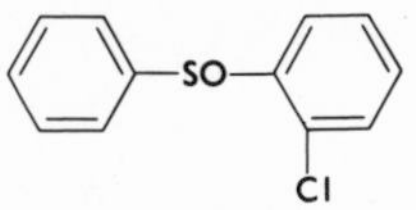 | 66 | 66 |
| 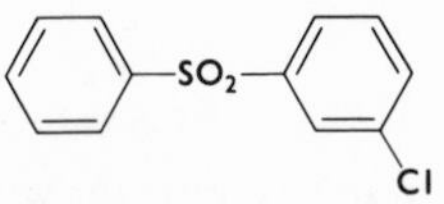 | 450 | 1 |
| 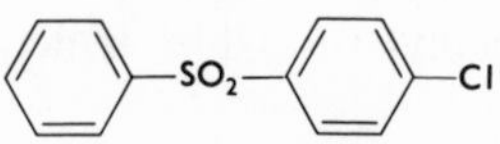 | 9,000 | 20 |
| 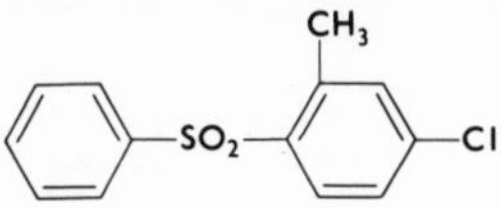 | 4,740 | 11 |
| 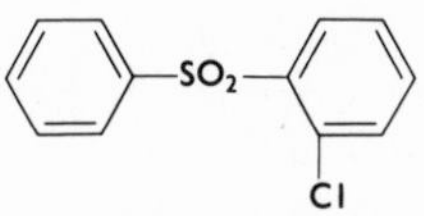 | 4,800 | 10 |

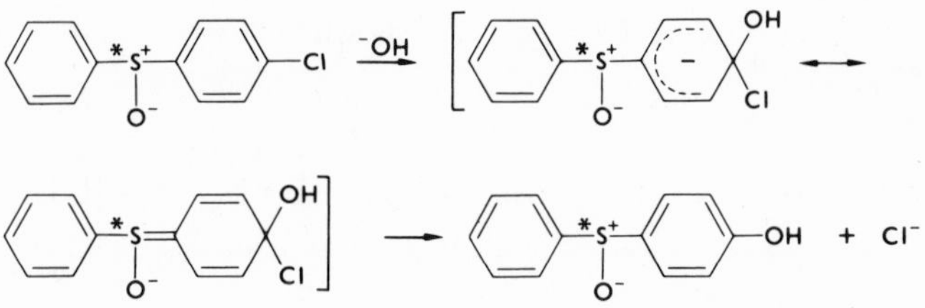

(8.113)

The fact that optically active $p$-benzenesulfinylphenol was found to retain its original optical activity after the hydrolysis of optically active $p$-chlorophenyl phenyl sulfoxide is also in keeping with the stereo-insensitivity of the $3d$-orbital resonance.[382] The fact that the rather small retardation of the rates by an *ortho* methyl group is due to the electron-releasing inductive effect, and certainly not because of the steric inhibition of resonance by *ortho* methyl groups, also supports the operation of a $3d$-orbital resonance effect, as shown before.

A retension of original optical activity was observed in the process of complete deuteration of aryl methyl sulfoxide with alkali ethoxide in deuteroethanol,[384] as shown below, evidently confirming the stereo-insensitivity of $3d$-orbital resonance of sulfinyl group:

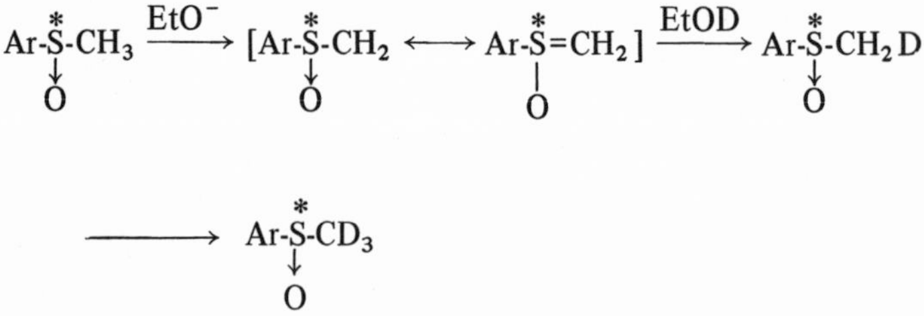

$$\overset{*}{\underset{\underset{O}{\downarrow}}{Ar\text{-}S}}\text{-}CH_3 \xrightarrow{EtO^-} [\overset{*}{\underset{\underset{O}{\downarrow}}{Ar\text{-}S}}\text{-}CH_2 \longleftrightarrow \overset{*}{\underset{\underset{O}{|}}{Ar\text{-}S}}=CH_2] \xrightarrow{EtOD} \overset{*}{\underset{\underset{O}{\downarrow}}{Ar\text{-}S}}\text{-}CH_2D$$

$$\longrightarrow \overset{*}{\underset{\underset{O}{\downarrow}}{Ar\text{-}S}}\text{-}CD_3 \qquad\qquad (8.114)$$

Only under very drastic conditions, such as in the treatment with alkali metal for a prolonged period, do aryl methyl sulfoxides undergo racemization.[385]

### 8.5.3.3  α-Sulfinyl Carbanions and Their Reactions

Because of the acidic nature of methylene proton $\alpha$ to a sulfinyl group ($pK_a$ of DMSO: $31 \cdot 3$),[386] an $\alpha$-sulfinyl carbanion is readily formed upon treatment with a strong base. A typical example is the formation of dimsyl anion by the treatment of DMSO either with sodium hydride or with sodium amide[387] (or with potassium $t$-butoxide,[388.5] in which case the equilibrium constant, $K$, for the reaction, $t$-BuOK + Me$_2$SO $\rightleftharpoons$ $t$-BuOH + MeSOCH$_2$K, was found to be $7.7 \times 10^{-3}$), as shown below:

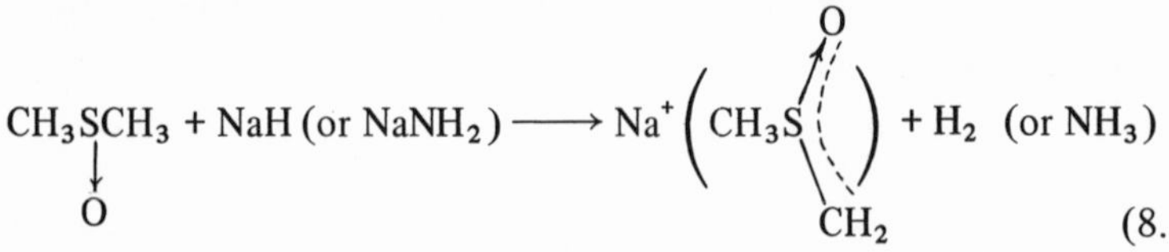

$$\underset{\underset{O}{\downarrow}}{CH_3SCH_3} + NaH \,(or\; NaNH_2) \longrightarrow Na^+ \left( CH_3\overset{\displaystyle O}{\underset{\displaystyle CH_2}{S}} \right) + H_2 \;(or\; NH_3)$$

$$\qquad\qquad\qquad\qquad\qquad\qquad\qquad\qquad\qquad\qquad (8.115)$$

Dimsyl sodium thus formed can react readily with various agents to afford useful chemical reagents. A few interesting reactions are shown below:

a) with carbonyl compounds[387,388,389,390,391]

$$CH_3SO\bar{C}H_2 + R_1\text{-}\overset{\displaystyle O}{\overset{\|}{C}}\text{-}R_2 \longrightarrow \underset{R_2}{\overset{R_1\ OH}{\underset{|}{\overset{\backslash\ |}{C}}}}\text{-}CH_2SOCH_3 \qquad (8.116)$$

b) with esters[388,393,394]

$$CH_3SO\bar{C}H_2 + R_1\text{-}\overset{\displaystyle O}{\overset{\|}{C}}\text{-}OR \longrightarrow R_1\text{-}\overset{\displaystyle O}{\overset{\|}{C}}\text{-}CH_2SOCH_3 \xrightarrow{\text{Al-Hg}} R_1\text{-}\overset{\displaystyle O}{\overset{\|}{C}}\text{-}CH_3$$

$$(8.117)$$

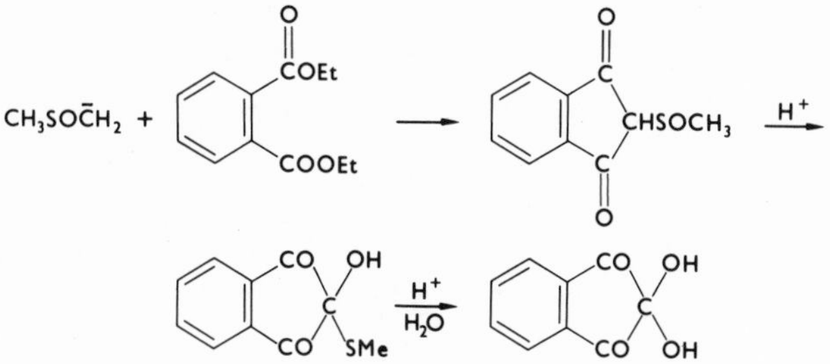

$$(8.118)$$

c) with imines[389]

$$CH_3SO\bar{C}H_2 + R_1R_2C{=}N\text{-}Ph \longrightarrow \underset{R_2\ NHPh}{\overset{R_1}{\underset{/|}{\overset{\backslash}{C}}}}\text{-}CH_2SOCH_3 \qquad (8.119)$$

d) with benzyne[387]

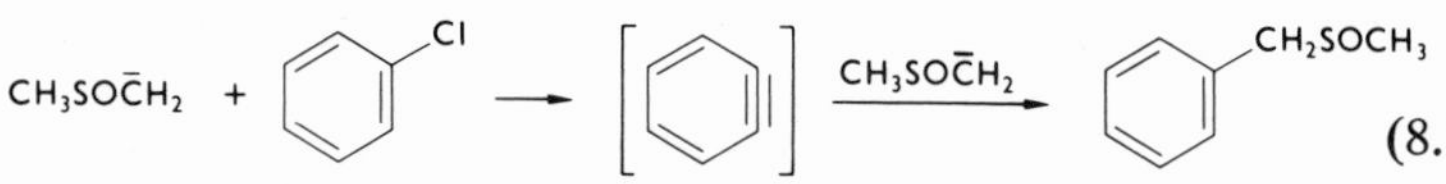

$$(8.120)$$

e) with carbenes[387]

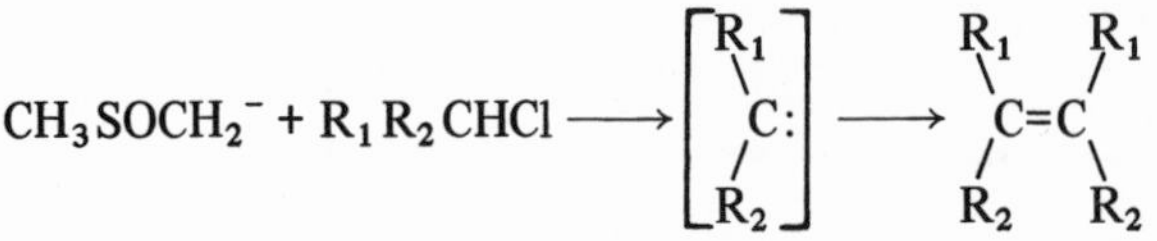

$$CH_3SOCH_2^- + R_1R_2CHCl \longrightarrow \begin{bmatrix} R_1 \\ C: \\ R_2 \end{bmatrix} \longrightarrow \begin{matrix} R_1 \quad R_1 \\ C=C \\ R_2 \quad R_2 \end{matrix} \qquad (8.121)$$

f) with olefins[395,396,397]

$$CH_3SOCH_2^- + CH_2=C(C_6H_5)_2 \rightleftharpoons CH_3SOCH_2CH_2^-C(C_6H_5)_2$$

$$(8.122)$$

$$\xrightarrow{DMSO} CH_3SOCH_2CH_2CH(C_6H_5)_2 \xrightarrow[Ei]{\Delta} CH_2=CH\text{-}CH(C_6H_5)_2$$

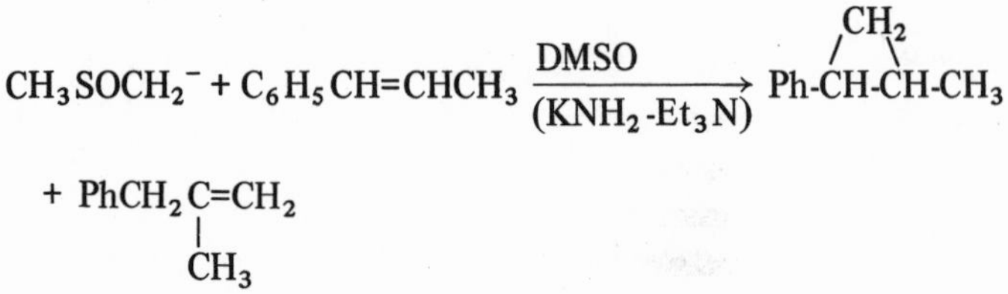

$$CH_3SOCH_2^- + C_6H_5CH=CHCH_3 \xrightarrow[(KNH_2\text{-}Et_3N)]{DMSO} Ph\text{-}\overset{\displaystyle CH_2}{\overset{/\quad\backslash}{CH\text{-}CH}}\text{-}CH_3 \qquad (8.123)$$

$$+ \underset{\underset{CH_3}{|}}{PhCH_2C}=CH_2$$

g) with acetylenes[398]

$$CH_3SOCH_2 \overline{\cdot} + C_6H_5C{\equiv}CC_6H_5 \longrightarrow CH_3SOCH_2\text{-}\overset{\overset{\displaystyle Ph}{|}}{C}=CHPh \qquad (8.124)$$

h) with aldehyde and diphenylmethane[391,399]

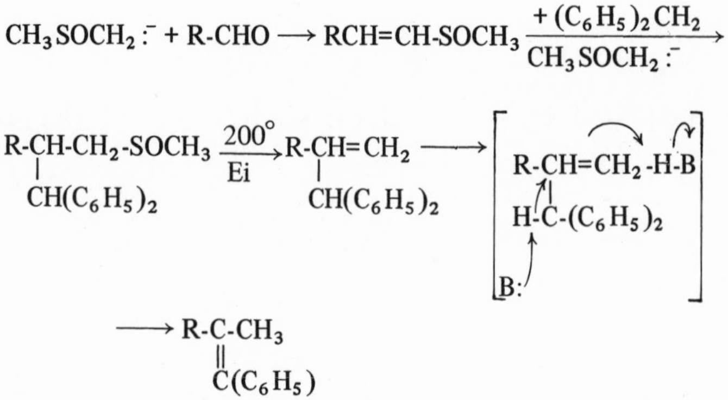

$$CH_3SOCH_2 \overline{\cdot} + R\text{-}CHO \longrightarrow RCH=CH\text{-}SOCH_3 \xrightarrow[CH_3SOCH_2\overline{\cdot}]{+ (C_6H_5)_2CH_2}$$

$$\underset{\underset{CH(C_6H_5)_2}{|}}{R\text{-}CH\text{-}CH_2\text{-}SOCH_3} \xrightarrow[Ei]{200°} \underset{\underset{CH(C_6H_5)_2}{|}}{R\text{-}CH=CH_2} \longrightarrow \begin{bmatrix} R\text{-}CH=CH_2\text{-}H\text{-}B \\ H\text{-}C\text{-}(C_6H_5)_2 \\ B: \end{bmatrix} \qquad (8.125)$$

$$\longrightarrow \underset{\underset{C(C_6H_5)}{||}}{R\text{-}C\text{-}CH_3}$$

i) with polycyclic aromatic hydrocarbons (*i.e.*, anthracene, phenanthrene, quinoline, acridine, etc.[400,401])

$$(8.126)$$

j) with alkyl halides or tosylates[402]

$$CH_3SOCH_2:^- + ROTs \xrightarrow{\text{DMSO}} R\text{-}CH_2SOCH_3 \xrightarrow{\text{Ni/H}_2} R\text{-}CH_3 \qquad (8.127)$$

Thus, dimsyl anion is now as useful as diethylmalonate anion in introducing a methylene linkage into an organic compound and is in great demand. To meet the demand, a convenient method of preparing dimsyl sodium that lasts more than two months in active form was developed by applying ultrasonic vibration during the reaction of DMSO and NaH in mineral oil.[403]

The dianion of dimethyl sulfoxide was suggested to be formed by the treatment of DMSO with alkali metal amide in liquid ammonia, since treatment of the solution with alkyl halide gave dealkylated sulfoxide while treatment with benzophenone gave the diol[404]:

$$(8.128)$$

$$\text{R-CCH}_2\text{SCH}_3 \overset{O}{\underset{}{\parallel}} \overset{O}{\underset{}{\uparrow}}$$

$$\xrightarrow{H^+} \quad R\text{-}\overset{O}{\overset{\parallel}{C}}\text{-}\underset{\underset{OH}{|}}{CH}\text{-}SCH_3 \xrightarrow{H_2O} R\text{-}\overset{O}{\overset{\parallel}{C}}\text{-}CHO + CH_3SH$$

$$\xrightarrow{LiAlH_4} \quad R\text{-}\underset{\underset{OH}{|}}{CH}\text{-}CH_2\text{-}\overset{O}{\overset{\uparrow}{S}}CH_3 \xrightarrow{-H_2O} RCH=CH\text{-}\overset{O}{\overset{\uparrow}{S}}CH_3$$

$$\xrightarrow{BrCH_2CO_2Et} \quad R\text{-}\overset{O}{\overset{\parallel}{C}}\text{-}\underset{\underset{CH_2CO_2Et}{|}}{CH}\overset{O}{\overset{\uparrow}{S}}CH \xrightarrow{Zn\text{-}HOAc} R\text{-}\overset{O}{\overset{\parallel}{C}}\text{-}CH_2CH_2\text{-}CO_2Et$$

$$\xrightarrow{CH_2=CH\text{-}CN} \quad R\text{-}\overset{O}{\overset{\parallel}{C}}\text{-}\underset{\underset{CH_2CH_2\text{-}CN}{|}}{CH}\text{-}SOCH_3$$

$$\xrightarrow[CO_2Et]{CH_2=CH\text{-}} \quad R\text{-}\overset{O}{\overset{\parallel}{C}}\text{-}\underset{\underset{CH_2CH_2\text{-}CO_2Et}{|}}{CH}\text{-}SOCH_3 \xrightarrow{Zn\text{-}HOAc} R\text{-}\overset{O}{\overset{\parallel}{C}}CH_2CH_2CH_2CO_2Et$$

$$\xrightarrow[ClC\text{-}OEt]{\overset{O}{\overset{\parallel}{}}} \quad R\text{-}C=\underset{\underset{OCO_2Et}{|}}{CH}SOMe$$

$$\xrightarrow{AcCl} \quad R\text{-}\underset{\underset{OAc}{|}}{C}=CH\text{-}SOCH_3 \xrightarrow{Zn\text{-}HOAc} R\text{-}\underset{\underset{OAc}{|}}{C}=CH_2$$

$$\xrightarrow{R'X} \quad R\text{-}\underset{\underset{OR'}{|}}{C}=CH\text{-}SOCH_3 \xrightarrow{Zn\text{-}HOAc} R\text{-}\underset{\underset{OR'}{|}}{C}=CH_2$$

$$(8.129)$$

$\beta$-Oxosulfoxides, formed in the reaction with esters are convenient starting compounds for the syntheses of a variety of derivatives, as shown above (eq. 8.129).[394,405-407] The facile reactions are mainly due to the strong acidic nature of the methylene protons sandwiched between the carbonyl and sulfinyl groups.

The initial step of the following unusual condensation reaction to form cyclopropyl sulfoxide is considered to involve the formation of an $\alpha$-sulfinyl carbanion[408]:

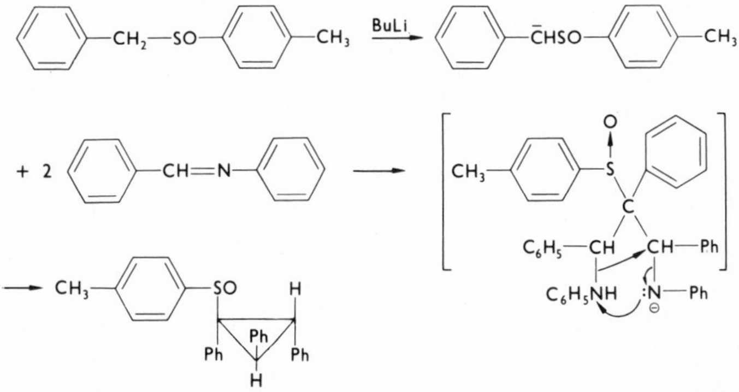

(8.130)

The formation of stilbene in the treatment of benzylic sulfoxides with phenyllithium is another reaction involving an $\alpha$-sulfinyl carbanion[409]:

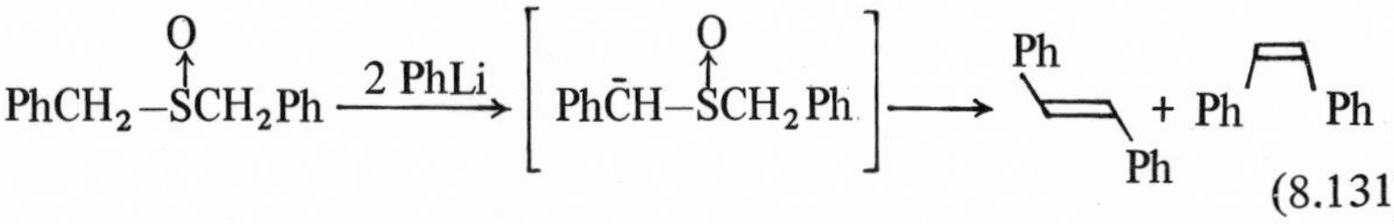

(8.131)

The electron-withdrawing effect of the sulfinyl group may be seen in the facile nucleophilic additions shown below[410,411]:

$$CH_2=CH-\underset{\underset{O}{\downarrow}}{S}-CH=CH_2 \xrightarrow[CH_3OH]{NaOCH_3} CH_3OCH_2CH_2\underset{\underset{O}{\downarrow}}{S}CH_2CH_2OCH_3$$

(8.132)

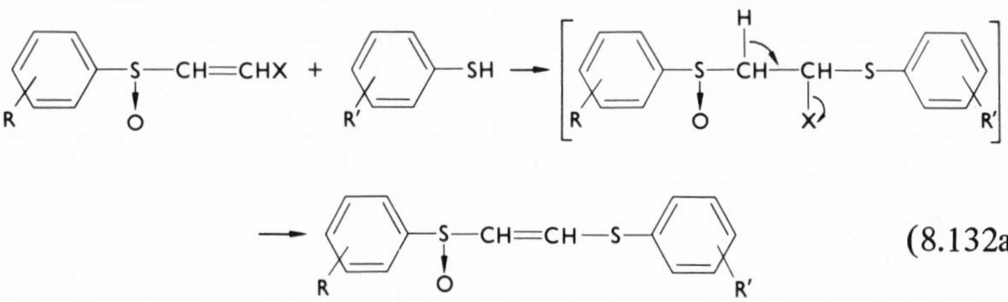

(8.132a)

An interesting example is the following reaction[412] in which isomerization precedes the addition (eq. 8.133). In the reaction with

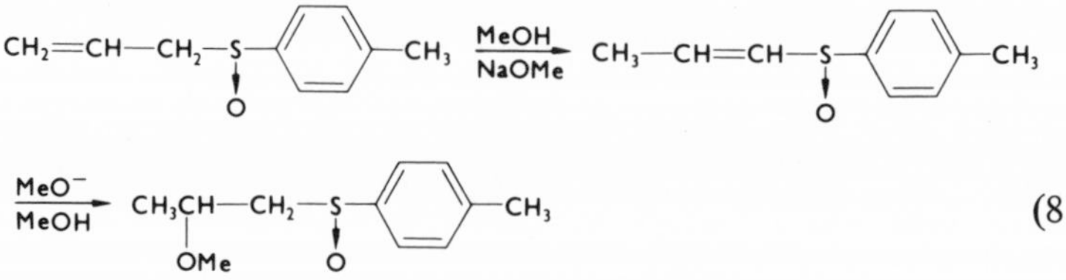

$$CH_2=CH-CH_2-\overset{\underset{O}{|}}{S}-\text{〈aryl〉}-CH_3 \xrightarrow{\underset{NaOMe}{MeOH}} CH_3-CH=CH-\overset{\underset{O}{|}}{S}-\text{〈aryl〉}-CH_3$$

$$\xrightarrow{\underset{MeOH}{MeO^-}} CH_3\underset{\underset{OMe}{|}}{CH}-CH_2-\overset{\underset{O}{|}}{S}-\text{〈aryl〉}-CH_3 \qquad (8.133)$$

piperidine, however, prior conversion to the sulfenate ester and subsequent nucleophilic replacement was observed to afford allyl alcohol and toluenesulfenylpiperidide.[412]

Vinyl sulfoxide is a dienophile, though not as good as vinyl sulfone, and reacts with cyclopentadiene[413] upon heating, $\alpha,\beta$-bis(sulfinyl)ethylene is a more reactive dienophile,[414] while $\beta$-sulfinylacrylic acid is a good dienophile which readily adds to cyclopentadiene.[415]

Though the methanesulfinyl group is electron-withdrawing, it is not a good leaving group, and hence the following elimination reaction (eq. 8.134) is believed to be nearly a E1cB-type reaction[416] on the basis of a large Hammett $\rho$ value, $-4.4$, and a small kinetic isotope effect for $\beta$-hydrogen removal, $k_H/k_D = 2.7$.

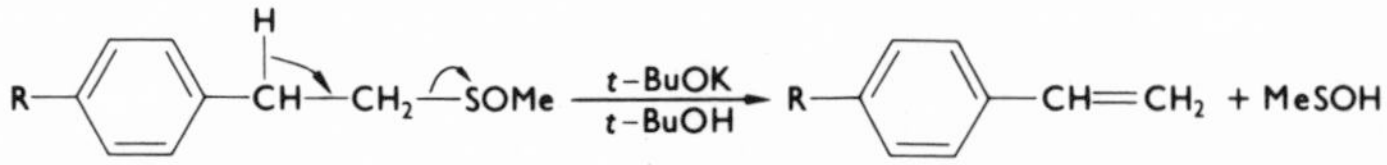

$$R-\text{〈aryl〉}-\overset{\overset{H}{|}}{CH}-CH_2-SOMe \xrightarrow{\underset{t-BuOH}{t-BuOK}} R-\text{〈aryl〉}-CH=CH_2 + MeSOH$$

$$(8.134)$$

The amount of 1,3-elimination to form cyclopropane increases with increasing electron-releasing property and bulk of the alkyl group in the following reaction[417]:

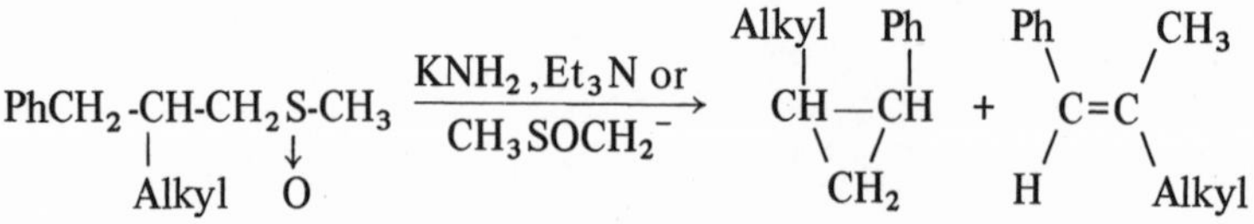

$$PhCH_2-\underset{\underset{Alkyl}{|}}{CH}-CH_2\underset{\underset{O}{\downarrow}}{S}-CH_3 \xrightarrow[CH_3SOCH_2^-]{KNH_2,Et_3N\text{ or}} \underset{\underset{CH_2}{\diagdown\diagup}}{\overset{\overset{Alkyl\quad Ph}{|\qquad|}}{CH-CH}} + \overset{\overset{Ph\quad CH_3}{\diagdown\;\diagup}}{\underset{\underset{H\qquad Alkyl}{\diagup\;\diagdown}}{C=C}}$$

$$trans$$

$$(8.135)$$

### 8.5.3.4 Hydrogen Iostopic Exchange and α-Halogenation

Because of the acidic nature of the methylene protons α- to a sulfinyl group, base-catalyzed hydrogen isotopic exchange of the α-sulfinyl protons proceeds very readily. However, there is a conformational preference in the deuteration.[166] Thus, one of the methylene protons of benzyl p-chlorophenyl sulfoxide undergoes H-D exchange faster than the other to result in stereo-selective deuteration,[418,419] however, the conformational preference depends very much on the solvent used.[420,421]

The stereochemistry of the base-catalyzed hydrogen isotope exchange of the conformationally rigid cis- and trans-4-phenyltetrahydrothiopyrane-1-oxides was studied in detail,[422] and it was revealed that the reaction is stereoselective in water and methanol-1-d, but nonstereoselective in t-butanol-1-d and DMSO-methanol mixture.

The order of proton acidity adjacent to the sulfinyl group is concluded to be (a) trans to the sulfinyl group and gauche to the sulfur lone pair, (b) gauche to the SO group and the sulfur lone pair, (c) gauche to the SO group and trans to the sulfur lone pair. A similar trend was observed in the following conformationally frozen sulfoxide in which the relative rates are $H_1 : H_2 : H_3 : H_4 = 1:1100:310:1250$.[423]

This is in keeping with the theoretical consideration that adjacent electron pairs and polar bonds tend to exist in the structure which maximizes their gauche interaction,[424] while the repulsion between the lone pair of the carbanion and the sulfoxide oxygen is minimized at the trans conformation.

α-Halogenation of alkyl sulfoxides is another interesting reaction associated with α-sulfinyl carbanions which has been studied extensively. The reaction with nitrosyl chloride in the presence of pyridine in chloroform (eq. 8.136),[425] the treatment of p-tolyl methyl sulfoxide with

$$R\text{-}\overset{O}{\underset{}{\overset{\uparrow}{S}}}\text{-}CH_2R' + NOCl \xrightarrow[\text{CHCl}_3]{\text{Pyridine}} R\text{-}\overset{O}{\underset{\underset{Cl}{|}}{\overset{\uparrow}{S}}}\text{-}CHR' \tag{8.136}$$

$$p\text{-}CH_3C_6H_4SOCH_3 + C_6H_5SO_2Cl \xrightarrow{\text{Pyridine}} p\text{-}CH_3C_6H_4SOCH_2Cl$$

$$\tag{8.137}$$

benzenesulfonyl chloride in the presence of pyridine (eq. 8.137),[426] the α-chlorination by t-butyl hypochlorite[427] and by sulfuryl chloride[428] of benzyl phenyl sulfoxide in the presence of potassium acetate, the α-chlorination of DMSO by N-chlorosuccinimide[429] are but a few examples. α-Chlorination of benzyl t-butyl sulfoxide,[421] phenyethyl phenyl sulfoxide[430] and thiolane-1-oxide was also achieved by treating them directly with chlorine in the presence of such a base as pyridine.

α-Bromination is also achieved similarly. Thus, p-tolyl methyl sulfoxide and thiolane-1-oxide are readily monobrominated with a mixture of bromine and N-bromosuccinimide in the presence of pyridine (eq. 8.138).[431] In the presence of pyridine, bromine alone also effects α-bromination similarly.[432]

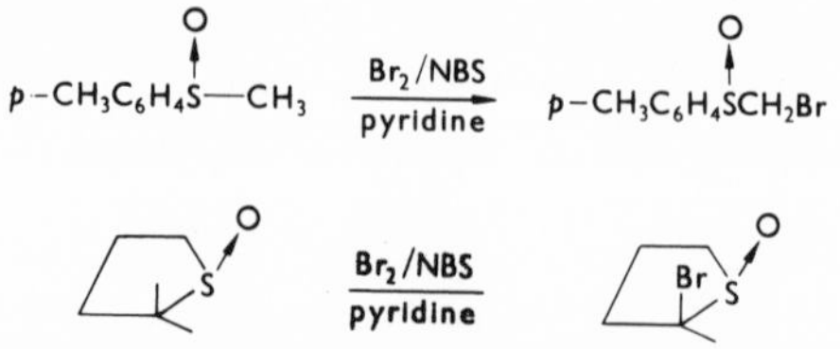

$$p\text{-}CH_3C_6H_4\overset{O}{\underset{}{\overset{\uparrow}{S}}}\text{—}CH_3 \xrightarrow[\text{pyridine}]{Br_2/NBS} p\text{-}CH_3C_6H_4SCH_2Br$$

$$\tag{8.138}$$

There is an interesting fluorination of DMSO, but the product is not α-fluoromethyl methyl sulfoxide but a moderately stable liquid of the tautomeric mixture shown below[433]:

α-Chlorination of alkyl sulfoxides is also stereoselective. For example, α-chlorination of both the *cis* and the *trans* isomers of 4-*p*-chlorophenyl-thiane-1-oxide gave only one isomer of the 2-chlorosulfoxide (eq. 8.139), while in the reactions of benzyl phenyl sulfoxide with NOCl, *p*-tosyl chloride, $SO_2Cl_2$ and iodobenzenedichloride both isomers (A:B) are formed in the ratios 90:10, 85:15 and 95:5, and in the reaction with *t*-BuOCl, the ratio is 30:70.

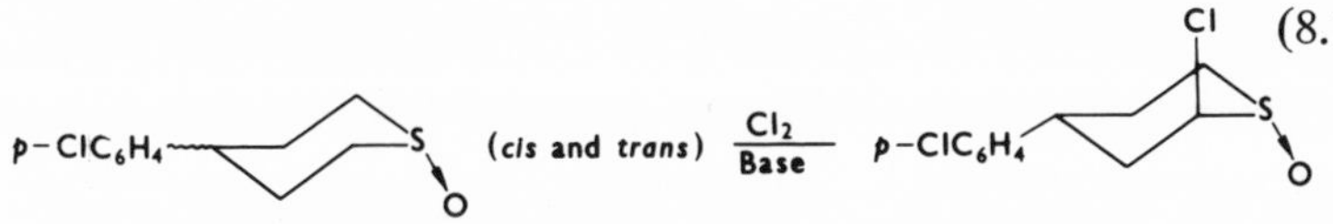

$$(8.139)$$

$$p\text{-ClC}_6\text{H}_4 \cdots \text{S}_O \quad (cis \text{ and } trans) \xrightarrow[\text{Base}]{\text{Cl}_2} p\text{-ClC}_6\text{H}_4 \cdots \text{S}_O$$

The reaction is presumed to proceed *via* the formation of a chloro-sulfoxonium salt which loses a proton during the rate-determining step to give a carbanion, as shown below:

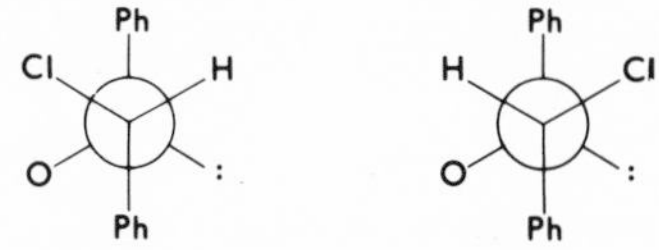

(A)        (B)

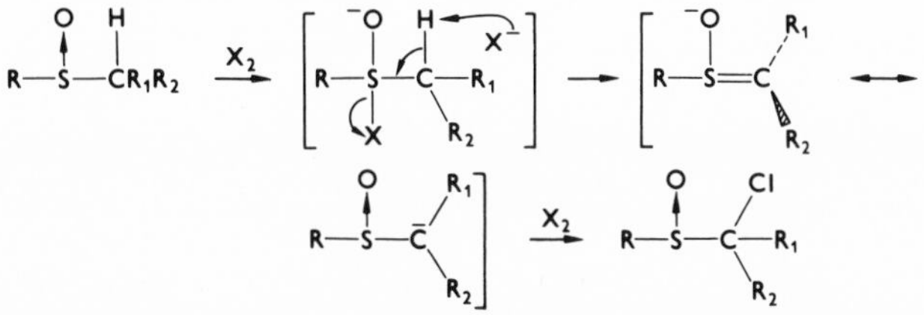

$$(8.140)$$

Evidence to support proton removal as being the rate-determining step is found in the large kinetic isotope effect, $k_H/k_D = 6$, while the ratio of the conformational isomers would be determined by the last product-forming step which is also responsible for steric control.

An interesting method of preparing chloromethyl sulfoxides is the following alkali cleavage reaction[434]:

$$\text{RSCCl=CHCl} \xrightarrow{5\%\text{NaOH}} \text{R-SCH}_2\text{Cl} \ (80\%) \qquad (8.141)$$
$$(R = p\text{-tolyl, benzyl, } t\text{-butyl})$$

α-Chlorosulfoxides thus formed have an active methylene and react with carbonyl compounds as shown below[435–437]:

$$\text{PhSCH}_2\text{Cl} + \text{N-BuLi} \xrightarrow[-78°]{\text{THF}} \text{Ph-SCHLi} \xrightarrow{R_1R_2\text{CO}}$$
$$\underset{\text{Cl}}{}$$

$$(8.142)$$

$$\text{Ph-S-CH-}\underset{\text{Cl}}{\overset{\text{OLi}}{\text{C}}}R_1R_2 \xrightarrow{\text{H}_2\text{O}} \text{Ph-S-CH-}\underset{\text{Cl}}{\overset{\text{OH}}{\text{C}}}R_1R_2 \xrightarrow[\text{MeOH-H}_2\text{O}]{\text{KOH}} \text{Ph-S-CH-CR}_1R_2$$

### 8.5.4 Other Reactions

The oxidation of sulfoxides to sulfones can be achieved by most of the reagents used for the oxidation of sulfides to sulfoxides. The most frequently used peracid oxidation involves a nucleophilic attack by sulfur on the peroxide oxygen, as in the odixation of sulfides. Evidence to support this may be found in the following oxidations of diaryl sulfoxides by substituted benzoic acids[438]:

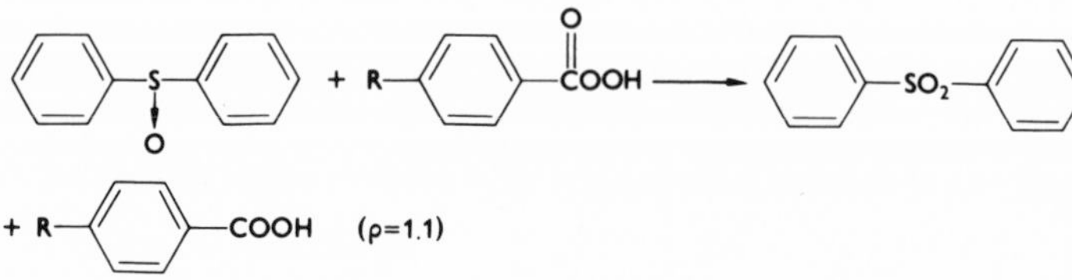

$$(8.143)$$

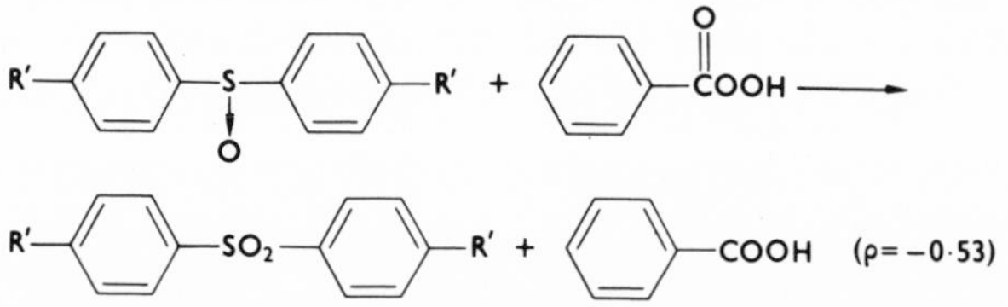

$$(8.144)$$

The Hammett $\rho$ value for the first reaction, 1.1, and for the second reaction, $-0.53$, are small, but in keeping with the mechanism. The presence of highly electron-withdrawing substituents and steric hindrance at the site of reactions tends to retard the rate. For example, bis(2-nitro-4-trifluoromethylphenyl) sulfoxide resists all attempts of oxidation with peracetic acid.[439]

The oxidation of p-tolyl methyl sulfoxide with substituted perbenzoic acid in both alkaline and acidic media also gave positive $\rho$ values,[440] while the alkaline oxidation was found to be more sensitive to steric hindrance than the acidic reaction.[441] the following scheme was suggested for the alkaline oxidation, where the addition complex undergoes cleavage at the rate-determining step[440,441]:

$$\text{PhCO}^-_3 + \overset{R}{\underset{R}{\diagdown}}S{\to}O \rightleftharpoons (\text{PhCOO-O-}\overset{R}{\underset{R}{\diagup}}S{-}O) \longrightarrow \text{PhCO}^-_2 + \text{RSO}_2\text{R}' \qquad (8.145)$$

The oxidation of dialkyl sulfoxides to sulfones is usually less facile than that of sulfides to sulfoxides. However, aryl alkyl and diaryl sulfides are relatively less reactive and are oxidized to the sulfoxides just as reluctantly, or sometimes more so, than sulfoxides are oxidized to sulfones. With sodium permanganate, even the oxidation of aliphatic sulfoxides to sulfones appears to proceed more readily than that of sulfides to sulfoxides. This was found to be the case in the oxidation by osmium tetroxide. Thus, dibenzyl sulfide was recovered intact under conditions in which the sulfoxide gave the sulfone in 95–100% yield.[442] Probably the nucleophilic attack by these oxidizing agents is easier on the trivalent sulfur atom of sulfoxides than on the divalent sulfur atom of sulfides.

Although *t*-butyl hypochlorite is considered as a selective oxidizing agent for sulfides to sulfoxides, sulfoxides are readily converted to sulfones by the same reagent in the presence of water,[443] *via* the formation of a chlorosulfoxonium intermediate.

Diphenyl sulfoxide is oxidized by pyridine N-oxide to the sulfone, however, dimethyl sulfoxide is mainly oxidized further to form methanesulfonic acid.[444] Nitric acid on its own is a selective oxidizing agent for sulfides to sulfoxides, and nitric acid in concentrated sulfuric acid (80–90 wt %) can oxidize sulfoxides to sulfones. For example, phenyl methyl and diphenyl sulfoxides are readily converted to the sulfones at room temperature.[445]

An interesting reaction is the intramolecular photochemical oxidation-reduction of 2-nitrophenyl phenyl sulfoxide shown in eq. 8.146.[446] Since the photoexcitation decreases the electron density on the sulfur atom while increasing that on the oxygen atoms of the nitro group, the while reaction is a kind of intramolecular nucleophilic attack of a nitro-oxygen atom on the sulfur atom.

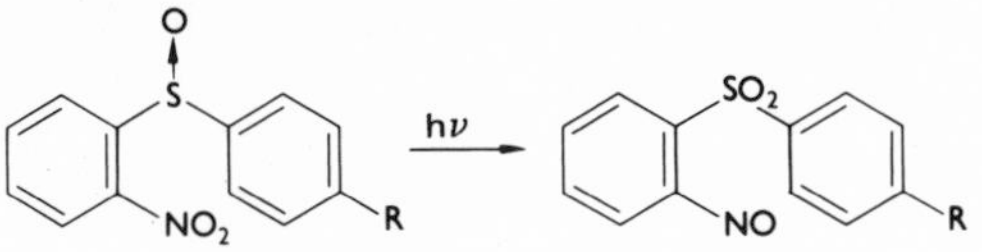

$$(8.146)$$

(R = H, Me,Cl and Ac)

Photolysis of sulfoxides sometimes gives the corresponding sulfine, which upon further photolysis gives the ketone,[447] as shown in eq. 8.147. Since the reaction is effected markedly by sensitization, it is believed to proceed through the triplet state.

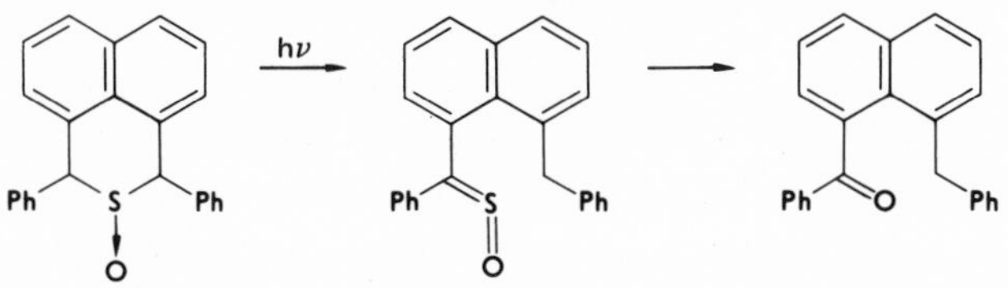

$$(8.147)$$

The following reaction may be a modification of the above process[448]:

$$\text{ArCOCH}_2\text{SOCH}_3 \xrightarrow{h\nu} \text{ArCOCH}_2\text{CH}_2\text{COAr} + \text{ArCOCH}_3$$

$$+ \text{ArCO}_2\text{H} + \text{CH}_3\text{SO}_2\text{SCH}_3 \tag{8.148}$$

## 8.6 CHEMICAL REACTIVITIES OF SULFILIMINES

### 8.6.1 Cleavage of S-N Bond

#### 8.6.1.1 Reduction

N-Sulfonylsulfilimines are readily reduced by tin and hydrochloric acid or by $\text{Na}_2\text{S}_2\text{O}_4$ to yield the corresponding sulfides and sulfonamides in high yields.[449] Catalytic reduction with paladium black is also effective in converting sulfilimines to sulfides.[450,451] Since these reductions do yield only sulfides without any further cleavage to form mercaptans, these procedures may be used for the purification of sulfides. N-Sulfonylsulfilimines are reduced to the sulfides by reaction with cyanide, benzenethiolate, thiocyanate ions, benzenethiol, and triphenylphosphine.[452] A well-studied reaction among these is that of N-p-toluenesulfonylsulfilimines with cyanide ion.[452,453] A kinetic study with substituted phenyl methyl N-tosylsulfilimine with cyanide ion in DHSO revealed that the reaction is of second order, first order with respect to both the sulfilimine and CN ion. While the Hammett $\rho$ value was found to be +0.92, the rate was markedly retarded by bulky *ortho* substituents.

Based on these observations, following mechanism has been postulated:

$$\tag{8.149}$$

Diaryl N-sulfonylsulfilimines are readily reduced to the sulfides by treatment with O,O-dialkyl dithiophosphoric acid. Treatment with either elemental sulfur or diaryl disulfides also reduces diaryl N-sulfilimines to the sulfides.[455] In the reaction of diphenyl N-p-toluenesulfonylsufilimine with diphenyl disulfide (eq. 8.150),[455] not only diphenyl sulfide but also N,N-bis(benzenesulfenyl)toluenesulfonamide are formed in good yields.

$$(8.150)$$

## 8.6.1.2 Hydrolysis and Other Nucleophilic Reactions

Earlier, Nicholet and Willard obtained the corresponding sulfonamide and an oily substance in the hydrolysis of diethyl N-$p$-toluenesulfonyl-sulfilimine which was believed to be diethyl sulfoxide because of a facile formation of diethyl sulfide upon reduction.[456] Mann and Pope, however, could not find any sulfoxide in the hydrolysis of several N-sulfonylsul-filimines with dilute hydrochloric acid.[457] A careful study of the hydrochloric acid-catalyzed hydrolysis of S,S-dibenzyl-N-tosylsulfilimine by Briscoe indicated the formation of tosylamide, benzyl alcohol, benzaldehyde, hydrogen sulfide and an oily sulfur-containing substance. The following mechanism, involving prior N protonation and subsequent hydrolysis, was suggested[458]:

$$(8.151)$$

$$C_6H_5CH_2OH + Ts\text{-}NH_2 + C_6H_5CH_2SOH \longrightarrow C_6H_5CHO + H_2S$$

Day and Cram found that optically active $p$-tolyl methyl N-$p$-tosylsulfilimine is converted to the corresponding sulfoxide of inverted configuration (optical purity: 96%) in 94% yield upon treatment with potassium hydroxide in methanol for one day at room temperature; whereas after hydrolysis with concentrated sulfuric or hydrochloric acid, they obtained the racemized sulfoxide.[459] They suggested the following mechanism involving both cleavage and formation of bonds at equatorial positions, analogous to the formation of sulfilimine from the sulfoxide and N-sulfinylsulfonamide[460]:

$$(8.152)$$

Similar inversions of configuration on the sulfur atom were observed in the hydrolysis of the following cyclic N-ethyl derivatives (eq. 8.153, 8.154).[461] However, these compounds are essentially sulfonium salts and the reactions do not represent those of a typical sulfilimine.

$$(8.153)$$

$$(8.154)$$

Formation of sulfoxides by hydrolysis appears to be rather limited. For example, alkaline treatment of cyclic sulfilimines leads to the predominant formation of rearranged products which are presumed to be formed by prior proton removal and subsequent nucleophilic attack by alkoxide (eq. 8.155).[462] Even with $p$-tolyl alkyl N-$p$-tosylsulfilimines, the predominant formation of $\alpha$-alkoxyalkyl sulfides was observed[462]:

$(n = 3, 4, 5; R = Me, Et, i\text{-}Pr, etc.)$

$$(8.155)$$

$$(8.156)$$

Unsubstituted free sulfilimines are quite unstable, but stay undecomposed in ethers, alcohols, and water in the absence of acid; however, in acidic conditions, they are known to decompose immediately to form the corresponding sulfoxides and ammonia[463]:

$$\begin{array}{c} R \\ \diagdown \\ S{\to}NH + H_2O \xrightarrow{\ OH^-\ } \end{array} \begin{array}{c} R \\ \diagdown \\ S{\to}O + NH_3 \\ \diagup \\ R' \end{array} \qquad (8.156a)$$

The reaction between sulfonylsulfilimines and phenylmagnesium bromide is similar to that for sulfoxides. From diphenyl N-tosylsulfilimine, biphenyl and diphenyl sulfide were obtained, apparently *via* the prior formation of triphenyl sulfonium ion, as shown in eq. 8.157,[464] while the reaction of phenyl methyl or phenyl benzyl N-p-tosylsulfilimine gives an α-phenylated sulfide as the major product (eq. 8.158).

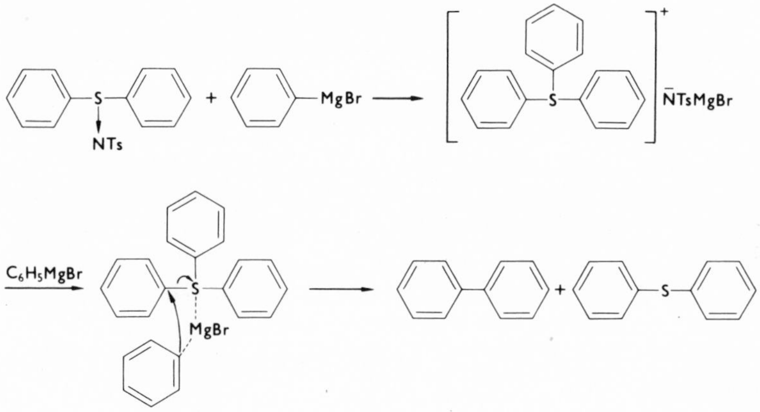

(8.157)

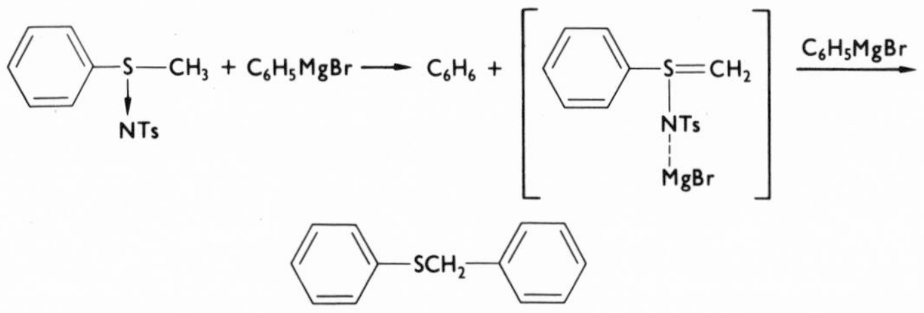

(8.158)

## 8.6.2 Basicity

### 8.6.2.1 $pK_a$ Values

The basicities of several N-sulfonylsulfilimines have been measured.[465,466] Some recently observed $pK_a$ values of a few representative sulfonylsulfilimines and the corresponding sulfoxides are listed in Table 8.7. Though the absolute $pK_a$ values of these compounds may be a matter of some controversy, the relative $pK_a$ values may be safely used for comparison and indicate that the basicity of sulfilimine nitrogen is not much different from that of sulfinyl oxygen.

**TABLE 8.7**

$pK_a$ Values of a Few Sulfoxides $\left(\begin{smallmatrix} R \\ R' \end{smallmatrix} SO\right)$ and Sulfilimines $\left(\begin{smallmatrix} R \\ R' \end{smallmatrix} S \to N\text{-}SO_2C_6H_4CH_3\text{-}p\right)$

| R | R' | $pK_a$ | |
|---|---|---|---|
|   |   | Sulfoxides | Sulfilimines |
| $CH_3$ | $CH_3$ | 0.911 | 0.573 |
| $CH_3$ | $C_6H_5$ | −0.488 | −1.69 |
| $CH_3$ | $p\text{-}CH_3OC_6H_4$ | −0.550 | −1.78* |
| $CH_3$ | $C_6H_5$ | −3.58 | −3.60 |

*Measured spectroscopically.[466]

### 8.6.2.2 Ei Reactions

In spite of the similar order of basicity, the sulfonylsulfilimines undergo Ei-type pyrolysis markedly more readily than the corresponding sulfoxides, as shown in Table 8.8.[467] The relatively small Hammett $\rho$ values, i.e., $\rho_x = +0.88$ and $\rho_y = -0.60$ and the substantial size of the kinetic isotope effect, 3.03, seem to indicate the reaction is an internal elimination (Ei) of an ideal type, in which both bond cleavage and bond formation take place simultaneously (eq. 8.159). The sulfenamide can be actually isolated in this reaction.

$$(8.159)$$

$$+ \text{Ph-S-NHTs} \xrightarrow{H_2O} \text{TsNH}_2 + [\text{PhS-OH}] \to \text{PhSH} + \text{PhSSPH}$$

**TABLE 8.8**

Rates of the Ei Reactions of N-Sulfonylsulfilimines $\left(X\!-\!\!\left\langle\bigcirc\right\rangle\!-\!\!\overset{\displaystyle S}{\underset{\displaystyle NSO_2}{|}}\!-\!C_2H_5\right)$ at 80°C $-\!\!\left\langle\bigcirc\right\rangle\!-\!Y$

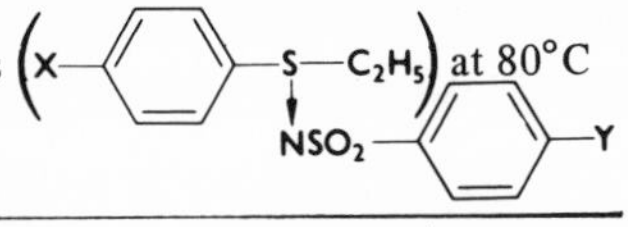

| X | Y | Solvent | $k$ (sec$^{-1}$) |
|---|---|---|---|
| H | CH$_3$ | benzene | 1.08 x 10$^{-5}$ |
| H | CH$_3$ | DMSO | 0.185 x 10$^{-5}$ |
| CH$_3$ | CH$_3$ | benzene | 0.707· x 10$^{-5}$ |
| CH$_3$O | CH$_3$ | benzene | 0.483 x 10$^{-5}$ |
| Br | CH$_3$ | benzene | 1.68 x 10$^{-5}$ |
| NO$_2$ | CH$_3$ | benzene | 4.89 x 10$^{-5}$ |
| H | H | benzene | 0.906 x 10$^{-5}$ |
| H | Br | benzene | 0.628 x 10$^{-5}$ |
| (Ph-S-CH$_2$CD$_3$) $\downarrow$ NTs | | benzene | 0.358 x 10$^{-5}$ |
| $p$-CH$_3$Ph-S-C$_2$H$_5$ $\downarrow$ O | | benzene | $<10^{-8}$ |
| $\overset{\displaystyle NH}{\underset{\displaystyle}{\parallel}}$ $p$-CH$_3$Ph-S-C$_2$H$_5$ $\downarrow$ O | | benzene | $<10^{-8}$ |

The stereochemistry of the Ei reaction was studied and is as shown below (eqs. 8.160, 8.161). The reaction is a stereospecific *cis* elimination, typical for Ei reactions.[468]

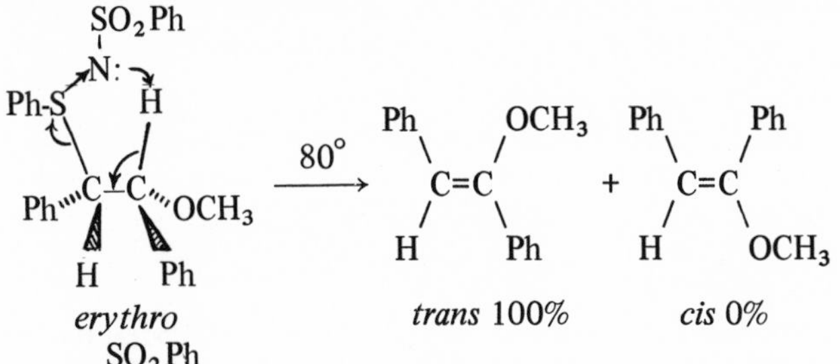

(8.160)

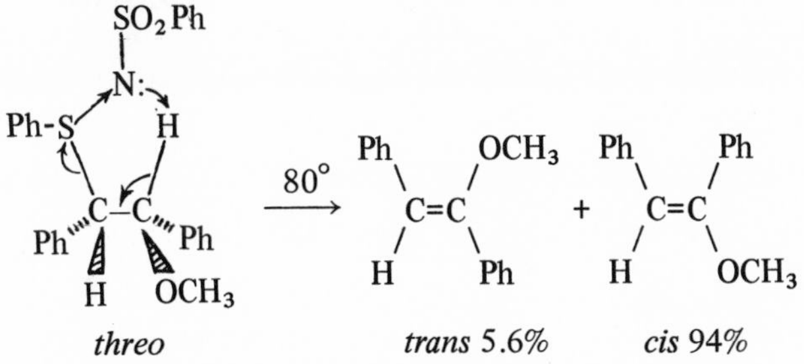

(8.161)

The relatively facile Ei reaction of N-sulfonylsulfilimines compared to that of the corresponding sulfoxides may be due to the shorter nonbonding distance between the $\beta$-hydrogen and the sulfinyl nitrogen atom (1.80Å) than that between the $\beta$-hydrogen and the sulfinyl oxygen (1.94Å). An unstable free sulfilimine, for example, diethylsulfilimine, also undergoes pyrolysis via an Ei process, together with other side reactions, as shown below[469,470]:

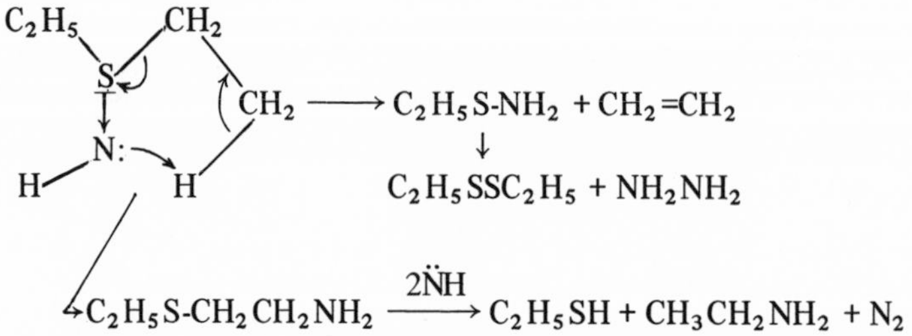

$$C_2H_5S\text{-}NH_2 + CH_2=CH_2$$

$$\downarrow$$

$$C_2H_5SSC_2H_5 + NH_2NH_2 \qquad (8.162)$$

$$C_2H_5S\text{-}CH_2CH_2NH_2 \xrightarrow{2\overset{..}{N}H} C_2H_5SH + CH_3CH_2NH_2 + N_2$$

However, pyrolysis of dimethyl sulfilimine appears to form nitrene and dimethyl sulfide and gives a mixture of methylamine, dimethyl sulfide, methane, nitrogen and ammonia.[469,470]

### 8.6.3 Oxidation

Free sulfilimines are readily oxidized to the corresponding sulfoximines by aqueous permanganate (eq. 8.163),[470] while the sulfones are obtained by oxidation with nitric acid.[470]

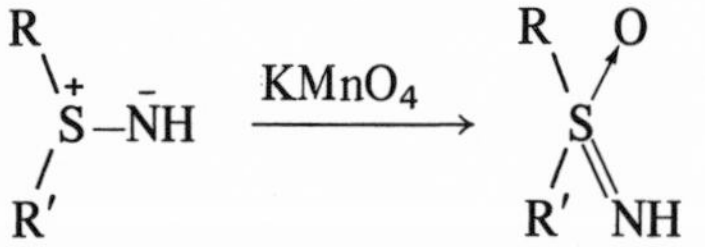

$$\begin{array}{c} R \\ \diagdown \overset{+}{S}\text{--}\overset{-}{NH} \\ \diagup \\ R' \end{array} \xrightarrow{KMnO_4} \begin{array}{c} R \quad O \\ \diagdown \diagup \\ S \\ \diagup \diagdown\diagdown \\ R' \quad NH \end{array} \qquad (8.163)$$

N-sulfonylsulfilimines can also be oxidized to the corresponding N-sulfonylsylfoximines by potassium permanganate[471,472] or m-chloroperbenzoic acid,[473] and subsequent treatment of the products with sulfuric acid gives free sulfoximines:

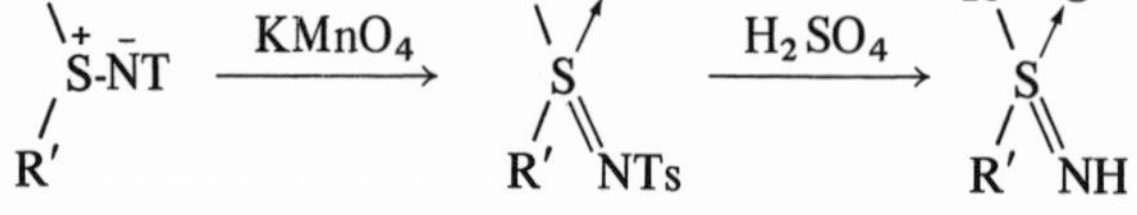

$$\begin{array}{c} R \\ \diagdown \overset{+}{S}\text{-}\overset{-}{NT} \\ \diagup \\ R' \end{array} \xrightarrow{KMnO_4} \begin{array}{c} R \quad O \\ \diagdown \diagup \\ S \\ \diagup \diagdown\diagdown \\ R' \quad NTs \end{array} \xrightarrow{H_2SO_4} \begin{array}{c} R \quad O \\ \diagdown \diagup \\ S \\ \diagup \diagdown\diagdown \\ R' \quad NH \end{array} \qquad (8.164)$$

The oxidation of optically active sulfilimines results in overall retention of configuration of the sulfoximines formed.[473,474] However, the oxidation of N-sulfonylsulfilimines is much more difficult than that of free sulfilimines, and the method cannot be applied for the preparation of diaryl N-sulfonylsulfoximines. The oxidation of N-sulfonylsulfilimines by hydrogen peroxide leads to the formation of the corresponding sulfones and sulfonamides.[475]

N-Sulfonylsulfilimines react with dialkyl sulfoxides in a manner somewhat similar to the Kornblum reaction with sulfoxides,[476] eventually giving rise to the formation of an unsymmetric disulfide and an aldehyde. The reaction is presumed to proceed as shown below:

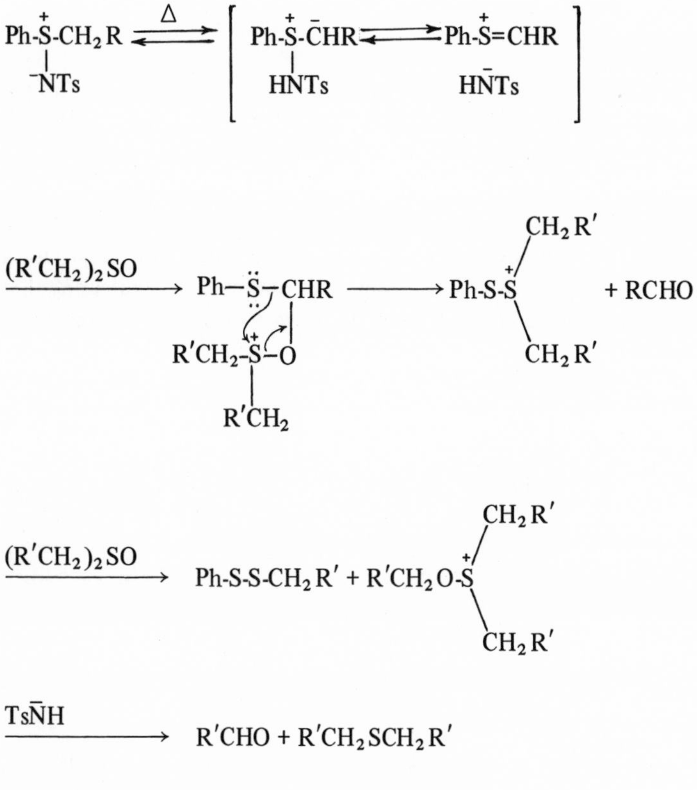

$$
\text{Ph-}\overset{+}{\underset{\text{-NTs}}{\text{S}}}\text{-CH}_2\text{R} \underset{\Delta}{\rightleftarrows} \left[ \text{Ph-}\overset{+}{\underset{\text{HNTs}}{\text{S}}}\text{-}\overset{-}{\text{C}}\text{HR} \rightleftarrows \text{Ph-}\overset{+}{\underset{\text{HNTs}}{\text{S}}}\text{=CHR} \right]
$$

$$
\xrightarrow{(\text{R}'\text{CH}_2)_2\text{SO}} \text{Ph-}\overset{\cdot\cdot}{\text{S}}\text{-CHR} \longrightarrow \text{Ph-S-}\overset{+}{\text{S}}\overset{\text{CH}_2\text{R}'}{\underset{\text{CH}_2\text{R}'}{}} + \text{RCHO}
$$

$$
\text{R}'\text{CH}_2\text{-}\overset{+}{\text{S}}\text{-O}
$$
$$
\text{R}'\text{CH}_2
$$

$$
\xrightarrow{(\text{R}'\text{CH}_2)_2\text{SO}} \text{Ph-S-S-CH}_2\text{R}' + \text{R}'\text{CH}_2\text{O-}\overset{+}{\text{S}}\overset{\text{CH}_2\text{R}'}{\underset{\text{CH}_2\text{R}'}{}}
$$

$$
\xrightarrow{\text{Ts}\overset{-}{\text{N}}\text{H}} \text{R}'\text{CHO} + \text{R}'\text{CH}_2\text{SCH}_2\text{R}' \tag{8.165}
$$

$\text{Ts} = p\text{-CH}_3\text{C}_6\text{H}_4\text{SO}_2 \, ; \, \text{R} = \text{H, Ph}; \, \text{R}' = \text{H, CH}_3$

### 8.6.4 Other Reactions

8.6.4.1 Isomerization

Ash, Challenger, *et al.* treated diallyl sulfide with chloramine-T and obtained crystalline diallyl N-*p*-tosylsulfilimine, which after 3-4 days, converted to an oily substance that turned out to be the isomerized product.[477,478] Since the isomerized product gave N-alkyl *p*-toluenesulfonamide upon hydrolysis, the isomerization was found to take place by the $S_N2'$-type allylic shift from S atom to N atom of the same molecule[477,478]:

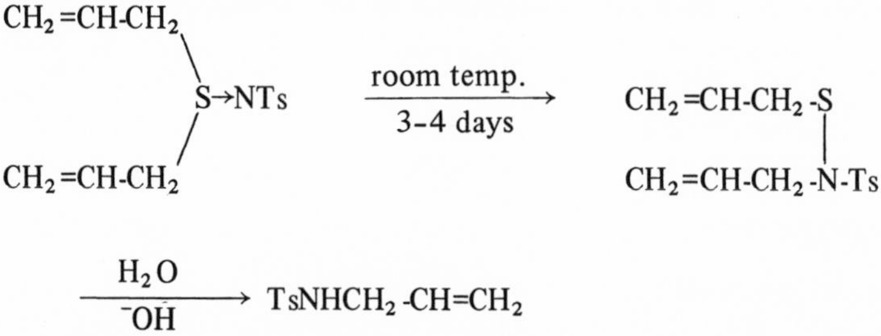

$$CH_2=CH-CH_2 \diagdown$$
$$\underset{CH_2=CH-CH_2 \diagup}{S \rightarrow NTs} \quad \xrightarrow[\text{3-4 days}]{\text{room temp.}} \quad \underset{CH_2=CH-CH_2-N\text{-Ts}}{CH_2=CH-CH_2-S}$$

$$\xrightarrow[^-OH]{H_2O} \quad TsNHCH_2\text{-}CH=CH_2 \qquad\qquad (8.166)$$

In the reaction between cinnamyl phenyl sulfide and chloramine-T, the corresponding sulfilimine was not isolated, but the rearranged N-phenylthio-N-1-phenylallylsulfonamide was obtained,[479] indicating that the isomerization of sulfilimine takes place *via* a cyclic $S_N2'$-type allylic rearrangement:

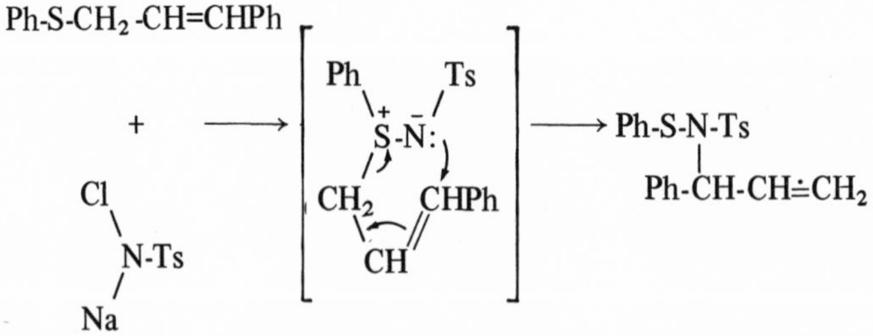

$$Ph\text{-}S\text{-}CH_2\text{-}CH=CHPh$$

$$+ \quad \longrightarrow \quad \begin{bmatrix} \underset{CH_2 \quad CHPh}{\overset{Ph \quad Ts}{\underset{\diagup}{\overset{\diagdown + \ - \diagup}{S\text{-}N:}}}} \\ CH \end{bmatrix} \quad \longrightarrow \quad \underset{Ph\text{-}CH\text{-}CH\doteq CH_2}{Ph\text{-}S\text{-}N\text{-}Ts} \qquad (8.167)$$

$$\underset{Na}{\overset{Cl}{\underset{\diagup}{\overset{\diagdown}{N\text{-}Ts}}}}$$

Similar isomerizations are known with other allylic amine oxides (eq. 8.168)[480,481,482,483] and sulfoxides (eq. 8.169),[484] which have similar semipolar linkages.

$$
\begin{array}{c}
\underset{\underset{\overset{|}{\text{O}}:}{\overset{\text{CH}_3\text{CH}_2}{|}}}{\text{C}_6\text{H}_5\text{-}\overset{+}{\text{N}}} \diagdown \Big( \overset{\text{CH}}{\underset{\text{CHR}}{\|}}
\end{array}
\longrightarrow
\text{C}_6\text{H}_5\text{-}\overset{\overset{\text{CH}_3}{|}}{\text{N}}\text{-O-}\overset{\overset{\text{R}}{|}}{\text{CH}}\text{-CH=CH}_2
\qquad (8.168)
$$

$$
\begin{array}{c}
\overset{..}{\underset{\overset{|}{\text{O}}:}{\text{Ar-}\overset{+}{\text{S}}}} \diagup^{\text{CH}_2} \diagdown \overset{\text{CH}}{\underset{\text{CH}_2}{\|}}
\end{array}
\longleftarrow
\text{Ar-S} \begin{array}{c} \text{CH}_2 \\ \| \\ \text{CH} \\ \diagdown \diagup \\ \text{CH}_2 \end{array} \overset{\text{O}}{\rightleftharpoons}
\begin{array}{c}
\overset{..}{\text{Ar-S}} \diagup^{\text{CH}_2} \diagdown \overset{\text{CH}}{\underset{\text{CH}_2}{\|}} \\ \downarrow \\ \text{O}
\end{array}
\qquad (8.169)
$$

The rate of isomerization of diallyl-N-arylsulfonylsulfilimine is substantially faster than that of a similar allyl aryl sulfoxide.[485]

Perhaps the longer S-N bond (as compared to the S-O bond) and hence the shorter nonbonding distance between the N atom and the terminal carbon of the allyl group (as compared to that between the sulfoxide oxygen and the same carbon) and the more positive character of the S atom of the sulfilimino group may be partly responsible for the facile isomerization of the allyl sulfilimines.

### 8.6.4.2 Others

An addition compound is obtained when $CO_2$ gas is passed into an etheral solution of a sulfilimine (eq. 8.170),[470] while introduction of carbon disulfide eventually reduces the sulfilimine to the sulfide (eq. 8.171).[470]

$$
\underset{\text{R}'}{\overset{\text{R}}{\diagup}}\overset{+}{\underset{}{\text{S}}}\text{-}\overset{-}{\text{NH}} + CO_2 \rightleftharpoons \underset{\text{R}}{\overset{\text{R}}{\diagup}}\overset{+}{\text{S}}\text{-NH-}\overset{\overset{\text{O}}{\|}}{\text{C}}\text{-O}^- \rightleftharpoons \underset{\text{R}}{\overset{\text{R}}{\diagup}}\overset{+}{\text{S}}\text{-}\overset{-}{\text{N}}\text{-}\overset{\overset{\text{O}}{\|}}{\text{C}}\text{-OH}
\qquad (8.170)
$$

$$
\underset{\text{R}'}{\overset{\text{R}}{\diagup}}\overset{+}{\text{S}}\text{-}\overset{-}{\text{NH}} + CS_2 \longrightarrow \underset{\text{R}'\text{S}-\text{C}=\text{S}}{\overset{\text{R}}{\diagdown}}\text{S}\text{-NH} \longrightarrow \underset{\overset{\|}{\text{S}}}{\overset{\overset{\text{C}}{\|}}{\text{NH}}} + \underset{\text{R}'}{\overset{\text{R}}{\diagdown}}\text{S=S} \longrightarrow \underset{\text{R}'}{\overset{\text{R}}{\diagdown}}\text{S} + \tfrac{1}{8}\text{S}_8
$$

$$
\qquad (8.171)
$$

N-Arylsulfilimines formed by the reaction between aromatic amines and DMSO in the presence of phosphorous pentoxide are unstable and undergo further rearrangement to afford $o$-methylthiomethylaniline derivatives[486]:

$$(8.172)$$

## 8.7 REFERENCES

1. For Reviews, see a) W.O. Ranky and D.C. Nelson "Dimethyl Sulfoxide" Chapter 17 in "Organic Sulfur Compounds" Edited by N. Kharasch, Pergamon Press, 1961 b) H. Uda "Chemistry of DMSO and related Compounds". *Yuki Gosei Kagaku-shi* **27**, 909–38 (1969) c) H.H. Szmant, "Chemistry of the Sulfoxide Group," Chapter 16 in "Organic Sulfur Compounds" edited by N. Kharasch, Pergamon Press. 1961 d) C.R. Johnson and J.C. Sharp, "The Chemistry of Sulfoxides", *Quart. Rep. Sulfur Chem.* **4**, No.1 (1969) e) S. Oae "Chemistry of Organosulfur Compounds" Kagakudojin (Kyoto) (1969) Chapter 6 "Sulfoxide."
2. For review, see S. Oae and K. Tsujihara, "Sulfilimines" *Kagaku*, **24**, 702–711 (1969).
3. a) T.P. Soboleva and E.F. Dynkkiev, *Khim, Pererab. Drev*, No.5,11 (1967); *Chem. Abstr.* **68**, 10975 (1967).
   b) S. Meszaros, *Period Polytech, Chem. Eng.* (Budapest) **13**, 79 (1969); *Chem. Abstr.* **72**, 66308z (1970)
4. J. G. Coma and V.G. Gerttula, U.S. 3,045,051 (1962).
   Chapter 2 in "Organic Sulfur Compounds" edited by N. Kharasch, Pergamon Press 1961.
6. a) S.R. Buc, H.B. Freyernucth and H.S. Schultz, U.S. 3,005,852 (1959), U.S. 3,006,963 (1961) b) N.V. Bogoslovskii, I.I. Lapkin, *Uch. Zap. Perm. Gos. Univ.* **141**, 300 (1966); *Chem. Abstr.* **69**, 2645 (1968) c) L.I. Denisova and V.A. Batyakina, *Khim. Farm. Zh.* **4**, 9 (1970); *Chem. Abstr.*, **73**, 120,241 (1970).
7. G. Modena and P.E. Todesco, *Boll. Sci. Fac. Chim. Ind. Bologna,* **23**, 31 (1965); *Chem. Abstr.,* **63**, 8168e (1965) b) R. Curci, R. DiPrete, J.D. Edwards and G. Modena, *Hydrogen-Bonded Solvent Syst. Proc. Symp* (1968), 303; *Chem. Abstr.,* **71**, 48904r (1969).
8. I.P. Gragerov and A.F. Levit, *Zh. Obshch. Khim,* **33**, 544 (1963); *Chem. Abstr.,* **59**, 1506h (1963).
9. L.J. Hughes, T.D. McMinn, Jr. and J.C. Burleson, U.S. 3,114,775 (1963); *Chem. Abstr.,* **60**, 6751a (1964).
10. G. Modena, and P.E. Todesco, *J. Chem. Soc.,* 4920 (1962).
11. F.G. Bordwell, and P.J. Boutan, *J. Amer. Chem. Soc.,* **79**, 717 (1957).
12. a) N. Marziano, A. Compagnini, P. Fiandaca, E. Maccarone and R. Passerini, *Ann. Chim.* (Rome) **58**, 59 (1968);*Chem. Abstr.* **69**, 18765 (1968).
    b) Y. Ogata and T. Kamei, *Tetrahedron,* **26**, 5667 (1970).
13. B. Ciocca and L. Canonica, *Gazz. Chim. Ital.* **76**, 113 (1946); *Chem. Abstr.* **40**, 7153 (1946).
14. C.C. Addison and J.C. Sheldon, *J.Chem. Soc.,* 2705 (1956).

15. R. D. Whitaker and C. L. Bennett, *Quart. J. Florida Acad. Sci.* **28**, 137 (1965); *Chem. Abstr.,* **63**, 9436d (1965).
16. S. Oae, N. Kunieda and W. Tagaki, *Chem. & Ind.* (London) 1790 (1965).
17. H.G. Hauthal, A. Onderka and W. Pritzkow, *J. prakt. Chem.,* **311**, 82 (1969); *Chem. Abstr.* **70**, 86796 (1969).
18. H.H. Szmant and R. Lapinski, *J. Amer. Chem. Soc.,* **80**, 6883 (1958).
19. D.E. Edwards and J.B. Stenlake, *J. Chem. Soc.,* 6338 (1958).
20. A. Schoberl and A. Wagner, "Methoden der Organischen Chemie" (Houben-Weyl) 4th Ed. George Thieme, Verlag, Stuttgart, Vol. IX p.221 (1955).
21. a) E. Arndt, and N. Bekir, *Chem. Ber.,* **63B**, 2390 (1930).
    b) A.H. Schlesinger and D.T. Mowry, *J. Amer. Chem. Soc.,* **73**, 2614 (1951).
    c) H. Kwart and R.K. Miller, *ibid.,* **78**, 5008 (1956).
22. a) T. Zincke and W. Frohneberg, *Chem. Ber.,* **43**, 834 (1909).
    b) Y. Ohnishi, K. Kikukawa, S. Oae, unpublished work.
23. W. Tagaki, K. Kikukawa, K. Ando and S. Oae, *Chem. & Ind.* (London) 1624 (1964).
24. a) R. Harville and S.F. Reed, Jr., *J. Org. Chem.,* **33**, 3976 (1968).
    b) D.L. Tuleen and P.J. Smith, *J. Tenn. Acad. Sci.,* **46**, 17 (1971); *Chem. Abstr.,* **74**, 87536 (1971).
25. J. Benes, *Collection Czeck. Chem. Commun.,* **28**, 1171 (1963).
26. W.D. Kingsbury and C.R. Johnson, *Chem. Commun.,* 365 (1969).
27. S. Oae, Y. Ohnishi, S. Kozuka and W. Tagaki, *Bull. Chem. Soc. Japan,* **39**, 364 (1966).
28. Y. Ohnishi, *M.S. Thesis,* OCU (1967).
29. U. Miotti, G. Modena and L. Sedea, *J. Chem. Soc.,* (B), 802 (1970).
30. T. Higuchi and K.H. Gensch *J. Amer. Chem. Soc.,* **88**, 3874, 5486 (1966).
31. K.H. Gensch, I.H. Pitman and T. Higuchi, *ibid.,* **90**, 2096 (1968).
32. W. Tagaki, M. Ochiai and S. Oae, *Tetrahedron Lett.,* 6131 (1968).
33. G. Allegra, G.E. Wilson, Jr., E. Benedetti, C. Pedone and R. Albert, *J. Amer. Chem. Soc.,* **92**, 4002 (1970).
34. a) A.H. Ford-Moore, *J. Chem. Soc.,* 2126 (1949).
    b) J.P.A. Castrillon and H.H. Szmant., *J Org. Chem.,* **32**, 976 (1967).
35. K.C. Schreiber and V.P. Fernandez, *J. Org. Chem.,* **26**, 2910 (1961).
36. a) G. Barbieri, M. Cinquini, S. Colonna and F. Montanari, *J. Chem. Soc.,* (C), 659 (1968).
    b) M. Cinquini, S. Colonna and F. Taddei, *Boll. Sci. Fac. Chim. Ind. Bologna,* **27**, 231 (1969); *Chem. Abstr.,* **72**, 89951 (1970).
37. M. Cinquini, S. Colonna and F. Montanari, *Chem. Commun.,* 607 (1969).
38. a) N.J. Leonard and C.R. Johnson, *J. Org. Chem.,* **27**, 282 (1962)
    b) C.R. Johnson and J.E. Keiser, *Org. Syn.,* **46**, 78 (1966)
    c) S.W. Weidman, R.A. Nicoletti, D.R. Peterson, C.R. Johnson, P.E. Rogers, 157th ACS Meeting (Minn).
39. P.S. Skell and M.F. Epstein, "Abstracts, 147th ACS Meeting, April 1964 Philadelphia" p. 26 N.
40. a) L. Skattebol, B. Boulette and S. Solomen, *J. Org. Chem.,* **32**, 3111 (1967).
    b) C. Walling and M.J. Mitz, *ibid.,* **32**, 1286 (1967).
    c) C.R. Johnson, J.J. Rigau, *J. Amer. Chem. Soc.,* **91**, 5398 (1969).
41. K. Kikukawa, W. Tagaki, N. Kunieda and S. Oae, *Bull. Chem. Soc. Japan,* **42**, 831 (1969).
42. E.J. Corey and M. Chaykovsky, *J. Amer. Chem. Soc.,* **84**, 866 (1962).
43. E.J. Corey and M. Chaykovsky, *ibid.,* **86**, 1639 (1964).
44. E.J. Corey and M. Chaykovsky, *ibid.,* **87**, 1345 (1965).
45, C. Walling and L. Bollyky, *J. Org. Chem.,* **29**, 2699 (1964).
46. I. Iwai and J. Ide, *Chem. Pharm. Bull.,* **13**, 663 (1965).
47. T.J. Broxton, Y.C. Mac, A.J. Parker and M. Ruane, *Aust. J. Chem.,* **19**, 521 (1966).
48. a) F. Effenberger and J. Daub, *Angew. Chem.* **76**, 435 (1964)
    b) F. Effenberger and J. Daub, *Chem. Ber.,* **102**, 104 (1969).
49. N.P. Volynskii, G.D. Gal'pern and U.V. Smalyaninov, *Neftekhimiya,* **1**, 473 (1961); **4**, 371 (1964).
50. V. Franzen, H.I. Joschek and C. Mertz, *Ann.,* **654**, 82 (1962).
51. A.G. Schultz and R.H. Schlessinger, *Chem. Commun.,* 748 (1970).

52. H. Reinheckel and D. Jahnke, *Chem. Ber.*, **99**, 1718 (1966).
53. S. Bast and K.K. Andersen, *J. Org. Chem.*, **33**, 846 (1968).
54. E.G. Miller, D.R. Rayner and K. Mislow, *J. Amer. Chem. Soc.* **88**, 3139 (1966).
55. N.S. Zefirov and F.A. Abdulvaleeva, *Vestn. Mosk. Univ. Khim*, **24**, 135 (1969); *Chem. Abstr.*, **71**, 112345 (1969).
56. S. Braverman and Y. Stabinsky, *Chem. Commun.*, 270 (1967); *Israel. J. Chem.*, **5**, 71 (1967).
57. P. Lepape, *Ann. Pharm. Fr.* **28**, 181 (1970) *Chem. Abstr.* **73**, 98541 (1970).
58. K.K. Andersen, *Tetrahedron Lett.*, 93 (1962).
59. a) K.K. Andersen, W. Gaffield, N.E. Papanikolaou, J.W. Foley, and R.I. Perkins, *J.Amer. Chem. Soc.*, **86**, 5637 (1964).
     b) K.K. Andersen, *J. Org. Chem.*, **29**, 1953 (1964).
     c) E.B. Fleischer, M. Axelrod, M. Green and K. Mislow, *J. Amer. Chem. Soc.*, **86**, 3395 (1964).
     d) P. Bickart, M. Axelrod, J. Jacobus and K. Mislow, *ibid.*, **89**, 697 (1967).
     e) K.K. Cheung, A. Kjaer, and G.A. Sim, *Chem. Commun.*, 100 (1965).
60. M. Kobayashi and M. Terao, *Bull. Chem. Soc. Japan*, **39**, 1343 (1966).
61. J. Jacobus and K. Mislow, *Chem. Commun.*, 253 (1968).
62. K. Balenovic, I. Bregovec, D. Francetic, I. Monkovic and V. Tomasic, *Chem. & Ind.* (London), 469 (1961).
63. A. Maccioni, *Boll. Sci. Fac. Chim. Ind. Bologna,* **23**, 41 (1965); *Chem. Abstr.* **63**, 8239f (1965).
64. a) K. Balenovic, N. Bregant and D. Francetic, *Tetrahedron Lett.*, 20 (1960).
     b) V. Folli, D. Iarossi, F. Montanari and G. Torre, *Boll. Sci. Fac. Chim. Ind. Bologna*, **25**, 159 (1967); *Chem. Abstr.*, **69**, 2588 (1968).
     c) K. Mislow, M.M. Green and M. Raban, *J. Amer. Chem. Soc.*, **87**, 2761 (1965).
65. A. Maccioni, F. Montanari, M. Secci and M. Tramontini, *Tetrahedron Lett.*, 607 (1961).
66. a) D. Iarossi, A. Pinetti, *Boll. Sci. Fac. Chim. Ind. Bologna*, **27**, 221 (1969); *Chem. Abstr.*, **72**, 131811 (1970).
     b) U. Folli, D. Iarossi and F. Montanari, *J. Chem. Soc.*, C, 1317, 1372 (1968).
67. R.M. Dodson, N. Newman and H.M. Tsuchiya, *J. Org. Chem.*, **27**, 2707 (1962).
68. B.J. Auret, D.R. Boyd, and H.B. Henbest, *J. Chem. Soc.*, (C), 2371, 2374 (1968); *Chem. Commun.* 66 (1966).
69. C.E. Holmlund, K.J. Sax, B.E. Nielsen, R.H. Hartman, R.H. Evans Jr. and R.H. Blank, *J. Org. Chem.*, **27**, 1468 (1962).
70. a) M. Balenovic and N. Bregant, *Chem & Ind.* (London), 1577 (1964).
     b) M. Mikolajczyk and M. Para, *Chem. Commun.*, 1192 (1969).
71. M. Kobayashi and A. Yabe, *Bull. Chem. Soc.*, *Japan* **40**, 224 (1967).
72. B.H. Nicolet and J. Willard, *Science*, **53**, 217 (1921).
73. F.G. Mann and W.J. Pope, *J. Chem. Soc.*, **121**, 1052 (1922).
74. V.G. Petrov, *J. Gen. Chem.* (USSR), **9**, 1635 (1939).
75. C.W. Todd, J.H. Fletcher and D.S. Tarbell, *J. Amer.Chem. Soc.*, **65**, 350 (1943).
76. M.V. Likhosherstov, *J. Gen. Chem.* (USSR), **17**, 1478 (1947).
77. T.P. Kawson, *J. Amer. Chem. Soc.*, **69**, 968 (1947).
78. M.A.McCall, D.S. Tarbell and M.A. Havill, *ibid.*, **73**, 4476 (1951).
79. M. Vecera and J. Petranek, *Chem. Listy*, **50**, 240 (1956).
80. G. Leandri and D. Spinelli, *Ann. Chim.*, **50**, 240 (1956).
81. A. Kucsman, I. Kapovits and M. Balla, *Tetrahedron*, **18**, 75 (1962).
82. J. Petranek, M. Vecera and M. Jurecek, *Collection Czeckoslov. Chem. Commun.*, **24**, 3637 (1959).
83. S.G. Clarke, J. Kenyon and H. Phillips, *J. Chem. Soc.*, 188 (1927).
84. C.R. Johnson and J.J. Rigau, *J. Org. Chem.*, **33**, 4340 (1968).
85. F. Ruff and A. Kucsman, *Acta Chim.* (Budapest), **62**, 437 (1969); **65**, 107 (1970).
86. K.K. Andersen, J. Bhattacharyya, and S.K. Mukhopadhyay., *J. Med. Chem.*, **13**, 759 (1970).
87. K. Tsujihara, N. Furukawa, K. Oae and S. Oae, *Bull. Chem. Soc. Japan*, **42**, 2631 (1969).
88. F.D. Marsh, U.S. P.3505401; *Chem. Abstr.* **72**, 132102 (1970).
89. A.J. Papa, *J. Org. Chem.*, **35**, 2837 (1970).
90. D.S. Tarbell and C. Weaver, *J. Amer. Chem. Soc.*, **63**, 2939 (1941).

91. G. Schulz and G. Kresze, *Angew. Chem.*, **75**, 1022 (1963). See also, T. Ohashi, K. Matsunaga, M. Okahara and S. Komori, *Synthesis*, 96 (1971).
92. A.G. Schering, Ger. P. 1229076 (1966).
93. J. Day and D.J. Cram, *J. Amer. Chem. Soc.*, **87**, 4398 (1965).
94. C. King, *J. Org. Chem.* **25**, 352 (1960).
95. J.J. Monagle, *ibid.*, **27**, 3851 (1962).
96. A. Appel and H. Rittersbacher, *Chem. Ber.*, **97**, 852 (1964).
97. J.B. O'Brien, *Chem. Abstr.*, **70**, 3189z.
98. R. Appel, W. Buchner and E. Gutle *Ann.*, **618**, 53 (1958).
99. R. Appel and W. Buchner, *Chem. Ber*, **95**, 849 (1962).
100. R. Appel and W. Buchner, *ibid.*, **95**, 855 (1962).
101. R. Appel and W. Buchner, *Angew. Chem.* **71**, 701 (1959).
102. R. Appel and W. Buchner, *Chem. Ber*, **95**, 2220 (1962).
    a) N. Furukawa, T. Omata, T. Yoshimura, T. Aida and S. Oae, *Tetrahedron Lett.*, 1619 (1972).
103. G.M. Phillips, J.S. Hunter and L.E. Sutton, *J. Chem. Soc.*, 146 (1945).
104. C.W.N. Cumper and S. Walker, *Trans. Faraday Soc.*, **52**, 193 (1956).
105. a) D. Barnard, J.M. Fabian and H.P. Koch, *J. Chem. Soc.*, 2442 (1949).
    b) H. Siebert, *Z.Anorg. Allg. Chem.*, **275**, 210, 223 (1954).
    c) A. Simon and H. Kriegsmann, *Z.Physik, Chem.* **204**, 369 (1955).
106. E.D. Amstutz, I.M. Hunsberger, and J.J. Chessick, *J. Amer. Chem. Soc.*, **73**, 1220 (1951)
107. a) S. Oae and C. Zalut, *J. Amer. Chem. Soc.*, **82**, 5359 (1960).
    b) S. Oae, M. Yoshihara and W. Tagaki, *Bull Chem. Soc. Japan*, **40**, 951 (1967).
    c) E.Z. Katsnelson, B.I. Ionin, and Ch. S. Frankovskii, *Zh. Org. Khim.*, **5**, 1099 (1969); *Chem. Abstr.*, **71**, 80544 (1969).
108. W. Moffitt, *Proc. Roy. Soc.*, **200**, 409 (1950).
109. G. Costa and P. Blasina, *Z. Physik. Chem.*, *Frankfurt* **4**, 24 (1955); *Chem. Abstr.*, **49** 8702i (1955).
110. a) A.I. Vogel, *J. Chem. Soc.*, 1820, 1833 (1948),
    b) C.C. Price and R.G. Gillis, *J. Amer. Chem. Soc.*, **75**, 4750 (1953).
111. a) S. Sugden, "The Parachor and Valency", Routledge & Kegan Paul, Ltd., London 1930, p. 115.
    b) A.L. Vogel and D.M. Cowan, *J. Chem. Soc.*, 16 (1943).
112. Y. Toshiyasu, Y. Taniguchi, K. Kimura, R. Fujishiro, M. Yoshihara, W. Tagaki and S. Oae, *Bull. Chem. Soc.*, *Japan* **42**, 1878 (1969).
113. a) G.L. Bendazzoli, F. Bernardi, P. Palmieri and C. Zauli, *J. Chem. Soc.*, (A), 2186 (1968).
    b) K. Ramaswamy, S. Swaminathan, *Indian J. Pure. Appl. Phys.* **7**, 807 (1969); *Chem. Abstr.*, **72**, 60929 (1970).
114. A. Kucsman, *Acta Chim. Acad. Sci. Hung.*, **3**, 47 (1953); *Chem. Abstr.* **47**, 10919e (1953).
115. a) A. Kucsman, I. Kapovits and F. Ruff, *Acta Chim. Acad. Sci. Hung.*, **40**, 95 (1964).
    b) A. Kucsman, F. Ruff and I. Kapovits, *Tetrahedron*, **22**, 1575 (1966).
116. K. Tsujihara, N. Furukawa and S. Oae, *Bull. Chem. Soc. Japan*, **43**, 2153 (1970).
117. A. Kalman, *Acta Crystallogr.*, **22**, 501 (1967).
118. H.P. Klug, *ibid.*, **24**, 792 (1968).
119. K.N. Trueblood and S.W. Mayer, *ibid.*, **9**, 628 (1956).
120. L. Pauling, "The Nature of the Chemical Bonds", Cornell University Press, Ithaca, N.Y., 1960, p.224.
121. H. Phillips, *J. Chem. Soc.*, 2552 (1925).
122. P.W.B. Harrison, J. Kenyon and H. Phillips, *J. Chem. Soc.*, 2079 (1926).
123. a) U. Folli, D. Iarossi and G. Torre, *Ric. Sci.*, **38**, 914 (1968). b) S. Oae, T. Kitao, Y. Kitaoka and S. Kawamura, *Bull. Chem. Soc. Japan*, **38**, 546 (1965).
124. a) M. Janczewski, *Rocz. Chem.*, **35**, 601 (1961); *Chem. Abstr.*, **56**, 5899b (1962).
    b) M. Janczewski, S. Dacka and J. Sack, *ibid.*, **36**, 1357 (1962); *Chem. Abstr.* **59**, 7445e (1963).
    c) M. Janczewski and W. Chamas, *ibid.*, **39**, 111 (1965); *Chem. Abstr.*, **63**, 4198d (1965).

d) M. Janczewski and M. Podgorski, *ibid.*, **40**, 145 (1966); *Chem. Abstr.*, **65**, 10541h (1966).
e) M. Janczewski and M. Podgorski, *ibid.*, **43**, 657, 683 (1969); *Chem. Abstr.* **71**, 38629, 70376 (1969).
125. a) S. Allenmark, C.E. Hagbert and O. Bohman, *Ark. Kemi.*, **30**, 269 (1969).
    b) S. Allenmark, O. Bohman and C.E. Hagbert, *ibid.*, **30**, 273 (1969).
    c) O. Bohman and S. Allenmark, *ibid.*, **31**, 299 (1969)
126. P. Karrer, N.J. Antia, and R. Schwyzer, *Helv. Chim. Acta.*, **34**, 1392 (1951).
127. a) A.C. Cope and E.A. Caress, *J. Amer. Chem. Soc.*, **88**, 1711 (1966).
    b) W. Kithing and C.J. Moore, *Inorg. Nucl. Chem. Lett.*, **4**, 691 (1968).
128. a) K.K. Andersen, *Tetrahedron Lett.*, **93**, (1962); see also K.K. Andersen, W. Gaffied, N.E. Papanikolaou, J.W. Foley and R.I. Perkins, *J. Amer. Chem. Soc.*, **86**, 5637 (1964); K.K. Andersen, *J. Org. Chem.*, **29**, 1953 (1964).
    b) K. Mislow, "On the Stereomutation of Sulfoxides", *Record Chem. Progress*, **28**. 217 (1967).
129. a) C.R. Johnson and W.O. Siegel, *Tetrahedron Lett.*, 1879 (1969), *J. Amer. Chem. Soc.*, **91**, 2796 (1969).
130. S. Allenmark, *Ark. Kemi.*, **26**, 73 (1966).
131. a) H.B. Henbest and S.A. Khan, *Proc. Chem. Soc.*, 56, (1964).
    b) C.R. Johnson and D.McCants, *J. Amer. Chem. Soc.*, **86**, 2935 (1964).
    c) J.C. Martin and J.J. Uebel, *ibid.*, **86**, 2936 (1964).
    d) J.B. Lambert and R.G. Keske, *J. Org. Chem.*, **31**, 3429 (1966).
132. A.L. Ternay, J. Herrmann, and S. Evans, *ibid.*, **34**, 940 (1969).
133. W. Amann and G. Kresze, *Tetrahedron Lett.*, 4909 (1968).
134. J.F. Carson and L.E. Boggs, and R.E. Lundin, *J. Org. Chem.*, **35**, 1594 (1970).
135. O. Bastiasen and H. Viervoll, *Acta. Chem. Scand.*, **2**, 702 (1948).
136. D. Dawish and G. Tourigny, *J. Amer. Chem. Soc.*, **88**, 4303 (1966).
137. a) D.R. Rayner, E.G. Miller, P.B. Bickart, A.J. Gordon and K. Mislow, *ibid.*, **88**, 3138 (1966).
    b) D.R. Rayner, A.J. Gordon and K. Mislow, *ibid.*, **90**, 4854 (1968).
138. E.G. Miller, D.R. Rayner, H.T. Thomas and K. Mislow, *ibid.*, **90**, 4861 (1968).
139. P. Bichart, F.W. Carson, J. Jacobus, E.G. Miller and K. Mislow, *ibid.*, **90**, 4869 (1968); see also D.J. Abbott and C.J.M. Stirling, *Chem. Commun.*, 165 (1968).
140. K. Mislow, M. Axelrod, D.R. Rayner, H. Gotthardt, L.M. Coyne and G.S. Hammond, *J. Amer. Chem. Soc.*, **87**, 4958 (1965).
141. R.S. Cooke and G.S. Hammond, *ibid.*, **90**, 2958 (1968).
142. M. Kishi and T. Komeno, *Tetrahedron Lett.*, 2641 (1971); see also A.G. Schultz and R.H. Schlessinger, *Chem. Commun.*, 1294 (1970).
143. B. Iselin, *Helv. Chim. Acta*, **44**, 61 (1961).
144. K. Mislow, T. Simmons, J.T. Melillo and A.L. Ternay Jr., *J. Amer. Chem. Soc.*, **86**, 1452 (1964); see also, D. Landini, F. Montanari, G. Modena, G. Scorrano, *Chem. Commun.*, 3 (1969).
145. W. Tagaki, K. Kikukawa, N. Kunieda and S. Oae, *Bull. Chem. Soc. Japan*, **39**, 614 (1966).
146. S. Oae, M. Yokoyama and M. Kise, *ibid.*, **41**, 1221 (1968).
147. S. Oae, T. Kitao and Y. Kitaoka, *Chem. & Ind.*, 291 (1961); *Bull. Chem. Soc. Japan*, **38**, 546 (1965).
148. S. Oae and N. Kunieda, *Bull Chem. Soc. Japan*, **41**, 696 (1968).
149. N. Kunieda and S. Oae, *ibid.*, **42**, 1324 (1969).
150. J. Day and D.J. Cram, *J. Amer. Chem. Soc.*, **87**, 4398 (1965).
151. N. Kunieda and S. Oae, *Bull. Chem. Soc. Japan*, **41**, 1025 (1968).
152. C.R. Johnson and D. McCants, Jr., *J. Amer. Chem. Soc.*, **86**, 2935 (1964).
153. a) S. Oae, N. Kunieda and W. Tagaki, *Chem. & Ind.*, 1790 (1965).
    b) N. Kunieda, K. Sakai and S. Oae, *Bull. Chem. Soc. Japan*, **42**, 1090 (1969).
154. a) S. Oae and M. Kise, *Tetrahedron Lett.*, 1409 (1967).
    b) S. Oae and M. Kise, *Bull. Chem. Soc. Japan*, **43**, 1416, 1421, 1804 (1970).
155. T. Cairns, G. Eglinton and D.T. Gibson, *Spectrochim. Acta.*, **20**, 31 (1964).
156. E. Jonsson, *Acta Chem. Scand.*, **21**, 1277 (1967).
157. E. Jonsson, *Tetrahedron Lett.*, 3675 (1967).
158. S. Allenmark, "Anchimerically Assisted Sulfoxide Reactions" in "Mechanisms of Reactions of Sulfur Compounds" Vol. 2 p. 177 (1968).
159. S. Allenmark and L.E. Hagberg, *Acta. Chem. Scand.*, **22**, 1694 (1968).

160. S. Allenmark and L.E. Hagberg, *ibid.*, **24**, 2225 (1970).
161. V. Folli, F. Montanari and G. Torre, *Tetrahedron Lett.*, 5037 (1966); see also M. Janczewski, and H. Maziarczyk, *Rocz. Chem.*, **42**, 657 (1968).
162. a) K. Mislow, M.M. Green, P. Laur, J.T. Melillo, T. Simmons and A.L. Ternay, Jr., *J. Amer. Chem. Soc.*, **87**, 1958 (1965).
    b) M. Axelrod, P. Bickart, M.L. Goldstein, M.M. Green, A. Kjaer and K. Mislow, *Tetrahedron Lett.*, 3249 (1968).
    c) E.A. Barnsley, *Tetrahedron*, **24**, 3747 (1968).
    d) R.M. Dodson, R.J. Cahill and V.C. Nelson, *Chem. Commun.*, 620 (1968).
    e) F.D. Saeva, D.R. Rayner and K. Mislow, *J. Amer. Chem. Soc.*, **90**, 4176 (1968).
163. a) M. Nishio and T. Ito, *Chem. Pharm. Bull.*, **13**, 1392 (1965).
    b) M. Nishio, *ibid.*, **15**, 1669 (1967).
    c) M. Nishio, *ibid.* **17**, 262, 274 (1969).
164. a) F. Taddei, *Congr. Conv. Simp. Sci.*, **11**, 106 (1967); *Boll. Sci. Fac. Chim. Ind. Bologna*, **26**, 19 (1968).
    b) F. Taddei, *J. Chem. Soc.*, (B), 653 (1970).
165. M. Nishio, *Chem. Commun.*, 562 (1968).
166. A. Rauk, E. Buncel, R.Y. Moir and S. Wolfe, *J. Amer. Chem. Soc.*, **87**, 5498 (1965).
167. E. Bullock, J.M.W. Scott and P.D. Golding, *Chem. Commun.*, 168 (1967).
168. M. Nishio, *ibid.*, 51, (1969); see also R.R. Fraser and Y.Y. Wigfield, *ibid.*, 1471 (1970).
169. a) W.H. Pirkle and S.D. Beare, *J. Amer. Chem. Soc.*, **90**, 6250 (1968).
    b) F.A.L. Anet, L.M. Sweeting, T.A. Whitney and D.J. Cram., *Tetrahedron Lett.*, 2617 (1968).
170. See for example, G. Binsch and G.R. Frenzen, *J. Amer. Chem. Soc.*, **91**, 3999 (1969).
171. H. Rheinboldt and E. Giesbrecht, *ibid.*, **68**, 973 (1946).
172. N. Marziano, and G. Montando, *Ann. Chim.* (Rome), **51**, 100 (1961).
173. S.G. Clarke, J. Kenyon and H. Phillips, *J. Chem. Soc.*, 188 (1927).
174. E.V. Bell and G.M. Bennett, *J. Chem. Soc.*, 1 (1930).
175. E.V. Bell and G.M. Bennett, *ibid.*, 1798 (1927), 86 (1928).
176. a) J. Day and D.J. Cram. *J. Amer. Chem. Soc.*, **87**, 4398 (1965).
    b) D.R. Rayner, D.M. von Schriltz, J. Day and D.J. Cram, *ibid.*, **90**, 2721 (1968).
    c) D.J. Cram, J. Day, D.R. Rayner, D.M. von Schriltz, D.J. Duchamp and D.C. Garwood, *ibid.*, **92**, 7369 (1970).
177. B.W. Christensen and A. Kjaer, *Chem. Commun.*, 934 (1969).
178. N. Furukawa, K. Harada, and S.Oae, *Tetrahedron Lett.*, 1377 (1972).
179. a) J.A. Smythe, *J. Chem. Soc.*, **95**, 349 (1909).
    b) M. Gayder and S. Swiles, *ibid.*, **97**, 2250 (1910).
    c) T.P. Hilditch, *Ber.*, **44**, 3588 (1911).
180. C.T. Chen and S-J. Yan, *Tetrahedron Lett.*, 3855 (1969).
181. O. Hinsberg, *Chem. Ber.*, **43**, 289 (1910).
182. E.N. Karaulova, and G.D. Gal'pern, *Zhur. Obshchei Khim.*, **29**, 3033 (1959).
183. D. Landini, F. Montanari, H. Hogeveen and G. Maccagnani, *Tetrahedron Lett.*, 2691 (1964).
184. G. Modena, G. Scorrano, D. Landini and F. Montanari, *Tetrahedron Lett.*, 3309 (1966).
185. J.H. Kruger, *J. Inorg. Chem.*, **5**, 132 (1966).
186. D. Landini, F. Montanari, G. Modena and G. Scorrano, *Chem. Commun.*, 86 (1968); *J.Amer. Chem. Soc.*, **92**, 7168 (1970).
187. R.A. Strecker and K.K. Andersen, *J. Org. Chem.*, **33**, 2234 (1968).
188. H. Hogeveen and F. Montanari, *Gazz. Chim. Ital.*, **94**, 176 (1964).
189. S. Allenmark, *Acta Chem. Scand.*, **20**, 910 (1966).
190. S. Allenmark *Ayk. Kemi*, **26**, 37 (1966); "Mechanisms of Reaction of Sulfur Compounds" Vol.2, 177 (1968).
191. F. Ostermayer and D.S. Tarbell, *J. Amer. Chem. Soc.*, **82**, 3752 (1960).
192. V.J. Traynelis and W.L. Hergenrother, *J. Org. Chem.*, **29**, 221 (1964).
193. B.R. James, F.T.T.Ng and G.L. Rempel, *Can. J. Chem.*, **47**, 4521 (1969).
194. H. Alper and E.C.H. Keung, *Tetrahedron Lett.*, 53 (1970).

195. a) T.H. Chan, A. Melnyk and D.N. Harpp, *Tetrahedron Lett.*, 201 (1969).
    b) A.I. Skobelina, I.U. Numanov, E.N. Karaulova, G.D. Gal'pern and T.S. Ovchinnkova, *Dokl. Ak'ad. Tadzh. SSR* 8 13 (1965); *Chem. Abstr.*, **64**, 12436g (1966).
    c) D.W. Chasar, *J. Org. Chem.*, **36**, 613 (1971).
196. C.N. Yiannios and J.Y. Karabinos, *J. Org. Chem.*, **28**, 3246 (1963).
    b) K. Balenovic and N. Bregant, *Chem. & Ind.* (London), 1577 (1964).
197. a) T.J. Wallace and J.J. Mahon, *J. Org. Chem.*, **30**, 1502 (1965).
    b) T.J. Wallace and H.A. Weiss, *Chem. & Ind.*, 1558 (1966).
198. A.A. Oswald and T.J. Wallace, Chapter 8, p. 213 in "Organic Sulfur Compounds" Vol. 2, edited by N. Kharasch and C.Y. Meyers, Pergamon Press, 1966.
199. H.C. Sorensen and L.L. Ingraham, *Arch. Biochem. Biophys.*, **134**, 214 (1969).
200. a) T.J. Wallace, *Chem. & Ind.* (London), 501 (1964).
    b) T.J. Wallace, *J. Amer. Chem. Soc.*, **86**, 2018 (1964)
    c) T.J. Wallace and J.J. Mahon, *ibid.*, **86**, 4099 (1964).
201. S. Oae and S. Kawamura, *Bull. Chem. Soc. Japan*, **36**, 163 (1963).
202. S. Kiso and S. Oae, *ibid.*, **40**, 1722 (1967).
203. W. Takagi, S. Kiso and S. Oae, *ibid.*, **38**, 414 (1965).
204. F.G. Bordwell and B.M. Pitt, *J. Amer. Chem. Soc.*, **77**, 572 (1966).
205. S. Searles, Jr., and H.R. Hays, *J. Org. Chem.* **23**, 2028 (1958).
206. H. Shimano, K. Ikura and S. Oae, unpublished work.
207. N.P. Volynskii, G.D. Gal'pern, and V.V. Smolyaninov, *Neftekhimiya* 1, 473 (1961).
208. S.K. Ray, R.A. Shaw and B.C. Smith, *Nature*, **196**, 372 (1962).
209. J.P.A. Castrillon and H.H. Szmant, *J. Org. Chem.*, **30**, 1338 (1965).
210. H.H. Szmant and O. Cox, *ibid.*, **31**, 1595 (1966).
211. J. Granoth, A. Kalir and Z. Pelah, *J. Chem. Soc.*, C, 2424 (1969).
212. R. Pummerer, *Chem. Ber.*, **43**, 1401 (1910).
213. L. Horner and P. Kaiser, *Ann.*, **626**, 19 (1959).
214. W.E. Parham and L.D. Edwards, *J. Org. Chem.*, **33**, 4150 (1968).
215. S. Oae, T. Kitao, S. Kawamura and Y. Kitaoka, *Tetrahedron*, **19**, 817 (1963).
216. S. Oae, T. Kitao and S. Kawamura, *ibid.*, **19**, 1783 (1963).
217. S. Oae, T. Kitao and Y. Kitaoka, *J. Amer. Chem. Soc.*, **84**, 3366 (1962).
218. F.G. Bordwell and B.M. Pitt, *ibid.*, **77**, 572 (1955).
219. a) S. Oae and M. Kise, *Tetrahedron Lett.*, 2261 (1968).
    b) M. Kise and S. Oae, *Bull. Chem. Soc. Japan*, **43**, 1426 (1970).
220. a) C.R. Johnson, J.C. Sharp and W.G. Phillips, *Tetrahedron Lett.*, 5299 (1967).
    b) C.R. Johnson and W.G. Phillips, *J. Amer. Chem. Soc.*, **91**, 682 (1969).
221. a) W.E. Parham and M.D. Bhavsar. *J. Org. Chem.*, **28**, 2686 (1963).
    b) W.E. Parham, L. Christensen, S.H. Groen and R.M. Dodson, *ibid.*, **29**, 2211 (1964).
222. R.H. Schlessinger and G.S. Ponticello, *Tetrahedron Lett.*, 4361 (1969).
223. M.P. Cava and G.E.M. Husbands, *J. Amer. Chem. Soc.*, 91, 3952 (1969).
224. W.J. Kenney, J.A. Walsh and D.A. Davenport, *ibid.*, **83**, 4019 (1961).
225. D.A. Davenport, D.B. Moss, J.E. Rhodes and J.A. Walsh, *J. Org. Chem.*, **34**, 3353 (1969).
226. T. Enkvist and C. Naupert, *Finska Kemistsamfundets Medd.*, **69**, 98 (1961); *Chem. Abstr.* 56, 4606c (1962).
227. D. Walker, *J. Org. Chem.*, **31**, 835 (1966).
228. T.L. Moore, *ibid.*, **32**, 2786 (1967).
229. a) H.D. Becker, *ibid.*, **29**, 1358 (1964).
    b) W.E. Parham and M.D. Bhavar, *ibid.*, **28**, 2686 (1963).
    c) H.D. Becker and G.A. Russell, *ibid.*, **28**, 1895 (1963).
    d) H.D. Becker, G.J. Mikol and G.A. Russell, *J. Amer. Chem. Soc.*, **85**, 3410 (1963).
    e) G.A. Russell and G.J. Mikol, *ibid.*, **88**, 5498 (1966).
    f) W.R. Sorenson, *J. Org. Chem.*, **24**, 978 (1959).
    g) E.N. Karaulova, G.D. Gal'pern, V.D. Nikitina, L. R. Barykina, I.V. Cherepanova, D.K. Zhestkov and F.V. Kozlova and G. Yu. Pek, *Neftekhimiya*, **10**, 559 (1970); *Chem. Abstr.*, **73**, 120009 (1970).
    h) C.T. Chen and S.J. Yan, *Bull. Inst. Chem. Acad. Sinica*, No 17, 75 (1970);

*Chem. Abstr.*, **73**, 76801 (1970).
230. R.B. Morin, D.O. Spry and R.A. Mueller, *Tetrahedron Lett.*, 849 (1969).
231. R.B. Morin and D.O. Spry, *Chem. Commun.*, 335 (1970).
232. W.L. Garkrecht, G.P. 2012955; *Chem. Abstr.*, **73**, 109781 (1970).
233. R.E. Boyle, *J. Org. Chem.*, **31**, 3880 (1966).
234. F.G. Bordwell and B.M. Pitt, *J. Amer. Chem. Soc.*, **77**, 572 (1955).
235. a) G.A. Russell and L.A. Ochrymowycz, *J. Org. Chem.*, **34**, 3618 (1969).
    b) G.A. Russell and L.A. Ochrymowycz, *ibid.*, **35**, 964, 2104 (1970).
    c) G.A. Russell and L.A. Ochrymowycz, *ibid.*, **35**, 3007 (1970).
236. R. Michelot and B. Tehoubar, *Bull. Soc. Chim. Fr.*, **1966**, 3039.
237. a) R. Oda and S. Takashima, *J. Chem. Soc. Japan*, **82**, 1423 (1961).
    b) R. Oda and K. Yamamoto, *ibid.*, **85**, 133 (1964).
    c) R. Oda and Y. Hayashi, *ibid.*, **87**, 291, 1110 (1966).
    d) Y. Hayashi and R. Oda, *J. Org. Chem.*, **32**, 457 (1967).
    e) M.G. Burdon and J.G. Moffatt, *J. Amer. Chem. Soc.*, **87**, 4656 (1965).
    f) K.E. Pfitzner, J.P. Marino and R.A. Olofson, *ibid.*, **87**, 4658 (1965).
    g) M.G. Burdon and J.G. Moffatt, *ibid.*, **88**, 5855 (1966).
    h) M.G. Burdon and J.G. Moffatt, *ibid.*, **89**, 4725 (1967).
    i) A.F. Cook and J.G. Moffatt, *ibid.*, **90**, 740 (1968).
    j) R.E. Harmon, C.V. Zenarosa and S.K. Gupta, *Tetrahedron Lett.*, 3781 (1969).
238. F. Goethals and P. de Radzitzky, *Bull. Soc. Chim. Belges.*, **73**, 546 (1964).
239. R.K. Blackwood, J.J. Beereboom, H.H. Rennhard, M. Schach von Wittenau and C.R. Stephens, *J. Amer. Chem. Soc.*, **85**, 3943 (1963).
240. G. Seitz, *Chem. Ber.*, **101**, 585 (1968).
241. C.F. Garbers, A.J.H. Labuschagne and D.F. Schneider, *Chem. Commun.*, 499 (1969).
242. D. Martin, H.J. Niclas and A. Weise, *Chem. Ber.*, **102**, 23 (1969).
243. D. Martin and H.J. Niclas, *Chem. Ber.*, **102**, 31 (1969).
244. J.G. Moffatt, *Quart. Rept. Sulfur Chemistry*, Vol. 3, No. 2, p. 95 (1968).
245. H. Potter, Am. Chem. Soc. Meeting, Cleveland, Ohio, April 11 (1960).
246. R. Oda and K. Yamamoto, *J. Org. Chem.*, **26**, 4679 (1961).
247. P. Manya, A. Sekera and P. Rumpf, *Tetrahedron*, **26**, 467 (1970) *C.R. Acad. Sci. Paris Sec.*, C, **264**, 1196 (1967).
248. S. Oae, *Quart. Rep. Sulfur Chemistry*, Vol. 5, 53–66 (1970).
249. S. Oae, T. Kitao and Y. Kitaoka, *Chem. & Ind.* (London), 291 (1961).
250. S. Oae, T. Kitao and Y. Kitaoka, *Bull. Chem. Soc. Japan*, **38**, 543 (1965).
251. R.J. Gillespie and R.C. Passerini, *J. Chem. Soc.*, 3850 (1956).
252. S. Oae, T. Kitao, Y, Kitaoka and S. Kawamura, *Bull. Chem. Soc. Japan*, **38**, 546 (1965).
253. S. Oae and N. Kunieda, *ibid.*, **41**, 696 (1968).
254. U. Schmidt, K. Kabitzke and K. Markau, *Angew. Chem.*, **72**, 708 (1960): see also U. Shmidt, *ibid.*, **76**, 629 (1964).
255. H.J. Shine, M. Rahman, H. Seeger and G.S. Wu, *J. Org. Chem.*, **32**, 1901 (1967).
256. H.J. Shine, "The Formation of Cations and Cation Radicals from Aromatic Sulfides and Sulfoxides" in "Organosulfur Chemistry", edited by M.J. Janssen, Intra Science (1967) Chapter 6, pp. 93–117.
257. N. Kunieda and S. Oae, *Bull. Chem. So. Japan*, **42**, 1324 (1969).
258. N. Kunieda and S. Oae, *ibid.*, **41**, 1025 (1968).
259. S. Oae, M. Yokoyama and M. Kise, *ibid.*, **41**, 1221 (1968).
260. K. Mislow, T. Simmons, J.T. Melillo, and A.L. Ternay, Jr., *J. Amer. Chem. Soc.*, **86**, 1452 (1964).
261. W. Tagaki, K. Kikukawa, N. Kunieda and S. Oae, *Bull. Chem. Soc. Japan*, **39**, 614 (1964).
262. H. Yoshida, T. Numata and S. Oae, *ibid.*, **44**, 2875 (1971).
263. a) D. Landini, F. Montanari, G. Modena and G. Scorrano, *Chem. Commun.*, 86 (1968).
    b) D. Landini, G. Modena, F. Montanari and G. Scorrano, *J. Amer. Chem. Soc.*, **92**, 7168 (1970).
264, N. Kunieda, K. Sakai and S. Oae, *Bull. Chem. Soc. Japan*, **42**, 1090 (1969).
265. S. Oae, and M. Kise, *Tetrahedron Lett.*, 1409 (1967); see also *Bull. Chem. Soc. Japan*, **43**, 1416 (1970).

266. M. Kise, and S. Oae, *Bull. Chem. Soc. Japan*, **43**, 1804 (1970).
267. E. Jonsson, *Acta Chem. Scand.*, **21**, 1277 (1967); *Tetrahedron Lett.*, 3675 (1967).
268. S. Oae, M. Kise, N. Furukawa and Y.H. Khim, *Tetrahedron Lett.*, 1415 (1967).
269. a) C.R. Johnson and W.G. Phillips, *Tetrahedron Lett.*, 2101 (1965) b) C.R. Johnson, *J. Amer. Chem. Soc.*, **85**, 1020 (1963).
270. a) J. Day and D.J. Cram, *J. Amer. Chem. Soc.*, **87**, 4398 (1965).
     b) D.R. Rayner, D.M. von Schriltz, J. Day and D.J. Cram, *ibid.*, **90**, 2721 (1968).
     c) D.J. Cram, J. Day, D.R. Rayner, D.M. von Schriltz, D.J. Duchamp and D.C. Garwood, *ibid.*, **92**, 7369 (1970).
271. S. Oae, M. Yokoyama, M. Kise and N. Furukawa, *Tetrahedron Lett.*, 4131 (1968).
272. B.W. Christensen and A. Kjaer, *Chem. Commun.*, 934 (1969).
273. M. Mikolajczyk, *Chem. & Ind.*, 2059 (1966).
274. T. Numata, K. Sakai, M. Kise, N. Kunieda and S. Oae, *International J. Sulfur Chem.*, A, **1**, 1 (1971).
275. T. Numata and S. Oae, *ibid.*, **1**, 6 (1971).
276. R. Kuhn and H. Tischmann, *Ann.*, **611**, 117 (1958).
277. S.G. Smith and S. Winstein, *Tetrahedron*, **3**, 317 (1958).
278. a) H.R. Nace and J.T. Monagle, *J. Org. Chem.*, **24**, 1792 (1959).
     b) J.M. Tien and I.M. Hunsbergen, *Chem & Ind.*, (London) 88 (1959).
279. N. Kornblum, J.W. Powers, G.J. Anderson, W.J. Jones, H.O. Larson, O. Levand and W.M. Weaver, *J. Amer. Chem. Soc.*, **79**, 6562 (1957).
280. N. Kornblum, W.J. Jones and G.J. Anderson, *ibid.*, **81**, 4113 (1959).
281. A.P. Johnson and A. Pelter, *J. Chem. Soc.*, 520 (1964).
282. R. Major and H. Hess, *J. Org. Chem.*, **23**, 1563 (1958).
283. D.N. Jones and M.R. Saeed, *J. Chem. Soc.*, 4657 (1963).
284. R. Iacona, H. Nace and A. Rowland, *J. Org. Chem.*, **29**, 3495 (1964).
285. R. Iacona and H. Nace, *ibid.*, **29**, 3498 (1964).
286. a) Y. Morisawa, and K. Tanabe, *Chem. Pharm. Bull.*, **17**, 1212 (1969).
     b) J. Doncet, D. Gagnaire, A. Robert, *C.R. Acad. Sci. Paris Sec.*, C, **268**, 1700 (1969).
     c) I. Lillien, *J. Org. Chem.*, **29**, 1631 (1964).
287. K. Torssell, *Tetrahedron Lett.*, 4445 (1966).
288. A.H. Fenselan and J.G. Moffatt, *J. Amer. Chem. Soc.*, **88**, 1762 (1966).
289. F.W. Swart and W.W. Epstein, *J. Org. Chem.*, **32**, 835 (1967).
290. D.H.R. Barton, B.J. Garner and R.G. Wightman, *J. Chem. Soc.*, 1855 (1964).
291. K.E. Pfitzner, and J.G. Moffatt, *J. Amer. Chem. Soc.*, **87**, 5670 (1965).
292. J.D. Albright and L. Goldman, *ibid.*, **89**, 2416 (1967).
293. K. Onodera, S. Hirano and K. Kashimura, *ibid.*, **87**, 4651 (1965).
294. T. Cohen and T. Tsuji, *J. Org. Chem.*, **26**, 1681 (1961).
295. J. Leitich and F. Wessely, *Monatsh*, **95**, 129 (1964).
296. W.H.W. Lunn, *J. Org. Chem.*, **30**, 2925 (1965).
297. V.J. Traynelis, W.L. Hergenrother, J.R. Livingston and J.A. Valicenti, *J. Org. Chem.*, **27**, 2377 (1962).
298. V.J. Traynelis and W.L. Hergenrother, *J. Amer. Chem. Soc.*, **86**, 298 (1964).
299. J.M. Tien, H.J. Tien and J.S. Ting, *Tetrahedron Lett.*, 1483 (1969).
300. W.H. Clement, T.J. Dangieri and R.W. Tuman, *Chem. & Ind.* (London) 755 (1969).
301. V.J. Traynelis, W.L. Hergenrother, H.T. Hanson and J.A. Valicenti, *J. Org. Chem*, **29**, 123 (1964).
302. S. Fujita, T. Hiyama and H. Nozaki, *Tetrahedron*, **26**, 4347 (1970).
303. D.J. Cram and A.C. Day, *J. Org. Chem.*, **31**, 1227 (1966).
304. M. Kise, T. Asari, N. Furukawa and S. Oae, *Chem. & Ind.* (London) 276 (1967).
305. C. Winterfeldt, *Chem. Ber.*, **98**, 1581 (1965).
306. M. Tamres and S. Searles, Jr., *J. Amer. Chem. Soc.*, **81**, 2100 (1959).
307. R.G. Laughlin, *J. Org. Chem.*, **25**, 864 (1960).
308. a) S.G. Terjesen and K. Sandved, *Kgl. Norske Videnskab. Selskabs. Fork.*, **10**, 117 (1937); *Kgl. Norske Videnskab. Selskabs. Skrifter*, No. 7, 20 (1938).
     b) E.M. Arnett, *Progr. Phys. Org. Chem.*, **1**, 223 (1963).
309. C.A. Streuli, *Anal. Chem.*, **30**, 997 (1958).

310. K.K. Andersen, W.H. Edmonds, J.B. Biasotti and R.A. Strecker, *J. Org. Chem.*, **31**, 2859 (1966).
311. P. Haake and R.D. Cook, *Tetrahedron Lett.*, 427 (1968).
312. K. Sakai, N. Kunieda and S. Oae, *Bull. Chem. Soc. Japan*, **42**, 1964 (1969).
313. N.C. Marziano, G. Cimino, U. Romano and R.C. Passerini, *Tetrahedron Lett.*, 2833 (1969); *Boll. Sedute Accad. Gioenia Sci. Natur. Catania* **9**, 589 (1969).
314. a) U. Quintily and G. Scorrano, *Chem. Commun.*, 260 (1971).
    b) D. Landini, G. Modena, G. Scorrano and F. Taddei, *J. Amer. Chem. Soc.*, **91**, 6703 (1969); see also *Boll Sci. Fac. Chim. Ind. Bologna*, **26**, 325 (1968).
315. J.F. Bunnett and F.P. Olsen, *Can. J. Chem.*, **44**, 1899 (1966).
316. R. Stewart and K. Yates, *J. Amer. Chem. Soc.*, **80**, 6355 (1958).
317. R.H. Figueroa, E. Roig and H.H. Szmant, *Spectrochem. Acta*, **22**, 1107 (1966).
318. S. Ghersetti, *Boll. Sci. Fac. Chim. Ind. Bologna*, **27**, 17 (1969).
319. V.F. Chesnokov, I.M. Bokhovkin, and I.V. Khazava, *Zh. Obshch. Khim.*, **39**, 500 (1969).
320. R.S. Drago, B. Wayland and R.L. Carlson, *J. Amer. Chem. Soc.*, **85**, 3125 (1963).
321. R. Reid and B. Muslin, *Trans. Illinois State. Acad. Sci.*, **59**, 48 (1966); *Chem. Abstr.* **65**, 601 (1966).
322. R. Klaebe, *Acta Chem. Scand.*, **18**, 999 (1964).
323. D.J. Pettitt, U.S.P.3,527,810·*Chem. Abstr.*.73, 109271x (1970).
324. R.N. Butler, J. Oakes and M.C.R. Symons, *J. Chem. Soc.*, (A) 1134 (1968).
325. P.W.N.M. van Leeuwen, and W.L. Groeneveld, *Rec. trav. chim. Pays-Bas*, **86**, 1217 (1967).
326. H.G. Langer, U.S. 3,471,250; *Chem. Abstr.*, **72**, 78404 (1970).
327. V.M. Marganian, J.E. Whisenhunt and J.C. Fanning, *J. Inorg. Nucl. Chem.*, **31**, 3775 (1969).
328. W. Kitching, C.J. Moore and D. Doddrell, *Aust. J. Chem.*, **22**, 1149 (1969).
329. Y.H. Shaw and N.C. Li, *Can. J. Chem.*, **48**, 2090 (1970).
330. I. Lindqvist and B. Fjellstroem, *Acta Chem. Scand.*, **14**, 2055 (1960).
331. W. Kitching, *Tetrahedron Lett.*, 3689 (1966).
332. A.J. Parker, *Quart. Revs.*, **16**, 163 (1962).
333. N. Kharasch and B.S. Thyagarajan, "The Chemistry of Dimethyl Sulfoxide for the Period 1961–1965", *Quart. Repts. Sulfur Chem.*, **1**, 1–91 (1966).
334. See W.S. MacGregor "Some Reactions and Solvent Properties of Dimethyl Sulfoxide", *ibid.*, **3**, 149 (1968).
335. a) D.J. Cram, B. Rickborn and G.R. Knox, *J. Amer. Chem. Soc.*, **82**, 6412 (1960).
    b) D.J. Cram, and M.R.V. Sahyum, *ibid.*, **85**, 1263 (1963).
    c) T.J. Wallace, J.E. Hofmann and A. Schriesheim, *ibid.*, **85**, 2739 (1963).
    d) J.E. Hofmann, T.J. Wallace and A. Schriesheim, *ibid.*, **86**, 1561 (1964).
    e) T.J. Wallace, H. Pobiner, J.E. Hofmann and A. Schriesheim, *J. Chem. Soc.*, 1271 (1965).
    f) C.H. Snyder and A.R. Soto, *J. Org. Chem.*, **29**, 742 (1964).
    g) A. Ledwith and Y. Shih-Lin, *Chem. & Ind.*, (London) 1867 (1964).
    h) N. Kornblum and H.W. Frazier, *J. Amer. Chem. Soc.*, **88**, 865 (1966).
    i) A.F. Cockerill, S. Rottschaefer and W.H. Saunders, Jr., *ibid.*, **89**, 901 (1967).
    j) T. Yoshida, Y. Yano and S. Oae, *Tetrahedron*, **27**, 5343 (1971).
336. a) D.J. Cram, "Fundamentals of Carbanion Chemistry" Academic Press, N.Y. (1965).
    b) D.J. Cram *et al.*, *J. Amer. Chem. Soc.*, **83**, 3678 (1961) and the subsequent papers of the series.
    c) C.C. Price and W.H. Snyder, *ibid*, **83**, 1773 (1961).
    d) L.W. Clark, *J. Phys. Chem.*, **60**, 825 (1956).
    e) S. Oae, K. Uneyama, W. Tagaki and I. Minamida, *Tetrahedron*, **24**, 5283 (1968).
    f) N. Kornblum, R.J. Berrigan and W.J. Le Noble, *J. Amer. Chem. Soc.*, **85**, 1141, 1148 (1963), **82**, 1257 (1960).
    g) E.C. Steiner and J.M. Gilbert, *ibid.*, **87**, 382 (1965).
    h) H.H. Szmant and M.N. Roman, *ibid.*, **88**, 4034 (1966).
    i) C.D. Ritchie and R. Uschold, *ibid.*, **86**, 4488 (1964).
    j) N. Kornblum, R. Seltzer and P. Haberfield, *ibid.*, **85**, 1148 (1963).

337. a) G.C. Finger and C.W. Kruse, *J. Amer. Chem. Soc.*, **78**, 6034 (1956).
b) G.C. Finger and L.D. Starr, *ibid.*, **81**, 2674 (1959).
c) M.D. Cava, R.L. Little and D.R. Napier, *ibid.*, **80**, 2257 (1958).
d) N. Kornblum and J.W. Powers, *J. Org. Chem.*, **22**, 455 (1957).
e) C.A. Kingsbury, *ibid.*, **29**, 3262 (1964).
f) W. Roberts and M.C. Whiting, *J. Chem. Soc.*, 1290 (1965).
g) C.H. Snyder and A.R. Soto, *J. Org. Chem.*, **30**, 673 (1965).
h) D.D. Roberts, *ibid.*, **31**, 4037 (1966).
i) E. Tommila, *Acta Chem. Scand.*, **20**, 923 (1966).
j) E. Tommila and M. Savolainen, *ibid.*, **20**, 946 (1966).
k) H. Bader, A.R. Hansen and F.J. McCarty, *J. Org. Chem.*, **31**, 2319 (1966).
l) J. Sfiras and A. Demeilliers, *Researches* (Paris) **15**, 83 (1966).
m) L.C. Dorman, *J. Org. Chem.*, **31**, 3666 (1966).
338. a) D.J. Cram, and R. Uyeda, *J. Amer. Chem. Soc.*, **86**, 5466 (1964).
b) D.C. Berndt an W.J. Adams, *J. Org. Chem.*, **31**, 976 (1966).
339. H.H. Szmant, *Quat. Repts. Sulfur Chem.*, **3**, 147 (1968).
340. a) N. Furukawa and M. Kise *Kagakukogyo* **17**, 225 (1966).
b) S. Oae, *Kagakukogyo*, **19**, 241 (1968); see also *Yukigoseikagaku Kyokaishi*, **27**, 793 (1969)
341. F.A. French, *Chem. Eng. News*, **44**, (No. 15) 48 (1966).
342. C.A. Kingsbury and D.J. Cram, *J. Amer. Chem. Soc.*, **82**, 1810 (1960).
343. D.R. Rayner, E.G. Miller, P. Bickart, A.J. Gordon and K. Mislow, *ibid.*, **88**, 3138 (1966).
344. S. Oae, K. Tsujihara and N. Furukawa, *Tetrahedron Lett.*, 2663 (1970).
345. D.W. Emerson, A.P. Craig and I.W. Potts, Jr., *J. Org. Chem.*, **32**, 102 (1967).
346. D.W. Emerson and T.J. Korniski, *ibid.*, **34**, 4115 (1969).
347. D.N. Jones and M.A. Saeed, *Proc. Chem. Soc.*, 81 (1964); D.N. Jones and M.J. Green, *J. Chem. Soc.*, C. 532 (1967).
348. J.L. Kice and J.D. Campbell, *J. Org. Chem.*, **32**, 1631 (1967).
349. I.D. Entwistle and R.A.W. Johnstone, *Chem. Commun.*, 29 (1965).
350. C. Walling and L. Bollyky, *J. Org. Chem.*, **29**, 2699 (1964).
351. A. Deljac, Z. Stefanac and K. Balenovic, *Tetrahedron, Suppl.*, **8**, 33 (1966).
352. W. Carruthers, I.D. Entwistle, R.A.W. Johnstone and B.J. Millard, *Chem. & Ind.* (London), 342 (1966).
353. D.G. Barnard-Smith and J.F. Ford, *Chem. Commun.*, 120 (1965).
354. J. Reid-Shelton and K.E. Davis, *J. Amer. Chem. Soc.*, **89**, 718 (1967).
355. G.E. Hartzell and J.N. Paige, *ibid.*, **88**, 2616 (1966).
356. G.E. Hartzell and J.N. Paige, *J. Org. Chem.*, **32**, 459 (1967).
357. K. Kondo, A. Negishi and N. Fukuyama, *Tetrahedron Lett.*, 2461 (1969).
358. R.D.G. Cooper, *J. Amer. Chem. Soc.*, **92**, 5010 (1970).
359. N.J. Leonard and C.R. Johnson, *ibid.*, **84**, 3701 (1962).
360. N.J. Leonard and W.L. Ripple, *J. Org. Chem.*, **28**, 1957 (1963).
361. C.R. Johnson and D. McCants Jr., *J. Amer. Chem. Soc.*, **86**, 2935 (1964) or J.C. Martin and J.J. Uebel, *ibid.*, **86**, 2936 (1964).
362. a) F. Montanari, R. Danieli, H. Hogeveen and G. Maccagnari, *Boll. Sci. Fac. Chim. Ind. Bologna*, **21**, 257 (1963); *Tetrahedron Lett.*, 2685 (1964).
b) M. Cinquini, S. Colonna and F. Montanari, *J. Chem. Soc.*, C, 572 (1970).
363. M. Cinquini, S. Colonna and F. Montanari, *Tetrahedron Lett.*, 3181 (1966); M. Cinquini and S. Colonna, *Boll. Sci. Fac. Chim. Ind. Bologna*, **27**, 157 (1969); *Chem. Abstr.*, **72**, 120734 (1970).
364. S. Oae, *J. Amer. Chem. Soc.*, **78**, 4030 (1956).
365. R.L. Shriner, H.C. Struck and W.J. Jorison, *ibid.*, **52**, 2060 (1930).
366. E.A. Fehnel and A.P. Paul, *ibid.*, **77**, 4241 (1955).
367. J.I. Brauman and N.J. Nelson, *ibid.*, **88**, 2332 (1966).
368. E. Buncel, E.A. Symons and A.W. Zabel, *Chem. Commun.*, 173 (1965).
369. M. Kobayashi, H. Minato, 24th Annual Meeting Japan Chem. Soc., Osaka, 1971, Spring.
370. F.G. Bordwell and G.D. Cooper, *J. Amer. Chem. Soc.*, **74**, 1058 (1952).
371. J.D. Roberts, E.A. McElhill and R. Armstrong, *ibid.*, **71**, 2923 (1949).
372. J.D. Roberts, R.A. Clement and J.J. Drysdale, *ibid.*, **73**, 2181 (1951).
373. C.C. Price and J.J. Hydock, *ibid.*, **74**, 1943 (1952).
374. F.G. Bordwell and P.J. Boutan, *ibid.*, **79**, 717 (1957).

375. H.H. Szmant and G. Suld, *ibid.*, **78**, 3400 (1956).
376. S. Oae and C. Zalut, *ibid.*, **82**, 5359 (1960).
377. S. Oae, M. Yoshihara and W. Tagaki, *Bull. Chem. Soc. Japan*, **40**, 959 (1967).
378. H. Kloosterziel and H.J. Backer, *Rec. Trav. Chim.*, **72**, 185 (1953); *Chem. Abstr.*, **48**, 2635 (1954).
379. E.L. Wehry, *J. Amer. Chem. Soc.*, **89**, 41 (1967).
380. G. Maccagnani and F. Tadei, *Bol. Sci. Fac. Chim. Ind. Bologna*, **23**, 381 (1965).
381. D.L. Hammick and R.B. Williams, *J. Chem. Soc.*, 211 (1938).
382. S. Oae and Y.H. Khim., *Bull. Chem. Soc., Japan*, **40**, 1716 (1967).
383. S. Oae and Y.H. Khim, *ibid.*, **42**, 1622 (1969).
384. Y.H. Khim, W. Tagaki, M. Kise, N. Furukawa and S. Oae, *ibid.*, **39**, 2556 (1966).
385. J. Jacobus and K. Mislow, *J. Amer. Chem. Soc.*, **89**, 5228 (1967).
386. D.J. Cram, "Fundamentals of Carbonion Chemistry", Academic Press, N.Y., 1965 p. 43.
387. E.J. Corey and M. Chaykovsky, *J. Amer. Chem. Soc.*, **84**, 866 (1962).
388. J.I. Brauman, J.A. Bryson, D.C. Kahl and N.J. Nelson, *J. Amer. Chem. Soc.*, **92**, 6679 (1970).
388.5 J.I. Brauman, J.A. Bryson, D.C. Kahl and N.J. Nelson, *J. Am. Chem. Soc.*, **92**, 6679 (1970).
389. unpublished result.
390. C. Walling and L. Bollyky, *J. Org. Chem.* **28**, 256 (1963).
391. G.A. Russell, G.E. Jansen, H.D. Becker and F.J. Swentowski, *J. Amer. Chem. Soc.*, **84**, 2652 (1962); see also P. Bravo, G. Gaudiano and P.P. Ponti, *Chem. & Ind.* (London), 253 (1971).
392. a) H.D. Becker, G.J. Mikol, and G.A. Russell, *J. Amer. Chem. Soc.*, **85**, 3410 (1963).
     b) M. Saquent and A. Thuillier, *Bull. Soc. Chim. France*, 1582 (1966).
393. a) E.J. Corey and M. Chaykovsky, *J. Amer. Chem. Soc.*, **86**, 1639, 1640 (1964).
     b) P.G. Gassman and G.D. Richmond, *J. Org. Chem.*, **31**, 2355 (1966).
394. G.A. Russell, E. Sabourin and G.J. Mikol, *J. Org. Chem.*, **31**, 2854 (1966).
395. P.A. Aragabright, J.E. Hofmann and A. Schriesheim, *ibid.*, **30**, 3233 (1965).
396. C. Walling and L. Bollyky, *ibid.*, **29**, 2699 (1964).
397. R. Baker and M.J. Spillett, *Chem. Commun.*, 757 (1966).
398. Japan P12, 894 (1967) *Chem. Abstr.*, **68**, 29426 (1968).
399. G.A. Russell and H.D. Becker, *J. Amer. Chem. Soc.*, **85**, 3406 (1963).
400. G.A. Russell and S.A. Weiner, *J. Org. Chem.*, **31**, 248 (1966).
401. H. Nozaki, Y. Yamamoto and R. Noyori, *Tetrahedron Lett.*, 1123 (1966).
402. T.J. Broxton, Y.C. Mac, A.J. Parker and M. Ruane, *Aust. J. Chem.*, **19**, 521 (1966).
403. K. Sjoberg, *Tetrahedron Lett.*, 6383 (1966).
404. H. Moskowitz, J. Blanc-Guenee and M. Miocque, *C.R. Acad. Sci. Paris, Sec.*, C, **267**, 898 (1968).
405. G.A. Russell and L.A. Ochrymowycz, *J. Org. Chem.*, **34**, 3624 (1969).
406. F. Bergmann and D. Diller, *Israel J. Chem.*, **7**, 57 (1969).
407. G.A. Russell and E.T. Sabourin, *J. Org. Chem.*, **34**, 2336 (1969).
408. A. Nudelman and D.J. Cram, *ibid.*, **34**, 3659 (1969).
409. R.H. Schlessinger, G.S. Ponticello, A.G. Schultz, I.S. Ponticello and J.M. Hoffman, *Tetrahedron Lett.*, 3963 (1968).
410. J.R. Alexander and H. McCombie, *J. Chem. Soc.*, 1913 (1931); see also *Chem. Abstr.* **56**, 9944 (1962).
411. F. Montanari and A. Negrini, *Gazz. Chim. Ital.*, **89**, 1543 (1959); *Chem. Abstr.*, **55**, 3494 (1961).
412. D.J. Abbott and C.J.M. Stirling, *J. Chem. Soc.*, C. 818 (1969).
413. E.N. Prilezhaeva, L.V. Tsymbal and M.F. Shostakovskii, *Doklady Akad. Nauk SSSR*, **138**, 1122 (1961); *Chem. Abstr.*, **55**, 24595 (1961).
414. E. Bertotti, C. Luciani and F. Montanari, *Gazz. Chim. Ital.*, **89**, 1564 (1959).
415. S. Ghersetti, H. Hogeveen, G. Maccagnani, F. Montanari and F. Taddei, *J. Chem. Soc.*, 3718 (1963).
416. R. Baker and M.J. Spillett, *J. Chem. Soc.*, B, 481 (1969).
417. R. Baker and M.J. Spillett, *ibid.*, B, 581 (1969).
418. M. Nishio, *Chem. Commun.* 562 (1968) 51 (1969).

419. a) J.E. Baldwin, R.E. Hackler and R.M. Scott, *ibid.*, 1415 (1969).
     b) S. Wolfe, A. Rauk, L.M. Tel and I.G. Csizmadia, *J. Chem. Soc.*, (B), 136 (1971).
420. T. Durst, R.R. Fraser, M.R. McClory, R.B. Swingle, R: Viau and Y.Y. Wigfield, *Can. J. Chem.*, **48**, 2148 (1970); T. Durst, R. Viau and M.R. McClory, *J. Amer. Chem. Soc.*, **93**, 3077 (1971).
421. M. Cinquini, S. Colonna, U. Folli and F. Montanari, *Boll. Sci. Fac. Chim. Ind. Bologna*, **27**, 203 (1969).
422. B.J. Hutchinson, K.K. Andersen and A.R. Katritzky, *J. Amer. Chem. Soc.*, **91**, 3839 (1969).
423. R.R. Fraser and F.J. Schuber, *Chem. Commun.*, 397 (1969).
424. S. Wolfe, 4th Symposium on Organic Sulfur in Venice June 1971.
425. R.N. Loeppky and D.C.K. Chang, *Tetrahedron Lett.*, 5415 (1968).
426. M. Hojo and Z. Yoshida, *J. Amer. Chem. Soc.*, **90**, 4496 (1968).
427. S. Iriuchijima and G. Tsuchihashi, *Tetrahedron Lett.*, 5259 (1969).
428. G. Tsuchihashi, K. Ogura, S. Iriuchijima, *Synthesis*, 89, (1971); see also K.C. Tin and T. Durst, *Tetrahedron Lett.*, 4643 (1970), T. Durst and K.C. Tin, *Can. J. Chem.*, **49**, 2374 (1971).
429. G. Tsuchihashi and S. Iriuchijima, *Bull. Chem. Soc. Japan*, **44**, 1726 (1971).
430. G. Tsuchihashi and S. Iriuchijima, *ibid.*, **43**, 2271 (1970).
431. S. Iriuchijima and G. Tsuchihashi, *Synthesis*, 588 (1970); see also D. Martin, A. Berger and R. Peschel, *J. prakt. Chem.* **312**, 683 (1970).
432. M. Cinquini and S. Colonna, *Boll. Sci. Foc. Chim. Ind. Bologna*, **27**, 201 (1969).
433. E.W. Lawless and G.J. Hennon, *Tetrahedron Lett.*, 6075 (1968).
434. M.S. Brown, *J. Org. Chem.*, **35**, 2831 (1970).
435. T. Durst, *J. Amer. Chem. Soc.*, **91**, 1034 (1969).
436. T. Durst and K.C. Tin, *Tetrahedron Lett.*, 2369 (1970).
437. D.F. Tavares, R.E. Estep and M. Blezard, *ibid.*, 2373 (1970).
438. S.A. Khan, M. Ashraf and A.B. Chughtai, *Sci. Res.* (Dacca, Pakistan) **4**, 135 (1967); *Chem. Abstr.* **68**, 77425d (1968).
439. H.H. Szmant, "Chemistry of Sulfoxide Group" Chapter 16 in "Org. Sulfur Compds", edited by N. Kharasch, Pergamon Press (1961).
440. R. Curci, A. Giovine and G. Modena, *Tetrahedron*, **22**, 1235 (1966).
441. R. Curci and G. Modena, *ibid.*, **22**, 1227 (1966).
442. H.B. Henbest and S.A. Khan, *Chem. Commun.*, 1036 (1968).
443. K. Kikukawa, W. Tagaki, N. Kunieda and S. Oae, *Bull. Chem. Soc. Japan*, **42**, 831 (1969).
444. M.E.C. Biffin, J. Miller and D.B. Paul, *Tetrahedron Lett.*, 1015 (1969).
445. N. Marziano, P. Fiandaca, E. Maccarone, U. Romano and R. Passerini, *ibid.*, 4785 (1967).
446. R. Tanikaga, Y. Higashino and A. Kaji, *ibid.*, 3273 (1970).
447. A.G. Schultz and R.H. Schlessinger, *Chem. Commun.*, 1483 (1969).
448. T. Shirafuji, Y. Yamamoto and H. Nozaki, *Tetrahedron Lett.*, 4097 (1969).
449. A.S.F. Ash, F. Challenger and D. Greenwood, *J. Chem. Soc.*, 1877 (1951).
450. M.A. McCall, D.S. Tarbell and M.A. Havill, *J. Amer. Chem. Soc.*, **73**, 4476 (1951).
451. D.S. Tarbell and M.A. McCall, *ibid.*, **74**, 48 (1952).
452. S. Oae, T. Aida, K. Tsujihara and N. Furukawa, *Tetrahedron Lett.*, 1145 (1971).
453. S. Oae, T. Aida, and N. Furukawa, *Int. J. Sulfur Chem.*, (A) (in press).
454. A. Nakanishi and S. Oae, *Chem. & Ind.*, (London) 960 (1971).
455. S. Oae and Y. Tsuchida, K. Tsujihara and N. Furukawa, *Bull. Chem. Soc. Japan*, **45**, 2856 (1972).
456. B.H. Nicolet and J. Willard, *Science*, **53**, 217 (1921).
457. F.G. Mann and W.J. Pope, *J. Chem. Soc.*, **121**, 1052 (1922).
458. P.A. Briscoe, Thesis Univ. ·of Leeds; cited in F. Challenger "The Properties of Some N-Chloramide (chloramines) and Sulfilimines "chapter 29, N.Kharasch. Ed. "Org. Sulfur Compds." Pergamon Press (1961).
459. J. Day and D.J. Cram. *J. Amer. Chem. Soc.*, **87**, 4398 (1965).
460. D.J. Cram, J. Day, D.R. Rayner, D.M. von Schriltz, D.J. Duchamp and D.C. Garwood, *ibid.*, **92**, 7369 (1970).
461. C.R. Johnson, J.J. Rigau, M. Haake, D. McCants, J.E. Keiser and A. Gertsema, *Tetrahedron Lett.*, 3719 (1968).

462. H. Kobayashi, N. Furukawa, T. Aida and S. Oae, *ibid.*, 3109 (1971).
463. R. Appel and W. Buechner, *Chem. Ber.*, **95**, 855 (1962).
464. a) N. Furukawa, T. Yoshimura and S. Oae, *Bull. Chem. Soc. Japan*, **45**, 2019 (1972)
    b) P.M. Manya, A. Sekera and P. Rumph, *Bull. Soc. Chim. Fr. 286 (1971).*
465. K.K. Andersen, W.H. Edmonds, J.B. Biasotti and R.A. Strecker, *J. Org. Chem.*, **31**, 2859 (1966).
466. K. Tsujihara, N. Furukawa and S. Oae, *Bull. Chem. Soc. Japan*, **43**, 2153 (1970).
467. S. Oae, K. Tsujihara and N. Furukawa, *Tetrahedron Lett.*, 2663 (1970).
468. K. Tsujihara, K. Harada, N. Furukawa and S. Oae, *Tetrahedron*, **27**, 6101 (1971).
469. R. Appel, W. Buechner and E. Guth, *Ann.*, **618**, 53 (1968).
470. R. Appel and W. Buechner, *Chem. Ber.*, **95**, 855 (1962).
471. H.R. Bentley, J.K. Whitehead, *J. Chem. Soc.*, 2081 (1950).
472. H.R. Bentley, E.E. McDermott and J.K. Whitehead, *Nature*, **165**, 735 (1950).
473. D.R. Rayner, D.M. von Schriltz, J. Day and D.J. Cram, *J. Amer. Chem. Soc.*, **90**, 2721 (1968).
474. G. Kresze and B. Wustrow, *Chem. Ber.*, **95**, 2652 (1962).
475. J. Holloway, J. Kenyon and H. Phillips, *J. Chem. Soc.*, 3000 (1928).
476. K. Tsujihara, T. Aida, N. Furukawa and S. Oae, *Tetrahedron Lett.*, 3415 (1970).
477. A.S.F. Ash, F. Challenger and D. Greenwood, *J. Chem. Soc.*, 1877 (1951).
478. A.S.F. Ash, F. Challenger, T.S. Stevens and J.L. Dunn, *ibid.*, 2792 (1952).
479. P.A. Briscoe, F. Challenger and P.S. Duckworth, *ibid.*, 1955 (1956).
480. J. Meisenheimer, *Chem. Ber.*, **52**, 1667 (1919).
481. J. Meisenheimer, H. Greeske and A. Willmersdorf, *ibid.*, **55**, 513 (1922).
482. R.F. Kleinschmidt and A.C. Cope, *J. Amer. Chem. Soc.*, **66**, 1929 (1944).
483. A.C. Cope and P.H. Towle, *ibid.*, **71**, 3931 (1949).
484. K. Mislow, *Chem. Progress*, **28**, 217 (1967).
485. J. Petranek and M. Vecera, *Coll. Czeck. Chem. Commun.*, **24**, 2191 (1959).
486. P. Claus and W. Vycudilik, *Tetrahedron Lett.*, 3607 (1968).

Chapter 9

# SULFONIUM SALTS

C.J.M. Stirling

*Department of Chemistry*

*University College of North Wales, Bangor LL57 2UW*

## 9.1 INTRODUCTION

The term "sulfonium salt" is applied to compounds which contain a tricoordinate sulfur atom bearing a positive charge on sulfur. Sulfonium salts therefore have the general structure: $ABC\overset{+}{S}\overset{-}{X}$. This is the most general description of sulfonium salts; this chapter will deal with those sulfonium salts in which at least two of the bonds to sulfur are from carbon. In most instances the third bond to sulfur will also be from carbon, but mention will be made of salts with bonds from sulfur to oxygen, sulfur, nitrogen, halogen, and hydrogen.

The chemistry of sulfonium salts has been briefly reviewed.[1] Aspects of bonding in sulfonium compounds have been reviewed by Price and Oae.[2]


## 9.2 PREPARATION

### 9.2.1 Alkylation of Sulfides

The sulfur atom in dialkyl sulfides is weakly nucleophilic and reaction with alkyl halides yields sulfonium salts directly[3]:

$$R_1 R_2 S + R_3 X \longrightarrow R_1 R_2 R_3 \overset{+}{S}\overset{-}{X} \qquad (9.1)$$

This simple reaction is promoted by the use of a polar solvent such as methanol because the transition state for the reaction is more polar than the starting materials and the rate of reaction decreases with chain length in the alkyl halide in accordance with general experience in nucleophilic substitution at $sp^3$ carbon. Methyl halides are the most reactive of the simple alkyl halides. In general, iodides are most reactive; in simple alkylation reactions, sulfides are less reactive than amines.[4]

Alkylation rates of a number of sulfides with a number of halides are listed by Reid[5]; many other alkylating agents can be applied to the formation of sulfonium salts from dialkyl sulfides. Dialkyl sulfates,[6] alkyl sulfonates,[7] including methyl fluorosulfonate[8] and nitrates[9] have been used.

For methionine, alkylation leads directly to the sulfonium salt[10] and not to the N-methyl derivative; the much greater basicity of nitrogen results in the amino-carboxyl zwitterion being formed and sulfur remains as the most nucleophilic centre.

When carbonium ions are generated in the presence of dialkyl sulfides, these are trapped with formation of sulfonium salt, and treatment of alcohols with, for example, strong acids in the presence of dialkyl sulfides leads to sulfonium sulfates directly.[11] Sulfonium salts are generated directly from a dialkyl sulfide in the presence of sulfuric acid. Thus treatment of dibenzyl sulfide with sulfuric acid gives the protonated sulfide which cleaves to give thiol and a stabilized carbonium ion. This ion, with the sulfide, gives the trialkyl sulfonium sulfate[12]:

$$PhCH_2SCH_2Ph + \overset{+}{H} \rightleftharpoons PhCH_2\overset{\overset{H}{|}}{\underset{+}{S}}CH_2Ph \tag{9.2}$$

$$(PhCH_2)_3\overset{+}{S} \ \overset{-}{HSO_4} \xleftarrow{PhCH_2SCH_2Ph} Ph\overset{+}{C}H_2 + HSCH_2Ph$$

Reaction of diazonium ions with dialkyl sulfides also produces sulfonium salts, this reaction undoubtedly occurring by way of preliminary decomposition of the diazonium ion to nitrogen and carbonium ion with subsequent nucleophilic attack by the sulfide upon the carbonium ion[13]:

$$\tag{9.3}$$

Formation of sulfonium salts can also result from nucleophilic attack of dialkyl sulfides upon cyclopropanes[14]:

$$\tag{9.4}$$

Related to this sequence is the formation[15] of a sulfonium salt by nucleophilic addition of a dialkyl sulfide to an electrophilic double bond in the presence of a proton donor. The sulfonium salt produced then has as counter ion that of the proton donor:

$$EtSCH_2CO_2H + CH_3CH=CHCO_2H \xrightarrow{HCl} Et\overset{+}{\underset{-}{S}}Cl \overset{\overset{CH_2CO_2H}{\diagup}}{\underset{\diagdown}{}} \tag{9.5}$$

$$CH(Me)CH_2CO_2H$$

Formation of bis-sulfonium salts by double alkylation of bis-sulfides has been achieved.[16]

Aryl sulfides and conjugated vinylic and acetylenic sulfides are difficult to alkylate with formation of sulfonium salt, because of the reduced nucleophilicity of the sulfur atom occasioned by $p\pi$ delocalization,[17] and the use of very reactive alkylating agents, particularly trialkyloxonium salts[18] and methyl fluorosulfonate[8] are effective for these substrates:

$$R_2S + R_3\overset{+}{O}\overset{-}{BF_4} \longrightarrow R_3\overset{+}{S}\overset{-}{BF_4} + R_2O \qquad (9.6)$$

Arylation of alkyl and aryl sulfides with diaryl iodonium salts is effective[19]:

$$Ph_2\overset{+}{I}\overset{-}{BF_4} + Ph_2S \longrightarrow Ph_3\overset{+}{S}\overset{-}{BF_4} + PhI \qquad (9.7)$$

Alkylation of alkyl sulfides with alkyl halides is subject to metal catalysis, which undoubtedly promotes heterolysis of the alkyl halide and raises reaction rate. Mercury, zinc, and ferric salts have been found[20] to be effective in this respect.

Formation of thiosulfonium salts $R\overset{+}{S}SRR'$ is achieved by alkylation of disulfides[21]:

$$MeSSMe + Me_3\overset{+}{O}\ \overset{-}{O_3}S \overbrace{\qquad}NO_2 \longrightarrow Me_2\overset{+}{S}-SMe\ \ \overset{-}{O_3}S\overbrace{\qquad}NO_2 \qquad (9.8)$$

### 9.2.2 Intramolecular Alkylation

Formation of cyclic sulfonium salts occurs when the sulfur atom and the leaving group are present in the same molecule. The formation of three-membered ring sulfonium salts (epi-sulfonium salts) has been extensively studied because of their formation as intermediates in addition of sulfinyl halides to olefins and in solvolysis of alkyl sulfides bearing $\beta$-leaving groups. These particular salts are considered separately.

Formation of many types of cyclic sulfonium salt by intramolecular alkylation has been observed. In formation of three-membered ring sulfonium salts, rates are dependent[22] upon the group attached to sulfur in the $\beta$-halogeno-alkyl sulfide as well as upon the groups attached to the carbon atoms linking the leaving group and the sulfur atom. Replacement of an alkyl group on sulfur by an aryl group gives approximately the same rate of cyclization in spite of the great reduction of the nucleophilicity of the sulfur atom. This is attributed to conjugation between the partially formed 3-membered ring in the cyclization transition state and the attached aryl group, which is, of course, not possible in the case of the alkyl sulfide. Likewise when aryl groups are attached to either the $\alpha$ or $\beta$ carbon atoms the resulting rate enhancement[22] is attributed to the same phenomenon, although substitution does in any event promote 3-membered ring formation.[23]

Intramolecular alkylation to give 4-membered ring sulfonium salts ($S_4$ participation) has been suggested to occur in a very favorable conformational system[24]; the rate of formation of a 5-membered ring sulfonium salt is greater than that of the corresponding 6-membered ring compound.[25, 26, 27] Formation of bridged cyclic sulfonium salts has been achieved[28] by intramolecular alkylation:

$$(9.9)$$

and[29]

$$(9.10)$$

Intramolecular alkylation of sulfides by carbonium ions is an equilibrium process in suitable solvents.[30] Thus the triarylmethanol **9.1** in liquid $SO_2$ gives the ion **9.2** as shown by nmr and uv spectra and, with two sulfur side chains **9.3**, nmr spectra at $10°$ show one coordinated and one free side chain, although at low concentrations some uncoordinated ion is present. At $55°C$, the different methyl and methylene signals coalesce showing rapid fluctuation of the tertiary carbon between the sulfur centers. The greater the degree of conjugative electron donation the more stable becomes the uncoordinated ion, and with one $p$-$NMe^2$ substituent, there is no evidence for sulfonium ion formation.

|   **9.1**   |   **9.2**   |   **9.3**   |

The isolation of an epi-sulfonium salt as a trinitrobenzenesulfonate[31] is noteworthy:

$$(9.11)$$

Neighboring group participation by sulfur in olefin bromination involves intramolecular formation of sulfonium salts[32] (*cf.* ref. 29):

$$(9.12)$$

### 9.2.3 Formation of Sulfonium Salts by Aromatic Electrophilic Substitution

The poor nucleophilicity of aryl sulfides has been referred to and most arylsulfonium salts have been prepared by methods other than alkylation at sulfur. Formation of a sulfur electrophile is conveniently achieved by protonation of a sulfoxide in the presence of, for example, sulfuric acid, and attack by this electrophilic species on an activated nucleus produces a sulfonium salt[33]:

$$(9.13)$$

This type of electrophilic substitution requires nuclei activated by oxygen or by sulfur atoms in this simple form. Formation of sulfonium salts from aromatic hydrocarbons and sulfoxides in the presence of Lewis acids such as aluminium trichloride has been achieved[34, 35]:

$$ArH + Ar_2SO + AlCl_3 \longrightarrow Ar_3\overset{+}{S} \, [HOAlCl_3]^-$$

$$(9.14)$$

In very similar sequences, treatment of benzene with diaryl sulfide dichlorides in the presence of aluminium chloride[36] and treatment of phenols with thionyl chloride yield [37] triarylsulfonium salts, in the latter instance with hydroxyl substituted nuclei. Electrophiles produced from protonated sulfoxides are probably involved in these reactions also. Formation of triarylsulfonium salts from treatment of aromatic hydrocarbons with thionyl chloride in the presence of aluminium chloride takes a similar course.[38]

Other references to arylsulfonium salts are given by Reid. [39]

### 9.2.4 Formation of Sulfonium Salts by Carbanion Addition to the Sulfinyl Group

This method has been mainly used for the rather inaccessible triaryl sulfonium salts. Treatment of, for example, diphenyl sulfoxide with phenylmagnesium bromide and working up in the presence of hydrobromic acid gives triphenylsulfonium bromide in 50% yield[40]:

$$Ph_2SO + ArMgBr \longrightarrow [Ph_2SAr]^+ [OMgBr]^- \xrightarrow{HBr} Ph_2\overset{+}{S}Ar\text{-}\overset{-}{Br} \quad (9.15)$$

### 9.2.5 Formation of Sulfonium Salts from Sulfides and Other Electrophiles

#### 9.2.5.1 Halosulfonium Salts

Halosulfonium salts of structure $R_2\overset{+}{S}X\ X^-$ are produced by nucleophilic attack of sulfides on halogens[41]; they are stable in the presence of Lewis acids:

$$Me_2S + SbCl_5 + Cl_2 \longrightarrow Me_2\overset{+}{S}\text{-}Cl\,(SbCl_6)^- \quad (9.16)$$

Sulfide dichlorides may be assigned the halosulfonium salt structure. They are generally unstable compounds and their only important reaction is the reversion reaction in which attack of the counter halide ion upon carbon adjacent to sulfonium sulfur occurs:

$$PhSCH_2Ph + Cl_2 \underset{}{\overset{fast}{\rightleftharpoons}} PhS\text{-}CH_2Ph \xrightarrow[AcOH]{slow} \quad (9.17)$$

$$PhCH_2Cl + PhCH_2OAc$$

The reversion reaction will be dealt with in greater detail below and for these sulfonium salts leads to "chlorinolysis" of the sulfide.[42, 43]

A halosulfonium salt is implicated in ring expansions of the type[44]:

$$(9.18)$$

### 9.2.5.2 Hydrosulfonium Salts

The simplest nucleophilic reaction of a sulfur compound with an electrophile is protonation of a sulfide. Hydrosulfonium salts are clearly intermediates in reactions in which a stabilized carbonium ion can be developed from the protonated sulfide as described earlier. Basicity of sulfides is considerably less than that of ethers or of amines, and reactions which depend upon initial protonation of the sulfide such as Zeisel-type dealkylations (below) are slow (eq. 9.19). [45]

$$(9.19)$$

It must be emphasized, however, that there is a large difference between hydrogen and carbon nucleophilicity: sulfur is a much more effective neighboring group than oxygen in, for example, the solvolysis of $\omega$-halogenoalkyl sulfides.[27]

### 9.2.5.3 Oxosulfonium Salts

Oxosulfonium salts are produced by oxygen protonation of sulfoxides and are probably the reactive electrophiles in formation of sulfonium salts by electrophilic substitution (above).

These intermediates are also probably involved in the epimerization of simple sulfoxides[46] and $\beta$-carboxy-sulfoxides[47] catalyzed by halide ion and protonic acids:

$$(9.20)$$

S-Hydroxy-S-phenylthiosulfonium salts, suggested as reaction intermediates, are obtained[48] by protonation of thiolsulfinic esters:

$$PhS\text{-}SOPh + \overset{+}{H} \;\rightleftharpoons\; PhS\text{-}\overset{+}{S}\text{-}Ph$$
$$\underset{OH}{|}$$

$$(9.21)$$

Alkoxysulfonium salts result from oxygen alkylation of sulfoxides. Displacement of alkoxy groups from such compounds in the course of borohydride reduction[49] has been investigated:

$$\underset{\underset{\text{OMe}}{|}}{\overset{+}{\text{PhSMe}}} \xrightarrow{\ ^-\text{BH}_4\ } \underset{\underset{\text{H}}{|}}{\overset{+}{\text{PhSMe}}} \xrightarrow{\ ^-\overset{+}{\text{H}}\ } \text{PhSMe} \qquad (9.22)$$

### 9.2.5.4 Iminosulfonium Salts

Reaction of sulfides with N-chloroamines yields iminodialkylsulfonium chlorides[50]; treatment of the iminosulfonium salt with a mild base gives the iminosulfuran (9.4):

$$\text{ClNHCO}_2\text{Et} + \text{Me}_2\text{S} \longrightarrow \text{Me}_2\overset{+}{\text{S}}\text{-NHCO}_2\text{Et} \ \ \bar{\text{Cl}}$$

$$\Big\downarrow \text{Et}_3\text{N} \qquad (9.23)$$

$$\text{Me}_2\text{S=N-CO}_2\text{Et} \longleftrightarrow \text{Me}_2\overset{+}{\text{S}}\text{-}\bar{\text{N}}\text{CO}_2\text{Et}$$

**9.4**

### 9.2.6 Oxidative Formation of Sulfonium Salts: Thiapyryllium Salts

Hydride ion transfer from a cyclic sulfide (9.5) can lead to formation [51] of thiapyryllium salt (9.6) when a suitably conjugated substrate is used:

$$(9.24)$$

**9.5**                                    **9.6**

Loss of hydride ion may, as in these cases, occur from a halosulfonium or other salt[52] in the presence of a suitable oxidizing agent, or may be effected by a stabilized carbonium ion such as triphenylmethyl[53]:

$$(9.25)$$

An interesting variation on these reactions is the cycloaddition of thiophosgene to 1,4-diphenylbutadiene yielding[54] the cyclic sulfide 9.7, dehydrogenated by excess thiophosgene to the thiapyryllium salt 9.8:

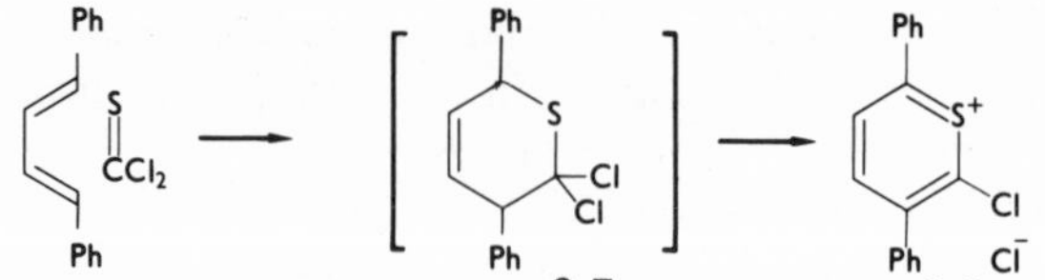

$$\text{(9.26)}$$

**9.7**                    **9.8**

Loss of nitrogen from the diazonium ion **9.9** yields[55] the thiapyryllium salt **9.10**, and rearrangement of the substituted thiophene **9.11** to a thiapyryllium ion has been observed in the mass spectrometer[55a]:

**9.9**              **9.10**              **9.11**

## 9.3 CHARACTERIZATION OF SULFONIUM SALTS

A wide variety of sulfonium salts, e.g., halides, crystallize well and do not show difficulties in characterization. It is often convenient, however, to change the common halide counter ion, for example, to picrate. This is simply achieved by treatment of the sulfonium halide with aqueous sodium picrate and the sulfonium picrate is precipitated. Methosulfates derived from alkylations with dimethyl sulfate also crystallize well.[56]

Other more complex counter ions have been employed in difficult cases. Episulfonium (thiiranium) salts, for example, have been characterized as their 2,4,6-trinitrobenzenesulfonates.[31]

Characterization of sulfonium salts as complexes with other metal salts has been commonly used and, in particular, complexes with salts of platinum, gold, and mercury have been applied.[57]

## 9.4 STRUCTURE AND STEREOCHEMISTRY OF SULFONIUM SALTS

Sulfonium salts are ionic and simple members of the series show high solubilities not only in water, but also in organic solvents such as chloroform.[58] Electrolytic dissociation in a range of solvents has been determined.[59]

Dissociation of sulfonium salts to ion pairs in non-polar solvents has been investigated by Hyne[60]; typically, trimethylsulfonium iodide ($10^{-3}$ M in EtOH) is present to the extent of 10% as ion-pairs.

Equivalence of C-S bonds is shown[61] by the formation of identical sulfonium salts in the reaction of methyl iodide with diethyl sulfide and of ethyl iodide with methyl ethyl sulfide, with subsequent characterization of the salts as complexes with heavy metal chlorides.

Sulfonium salt stereochemistry was pioneered by Pope and Peachey,[62] who resolved the ion **9.12** into enantiomeric forms as the

$$MeEt\overset{+}{S}CH_2CO_2HX^-$$

**9.12**

(+)-camphorsulfonate. Treatment of methyl ethyl sulfide with (+)-menthyl bromoacetate gives directly a diastereoisomeric mixture of sulfonium salts[63]:

$$MeSEt + BrCH_2CO_2\text{-(+)-menthyl} \rightarrow MeEt\overset{+}{S}\text{-}CH_2CO_2\text{-(+)-menthyl } \overset{-}{Br}$$

(9.27)

Likewise the bis-sulfonium salt $MeEt\overset{+}{S}CH_2CH_2\overset{+}{S}EtMe$ can be isolated[64] in diastereoisomeric forms, again demonstrating the pyramidal configuration of the sulfur atom.

The Raman spectrum[65] of trimethylsulfonium bromide suggests a pyramidal structure of the ion with sulfur at the apex, and X-ray crystallography[66] of tetrahydrothiophen bromosulfonium bromide **9.13** shows that the sulfur atom is pyramidal:

**9.13**

These observations are consistent with the idea that in sulfonium salts $3p$ bonding orbitals are used producing pyramidal $3p$ geometry, the unshared electron pair on sulfur remaining in the $3s$ orbital.

Pyramidal inversion in sulfonium salts is analogous to that found[67] for other sulfur compounds such as sulfoxides.

It has been concluded[68] that racemization of $t$-butylethylmethyl-sulfonium perchlorate involves neither heterolysis nor reversion because racemization occurs faster than solvolysis in a variety of solvents. The reversion reaction cannot be involved because the counter ion is the non-nucleophilic perchlorate ion and, while replacement of hydrogen atoms in a $t$-butyl group by methoxy and by methyl groups produces the expected decrease and increase respectively in solvolysis rate, the rates of racemization are affected in the opposite sense. Pyramidal inversion of l-adamantylethylmethylsulfonium perchlorate is also known to occur.[69]

Cyclic sulfonium salts have been partially resolved and because racemization in acetic acid at $100°C$ was much slower than for acyclic sulfonium salts, pyramidal inversion is suggested.[69a]

**TABLE 9.1**

Comparison of Ultraviolet Spectra of Benzene Derivatives Containing Sulfonium and Other Substituents

| R | R' | X | Ph-X $\lambda_{max}$ (log $\epsilon$) | Ph-X $\lambda'_{max}$ (log $\epsilon$) | HO(R,R') $\lambda_{max}$ (log $\epsilon$) | HO(R,R') $\lambda'_{max}$ (log $\epsilon$) | $\bar{O}$(R,R') $\lambda_{max}$ (log $\epsilon$) | $\bar{O}$(R,R') $\lambda'_{max}$ (log $\epsilon$) | HO(R) $\lambda_{max}$ (log $\epsilon$) | HO(R) $\lambda'_{max}$ (log $\epsilon$) | $\bar{O}$(R) $\lambda_{max}$ (log $\epsilon$) | $\bar{O}$(R) $\lambda'_{max}$ (log $\epsilon$) |
|---|---|---|---|---|---|---|---|---|---|---|---|---|
| H | H | H | 203.5, 3.87 | 254, 2.31 | 210.5, 3.79 | 270, 3.16 | 235, 3.97 | 287, 3.41 | 210.5, 3.79 | 270, 3.16 | 235, 3.97 | 287, 3.41 |
| H | H | $\overset{+}{S}Me_2\,\bar{Cl}$ | 220, 3.94 | 265, 3.04 | 242, 4.03 | 264, 3.06 | 269, 4.29 | — | 216, 4.1 | 286, 3.27 | 218.5, 4.34 | 312, 3.41 |
| H | H | $\overset{+}{N}Me_3\,\bar{Cl}$ | 203, 3.87 | 254, 2.31 | 221, 3.93 | 269, 3.33 | 244, 4.15 | 284, 3.60 | 220, 3.85 | 272, 3.36 | 238, 4.06 | 293, 3.60 |
| H | H | $MeSO_2$ | 217, 3.83 | 264, 2.99 | 239, 4.25 | 268, 3.22 | 268, 4.31 | — | 223, 3.82 | 286, 3.50 | 248, 3.92 | 314, 3.60 |
| Me | H | $\overset{+}{S}Me_2\,\bar{Cl}$ | — | | 245, 4.00 | 271, 3.46 | 274, 4.11 | — | 215, 4.06 | 295, 3.54 | 218, 4.25 | 324, 3.58 |
| Me | H | $\overset{+}{N}Me_3\,\bar{Cl}$ | — | | 221, 3.86 | 271, 3.08 | 243, 4.04 | 289, 3.37 | 224, 4.01 | 277, 3.41 | 243, 4.08 | 298, 3.55 |
| Br | Br | H | — | | 218, 4.02 | 280, 3.33 | 242, 3.72 | 298, 3.51 | | | | |
| Br | Br | $\overset{+}{S}Me_2\,\bar{Cl}O_4$ | | | 259, 3.85 | 293, 3.30 | 287, 4.14 | | | | | |

## 9.5  ULTRAVIOLET SPECTRA OF SULFONIUM SALTS

Sulfonium substituents cause[70] marked bathochromic shifts on the spectra of aromatic compounds. This is in contrast to ammonium substituents where no such effect is observed. Further, when other conjugative substituents are present in the nucleus, further bathochromic shifts are observed which can be attributed to the ability of the sulfonium group to conjugate with the nucleus and with electron-releasing conjugative substituents in the excited state. Relevant data are in Table 9.1.

Another important feature of the uv spectral properties of aromatic sulfonium salts is that conjugation with sulfonium substituents is not appreciably inhibited by large *ortho* groups which are known to inhibit $p\pi$ resonance sterically in compounds such as aromatic tertiary amines. It is concluded that overlap between the $\pi$ orbital of the benzene nucleus and $d$ orbitals on sulfur permits the sulfonium substituent to accept electrons conjugatively without stereoelectronic restriction.

A bathochromic shift due to the insertion of a sulfonium substituent is also found with *meta*-substituted derivatives. In these instances direct conjugation is clearly impossible, and it has been suggested[71] that an electron-accepting resonance interaction with the *meta* substituent is possible in the excited state and involves canonical structures such as:

(9.28)

## 9.6  REACTIONS OF SULFONIUM SALTS

### 9.6.1  Electronic Effects of the Sulfonium Substituent

Before reactions of sulfonium salts are discussed, the effect of the sulfonium group on its molecular environment must be considered. Because of the positive charge on sulfur, the sulfonium substituent produces a strong dipole with resultant weakening[72] of the bond between carbon and sulfur. Polarization of the bond which attaches the sulfonium substituent to the rest of the molecule and the ability of the sulfur atom to accept the electron pair of this bond confer leaving-group ability on the sulfonium substituent, and we shall see many examples of reactions in which the sulfonium group is displaced with formation of a sulfide. In this respect the sulfonium group is a better leaving group from $sp^3$ carbon than an ammonium group.[73]

A positive sulfonium pole attached to an aromatic nucleus has the expected profound effect on the reactivity of that nucleus towards electrophiles. In addition, the ionic nature of sulfonium salts renders their reactivity much more subject to solvent effects than is the case with electrically neutral substrates. The electron-accepting conjugative property of the sulfonium group on the ultraviolet spectra of compounds has already been alluded to and the effect is also seen in the stabilization by sulfonium substituents of the conjugate bases of Brønsted acids of various types. The most striking instances are found in the stabilization of carbanions adjacent to a sulfonium substituent; these species are the sulfonium ylides which have been the subject of much recent work. A great many other reactions which involve either carbanion or carbanionic intermediates are profoundly affected by the presence of sulfonium substituents in the substrate. The effect of the sulfonium group may be purely inductive or may be a combination of inductive and special resonance factors. The present controversy in this area is a measure of the lack of true understanding of the forces which are operative.

### 9.6.2 Effect of the Sulfonium Group in Anion Stabilization

In connection with the problem of whether or not sulfur is able to expand its valency shell and behave as an electron-accepting conjugative substituent ($d$-orbital resonance), much comparison has been made of the effects of ammonium and sulfonium substituents upon the dissociation constants of carboxylic acids and phenols. The reasoning is based on the conclusion that for nitrogen, $d$ orbitals are not available for electron-acceptive conjugation and that therefore any significant differences between the effects of ammonium and sulfonium substituents on acid dissociation reveal any tendency on the part of sulfur to expand its valency shell.

For benzoic acid, there is no direct conjugation between the 'onium substituent and the carboxylate group. The data show that the difference in dissociation constant between the *para* and *meta* ammonium salts and

**TABLE 9.2**
Dissociation Constants of Benzoic Acids and Phenols Bearing 'Onium Substituents

| Substance | Ref. | Solvent | $pK_a$ | $\sigma$ | $\sigma_m - \sigma_p$ |
|-----------|------|---------|--------|----------|------------------------|
| $p$-$(CH_3)_3\overset{+}{N}C_6H_4COOH$ | 74 | 50% EtOH | 4.42 | 0.88 | |
| $m$-$(CH_3)_3\overset{+}{N}C_6H_4COOH$ | 74 | 50% EtOH | 4.22 | 1.02 | +0.14 |
| $p$-$(CH_3)_2\overset{+}{S}C_6H_4COOH$ | 70 | $H_2O$ | 3.27 | 0.90 | |
| $m$-$(CH_3)_2\overset{+}{S}C_6H_4COOH$ | 70 | $H_2O$ | 3.22 | 1.00 | +0.10 |
| $p$-$(CH_3)_3\overset{+}{N}C_6H_4OH$ | 70,75 | $H_2O$ | 8.35 | 0.76 | |
| $m$-$(CH_3)_3\overset{+}{N}C_6H_4OH$ | 70,75 | $H_2O$ | 8.03 | 0.84 | +0.08 |
| $p$-$(CH_3)_2\overset{+}{S}C_6H_4OH$ | 70,75 | $H_2O$ | 7.30 | 1.16 | |
| $m$-$(CH_3)_2\overset{+}{S}C_6H_4OH$ | 70,75 | $H_2O$ | 7.67 | 1.00 | −0.16 |

the *para* and *meta* sulfonium salts is similar and that in each case the *meta* acid is the stronger. This is consistent with the separation dependence of the operation of the inductive effect of the 'onium pole in stabilizing the carboxylate ion. By contrast, for the dissociation of phenols, there is the possibility of direct conjugation between the oxygen atom of the phenoxide ion and a *p* substituent, an effect which renders the use of simple Hammett sigma values inapplicable for substituents in which direct conjugation is possible. It can be seen that for ammonium salts, the *meta* derivative is again the stronger acid, and again the greater inductive effect of the *meta* than the *para* substituent can be considered to be responsible. The difference between $\sigma_p - \sigma_m$ is negative. For the sulfonium substituted phenols, however, the *p* derivative is the stronger acid and $\sigma_p - \sigma_m$ is positive. A reasonable explanation for this observation is that the electron accepting *conjugative* effect of the sulfonium pole is now appreciable. This can operate so as to stabilize the phenoxide ion only when the sulfonium substituent is *para* to oxygen. Its effect is sufficient to dominate the greater inductive effect of the closer *meta* substituent. The effect can be represented:

It is important that one of the canonical forms does not involve charge separation.

Further evidence that *d*-orbital resonance[76] is involved in the stabilization of phenoxide ions bearing conjugated sulfonium groups comes from the pattern of dissociation constants of phenols bearing *p*-dimethylsulphonium substituents flanked by bulky groups (Table 9.3). The effect of bulky groups is undoubtedly to force the substituent out of the plane of the nucleus, and for the familiar $p\pi$ resonance involved with substituents such as dimethylamino, the effect of this enforced non-planarity is to inhibit resonance interaction between nucleus and substituent. Orbital overlap is severely reduced and the stereoelectronic requirements are therefore quite strict. For a resonance interaction in which *d* orbitals are involved, however, the spatial distribution of the *d* orbitals renders them less sensitive to deviation from precise alignment, and steric inhibition of resonance is not observed.

Stabilization of carbanions by adjacent sulfonium groups has been much studied, particularly in connection with ylide chemistry (below). Some dissociation constants of sulfonium salts are compared with other 'onium salts (Table 9.4).

TABLE 9.3

$pK_a$ Data for 'Onium-Substituted Phenols[71]

|  | $pK_a$ | $\sigma$ | $\Sigma\sigma$ |
|---|---|---|---|
| 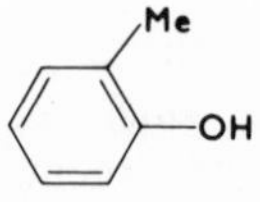 | 10.02 | −0.03 | |
| 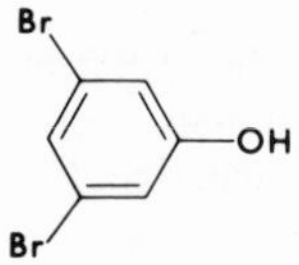 | 8.01 | 0.85 | |
| 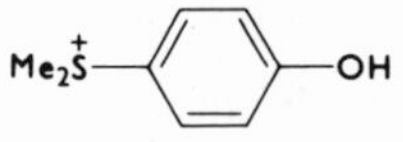 | 7.30 | 1.16 | |
| 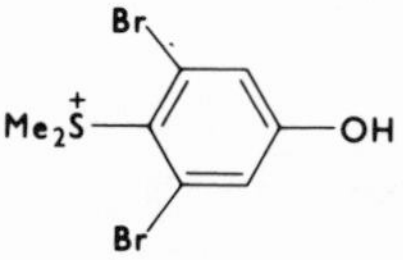 | 5.30 | 2.03 | 2.04 |
| 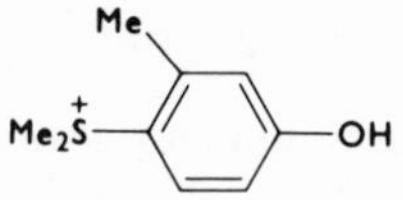 | 7.60 | 1.03 | 1.13 |

TABLE 9.4

$pK_a$ Data for Sulfonium Salts (Water at 25°C)

| Salt | $pK_a$ | ref. | Salt | $pK_a$ | ref. |
|---|---|---|---|---|---|
| $Me_2\overset{+}{S}CH_2COPh$ | 8.25 | 77 | $Me_3\overset{+}{P}CH_2COPh$ | 9.2 | 77 |
| $Me_2\overset{+}{S}CH_2CH_2OPh$ | 18.9 | 78, 79 | $(n\text{-}Bu)_3\text{9-fluorenyl-}\overset{+}{P}$ | 8.0 | 80 |
| $Me_2\text{-9-fluorenyl-}\overset{+}{S}$ | 7.3 | 80 | $Ph_3\text{-9-fluorenyl-}\overset{+}{As}$ | 7.8 | 81 |
| $Me_3\text{-9-fluorenyl-}\overset{+}{N}$ * | ~24 | 82 | $Ph_3\text{-9-fluorenyl-}\overset{+}{P}$ | 7.5 | 80 |

*Generation of the ylide required treatment with ethereal PhLi.

It is notable that the perphenyl 'onium salts are much more acidic than the permethyl 'onium salts, and this may reasonably be ascribed to the differential inductive effect between methyl and phenyl groups. Electron

**TABLE 9.5**

Deuterium-Hydrogen Exchange in 'Onium Salts

| Substrate | $\Delta H^{\ddagger}$ (kcal/mol$^{-1}$) | $\Delta S^{\ddagger}$ (cal/°/mol$^{-1}$) |
|---|---|---|
| $(CH_3)_3\overset{+}{S}\overset{-}{I}$ | 22.4 | $-1 \pm 2$ |
| $(CH_3)_4\overset{+}{P}\overset{-}{I}$ | $25.6 \pm 0.2$ | $+4 \pm 1$ |
| $(CH_3)_4\overset{+}{N}\overset{-}{I}$ | $33.2 \pm 0.6$ | $-15 \pm 2$ |

(catalyst: $D\overline{O}$ in $D_2O$)

withdrawal from the heteroatom causes contraction of the $d$ orbitals with better resulting overlap with the adjacent $2p$ orbital on carbon.

Deuterium-hydrogen exchange rates in 'onium salts have been used to assess their kinetic acidities. Data obtained by Doering and Hoffman[83] are given in Table 9.5.

Both enthalpies and entropies of activation for deuterium-hydrogen exchange favor the second-row 'onium salts. The large decrease in enthalpy of activation is to be attributed to $d$-orbital resonance stabilization of the carbanionic transition state for deuterium-hydrogen exchange, and the small entropies of activation for the second-row elements are consistent with the formation of a species in which charge separation and hence the degree of solvation in the transition state is lower than that in the starting materials. By contrast, in the ammonium salt, an additional center requiring solvation is produced with an attendant decrease in solvent entropy.

Dissociation constants (Table 9.6) of methylene bis-sulfonium salts have recently been determined,[84] and show that as the size of the alkyl substituents on sulfur increases the acidity of the salts increases, contrary to expectations based on polar effects in other systems. It is suggested that the observed differences are to be accounted for by solvation. With small alkyl groups on sulfur, salts are well solvated and formation of the conjugate base occurs with loss of solvation energy because of the ylide

**TABLE 9.6**

$pK_a$ ($H_2O$) for Methylene Bis-Dialkylsulfonium Fluoroborates, $RR'\overset{+}{S}CH_2\overset{+}{S}RR'(B\overline{F}_4)_2$

| Compound | $pK_a$ |
|---|---|
| R = R' = Me | 9.19 |
| R = Me;  R' = Et | 8.51 |
| R = Me;  R' = i-Pr | 7.76 |
| R = Et;  R = t-Bu | 7.14 |

character of the conjugate base. This effect is greater than in the case of the sulfonium salts containing more highly branched alkyl groups which are less solvated and hence lose less solvation energy when the conjugate base is formed.

In spite of the limited amount of accurate $pK_a$ data for sulfonium salts, it is clear that the sulfonium group exercises a considerable stabilizing influence on an adjacent carbanion, that the effect is comparable with that found in other second-row elements capable of expanding their valency shells, such as phosphorus and arsenic, and is very much greater than for the corresponding ammonium salts. Johnson[85] gives a list of known sulfonium ylides with the conditions for their preparation from the conjugate acids. In those cases in which the $\alpha$-carbon atom also has attached to another group capable of stabilizing a carbanion such as phenyl, carboalkoxy, or acyl, the conditions required to form the ylide, e.g., treatment with aqueous sodium hydroxide, indicate that the parent sulfonium salts have $pK_a$ values in the range 12–16.

Another interesting aspect of the stabilization of carbanions by adjacent sulfonium centers has arisen from studies of the biological properties of thiamine, which may well be due to the stability of the carbanion resulting from the removal of proton from $C_2$ of the thiazolium portion of the molecule[86]:

$$(9.29)$$

It has been shown[87] that 3,4-dimethylthiazolium bromide exchanges hydrogen at $C_2$ for deuterium in a neutral medium. Although imidazolium ions and oxazolium ions also exchange[88] hydrogen at $C_2$, the relative rates of these exchanges indicate that such exchange is much preferred in the sulfur system.

Another simple piece of evidence[89] for greater carbanion stabilization by sulfonium groups than by ammonium groups is the observation for the pyruvic acids, $R_X\overset{+}{M}CH_2COCO_2H$, that the ammonium derivative (M = N) is monobasic while the sulfonium derivative (M = S) is dibasic and the methyl ester has a $pK_a$ of 5.5.

Further comparison[90] may be found in the decarboxylation of acetic acids bearing an $\alpha$-'onium substitutent. The sulfonium salt,

$C_{12}H_{25}-\overset{+}{S}MeCH_2CO_2H\overline{O}Ts$ rapidly loses carbon dioxide to give the simple sulfonium salt, $C_{12}H_{25}\overset{+}{S}Me_2\overline{O}Ts$, whereas the corresponding ammonium betaine does not decarboxylate so readily.

Sulfonium ylides, the carbanions produced by proton removal adjacent to the sulfonium group, have been studied intensively in recent years and a later section of this chapter is devoted specifically to them.

### 9.6.3 The Sulfonium Group as Leaving Group in Displacement Reactions

The presence of the sulfonium pole causes strong polarization of the carbon-sulfur bond, and we may expect that nucleophiles will be encouraged to attack at carbon with displacement of the sulfonium group. Further, under suitable circumstances, dissociation of sulfonium salts to sulfide and carbonium ion may be observed. Because of the presence of a positive pole, sulfonium salts are highly solvated, and reactions in which they are involved are very solvent dependent.

In principle, the simplest type of reaction involving a sulfonium group is reversible dissociation to carbonium ion and sulfide. This type of reaction has been observed by Darwish and his collaborators[91] in a study of the racemization of benzylethylmethylsulfonium perchlorates (Table 9.7).

Results for the first three sulfonium salts are consistent with a scheme in which racemization is independent of solvolysis and involves pyramidal inversion at sulfur. Electron-withdrawing groups such as $p$-nitro have negligible effect on inversion rate. For the $p$-methoxy substituted compound, however, solvolysis is very much more rapid, consistent with the formation of a stabilized $p$-methoxy-benzylcarbonium ion. The increase in racemization, although not paralleled closely by the increase in solvolysis rate, cannot be accounted for in terms of a pyramidal inversion mechanism. The lower racemization rate is, of course, consistent with neutral molecule-ion return from species with different degrees of association.

**TABLE 9.7**

Solvolysis and Racemization in Sulfonium Salts $R\overset{+}{S}EtMe$ $ClO_4^-$

| R | $k_{rel.}$ solvolysis | $k_{rel.}$ racemization |
|---|---|---|
| $PhCH_2$ | 1 | 1 |
| $p$-$NO_2C_6H_4CH_2$ | 0.2 | 0.99 |
| $PhCOCH_2$ | 0.03 | 0.6 |
| $p$-$MeOC_6H_4CH_2$ | $10^3$ | 15 |

### 9.6.4 Reversion in Sulfonium Salts

This term will be used to indicate the reverse of the reaction by which sulfonium salts are formed from sulfides and alkylating agents. In the simple case,

$$R_2S + R'X \rightleftharpoons R_2\overset{+}{S}R'\bar{X} \qquad (9.30)$$

pressure causes the equilibrium to move to the right and increase of temperature to the left. The latter change corresponds to the free energy change associated with the entropy term. Reversion is characteristic of simple alkylsulfonium salts and frequently leads to unexpected products from the alkylation of sulfides. Alkylation of bis-2-methylpropyl sulfide with methyl iodide,[92] for example, gives trimethyl sulfonium iodide by a sequence of alkylation and reversion. Kinetics of reversion in trialkyl-sulfonium salts have been investigated,[93, 94] and the reversion reaction is no exception to the majority of sulfonium salt reactions in being very sensitive to solvents because of the charged nature of the substrate (eq. 9.31).

Racemization involving nucleophilic displacement of the sulfonium group and subsequent regeneration of the sulfonium salt has been suggested [95,96] to occur in phenacylsulfonium salts.

$$
\underset{\substack{| \\ p\text{-Tolyl} \\ \textbf{9.14}}}{\overset{\substack{(CH_2)_n \\ CH_2 \qquad CH_2 \\ \overset{+}{S} \qquad Br^-}}{}} \rightleftharpoons p\text{-Tolyl-SCH}_2(CH_2)_n CH_2 Br \qquad (9.31)
$$

In a recent study of reversion[97] in cyclic sulfonium salts, it was found for the salts (9.14) (n = 2 and 3) that the reactions were 40 and 136 times respectively more rapid in $t$-butanol than in ethanol because the less polar solvent solvates the sulfonium salt poorly, and hence the rate of reversion increases. Furthermore, reversion for the 5-membered ring (n = 2) in ethanol was 75 times more rapid than for the 6-membered ring.

Degrees of selectivity in the reversion reaction are quite obvious in that for an aryldimethylsulfonium salt,

$$p\text{-Tolyl-}\overset{+}{S}(Me)_2 \cdot \bar{O}SO_2 OMe \rightleftharpoons p\text{-Tolyl-SMe} + Me_2SO_4 \qquad (9.32)$$

displacement on $sp^3$ carbon is very much more rapid than at $sp^2$ (aromatic carbon) and reversion is unidirectional.[98]

Reversion in arylsulfonium halides albeit slow has, however, been observed[35]:

$$Ar_3 \overset{+}{S} \, Cl^- \; \underset{\Longleftarrow}{\xrightarrow{250°C}} \; Ar_2 S \; + \; ArCl \qquad (9.33)$$

## 9.6.5 Reactions with Other Nucleophiles

Reactions of sulfonium salts with nucleophiles figured in the classical development of our understanding of displacement processes in aliphatic systems and this aspect of their chemistry have been reviewed.[99]

An interesting general aspect of the reactivity of sulfonium salts is that the mechanistic changeover from $S_N 1$ processes, which is favored by low carbon nucleophilicity in the nucleophile and increasing solvent polarity, tends to occur more readily with sulfonium salts than with alkyl halides.

The role of solvent effects in nucleophilic displacements at carbon in sulfonium salts has already been referred to in the reversion reaction and very large effects are found with other nucleophiles even when the solvent may be regarded as undergoing a relatively small change. These very large effects are encountered when the nucleophile and substrate have opposite charges so that the transition state is much less solvated than the reactants. By contrast, as can be seen in Table 9.8, when the nucleophile attacking sulfonium salt is a neutral nucleophile, then the solvent effects are less important.

**TABLE 9.8**

Solvent Effects in Nucleophilic Substitutions

| Example | Rate constant | Volume: % $H_2O$ in aqueous EtOH | | | | $t°$ |
|---------|---------------|------|------|------|------|------|
|         |               | 0 | 20 | 40 | 100 | |
| Bimolecular mechanism[†] | | | | | | |
| $i$-PrBr + OH$^-$ | $10^5 \, k_2$ | 6.0 | 4.9 | 3.0 | – | 55 |
| $Me_3\overset{+}{S}$ + OH$^-$ | $10^4 \, k_2$ | 7240 | 178 | 5.1 | 0.37 | 100 |
| $Me_3\overset{+}{S}$ + NMe$_3$ | $10^5 \, k_2$ | 6.67 | – | – | 0.65 | 45 |
| Unimolecular mechanism* | | | | | | |
| $t$-BuCl | $10^6 \, k_1$ | – | 9.14 | 126 | <1294 | 25 |
| $t$-Bu$\overset{+}{S}$(Me)$_2$ | $10^5 \, k_1$ | 1.90 | 1.24 | – | 0.60 | 50 |

*Units: sec$^{-1}$.
†Units: M$^{-1}$ sec$^{-1}$.

In the case of $S_N 1$ reactions of sulfonium salts, dissociation involves dispersion of charges in the transition state and results in only a small decrease of velocity constant with change to a less polar solvent.

It has been demonstrated[100] that bimolecular displacement of dimethyl sulfide from dimethyl α-methylbenzylsulfonium ion by azide ion occurs with inversion of configuration at carbon.

Alkaline hydrolysis of sulfonium salts is subject to competition from elimination, which is dealt with below. Just as for reactions of ammonium salts with basic nucleophiles, the inductive effect of an 'onium group promotes β-elimination because of the acidifying effect on β protons. 'Onium-leaving groups do not leave as readily as halide ions and because elimination reactions are markedly less sensitive to leaving group differential than $S_N 2$ displacements (below), elimination is often the major reaction path with this type of substrate. Hydrolysis of aryldimethylsulfonium salts, however, gives[101] selective formation of the aryl methyl sulfide in high yield under alkaline conditions.

Neutral and alkaline hydrolyses of trimethyl- and benzylmethylphenylsulfonium salts have been studied by Swain and his colleagues.[102] They find that, in water, the salts solvolyze by first-order processes when the counter ion is non-nucleophilic, e.g., perchlorate. Ethanolysis is about 20 times as rapid as hydrolysis, the small difference being consistent with solvent effects discussed above. For the benzylsulfonium salts, the data show that when electron-withdrawing substituents are present in the benzylic group, the transition state charge separation, calculated from the negative salt effect, decreases. This is in qualitative agreement with predictions that, for reactions of this type, electron-withdrawal at the benzyl group should result in contraction of the O-C and C-S bonds with consequent reduction in charge separation.

Reactions of triarylsulfonium salts with methoxide ion have recently been studied.[103] The chief products are the aryl methyl ether and the diaryl sulfide, together with some of the hydrocarbon, ArH:

$$Ar_3 \overset{+}{S} + \bar{O}Me \longrightarrow ArH + ArOMe + ArSAr$$

$$Ar_3 \overset{+}{S} + \bar{O}Me \rightleftharpoons Ar_3S\text{-}OMe \rightleftharpoons [Ar_3S \cdot + \cdot OMe]_{cage}$$
$$\downarrow$$
$$Ar_2 S + ArOMe \qquad\qquad (9.34)$$

and $Ar_3 S \cdot \longrightarrow Ar_2 S + Ar \cdot$

$$Ar \cdot + MeOH \longrightarrow ArH + \cdot CH_2 OH$$

It is suggested that the first stage is the formation of the neutral tetracoordinate species, which then undergoes homolytic decomposition to give radicals in a cage. These then break down to give the diaryl sulfide and the ether. The hydrocarbon arises by homolysis of the triaryl sulfide radical to diaryl sulfide and an aryl radical, which then abstracts a hydrogen atom from methanol with the formation of the hydroxy methyl radical. This subsequently turns up in the products as formaldehyde, *etc.* Formation of tight ion pairs in this type of system has been referred to earlier.

Reaction of triarylsulfonium bromide with ethoxide ion, however, has been suggested—on the basis of the extremely large solvent effect—to proceed *via* an ionic path that involves the nucleophilic attack of ethoxide ion on sulfur as shown below[103a]:

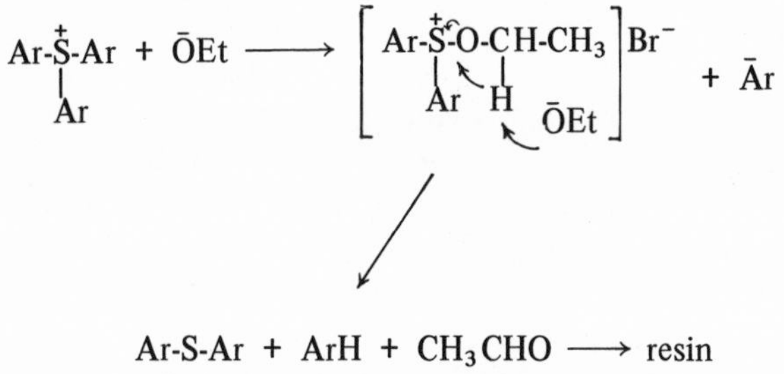

$$
\text{Ar-}\overset{+}{\underset{\text{Ar}}{\text{S}}}\text{-Ar} + \bar{\text{O}}\text{Et} \longrightarrow \left[ \overset{+}{\underset{\text{Ar}}{\text{Ar-S-O-CH-CH}_3}} \; \overset{}{\underset{\bar{\text{O}}\text{Et}}{\text{H}}} \right] \text{Br}^- + \bar{\text{Ar}}
$$

$$(9.35)$$

$$
\text{Ar-S-Ar} + \text{ArH} + \text{CH}_3\text{CHO} \longrightarrow \text{resin}
$$

Alkylation of carboxylate ions by sulfonium salts has been observed[104]; trimethylsulfonium iodide with silver mesitoate gives the methyl ester in hot methanol:

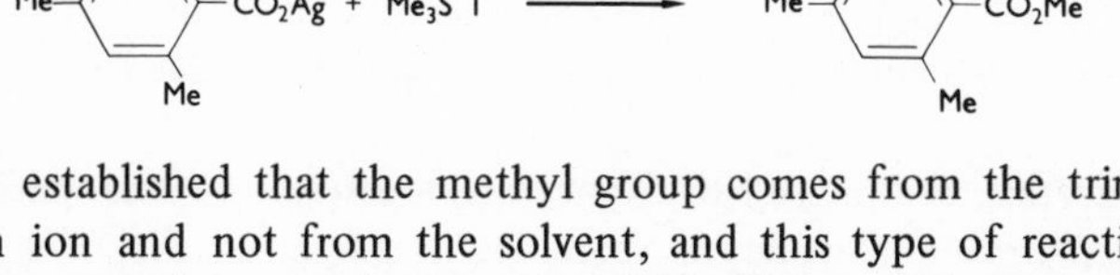

$$(9.36)$$

It was established that the methyl group comes from the trimethylsulfonium ion and not from the solvent, and this type of reaction has recently been used in attachment of peptide chains to resins in connection with solid-phase peptide synthesis.[105] The resin is synthesized with benzyldimethylsulfonium groups and nucleophilic attack by carboxylate ion is preferentially at the benzyl group with displacement of dimethyl sulfide. About 10% of the reaction, however, involves attack at the methyl group giving the (irrelevant) methyl ester of the peptide acid.

### 9.6.6 Chlorinolysis of Sulfides

Formation of halosulfonium salts of the type $R_2SX^+X^-$ by nucleophilic attack of sulfides on halogen molecules[106] has been referred to earlier. Halosulfonium salts are unstable particularly with respect to nucleophilic attack by the counter ion, which results in cleavage of the original sulfide.[107] A nice example of this is shown in the halogenation of thioacetals,[108] when nucleophilic attack to form sulfenyl halide and α-chlorosulfide results (eq. 9.37).

$$(MeS)_2CH_2 + Cl_2 \longrightarrow CH_2 \overset{+}{S} \underset{Me}{\overset{Cl}{\diagdown}} \longrightarrow MeSCH_2Cl + MeSCl \quad (9.37)$$

Benzylic sulfides are particularly prone to this type of reaction because of the ready displacement of sulfonium groups from benzylic sites. Studies of salt effects reveal the role of ion-pair equilibria in these systems.[42] Use of optically active α-ethylbenzyl phenyl sulfide shows[43] that for chlorinolysis in acetic acid, the acetate and chloride products arise from *different* species involved in ion-pair solvent equilibria.

### 9.6.7 Hydrogen Species in Sulfonium Displacements

As mentioned earlier, sulfides are weakly basic and small amounts of protonated sulfide exist at equilibrium in strongly acidic conditions. The obvious comparison is with the corresponding oxygen analogues, and thioethers are much less subject to Zeisel-type cleavage in the presence of strong acids possessing nucleophilic anions such as hydrogen iodide. It is clear that this is mainly due to the less favorable protonation equilibrium rather than to the slower displacement reaction. Cleavage of aryl thioethers[109] is very much accelerated by groups such as p-MeO in the aromatic nucleus. These raise electron density at sulfur and increase the equilibrium concentration of protonated species involved in the cleavage reaction.

Thioacetals, as opposed to oxyacetals, are stable to acids and normally require heavy metal catalysis (e.g., $Hg^{++}$) for cleavage.[110, 111] Note the obvious point that nucleophilic attack on protonated sulfide involves competition between deprotonation (nucleophilic attack on hydrogen) and dealkylation (nucleophilic attack on carbon).

Hydride ion attack on sulfonium salts appears to occur in the expected manner. Lithium aluminium hydride reduction of trimethylsulfonium bromide gives[112] methane and dimethyl sulfide. With alkoxysulfonium salts, however, deuterium labeling experiments show[113] that attack of borohydride ion results in displacement of the alkoxy group from sulfur and formation of a protonated sulfide:

$$
\underset{\underset{H}{\mid}}{Ph\overset{+}{S}Me} \quad (OMe) + \bar{B}H_4 \longrightarrow PhS\overset{+}{M}e \longrightarrow PhSMe \qquad (9.38)
$$

### 9.6.8 Addition to Alkenes

Addition of thiosulfonium salts to double bonds yielding $\beta$-thioalkylsulfonium salts has been reported.[114] In this case, the high reactivity of thiosulfonium salts renders them susceptible to nucleophilic attack by a carbon-carbon double bond and this is confirmed by other studies[115] of $Me_2\overset{+}{S}$ as a leaving group from sulfur:

$$
MeS\overset{+}{S}Me_2\bar{X} \quad + \quad \text{(cyclooctadiene)} \longrightarrow \text{(ring with } SMe \text{ and } \overset{+}{S}Me_2\ \bar{X}) \qquad (9.39)
$$

$$(\bar{X} = 2,4,6\text{-trinitrobenzenesulfonate ion})$$

### 9.6.9 The Sulfonium Group in Elimination Reactions

#### 9.6.9.1 As Leaving Group

The most familiar role is that in which the sulfonium group acts as a leaving group in 1,2-eliminations. Sulfonium compounds played an important part in defining the fundamental ideas on transition states in elimination reactions, particularly associated with the Ingold school.[116]

Reference has already been made to heterolysis of sulfonium salts with formation of stabilized carbonium ions. This process occurs in 1,2-eliminations which have the unimolecular (E1) mechanism (eq. 9.40).[117-121]

In the unimolecular heterolysis of sulfonium salts there is, of course, competition to give products of substitution and elimination. Solvolysis of $t$-butyl dimethylsulfonium ion in 80% aqueous ethanol gives 30% of product by elimination, and the rate of the reaction is roughly one eighth of that with $t$-butyl chloride which also gives 30% of elimination product.

$$
\underset{\underset{\mid}{\mid}}{-C}\underset{\underset{\mid}{\mid}}{\overset{H}{-}}C-\overset{+}{S}R_2 \rightleftarrows \underset{\mid}{-}C\overset{H}{\underset{\mid}{-}}C\diagup + R_2S \xrightarrow{fast} H^+ + \diagup C=C\diagdown \qquad (9.40)
$$

In sulfonium salts which possess hydrogen on carbon $\beta$ to sulfur and which are not disposed to undergo unimolecular heterolysis (i.e., solvent and/or structure not favorable), nucleophilic attack on the salt can cause either substitution, as previously described, or 1,2-elimination:

$$
\begin{array}{c}
\text{H} \\
| \quad | \quad + \\
-\text{C}-\text{C}-\overset{+}{\text{S}}\text{R}_2 + \text{Nu:} \\
| \quad |
\end{array}
\left\{
\begin{array}{l}
\text{NuH} + \text{C}=\text{C} + \text{R}_2\text{S} \\[1em]
\text{H} \\
| \quad | \\
-\text{C}-\text{C}-\text{Nu} + \text{R}_2\text{S} \\
| \quad |
\end{array}
\right.
\quad (9.41)
$$

The choice between these two modes of reaction depends upon the relationship between proton and carbon nucleophilicity for the nucleophile in question. We have already seen that the reversion reaction, usually involving a weakly basic counter ion such as iodide, gives substitution product exclusively. When, however, the nucleophile is strongly basic, 1,2-elimination appears. Thus triethylsulfonium hydroxide decomposes on heating to diethyl sulfide, ethylene, and water, a reaction analogous to the behavior of quaternary ammonium hydroxides.

Kinetic investiation[122] of bimolecular reactions of sulfonium ions with ethoxide ions revealed (Table 9.9) a number of important aspects of elimination reactions in which the leaving group is a rather poor one and the $\beta$ proton is appreciably acidic.

**TABLE 9.9**

Bimolecular Elimination in Sulfonium Salts and Alkyl Bromides

| Substituent | $\overset{\beta\quad\alpha}{CH_3CH_2\overset{+}{S}(Me)_2}$ $k_{rel.}$ | $\overset{\beta\quad\alpha}{CH_3CH_2Br}$ $k_{rel.}$ |
|---|---|---|
| none | 1 | 1 |
| $\alpha$-Me | 11.4 | 2.5 |
| $\beta$-Me | 0.5 | 3.3 |
| $\beta$-Et | 0.3 | 2.7 |
| $\beta,\beta$-Me$_2$ | 0.25 | 5.4 |
| $\alpha,\beta$-Me$_2$ | 5.8 | 1.0 |
| $\alpha,\alpha$-Me$_2$ | 65 | 27 |
| $\beta$-Ph | 430 | 350 |
| $\alpha$-Ph | 100 | 50 |

(Reactions with sodium ethoxide in ethanol)

As can be seen in Table 9.9, all α substituents accelerate elimination in both sulfonium salt and alkyl bromide. This effect can reasonably be ascribed to conjugation of the α substituent with the partially formed double bond in the transition state. On the contrary, however, when alkyl groups are placed on the β-carbon atom, elimination in alkyl halides is accelerated as before, but for sulfonium salts β-alkylation has an adverse effect on rate. This is due to the reduction in acidity of the β-C-H bond which is caused by β-alkylation, consistent with the view that elimination in sulfonium salts has a much more carbanionic type of transition state than for alkyl halides. This situation is occasioned by the powerful inductive effect of the 'onium pole which renders the β-C-H bond relatively more acidic than in an alkyl halide. This, coupled with the fact that the 'onium group is a poorer leaving group than halogen, leads to a transition state with a larger degree of β-C-H bond extension than for the alkyl halide and a smaller degree of extension of the bond to the leaving group. When the β substituent is phenyl, however, the effect is to accelerate both types of substrate. In the sulfonium salt, conjugative stabilization of the carbanionic center produced at $C_\beta$ is evidently important.

The role of the sulfonium substituent in stabilizing adjacent carbanionic centers has already been discussed and the effect of this substituent in elimination reactions is understandable. Conjugative effects operate in transition states for elimination from 'onium salts just as in alkyl halides, but the substantial inductive effect of the sulfonium pole is superimposed. This makes elimination reactions in sulfonium salts very sensitive to substituents which either stabilize or destabilize the carbanionic center developed at $C_\beta$. For simple alkyl sulfonium salts, this effect determines the orientation of elimination. Elimination will occur less readily from alkylated β-carbon atoms ("Hofmann control"), which are preferred in eliminations from alkyl halides.

More recent work on reactions in which the sulfonium group is the leaving group has concentrated on the interrelationship of the sulfonium and other leaving groups in the hypothesis of variable transition states in elimination reactions.[123] Particular attention has been paid to the 2-phenylethyl system because of the possibility of examining transition states by the application of linear free energy treatments. Data are given in Table 9.10.

As the leaving group becomes increasingly poor, carbonionic character at $C_\beta$ increases (increasingly positive value of $\rho$) and the isotope effect decreases. These trends point to the increasing degree of β-C-H bond cleavage which is required to expel the leaving group.

**TABLE 9.10**

E2 Reactions of $\beta$-Arylethyl Compounds (ArCH$_2$CH$_2$X)

| Leaving Group (X) | $k_{rel.}$ | $\rho$ [a] | $k_H/k_D$ | Leaving Group Isotope Effect |
|---|---|---|---|---|
| Br | 530 | +2.14 | 7.1 | – |
| $\overset{+}{S}Me_2$ | 1[b,c] | +2.75 | 5.1[b] | ~35% of theoretical maximum[d] |
| $\overset{+}{N}Me_3$ | 0.02[b] | +3.77 | 3.0[b] | ~30% of theoretical maximum[d] |
| | | (NaOH-EtOH at 30°C) | | |

[a]Ref. 124. [b]Ref. 125. [c]Value corrected for special entropy effect. [d]Ref. 126.

Note that the leaving group isotope effects do not entirely fall in with this picture as it appears clear that the sulfur leaving group isotope effect is at least comparable in magnitude with the nitrogen leaving group isotope effect.

The solvent isotope effect observed for elimination when $^-$OH in H$_2$O is replaced by $^-$OD in D$_2$O in the 2-phenylethyl system has been measured.[127] For bromide and sulfonium leaving groups the value is 1.57 at 80°C, and for the ammonium group it is 1.79. The conclusion is that bonding between the $\beta$ proton and the base is well advanced and there seems no doubt that the degree of C-H bond cleavage is greater in the case of the ammonium compound than for the sulfonium compound. This is in agreement with the general view that the poorer the leaving group, the more the transition state shifts towards $\beta$-carbanionic character. Experiments by Bunnett and Bacciochi[128] assess the degree of bonding of base to proton at the transition state by measuring the EtS$^-$/MeO$^-$ rate ratio, which changes from 6.5 with chloride as leaving group through 0.8 for $\overset{+}{S}Me_2$ to 0.05 with SO$_2$Me. Note again the central position of the sulfonium ion in the leaving-group range.

Although the amount of reported work is limited, eliminations involving sulfonium leaving groups appear to be subject to the same spatial restraints as reactions with alkyl halides. Thus Cristol and Stermitz[129] found that both *cis* and *trans* 1-dimethylsulfonio-2-phenylcyclohexane afforded 1-phenylcyclohexene as the dominant elimination product with potassium hydroxide in ethanol:

$$(9.42)$$

The *cis* isomer can react by *anti*-elimination and yields the conjugated alkene (72%) together with non-conjugated alkene (1%) and substitution product (7.5%). The *trans* isomer, however, can form the conjugated alkene only by *syn*-elimination. This product is obtained in 22% yield together with the non-conjugated isomer (2%) and substitution product (61%). An $\alpha'$, $\beta$-elimination mechanism for the *trans* isomer was regarded at that time as being untenable when applied to simple 1,2-eliminations. An alternative was formation of an ylide which would allow pre-isomerization of the *trans* isomer and normal *anti*-elimination. Deuterium labeling experiments to distinguish between these possibilities were not carried out, but evidence for $\alpha'$, $\beta$-elimination in another system has been found by Franzen and Mertz,[130] who treated $\alpha$-perdeuterotriethylsulfonium bromide with sodium triphenylmethide in ether. The products were triphenylmethane, ethylene, and diethyl sulfide. Of the triphenylmethane obtained, 75% contained deuterium, showing preferential removal of $\alpha$-deuterons. The labeling pattern in the olefin obtained showed that $\alpha'$, $\beta$-elimination *via* the ylide (9.15) was a probable pathway:

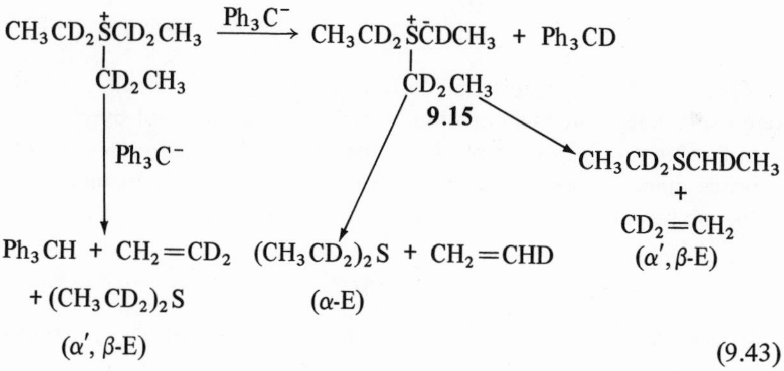

$$
\underset{\substack{|\\ CD_2CH_3}}{CH_3CD_2\overset{+}{S}CD_2CH_3} \xrightarrow{\ Ph_3C^-\ } \underset{\substack{|\\ CD_2CH_3 \\ \mathbf{9.15}}}{CH_3CD_2\overset{+}{\underset{\cdot\cdot}{S}}CDCH_3} + Ph_3CD
$$

$$
\begin{array}{l}
Ph_3CH + CH_2{=}CD_2 \quad (CH_3CD_2)_2S + CH_2{=}CHD \\
+ (CH_3CD_2)_2S \qquad\qquad\qquad (\alpha\text{-E}) \\
(\alpha', \beta\text{-E})
\end{array}
\qquad
\begin{array}{l}
CH_3CD_2SCHDCH_3 \\
+ \\
CD_2{=}CH_2 \\
(\alpha', \beta\text{-E})
\end{array}
$$

$$(9.43)$$

As will be seen below, formation of ylides by attack of arynes on sulfides occurs readily, but when these contain $\beta$-hydrogen atoms, intramolecular proton transfer occurs and the products are alkene and aryl alkyl sulfide[131]:

$$\text{(9.44)}$$

Related fragmentation is observed[132] in reactions of cyclic sulfonium salts with strong bases:

$$\text{(9.45)}$$

Base-induced reaction of alkoxysulfonium salts to produce sulfide and carbonyl compound involves a cyclic elimination with the sulfonium as departing group. The mechanism is indicated[133] by specific labeling:

$$\text{(9.46)}$$

The preceding examples of sulfonium eliminations involve unactivated substrates, but $\beta$-elimination is very much accelerated by electron withdrawing groups. A simple synthetic route[134] to $\alpha$-acetylenic acids depends upon successive elimination of sulfonium substituents from thioketals bearing conjugative groups such as carboalkoxy, e.g.,

$$PhC(SMe)_2CH_2CO_2Et \xrightarrow{Me_2SO_4} Ph\overset{+}{C}=CHCO_2Et \quad NaOH/0° \tag{9.47}$$
$$\searrow PhC\equiv CCO_2Et \quad (80\%)$$

### 9.6.9.2 As Activating Group

In pioneering work, Rydon[135,136] showed that the sulfonium group powerfully accelerated reactions in which a leaving group $\beta$ to the sulfonium group was eliminated under basic conditions. For example, esters bearing a $\beta$-sulfonium substituent in the alkoxy group decomposed readily to form carboxylate ion, acetylene, and dialkyl sulfide under very mildly basic conditions:

$$Me_2 \overset{+}{S}CH_2 CH_2 OCOPh \xrightarrow{^-OH} Me_2 \overset{+}{S}CH=CH_2 + {}^-OCOPh$$

$$^-OH \searrow \qquad\qquad (9.48)$$

$$Me_2 S + CH{\equiv}CH$$

More recently, this type of very rapid elimination reaction has been exploited by Rydon and his collaborators in devising protecting groups for carboxyl and amino groups in peptide synthesis[137]:

$$P\text{-NHCHR.}CO_2 H + ClCH_2 CH_2 SMe \xrightarrow{Et_3 N}$$

$$P\text{-NHCHR.}CO_2 CH_2 CH_2 SMe \xrightarrow[\text{(ii) couple}]{\text{(i) remove N-protecting group}}$$

$$P'\text{-NHCHR'-CONHCHR-}CO_2 CH_2 CH_2 SMe \xrightarrow{MeI} \qquad (9.49)$$

$$P'\text{-NHCHR'-CONHCHR-}CO_2 CH_2 CH_2 \overset{+}{S}Me_2 \bar{I} \xrightarrow{\text{aqueous}} {\text{NaOH}}$$

$$P'\text{-NHCHR'-CONHCHR-}CO_2 H + CH_2 {=} CH\overset{+}{S}Me_2 \bar{I}$$

The activating effect of the sulfonium group in eliminations has received quantitative study recently by Crosby and Stirling[138] in the system:

$$X\text{-}CH_2 CH_2 \text{-}OPh \xrightarrow{Et\bar{O}/EtOH} X\text{-}CH{=}CH_2 + \bar{O}Ph \qquad (9.50)$$

The dimethylsulfonio substituent ($X = Me_2\overset{+}{S}$) is found to cause very rapid elimination of phenoxide under basic conditions, and the rates of elimination with this substituent compared with others are shown in Table 9.11.

TABLE 9.11

Activated Eliminations of Phenoxide from $\beta$-Substituted Phenoxyethanes[138]

$$X\text{-}CH_2CH_2\text{-}OPh \xrightarrow{\text{base}} X\text{-}CH_2{=}CH_2 + \bar{O}Ph$$

| Substituent X | $k_{rel}.Et\bar{O}/EtOH$ | $k_{rel}.\bar{O}H/H_2O$ |
|---|---|---|
| $Ph_3\overset{+}{P}$ | $1.2 \times 10^3$ | $6.4 \times 10^{-1}$ |
| $Me_2\overset{+}{S}$ | $6.5$ | $1.6 \times 10^{-2}$ |
| $CH_3CO$ | $1$ | $1$ |
| $CO_2Et$ | $1.9 \times 10^{-3}$ | $4.9 \times 10^{-3}$ |
| $Me_3\overset{+}{N}$ | $9.7 \times 10^{-9}$ | $-$ |

$(k_{EtOH/H_2O}$ for $X = CH_3CO$ is 10)

It has been shown that for these eliminations, reversible formation of the $\beta$-carbanion is then followed by loss of the leaving group in a slower step (E1cB mechanism). Two important points emerge from the data: (1) As there is evidence from other sources[139] that the observed elimination rate is very largely dependent upon the dissociation constant of the appropriate carbon acid, the scale of relative rates is also a scale of acidites in the appropriate solvent. The great effectiveness of the second-row 'onium salts in promoting dissociation of the adjacent C-H bond is in contrast to the effect of the first-row ammonium substituent. It does not seem likely that this difference is due to a difference in inductive effects of first- and second-row substituents as this should produce results in the reverse order. Stabilization of the carbanion by involvement of the $d$ orbitals of the second row elements appears probable. (2) Change of solvent has a pronounced effect on elimination rate. For the ester (X = $CO_2Et$), transfer from ethanol to water changes the rate constant by less than one power of 10. By contrast, for the sulfonium salt (X = $Me_2\overset{+}{S}$), the rate constant decreases by more than $10^3$. In this system, ionization converts a positively charged substrate to a neutral one and causes loss of solvation, which is undoubtedly responsible for the effect on rates. The phosphonium compound is affected in a similar way.

In connection with earlier studies of the involvement of sulfur $d$ orbitals in carbanion stabilization, Doering and Schreiber[140] compared the reactions of 2-bromoethyldimethylsulfonium bromide and 2-bromoethyltrimethylammonium bromide with aqueous sodium hydroxide. For the sulfonium salt, they found that rapid elimination to give the vinyl sulfonium salt occurred, whereas with the ammonium salt,

slow nucleophilic displacement of bromine occurred to give the
$\beta$-hydroxyethyl 'onium salt. The difference in behavior was ascribed to the
ability of the sulfonium substituent to raise the dissociation constant, of
the $\beta$-C-H bond and hence cause a preference of elimination *versus*
substitution. This is entirely in accord with the later work of Crosby and
Stirling,[138] and a further interesting comparison is possible between the
results for the bromides and the phenyl ethers. Replacement of bromine
by a phenoxy substituent produces a decrease in rate constant of only
about $10^4$, in marked contrast to the corresponding change for an $S_N2$
displacement in which phenoxide is virtually unkown as a leaving group.

### 9.6.10 The Sulfonium Group in Nucleophilic Additions to Carbon-Carbon Multiple Bonds

The ability of the sulfonium group to stabilize an adjacent carbanion is
also revealed in its effect in nucleophilic addition reactions in which the
transition state is either carbonionic or an actual carbanion:

$$X-C=C\underset{|}{\overset{/}{\diagdown}} + Nu: \rightleftharpoons X-\bar{C}-\overset{|}{\underset{|}{C}}-Nu \underset{B:}{\overset{SH}{\rightleftharpoons}} X-\overset{H}{\underset{|}{C}}-\overset{|}{\underset{|}{C}}-Nu \qquad (9.51)$$

This area was again pioneered by Doering,[140] who compared the
reactivities of vinylsulfonium salts and vinylammonium salts in
nucleophilic addition. Dimethylvinylsulfonium bromide was prepared by
treatment of 2-bromoethyldimethylsulfonium bromide with silver oxide,
rapid neutralization being required to prevent nucleophilic addition of the
solvent. Addition of a number of nucleophiles occurred readily:

$$
\begin{array}{c}
Me_2\overset{+}{S}CH=CH_2 \\[2ex]
\diagup \quad H\bar{O} \Big| Et\bar{O} \quad \bar{C}H(CO_2Et)_2 \diagdown \qquad (9.52) \\[2ex]
Me_2\overset{+}{S}CH_2CH_2OH \quad Me_2\overset{+}{S}CH_2CH_2OEt \quad Me_2\overset{+}{S}CH_2CH_2CH(CO_2Et)_2
\end{array}
$$

Recently there has been much interest in this type of reaction,
particularly with respect to the subsequent reactions of the adducts
formed from nucleophiles and $\alpha, \beta$-unsaturated sulfonium salts.

Gosselck and his collaborators[134, 141] have studied addition of carbon nucleophiles to simple $\alpha$, $\beta$-unsaturated sulfonium salts which yield cyclopropanes. These salts are readily obtained, for example,[142] by nucleophilic addition of the conjugate base of a $\alpha$-alkylthioester to an aromatic aldehyde with subsequent alkylation of the resulting $\alpha$, $\beta$-unsaturated sulfide:

$$\text{ArCHO} + \text{MeSCH}_2\text{CO}_2\text{Et} \xrightarrow{\text{NaOEt}} \text{ArCH=C}\underset{\text{SMe}}{\overset{\text{CO}_2\text{Et}}{\diagdown}} \qquad (9.53)$$

$$\text{ArCH=C}\underset{\overset{+}{\text{SMe}_2}}{\overset{\text{CO}_2\text{Et}}{\diagdown}} \qquad\qquad \text{Me}_2\text{SO}_4$$

Addition of the conjugate base of malononitrile to sulfonium salts of the type, $\text{Me}_2\overset{+}{\text{S}}\text{CR}_1=\text{CHR}_2$, gives cyclopropanes following proton transfer in the ylide so produced[143]:

$$\text{Me}_2\overset{+}{\text{S}}\text{-C}=\text{CHPh} + \bar{\text{C}}\text{H(CN)}_2 \longrightarrow \text{Me}_2\overset{+}{\text{S}}\text{-}\bar{\text{C}}\text{-CHPh} \qquad (9.54)$$

Addition of hydroxyl ion to $\alpha$-phenacyl-$\alpha$,$\beta$-unsaturated sulfonium salts in which the intermediate ylide is highly stabilized has been observed[144]:

$$\qquad (9.55)$$

Proton transfer in the stabilized ylide is followed by re-elimination in a different direction giving aromatic aldehyde and the phenacylsulfonium ylide.

Addition of a phenacylsulfonium ylide to an $\alpha,\beta$-unsaturated sulfonium salt yields an ylide with a $\gamma$-sulfonium group. Proton transfer and subsequent intramolecular nucleophilic displacement yields the 1,1-disubstituted cyclopropane[145]:

$$CH_2{=}CH\text{-}\overset{+}{S}Me_2\,\bar{B}r \;+\; RCO\bar{C}H\overset{+}{S}Me_2 \longrightarrow RCOCH\overset{+}{S}Me_2$$

$$(9.56)$$

In the case of $\alpha$-phenacyl-$\beta$-aryl-$\alpha,\beta$-unsaturated sulfonium salts, addition of ethoxide ion yields a stabilized ylide whose conjugate acid can be directly isolated as its hydrogen sulfate. Hydrolysis gives phenacylsulfonium salt and aromatic aldehyde:

$$Me_2\overset{+}{S}\text{-}CCO\text{-}R \;+\; \bar{O}Et \longrightarrow Me_2\overset{+}{S}\bar{C}COR \xrightarrow{H_2O} Me_2\overset{+}{S}CH_2COR \;+\; ArCHO$$
$$\underset{CHAr}{\overset{\parallel}{}} \qquad\qquad \underset{EtOCHAr}{\overset{|}{}} \qquad\qquad\qquad\qquad\qquad (9.57)$$

Hydroxide ion adds to the bis-phenyl salt **(9.16)**, and intramolecular nucleophilic displacement by oxygen then yields *trans*-stilbene oxide; amide ion, a much stronger base, causes elimination of the sulfonium group with formation of tolan[146]:

$$(9.58)$$

Additions to allenic sulfonium salts have been studied by Stirling and his collaborators,[147] particularly with respect to the subsequent reaction to the adducts formed. Allenyldimethylsulfonium bromide is very readily prepared from propargyl bromide and dimethyl sulfide:

$$Me_2S + BrCH_2C{\equiv}CH \xrightarrow{CH_3CN} Me_2\overset{+}{S}CH_2C{\equiv}CH$$

$$\underset{Me_2\overset{+}{S}\text{-}CH_2\text{-}\underset{\underset{}{|}}{C}{=}CH_2}{\overset{\overset{OCOR}{|}}{}} \xleftarrow{R\bar{C}O_2} Me_2\overset{+}{S}\text{-}CH{=}C{=}CH_2 \quad \Big\downarrow MeOH$$

$$\overset{OCOR}{\underset{Me_2\overset{+}{S}\text{-}CH_2\text{-}\underset{|}{C}{=}CH_2}{|}} \tag{9.59}$$

$$+ Me_2\overset{+}{S}CH_2COCH_3$$

Isomerization of the propargylic sulfonium salt to the allenic sulfonium salt occurs spontaneously in methanol and very rapidly in the presence of, for example, triethylamine. The conjugated double bond is very electrophilic and even nucleophiles as weak as carboxylate ions add to give adducts.[147] These adducts are activated vinyl esters and react readily with amines to give acylamides and the sulfonium ketone; the overall process is an extremely simple way of converting the anion of an acid to an amide. Carboxylic esters can be obtained by using alkoxide ion as nucleophile. Note that the addition of carboxylate ion is much more rapid than any other reaction, but that in the presence of excess of carboxylate ion, dealkylation of the sulfonium salt can occur with formation of the alkyl ester of the carboxylic acid. It is also notable that the corresponding propargylic ammonium salt does not isomerize to the allenic isomer and no products of nucleophilic addition to this 'onium salt can be obtained:

$$Me_3\overset{+}{N}\text{-}CH_2C{\equiv}CH \xrightarrow[\;\;]{Et_3N} \!\!\!\!\!\!\!\! \text{||} \quad Me_3\overset{+}{N}CH{=}C{=}CH_2 \tag{9.60}$$

Addition of nucleophiles occurs[148] at the central carbon atom of the allenic system, and this reaction can be followed by a number of alternatives, depending upon the nature of the nucleophile. Particularly notable for its simplicity is a new general furan synthesis involving addition of enolate ions[149]:

For allenic sulfonium salts with a cyclic group attached to sulfur, intramolecular nucleophilic ring opening can occur with the formation[150] of medium ring sulfides:

$$(9.62)$$

It is clear that this type of reaction can be extended to give a wide variety of unusual sulfur-containing compounds from very simple starting materials.

Addition of methoxide to a vinylmethoxysulfonium salt produces an ylide which is considered to dissociate to alkoxide ion and α-alkylthiocarbonium ion. This, on reaction with solvent, gives the α-alkoxy sulfide in a Pummerer-type sequence[133]:

$$\rightarrow Ph\overset{+}{S}=CH\text{-}CH_2OMe \xrightarrow{\ MeOH\ } PhSCH\text{-}CH_2OMe \qquad (9.63)$$

### 9.6.11 Rearrangements of Sulfonium Salts

The overwhelming majority of sulfonium salt rearrangements which have so far been examined occur under basic conditions and almost certainly involve ylide intermediates. This type of reaction will, therefore, be discussed under the heading of ylides (below) and comments on concerted reactions will be included in that section.

### 9.6.12 The Sulfonium Group in Electrophilic Aromatic Substitution[151]

The effect of the sulfonium group on an aromatic nucleus is to reduce electron density in the nucleus and hence to deactivate it towards electrophilic substitution (−I effect). The *meta* isomer is always predominantly obtained. Thus mono-nitration of dimethylphenylsulfonium ion was claimed to yield *meta* derivative exclusively.[152]

Insertion of a methylene group between the 'onium substituent and the nucleus decreases the proportion of *meta* substitution to 52%, while for the corresponding ammonium and phosphonium ions, the proportions are 80% and 10% respectively. Further, while dimethylphenylselenonium ion gives 100% *meta* derivative in mono-nitration, benzyldimethylselenonium ion gives only 16% of the total substitution at the *meta* position, which is consistent with the improvement in nuclear screening for selenium as compared with sulfur.

More recently it has been shown that small amounts of *ortho* (3.6%) and *para* (6.0%) isomers are produced in the nitration of dimethylphenylsulfonium ion.[153] A resonance interaction of the sulfonium ion with the nucleus is suggested to operate, but it is not clear whether this is a $\pi$ $(d-p)$ and/or a $\pi$ $(p-p)$ interaction.

Silver ion-catalyzed nuclear halogenation of sulfonium salts gives considerably higher *ortho/para* ratios than nitration. Chlorination of dimethylphenylsulfonium ion gives,[154] for example, 15% *ortho*, 80% *meta*, and 5% *para* substitution. Involvement of a cyclic intermediate has been suggested to account for this difference.

## 9.7 EPI-SULFONIUM SALTS (THIIRANIUM SALTS)

These compounds deserve special mention because of their frequent occurrence as reaction intermediates, particularly in the addition of sulfenyl halides to olefins[155] and in the solvolysis of halides or sulfonates bearing $\beta$-thio groups.[27] Their chemistry has been recently reviewed.[156]

In spite of the group's high reactivity, a member of the episulfonium salt series has been isolated[31] from treatment of 2-bromocyclooctyl butyl sulfide with silver 2,4,6-trinitrobenzenesulfonate. Formation of the cyclooctyl episulfonium ion occurs directly.

Very rapid hydrolysis of $\beta$-chloroalkyl sulfides has been attributed to the intermediate formation of episulfonium ions in which the $\beta$-sulfur atom participates as a neighboring group. Much attention[157] has been paid to the hydrolysis of mustard gas (bis-2-chloroethyl sulfide) which readily cyclizes to the ion (9.17):

$$ClCH_2CH_2SCH_2CH_2Cl \longrightarrow ClCH_2CH_2\overset{+}{S}\underset{\textbf{9.17}}{\triangle} \quad \bar{Cl} \qquad (9.64)$$

Neighboring group participation by sulfur leads to interesting comparisons in reactivity. Thus solvolysis of 2-chloroethyl phenyl sulfide in aqueous dioxane occurs 600 times more rapidly than for 3-chloropropyl phenyl sulfide and 5 times more rapidly than 4-chlorobutyl phenyl sulfide. By contrast, when the group attached to sulfur is ethyl instead of phenyl, the 4-chlorobutyl sulfide is very much more reactive than the 2-chloroethyl sulfide.[27] Stirling and his collaborators have discussed this phenomenon and have invoked specific conjugative effects with 3-membered rings.[22,25]

The easy formation of episulfonium salts is also responsible for the rapid reactions involving ring expansions and contractions which have been observed[158] with cyclic sulfides:

$$(9.65)$$

$$(9.66)$$

Similarly, the rearrangement of bromoalkyl sulfides may be attributed to[159] intermediate formation of episulfonium ions.

The role of episulfonium ions as intermediates in the electrophilic addition of sulfenyl halides to olefins is well established. Addition of 2,4-dinitrobenzenesulfenyl chloride to simple alkenes occurs predominantly with Markovnikov orientation[160] indicating the electrophile to be $ArS^+$, and, in reactions with styrene, only the Markovnikov product is obtained. The Hammett $\rho$ value for 2,4-dinitrobenzenesulfenyl chloride additions[161] in acetic acid is $-2.20$ and rates of addition are strongly accelerated[162] in polar solvents. A typical rate ratio for carbon tetrachloride:nitrobenzene is 1:3000.

Importantly, additions to *cis-trans* pairs of olefins are stereospecifically *trans*,[163] leading to formulation of the episulfonium intermediate in a way very similar to the proposals of bromonium ion intermediates in alkene halogenation. *Trans* stereochemistry is, of course, dictated by rearside attack on the 3-membered ring with displacement of the sulfonium group:

$$(9.67)$$

Kinetic data consistent with the involvement of the sulfonium intermediate have been obtained by Brown and Hogg,[164] who showed that rates of addition of arenesulfenyl chlorides to cyclohexene require, for correlation by a linear free energy relationship, use of Hammett $\sigma^+$ values, implying enhanced mesomeric release from *para* substituents. This observation is in accord with formation of a thiiranium ion rather than an open carbonium ion.

It seems probable that episulfonium salts are also involved in the addition of sulfenyl halides to acetylenes. Acetylenes are less reactive in electrophilic addition reactions, but alkylation increases reactivity much more than for the corresponding alkenes because nucleophilicity is increased without adverse steric requirements.[165] Reactivities[166] of alkenes are much less sensitive to structure in MeSCl additions than for additions of chlorine. This suggests that transition states for the former system have the positive charge localized on sulfur.

Substituted episulfonium compounds are produced by addition of alkanesulfenyl chlorides to alkenes of the structure $RCH{=}CH_2$, where R = vinyl, COCl, $CO_2Me$, CN, $SO_2R$, *etc.* Orientation in the final product is determined by the ease of nucleophilic attack adjacent to the substituent.[166] This occurs $\alpha$ with R = COR, but $\beta$ with R = $SO_2Me$, and $\alpha$ attack of chloride on the episulfonium ion parallels the solvolytic reactivity of the chloride, $RCH_2Cl$, which is dependent upon interaction between nucleophile and substituent.

Charge distribution in episulfonium ions has been discussed[167] in some detail in connection with sulfenyl chloride-alkene additions. Sensitivity to steric effects in ring cleavage of episulfonium ions produced by addition of alkanesulfenyl halides to simple alkenes leads to kinetic anti-Markovnikov orientation. Subsequent isomerization to the Markovnikov adducts, however, occurs readily. This investigation is a timely reminder of the dangers of over-generalization from previously collected data mainly derived from reactions with 2,4-dinitrobenzenesulfenyl halides.

All of these reactions have the common feature that the episulfonium ion is extremely susceptible to nucleophilic substitution, as ring strain is relieved, and the sulfonium group is a moderately good leaving group.

## 9.8 SULFONIUM YLIDES

There has been a great deal of recent interest in this aspect of the chemistry of sulfonium salts because of the importance of sulfonium ylides in a number of novel synthetic procedures and because of their probable role in biogenetic sequences.

Sulfonium ylides have the general structure $R_1R_2\overset{+}{S}\overset{-}{C}R_3R_4$. Their chemistry has been reviewed by Johnson[168] and by Lowe.[169]

Dimethylsulfonium fluorenylide[170] is readily obtained by proton transfer from the conjugate acid. In formation of an exceptionally stabilized sulfonium ylide of this sort, and also for others such as phenacyl ylides (below), mild bases suffice for proton transfer. More reactive (and more basic) ylides are obtained from simple sulfonium salts using bases such as phenyl lithium or methylsulfinyl carbanion[171]:

$$
(CH_3)_3\overset{+}{S}\overset{-}{I} \quad \underset{\substack{BuLi \\ THF}}{\overset{CH_3SO\overset{-}{C}H_2/(CH_3)_2SO}{\Longleftrightarrow}} \quad (CH_3)_2\overset{+}{S}\text{-}\overset{-}{C}H_2 \qquad (9.68)
$$

Proton transfer from sulfonium salts is the most common method for obtaining ylides. For those in which the carbanion is additionally stabilized by, for example, carbonyl groups, the reaction of sulfoxides with active methylene compounds gives[172] sulfonium ylides directly:

$$
PhSOCH_3 + CH_2(COMe)_2 \quad \xrightarrow[100°C]{Ac_2O} \quad \underset{Ph}{\overset{Me}{\diagdown}}\overset{+}{S}\text{-}\overset{-}{C}(COMe)_2 \qquad (9.69)
$$

$$
Et_2SO + CH_2(CN)_2 \quad \xrightarrow[H_3PO_4]{DCC} \quad Et_2\overset{+}{S}\text{-}\overset{-}{C}(CN)_2 \qquad (9.70)
$$

Reactions of carbenes with sulfides also yields[173] ylides directly and this is an especially simple method when the carbene is generated photolytically from a diazo compound:

$$
Et_2S + N_2C(CO_2Me)_2 \quad \xrightarrow{h\nu} \quad Et_2\overset{+}{S}\text{-}\overset{-}{C}(CO_2Me)_2 \qquad (9.71)
$$

The exceptionally stabilized cyclopentadienyl ylide (9.18) is produced[174] directly from the diazonium compound and diphenyl disulfide:

$$(9.72)$$

In a rather similar way, direct reaction of benzyne with sulfides yields arylalkylsulfonium ylides (see eq. 9.84).[175]

Basicity of ylides has been discussed earlier in connection with the acidity of sulfonium salts and will not be further discussed here.

The interesting question of ylide racemization has been investigated[176] for methylethylphenacylsulfonium ylide, $\text{PhCO\overset{-}{C}H\overset{+}{S}MeEt}$. Pyramidal inversion occurs without decomposition 200 times faster than for the conjugate acid. This rate difference may be attributed to the greater electronegativity of the $\text{PhCOCH}_2$ group than the $\text{PhCOCH}^-$ group, and perhaps also the greater extent of $(p-d)$ $\pi$-bonding in the ylide. The rather easy racemization of the ylide relative to sulfoxides is surprising. It is suggested[176] that electronic repulsion in the inversion transition state is less for the ylide, and that $(p-d)$ $\pi$-bonding is more effective for the pyramidal sulfoxide ground state.

### 9.8.1 Reactions of Sulfonium Ylides

Sulfonium ylides show the reactivity typical of carbanions towards carbonyl groups and towards electrophilic alkenes. The key to the understanding of these reactions is the fact that the intermediate formed by addition is an anion bearing a $\gamma$-leaving group (the sulfonium group):

$$\overset{+}{\underset{|}{S}}{-}C\overset{|}{\underset{|}{}} + X{=}Y \longrightarrow \overset{+}{\underset{|}{S}}{-}C\overset{\overset{\bar{Y}}{\diagdown}X}{\diagup} \longrightarrow \overset{|}{\underset{|}{S}}{:} + \overset{Y{-}X}{\underset{\diagup\diagdown}{C}} \qquad (9.73)$$

Addition to ketones of a simple ylide, such as dimethylsulfonium methyl ylide, yields epoxides in nearly quantitative yields[171]:

$$\text{Me}_2\overset{+}{\underset{-}{S}}\text{CH}_2 + \text{PhCHO} \longrightarrow \underset{\underset{\text{(75\%)}}{\overset{\diagdown}{\text{CH}_2}\diagup}}{\text{PhCH}{-}{-}\text{O}} + \text{Me}_2\text{S} \qquad (9.74)$$

Note that this type of reaction differs from the reaction of the corresponding phosphonium ylides in which olefins are produced. The difference in reactivity is to be ascribed to the fact that the sulfonium group is a much better leaving group than the phosphonium group in intramolecular displacement at $sp^3$ carbon.

Most epoxide-forming reactions have been run with simple ylides and very strong bases but it has been shown recently that good yields of epoxides can be obtained[177] very simply from, for example, trimethylsulfonium salts with aromatic aldehydes and as weak a base as aqueous sodium hydroxide:

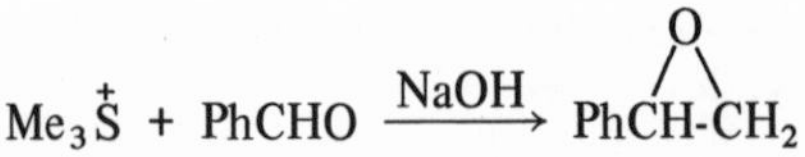

$$Me_3\overset{+}{S} + PhCHO \xrightarrow{NaOH} Ph\overset{O}{\overset{/\backslash}{CH\text{-}CH_2}} \qquad (9.75)$$

Reaction of cyclopropylsulfonium ylides with ketones generates α-cyclopropylepoxides, which rearrange to cyclobutanones. When the ketone is cyclic, "spiro-annelation" results[178] :

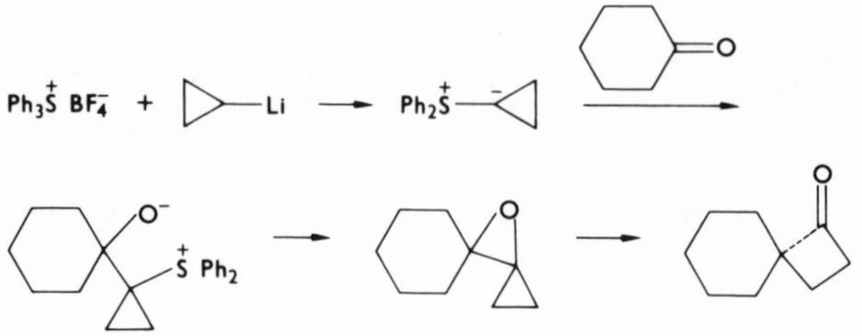

$$Ph_3\overset{+}{S}\ BF_4^- + \triangleright\!\!-Li \longrightarrow Ph_2\overset{+}{S}\!\!-\!\!\triangleleft \longrightarrow$$

(9.76)

Reaction of sulfonium ylides with aziridines gives[179] azabicyclobutanes:

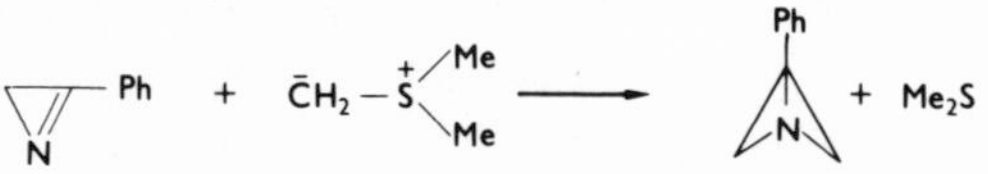

$$\underset{N}{\triangle\!\!/}\!\!-Ph + \bar{C}H_2-\overset{+}{S}\overset{Me}{\underset{Me}{\big\langle}} \longrightarrow \underset{(60\%)}{\overset{Ph}{\underset{N}{\triangle}}} + Me_2S \qquad (9.77)$$

Addition to electrophilic olefins yields[180] cyclopropanes by a process directly analogous to the epoxide synthesis. Once again the second stage of the reaction is a 1,3-intramolecular displacement:

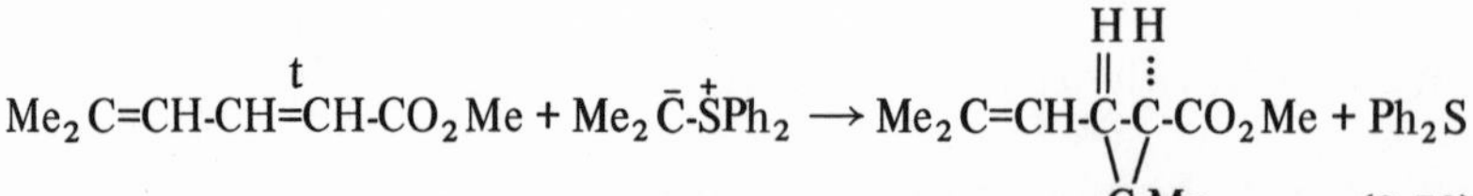

$$Me_2C=CH\text{-}\overset{t}{CH}=CH\text{-}CO_2Me + Me_2\bar{C}\text{-}\overset{+}{S}Ph_2 \rightarrow Me_2C=CH\text{-}\overset{\overset{H}{\|}\,\overset{H}{:}}{C\text{-}C}\text{-}CO_2Me + Ph_2S$$
$$\underset{C\text{-}Me_2}{\overset{\backslash/}{}} \qquad (9.78)$$

(±-methyl *trans*-chrysanthemate)

Formation of a cyclopropane by photolysis of the same type of ylide gives phenacylcarbene which then gives[181] the normal addition reaction with olefins:

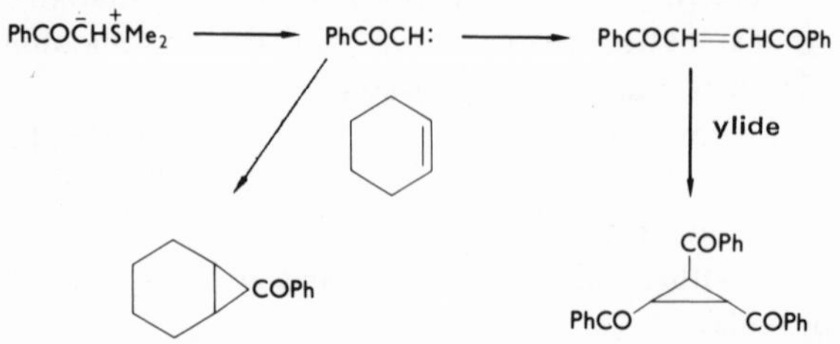

$$PhCO\bar{C}H\overset{+}{S}Me_2 \longrightarrow PhCOCH: \longrightarrow PhCOCH=CHCOPh$$

ylide

(9.79)

### 9.8.2 Rearrangement

This is another common reaction type in sulfonium ylide chemistry and a number of variations have been investigated.

The simplest is the Stevens rearrangement[182] in which a base removes the most acidic proton and the resulting anion performs an intramolecular nucleophilic displacement:

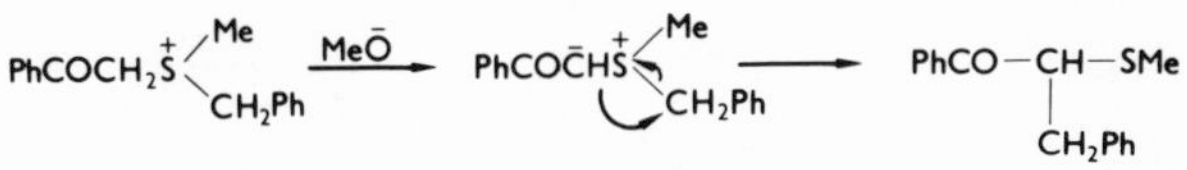

$$PhCOCH_2\overset{+}{S}\diagup^{Me}_{\diagdown CH_2Ph} \quad \xrightarrow{Me\bar{O}} \quad PhCO\bar{C}HS\diagup^{Me}_{\diagdown CH_2Ph} \quad \longrightarrow \quad PhCO-\underset{CH_2Ph}{CH}-SMe \qquad (9.80)$$

A related rearrangement (Sommelet) occurs in sulfonium ylides which bear aromatic nuclei. Treatment of benzyldimethylsulfonium salts with potassium amide in liquid ammonia gives[183, 184] o-methylbenzyl sulfide:

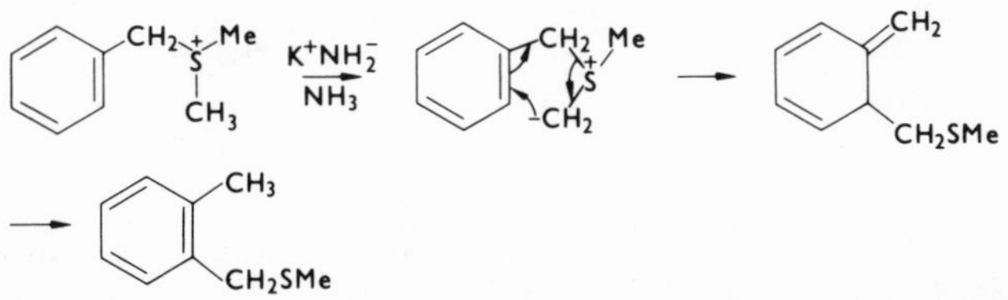

$$(9.81)$$

Both rearrangement types can, however, occur simultaneously as shown in the reaction of dibenzylmethylsulfonium ion with base. This gives a mixture of Stevens rearrangement product and Sommelet rearrangement product, the amount of the latter increasing with medium basicity:

(Sommelet) (9.82)

(Stevens) (9.83)

Recently, chemically induced nuclear spin polarization has been observed[185] in the Stevens rearrangement of the ylide produced from reaction of dibenzyl sulfide and benzyne. This suggests a radical recombination mechanism:

$$\text{(9.84)}$$

Rearrangement of phenacyldimethylsulfonium ylide gives [186] the enol ether **(9.19)**:

$$\text{(9.85)}$$

**9.19**

The reaction is best viewed[187] as a sigmatropic rearrangement in which the covalent formulation of the ylide is involved.

Attention has recently been paid to rearrangement of allylic sulfonium ylides in connection with their possible involvement in biogenetic sequences.[188, 189] Treatment of diallylsulfonium salts with bases readily gives rearranged sulfides[190],

$$\text{(9.86)}$$

and this rearrangement is of the same carbocyclic type as the Sommelet rearrangement discussed above. Baldwin[191] has rationalized several sulfonium rearrangements on the common basis of a 5-center electrocyclic process:

$$\text{(9.87)}$$

The rearrangement described above leads to head-to-tail coupling of allylic units; for an unsymmetrical allylic sulfide, alkylation and ylide rearrangement gives tail-to-tail coupling:[192] and provides a model for enzymatic coupling of farnesol pyrophosphate. Rearrangement of $\beta,\gamma$-acetylenic sulfonium salts similarly gives allenic sulfides[193]:

$$\text{Et} - \overset{+}{\underset{}{S}} \left\langle \begin{array}{c} \equiv - \text{Ph} \\ \equiv - \text{Ph} \end{array} \right. \longrightarrow \text{Et} - \text{S} - \overset{\displaystyle{/\!/}^{Ph}}{\underset{\displaystyle{\underset{Ph}{\|}}}{}} \tag{9.88}$$

## 9.9 SULFOXONIUM YLIDES

Methylation of dimethyl sulfoxide with methyl iodide gives[194] trimethylsulfoxonium iodide, an exceptional reaction in that other sulfoxides and halides give products of O-alkylation:

$$\text{Me}_2\text{SO} + \text{MeI} \longrightarrow \text{Me}_3\overset{+}{\text{S}}\text{O}\bar{\text{I}} \tag{9.89}$$

This sulfonium salt underwent deuterium-hydrogen exchange in $D_2O$ one hundred times faster than trimethylsulfonium ion, indicating that the ylide was probably much more stable. This has been confirmed in subsequent experimental work, mainly by Corey and his collaborators,[171] who prepared it by treatment of the trimethyloxysulfonium iodide with sodium hydride in tetrahydrofuran:

$$\text{Me}_3\overset{+}{\text{S}}\text{O} + \text{NaH} \xrightarrow{\text{THF}} \text{Me}_2\overset{+}{\text{S}}\text{O}\bar{\text{C}}\text{H}_2 \tag{9.90}$$

Sulfoxonium ylides stabilized by $\alpha$-keto and $\alpha$-alkoxycarbonyl groups have been obtained by silver or copper-catalyzed decompositions of diazoesters and ketones in dimethyl sulfoxide[195]:

$$\text{NO}_2\text{—}\!\!\left\langle\bigcirc\right\rangle\!\!\text{—}\overset{\underset{\displaystyle\|}{}}{\underset{\text{N}_2}{\text{C}}}\!\text{—CO}_2\text{Et} + (\text{CH}_3)_2\text{SO} \xrightarrow[120°]{\text{Ag}_2\text{O}} \text{NO}_2\text{—}\!\!\left\langle\bigcirc\right\rangle\!\!\text{—}\overset{\displaystyle{\overset{}{\text{C}}\text{—CO}_2\text{Et}}}{\underset{\displaystyle{\underset{Me}{\overset{S^+}{|}}\underset{\displaystyle\underset{O}{\|}}{}Me}}{}} \tag{9.91}$$

Reaction of sulfoxonium ylides with ketones gives oxirans in exactly the same way as the corresponding reactions with sulfonium ylides. The important difference was found with $\alpha,\beta$-unsaturated ketones when cyclopropanes are formed. This is in contrast to the corresponding reactions with sulfonium ylides which give $\alpha,\beta$-unsaturated epoxides preferentially.[171]

Trimethylsulfoxonium ylide alkylates active hydrogen compounds. β-Naphthol is converted into β-naphthyl methyl ether and p-nitrobenzoic acid into its methyl ester.[196] New ylides are produced with acylating agents such as acid chlorides, anhydrides, and ketenes[197]:

$$Me_2^+\overset{\overset{\displaystyle O}{\|}}{S}\text{-}\bar{C}H_2 + PhCOCl \longrightarrow PhCO\bar{C}H\overset{+}{S}(CH_3)_2 + (CH_3)_3\overset{+}{S}OCl^- \quad (9.92)$$

Photolysis of the β-ketosulfoxonium ylide in an alcohol produces a chain-extended ester in a sequence analogous to the Arndt-Eistert reaction:

$$RCO\bar{C}\overset{\overset{\displaystyle O}{\|}}{S}Me_2^+ \xrightarrow[R'OH]{h\nu} RCH_2CO_2R' + Me_2SO \quad (9.93)$$

Sulfoxonium ylides with acetylenic ketones yield[198] thiobenzene monoxide derivatives:

$$(9.94)$$

The same type of compound is obtained from acyl cyclohexanones[199]:

$$(9.95)$$

A number of other heterocyclic systems are produced from the addition reactions of sulfoxonium ylides. With benzonitrile, the nitrogen-sulfur heterocycle (9.20) is obtained[200]:

$$Me_2\overset{+}{S}-CH_3\bar{I} + PhC\equiv N \xrightarrow{NaH} \left[ Me_2\overset{O}{\overset{\|}{S}}\!=\!CH-\overset{Ph}{\overset{|}{C}}\!=\!NH_2 \right]^{+} \bar{I} \longrightarrow \quad (9.96)$$

**9.20**

Addition to nitrile oxides[201] and nitrile imines[202] gives the 5-membered heterocycles:

$$(9.97)$$

Much of the significant work on sulfonium salts and their derivatives is of very recent origin. The versatility of the sulfonium group as carbanion stabilizer and leaving group in displacement reactions renders sulfonium salts particularly useful tools in certain areas of organic synthesis. Much ingenuity has been shown in the investigation of the potential of this interesting class of compound and it is certain that this area is a rich vein for future discovery in the field of sulfur chemistry.

### 9.10 REFERENCES

1. J. Goerdeler, "Houben-Weyl, Methoden der organischen Chemie," Vol. 9, Springer Verlag.
2. C.C. Price and S. Oae, "Sulfur Bonding," Ronald Press, New York, 1962.
3. Ref. 1, pp. 175-183.
4. R.G. Pearson, H. Sobel and J. Songstad, *J. Amer. Chem. Soc.*, **90**, 319 (1968).
5. E.E. Reid, "Organic Chemistry of Bivalent Sulfur," Vol. II, Chemical Publishing Co. Inc, New York, 1960, p. 68.
6. F.E. Ray and J. L. Farmer, *J. Org. Chem.*, **8**, 391 (1943).
7. C.G. Swain and E.R. Thornton, *J. Amer. Chem. Soc.*, **83**, 4033 (1961).
8. M.G. Ahmed, R.W. Alder, G.H. James, M.L. Sinnott, and M.C. Whiting, *Chem. Commun.*, 1533 (1968).
9. F.E. Ray and G.J. Szasz, *J. Org. Chem.*, **8**, 121 (1943).
10. T.F. Lavine, N.F. Floyd and M.S. Cammaroti, *J. Biol. Chem.*, **207**, 107 (1954).
11. T.W. Milligan and B.C. Minor, *J. Org. Chem.*, **28**, 235 (1963).
12. O. Haas and G. Dougherty, *J. Amer. Chem. Soc.*, **65**, 1238 (1943).

13. A. Ginsberg and J. Goerdeler, *Chem. Ber.,* **94**, 2043 (1961).
14. N.F. Blau and C.G. Stuckwisch, *J. Org. Chem.,* **27**, 370 (1962).
15. A. Schöberl and G. Lange, *Annales*, **599**, 140 (1956).
16. Ref. 5, p. 70.
17 H.J. Boonstra and L. Brandsma, unpublished results quoted in "The Chemistry of the Ether Linkage," Ed. S. Patai Interscience, New York, 1967, p. 152.
18. H. Meerwein, *J. prakt. Chem.,* **147**, 257 (1937).
19. L.G. Markarowa and A.N. Nesmeyanov, *Izv. Akad. Nauk SSSR,* 617 (1945) (*Chem. Abstr.,* **40**, 4686 (1946)).
20. S. Smiles, *J. Chem. Soc.,* **77**, 160 (1900).
21. J.L. Kice, C.G. Venier and L. Heasley, *J. Amer. Chem. Soc.,* **89**, 3557 (1967).
22. R. Bird and C.J.M. Stirling, *J. Chem. Soc.,* 1221 (1973).
23. N. Allinger and V. Zalkow, *J. Org. Chem.,* **25**, 701 (1960).
24. D.N. Jones, M.J. Green, M.A. Saeed, and R.D. Whitehouse, *J. Chem. Soc. C,* 1362 (1968).
25. A.C. Knipe and C.J.M. Stirling, *J. Chem. Soc.* (B), 67 (1968).
26. K.D. Gunermann, *Angew. Chem. Internat. Edn.,* **2**, 674 (1963).
27. B. Capon, *Quart. Rev. Chem. Soc.,* **18**, 45 (1964).
28. V. Prelog and E. Cerkovnikov, *Annales*, **537**, 214 (1939); V. Prelog and D. Kohlbach, *Chem. Ber.,* **72**, 672 (1939).
29. P. Wilder and L.A. Feliu-Otero, *J. Org. Chem.,* **30**, 2560 (1965).
30. R. Breslow, S. Garratt, L. Kaplan, and D. La Follette, *J. Amer. Chem. Soc.,* **90**, 4051, 4056 (1968).
31. D.J. Pettitt and G.K. Helmkamp, *J. Org. Chem.,* **29**, 2702 (1964).
32. P. Wilder and L.A. Feliu-Otero, *J. Org. Chem.,* **31**, 4264 (1966).
33. F. Krollpfeiffer and W. Hahn, *Chem. Ber.,* **86**, 1049 (1953).
34. C. Courtot and Tse-Yei-Tung, *C.R. Acad. Sci., Ser. C,* **197**, 1227 (1933).
35. G.H. Wiegand and W.E. McEwen, *J. Org. Chem.,* **33**, 2671 (1968).
36. G. Dougherty and P.D. Hammond, *J. Amer. Chem. Soc.,* **61**, 80 (1939).
37. D. Libermann, *C.R. Acad. Sci, Ser.C,* **197**, 921 (1933).
38. C. Courtot and Tse-Yei-Tung, *C.R. Acad. Sci., Ser.C,* **200**, 1541 (1935).
39. Ref. 5, Chapter 2.
40. B.S. Wildi, S.W. Taylor, and H.A. Potratz, *J. Amer. Chem. Soc.,* **73**, 1965 (1951).
41. H. Meerwein, K.F. Zenner, and R. Gipp, *Annales*, **688**, 67 (1965).
42. H. Kwart, R.W. Body, and D.M. Hoffman, *Chem. Commun.,* 765 (1967).
43. H. Kwart and P.R. Strilko, *Chem. Commun.,* 767 (1967).
44. G.E. Wilson, *J. Amer. Chem. Soc.,* **87**, 3785 (1965).
45. C.M. Suter and H.L. Hansen, *J. Amer. Chem. Soc.,* **54**, 4100 (1932).
46. D. Landini, F. Montanari, G. Modena, and G. Scorrano, *Chem. Commun.,* 86 (1968).
47. S. Allenmark and C.E. Hagberg, *Acta Chem. Scand.,* **22**, 1461, 1694 (1968).
48. G.K. Helmkamp, H.N. Cassey, B.A. Olsen, and D.J. Pettitt, *J. Org. Chem.,* **30**, 933 (1965).
49. C.R. Johnson and W.G. Phillips, *J. Org. Chem.,* **32**, 3234 (1967).
50. G.F. Whitfield, H.S. Beilan, D. Saika, and D. Swern, *Tetrahedron Lett.,* 3543 (1970), and references cited.
51. A. Lüttringhaus and N. Engelhard, *Chem. Ber.,* **93**, 1525 (1960).
52. H.J. Shine and L. Hughes, *J. Org. Chem.,* **31**, 3142 (1966).
53. T.E. Young and P.H. Scott, *J. Org. Chem.,* **30**, 3613 (1965).
54. G. Laban and R. Mayer, *Z. Chem.,* **7**, 227 (1967).
55. R. Pettit, *Tetrahedron Lett.,* 11 (1960).
55a.V. Hanus and V. Cermak, *Collect. Czech. Chem. Commun.,* **24**, 1602 (1959).
56. V. Prelog, H. Hahn, H. Brauchli, and H.C. Beyermann, *Helv. Chim. Acta,* **27**, 1209 (1944).
57. Ref.1, p.187.
58. Ref. 5, p.70.
59. Ref.5, p.71.
60. J.B. Hyne, *Canad. J. Chem.,* **39**, 1207 (1961).
61. H. Klinger and A. Maassen, *Annales*, **243**, 193 (1888).
62. W.J. Pope and S.J. Peachey, *J. Chem. Soc.,* **77**, 1072 (1900).
63. S. Smiles, *J. Chem. Soc.,* **87**, 450 (1905).

64. E. Wedekind, *Chem. Ber.*, **58**, 2510 (1925).
65. H. Siebert, *Z. Anorg. Chem.*, **271**, 65 (1952).
66. G. Allegra, G.E. Wilson, E. Benedetti, C. Pedone, and R. Albert, *J. Amer. Chem. Soc.*, **4002** (1970).
67. D.R. Rainer, E.G. Miller, P. Bickhart, A.J. Gordon and K. Mislow, *J. Amer. Chem. Soc.*, **88**, 3138 (1966).
68. D. Darwish and G. Tourigny, *J. Amer. Chem. Soc.*, **88**, 4304 (1966).
69. R. Scartazzini and K. Mislow, *Tetrahedron Lett.*, 2719 (1967).
69a. A. Garbesi, N. Corsi, and A. Fava, *Helv. Chim. Acta*, **53**, 1499 (1970).
70. F.G. Bordwell and P.S. Boutan, *J. Amer. Chem. Soc.*, **78**, 87 (1956).
71. S. Oae and C.C. Price, *J. Amer. Chem. Soc.*, **80**, 4938 (1958) and references cited.
72. L. Pauling, "The Nature of the Chemical Bond," 3rd Ed., Cornell University Press, Ithaca, N.Y. 1960.
73. Ref. 99 Section **VII**, 24b.
74. J.D. Roberts, R.D. Clement and J.J. Drysdale, *J. Amer. Chem. Soc.*, **73**, 2181 (1951).
75. S. Oae and C.C. Price, *ibid.*, **80**, 3425 (1958).
76. G. Cilento, *Chem. Rev.*, **60**, 147 (1960).
77. G. Aksnes and J. Songstad, *Acta Chem. Scand.*, **18**, 655 (1964).
78. J. Crosby, Ph.D. Thesis, London University, 1968.
79. J. Crosby and C.J.M. Stirling, *J. Chem. Soc. B*, 671 (1970).
80. A.W. Johnson and R.B. Lacount, *J. Amer. Chem. Soc.*, **83**, 417 (1961).
81. A.W. Johnson and R.B. Lacount, *Tetrahedron*, **9**, 130 (1960).
82. G. Wittig and G.B. Felletschin, *Annales*, **555**, 133 (1944).
83. W. von E. Doering and A.K. Hoffman, *J. Amer. Chem. Soc.*, **77**, 521 (1955).
84. C.P. Lillya and E.F. Miller, *Tetrahedron Lett.*, 1281 (1968).
85. A.W. Johnson, "Ylid Chemistry," Academic Press, New York 1966, p. 315.
86. F.G. White and L.L. Ingraham, *J. Amer. Chem. Soc.*, **84**, 3109 (1962).
87. R. Breslow, *J. Amer. Chem. Soc.*, **80**, 3179 (1958).
88. P. Haake and W.B. Miller, *J. Amer. Chem. Soc.*, **85**, 4044 (1963).
89. N.F. Blau and C.G. Stuckwisch, *J. Org. Chem.*, **22**, 82 (1957).
90. D.M. Burness, *J. Org. Chem.* **24**, 849 (1959).
91. D. Darwich, Sai Hong Hui and R. Tomilson, *J. Amer. Chem. Soc.*, **90**, 5631 (1968).
92. S.R. Reymenn, *Chem. Ber.*, **7**, 1288 (1874).
93. H.A. Taylor and W.C.M. Lewis, *J. Chem. Soc.*, 665 (1922).
94. R.F. Corran, *Trans. Faraday Soc.*, **23**, 605 (1927).
95. M.P. Balfe, J. Kenyon and H. Phillips, *J. Chem. Soc.*, 2554 (1930).
96. J.F. Kincaid and F.C. Henriques, *J. Amer. Chem. Soc.*, **62**, 1474 (1940).
97. A.C. Knipe and C.J.M. Stirling, *J. Chem. Soc. B*, 1218 (1968).
98. K. von Auers, *Chem. Ber.*, **53**, 2285 (1920).
99. C.K. Ingold, "Structure and Mechanisms in Organic Chemistry," Bell, London, 1953, Chapter 7.
100. Ref. 99, p. 380.
101. F. Krollpfeiffer, H. Hartmann and F. Schmidt, *Annales*, **563**, 15 (1949).
102. C.G. Swain, W.D. Burrows and B.J. Schowen, *J. Org. Chem.*, **33**, 2534 (1968).
103. J. Knapczyk and W.E. McEwen, *J. Amer. Chem. Soc.*, **91**, 145 (1969).
103a. S. Oae and Y.H. Khim, *Bull. Chem. Soc., Japan*, **42**, 3528 (1969).
104. T.R. Lewis and S. Archer, *J. Amer. Chem. Soc.*, **73**, 2109 (1951).
105. L.C. Dorman and J. Love, *J. Org. Chem.*, **35**, 158 (1969).
106. D.S. Tarbell and D.P. Harnish, *Chem. Rev.*, **49**, 17 (1951).
107. H. Kwart and L.J. Miller, *J. Amer. Chem. Soc.*, **80**, 884 (1958), and references cited.
108. H. Böhme and H.J. Grau, *Annales*, **577**, 68 (1952).
109. J. Gierer and B. Alfredsson, *Chem. Ber.*, **90**, 1240 (1957).
110. R.A. Baxter, G.T. Newbold, and F.S. Spring, *J. Chem. Soc.*, 370 (1947).
111. B. Holmberg, *J. prakt. Chem.*, **135**, 57 (1932).
112. S. Asperger, D. Stevanovi, D. Hegedic, D. Pavlovic, and L. Klasinc, *J. Org. Chem.*, **34**, 2526 (1968).
113. C.R. Johnson and W.G. Phillips, *J. Org. Chem.*, **32**. 3233 (1967).
114. B.A. Olsen and D.J. Pettitt, *J. Org. Chem.*, **30**, 676 (1965).

115. J.L. Kice and N.A. Favstritsky, *J. Amer. Chem. Soc.*, **91**, 1751 (1969).
116. Ref. 99, Chapter 8.
117. K.A. Cooper, E.D. Hughes, C.K. Ingold, and B.J. Macnulty, *J. Chem. Soc.*, 2038 (1940).
118. K.A. Cooper, M.L. Dhar, E.D. Hughes, C.K. Ingold, B.J. Macnulty and L.I. Woolf, *ibid.*, 2043 (1940).
119. K.A. Cooper, E.D. Hughes, C.K. Ingold, G.A. Maw and B.J. Macnulty, *ibid.*, 2049 (1940).
120. E.D. Hughes, C.K. Ingold and L.I. Woolf, *ibid.*, 2084 (1940).
121. E.D. Hughes, C.K. Ingold and A.M. Mandour, *ibid.*, 2090 (1940).
122. Ref. 99, Section VIII-31e.
123. J.F. Bunnett, "Survey of Progress in Chem," Vol. 5, 53 (1969).
124. W.H. Saunders and D.H. Edison, *J. Amer. Chem. Soc.*, **82**, 138 (1960).
125. W.H. Saunders In "Alkenes," Ed. S. Patai, Interscience, New York, 1965.
126. W.H. Saunders, A.F. Cockerill, S. Asberger, L. Klasinc, and D. Stefanovic, *J. Amer. Chem. Soc.*, **88**, 848 (1966).
127. L.J. Steffa and E.R. Thornton, *J. Amer. Chem. Soc.*, **89**, 6149 (1967).
128. J.F. Bunnett and E. Bacciochi, *J. Org. Chem.*, **32**, 11 (1967).
129. S.J. Cristol and F.R. Stermitz, *J. Amer. Chem. Soc.*, **82**, 4692 (1960).
130. V. Franzen and C. Mertz, *Chem. Ber.*, **93**, 2819 (1960).
131. V. Franzen, H.I. Joschek and C. Mertz, *Annales*, **654**, 82 (1962).
132. F. Weygand and H. Daniel, *Chem. Ber.*, **94**, 3145 (1961).
133. C.R. Johnson and W.G. Phillips, *J. Org. Chem.*, **32**, 1926 (1967).
134. J. Gosselck, L. Biress, H. Schenk, and G. Schmidt, *Angew. Chem. Internat. Edn.*, **4**, 1080 (1965).
135. C.W. Crane and H.N. Rydon, *J. Chem. Soc.*, 766 (1947).
136. P. Mamalis and H.N. Rydon, *J. Chem. Soc.*, 1049 (1956).
137. M.J.S.A. Amaral, G.J. Barratt, H.N. Rydon and J.E. Willett, *J. Chem. Soc. C*, 807 (1966).
138. J. Crosby and C.J.M. Stirling, *J. Chem. Soc.* (B), 679 (1970).
139. J. Crosby, R.P. Redman and C.J.M. Stirling, unpublished work.
140. W. von E. Doering and K.C. Schreiber, *J. Amer. Chem. Soc.*, 77, 514 (1955).
141. J. Gosselck, L. Beress, H. Schenk and G. Schmidt, *Angew. Chem. Internat. Edn.*, 5, 596 (1966).
142. J. Gosselck and G. Schmidt, *Tetrahedron Lett.*, 2615 (1969).
143. J. Gosselck, H. Ahlbrecht, F. Dost, H. Schenk and G. Schmidt, *Tetrahedron Lett.*, 995 (1968).
144. G. Schmidt and J. Gosselck, *Tetrahedron Lett.*, 2623 (1969).
145. G. Schmidt and J. Gosselck, *Tetrahedron Lett.*, 3445 (1969).
146. J. Gosselck, G. Schmidt, L. Beress and H. Schenk, *Tetrahedron Lett.*, 331 (1968).
147. G.D. Appleyard and C.J.M. Stirling, *J. Chem. Soc. C*, 1904 (1969).
148. P.D. Howes and C.J.M. Stirling, *J. Chem. Soc.*, 59 (1973).
149. J.W. Batty, P.D. Howes and C.J.M. Stirling, *J. Chem. Soc.*, 65 (1973).
150. J.W. Batty, P.D. Howes and C.J.M. Stirling, to be published.
151. R.O.C. Norman and R. Taylor, "Electrophilic Substitution in Benzenoid Compounds," Elsevier, Amsterdam, 1965.
152. J.W. Baker and W.G. Moffitt, *J. Chem. Soc.*, 1722 (1930).
153. H.M. Gilow and G.L. Walker, *J. Org. Chem.*, **32**, 2580 (1967).
154. H.M. Gilow, R.B. Camp and E.C. Clifton, *J. Org. Chem.*, **33**, 230 (1968).
155. N. Kharasch, Z.S. Ariyan and A.J. Havlik, *Quarterly Reports on Sulfur Chemistry*, **1**, 93 (1966), and references cited.
156. W.H. Mueller, *Angew. Chem. Internat. Edn.*, **8**, 482 (1969).
157. P.D. Bartlett and G.C. Swain, *J. Amer. Chem. Soc.*, **71**, 1406 (1949).
158. L. Brandsma and J.F. Arens, "Chemistry of the Ether Linkage," Ed. S. Patai, Interscience, New York, 1967, p. 593.
159. Ref. 158, pp. 569-70.
160. N. Kharasch and C.M. Buess, *J. Amer. Chem. Soc.*, **71**, 2724 (1949).
161. W.L. Orr and N. Kharasch, *J. Amer. Chem. Soc.*, **78**, 1201 (1956).
162. D.R. Hogg and N. Kharasch, *J. Amer. Chem. Soc.*, **78**, 2728 (1956).
163. A.J. Havlik and N. Kharasch, *J. Amer. Chem. Soc.*, **78**, 1207 (1956).
164. C. Brown and D.R. Hogg, *Chem. Commun.*, 357 (1965).

165. W.A. Thaler, *J. Org. Chem.*, 34, 871(1969).
166. W. Thaler, W.H. Mueller, and P.E. Butler, *J. Amer. Chem. Soc.*, 90, 2069 (1968).
167. W.H. Mueller and P.E. Butler, *J. Amer. Chem. Soc.*, 90, 2075 (1968).
168. A.W. Johnson, "Ylid Chemistry," Academic Press, New York, 1966.
169. P.A. Lowe, *Chem. Ind.* (London), 1070 (1970).
170. C.K. Ingold and J.A. Jessop, *J. Chem. Soc.*, 137 (1930).
171. E.J. Corey and M. Chaykovsky, *J. Amer. Chem. Soc.*, 87, 1353 (1965).
172. H. Nozaki, *Tetrahedron,* 23, 4279 (1969); A.F. Cook and J.G. Moffatt, *J. Amer. Chem. Soc.*, 90, 740 (1968).
173. W. Ando, T. Yagihara, S. Tozune, and T. Migita, *J. Amer. Chem. Soc.*, 91, 2786 (1969).
174. D. Lloyd and M.I.C. Singer, *Chem. & Ind.* (London), 118 (1967).
175. B. Franzen, H.J. Schmidt and C. Mertz, *Chem. Ber.*, 94, 2942 (1961).
176. D. Darwish and R.L. Tomlinson, *J. Amer. Chem. Soc.*, 5938 (1968).
177. M.J. Hatch, *J. Org. Chem.*, 34, 2133 (1969).
178. B.M. Trost, R. LaRochelle, and M.J. Bogdanowicz, *Tetrahedron Lett.*, 3449 (1970).
179. A.G. Hortmann and D.A. Robertson, *J. Amer. Chem. Soc.*, 89, 5974 (1967).
180. E.J. Corey and M. Jautelat, *J. Amer. Chem. Soc.*, 89, 3912 (1967).
181. B.M. Trost, *J. Amer. Chem. Soc.*, 88, 1587 (1966).
182. T. Thompson and T.S. Stevens, *J. Chem. Soc.*, 69 (1932).
183. L.A. Pinck and G.E. Hilbert, *J. Amer. Chem. Soc.*, 68, 751 (1946).
184. C.R. Hauser, S.W. Kantor and W.R. Brasen, *J. Amer. Chem. Soc.*, 75, 2660 (1953).
185. Y. Hayashi and R. Oda, *Tetrahedron Lett.*, 5381 (1968).
186. K.W. Ratts and A.N. Yao, *J. Org. Chem.*, 33, 70 (1968).
187. G.M. Blackburn, W.D. Ollis, J.D. Plackett, C. Smith and I.O. Sutherland, *Chem. Commun.*, 186 (1968).
188. J.W. Cornforth, R.H. Cornforth, C. Donninger and G. Popjak, *Proc. Roy. Soc.*, B, 163, 492 (1966).
189. R.B. Clayton, *Quart. Rev. Chem. Soc.*, 19, 168 (1965).
190. J.E. Baldwin, R.E. Hackler and D.P. Kelly, *Chem. Commun,* 537 (1968), and references cited.
191. J.E. Baldwin, R.E. Hackler and D.P. Kelly, *Chem. Commun.*, 538 (1968) and references cited.
192. J.E. Baldwin, R.E. Hackler and D.P. Kelly, *J. Amer. Chem. Soc.*, 90, 4758 (1968).
193. J.E. Baldwin, R.E. Hackler and D.P. Kelly, *Chem. Commun.*, 1083 (1968).
194. R. Kuhn and H. Trischmann, *Annales*, 611, 117 (1958).
195. F. Dost and J. Gosselck, *Tetrahedron Lett.*, 5091 (1970).
196. H. Metzger, H. Konig and K. Seelert, *Tetrahedron Lett.*, 867 (1964).
197. E.J. Corey and M. Chaykovsky, *J. Amer. Chem. Soc.*, 86, 1640 (1964).
198. A.J. Hortman, *J. Amer. Chem. Soc.*, 87, 4972 (1965).
199. B. Holt, J. Howard and P.A. Lowe, *Tetrahedron Lett.*, 4937 (1969).
200. H. Konig, H. Metzger and K. Seelert, *Chem. Ber.*, 98, 3724 (1965).
201. A. Umani-Ronchi, P. Bravo and G. Gaudiano, *Tetrahedron Lett.*, 3477 (1966).
202. G. Gaudiano, A. Umani-Ronchi, P. Bravo and M. Acampora, *Tetrahedron Lett.*, 107 (1967) and references cited.

*Chapter 10*

# SULFONES AND SULFOXIMINES

William E. Truce, Thomas C. Klingler, and William W. Brand
*Department of Chemistry, Purdue University*
*West Lafayette, Indiana 47907, USA*

## 10.1 PHYSICAL PROPERTIES

Sulfones are compounds in which the sulfur atom is bonded to two carbon atoms and two terminal oxygens in a tetrahedral arrangement, which is easily deformable.[1] The sulfonyl moiety probably is a hydrid of the forms shown below.[4]

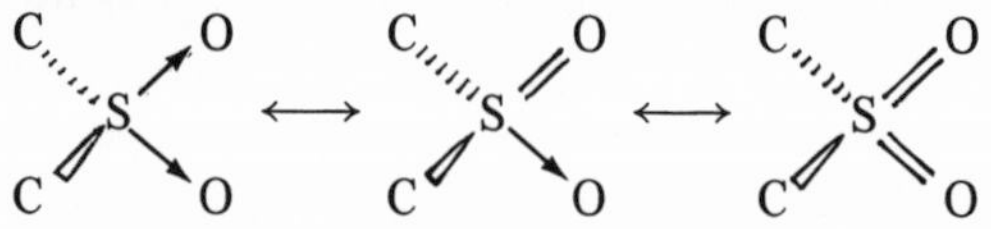

The sulfur-oxygen bonds are polar giving rise to a large dipole moment relative to the analogous ketones ($CH_3SO_2CH_3$, 3.22D; $CH_3COCH_3$, 2.88D). This polarity manifests itself in the physical properties of sulfones. They occur as either solids or high boiling liquids. Melting points generally exceed those of the analogous ketones by $60-120°$ (see Table 8.1). Their ability to hydrogen bond with alcohols and other proton donors is well documented,[3] although such bonding is less pronounced than for the corresponding sulfoxides. This may be attributable to the enhanced formal charge residing on the sulfonyl sulfur.

With the exception of the small ring compounds, sulfones are generally quite stable at elevated temperatures. Di-*p*-tolyl sulfone, for example, can be distilled without decomposition at $405°$.[1] While sulfones are soluble in most organic solvents, only those of low molecular weight show any appreciable solubility in water.[1]

The sulfonyl group exhibits two intense characteristic bands in the infrared due to the symmetric and asymmetric stretching modes. These occur at $1300\sim1350\,cm^{-1}$ and $1120\sim1160\,cm^{-1}$ respectively.[3,4] *ortho*-Substituted diaryl sulfones commonly contain bands at both 1150 and 1130 $cm^{-1}$ as well as at 1307 $cm^{-1}$.[3] The frequencies of these absorptions are relatively insensitive to adjacent olefinic linkages with the exception of thiirene dioxides in which the bands occur at 1253 and 1149 $cm^{-1}$.[5]

Dialkyl sulfones are completely transparent in the ultraviolet region down to 200 nm.[3,4] Bis-(sulfonyl)methanes in basic solution do show some absorption; this is no doubt due to the anion.[3] Vinyl sulfones give only weak absorption relative to vinyl sulfides and ketones and only slight interactions with the aromatic ring in aryl sulfones are indicated when uv is used.[3,4]

**TABLE 10.1**

Comparison of Melting Points (°C) of Representative
Sulfones and Ketones (R-X-R')

| R | R' | X = $SO_2$[a] | X = $CO$[b] | mp difference |
|---|---|---|---|---|
| $CH_3$ | $CH_3$ | 109 | -95 | 204 |
| $CH_3$ | $C_2H_5$ | 36 | -87 | 123 |
| $CH_3$ | $(CH_3)_3C$ | 79 | -53 | 132 |
| $C_2H_5$ | $C_2H_5$ | 74 | -42 | 116 |
| $CH_3$ | $C_6H_5CH_2$ | 127 | 27 | 100 |
| $C_6H_5CH_2$ | $C_6H_5CH_2$ | 150 | 35 | 115 |
| $C_6H_5$ | $CH_3$ | 88 | 20 | 68 |
| $C_6H_5$ | $C_2H_5$ | 41 | 19 | 22 |
| $C_6H_5$ | $C_6H_5$ | 128 | 48 | 80 |
| $C_6H_5$ | $p\text{-}CH_3C_6H_4$ | 125 | 60 | 85 |
| $p\text{-}CH_3C_6H_4$ | $p\text{-}CH_3C_6H_4$ | 158 | 95 | 63 |
| $\alpha\text{-}C_{10}H_7$ | $CH_3$ | 103 | 34 | 69 |
| $\beta\text{-}C_{10}H_7$ | $CH_3$ | 142 | 56 | 86 |

[a] Data taken from reference 1.
[b] Data taken from reference 2.

## 10.2 PREPARATIONS OF SULFONES

### 10.2.1 Oxidations of Sulfides and Sulfoxides

Of the many methods available for the preparation of sulfones, the most widely used approach involves direct oxidation of sulfides and sulfoxides. A variety of oxidizing agents has been utilized to effect these transformations, including hydrogen peroxide, peracids, hydroperoxides, chlorine, positive halogen reagents, nitric acid, oxides of nitrogen, oxygen, ozone, metal oxides, and electrolytic methods. An extensive tabulation of experimental results for these processes have been compiled elsewhere.[1]

$$C_6H_5CH_2SC_2H_5 \xrightarrow[\substack{HOAc \\ 80°}]{H_2O_2} C_6H_5CH_2SO_2C_2H_5 \qquad (10.1)$$
$$87\%$$

By far, the most common oxidant is hydrogen peroxide, generally in acetic acid (eq. 10.1). This method generates the sulfone in excellent yield and can be carried out in the presence of other functional groups such as hydroxyl.[6] The oxidation proceeds through a sulfoxide intermediate, at which point the oxidation may be stopped by the presence of perchloric acid in the reaction medium. This cessation of oxidation is attributable to the intervention of a sulfoxide-perchlorate complex.[7] Peroxide oxidation may be carried out in non-acidic media in the presence of catalytic amounts of tungsten, molybdenum, or vanadium salts.[8]

Oxidation of sulfides and sulfoxides by peracids may proceed by either of two mechanisms (eqs. 10.2 and 10.3). The mode of oxidation is controlled largely by the pH of the reaction mixture,[9] whereas solvent effects are minimal.[10] A Hammet study of oxidations to produce diaryl sulfones in aqueous solution has shown a negative $\rho$ value below pH 7, and a positive $\rho$ value above pH 10.[11] In such transformations alkyl sulfides are generally oxidized with greater facility than aryl sulfides.[12] The oxidation of $\alpha$-halo sulfides to $\alpha$-halo sulfones with peracid (e.g., *m*-chloroperbenzoic acid) in a nonsolvolizing medium is the method of choice for the preparation of these compounds, as extensive decomposition of the sulfide occurs under other conditions.[13]

$$C_6H_5CO_3^- + R_2SO \longrightarrow C_6H_5\overset{O}{\overset{\|}{C}}O\!-\!O\!-\!\bar{S}\!-\!\bar{O} \longrightarrow R_2SO_2 + C_6H_5CO_2^-$$
$$\underset{R \quad R}{\phantom{xxx}}$$

$$(10.2)$$

$$C_6H_5\text{-}\overset{O}{\overset{\|}{C}}\quad O\quad :\!\overset{}{\underset{R}{S}}\!-\!R \longrightarrow R_2SO_2 + C_6H_5CO_2H \qquad (10.3)$$

Generally, organic hydroperoxides are used for the preparation of sulfoxides from sulfides. Sulfones may be obtained in the presence of transition metal compounds, such as molybdovanadic acid, in catalytic amounts.[14] Under these conditions sulfones are produced in good yield at reaction temperatures above 55°. Below 55° sulfoxides are obtained as the only products. Unsaturated sulfides may be oxidized without affecting the olefinic linkage.[15]

A number of halogen compounds are capable of oxidizing sulfides and sulfoxides to sulfones. Aryl sulfides are converted to sulfones by the action of $t$-butyl hypochlorite. Alkyl sulfides, in contrast, are oxidized to sulfoxides only.[16] A more general reagent is sodium hypochlorite, which generates sulfones from both alkyl and aryl sulfides.[17] At low temperature N-chlorobenzotriazole converts sulfoxides to sulfones in high yields.[18]

Chlorine in aqueous methanol converts sulfoxides to sulfones in moderate yields. This method suffers from the fact that cleavage may occur (eq. 10.4). Diphenyl sulfone was prepared from the sulfoxide in this manner in 42% yield.[19]

$$(C_6H_5CH_2)_2SO \xrightarrow[\text{aq. CH}_3\text{OH}]{\text{Cl}_2} \underset{40\%}{(C_6H_5CH_2)_2SO_2} + \underset{19\%}{C_6H_5CH_2SO_2Cl}$$

$$(10.4)$$

Utilization of stoichiometric amounts of iodobenzene dichloride permits selective oxidation of sulfides to sulfones (eq. 10.5) or to sulfoxides (eq. 10.6), depending on the amount used. The presence of electron-withdrawing substituents in aryl sulfides may terminate oxidation at the sulfoxide stage (eq. 10.6). Benzylic positions are chlorinated, giving rise to $\alpha$-chloro sulfoxides from benzylic sulfides.[20]

$$(C_6H_5)_2S + 2C_6H_5ICl_2 \xrightarrow[C_5H_5N]{H_2O} \underset{100\%}{(C_6H_5)_2SO_2} \qquad (10.5)$$

$$(10.6)$$

Oxidations of organic substrates with nitric acid are well known and have been applied to sulfides. Heating the sulfide in concentrated nitric acid above $120°$ gives a good yield of sulfone. The volatility of low boiling sulfides, an apparent drawback, is diminished due to protonation of the sulfide to the sulfonium salt.[21] In nitroethane solvent, aryl sulfides bearing strongly electron-releasing substituents may be cleaved to the nitroarene and the corresponding sulfonic acid.[22]

Ozonolysis of sulfides at low temperature proceeds in distinct steps to produce first sulfoxide and then sulfone. Oxidation may be stopped at the sulfoxide stage.[23] Alkyl sulfides are oxidized more readily than aryl sulfides, the latter frequently yielding mixtures of both oxidation states even with excess ozone.[23] Ozonization of olefinic sulfides proceeds with preferential attack at the unsaturated linkage,[24] while hydroxy sulfides are easily converted to hydroxy sulfones.[25]

Molecular oxygen is an effective oxidant in the presence of rhodium and iridium salts[26] or vandium metal.[27] Alkyl sulfoxides are also converted to sulfones photochemically by the action of oxygen in the presence of a sensitizer.[28]

Various metallic oxides have been used for the preparation of sulfones. Permanganate in acidic or basic media oxidizes both sulfides and sulfoxides to sulfones in good yields.[29] Under neutral conditions permanganate or osmium tetroxide will oxidize sulfoxides, but not sulfides to sulfones.[30] Sodium dichromate[31] and ruthenium tetroxide[32] similarly oxidize both sulfides and sulfoxides to sulfones. Chromic acid has been employed for preparation of sulfones from sulfides; however, yields are frequently inferior to those obtained *via* permanganate oxidation. Use of selenium dioxide also has been reported.[33]

Finally, electrochemical oxidation of sulfides to sulfones often proceeds in good yield. Yields are frequently enhanced by the presence of salts or oxides of tungsten, vanadium, molybdenum, or selenium.[34] Under anhydrous conditions the oxidation may proceed with cleavage and further oxidation to sulfonic acid salts.[35]

### 10.2.2 Aromatic Sulfonylation

A widely used method for the preparation of aryl sulfones involves the reaction between an arene and a sulfonyl halide in the presence of a suitable Lewis acid (eqs. 10.7 and 10.8).[36] Although sulfonyl bromides and fluorides have been used in the reaction, sulfonyl chlorides are most commonly employed. The sulfonylation proceeds in good yield; arenesulfonyl chlorides usually give slightly better results than alkyl analogues. Methanesulfonyl chloride with methoxy-substituted aromatics, such as hydroquinone dimethyl ether, gives predominantly methanesulfonate esters, although sulfones can be obtained from arenesulfonyl chlorides and the same methyl esters.

$$CH_3SO_2Cl + C_6H_6 \xrightarrow{AlCl_3} CH_3SO_2C_6H_5 \qquad (10.7)$$
$$80\%$$

$$C_6H_5SO_2Cl + C_6H_6 \xrightarrow[CH_2Cl_2]{AlCl_3} \underset{90\%}{(C_6H_5)_2SO_2} \qquad (10.8)$$

The reactivity of a substituted arenesulfonyl chloride is influenced by the nature of the substituent. Electron-releasing groups *para* to the $SO_2Cl$ moiety increase the rate of reaction, whereas electron-withdrawing substituents have the opposite effect; however, even with *p*-nitro substituents, reaction is still possible. Heterocyclic sulfonyl chlorides have also been used successfully.

Substituents on the aromatic hydrocarbon exert orienting effects similar to those observed in Friedel-Crafts acylations, although the selectivity is considerably lower with sulfonylations. Strongly deactivating groups such as nitro or nitrile prevent reaction from occurring. Thiophene and activated thiophene nuclei are too reactive under sulfonylation conditions and extensive decomposition results. Deactivated thiophene rings (e.g., 2,5-dichlorothiophene) can be sulfonylated without extensive decomposition. Fused ring aromatics are also sulfonylated in reduced yields with considerable tar formation when the reactions are catalyzed by $AlCl_3$. More suitable catalysts for such systems are $SnCl_4$ or $ZnCl_2$, which enhance sulfonylation by diminishing undesirable side reactions.

The initial step in the mechanism (eq., 10.9) for $AlCl_3$-catalyzed reactions involves formation of a complex between the catalyst and the sulfonyl chloride. A similar complex exists between the sulfone product and $AlCl_3$, which in fact is stronger than the initial complex. For this reason a full equivalent of Lewis acid is required. Addition of a slight excess of catalyst frequently induces a large rate enhancement of sulfone formation. Other catalysts that have been used include $FeCl_3$, $SbCl_5$ $AlBr_3$, $ZnCl_4$, $AnCl_2$, $GaCl_3$ and $AgClO_4$.

$$RSO_2Cl + AlCl_3 \rightleftharpoons RSO_2Cl \cdot AlCl_3 \qquad (10.9)$$

$$RSO_2Cl \cdot AlCl_3 \rightleftharpoons RSO_2^+ + AlCl_4^- \qquad (10.10)$$

$$RSO_2^+ + AlCl_4^- + ArH \longrightarrow ArSO_2 \ R \cdot AlCl_3 + HCl \qquad (10.11)$$

Following complexation, ionization to sulfonylium ion and $AlCl_4^-$ occurs (eq. 10.10). Subsequently, the aromatic hydrocarbon is attacked to produce sulfone complex and HCl (eq. 10.11).[37] The rate-determining step is dependent upon the reactivity of the aromatic hydrocarbon. Where ArH is at least as reactive as toluene, eq. 10.10 becomes rate-determining, and the reaction proceeding independently of (ArH); where ArH is less reactive than benzene, the product-forming step (eq. 10.11) becomes rate-determining, and the reaction exhibits first-order dependence in ArH, $RSO_2Cl$, and Lewis acid.

The course of reaction is frequently solvent and temperature dependent. Generally the reaction is carried out with ArH acting as both solvent and reactant or in nitrobenzene or carbon disulfide in which the addition complex (eq. 10.9) is soluble. Nitrobenzene also complexes with $AlCl_3$ requiring the modification of eq. 10.9 such that the solvent complex, rather than free $AlCl_3$, reacts with $RSO_2Cl$. Other solvents such as chlorinated hydrocarbons and nitromethane have also been used.

The sulfonylation of naphthalene, as shown in eq. 10.12, has been reported to show a solvent effect on product orientation. In dichloroethylene at 25° the preponderant attack occurs at the α-position, while at elevated temperature in nitrobenzene the β-isomer predominates. This comparison may, however, be spurious in that the α-isomer is known to isomerize to β in the presence of HCl at high temperature.[38]

| | |
|---|---|
| $C_2H_4Cl_2$  80% | 20% |
| $C_6H_5NO_2$  7% | 93% |

(10.12)

A marked temperature effect was observed in the $ZnCl_2$-catalyzed reaction between *p*-nitrobenzenesulfonyl chloride and resorcinol dimethyl ether (eq. 10.13). Cleavage occurs at 140°, while sulfonylation occurs at 125°.

(10.13)

Sulfones are frequently by-products in aromatic sulfonation reactions.[39] By removing the water produced, either as an azeotrope (eq. 10.14)[40] or by direct distillation (eq. 10.15),[41] the sulfone is obtained as the major product.

$$C_6H_6 + H_2SO_4 \text{ (conc.)} \xrightarrow[(-H_2O)]{\Delta} (C_6H_5)_2SO_2 + C_6H_5SO_3H$$

$$80\% \qquad\qquad 15\% \quad (10.14)$$

An analogous method of preparation of aryl sulfones involves treatment of an aromatic hydrocarbon with sulfuric acid and dimethyl pyrosulfate (eq. 10.16).[42] Possibly reaction proceeds through an $ArSO_2OSO_2OCH_3$ intermediate.

Unsymmetrical aryl sulfones may be prepared by reaction of an arenesulfonic acid with an aromatic substrate above 150°. The reaction is facilitated by removal of water, as by addition of excess $P_4O_{10}$[43] or by passing a gaseous hydrocarbon stream through the hot reaction mixture.[44] In such reactions at elevated temperatures trans-sulfonation may intervene, as shown in eq. 10.17, to reduce the yield of the desired product.[45] Similar behavior is shown (eq. 10.18) in the reaction of naphthalensulfonic acids in polyphosphoric acid to form symmetrical sulfones even in the absence of additional hydrocarbon.[46] Addition of dimethyl pyrosulfate to a hydrocarbon and sulfonic acid (eq. 10.19) effects sulfone formation at reduced temperatures.[47]

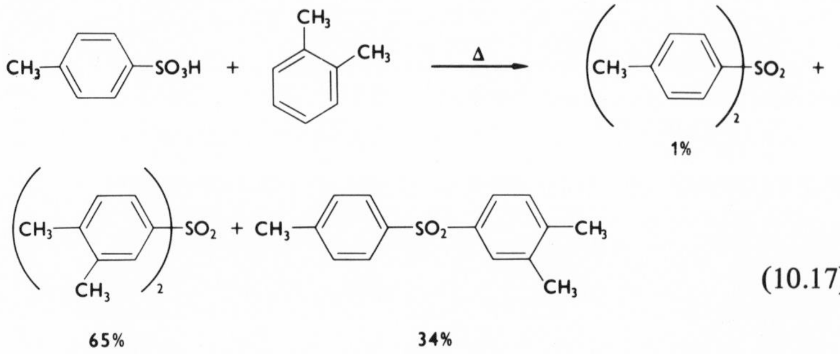

(10.17)

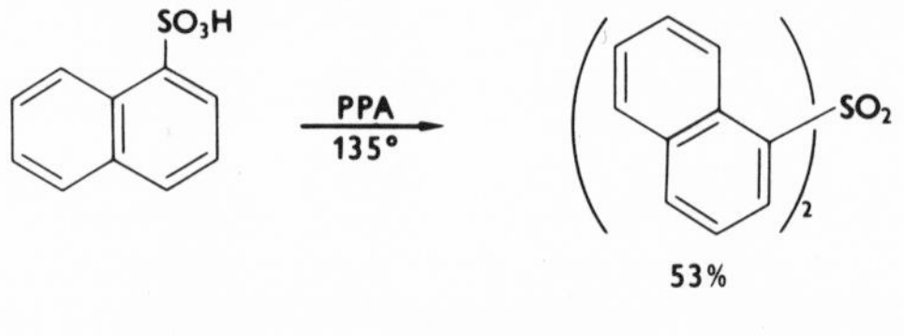

(10.18)

(10.19)

### 10.2.3 Synthesis *via* Sulfinates

Salts of sulfinic acids readily displace halide ion from primary alkyl halides to form sulfones (eq. 10.20). Hydroxylic solvents such as methanol or ethanol are commonly used.[48] Higher boiling solvents such as dipropylene glycol often give excellent yields, but may produce product purification problems.[49] Since the advent of solvents known to facilitate nucleophilic displacements, such solvents as glyme[49] and DMF[50] have been used. Hard alkylating agents, such as dimethyl sulfate or diazomethane, produce predominantly sulfinate esters.[50]

$$p\text{-CH}_3\text{C}_6\text{H}_4\text{SO}_2\text{Na} + \text{CH}_3\text{I} \longrightarrow p\text{-CH}_3\text{C}_6\text{H}_4\text{SO}_2\text{CH}_3 \qquad (10.20)^{48}$$

85%

Sulfinate esters may be induced to rearrange to sulfones by heating or under acidic conditions (eq. 10.21). Such rearrangements are facilitated by groups which stabilize the intermediate carbonium ion. As a result, methyl or phenyl sulfinate esters tend to be stable under the rearrangement conditions.[51]

$$(10.21)$$

Sulfinate displacements are also known to occur at aromatic centers activated toward nucleophilic attack[52,53]:

$$(10.22)^{53}$$

$$(10.23)^{52}$$

$$(10.24)^{53}$$

An oxidative sulfonylation of 4-aminophenol with a sodium sulfinate has been carried out by enzymatic catalysis.[54]

$$(10.25)$$

As shown in eq. 10.26, diaryl ethers bearing an *ortho* sulfinate group in one ring and an electron-withdrawing substituent in the other may be induced to rearrange to aryl sulfones. This may be considered a reverse Smiles rearrangement as the course of reaction is reversed under basic conditions. For rearrangement to occur, a buffered solution is required such that the sulfinic acid is ionized but not the phenolic product.

$$\text{(10.26)}$$

Alkyl sulfones may be prepared in moderate to good yields by the addition of sulfinate salts to Michael olefins, as shown in eq. 10.27.[56] The activating group (Y) may be carboxylate, amide, nitrile, keto, aldehyde, or sulfonyl. Both alkyl and aryl sulfinates may be used with comparable results. The replacement of the trimethylammonium group by sulfinate in eq. 10.28 has been described as a displacement,[52] although another plausible mechanistic pathway would appear to be an elimination-addition sequence.

$$\text{RSO}_2\text{Na} + \overset{\text{Y}}{\underset{\text{H}}{\text{C}=\text{C}}} \quad \xrightarrow[\text{NaH}_2\text{PO}_3]{\text{H}_2\text{O}} \quad -\overset{|}{\underset{\text{R-SO}_2}{\text{C}}}\text{-CH}_2\text{Y} \qquad \text{(10.27)}$$

$$\text{ArSO}_2\text{Na} + \overset{\text{O}}{\overset{\|}{\text{ArC}}}\text{-CH}_2\text{CH}_2\overset{+}{\text{N}}(\text{CH}_3)_2 \longrightarrow \overset{\text{O}}{\overset{\|}{\text{ArC}}}\text{CH}_2\text{CH}_2\text{SO}_2\text{Ar} \qquad \text{(10.28)}$$

Addition of sulfinic acids to aldehydes yields α-hydroxy sulfones (eqs. 10.29 and 10.30).[57] These adducts are crystalline compounds which revert to starting materials on standing. They may be converted to α-aminosulfones by condensation with amines or hydroxylamine. Primary amines (eq. 10.31) and ammonium hydroxide (eq. 10.32) form disulfones. Aromatic (eq. 10.33) and secondary amines incorporate only one sulfonyl unit.[57]

$$\text{(10.29)}$$

$$n\text{-C}_4\text{H}_9\text{SO}_2\text{H} + \text{C}_3\text{H}_7\text{CHO} \longrightarrow n\text{-C}_4\text{H}_9\text{SO}_2\underset{\text{OH}}{\overset{|}{\text{CHC}_3\text{H}_7}} \qquad \text{(10.30)}$$

$$2\,C_5H_{11}SO_2CH_2OH + CH_3NH_2 \longrightarrow (C_5H_{11}SO_2CH_2)_2NCH_3$$

$$(10.31)$$

$$2\,C_5H_{11}SO_2CH_2OH + NH_4OH \longrightarrow (C_5H_{11}SO_2CH_2)_2NH$$

$$(10.32)$$

$$C_5H_{11}SO_2CH_2OH + C_6H_5NH_2 \longrightarrow C_5H_{11}SO_2CH_2NHC_6H_5$$

$$(10.33)$$

### 10.2.4 Additons of Sulfonyl Halides to Olefins and Acetylenes

In the presence of light or benzoyl peroxide, sulfonyl halides add by a free radical mechanism to olefins to form $\beta$-halosulfones in excellent yields.[58] Sulfonyl chlorides add readily to unactivated olefins, but with olefins such as styrene only polymeric products result. This difficulty may be overcome by carrying the addition out in the presence of catalytic amounts of cupric chloride and triethylamine hydrochloride.[59]

$$RSO_2Cl \xrightarrow{h\nu} RSO_2\cdot + Cl\cdot \qquad (10.34)$$

$$RSO_2\cdot + C_6H_5CH=CH_2 \rightleftarrows RSO_2CH_2\overset{\cdot}{C}HC_6H_5 \qquad (10.35)$$
$$\mathbf{10.1}$$

$$\mathbf{10.1} + RSO_2Cl \longrightarrow RSO_2CH_2\underset{\underset{Cl}{|}}{C}HC_6H_5 + RSO_2\cdot \qquad (10.36)$$
$$\mathbf{10.2}$$

$$\mathbf{10.1} + xC_6H_5CH=CH_2 \longrightarrow \text{Polymer} \qquad (10.37)$$

$$\mathbf{10.1} + CuCl_2 \longrightarrow \mathbf{10.2} + CuCl \qquad (10.38)$$

$$RSO_2Cl + CuCl \longrightarrow RSO_2\cdot + CuCl_2 \qquad (10.39)$$

The radical intermediate (10.1) is intercepted by the cupric salt and undergoes ligand transfer (eq. 10.38) faster than polymerization (eq. 10.37), which predominates in the absence of cupric salt.[59] The copper-catalyzed addition proceeds well with both activated and unactivated olefins[59,60]:

$$CH_3CH_2CH=CH_2 \ + \ C_6H_5SO_2Cl \ \xrightarrow{\begin{array}{c} CuCl_2(95-100°) \\ \hline Et_3N.HCl \ CH_3CN \end{array}}$$

$$(10.40)$$

$$\begin{array}{c} CH_3CH_2CHCH_2SO_2C_6H_5 \\ | \qquad 86\% \\ Cl \end{array}$$

Whereas sulfonyl chlorides in the absence of cupric chloride cause polymerization of styrene, sulfonyl bromides and iodides add cleanly to give 1:1 adducts.[58] Further studies of sulfonyl halide additions to acrylonitrile have shown only iodide capable of undergoing chain transfer rapidly enough to compete with polymerization. Therefore the enhanced reactivity of sulfonyl iodides and bromides permits radical addition to olefins under considerably milder conditions than with sulfonyl chlorides. These adducts, once formed, are readily dehydrohalogenated with base, constituting a valuable synthesis of vinyl sulfones (see Sect. 10.3.7).

Conjugated dienes undergo 1,4-addition of sulfonyl chlorides in the presence of $CuCl_2$ to give $\delta$-chloro-$\beta,\gamma$-unsaturated sulfones (eq. 10.41).[60] The reaction has been extended to bis-sulfonyl chlorides and to unconjugated dienes.[61]

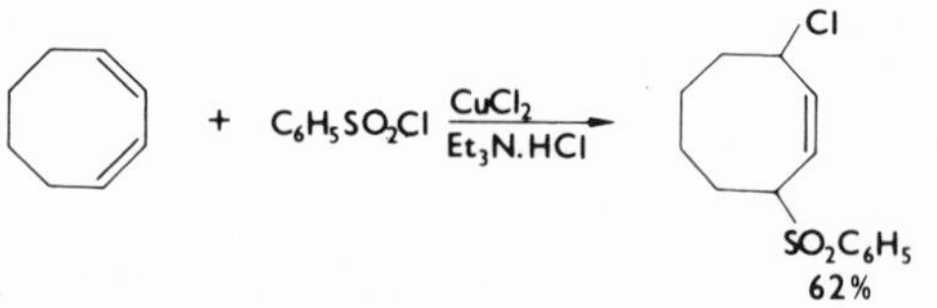

$$(10.41)$$

Benzenesulfonyl chloride adds to phenylacetylene to produce the $\beta$-chlorovinyl sulfone in 45% yield.[60] The sulfonyl iodides are more reactive toward acetylenes, generally resulting in better yields of $\beta$-halovinyl sulfones[62,63]:

$$C_6H_5C{\equiv}CH \ + \ RSO_2I \ \xrightarrow[\Delta]{light} \ \begin{array}{c} C_6H_5 \qquad SO_2R \\ \diagdown \quad \diagup \\ C=C \\ \diagup \quad \diagdown \\ I \qquad H \end{array}$$

$$(10.42)$$

$$
\begin{array}{ll}
R = CH_3 & 73\% \\
R = C_2H_5 & 80\% \\
R = p\text{-}C_7H_7 & 87\%
\end{array}
$$

The reaction may be induced thermally, but is catalyzed by light. Similarly sulfonyl iodides add to allenes to give 1:1 adducts. The direction of addition is apparently controlled by the substituents present on the allene substrate[63]:

$$H_2C=C=CH_2 + ArSO_2I \xrightarrow[Et_2O]{hv} CH_2=C-CH_2SO_2Ar + CH_2=C(SO_2Ar)CH_2SO_2Ar$$

$$\underset{31\%}{\overset{|}{\underset{I}{|}}} \qquad 13\% \qquad (10.43)$$

$$C_6H_5CH=C=CH_2 + CH_3-\!\!\!\bigcirc\!\!\!-SO_2I \xrightarrow[Ether]{hv} \underset{H}{\overset{C_6H_5}{>}}C=C\underset{SO_2Ar}{\overset{CH_2I}{<}} \quad (10.44)$$

$$80\%$$

## 10.2.5 Reactions of Sulfonic Acid Derivatives with Organometallic Reagents

Sulfonate esters react with Grignard reagents to produce sulfones.[64] Aryl arenesulfonates and aryl or alkyl alkanesulfonates undergo this transformation. On the other hand, aryl Grignards (eq. 10.45) give considerably better results than the aliphatic reagents. For this reason, the method is limited to the preparation of diaryl and aryl alkyl sulfones.[64]

$$CH_3-\!\!\!\bigcirc\!\!\!-SO_2OC_6H_5 + CH_3-\!\!\!\bigcirc\!\!\!-MgBr \xrightarrow[35°]{Et_2O} \left(CH_3-\!\!\!\bigcirc\!\!\!-SO_2\right)_2 \quad (10.45)$$

$$45\%$$

At low temperatures, arenesulfonyl chlorides also interact with Grignard reagents to form sulfones in low yields (eq. 10.46).[65] This mode of sulfone synthesis is very limited, as sulfoxides are frequently produced as the only products.[66] Sulfones are reported to arise in low yields from similar reactions of organocadmiums[67] or organomercury[68] compounds.

$$CH_3-\!\!\!\bigcirc\!\!\!-SO_2Cl + C_6H_5MgBr \xrightarrow[-5°]{Et_2O} CH_3-\!\!\!\bigcirc\!\!\!-SO_2C_6H_5 + C_6H_5Cl$$

$$\qquad\qquad\qquad\qquad 33\% \qquad\qquad 11\%$$

$$+ CH_3-\!\!\!\bigcirc\!\!\!-SO_2$$

$$11\% \qquad\qquad (10.46)[65]$$

Sulfonyl fluorides are superior to the chlorides for preparing sulfones by treatment with a Grignard reagent or organolithium reagent.[69] With excess organometallics, subsequent metallation and reaction with additional sulfonyl fluoride can be used for preparing β-disulfones[70]:

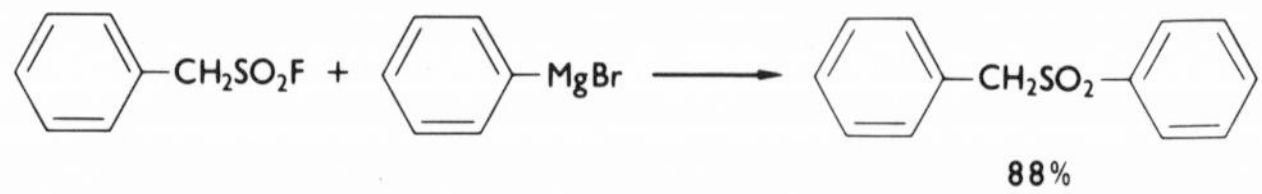

$$\text{(10.47)}$$

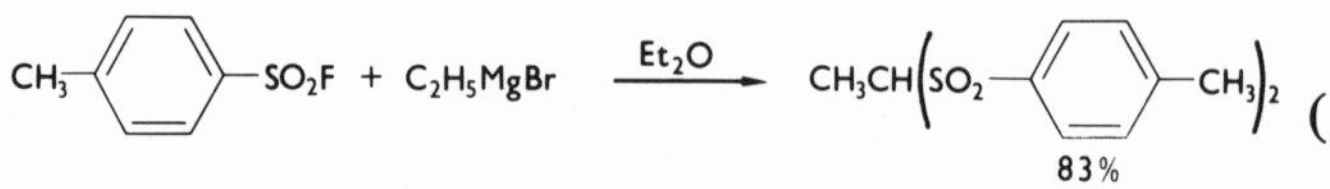

88%

$$\text{(10.48)}$$

83%

Although thionyl chloride reacts readily with organometallic reagents to form sulfoxides, the analogous reaction with sulfuryl chloride can be used to make sulfonyl chlorides[71,72] (the corresponding sulfide and sulfoxide being possible by-products),[71] but is not a useful approach to sulfones.

### 10.2.6 Fries and Related Rearrangements

In the presence of a Lewis acid, aryl sulfonates will rearrange to hydroxyaryl sulfones (eq. 10.49). This is an extension of the more extensively investigated Fries rearrangement of carboxylic esters.[36]

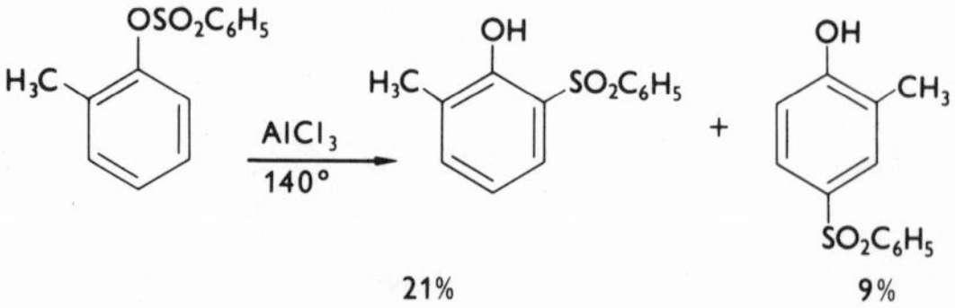

$$\text{(10.49)}^{73}$$

21%                        9%

Migration of the sulfonyl moiety occurs predominantly to the *ortho* position, although some *para* isomer is also formed. As in the rearrangement of carboxylic esters, use of solvents favoring charge separation, such as nitrobenzene, will increase the proportion of *para* isomer. The reaction is usually carried out without solvent by heating to $120°-160°$. Other catalysts which have been used are $ZnCl_2$ and HF. Yields are low in most cases, in the order of 10–40%.[36] The rearrangement may also be carried out in the absence of catalyst under photolytic conditions (eqs. 10.50 and 10.51).

$$(10.50)^{74}$$

$$(10.51)$$

N,N-Diaryl and N-aryl-N-alkyl arylsulfonamides on treatment with hot, concentrated sulfuric acid undergo a rearrangement analogous to the Fries rearrangement to give *o*-aminodiaryl sulfones (eq. 10.52).[75] Under the same conditions methanesulfonamides are hydrolyzed, which is also true of primary anilides in general. Sulfonamides of N-alkylanilines undergo rearrangement more easily than those of diarylamines. Rearrangement is facilitated by electron-donating substituents on the N-aryl group and retarded by electron-withdrawing substituents.

$$(10.52)$$

The derivatives of primary anilides will rearrange in the presence of ZnCl$_2$ and HCl to give predominantly the *para* isomer. If the *para* position is blocked, *ortho* substitution occurs, as in eq. 10.53. Under these conditions an additional N-alkyl group will cause *ortho* substitution even in the absence of the *p*-blocking group; the reason for this change in orientation is unclear.[76]

$$(10.53)$$

An arylsulfonamide intermediate is believed to be involved in the formation of sulfones from arenesulfinic acids and diazobenzene or phenazine[77]:

$$(10.54)$$

### 10.2.7 Small Ring Sulfones

Thietane and thiete dioxides are available *via* cycloaddition reactions of sulfenes ($\diagdown$C=SO$_2$) to electron-rich multiple bonds[78] as found in enamines (eq. 10.55), ketene acetals (eq. 10.55), ketene O,N-acetals (eq. 10.56) and ynamines (eq. 10.57).[79,80]

$$(10.55)$$

$$(10.56)$$

$$(10.57)$$

Ar = *p* − tolyl

Thiirane dioxides (episulfones) are available through sulfene precursors.[79] Symmetrical episulfones may be obtained either by reaction of diazoalkanes with sulfur dioxide (eq. 10.58)[81] or with an appropriate sulfene (eq. 10.59)[79] The unsymmetrical isomers may be prepared as in eq. 10.59, where R $\neq$ R′ or by addition of phosphorus ylides to sulfene (eq. 10.60).[83]

$$C_6H_5CHN_2 + SO_2 \xrightarrow{-20°} C_6H_5\diagdown\!\!\!\diagup C_6H_5 \qquad\qquad (10.58)[82]$$
$$SO_2$$

$$RCH_2SO_2Cl + R'CHN_2 \xrightarrow{NEt_3} R\diagdown\!\!\!\diagup R' \qquad\qquad (10.59)$$
$$SO_2$$

$$(C_6H_5)_3\overset{+}{P}-\overset{-}{C} \quad + \quad CH_3SO_2Cl \xrightarrow{NEt_3} CH_2 \diagdown\!\!\!\diagup{}^{SO_2}\!\!\!C$$

$$(10.60)$$

$$(C_6H_5)_3P - CHSO_2CH$$

Thiirene dioxides, the unsaturated analogues of episulfones, are prepared under modified Ramberg-Bäcklund treatment of $\alpha,\alpha'$-dibromsulfones (eq. 10.108)[5] or dehydrohalogenation of halothiirane dioxides (eq. 10.109).[84]

## 10.2.8 Miscellaneous Preparations

In an extension of the Diels-Alder reaction, sulfur dioxide will add in a concerted fashion to substituted 1,3-butadienes to produce cyclic $\beta,\gamma$-unsaturated sulfones.[85] The cycloaddition occurs by a stereospecific disrotatory process (eq. 10.61).[86] Sulfur dioxide will also add to simple Michael olefins in the presence of formic acid and trimethylamine (eq. 10.62).[87]

$$(10.61)$$

$$CH_2{=}CHCO_2Et + SO_2 \xrightarrow[NMe_3]{HCO_2H,\ \Delta} SO_2(CH_2CH_2CO_2Et)_2 \qquad (10.62)$$

$$60\%$$

Condensations of sulfonyl phosphorous ylides with aldehydes have been reported to give good yields of vinyl sulfones[88] :

$$C_2H_5SO_2\overset{-}{C}H\overset{+}{-}PO(OEt)_2 \ + \ Cl_3CC\overset{O}{\underset{H}{\diagdown}} \ \longrightarrow \ C_2H_5SO_2CH{=}CHCCl_3$$

$$76\% \qquad (10.63)$$

Low yields of sulfones have been obtained by the thermal decomposition of benzoyl peroxide in the presence of sulfur dioxide and an aromatic hydrocarbon *via* a radical mechanism[89] :

$$CH_3{-}\langle\!\!\!\!\!\bigcirc\!\!\!\!\!\rangle{-}CH_3 + (C_6H_5\overset{O}{\overset{\|}{C}}{-}O{-})_2 \ \xrightarrow[\Delta]{SO_2} \ CH_3 {-}\langle\!\!\!\!\!\bigcirc\!\!\!\!\!\rangle{-}CH_2SO_2C_6H_5$$

$$18\%$$

$$(10.64)$$

## 10.3 REACTIONS OF SULFONES

Most of the reactions of sulfones are base-catalyzed (related to the strong electron-attracting character of the sulfonyl group and to the stability of the sulfinate ion) with only a few reactions readily occurring under acidic or thermal conditions. Carbanionic reactions predominate, and electron deficiency adjacent to the sulfonyl group is unfavored as is evidenced by the facts that $\alpha$-keto sulfones are unknown, that electrophilic additions to vinyl sulfones are difficult, that halogens $\alpha$ to a sulfonyl group are inert toward Lewis acids, and that radical halogenation of sulfolane goes only at positions remote from the sulfonyl group (eq. 10.64a).[90] Furthermore, the sulfonyl group, like the nitro group, is deactivating and *m*-directing

$$\langle\underset{O^{\diagup S}{\diagdown}O}{\ }\rangle + Cl_2 \ \xrightarrow[\Delta,\ h\nu]{CCl_4} \ \langle\underset{O^{\diagup S}{\diagdown}O}{\overset{Cl}{\ }}\rangle + \langle\underset{O^{\diagup S}{\diagdown}O}{\overset{Cl\quad Cl}{\ }}\rangle + \langle\underset{O^{\diagup S}{\diagdown}O}{\overset{Cl\ \ Cl}{\ \ \diagdown Cl}}\rangle + HCl$$

$$(10.64a)$$

### 10.3.1  Acidity of Sulfones and Nature of the Carbanions

The strong electronegative character of the sulfonyl group markedly enhances the acidity of $\alpha$-hydrogens. Metallation of sulfones was reported as early as 1935.[91] Much of the recent chemistry of sulfones has involved the intermediacy of sulfone-stabilized carbanions from metallation with strong bases. Commonly-used metallating agents have been Grignard reagents,[92] potassium *t*-butoxide,[93] *n*-butyllithium,[94] metal hydrides,[95] and alkali metal amides.[96,97]

Metallation may occur at positions other than at an α-carbon. Diaryl sulfones may be metallated on the aromatic nucleus, or at a benzylic position, the latter being the basis for the Truce-Smiles rearrangement. It has been proposed that the reactivity of sulfones toward proton abstraction follows the order[98]:

$$-SO_2\overset{|}{\underset{|}{C}}-H \quad > \quad -SO_2-\underset{}{\bigcirc}\overset{CH_2-H}{} \quad > \quad -SO_2-\overset{H}{\bigcirc}$$

This order is suggested by eqs. 10.65 and 10.66. Metallation on the aromatic nucleus is carried out with *n*-butyllithium, and occurs at the *ortho* positions.[99,100] The latter metallation may occur predominantly at an *ortho* position even with an alkly group *meta* or *para* to the sulfonyl, as in eq. 10.67.[101] This is in contrast to the other bases mentioned above, which act in accordance with eq. 10.66.[96] Initial coordination between the sulfonyl group and the alkyllithium reagent, directing metallation to the *ortho* position, has been proposed to account for this seemingly anomolous behavior.[101]

$$CH_3-\bigcirc-SO_2CH_2C_3H_7 \xrightarrow[\substack{2\ CO_2 \\ 3\ H_3O^+}]{1\ BuLi} CH_3-\bigcirc-SO_2\underset{\underset{CO_2H}{|}}{C}HC_3H_7 \qquad (10.65)$$

$$CH_3-\bigcirc-SO_2CH_3 \xrightarrow[\substack{2\ C_6H_5CH_2Cl}]{1\ NaNH_2} C_6H_5CH_2CH_2-\bigcirc-SO_2C_6H_5 \qquad (10.66)$$

$$CH-\bigcirc-SO_2C_6H_5 \xrightarrow[\substack{2\ CO_2 \\ 3\ H_3O^+}]{1\ BuLi} CH_3-\overset{CO_2H}{\bigcirc}-SO_2C_6H_5 + p-CH_3C_6H_4SO-\overset{CO_2H}{\bigcirc}$$

$$(10.67)$$

The acidity of hydrogens α to the sulfonyl group has been compared with that of compounds, containing other electron-withdrawing substituents, such as ketones, nitroalkanes, and nitriles.[102,103] The studies of sulfonyl systems have been directed towards (a) the mode of stabilization at the ensuing carbanionic center and (b) the configuration and conformation of the adjacent carbanion.

**TABLE 10.2**[102]

$$ZSO_2CHXY \xrightarrow[25°]{DMSO} [ZSO_2CXY]^-$$

| X | Y | Z | $pK_a$ |
|---|---|---|---|
| H | H | $CH_3$ | 28.5 |
| H | H | $C_6H_5$ | 27 |
| H | $CH_3$ | $C_6H_5$ | 29 |
| H | $C_6H_5$ | $CH_2C_6H_5$ | 22 |
| $CH_3$ | $C_6H_5$ | $CH(CH_3)C_6H_5$ | 23.5 |

The possibility of resonance stabilization by $2p$-$3d$ overlap sets the sulfonyl system apart from the nitro and carbonyl systems. The degree to which this occurs, if at all, has been the subject of much debate. When the acidities of a number of sulfones were measured (Table 10.2),[102] the following substituent effects were observed: Replacement of an $\alpha$-hydrogen by a methyl group (X or Y = $CH_3$) increases the $pK_a$ by 1.5 to 2 units. Replacement of an $\alpha$-hydrogen by a phenyl group (X or Y = $C_6H_5$) results in a lowering of the $pK_a$ by 6.5 units. Changing Z (Table 10.2) from methyl to phenyl lowers the $pK_a$ by 1.5 units. The increase in $pK_a$ resulting from an $\alpha$-methyl substituent stands in contrast to a decrease of 1.7 $pK_a$ units in going from nitromethane ($pK_a$ = 10.2) to nitroethane ($pK_a$ = 8.5).[104] This has been rationalized as being due to the degree of resonance stabilization in the conjugate base.[102] For the nitronate anion, the *aci* form (**10.4**) makes a considerable contribution to the resonance hybrid, probably because of comparable C=N and N=O bond energies (147 and 145 kcal/mole). $\alpha$-Methyl substituents should then stabilize the C=N in a manner analogous to that observed for olefins. With the sulfonyl carbanion, the conjugative stabilization is diminished due to poor $2p$-$3d$ overlap (**10.6**), thereby resulting in only a very small stabilization of the anion due to any increase in stability of the C=S bond in structure **10.6**. This effect, if any, is completely overwhelmed by the inductive destabilization of **10.5** by an alkyl substituent.

$$R\overset{-}{C}H\text{-}NO_2 \longleftrightarrow RCH\text{=}NO_2^-$$

**10.3**                    **10.4**

$$RCH\text{-}SO_2CH_3 \longleftrightarrow [RCH\text{=}SO_2CH_3]^-$$

**10.5**                    **10.6**

The large stabilizing inductive effect of a sulfonyl group makes investigation of the degree of resonance stabilization resulting from $2p$-$3d$ overlap difficult. In a study of the $pK_a$ values of substituted phenols, anilinum ions, and benzoic acids, Bordwell and Cooper found that a significantly larger Hammett sigma constant was needed to express the effect of a $p$-$CH_3SO_2$ substituent relative to that of the *meta* substituent in the phenols and benzoic acids (e.g., **10.7** and **10.8**). This difference was taken as a measure of the resonance contribution of the sulfonyl group.[105] This and other studies[106] seem to indicate the resonance and

<table>
<tr><td align="center">OH<br>⬡<br>SO₂CH₃</td><td align="center">CO₂H<br>⬡<br>SO₂CH₃</td></tr>
<tr><td align="center">**10.7**</td><td align="center">**10.8**</td></tr>
<tr><td align="center">$\sigma_p = 0.98$</td><td align="center">$\sigma_p = 0.72$</td></tr>
<tr><td align="center">$\sigma_m = 0.70$</td><td align="center">$\sigma_m = 0.65$</td></tr>
</table>

inductive effects in these systems are not completely separable.[107]

Base-catalyzed hydrogen-deuterium exchange at an asymmetric carbon adjacent to a sulfonyl group has been found to proceed with a larger rate of exchange ($k_e$) than rate of racemization ($k_\alpha$). For 2-octyl phenyl sulfone (**10.9**), $k_e/k_\alpha$ was found to vary from 10 to 1980, depending upon solvent and conditions.[108,109] Similar behavior has been observed in other systems containing two unsubstituted oxygens bound to a second-row element.[110] Such behavior has led to questions concerning the configuration and conformation of these carbanions.

$$\begin{array}{c} CH_3 \\ | \; {}^* \\ C_6H_5SO_2\text{-}C\text{-}H \\ | \\ C_6H_{13}\text{-}n \end{array}$$

**10.9**

Two rationales to account for the maintenance of asymmetry in $\alpha$-sulfonyl carbanions have been proposed[107,111]: (1) the carbanion exists in pyramidal configuration **10.10, 10.11,** or **10.12** in which there exists a high barrier to inversion; or (2) the carbanion is planar with a high barrier to rotation (**10.13** or **10.14**). Of conformations **10.13** and **10.14**,

**10.10**              **10.11**              **10.12**

**10.13**              **10.14**

only **10.13** is dissymmetric and proton transfer would have to occur at only one lobe of the $p$ orbital. In the reverse aldol fragmentation of **10.15**, the product **10.16** demonstrates that cleavage has occurred with a high degree of inversion at the asymmetric center (*).[111] Protonation has therefore occurred from the direction *syn* to the sulfonyl oxygens. Decarboxylation of optically-active sulfonyl acetic acids leads to similar results.[112] The stereochemical considerations of the Ramberg-Bäcklund reaction also indicate **10.12** or **10.13** as the more stable form of the anion.[113,114] Initial attempts to determine the preferred conformation and configuration of the sulfonyl carbanion[107–114] were ambiguous in that the results were consistent with either **10.12** or **10.13**.[107] Nonempirical molecular orbital calculations on the hydrogen methyl sulfonyl carbanion ($HSO_2CH_2^-$) indicate that while **10.13** is more stable than **10.14** by 2.5 kcal/mole, **10.12** is more stable than **10.13** or **10.14** by 2.5 and 5.0 kcal/mole, respectively.[115] In addition, these calculations appear to show no significant contribution to the structure of the carbanion from the $d$ orbitals of sulfur. These authors state that a minimal basis set of Gaussian-type functions were used and more refined calculations are in progress. A subsequent study of the rates of exchange of a conformationally fixed sulfone showed qualitative agreement with the M.O. calculations, although the calculated energy differences were thought to be too large.[116]

R (+) **10.15**                              S (−) **10.16**                    (10.68)

The question as to the hybridization of the $\alpha$-sulfonyl carbanion has yet to be resolved. Considerable evidence has been amassed both for and against the pyramidal configuration; however no completely unambiguous proof of structure has yet been put forward.[107]

### 10.3.2 Alkylations, Acylations, Halogenations, Nitrations, and Related Reactions of Sulfone-Stabilized Carbanions

Sulfone-stabilized carbanions are of interest not only because of their potential asymmetry, but also because of their use as synthetic intermediates in preparing more complex molecules.

Reaction of sulfonyl carbanions with aldehydes or ketones results in the formation of $\beta$-hydroxy sulfones.[92-94,97,117-120] The carbanion is normally generated prior to introduction of the carbonyl reagent to avoid reaction of the metallating reagent with the carbonyl group (eq. 10.70).[94] While such additions of alkyl sulfones proceed readily with most ketones and aldehydes, benzylic sulfones are unreactive except when the metal atom is Mg.[97,119,121] Apparently with the benzylic system, **10.18** is less thermodynamically stable than **10.17** when M is Li, Na, or K and the equilibrium (eq. 10.71) lies far to the left. If M = MgBr, the reaction proceeds, presumably because of the increased chelation of Mg with the oxygen relative to the alkali metal salts, thus shifting the equilibrium towards **10.18**. The reversibility of the reaction (eq. 10.72) has been used to induce reverse aldol-type fragmentation of a number of $\beta$-hydroxy sulfones.[122-125]

$$RSO_2\overset{|}{\underset{|}{C}}{}^- + \overset{\backslash}{\underset{/}{C}}{=}O \xrightarrow{\text{H}^+} RSO_2\overset{|}{\underset{|}{C}}{}\text{-}\overset{\overset{\text{OH}}{|}}{\underset{|}{C}} \qquad (10.69)$$

$$CH_3SO_2CH_3 \xrightarrow{n\text{-BuLi}} [CH_3SO_2CH_2^-\,Li^+]$$

$$\xrightarrow[C_6H_5CHO]{} CH_3SO_2CH_2\overset{\overset{\text{OH}}{|}}{C}HC_6H_5 \qquad (10.70)$$

$$85\%$$

$$ArSO_2\overset{\overset{\text{M}}{|}}{C}HC_6H_5 + \overset{\backslash}{\underset{/}{C}}{=}O \rightleftarrows ArSO_2CH\text{-}\overset{\overset{\text{OM}}{|}}{\underset{\underset{C_6H_5}{|}}{C}}{-} \qquad (10.71)$$

$$\textbf{10.17} \qquad\qquad\qquad \textbf{10.18}$$

$$\underset{\underset{CH_3\ H}{|\quad|}}{\overset{\underset{Cl\quad OH}{|\quad|}}{CH_3SO_2C-CC_6H_5}} \xrightarrow[\text{aq. DME, }\Delta]{\text{KOH}} C_6H_5CHO + CH_3SO_2\overset{\underset{|}{Cl}}{\underset{|}{C}}HCH_3$$

$$(10.72)$$

$$ArSO_2CH_2CH{=}CH_2 \xrightarrow[H_2O]{OH^-} [ArSO_2CH_2\overset{\underset{|}{OH}}{C}HCH_3]$$

$$ArSO_2CH_3 + CH_3\overset{O}{\underset{H}{C}}$$

$$(10.73)$$

Unlike many organometallic reagents, additions of sulfonyl carbanions to vinyl ketones and esters do not usually lead to 1,4-adducts. Only 1,2-addition products are normally observed.[126-128]

β-Keto sulfones can be prepared by reaction of sulfonyl carbanions with acid chlorides[91,129,130] or carboxylic esters,[93,131,132] or by oxidation of the corresponding β-hydroxy sulfones.[92,120] The usefulness of such reactions is illustrated in eq. 10.75 in which the tricyclic hydroxy ketone **10.19** was synthesized from a lactone,[133] and in similar reactions with phthalate esters.[134]

$$CH_3SO_2CH_3 \xrightarrow[\substack{2.\ C_6H_5CO_2CH_3 \\ 3.\ H^+}]{1.\ NaH} CH_3SO_2CH_2\overset{\overset{O}{\|}}{C}C_6H_5 + CH_3OH$$

$$87\%$$

$$(10.74)$$

$$(10.75)$$

Sulfonyl carbanions are readily alkylated by primary alkylhalides.[92-94,97,119,120,129] Reacting these anions with nitrate esters gives α-nitro sulfones.[135] Carbonation followed by acidification produces sulfonylacetic acids in good yield.[91,98,129,136,137]

$$C_6H_5SO_2CH_2C_6H_5 \xrightarrow[\substack{\text{2. EtONO}_2 \\ \text{3. H}^+}]{\text{1. } t\text{-BuOK}} C_6H_5SO_2\overset{\overset{\displaystyle NO_2}{|}}{C}HC_6H_5 \; + \; EtOH$$

$$(10.76)$$

$$p\text{-CH}_3C_6H_4SO_2\overset{\overset{\displaystyle MgBr}{|}}{C}HCH(C_6H_5)_2 \xrightarrow[\text{2. H}^+]{\text{1. CO}_2} p\text{-CH}_3C_6H_4SO_2\overset{\overset{\displaystyle CO_2H}{|}}{C}HCH(C_6H_5)_2$$

$$(10.77)$$

Polycyclic aromatics have been methylated by the anion of phenyl methyl sulfone as shown[138]:

$$(10.78)$$

Halogenations of sulfonyl carbanions are possible,[139] but $\alpha$-halo sulfones are more accessible by other methods (see Sect. 10.2.1). Condensation of sulfones with $CCl_4$ in the presence of potassium hydroxide gives rise to complete chlorination at the $\alpha$-positions (eq. 10.79). If hydrogens are present on both $\alpha$-positions, chlorination is followed by the production of olefins *via* the Ramberg-Bäcklund reaction along with some products resulting from dichlorocarbene additions as shown in eq. 10.80.[140]

$$C_6H_5SO_2CH_2C_6H_5 \xrightarrow[\text{CCl}_4, \text{H}_2\text{O}]{\text{KOH}, t\text{-BuOH}} \underset{100\%}{C_6H_5SO_2CCl_2C_6H_5} \qquad (10.79)$$

$$(10.80)$$

Although vinyl sulfones have been extensively studied, there has been only one report to date of epoxidation of these compounds to yield $\alpha,\beta$-epoxy sulfones in a base-catalyzed reaction.[141] These compounds are accessible, however, by Darzens-type condensations of carbonyl compounds with $\alpha$-halo sulfones.[142] The product in this reaction is the one in which the more bulky $\beta$-substituent ($R_2$ or $R_3$ in eq. 10.82) is *trans* to the sulfonyl group. Under similar conditions a diastereomeric mixture of **10.21** produced a mixture of *cis* and *trans* epoxides **10.22**.[125] In some cases, the epoxy sulfone undergoes further base-catalyzed rearrangement to the $\beta$-keto sulfone[125,143] *via* the enolate, **10.23**. In the reaction of *p*-tolyl $\alpha$-chloromethyl sulfone with acetophenone, the isolation of the sulfonyl aldehyde, **10.24**, is anomolous in that such an enolate intermediate is not possible.[142]

$$p\text{-CH}_3\text{C}_6\text{H}_4\text{SO}_2\text{CH}_2\text{Cl} + \text{C}_6\text{H}_5\text{CHO} \xrightarrow[\text{$t$-BuOH, 10-15$^\circ$}]{\text{$t$-BuOK}}$$

$$p\text{-CH}_3\text{C}_6\text{H}_4\text{SO}_2 \overset{\displaystyle\text{O}}{\underset{\displaystyle\text{H}}{\overset{\diagup\ \diagdown}{\text{C}-\text{C}}}} \underset{\displaystyle\text{C}_6\text{H}_5}{\text{H}}$$

95%

(10.81)

$$\text{RSO}_2\overset{\displaystyle}{\underset{\displaystyle\text{Cl}}{\text{CR}_1}} + \text{R}_2\text{R}_3\text{C}{=}\text{O} \rightleftharpoons \text{RSO}_2\overset{\displaystyle\text{R}_1}{\underset{\displaystyle\text{Cl}}{\text{C}}}\overset{\displaystyle\text{O}^-}{\underset{\displaystyle\text{R}_3}{\text{C}}}{-}\text{R}_2 \longrightarrow$$

**10.20**

$$\text{RSO}_2\overset{\displaystyle\text{O}}{\underset{\displaystyle\text{R}_1}{\overset{\diagup\ \diagdown}{\text{C}-\text{C}}}}\text{R}_2\text{R}_3 \qquad (10.82)$$

$$\text{CH}_3\text{SO}_2\overset{\displaystyle\text{CH}_3}{\underset{\displaystyle\text{Cl}}{\text{C}}}{-}\overset{\displaystyle\text{OH}}{\text{CHC}_6\text{H}_5} \xrightarrow[\text{DMF, 0}^\circ]{\text{$t$-BuOK}} \text{CH}_3\text{SO}_2\overset{\displaystyle\text{O}}{\underset{\displaystyle\text{CH}_3}{\overset{\diagup\ \diagdown}{\text{C}-\text{CHC}_6\text{H}_5}}} \qquad (10.83)$$

70%

**10.21**                                    **10.22**

$$RSO_2 \overset{O}{\underset{/ \backslash}{CH-CHR'}} \xrightarrow{B^-} [RSO_2 CH \overset{O^-}{=} \overset{|}{C}R'] \longrightarrow RSO_2 CH_2 \overset{O}{\underset{||}{C}}R' \qquad (10.84)$$

$$\underset{\textbf{10.23}}{}$$

$R = C_6H_5, CH_3$

$R' = C_6H_5, p\text{-}O_2NC_6H_4$

$$p\text{-}CH_3C_6H_4SO_2CH_2Cl + C_6H_5\overset{O}{\underset{||}{C}}CH_3 \xrightarrow{t\text{-BuOK}} p\text{-}CH_3C_6H_4SO_2\overset{CH_3}{\underset{|}{C}}\text{-CHO}$$
$$\underset{C_6H_5}{|}$$

$$\textbf{10.24}$$

$$(10.85)$$

### 10.3.3 Cyclizations of Halosulfones

Treatment of $\gamma$-chloro sulfones with base results in conversion to the cyclopropyl sulfones in good yield. The base/solvent systems that have been used in this reaction include $t$-BuOK/$t$-BuOH,[144,145] NaNH$_2$/ Glyme,[146] and $n$-BuLi/THF.[147] Bases as weak as pyridine, sodium carbonate, or sodium acetate can effect cyclization of 1,1-bis(sulfonyl)-3-haloalkanes (eq. 10.87).[148]

$$RSO_2CH_2CH_2CH_2Cl \xrightarrow[\text{Glyme}]{NaNH_2} RSO_2\text{--}\triangleleft \qquad (10.86)$$

$R = $ aryl or alkyl                      60–98%

$$(EtSO_2)_2CHCH_2CH_2Cl \xrightarrow[\text{H}_2\text{O}]{Na_2CO_3} \overset{SO_2Et}{\underset{SO_2Et}{\triangleright}} \qquad (10.87)$$

Hammett treatment of a study of substituent effects on the rate of cyclization of aryl 3-chloropropyl sulfones resulted in a reaction constant, $\rho$, of $+2.32$.[149] This and other evidence was taken to indicate a stepwise process (path a) rather than a concerted one (path b).

$$ArSO_2CH_2CH_2CH_2CH_2Cl \quad \begin{array}{c} ArSO_2\overset{-}{C}H \quad CH_2-Cl \\ a \parallel \quad \quad \diagdown / \\ \quad \quad B^-H \quad CH_2 \quad \quad ArSO_2-\triangleleft \\ b \diagdown \quad \quad \mid^- \\ \quad \quad ArSO_2\overset{}{C}H\overset{}{C}H_2-Cl \\ \quad \quad \diagdown / \\ \quad \quad CH_2 \end{array}$$

(10.88)

In order for the reaction to proceed, the system must be able to assume a semi-W form transition state,[150] as was shown in the closure of the tricyclo[2.2.1.0$^{2,6}$]heptane ring system, **10.25**, to the quadracyclo [2.2.1.0.$^{2,6}$0$^{3,5}$]heptane system, **10.26**.[151] Treatment of *exo*-bromide, **10.25**, with base resulted in the rapid (complete after 5 minutes) formation of **10.26**, while the corresponding *endo*-bromide under the same conditions gave only recovery of starting material after 7 hours.

$$\text{Br}\diagup\!\!\!\!\diagup\text{SO}_2\text{C}_6\text{H}_5 \quad \xrightarrow[\text{DMSO}]{t\text{-BuOK}} \quad \diagup\!\!\!\!\diagup\text{SO}_2\text{C}_6\text{H}_5 \quad 75\%$$

(10.89)

**10.25**                          **10.26**

Cyclobutyl and cyclopentyl aryl sulfones can also be prepared by the analogous cyclization of δ- and ε-chlorosulfones.[152] When the possibility of forming rings of differing sizes exists, the product of cyclization is controlled by ring size, and not by the acidity of the potential carbanion sites as in sulfides.[146,152]

$$\left[\text{SO}_2\right]\!\!-\!\!C_6H_5 \quad \times\!\!- \quad C_6H_5CH_2SO_2(CH_2)_3Cl \quad \longrightarrow \quad C_6H_5CH_2SO_2-\triangleleft$$

(10.90)

$$\left[\text{S}\right]\!\!-\!\!C_6H_5 \quad \longleftarrow \quad C_6H_5CH_2S(CH_2)_3Cl \quad \times\!\!- \quad C_6H_5CH_2-S-\triangleleft$$
95%

(10.91)

$$\left[\text{SO}_2\right] \quad \longleftarrow \quad C_6H_5CH_2SO_2(CH_2)_4Cl \quad \times\!\!- \quad C_6H_5CH_2SO_2-\square$$
60%

(10.92)

$$\left[\text{SO}_2\right]\!\!-\!\!C_6H_5 \quad \times\!\!- \quad C_6H_5CH_2SO_2(CH_2)_4Cl \quad \longrightarrow \quad C_6H_5CH_2SO_2-\pentagon$$
79%

(10.93)

### 10.3.4 The Ramberg-Bäcklund Reaction

In view of the limited reactivity of α-halosulfones to intermolecular nucleophilic displacement,[153] the report by Ramberg and Bäcklund in 1940[154] that α-haloethyl sulfones, when treated with $2N$ KOH, underwent facile release of halide ion and generation of 2-butene was of considerable interest. The scope and mechanism of this rearrangement have since been extensively explored.

$$CH_3CH_2SO_2\underset{\underset{X}{|}}{C}HCH_3 \xrightarrow[H_2O]{KOH} CH_3CH=CHCH_3 + KBr + K_2SO_3 \quad (10.94)$$

While the significance of the Ramberg-Bäcklund reaction would appear to be limited by the wide variety of methods available for olefin synthesis, a number of important applications have been developed. The conversion of α-chlorosulfone **(10.27)** to the cyclobutene derivative of tetrahydronaphthalene **(10.28)** shown in eq. 10.95[155] and the preparation of $\Delta^{1,5}$-bicyclo[3.3.0]octene **(10.30)** from 1-chloro-9-thiabicyclo[3.3.1] nonane 9,9-dioxide **(10.29)**[72] in eq. 10.96 demonstrate its usefulness as a route to fused ring carbocyclics. Deuterated olefins are obtained when the reaction of **10.31** is run in deuterated solvents, as illustrated in eq. 10.97, which is the first reported preparation of 9,10-dideuteriophenanthrene.[156] By the sequence shown in eq. 10.98, olefinic chains may be elongated.[157] Dihalosulfones are converted to ethylene-sulfonic acids or esters and/or acetylenes, depending on the reaction conditions employed (eqs. 10.99[158] and 10.100[159]). α,α,α-Trichloromethyl sulfones may be converted to chloroalkene-sulfonic acids (eq. 10.101), β-dichloromethyl sulfonic acids, and dichloroolefins under the reaction conditions.[160] The

$$(10.95)$$

54%

10.27      10.28

$$(10.96)$$

75%

10.29      10.30

$$\xrightarrow[\substack{D_2O \\ \text{Dioxane}}]{\text{NaOD}} \qquad (10.97)$$

**10.31**                                      92%

$$\underset{/}{\overset{\backslash}{C}}=\underset{\backslash}{\overset{/}{C}} \xrightarrow[2.\ H_2O]{1.H_2S,\, h\nu} -CH\text{-}\underset{|}{\overset{|}{C}}SH \xrightarrow[4.\ [O]]{3.\ CH_2O,\ HCl} -CH_2CH_2SO_2CH_2Cl$$

$$RCH_2CH=CH_2 \xleftarrow{\hspace{2cm}} \Big\vert\ 5.\ B^- \quad (10.98)$$

$$\left(\begin{array}{c} C_6H_5CH \\ | \\ Br_2 \end{array}\right)SO_2 \xrightarrow[H_2O]{NaOH} \underset{(13\%)}{C_6H_5C=CC_6H_5} +$$

$$\underset{H}{\overset{C_6H_5}{\diagdown}}C=C\underset{SO_3H}{\overset{C_6H_5}{\diagup}} + \text{Vinyl Bromides} \quad (10.99)$$
$$79\% \qquad\qquad (2\%)$$

$$n\text{-}C_5H_{11}CH_2SO_2CHCl_2 \xrightarrow[t\text{-BuOH}]{t\text{-BuOK}} n\text{-}C_5H_{11}C{\equiv}CH \quad (10.100)$$
$$46\%$$

$$C_6H_5(CH_2)_3SO_2CCl_3 \xrightarrow[H_2O]{NaOH} C_6H_5(CH_2)_2\underset{\underset{SO_3H}{|}}{C}=CHCl +$$
$$(66\%) \qquad\qquad (10.101)$$

$$C_6H_5(CH_2)_2C{\equiv}CH + \text{mono- and dichloroolefins}$$
$$(1\%)$$

gross mechanistic pathway of this reaction involves the pre-equilibrium formation of the sulfone-stabilized carbanion (eq. 10.102), formation of an episulfone intermediate *via* a 1,3-intramolecular displacement of halide (eq. 10.103), and loss of $SO_2$ to yield the olefin (eq. 10.104).

$$\underset{R_1R_2\overset{|}{C}\text{-}SO_2\overset{|}{C}R_3R_4}{\overset{H\quad\ X}{}} + B^- \xrightleftharpoons[k_{-1}]{k_1} R_1R_2\overset{-}{C}SO_2\overset{\overset{X}{|}}{C}R_3R_4 + BH$$

$$(10.102)$$

$$R_1 R_2 \bar{C}\text{-}SO_2\text{-}CR_3 R_4 \overset{k_2}{\longrightarrow} R_1 R_2 C\underset{\underset{\substack{S \\ / \backslash \\ O \quad O}}{\backslash /}}{-} CR_3 R_4 \qquad (10.103)$$

$$R_1 R_2 C\underset{\underset{\substack{S \\ / \backslash \\ O \quad O}}{\backslash /}}{-} CR_3 R_4 \overset{k_3}{\longrightarrow} R_1 R_2 C{=}CR_3 R_4 \qquad (10.104)$$

The rapid reversible carbanion formation has been demonstrated by treatment of $\alpha$-bromobenzyl benzyl sulfone with $NaOCH_3$ and $CH_3OD$. After one half-life, examination of recovered starting material revealed complete deuterium incorporation at the $\alpha$ positions.[161] Similar results have been obtained for $\alpha$-haloalkyl sulfones[162,163] and benzyl $\alpha$-chlorobenzyl sulfones.[156] These observations necessitate that $k_{-1} \gg k_2$, as the rate of exchange is much greater than the rate for loss of halide ion.[161] The participation of the equilibrium in the reaction pathway is also implicated by the observation of a leaving–group effect (Br/Cl rate ratio = 620 at $0°$),[161] which is significantly larger than that normally observed for intermolecular displacements.[164] Thiirane dioxide intermediates (episulfones), which are available by other methods,[81,165] have been shown to lose $SO_2$ under these conditons at a much greater rate than the rate for loss of halide ion. Therefore, the overall reaction rate ($k_{obs}$) can be approximated by the equation $k_{obs} = K_{eq} \cdot k_2$ where $K_{eq} = k_1/k_{-1}$.[161] Formation of the episulfone from a carbene or zwitterion has been considered,[157,167] but the above mechanism seems most reasonable in light of experimental observations.

Interestingly, the Ramberg-Bäcklund reaction leads to predominant formation of the less thermodynamically stable *cis* olefin (Table 10.3). Since the *cis* and *trans* episulfones are known to decompose stereospecifically to the *cis* and *trans* olefins, respectively,[162,163,82] the product isomer ratios must be determined by the ratio for formation of the *cis* and *trans* episulfones. That the 1,3-displacement reaction occurs by a semi-W transition state[150] with double inversion has been shown by the rearrangement of **10.29**,[72] in which the centers are rigidly held in such a conformation, and by the reaction of *meso* **10.32** and *d,l*-$\alpha$-bromobenzyl sulfone, **10.33**, with triphenylphosphine.[114] Compound **10.32** gives rise to 95% *cis* stilbene and 5% *trans*, while **10.33** yields 96% *trans* and only 4% *cis*, (eqs. 10.105 and 10.106).

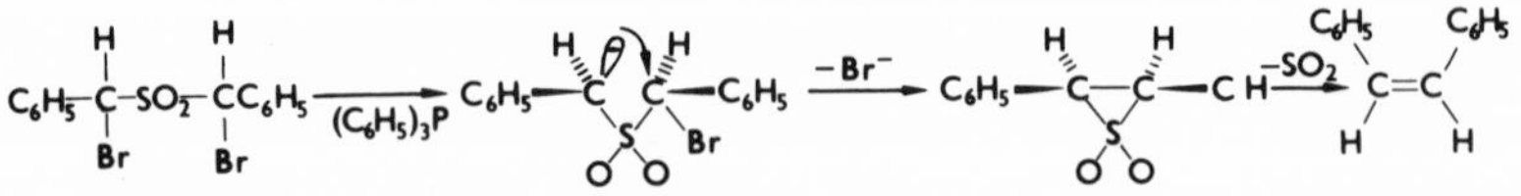

**10.32**                                                                          (10.105)

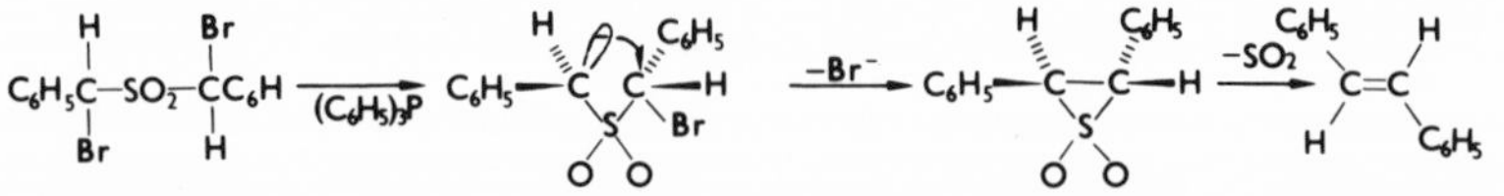

**10.33**

                                                                                   (10.106)

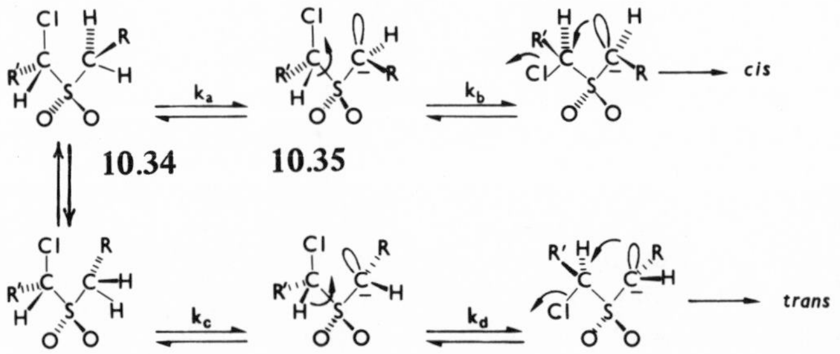

**10.34      10.35**

                                                                                   (10.107)

**10.36              10.37**

The preferred formation of *cis* episulfone is accounted for as shown in eq. 10.107. On the basis of steric effects, conformation **10.34** would be more stable than **10.36** and, being of lower energy, make a proportionately larger contribution to the rotomer population. The rate of proton abstraction ($k_c$) from **10.36** should be retarded in relation to $k_a$ as a result of increased steric compression in **10.37** as compared to **10.35**. Rotation of the adjacent C-S bond to attain the semi-W conformation should result in $k_b$ being larger than $k_d$ as $R'$ must rotate past the sulfonyl grouping in **10.37** but only past hydrogen in **10.35**. Therefore the additive effect of these equilibria should lead to predominant formation of *cis* episulfone.[166] As the bulk of R (eq. 10.107) increases, however, the reaction becomes less stereoselective (see entries 1,4,5,6 and 7 in Table 10.3).

In the case of dihalosulfones, the reaction proceeds through a thiirene dioxide intermediate. These have not been isolated from the reactions with hydroxide or alkoxide ion; however, 2,3-diphenylthiirene 1,1-dioxide has been obtained by treatment of a mixture of **10.32** and **10.33** with triethylamine (eq. 10.108).[5] Methyl thiirene dioxide has been obtained by the condensation of chloromethyl sulfene with diazomethane (eq. 10.109).[84]

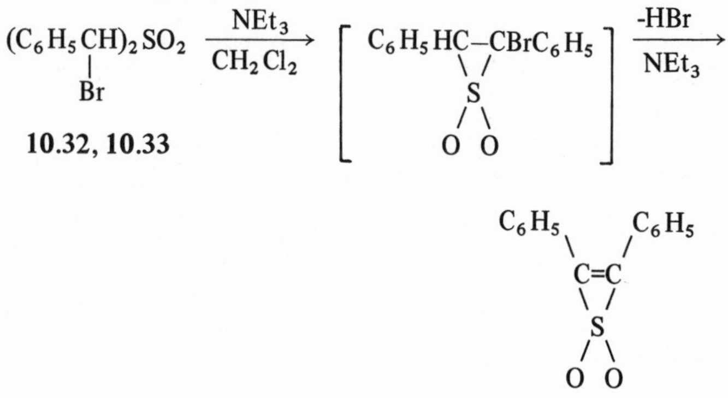

$$(C_6H_5CH)_2SO_2 \xrightarrow[\underset{Br}{}]{NEt_3 \atop CH_2Cl_2} \left[ \begin{array}{c} C_6H_5HC{-}CBrC_6H_5 \\ \diagdown \diagup \\ S \\ \diagup \diagdown \\ O \quad O \end{array} \right] \xrightarrow[NEt_3]{-HBr}$$

**10.32, 10.33**

$$\begin{array}{c} C_6H_5 \diagdown \qquad \diagup C_6H_5 \\ C{=}C \\ \diagdown \diagup \\ S \\ \diagup \diagdown \\ O \quad O \end{array}$$

(10.108)

### TABLE 10.3

$$RCH_2SO_2CHClR' \xrightarrow{base} RCH{=}CHR'$$

|  | R | R' | Product | % cis | % trans | Ref. |
|---|---|---|---|---|---|---|
| 1[a] | $CH_3$ | $CH_3$ | 2-butene | 78.8 | 21.2 | 163a |
| 2[b] | $CH_3$ | $CH_3$ | 2-butene | 78.1 | 21.9 | 163a |
| 3[c] | $CH_3$ | $CH_3$ | 2-butene | 22.6 | 77.4 | 163a |
| 4[b] | $C_2H_5$ | $CH_3$ | 2-pentene | 68.5 | 34.3 | 163a |
| 5[b] | $CH_3$ | $C_2H_5$ | 2-pentene | 71.3 | 28.7 | 163a |
| 6[a] | $CH_3$ | $CH(CH_3)_2$ | 4-methyl-2-pentene | 59.4 | 40.6 | 163b |
| 7[a] | $CH(CH_3)_2$ | $CH_3$ | 4-methyl-2-pentene | 51.0 | 49.0 | 163b |
| 8[d] | $C_6H_5$ | $C_6H_5$ | stilbene | 0 | 100 | 163a |

[a]Base: 2N NaOH, 100°.  [c]Base: 1M t-BuOK in t-BuOH, 93°.
[b]Base: 2N KOH, 100°.  [d]Base: 2N NaOH in aqueous dioxane, 100°.

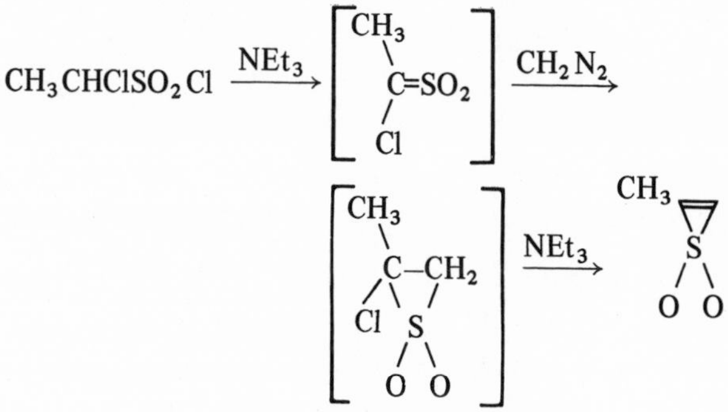

$$CH_3CHClSO_2Cl \xrightarrow{NEt_3} \left[ \begin{array}{c} CH_3 \\ \diagdown \\ C{=}SO_2 \\ \diagup \\ Cl \end{array} \right] \xrightarrow{CH_2N_2}$$

$$\left[ \begin{array}{c} CH_3 \\ \diagdown \\ C{-}CH_2 \\ \diagup \diagdown \diagup \\ Cl \quad S \\ \diagup \diagdown \\ O \quad O \end{array} \right] \xrightarrow{NEt_3} \begin{array}{c} CH_3 \\ \diagup\!\!\triangledown \\ S \\ \diagup \diagdown \\ O \quad O \end{array}$$

(10.109)

These unsaturated isomers are considerably more stable than their saturated analogues. This may be attributed to the increased stabilization of attaining Hückel aromaticity.

It should be noted (Table 10.3) that while *cis* olefins are generally the predominant products, the more stable *trans* isomers may be obtained by using a stronger base in a less polar medium (i.e., *t*-BuOK/*t*-BuOH). Benzylic sulfones are converted exclusively to *trans* olefins even by hydroxide ion. This results from epimerization of the *cis* episulfones rather than by post isomerization of the olefin. When **10.38** was heated with NaOH in $D_2O$, less than 0.5% of *trans*-2-butene was isolated with only 5% deuterium incorporation. When exposed to *t*-BuOK in *t*-BuOH, **10.38** was converted to a mixture of 19.4% *cis* and 80.6% *trans*-2-butene, which exhibited essentially complete deuterium incorporation.[162,163] This is nearly the same isomer ratio obtained from α-chloroethyl ethyl sulfone

and *t*-BuOK. The solvent plays an important role in the reaction as evidenced by failure of **10.27** to react with NaOD in $D_2O$, although deuterium incorporation was complete. Employing the less polar solvent THF, **10.27** was cleanly converted to **10.28**.[155]

The mechanism by which $SO_2$ is expelled from the episulfones is still open to question. In the absence of base, the rate of decomposition shows a marked increase with increasing solvent polarity. In the presence of alkoxide ion, the decomposition is second order, exhibiting first-order dependence on both episulfone and base. Nonbasic decomposition has been proposed to proceed *via* a singlet diradical intermediate, **10.39**, in which expulsion of $SO_2$ occurs faster than bond rotation. Basic catalysis is attributed to direct attack on the sulfonyl grouping.[82] Thermal decomposition of the episulfones by a concerted elimination in accordance with the Hoffmann-Woodward rules[168,169] is symmetry forbidden.[166] On the basis of the products arising from the thermal decomposition of dibenzoyl stilbene episulfone, **10.40**, an initial rearrangement to the 1,3,2-dioxathiolane, **10.41**, followed by concerted fragmentation has been postulated, as shown in eq. 10.114.[170]

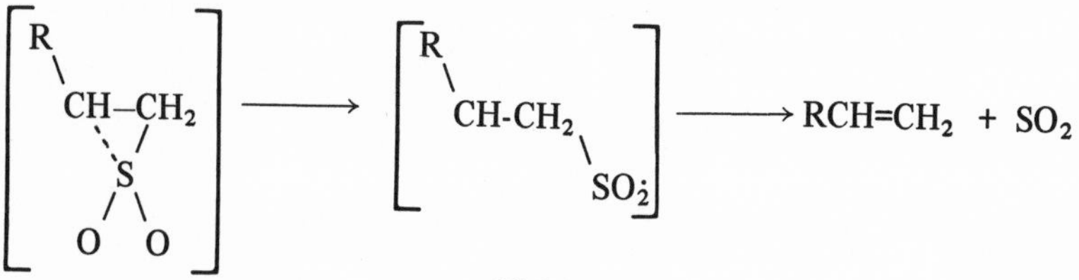

**10.39** (10.112)

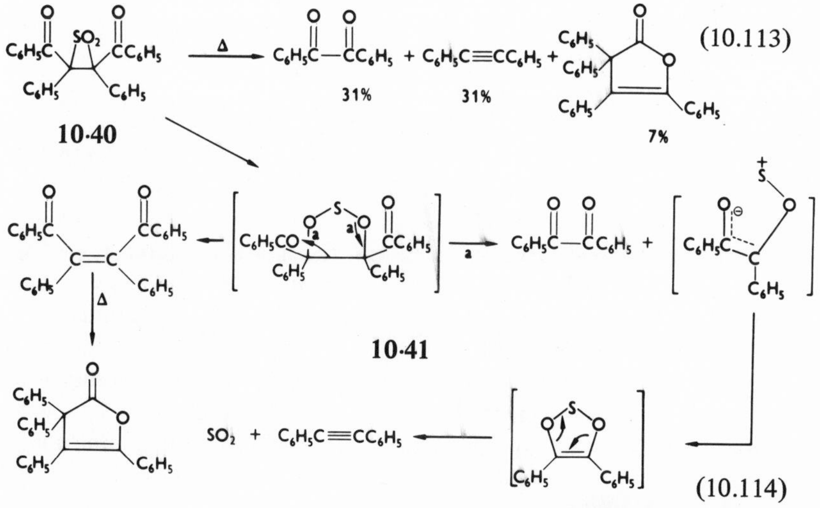

**10-40** (10.113)

**10-41** (10.114)

### 10.3.5 The Smiles Rearrangement

The Smiles rearrangement involves an intramolecular displacement at an aromatic ring by a nucleophilic center attached to the group being displaced through two or three atoms. The rearrangement is normally catalyzed by base, but a few exceptions to this are known.[55]

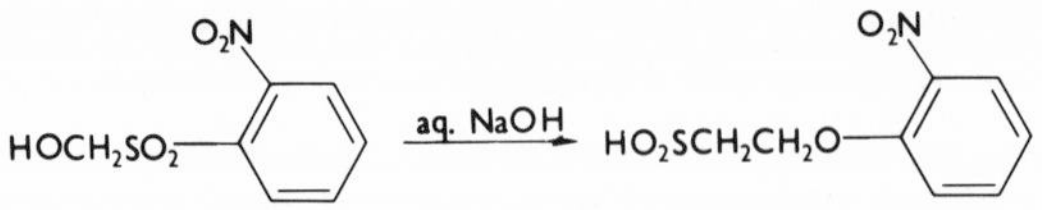

$HOCH_2SO_2$—⟨aryl, $O_2N$⟩ $\xrightarrow{\text{aq. NaOH}}$ $HO_2SCH_2CH_2O$—⟨aryl, $O_2N$⟩ $(10.115)^{171}$

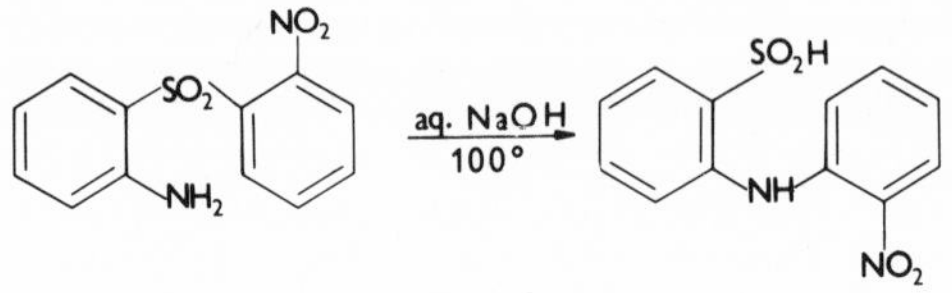

⟨aryl $SO_2$–aryl $NO_2$, $NH_2$⟩ $\xrightarrow[100°]{\text{aq. NaOH}}$ ⟨aryl $SO_2H$–NH–aryl $NO_2$⟩ $(10.116)^{172}$

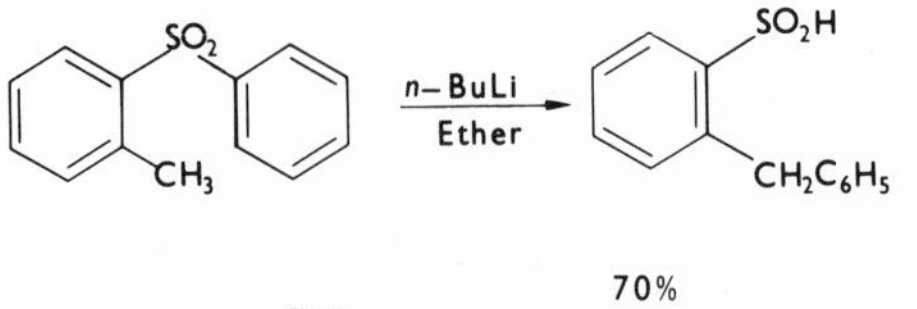

⟨aryl $SO_2$–aryl, $CH_3$⟩ $\xrightarrow[\text{Ether}]{n-\text{BuLi}}$ ⟨aryl $SO_2H$, $CH_2C_6H_5$⟩ $(10.117)^{173}$

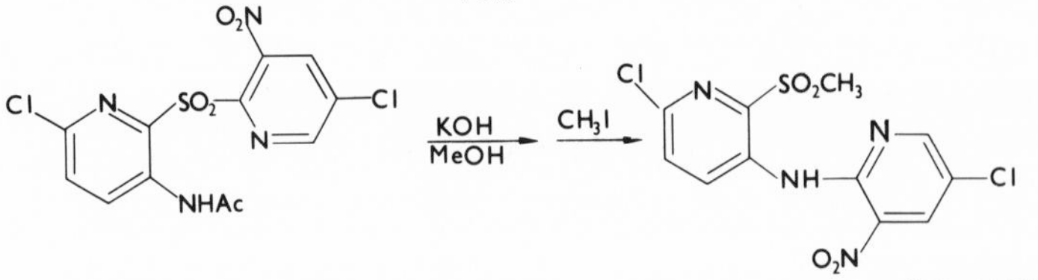

70%

⟨pyridine $Cl$, $O_2N$, $SO_2$–pyridine $Cl$, NHAc⟩ $\xrightarrow[\text{MeOH}]{\text{KOH}}$ $\xrightarrow{CH_3I}$ ⟨pyridine $Cl$, $SO_2CH_3$, NH–pyridine $Cl$, $O_2N$⟩ $(10.118)^{174}$

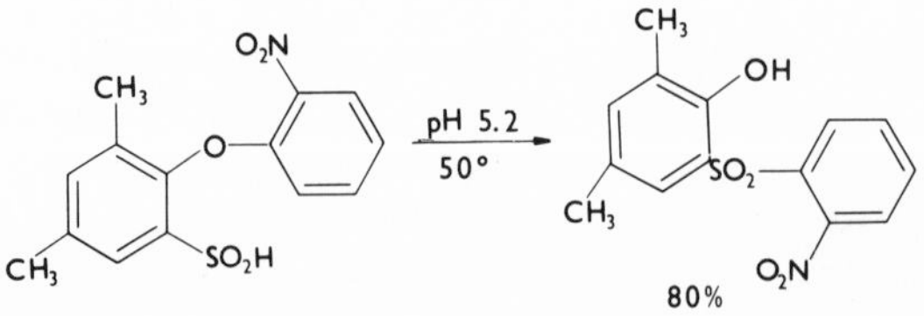

⟨aryl $CH_3$, $CH_3$, $SO_2H$, O–aryl $O_2N$⟩ $\xrightarrow[50°]{\text{pH 5.2}}$ ⟨aryl $CH_3$, $CH_3$, OH, $SO_2$–aryl $O_2N$⟩ $(10.119)^{175}$

80%

Diaryl and alkyl aryl sulfones are the most common types of compounds to undergo this rearrangement, although many other substrates, such as ethers,[176,177] sulfoxides,[178,179] sulfides,[172] sulfonamides,[180] carboxylic esters,[181] sulfonic esters,[181] iodonium compounds,[182] and phosphonium compounds,[183] have all been found to undergo like rearrangement.[55] The nucleophilic function may be a nitrogen, sulfur (-S⁻ or in some cases -SO₂⁻), oxygen (including carboxylate with iodonium compounds), or a carbanion. The rearrangement may also be catalyzed by acid if one or both aromatic rings are pyridine.[184] There are at least two mechanisms possible for this reaction:

$$\textbf{10.42} \qquad \textbf{10.43} \qquad (10.120)$$

$$\textbf{10.44} \qquad (10.121)$$

In eq. 10.120, the nucleophilic function, Y⁻, attacks the ring, leading to intermediate **10.43** in which the negative charge is delocalized around the aromatic ring. Subsequent expulsion of X⁻ leads to rearranged product. This mechanism should be favored by the presence of an electron-attracting group in conjugation with the anion.

The second mechanism (eq. 10.121) involves a direct displacement of X⁻ by Y⁻. In the transiton state, **10.44**, the aromatic π cloud is not disturbed. In both mechanisms, if YH has a free pair of electrons, as with -ṄH, the loss of a proton may also be concerted with nucleophilic attack, or may occur after the rearrangement has taken place.

An electron-withdrawing activating group is usually required for rearrangement to occur. An o- or p-nitro group is most commonly used for activation, but some cases are known in which a p-sulfonyl,[185] or an o-halogen substituent,[186,187] is sufficient to allow rearrangement. The first reported example of a Smiles rearrangement,[188] that of bis(2-hydroxy-l-napthyl)sulfide, had only the o-hydroxyl group for activation. Systems not generally requiring activation are the iodonium and phosphonium compounds, and o-methyl diaryl sulfones.

The nucleophilicity of Y⁻ and the nature of the leaving group, X⁻, play an important role in determining whether rearrangement will take place. These factors, in turn, are affected by substituents on the aromatic ring connecting X and Y in diaryl systems. If X⁻ is a good leaving group, or if Y⁻ is a strong nucleophile, rearrangement occurs readily. As X⁻ becomes a poorer leaving group,[172,189] or Y⁻ becomes a weaker nucleophile,[172] rearrangement becomes more difficult. For example, when N-substituted o-aminodiphenyl sulfides were treated with base, the N-methyl and unsubstituted derivatives were not acidic enough to rearrange; the N-acylamines underwent Smiles rearrangement, and the N-picryl- and

N-sulfonylamines formed anions, but were too weakly nucleophilic to rearrange.[172] Similar results were obtained with the corresponding sulfones,[172] except that the unsubstituted amine also rearranged when the leaving group was $SO_2^-$, a better leaving group.

Substituents in the ring connecting X and Y can have a great effect upon the rearrangement. An electron-withdrawing substituent will, by induction or resonance, stabilize the anion, $Y^-$, therefore making YH more acidic. It will also stabilize the developing charge on the leaving group, $X^-$, thus promoting rearrangement.[172] The relative importance of these two effects will depend upon the X and Y groups as well as the position of the substituent on the ring,[190] but both work to favor rearrangement. However, if the system is one which is already favorable for rearrangement, a substituent such as nitro that is *ortho* or *para* to $Y^-$ may decrease its nucleophilicity by charge delocalization to the point where rearrangement does not occur.[181] The opposite situation would occur with an electron–releasing substituent.

The presence of a second *ortho* group on the ring connecting X and Y was shown to cause an increase in the rate of rearrangement.[191] This was originally attributed to an inductive effect, but strong evidence has been presented to indicate that this is, in fact, a steric effect.[192] The steric acceleration is thought to be due to the different conformations possible:

|        10.45        |        10.46        |        10.47        |

Conformation **10.45** is the one needed for rearrangement. Other conformations are illustrated by **10.46** and **10.47**. If R is larger than hydrogen, the relative amount of **10.46** is decreased, thus increasing the population of **10.45**. If any $o'$-substituents are present in the migrating ring as activating groups, conformation **10.47** would now have the $o'$-group opposed to one of the *ortho* groups on the ring. The overall effect is to make **10.45** more favorable, thus raising the ground state of the system, and lowering the free energy barrier to rearrangement.

A synthetically useful variation of the Smiles rearrangement leads to the formation of a fused ring system. In these reactions, rearrangement is followed by displacement of an *o*-substituent by $X^-$, or displacement of X by a nucleophilic *o*-substituent. Phenazines, phenoxazines, phenothiazines, azaphenothiazines, and dipyridothiazines have been synthesized in this way.[55]

$$(10.122)^{193}$$

$$(10.123)^{55}$$

The rearrangement of *o*-methyldiaryl sulfones with strong bases represents the only case, besides iodonium and phosphonium compounds, in which a Smiles rearrangement occurs without additional activating groups. Treatment of these sulfones with *n*-butyllithium in ether, or potassium *t*-butoxide in DMSO gives the *o*-benzylbenzenesulfinic acid. products in high yield.[101]

$$(10.124)^{194}$$

The mesityl naphthyl sulfones are an exception to this, in which *n*-butyllithium in ether and potassium *t*-butoxide in DMSO lead to different products. Mesityl 1-naphthyl sulfone, when treated with *n*-butyllithium in ether, gave the expected Truce-Smiles product, 2-(1′-naphthylmethyl)-4,6-dimethylbenzenesulfinic acid (**10.48**); however, when the same sulfone was treated with potassium *t*-butoxide in DMSO, the product was 2-(2′-napthylmethyl)-4,6-dimethylbenzenesulfinic acid (**10.49**).[195] An addition, β-elimination mechanism has been proposed,[195] and one of the intermediates, **10.50**, in this mechanism has been isolated.[196]

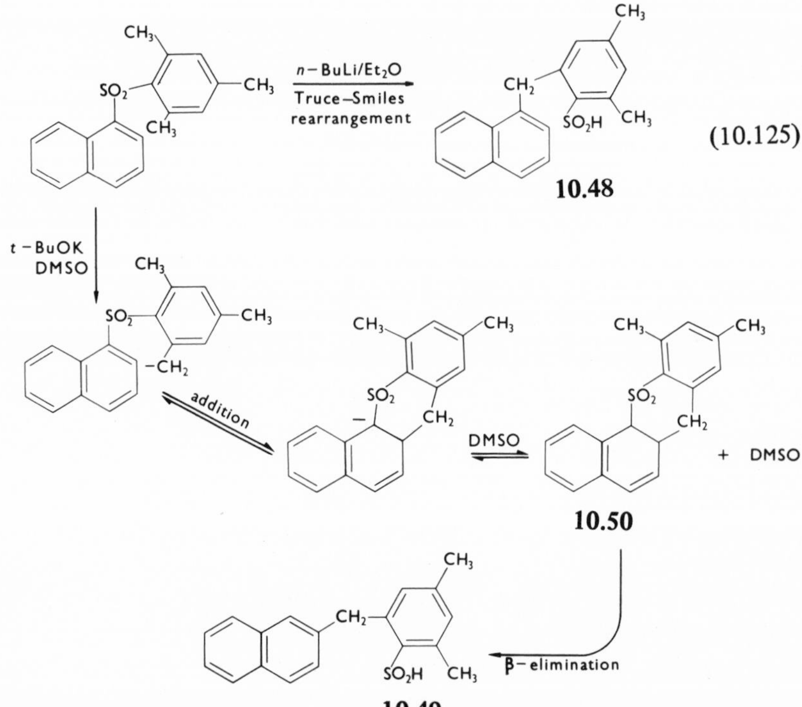

(10.125)

**10.48**

**10.50**

**10.49**

Compounds analogous to **10.50** have also been isolated from *o*-methyl-diaryl sulfones, which give a normal Truce-Smiles rearrangement.[196-199] Mesityl *m*-tolyl sulfone, on treatment with *n*-butyllithium in ether at 0° for a short time followed by quenching with $CO_2$ and decarboxylation of the resulting carboxylate, gave as a product 1,5,7-trimethyl-4a,9a-dihydro-thioxanthene 10,10-dioxide (**10.51**), analogous to **10.50**. Compound **10.51**, when treated under normal Truce-Smiles conditions, gave the same rearrangement product, 2-(3′-methylbenzyl)-4,6-dimethylbenzenesulfinic acid (**10.52**), as did mesityl *m*-tolyl sulfone under identical conditions. However, treatment of **10.51** with sodium ethoxide in boiling ethanol gave 2-(2′-methylbenzyl)-4,6-dimethylbenzenesulfinic acid (**10.53**).[195]

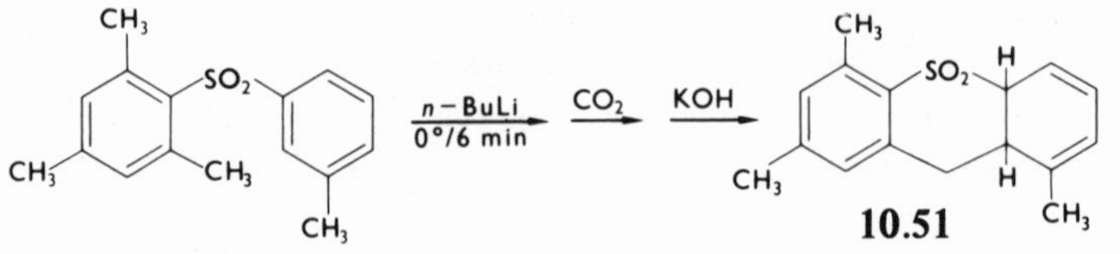

10.126)

**10.51**

10.51 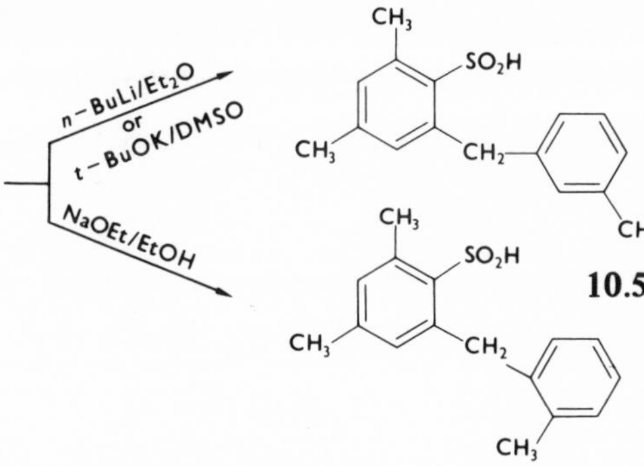


(10.127)

10.52

(10.128)

10.53

### 10.3.6 Elimination Reactions of Sulfones

$\beta$-Elimination reactions from $\beta$-substituted sulfones are influenced by the enhanced acidity of the $\alpha$-hydrogen, and by steric interactions in the transition state. In general, base-catalyzed eliminations in these systems proceed preferentially by a *trans* process, although *cis* elimination is preferred over removal of an unactivated $\gamma$-proton resulting in a *trans* elimination to a $\beta,\gamma$-unsaturated sulfone. This is illustrated in eq. 10.129 in which elimination of $p$-toluenesulfonate from *trans*-$\beta$-tosyl-cyclohexyl (**10.54**) and -cyclopentyl (**10.55**) aryl sulfones by hydroxide ion[200] or triethylamine[201] proceeds to generate the $\alpha,\beta$-unsaturated sulfones. *Trans* elimination would produce $\beta,\gamma$-unsaturated sulfones.

$$p\text{-}CH_3C_6H_4SO_2\text{-}\underset{\underset{H}{|}}{C}\overset{(CH_2)_n}{\overline{\qquad\qquad}}\underset{\underset{OTs}{|}}{C}\text{-}H \xrightarrow[\text{50\% aq. dioxane}]{Et_3N} p\text{-}CH_3C_6H_4SO_2\text{-}C\overset{(CH_2)_n}{=\!=\!=}C\text{-}H$$

**10.54** (n = 4)
**10.55** (n = 3)

(10.129)

The acyclic 3-$p$-tolylsulfonyl-2-butanol derivatives, **10.56** and **10.57**, are converted cleanly to the respective olefins resulting from a *trans* elimination.[201] These results can be rationalized by a concerted E2 elimination, although a recent study[202] has led to the postulation of an E2cB mechanism. Evidence supporting such a mechanism arose from a study of a series of aryl sulfones (**10.58**), in which Brønsted plot and Hammett study of aryl substituents revealed considerable carbanion character in the transition state. The leaving-group effect was likewise found to be relatively small.[203]

$$\begin{array}{c} \text{BsO} \quad \text{CH}_3 \\ \quad | \qquad \vdots \\ \text{H''C}-\text{C} \quad \text{SO}_2\text{C}_6\text{H}_4\text{CH}_3\text{-}p \\ \quad / \quad | \\ \text{CH}_3 \quad \text{H} \end{array} \xrightarrow{\text{Et}_3\text{N}} \begin{array}{c} \text{H} \quad\quad \text{CH}_3 \\ \quad \backslash \quad\quad / \\ \quad \text{C}=\text{C} \\ \quad / \quad\quad \backslash \\ \text{CH}_3 \quad \text{SO}_2\text{C}_6\text{H}_4\text{CH}_3\text{-}p \end{array} \qquad (10.130)$$

*threo*       **10.56**

$$\begin{array}{c} \text{BsO} \quad\quad \text{SO}_2\text{C}_6\text{H}_4\text{CH}_3\text{-}p \\ \quad \backslash \qquad \vdots \\ \text{H}\text{''''''}\,\text{C}-\text{C}-\text{CH}_3 \\ \qquad \diagup \quad | \\ \quad \text{CH}_3 \quad \text{H} \end{array} \xrightarrow{\text{Et}_3\text{N}} \begin{array}{c} \text{H} \quad\quad \text{SO}_2\text{C}_6\text{H}_4\text{CH}_3\text{-}p \\ \quad \backslash \quad / \\ \quad \text{C}=\text{C} \\ \quad / \quad\quad \backslash \\ \text{CH}_3 \quad\quad \text{CH}_3 \end{array} \qquad (10.131)$$

*erythro*  **10.57**

$$\text{ArSO}_2\text{CH}_2\text{CH}_2\text{X} \xrightarrow[\text{CH}_3\text{CN}]{\text{R}_3\text{N}} \text{ArSO}_2\text{CH}=\text{CH}_2 \qquad (10.132)$$

**10.58**                    X = Cl, OTs, Br

Dehydrohalogenation of the *erythro* and *threo* sulfonyl 1,2-diphenylethanes, **10.59** and **10.61**, gives rise to a stereoconvergent elimination in which both **10.59** and **10.61** produce the thermodynamically more stable isomer **10.60**.[204] Steric compression in the transition state between the bulky sulfonyl[205] group and the β-phenyl of **10.61** causes the elimination to proceed by an ElcB mechanism rather than an E2 mechanism as does **10.59**.

$$\begin{array}{c} \text{H} \quad\quad \text{C}_6\text{H}_5 \\ \quad \backslash \quad \vdots \\ \quad \text{C}-\text{C}\text{—H} \\ \diagup \quad\quad \backslash \\ \text{C}_6\text{H}_5 \quad\quad \text{Cl} \\ \diagup \\ p\text{-CH}_3\text{C}_6\text{H}_4\text{SO}_2 \\ \textbf{10.59} \end{array} \xrightarrow[\text{DMSO}]{t\text{-BuOK}} \begin{array}{c} \text{C}_6\text{H}_5 \quad \text{CH}_6\text{H}_5 \\ \quad \backslash \quad / \\ \quad \text{C}=\text{C} \\ \quad / \quad\quad \backslash \\ p\text{-CH}_3\text{C}_6\text{H}_4\text{SO}_2 \quad \text{H} \\ \textbf{10.60} \end{array} \qquad (10.133)$$

$$\begin{array}{c} t\text{-BuOK} \\ \text{DMSO} \end{array}$$

$$\begin{array}{c} \text{H} \quad \text{H} \\ \quad \backslash \quad \vdots\; \text{C}_6\text{H}_5 \\ \quad \text{C}-\text{C} \diagup \\ \text{C}_6\text{H}_5 \diagup \quad \backslash \\ \qquad\quad \text{Cl} \\ p\text{-CH}_3\text{C}_6\text{H}_4\text{SO}_2 \\ \textbf{10.61} \end{array} \xrightarrow[\text{DMSO}]{t\text{-BuOK}} \;\;\bcancel{\quad}\;\; \begin{array}{c} \text{C}_6\text{H}_5 \quad \text{H} \\ \quad \backslash \quad / \\ \quad \text{C}=\text{C} \\ \quad / \quad\quad \backslash \\ p\text{-CH}_3\text{C}_6\text{H}_4\text{SO}_2 \quad \text{C}_6\text{H}_5 \end{array}$$

$$(10.134)$$

A similar effect operates in the pyrolysis of xanthate derivatives of β-hydroxy sulfones.[206] Pyrolysis of **10.62** proceeds *via* a first-order *trans* elimination rather than the normal *cis* elimination. Similarly, the xanthate derivative of **10.57** undergoes pyrolysis by 80% *trans* and 20% *cis* elimination, although the *threo* isomer, **10.56**, proceeds with 95% *cis* elimination. This has been explained in terms of severe steric interactions in the transition state for a *cis* elimination from **10.57**.

$$\begin{array}{c} \includegraphics{} \end{array} \quad SO_2C_6H_4CH_3-p \xrightarrow{\ 210°\ } \begin{array}{c} \includegraphics{} \end{array} - SO_2C_6H_4CH_3-p \qquad (10.135)$$

$OCS_2CH_3$

**10.62**

If the β-substituent of a sulfone is a poorer leaving group than sulfinate, the sulfonyl group may be lost in an elimination reaction. Treatment of **10.63** with alumina leads to the elimination of α-toluenesulfinic acid to yield β-nitrostyrene.[207] 1,4-Eliminations from 3-sulfolenes can be used to prepare dienic sulfinates and, subsequently, dienic sulfones (eq. 10.137).[208] Treatment of **10.64** with amylsodium leads to aryl migration to form **10.65** followed by loss of sulfinate.[209]

$$C_6H_5CH_2SO_2CHCH_2NO_2 \xrightarrow{Al_2O_3} C_6H_5CH=CHNO_2 +$$

$$\underset{\displaystyle C_6H_5}{|} \qquad\qquad\qquad [C_6H_5CH_2SO_2^- \cdot Al_2O_3H^+]$$

$$\qquad\qquad\qquad\qquad\qquad\qquad\qquad\qquad (10.136)$$

**10.63**

$$\begin{array}{c} \includegraphics{} \end{array}\!\!R \xrightarrow[\Delta]{EtMgBr} RCH=CH\text{-}CH=CHSO_2MgBr \xrightarrow{C_6H_5CH_2Cl}$$

$$O\quad O$$

$$\qquad\qquad\qquad\qquad RCH=CHCH=CHSO_2CH_2C_6H_5$$

$$\qquad\qquad\qquad\qquad\qquad\qquad\qquad (10.137)$$

$$(C_6H_5)_3CCH_2SO_2C_6H_5 \xrightarrow{Amyl\ Na} [(C_6H_5)_2\overset{\curvearrowright}{C}\text{-}\overset{-}{C}HSO_2C_6H_5] \longrightarrow$$

$$\textbf{10.64}\qquad\qquad\qquad \textbf{10.65}\quad \underset{\displaystyle C_6H_5}{|}$$

$$(C_6H_5)_2C=CHC_6H_5 + C_6H_5SO_2Na$$

$$\qquad\qquad\qquad\qquad\qquad\qquad (10.138)$$

Sulfinate was used as the leaving group in the 1,3-displacement reaction of **10.66** during the preparation of the dicyclopropyl ketone (**10.67**).[210]

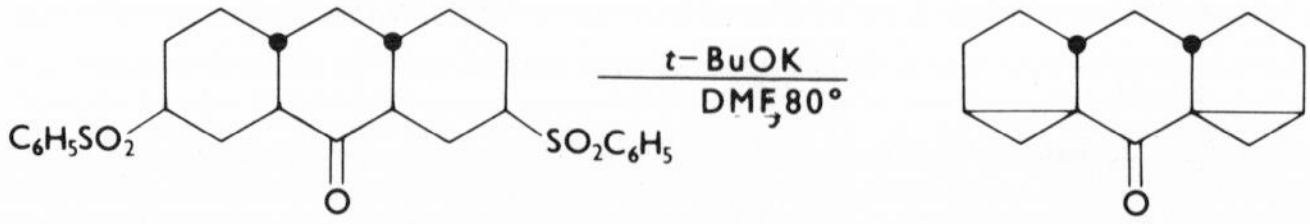

$$
\text{10.66} \xrightarrow[\text{DMF, 80°}]{t-\text{BuOK}} \text{10.67} \qquad (10.139)
$$

**10.66**                                             [34%] **10.67**

Similarly, the carbanion from 9-phenylsulfonylfluorene reacts with nitrosobenzene to give N-phenylfluorenone ketoxime, presumably *via* the oxazirane, **10.68**, which arises from displacement by oxygen of the phenylsulfinate.[211]

(10.140)

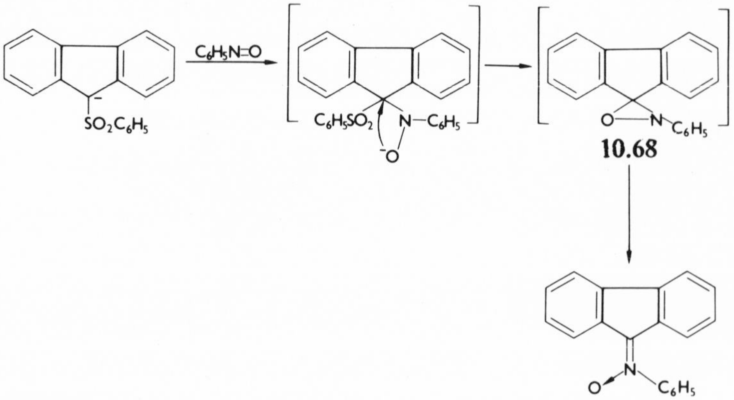

**10.68**

Still another interesting example of displacement of sulfinate is involved in the synthesis of vicinal dinitro compounds,[211a] i.e.,

$$\text{ArSO}_2\text{Na} + (\text{CH}_3)_2\text{C(NO}_2)\text{Br} \longrightarrow$$

$$\text{ArSO}_2\text{C(CH}_3)_2\text{NO}_2 \xrightarrow{\text{LiC(CH}_3)_2\text{NO}_2} [\text{O}_2\text{NC(CH}_3)_2]_2$$

### 10.3.7  Reductions of Sulfones

Sulfones exhibit a marked stability toward reducing agents. Alkyl and alkyl aryl sulfones are not reduced by zinc in mineral acid, zinc dust or fuming hydriodic acid.[212] $\alpha,\beta$-Dihalo sulfones can be reduced to their respective sulfides by passing hydrogen sulfide into a solution of the sulfone in acetic acid containing hydrogen bromide.[213] Thiophene-1,1-dioxides are reduced to thiophenes with zinc and hydrochloric acid in acetic acid solvent.[214,215]

$$(10.141)$$

The only general method for the reduction of sulfones to sulfides involves refluxing with lithium aluminium hydride in ether solvent.[215] Reduction of **10.69** with lithium aluminium hydride proceeds slowly to form the dihydrobenzothiophene, **10.72**. That the double bond is reduced prior to reduction of the sulfonyl group has been demonstrated by the failure of the corresponding vinyl sulfide, **10.71**, to undergo reduction (eq. 10.142).[215] Thietane dioxides and tetrahydrothiophene dioxides are easily reduced at 35°, while tetrahydrothiopyran dioxide and acyclic sulfones require higher temperature (92°).[215] The low reactivity of acyclic sulfones has been used in the hydride reduction of vinyl sulfone, **10.73**, to the saturated analogue, **10.74**.[216]

$$(10.142)$$

$$(n\text{-Bu})_2 SO_2 \xrightarrow[n\text{-BuOEt, }92°]{\text{LiAlH}_4} (n\text{-Bu})_2 S \qquad (10.143)$$

$$73\%$$

$$CH_3 SO_2 CH{=}CHC_6 H_5 \xrightarrow{\text{LiAlH}_4} CH_3 SO_2 CH_2 CH_2 C_6 H_5 \qquad (10.144)$$

$$55\%$$

**10.73**                **10.74**

Reduction of sulfones with alkali metals in ammonia or methylamine results in carbon-sulfur bond cleavage to form a hydrocarbon and sulfinate or mercaptan. Lithium metal in methylamine cleaves all sulfones,[217,218] while sodium in the same solvent cleaves aryl sulfones but does not react with alkyl or alkyl aryl sulfones. The reduction product from aryl sulfones is dependent upon the metal used, as is indicated in eq. 10.145. Lithium in methylamine gives the hydrocarbon and the thiophenol, but sodium in methylamine gives the sulfinate instead of the thiophenol.[217]

$$
(p\text{-}CH_3C_6H_4)_2SO_2
\begin{cases}
\xrightarrow[CH_3NH_2]{Li} & p\text{-}CH_3C_6H_4SH + CH_3C_6H_5 \\
& 100\% \\
\xrightarrow[CH_3NH_2]{Na} & p\text{-}CH_3C_6H_4SO_2Na + CH_3C_6H_5 \\
& 85\%
\end{cases}
\tag{10.145}
$$

The direction of cleavage of unsymmetrical sulfones by lithium in methylamine has only in some cases been clearly worked out (Table 10.4). With alkyl aryl sulfones (**10.75**), the direction of cleavage is determined by the alkyl group, R. If R is a primary alkyl,[218] cleavage occurs at the aryl-sulfur bond (A), resulting in formation of the alkyl sulfinate as the major product. A mixture of cleavage products resulting from cleavage at either A or B is obtained with secondary alkyl substituents, although cleavage of the aryl-sulfur bond predominates.[146,219] If R is a tertiary alkyl, cleavage of the alkyl-sulfur bond (B) occurs almost exclusively.[219] There appear to be no obvious trends in the direction of cleavage of alkyl sulfones except with methyl or cycloalkyl sulfones, in which case anionic stability[219] and ring size[218] appear to be the predominant factors.

$$
Ar\overset{A}{-\!\!-}SO_2\overset{B}{-\!\!-}R
$$

**10.75**

**TABLE 10.4**

$$RSO_2R' \xrightarrow[\text{CH}_3\text{NH}_2]{\text{Li}} RH + R'H + RSO_2Li + R'SO_2Li$$

| R | R' | % Yield | Mole % | | | | Ref. |
|---|----|---------|--------|--|--|--|------|
| | | | RH | R'H | RSO$_2$Li | R'SO$_2$Li | |
| $C_6H_5$ | $n\text{-}C_{10}H_{21}$ | 95 | $-^a$ | 0 | 0 | 100 | 217 |
| $C_6H_5$ | $CH(CH_3)C_6H_{13}$ | 83 | 61 | 39$^b$ | $-^a$ | $-^a$ | 219 |
| $C_6H_5$ | $C(CH_3)_2C_3H_7$ | 59 | 0 | 100 | $-^a$ | $-^a$ | 219 |
| $n\text{-}C_8H_{17}$ | $CH_3$ | 86 | $-^a$ | $-^a$ | >86 | 0 | 219 |
| $n\text{-}C_6H_{13}$ | $C(CH_3)_2C_3H_7$ | 72 | 44 | 56 | $-^a$ | $-^a$ | 219 |
| $(CH_2)_4CH$ | $(CH_2)_5CH$ | 96 | 83 | 17 | $-^a$ | $-^a$ | 218 |
| $(CH_2)_4CH$ | $n\text{-}C_7H_{15}$ | 78 | 70 | 30 | $-^a$ | $-^a$ | 218 |

$^a$Not isolated.
$^b$$n$-Octane: 10%; cyclohexene: 29%.

Sodium amalgum in boiling ethanol cleaves aryl and alkyl aryl sulfones to aromatic sulfinic acids and aryl or alkyl hydrocarbons.[220] Sodium in ammonia has been found to be a superior method for the cleavage of S-benzyl-L-cysteine dioxides[221] in the synthesis of cysteine sulfinates:

$$C_6H_5CH_2SO_2CH_2\underset{\underset{NH_2}{|}}{C}HCO_2H \xrightarrow[\text{NH}_3]{\text{Na}} C_6H_5CH_3 +$$

$$NaO_2SCH_2\underset{\underset{NH_2}{|}}{C}HCO_2H \qquad (10.146)$$

Polarographic reduction of aryl and alkyl aryl sulfones proceed readily to give the sulfinate and hydrocarbon,[222-226] and can be performed on a preparative scale.[227]

$$(10.147)$$

Aryl and alkyl aryl sulfones undergo cleavage with hot sodium amide in piperidine to produce an N-arylpiperidine and a sulfinic acid.[228] Although the analogous reaction with aryl halides proceeds *via* an aryne intermediate, this cleavage has been shown to occur by direct displacement with no *ortho* substitution.[229]

$$ArSO_2R \; + \; \underset{}{\bigcirc}NH \; + \; NaNH_2 \; \xrightarrow{\Delta} \; \underset{}{\bigcirc}N\!-\!Ar \; + \; RSO_2 \quad\quad (10.148)$$

Cleavage of aryl sulfones has also been reported with aqueous potassium hydroxide at 200°,[230] by chlorine at 130°,[231] and by phosphorous pentachloride.[1,232]

The ready availability of $\beta$-keto sulfones would make them useful synthetic intermediates *via* subsequent cleavage of such systems. Aluminum amalgum in aqueous tetrahydrofuran cleaves the carbon-sulfur bond, producing an alkyl ketone, as in eq. 10.149. With $\alpha$-aroyl sulfones, zinc in acetic acid often gives better yields of the carbon-sulfur bond cleavage product than does the analogous aluminum amalgum reduction.[133] An attempted Wolff-Kishner reduction of a $\beta$-keto sulfone resulted instead in cleavage to the methyl sulfone (eq. 10.150),[233] a reaction common to these systems under basic conditions (eq. 10.151).[234]

$$(10.149)$$

$$89\%$$

$$(CH_3)_3C\overset{O}{\overset{\|}{C}}CH_2SO_2Ar \xrightarrow[120°]{N_2H_4\cdot H_2O} \xrightarrow[200°]{KOH} CH_3SO_2Ar \quad\quad (10.150)$$
$$76\%$$

$$C_6H_5CH_2\overset{O}{\overset{\|}{C}}CHSO_2CH_2C_6H_5 \xrightarrow{\underset{H}{\overset{\bigcirc\!\!\!N}{}}} C_6H_5CH_2\overset{O}{\overset{\|}{C}}N\!\!\bigcirc \; + \; C_6H_5CH_2SO_2CH_2C_6H_5$$
$$\underset{C_6H_5}{} \quad\quad\quad\quad\quad\quad\quad\quad\quad\quad\quad\quad 100\% \;\; (10.151)$$

Most sulfones are completely desulfurized by the action of Raney nickel in alcohol[235] or of nickel aluminum alloy in aqueous sodium hydroxide.[134] This reduction, unlike that of sulfides and sulfoxides, proceeds with retention of optical activity when the sulfonyl group is attached to a chiral center. Whether the reaction proceeds with net inversion or retention, however, is dependent upon the conditions employed.[236]

$$CH_2=CHCH_2SO_2CH_2CH_2OH \xrightarrow[EtOH, \Delta]{Ni(R)} \begin{array}{c} CH_3 \\ | \\ C_6H_5\text{-}CH\text{-}CONH_2 \\ * \end{array}$$

inversion, 65% optical purity

$$\begin{array}{c} CH_3 \\ | \\ C_6H_5\text{-}C\text{-}CONH_2 \\ | \\ SO_2C_6H_5 \end{array}$$

(10.152)

$$\xrightarrow[acetone]{Ni(R)} \begin{array}{c} CH_3 \\ | \\ C_6H_5\text{-}CH\text{-}CONH_2 \\ * \end{array}$$

retention, 67% optical purity

## 10.3.8 Pyrolysis of Sulfones

Sulfones in general are thermally very stable compounds.[237] Bis-1,4-(butylsulfonyl)butane at 275° decomposes to the extent of only 0.8% during one hour to give sulfur dioxide, 1-butene, and cis-2-butene.[238] A correlation has been observed between the number of β-hydrogens and thermal stability in dialkyl and alkyl aryl sulfones.[239] Alkyl aryl sulfones are more stable than dialkyl sulfones, and stability of the latter decreases in the order, $1° > 2° > 3°$.[239] Diaryl sulfones give only a small amount of decomposition even at temperatures as high as 800°.[240]

In contrast, allylic and benzylic sulfones are more prone to decompose. Allylic compounds between 150° and 400° decompose with rearrangement and extrusion of $SO_2$. This pyrolysis leaves many functional groups such as cyano, keto, halo, and carboxylic ester unchanged.[241]

$$\begin{array}{c} CH_2=CHCHSO_2CH_2C_6H_5 \\ | \\ CH_3 \end{array} \xrightarrow[50 \text{ mm Hg}]{210°} \begin{array}{c} CH_3CH=CHCH_2CH_2C_6H_5 \\ 35\% \end{array}$$

(10.153)

$$\begin{array}{c} O \\ \| \\ CH_2=CCH_2SO_2CH_2CH_2OCCH_3 \\ | \\ Cl \end{array} \xrightarrow[100 \text{ mm Hg}]{170\text{-}190°} \begin{array}{c} CH_3CH=CHCH_2CH_2C_6H_5 \\ 56\% \end{array}$$

(10.154)

$$\begin{array}{c} CH_2=CCH_2SO_2CH_3 \\ | \\ CN \end{array} \xrightarrow[90 \text{ mm Hg}]{190\text{-}240°} \begin{array}{c} CH_2=CCH_2CH_3 \\ | \\ CN \end{array}$$

(10.155)

The mechanism of this reaction is thought to be dependent upon temperature. At lower temperatures a cyclic, concerted mechanism (eq. 10.156) is thought to be operative, while at higher temperatures dissociation into $SO_2$ and alkyl radicals is thought to be followed by radical recombination (eq. 10.157).[241]

$$
\begin{array}{ccc}
\underset{CH_2}{\overset{R}{\underset{\diagdown}{\overset{\diagup}{CH-CH}}}}\!\!\!\!\diagdown_{SO_2} & \underset{CH_2}{\overset{R}{\overset{\diagup}{CH{\cdots}CH}}}\cdots SO_2 & \underset{CH_2}{\overset{CHR}{\overset{\diagup\!\!\diagup}{CH}}} \\
\quad\;\; \diagup \qquad\qquad\quad\quad \cdots\;\quad\;\cdots & & \diagdown \\
\quad\;\; CR_3 \qquad\qquad\qquad CR_3 & & CR_3
\end{array}
$$

$$
\overset{R}{\underset{CR_3}{\overset{\diagup}{\underset{\diagup}{\underset{CH_2}{\overset{\diagup}{CH-CH}}}}\diagdown SO_2}} \longrightarrow \overset{R}{\underset{\cdot CR_3}{\overset{\diagup}{\underset{CH_2^{\cdot}}{\overset{\diagup}{CH{\cdots}CH}}}\cdot SO_2}} \longrightarrow \overset{CHR}{\underset{CR_3}{\overset{\diagup\diagup}{\underset{CH_2}{\overset{\diagup}{CH}}}}} + SO_2
$$

$$\text{(10.156)}$$

$$CH_2{=}CH\text{-}CH_2 SO_2 R \longrightarrow CH_2{=}CH\text{-}CH_2 + SO_2 + R \longrightarrow$$

$$CH_2{=}CH\text{-}CH_2 R + SO_2 \qquad\qquad \text{(10.157)}$$

Evidence for these two mechanisms has been obtained by observing the products of pyrolysis of substituted allyl alkyl sulfones[241]:

$$
\underset{CH_2}{\overset{CHCH_3}{\overset{\diagup\diagdown}{\underset{\diagup}{CH\;\;SO_2}}}}\underset{CH_2 C_6 H_5}{} \xrightarrow{210^\circ} \underset{\underset{CH_2 C_6 H_5}{}}{\overset{CHCH_3}{\underset{CH_2}{\overset{\diagup\diagup}{\underset{\diagdown}{CH}}}}} + SO_2 \qquad \text{(10.158)}
$$

$$\overset{}{32\%}$$

$$
CH_2{=}CHCH_2 SO_2 \overset{CH_3}{\underset{CH_3}{\overset{\mid}{\underset{\mid}{C}}}}CH_2 CH_3 \xrightarrow{350^\circ} CH_2{=}CHCH_3
$$

$$24\%$$

$$+ \; CH_2{=}CHCH_2 CH_2 CH{=}CH_2 \; +$$

$$8\% \qquad\qquad \text{(10.159)}$$

$$+ \ CH_2{=}\underset{\underset{\displaystyle CH_3}{|}}{C}CH_2CH_3 \ + \ CH_3\underset{\underset{\displaystyle CH_3}{|}}{\overset{\overset{\displaystyle CH_3}{|}}{C}}CH_2CH_3 \ + \ CH_2{=}CHCH_2\underset{\underset{\displaystyle CH_3}{|}}{\overset{\overset{\displaystyle CH_3}{|}}{C}}CH_2CH_3$$

$$\qquad\quad 16\% \qquad\qquad\quad 4\% \qquad\qquad\qquad 30\%$$

$$(10.159 \ cont'd)$$

Benzyl sulfones lose $SO_2$ on heating at reduced pressure to form 1,2-diarylethanes.[242]

$$(10.160)$$

This reaction probably proceeds by formation and recombination of benzyl radicals. In 1903, benzyl sulfone was reported to decompose thermally to give stilbene and toluene.[243] This has since been rationalized in terms of hydrogen atom abstraction from the 1,2-diphenylethane by benzyl radicals:[241]

$$C_6H_5CH_2SO_2CH_2C_6H_5 \ \xrightarrow{290°} \ C_6H_5CH{=}CHC_6H_5 \ + \ C_6H_5CH_3 \ + \ SO_2$$

$$SO_2 \ + \ 2C_6H_5CH_2{\cdot} \ \longrightarrow \ C_6H_5CH_2CH_2C_6H_5 \ \xrightarrow{\phantom{aaa}} C_6H_5CH_2{\cdot}$$

$$(10.161)$$

$o$-Methyldiaryl sulfones lose $SO_2$ on strong heating to give diarylmethanes. This reaction was suggested to be a Truce-Smiles rearrangement with the sulfonyl oxygen acting as the base.[244]

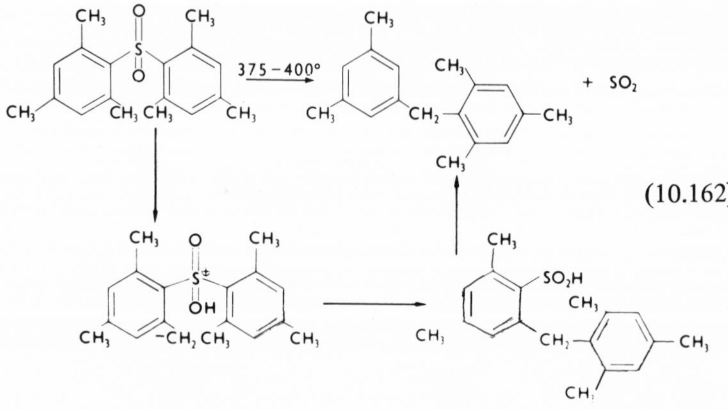

$$(10.162)$$

Some cyclic sulfones are more prone to pyrolytic decomposition than their acyclic analogues. Thermolysis of episulfones to $SO_2$ and olefin has been discussed in the section on the Ramberg-Bäcklund reaction (Sect. 10.3.4).

Thietane 1,1-dioxides react in a similar manner giving $SO_2$ and cyclopropanes. In the reaction of 2,4-diphenylthietane 1,1-dioxide (**10.76**), the ratio of *cis*- to *trans*-1,2-diphenylcyclopropane seemed to be independent of whether *cis*- or *trans*-thiethane dioxide was used as a starting material.[245]

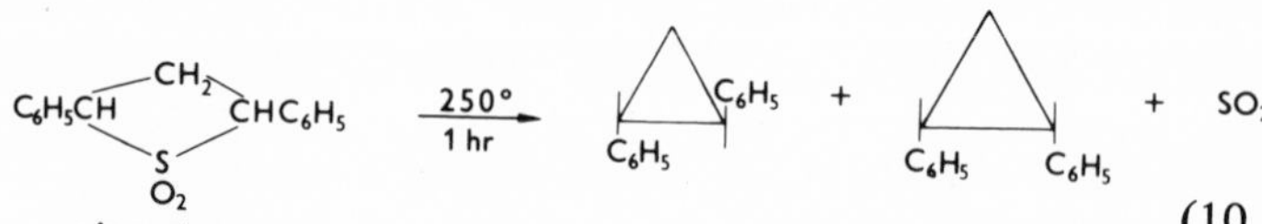

$$(10.163)$$

8 : 1

**10.76**

2,5-Dihydrothiophene 1,1-dioxides readily undergo a retro Diels-Alder reaction, giving $SO_2$ and a butadiene with greater than 99% stereospecificity *via* a concerted, disrotatory elimination.[85,246,247]

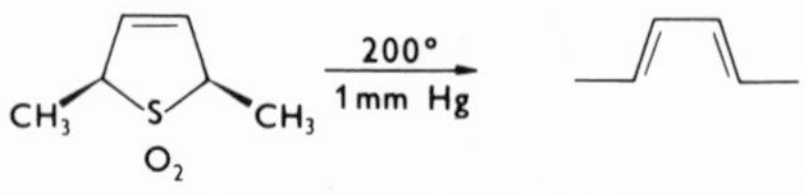

$$(10.164)$$

In a similar manner, pyrolysis of the 2,7-dimethyl-2,7-dihydrothiepin 1,1-dioxides have been shown to eliminate $SO_2$ *via* a stereospecific *trans* concerted elimination.[248]

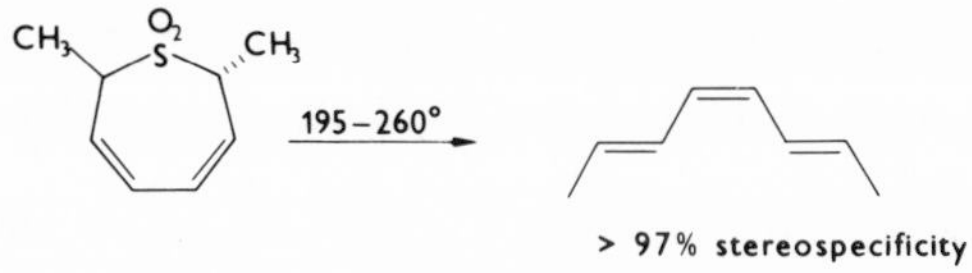

$$\text{(10.165)}$$

> 97% stereospecificity

The 1,3-dihydroisothianaphthenes, **10.77**, can be used to prepare benzocyclobutenes.[249-251] With proper temperature control of the reaction, dibenzo-1,5-cyclooctadiene, **10.78**, can be made the major product.[249]

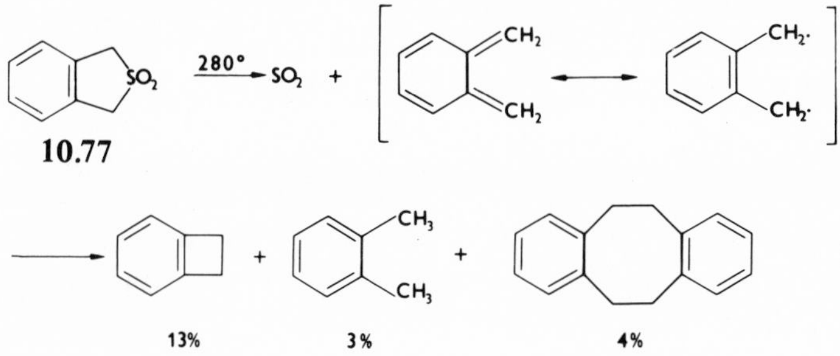

**10.77**

$$\text{(10.166)}$$

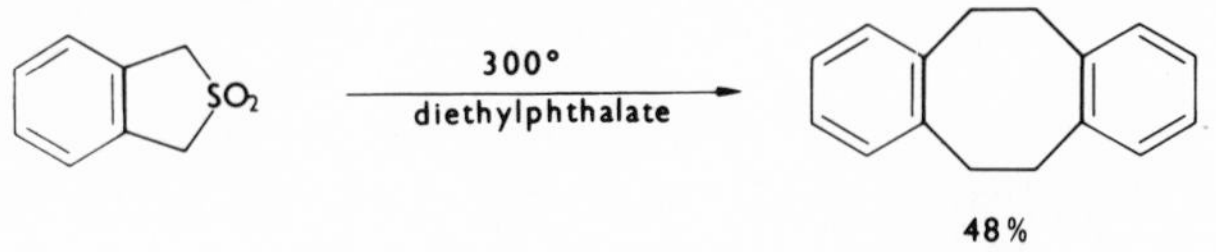

13%        3%        4%

$$\text{(10.167)}$$

48%
**10.78**

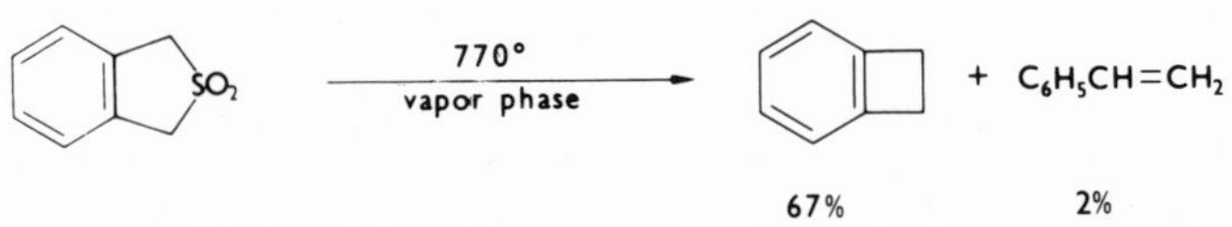

$$\text{(10.168)}$$

67%        2%

Benzothiophene dimerizes with the loss of $SO_2$ at 200°, with or without solvent.[252]

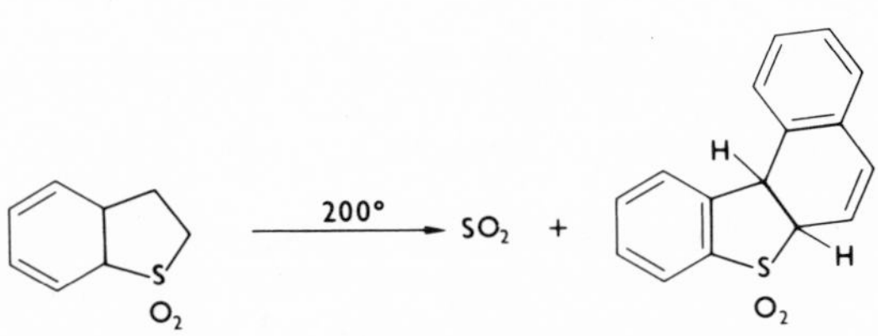

$$\text{(10.169)}$$

Several studies have been made on the pyrolysis of polysulfones.[238,253-255] Polysulfones from polymerization of $SO_2$ and olefins decompose at about 200° to give back the monomers.[253,254] At higher temperatures more complex product mixtures result.[254] Polymers with more than two carbons between the sulfonyl moieties give less than 10% weight loss up to 300°,[238,255] producing olefinic products probably *via* radical abstraction of a $\beta$-proton giving an olefin, $SO_2$ and an alkyl radical which can propagate the chain reaction.[238]

### 10.3.9 Reactions of Vinyl Sulfones

Eliminations of substituents $\beta$ to the sulfonyl group result in the formation of an $\alpha,\beta$-unsaturated sulfone in preference to the $\beta,\gamma$-isomer (see Sect. 10.3.6). Base-catalyzed isomerization of these products indicates that they arise from kinetic rather than thermodynamic control.[256-258] With the exception of *p*-tolyl propenyl sulfone, the $\beta,\gamma$-isomers are considerably more stable than the $\alpha,\beta$-unsaturated sulfones. Equilibrium of the pentenyl methyl sulfones (**10.79**, n = 2, R = $CH_3$) produced an equilibrium mixture containing > 99% of **10.79b**.[256] Similar treatment of butenyl butyl sulfone (**10.79**, n = 1, R = *n*-$C_4H_9$) gave a mixture containing the $\beta,\gamma$-isomer to the extent of 84%.[258]

$$CH_3(CH_2)nCH=CHSO_2 R \rightleftharpoons CH_3(CH_2)_{n-1} CH=CHCH_2SO_2 R \quad (10.170)$$
$$\text{10.79a} \qquad\qquad\qquad\qquad \text{10.79b}$$

The relative stabilities of unsaturated cyclic sulfones is related to ring size and alkyl substituents. The isomers of butadiene sulfone (**10.80a** and **10.81a**) are present in an equilibrium mixture to the extent of 42% and 58%, respectively.[257] However, a methyl group at position 3 (**10.80b** and **10.81b**) enhances the stability of the $\alpha,\beta$-unsaturated isomer so that it is present at equilibrium to the extent of 86%.[258] In contrast, thiacyclohex-3-ene 1,1-dioxide, **10.82**, predominates over the $\alpha,\beta$-isomer to the extent of 97 to 3 at equilibrium.[258]

$$(10.171)$$

a. R = H
b. R = $CH_3$

**10.80**        **10.81**

3                97

$$(10.172)$$

**10.82**

On the basis of thermodynamic parameters determined from thermochemical data,[259,260] it has been concluded that, rather than being stabilized by conjugation with the sulfonyl group, the double bond is destabilized by the sulfonyl group by approximately 4 kcal/mole. This destabilization is probably a result of the large inductive effect of the sulfonyl group, with some inductive stabilization by alkyl groups.[259]

This large inductive effect of the sulfonyl group also greatly influences the reactivity of the olefin. The vinyl hydrogen $\alpha$ to sulfonyl is enhanced in acidity,[257] and electrophilic additions to the double bond are greatly retarded. Normal addition of hydrogen halide to compounds **10.80** and **10.81** failed, although zinc chloride or bromide can be used to produce a moderate yield of $\beta$-halo sulfone.[261] The retarding effect of the sulfonyl group towards electrophilic addition is clearly illustrated by the rate of addition of bromine to the isomers of butenyl butyl sulfone (Table 10.5).[262]

**TABLE 10.5[262]**

Rate of Addition of $Br_2$ to $n\text{-}C_4H_9SO_2R$ in Acetic Acid at 25°

| R | $k_{add'n} \times 10^2$ ($\ell \cdot mol^{-1} \cdot min^{-1}$) |
|---|---|
| -CH=CHCH$_2$CH$_3$ | 0.26 |
| -CH$_2$CH=CHCH$_3$ | 214.2 |
| -CH$_2$CH$_2$CH=CH$_2$ | 16,670 |
| -CH$_2$CH$_2$CH$_2$CH=CH$_2$ | 102,600 |

The double bonds of vinyl sulfones are activated towards nucleophilic addition. Nucleophiles such as thiolate,[91] sodium diethyl malonate,[91] phenylmagnesium bromide,[91] hydroxide,[1] phenoxide,[1] and amines[1] have been observed to add to vinyl sulfones. Base-catalyzed epoxidation has also been reported.[140]

$\beta$-Substituted saturated sulfones may undergo nucleophilic displacement of the $\beta$-substituent by amines of alkoxide.[263,264] Although the resulting products may be viewed as arising *via* direct displacement, an elimination-addition mechanism (eq. 10.174) is more plausible.[264]

$$C_6H_5CH=CHSO_2C_6H_4CH_3\text{-}p$$

$$\xrightarrow{p\text{-}CH_3C_6H_4S^-} \quad C_6H_5\underset{\underset{SAr}{|}}{CH}CH_2SO_2Ar$$

$$\xrightarrow{NaCH(CO_2Et)_2} \quad C_6H_5\underset{\underset{CH(CO_2Et)_2}{|}}{CH}CH_2SO_2Ar$$
$$100\%$$

$$\xrightarrow[2.\,H_2O]{1.\,C_6H_5MgBr} \quad (C_6H_5)_2CHCH_2SO_2Ar$$
$$50\%$$

$$\xrightarrow[OH^-]{H_2O_2} \quad C_6H_5\overset{}{\cdots}\underset{H}{\overset{}{C}}\overset{O}{\underset{SO_2Ar}{\overset{}{C}}}\cdots H$$

$$(10.173)$$

$$ArSO_2CH_2CH_2Y + Z \longrightarrow [ArSO_2CH=CH_2] \longrightarrow ArSO_2CH_2CH_2Z$$

$$(10.174)$$

$$Y = SO_2Ar,\ \text{halogen, OR, OAr}$$
$$Z = RO^-,\ R_2NH$$

Vinyl sulfones also readily undergo certain cycloaddition reactions. For example, nitrones cycloadd to vinyl sulfones in a concerted process to give isoxazolidines, **10.83**.[265] They are also excellent dienophiles in the Diels-Alder reaction, as shown in eq. 10.176.[252, 264-268]

$$\underset{H}{\overset{C_6H_5}{\diagdown}}C=\overset{O}{\underset{CH_3}{\overset{\|}{N}}} + \underset{H}{\overset{C_6H_5SO_2}{\diagdown}}C=C\overset{H}{\underset{C_6H_3}{\diagup}} \xrightarrow[toluene]{100°} \quad$$

$$(10.175)$$

$$53\%$$
$$\textbf{10.83}$$

$$RSO_2CH=CH_2 + \quad \bigg\backslash\!\!\!\diagup \quad \longrightarrow \quad$$

$$(10.176)$$

## 10.3.10 α-Diazo Sulfones

α-Diazosulfones have been known only since 1962.[269] They can be prepared by nitrosation of N-sulfonylmethyl urethanes, followed by treatment of the N-nitroso compound with base (eq. 10.178).[270-272] α-Diazo-β-keto sulfones,[273] or bis-sulfonyl diazomethane[274] is prepared from the corresponding β-keto sulfone (eq. 10.179) or disulfone and p-toluenesulfonyl azide in the presence of a base. The α-diazo-β-keto sulfones can subsequently be cleaved by triethylamine in dry methanol to give α-diazo sulfones (eq. 10.179).[275] Bis-sulfonyl diazomethanes may also be prepared from oxidation of the corresponding hydrazone, which is prepared from $Br_2C=N-N=CBr_2$ and the sulfinate salt (eq. 10.180).[276]

$$RSO_2H + CH_2O + H_2NCO_2Et \xrightarrow{\text{rt}} RSO_2CH_2NHCO_2Et \qquad (10.177)$$

$$RSO_2CH_2NHCO_2Et \xrightarrow{\text{NOCl}} \xrightarrow{\text{base}} RSO_2CHN_2 \qquad (10.178)$$

$$ArSO_2CH_2\overset{\overset{\displaystyle O}{\|}}{C}R + TsN_3 \xrightarrow{\text{base}} ArSO_2\overset{\overset{\displaystyle O}{\|}}{\underset{\underset{\displaystyle N_2}{\|}}{C}}CR \xrightarrow[\text{Et}_3\text{N}]{\text{MeOH}} ArSO_2CHN_2$$

$$(10.179)$$

$$(Br_2C=N)_2 + C_6H_5SO_2Na \longrightarrow (C_6H_5SO_2)_2C=N-NH_2$$
$$\xrightarrow{[O]} (C_6H_5SO_2)_2CN_2 \qquad (10.180)$$

α-Diazo sulfones are hydrolyzed in aqueous acid to give the α-hydroxy sulfones. These readily dissociate to the respective aldehyde and sulfinic acid.[277,278] If a nucleophile such as chloride ion is present, an α-substituted sulfone is formed.[278] Anhydrous HCl gives practically quantitative conversion to the α-chloro sulfone.[272]

$$C_6H_5CH_2SO_2CHN_2 \xrightarrow[\text{H}_2\text{O-dioxane}]{\text{HClO}^4} C_6H_5CH_2SO_2CH_2OH$$

$$\rightleftharpoons C_6H_5CH_2SO_2H + CH_2O$$

$$(10.181)$$

$$C_6H_5CH_2SO_2CHN_2 \xrightarrow[\text{H}_2\text{O-dioxane}]{\text{HCl}} C_6H_5CH_2SO_2CH_2Cl \quad (10.182)^{278}$$
$$91\%$$

Perchlorate or sulfonate esters can be isolated from reaction with perchloric or *p*-toluenesulfonic acid, respectively, in an inert solvent (eqs. 10.183 and 10.184).[279] α-Acylamido sulfones arise from reaction in acetonitrile or propionitrile.[279]

$$p\text{-}CH_3C_6H_4SO_2CHN_2 \xrightarrow[\text{ClCH}_2\text{CH}_2\text{Cl}]{70\% \text{ aq. HClO}_4} p\text{-}CH_3C_6H_4SO_2CH_2OClO_3$$
$$49\%$$
$$(10.183)$$
$$+ p\text{-}CH_3C_6H_4SO_2CH_2OH$$
$$10\%$$

$$p\text{-}CH_3C_6H_4SO_2CHN_2 \xrightarrow[\text{ClCH}_2\text{CH}_2\text{Cl}]{\text{TsOH}} p\text{-}CH_3C_6H_4SO_2CH_2OTs$$
$$72\%$$

Ts = *p*-toluenesulfonyl                                              (10.184)

$$p\text{-}CH_3C_6H_4SO_2CHN_2 \xrightarrow[\text{CH}_3\text{CN}]{70\% \text{ HClO}_4} ArSO_2CH_2\overset{\overset{\displaystyle O}{\displaystyle \|}}{N}HCCH_3 +$$
$$52\% \qquad (10.185)$$

$$ArSO_2CH_2OClO_3 + ArSO_2CH_2OH$$
$$13\% \qquad\qquad 13\%$$

If the α-diazo sulfone contains an internal nucleophile in the β′ position, cyclization will take place in the presence of $HBF_3OH$ when the reaction is run in the absence of any external nucleophile.[280]

Triphenylphosphine forms a 1:1 addition compound with α-diazo sulfones in which the nitrogen is retained, making these derivatives useful for structural characterization.[270,272] In some cases, these addition compounds are spontaneously hydrolyzed to triphenylphosphine oxide plus the hydrazone of an α-sulfonyl aldehyde.[275]

$$\text{(10.186)}$$

$$\text{(10.187)}$$

$$C_6H_5CH_2SO_2CHN_2 \xrightarrow{(C_6H_5)_3P} C_6H_5CH_2SO_2CH=N-N=P(C_6H_5)_3$$

$$93\% \qquad \text{(10.188)}$$

$$\text{(10.189)}$$

Like other diazo compounds, α-diazo sulfones will decompose photolytically to nitrogen plus the α-sulfonyl carbene. The carbene will add to an olefin to form a cyclopropane,[281] or to an alcohol to form an α-sulfonyl ether.[276,281]

$$p\text{-}CH_3OC_6H_4SO_2CHN_2 + CH_2=C\begin{array}{c}CH_3\\ \\CH_3\end{array} \xrightarrow{h\nu} p\text{-}CH_3OC_6H_4SO_2CH\begin{array}{c}CH_2\\ | \\ C-CH_3\\ CH_3\end{array}$$

$$\text{(10.190)}$$

$$p\text{-}CH_3OC_6H_4SO_2CHN_2 \xrightarrow[CH_3OH]{h\nu} p\text{-}CH_3OC_6H_4SO_2CH_2OCH_3$$

$$\text{(10.191)}$$

A few cases have been observed in which, in analogy with the Wolff rearrangement, the carbene undergoes rearrangement to form in low yield an α-sulfonyl ketene from an α-diazo-β-keto sulfone,[282] or a sulfene from an unsubstituted α-diazo sulfone.[283]

$$p\text{-}CH_3C_6H_4SO_2\overset{\overset{\displaystyle N_2}{\|}}{\underset{\underset{\displaystyle O}{\|}}{C}}CCH_3 \xrightarrow[hv]{EtOH} [p\text{-}CH_3C_6H_4SO_2\underset{\underset{\displaystyle CH_3}{|}}{C}=C=O] \longrightarrow$$

$$(10.192)$$

$$p\text{-}CH_3C_6H_4SO_2\underset{\underset{\displaystyle CH_3}{|}}{CH}\overset{\overset{\displaystyle O}{\|}}{C}\text{-}OEt$$
$$20\%$$

$$p\text{-}CH_3OC_6H_4SO_2CHN_2 \xrightarrow[hv\ -10°]{CH_3OH}$$

$$[p\text{-}CH_3OC_6H_4SO_2\ddot{C}H] \longrightarrow [p\text{-}CH_3OC_6H_4CH{=}SO_2]$$

$$\downarrow \qquad\qquad\qquad\qquad\qquad \downarrow$$

$$p\text{-}CH_3OC_6H_4SO_2CH_2OCH_3 + p\text{-}CH_3OC_6H_4CH_2SO_2OCH_3 \qquad (10.193)$$
$$78\% \qquad\qquad\qquad 12\%$$

## 10.4 SULFOXIMINES

Sulfoximines are compounds of sulfur in the same oxidation state as sulfones, but with one oxygen replaced by an =NH group (10.84). These compounds are also known as sulfone imines, a name which would eliminate the ambiguity concerning their oxidation state relative to sulfoxides and sulfones.[284]

Sulfoximines are stable compounds, similar in physical properties to sulfoxides, the sulfoximines being more basic.[285] Like sulfoxides, sulfoximines may possess an asymmetric center at sulfur.[286] These have been correlated with optically active sulfoxides and sulfilimines.[287]

Various sulfoximines have been shown to possess physiological activity.[288]

$$\begin{array}{cc} R_1 & O \\ \diagdown & \diagup\!\!\!\diagup \\ & S \\ \diagup & \diagdown\!\!\!\diagdown \\ R_2 & NH \end{array}$$

**10.84**

## 10.4.1 Preparations of Sulfoximines

The first sulfoximine was prepared by treatment of zein or wheat flour with $NCl_3$, giving methionine sulfoximine,[288] a reaction which subsequently was shown to occur with methionine sulfoxide but not with methionine.[289]

A better preparation of sulfoximines from sulfoxides involves the use of hydrazoic acid in sulfuric acid (eq. 10.194).[285] The yields in this reaction are generally 20–50%. Sulfides with an excess of hydrazoic acid will give the sulfoximine directly, although in some cases hydrolysis to the sulfone occurs.[285,290]

$$(C_2H_5)_2SO + HN_3 \xrightarrow[\text{CHCl}_3]{\text{H}_2\text{SO}_4} \underset{49\%}{(C_2H_5)_2S\overset{O}{\underset{NH}{\lVert}}} \qquad (10.194)$$

$$\underset{HO_2C}{\overset{H_2N}{\diagdown}}CHCH_2CH_2SCH_3 + \qquad (10.195)$$

$$\text{excess } HN_3 \xrightarrow[\text{CHCl}_3]{\text{con. H}_2\text{SO}_4} \underset{HO_2C}{\overset{H_2N}{\diagdown}}CHCH_2CH_2\overset{O}{\underset{NH}{\overset{\lVert}{\underset{\lVert}{S}}}}CH_3$$

$$27\%$$

$$(CH_3)_2S + \text{excess } HN_3 \xrightarrow[\text{CHCl}_3]{\text{con. H}_2\text{SO}_4} (CH_3)_2SO + (CH_3)_2SO_2$$

$$(10.196)$$

Alternatively, N-substituted sulfilimines may be oxidized to sulfoximines with $m$-chloroperbenzoic acid,[287] or with $KMnO_4$[291-293] in good yield. These oxidations proceed with retention of configuration, giving products of greater than 95% optical purity when optically pure sulfilimines are used.[287,291] The sulfoximine group is resistant to further reaction with the oxidizing agent, so that other groups in the molecule can be oxidized.[292]

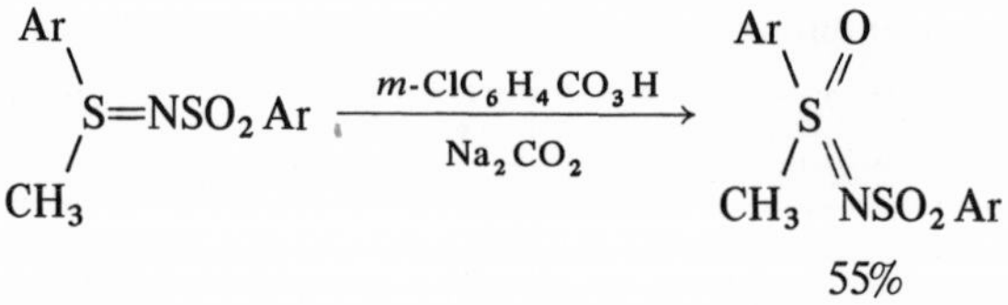

$$\underset{\underset{CH_3}{|}}{\overset{\overset{Ar}{|}}{S}}=NSO_2\,Ar \xrightarrow[\text{Na}_2\text{CO}_2]{m\text{-ClC}_6\text{H}_4\text{CO}_3\text{H}} \underset{\underset{CH_3\quad NSO_2\,Ar}{|}}{\overset{\overset{Ar\quad O}{|\;\;\;\nparallel}}{S}} \qquad (10.197)$$

$$55\%$$

Ar = *p*-tolyl

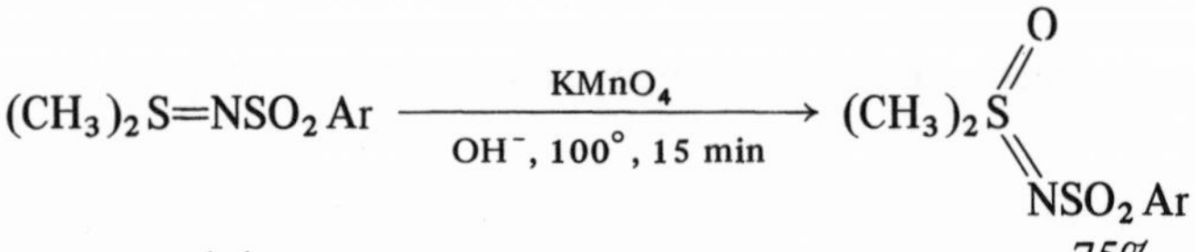

$$(CH_3)_2\,S=NSO_2\,Ar \xrightarrow[\text{OH}^-,\,100°,\,15\text{ min}]{KMnO_4} (CH_3)_2\underset{NSO_2\,Ar}{\overset{O}{S}} \qquad (10.198)$$

Ar = *p*-tolyl                                                            75%

$$(CH_3)_2\,\underset{NSO_2}{\overset{O}{S}}\!\!-\!\!\underbrace{\phantom{xxx}}\!\!-\!CH_3 \xrightarrow{\;KMnO_4\;} (CH_3)_2\,\underset{NSO_2}{\overset{O}{S}}\!\!-\!\!\underbrace{\phantom{xxx}}\!\!-\!CO_2H \qquad (10.199)$$

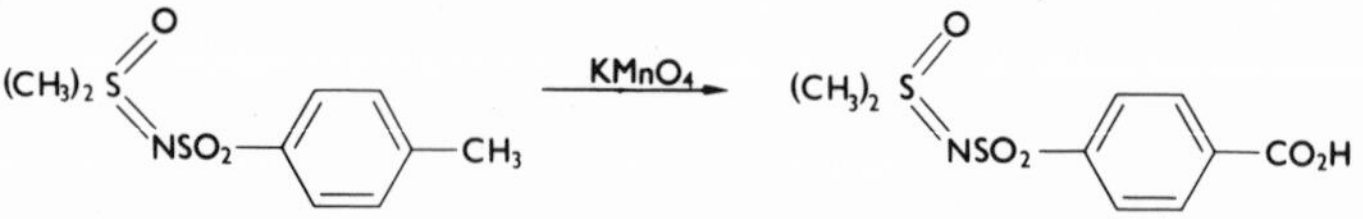

Sulfides can also be converted to sulfoximines with sulfonyl or acyl azides, either photolytically or pyrolytically.[287,294-297] A nitrene is the probable intermediate. In the copper-catalyzed reaction (eq. 10.203),[287,291] greater than 90% optical purity is maintained.

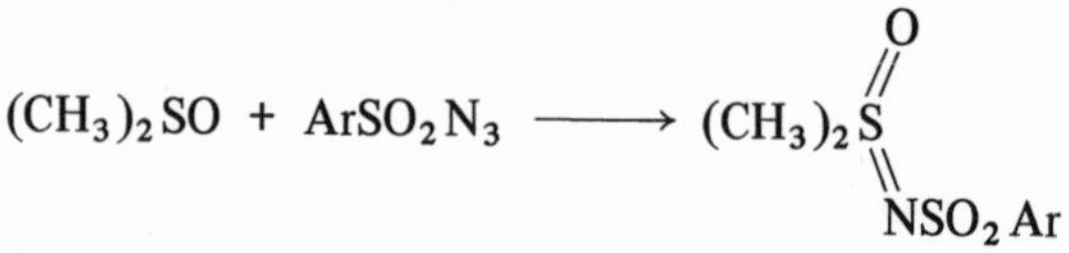

$$(CH_3)_2\,SO + ArSO_2N_3 \longrightarrow (CH_3)_2\underset{NSO_2\,Ar}{\overset{O}{S}} \qquad (10.200)^{[294]}$$

Ar = *p*-tolyl

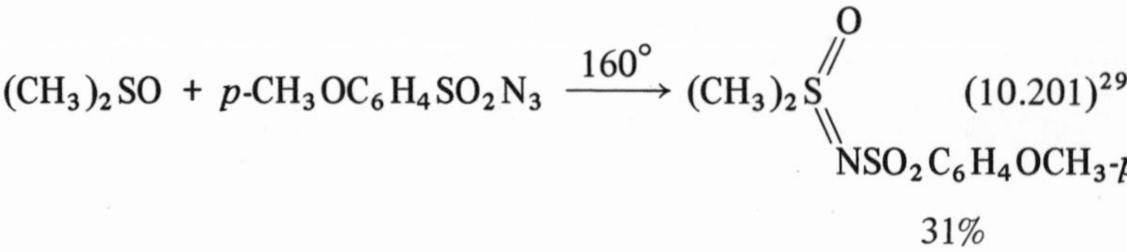

$$(CH_3)_2\,SO + p\text{-CH}_3\text{OC}_6\text{H}_4\text{SO}_2\text{N}_3 \xrightarrow{160°} (CH_3)_2\underset{NSO_2\,C_6H_4OCH_3\text{-}p}{\overset{O}{S}} \qquad (10.201)^{[294]}$$

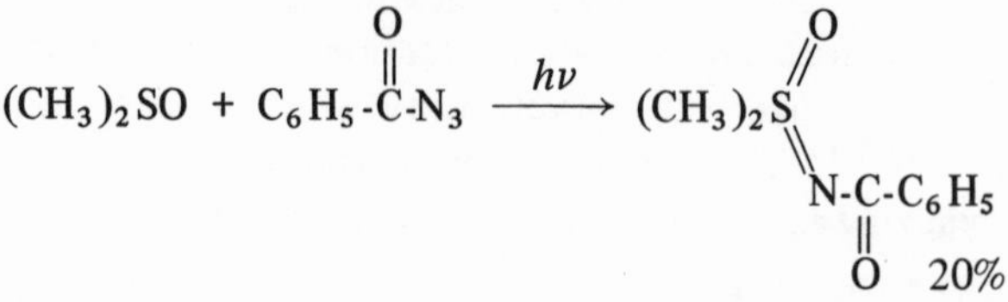

$$31\%$$

$$(CH_3)_2\,SO + C_6H_5\text{-}\overset{\overset{O}{\parallel}}{C}\text{-}N_3 \xrightarrow{h\nu} (CH_3)_2\underset{\underset{O\quad\;\;}{\overset{}{N\text{-}C\text{-}C_6H_5}}}{\overset{O}{S}} \qquad (10.202)^{[295]}$$

$$20\%$$

$$\underset{\substack{\diagup \\ Ar}}{\overset{\substack{CH_3 \\ \diagdown}}{S}}O \;+\; ArSO_2N_3 \;\xrightarrow[\Delta]{Cu}\; \underset{\substack{\diagup \quad \diagdown \\ Ar \quad NSO_2Ar}}{\overset{\substack{CH_3 \quad O \\ \diagdown \quad \diagup}}{S}} \qquad\qquad (10.203)^{296}$$

$$65\%$$

Ar = p-tolyl

$$(CH_3)_2SO \;+\; \text{chloroamine-T} \;\xrightarrow[80°]{Cu}\; (CH_3)_2\underset{NSO_2Ar}{\overset{O}{S}} \qquad (10.204)^{297}$$

$$80\%$$

A similar reaction, which is thought to go through an acyl nitrene, involves thermolysis of $\Delta^2$-1,4,2-dioxazolinones.[298]

$$(CH_3)_2SO \;+\; \underset{\substack{O \quad O}}{\overset{N-O}{Ar-C \diagdown \quad \diagup C}} \;\xrightarrow{150°}\; (CH_3)_2\underset{\underset{O}{N-C-Ar}}{\overset{O}{S}} \;+\; C_6H_5NHCNHC_6H_5$$

$$(10.205)$$

| | | |
|---|---|---|
| Ar = $C_6H_5$ | 51% | 18% |
| Ar = $p$-$O_2NC_6H_4$ | 90% | 0% |

Cyclic aminonitrenes, prepared by oxidation of the corresponding hydrazines with lead tetraacetate, react with dimethyl sulfoxide to give N-amino sulfoximines. This reaction can be reversed photolytically, giving the sulfoxide and nitrene.[299]

$$(10.206)$$

$$75\%$$

Sulfoximines are sometimes formed in low yield as a side product of sulfone diimine preparation from sulfides and chloramine.[284]

$$R\text{-S-CH}_3 \xrightarrow[\text{(CH}_3)_2\text{CHOH}]{\text{ClNH}_2} \underset{\substack{\diagup \; \diagdown \\ \text{CH}_3 \;\; \text{NH}}}{\overset{\substack{\text{R} \quad \text{NH} \\ \diagdown \; /\!\!/}}{\text{S}}} + \underset{\substack{\diagup \; \diagdown \\ \text{CH}_3 \;\; \text{NH}}}{\overset{\substack{\text{R} \quad \text{O} \\ \diagdown \; /\!\!/}}{\text{S}}} \qquad (10.207)$$

R = dodecyl

$\qquad\qquad\qquad$ 44%$\qquad$ 6%

### 10.4.2 Reactions of Sulfoximines

Sulfoximines can be converted to sulfones by treatment with nitrous acid,[285] hydrogen peroxide,[289,300] or by prolonged hydrolysis in hot concentrated acid or base.[300]

Reduction to the sulfoxide occurs in low yield with complete retention of optical activity in the presence of one equivalent of nitrosyl hexafluorophosphate in nitromethane.[287]

$$\underset{\substack{\diagup \; \diagdown \\ \text{Ar} \;\; \text{NH}}}{\overset{\substack{\text{CH}_3 \;\; \text{O} \\ \diagdown \; /\!\!/}}{\text{S}}} \xrightarrow[\text{CH}_3\text{NO}_2]{\text{NO}^+\text{PF}_6^-} \underset{\text{Ar}}{\overset{\text{CH}_3}{\diagdown}} \text{SO} \qquad (10.208)$$

Ar = *p*-tolyl $\qquad\qquad\qquad$ 20%

The imine group is amphoteric, thus allowing a variety of condensations to take place at the nitrogen atom of sulfoximines. Acylation and sulfonylation occur readily (eqs. 10.209 and 10.210). Deacylation, likewise, goes in high yield with concentrated sulfuric acid at room temperature.[287]

$$(\text{CH}_3)_2\overset{\substack{\text{O} \\ /\!\!/}}{\underset{\backslash\backslash}{\text{S}}} + \text{C}_6\text{H}_5\text{-C-Cl} \xrightarrow{\text{pyridine}} (\text{CH}_3)_2\overset{\substack{\text{O} \\ /\!\!/}}{\underset{\backslash\backslash}{\text{S}}} \qquad (10.209)^{[294]}$$

$$\text{NH} \qquad\qquad\qquad\qquad \underset{\substack{\text{O} \quad 74\%}}{\overset{}{\text{N-C-C}_6\text{H}_5}}$$

$$\underset{\substack{\diagup \; \diagdown \\ \text{Ar} \;\; \text{NH}}}{\overset{\substack{\text{CH}_3 \;\; \text{O} \\ \diagdown \; /\!\!/}}{\text{S}}} \xrightarrow[]{\text{Na}} \xrightarrow{\text{ArSO}_2\text{Cl}} \underset{\substack{\diagup \; \diagdown \\ \text{Ar} \;\; \text{NSO}_2\text{Ar}}}{\overset{\substack{\text{CH}_3 \;\; \text{O} \\ \diagdown \; /\!\!/}}{\text{S}}} \qquad (10.210)^{[287]}$$

$$\qquad\qquad\qquad\qquad\qquad\qquad 62\%$$

The imine group can be metallated with sodium[285] or with *n*-butyllithium,[301] and can be N-halogenated with sodium hypochlorite[287] or bromine in ether.[302]

Metallated sulfoximines condense with trialkyl[301] or triaryl[302] phosphine dihalides to form N-(trialkyl- or triarylphosphoranylidene) dialkyl sulfone iminium halides.[301] These can be hydrolyzed to the corresponding phosphine oxide and sulfoximine in dilute acid or base.[301] Similarly, N-halo sulfoximines react with triphenylphosphine or dimethyl sulfide to form an addition salt.[302]

$$(CH_3)_2\overset{\overset{O}{\parallel}}{\underset{\parallel}{S}}_{NH} \xrightarrow{n\text{-BuLi}} \xrightarrow{R(CH_3)_2PCl_2} (CH_3)_2\overset{\overset{O}{\parallel}}{\underset{\parallel}{S}}_{{}^+N=P(CH_3)_2R} \quad Cl^-$$
R = dodecyl
63%

$$(10.211)$$

$$(CH_3)_2\overset{\overset{O}{\parallel}}{\underset{\parallel}{S}}_{NH} \xrightarrow{Br_2} \xrightarrow{P(C_6H_5)_3} (CH_3)_2\overset{\overset{O}{\parallel}}{\underset{\parallel}{S}}_{{}^+N=P(C_6H_5)_3} \quad Br^-$$
36%

$$+ \quad (CH_3)_2\overset{\overset{O}{\parallel}}{\underset{\parallel}{S}}_{{}^+NH_2} \quad Br^- \qquad (10.212)$$
27%

$$(CH_3)_2\overset{\overset{O}{\parallel}}{\underset{\parallel}{S}}_{NH} \xrightarrow{Br_2} \xrightarrow{(CH_3)_2S} (CH_3)_2\overset{\overset{O}{\parallel}}{\underset{\parallel}{S}}_{{}^+N=S(CH_3)_2} \quad Br^- \qquad (10.213)$$
17.5%

The sulfoximine N-H also adds to alkyl and alkylsulfonyl isocyanates, alkyl and acyl isothiocyanates, isocyanic acid, and phosgene in good yield.[303]

$$(CH_3)_2\overset{\displaystyle O}{\underset{\displaystyle NH}{S}} \;+\; C_2H_5N=C=O \;\longrightarrow\; (CH_3)_2\overset{\displaystyle O}{\underset{\displaystyle \underset{\overset{\|}{O}\quad 85\%}{N\text{-}C\text{-}NHC_2H_5}}{S}} \qquad (10.214)$$

$$(CH_3)_2\overset{\displaystyle O}{\underset{\displaystyle NH}{S}} \;+\; C_6H_5N=C=S \;\longrightarrow\; (CH_3)_2\overset{\displaystyle O}{\underset{\displaystyle \underset{\overset{\|}{S}\quad 85\%}{N\text{-}C\text{-}NHC_6H_5}}{S}} \qquad (10.215)$$

$$(CH_3)_2\overset{\displaystyle O}{\underset{\displaystyle NH}{S}} \;+\; COCl_2 \;\xrightarrow{\;Et_3N\;}\; (CH_3)_2\overset{\displaystyle O}{S}\underset{\displaystyle \underset{\overset{\|}{O}}{N\text{-}C\text{-}N}}{}\overset{\displaystyle O}{S}(CH_3)_2 \qquad 45\% \qquad (10.216)$$

Sulfoximines are readily alkylated with trimethyloxonium fluoroborate to give, after basic workup, N-methyl sulfoximines.[304,305] Electrophilic olefins with a catalytic amount of sodium hydride also give N-alkylation.[304] A second alkyl group can be attached to the nitrogen by means of trimethyloxonium fluoroborate, thus giving a nitrogen analogue of a trialkylsulfoxonium salt.[304,305]

$$\underset{CH_3\;\;NH}{\overset{C_6H_5\;\;O}{S}} \;+\; CH_2{=}CHCO_2CH_3 \;\xrightarrow{\;Cat.\ NaH\;}\; \underset{CH_3\;\;NCH_2CH_2CO_2CH_3}{\overset{C_6H_5\;\;O}{S}} \qquad (10.217)$$

$$\underset{CH_3\;\;NCH_3}{\overset{C_6H_5\;\;O}{S}} \;+\; (CH_3)_3O^+BF_4^- \;\longrightarrow\; \underset{CH_3\;\;N(CH_3)_2}{\overset{C_6H_5\;\;O}{S^+}} \qquad BF_4^- \qquad (10.218)$$

The N,N-dialkyl salts react with methoxide ion to give a mixture of sulfone, sulfinamide and methyl sulfonate.[305] Sodium hydride with these salts, however, produces ylides similar to sulfoxonium ylides. These ylides can be C-acylated with acyl halides, or will add to isocyanates. They also react with aldehydes, ketones, imines, or electrophilic olefins to form the corresponding epozides, aziridines or cyclopropanes in good yield.[305-307]

$$\underset{\substack{| \\ CH_3}}{\overset{\substack{C_6H_5 \quad O \\ \diagdown \quad \diagup\diagup}}{S^+}} BF_4^- \xrightarrow{CH_3O^-} C_6H_5\text{-}\underset{\substack{\| \\ }}{\overset{\substack{O \\ \|}}{S}}\text{-}N(CH_3)_2 + C_6H_5\text{-}\underset{\substack{\| \\ O}}{\overset{\substack{O \\ \|}}{S}}\text{-}OCH_3$$

$$+ \ C_6H_5SO_2CH_3$$

$$(10.219)$$

$$\underset{\substack{| \quad \backslash \\ CH_3 \ N(CH_3)_2}}{\overset{\substack{C_6H_5 \ O \\ \backslash \ \diagup\diagup}}{S^+}} BF_4^- \xrightarrow[\text{DMSO}]{\text{NaH}} \underset{\substack{| \quad \backslash \\ {}^-CH_2 \ N(CH_3)_2}}{\overset{\substack{C_6H_5 \ O \\ \backslash \ \diagup\diagup}}{S^+}} + \ H_2 \qquad (10.220)$$

$$\underset{\substack{{}^-CH_2 \ N(CH_3)_2}}{\overset{\substack{C_6H_5 \ O \\ \backslash\diagup\diagup}}{S^+}} + \ C_6H_5\overset{\substack{O \\ \|}}{C}\text{-}Cl \longrightarrow \underset{\substack{{}^-CH \quad N(CH_3)_2 \\ | \\ C_6H_5\text{-}C \\ \diagdown \\ O}}{\overset{\substack{C_6H_5 \ O \\ \diagdown\diagup}}{S^+}} \qquad (10.221)$$

$$70\%$$

$$\underset{\substack{{}^-CH_2 \ N(CH_3)_2}}{\overset{\substack{C_6H_5 \ O \\ \backslash\diagup\diagup}}{S^+}} + \ C_6H_5NCO \longrightarrow \underset{\substack{{}^-CH \quad N(CH_3)_2 \\ | \\ C_6H_5\text{-}NH\text{-}C \\ \diagdown \\ O}}{\overset{\substack{C_6H_5 \ O \\ \diagdown\diagup}}{S^+}} \qquad (10.222)$$

$$\underset{\substack{{}^-CH_2 \quad N(CH_3)_2}}{\overset{\substack{C_6H_5 \quad O \\ \diagdown\quad\diagup\diagup}}{S^{+\cdot}}} + \ \underset{\substack{H \quad COC_6H_5}}{\overset{\substack{C_6H_5CO \quad H}}{C=C}} \longrightarrow \underset{\substack{H \quad 80\% \quad COC_6H_5}}{\overset{\substack{C_6H_5CO \quad H}}{\triangle}}$$

$$(10.223)$$

$$\underset{\substack{\displaystyle\overset{|}{C}\text{H}_3 \\ \displaystyle\overset{|}{C}\text{H}_3 }}{\underset{\displaystyle\overset{|}{\text{–C}}\quad\text{N(CH}_3)_2}{\overset{\displaystyle\text{C}_6\text{H}_5\diagdown\diagup\text{O}}{\text{S}^+}}} \quad + \quad \underset{\substack{\text{H}\quad\text{COC}_6\text{H}_5}}{\overset{\text{C}_6\text{H}_5\quad\text{H}}{\text{C=C}}} \quad\longrightarrow\quad \underset{\substack{\text{H}\qquad\qquad\text{COC}_6\text{H}_5 \\ 60\%}}{\overset{\text{CH}_3\qquad\text{CH}_3}{\text{C}_6\text{H}_5 \diagup\!\!\!\!\diagdown\text{H}}}$$

$$(10.224)$$

$$\underset{\substack{\displaystyle\overset{|}{}\quad\diagdown \\ \text{–CH}_2\quad\text{N(CH}_3)_2}}{\overset{\text{C}_6\text{H}_5\diagdown\diagup\text{O}}{\text{S}^+}} \quad + \quad p\text{-ClC}_6\text{H}_4\text{CHO} \quad\longrightarrow\quad \underset{60\%}{p\text{-ClC}_6\text{H}_4\text{CH-CH}_2}\overset{\text{O}}{\diagup\diagdown} \qquad (10.225)$$

These ylides are more stable and prepared under milder conditions than the corresponding sulfoxonium ylides, and are thought to be more generally useful than the latter. Optically active ylides are also possible, and may make asymmetric induction possible in cyclopropane synthesis.[305]

## 10.5 REFERENCES

1. C.M. Sutter, "The Organic Chemistry of Sulfur," J. Wiley and Sons, Inc., 1944, p. 683.
2. R.C. Weast (Editor), "Handbook of Chemistry and Physics," 49th Ed., Chemical Rubber Publishing Co., 1968.
3. C.C. Price and S. Oae, "Sulfur Bonding," The Ronald Press Company, New York, 1962, pp. 61–128.
4. N. Kharasch (Editor), "Organic Sulfur Compounds," Vol. 1, Pergamon Press, New York, 1961, pp. 47–74.
5. L.A. Carpino and L.V. McAdams, III, *J. Amer. Chem. Soc.*, 87, 5804 (1965); *ibid.*, 93, in press (1971).
6. M. Portelli and B. Soranza, *Ann. Chim.* (Rome), 52, 1280 (1962); *Chem. Abstr.*, 59, 493 (1963).
7. E.N. Karaulova, G.D. Gal'pern, and T.A. Bardina, *Dokl. Akad. Nauk SSSR*, 173, 104 (1967); *Chem. Abstr.*, 67, 53578 (1967).
8. H.S. Schultz, H.B. Freyermuth, and S.R. Buc, *J. Org. Chem.*, 28, 1140 (1963).
9. R. Curci, A. Giovine, and G. Modena, *Tetrahedron*, 22, 1235 (1966).
10. G. Modena and P.E. Todesco, *Boll. Sci. Fac. Chim. Ind. Bologna*, 23, 31 (1965); *Chem. Abstr.*, 63, 8168 (1965).
11. R. Curci and G. Modena, *Tetrahedron Lett.*, 863 (1965).
12. A. Greco, G. Modena, and P.E. Todesco, *Gazz. Chim. Ital.*, 90, 671 (1960); *Chem. Abstr.*, 55, 16510 (1961).
13. R. Sowada, *Zeit. Chem.*, 8, 361 (1968).
14. I. Seree de Roch and P. Menguy, Fr. 1,540,284, Sept. 27 (1968); *Chem. Abstr.*, 71, 80944 (1969).
15. L. Kuhnen, *Angew. Chem., Int. Ed. Eng.*, 5, 893 (1966).
16. L. Skatteboel, B. Boulette, and S. Solomon, *J. Org. Chem.*, 32, 3111 (1967).
17. A.E. Wood and E.G. Travis, *J. Amer. Chem. Soc.*, 50, 1226 (1928).
18. W.D. Kingsbury and C.R. Johnson, *Chem. Commun.*, 365 (1969).
19. V.I. Dronov, A.U. Baisheva, A.E. Pototskaya, and L.M. Soskova, *Khim. Seraorgon. Soedin., Soderzhasch Neft. Nefteprod., Akad. Nauk SSSR, Bashkirsh.*

*Filial*, 7, 40 (1964); *Chem. Abstr.*, **63**, 4235 (1965).
20. G. Barberi, M. Cinquini, S. Colonna, and F. Montanari, *J. Chem. Soc.*, C, 659 (1968).
21. D.W. Goheen and C.F. Bennett, *J. Org. Chem.*, **26**, 1331 (1961).
22. A. Compagnini, M. Santagati, N. Marziano, and R. Passerini, *Biol. Sedute Accad. Gioenia Sci. Natur. Catania*, [4] **9**, 585 (1969); *Chem. Abstr.*, **72**, 43038 (1970).
23. A. Maggiolo and Blair, *Advances in Chem. Ser.* No. **21**, 200 (1959).
24. D. Barnard, *J. Chem. Soc.*, 4547 (1957).
25. L.J. Hughes, T.D. McMinn, Jr., and J.C. Burleson, U.S. 3,114,775, Dec. 17. 1963; *Chem. Abstr.*, **60**, 6751 (1964).
26. J. Trocha-Grimshaw and H.B. Henbest, *Chem. Commun.*, 1035 (1968).
27. P.S. Makoveev and A.V. Mashkina, *Neftekhimiya*, 302 (1969); *Chem. Abstr.*, **71**, 38195 (1969).
28. G.O. Schenck and C.H. Krauch, *Chem. Ber.*, **96**, 517 (1963).
29. K.B. Wiberg, "Oxidation in Organic Chemistry," Vol. 5A, Academic Press, 1965, p. 63.
30. H.B. Henbest and S.A. Khan, *Chem. Commun.*, 1036 (1968).
31. H.E. Armstrong, *Chem. Ber.*, **7**, 407 (1874).
32. C. Djerassi and R.R. Engle, *J. Amer. Chem. Soc.*, **75**, 3838 (1953).
33. N.M. Mel'nikov, *Uspekhi Khim.*, **5**, 443 (1936); *Chem. Abstr.*, **30**, 5180 (1936).
34. C.F. Bennett and D.W. Goheen, U.S. 3,418,224, Dec. 24, 1968; *Chem. Abstr.*, **70**, 43434 (1969).
35. P.T. Cottrell and C.K. Mann, *J. Electrochem. Soc.*, **116**, 1499 (1969).
36. G. Olah (Editor), "Friedel-Crafts and Related Reactions," Vol. 3, part 2, Interscience Publishers, New York, 1964, pp. 1319-1354.
37. F.R. Jensen and H.C. Brown, *J. Amer. Chem. Soc.*, **80**, 4042 (1958).
38. V.A. Koptyug, T.N. Gerasimova, and V.N. Vorozhtsov, Jr., *Khim Nauka, i Prom.*, **4**, 414 (1959); *Chem. Abstr.*, **54**, 438 (1960).
39. W.H.C. Rueggeberg, T.W. Sauls, and S.L. Norwood, *J. Org. Chem.*, **20**, 455 (1955).
40. G. Fouque and J. Lacroix, *Bull. Soc. Chim.*, **33**, 180 (1923); *Chem. Abstr.*, **17**, 1958 (1923).
41. V.V. Kozlov, T.I. Vol'fson, N.A. Kozlova, and G.S. Tubyanskaya, *J. Gen. Chem.* (USSR) (Eng. tr.), **32**, 3373 (1962).
42. R. Joly, R. Bucourt, and J. Mathieu, *Rec. Trav. Chim. Pays-Bas*, **78**, 527 (1959); *Chem. Abstr.*, **54**, 4449 (1960).
43. A. Michael and A. Adair, *Chem. Ber.*, **10**, 583 (1877).
44. H. Meyer, *Annales*, **433**, 327 (1923).
45. H. Drews, S.M. Meyerson, and E.K. Fields, *Angew. Chem.*, **72**, 493 (1960).
46. D.A. Denton and H. Suschitzky, *J. Chem. Soc.*, 4741 (1963).
47. L. Velluz, R. Joly, and R. Bucourt, *Compt. Rend.*, **248**, 114 (1959); *Chem. Abstr.*, **53**, 17946 (1959).
48. P. Oxley, M.W. Partridge, T.D. Robson, and W.F. Short, *J. Chem. Soc.*, 763 (1946).
49. J.M. Kauffman, *J. Chem. Eng. Data*, **14**, 498 (1969).
50. J.S. Meek and J.S. Fowler, *J. Org. Chem.*, **33**, 3422 (1968).
51. A.H. Wragg, J.S. McFadyen, and T.S. Stevens, *J. Chem. Soc.*, 3603 (1958).
52. O.R. Hansen and R. Hammer, *Acta Chem. Scand.*, **7**, 1331 (1953).
53. O.R. Roblin, Jr., J.H. Williams, and G.W. Anderson, *J. Amer. Chem. Soc.*, **63**, 1930 (1941).
54. K. Bailey, B.R. Brown, and B. Chalmers, *Chem. Commun.*, 618 (1967).
55. W.E. Truce, E.M. Kreider, and W.W. Brand, *Org. Reactions*, **18**, 99 (1970).
56. O. Achmatowicz and J. Michalski, *Roczniki Chem.*, **30**, 243 (1956); *Chem. Abstr.*, **51**, 1064 (1957).
57. H. Bredereck and E. Bäder, *Chem. Ber.*, **87**, 129 (1954).
58. J.H. McNamara and P.S. Skell, "Abstracts of Papers of the 135th National Meeting of the American Chemical Society, 1959," 13L.
59. M. Asscher and D. Vofsi, *J. Chem. Soc.*, 4962 (1964).
60. C.T. Goralski, Ph.D. Thesis, Purdue University, 1969.
61. M. Asscher, D. Vofsi, and A. Katchalsky, Belg. 654,544 (1965); *Chem. Abstr.*, **65**, P 5404 (1966).

62. W.E. Truce and G.C. Wolf, *Chem. Commun.*, 150 (1969).
63. G.C. Wolf, Ph.D. Thesis, Purdue University, 1970.
64. H. Gilman, N.J. Beaber, and C.H. Myers, *J. Amer. Chem. Soc.*, 47, 2047 (1925).
65. H. Burton and W.A. Davy, *J. Chem. Soc.*, 528 (1948).
66. H. Gilman and R.E. Fothergill, *J. Amer. Chem. Soc.*, 51, 3501 (1929).
67. H. Gilman and J.F. Nelson, *Rec. Trav. Chim. Pays-Bas*, 55, 518 (1963).
68. F. Whitmore and N. Thurman, *J. Amer. Chem. Soc.*, 45, 1068 (1923).
69. Y. Shirota, T. Nagai, and N. Tokura, *Tetrahedron*, 25, 3193 (1969).
70. H. Fukuda, F.J. Frank, and W.E. Truce, *J. Org. Chem.*, 28, 1420 (1963).
71. B. Oddo, *Gazz. Chim. Ital.* 41, I, 11 (1911); *Chem. Abstr.*, 5, 2635 (1911).
72. L.A. Paquette and R.B. Houser, *J. Amer. Chem. Soc.*, 91, 3870 (1969).
73. J.L. Stratenus and E. Havinga, *Rec. Trav. Chim. Pays-Bas*, 85, 434 (1966).
74. B.K. Snell, *J. Chem. Soc.*, C, 2367 (1968).
75. S. Searles and S. Nukina, *Chem. Rev.*, 59, 1077 (1959).
76. T.N. Gerasimova, V.A. Bushmelev, and V.A. Koptyug, *Zh. Org. Khim.*, 1, 1667 (1965) [Eng. tr., p. 1690].
77. W. Bradley and J.D. Hannon, *Chem. Ind.* (London), 540 (1959).
78. W.E. Truce and L.K. Liu, *Mech. React. Sulfur Cmpds.*, 4 (1970).
79. G. Opitz, *Angew. Chem. Internat. Ed. Engl.*, 6, 107 (1967).
80. W.E. Truce, R.H. Bavry, and P.S. Bailey, Jr., *Tetrahedron Lett.*, 5651 (1968).
81. H. Staudinger and F. Pfenninger, *Chem. Ber.*, 49, 1941 (1916).
82. F.G. Bordwell, J.M. Williams, Jr., E.B. Hoyt, Jr., and B.B. Jarvis, *J. Amer. Chem. Soc.*, 90, 429 (1968).
83. Y. Ito, M. Okano and R. Oda, *Tetrahedron*, 23, 2137 (1967).
84. L.A. Carpino and R.H. Rynbrandt, *J. Amer. Chem. Soc.*, 88, 5682 (1966); L.A. Carpino, R.H. Rynbrandt, and J.W. Spiewak, *ibid*, 93, in press (1971).
85. J. Hamer, "1,4-Cycloaddition Reactions," Academic Press Inc., New York, 1967, p. 13.
86. W.L. Mock, *J. Amer. Chem. Soc.*, 88, 2857 (1966).
87. K. Wagner, Ger. 1,222,048, Aug. 4, 1966; H.W. Gibson and D.A. McKenzie, *J. Org. Chem.*, 35, 2994 (1970).
88. I.C. Popoff, J.L. Dever, and R. Gordon, *J. Org. Chem.*, 34, 1128 (1969).
89. C.M.M. da Silva Correa, A.S. Lindsay, and W.A. Waters, *J. Chem. Soc.*, C, 1872 (1968).
90. V.I. Dronov and V.A. Snegotskaya, *Khim. Seraorg. Soedin., Soderzh. Neftyakh Nefteprod.*, 8, 133 (1968); *Chem. Abstr.*, 71, 81066 (1969).
91. E.P. Kohler and H.A. Potter, *J. Amer. Chem. Soc.*, 57, 1316 (1935).
92. L. Field and J.W. McFarland, *ibid.*, 75, 5582 (1953).
93. H.D. Becker and G.A. Russell, *J. Org. Chem.*, 28, 1896 (1963).
94. W.E. Truce and K.R. Buser, *J. Amer. Chem. Soc.*, 76, 3577 (1954).
95. E.J. Corey and M. Chaykovsky, *ibid.*, 76, 5377 (1954).
96. G.P. Crowther and C.R. Hauser, *J. Org. Chem.*, 33, 2228 (1968).
97. D.F. Tavares and P.F. Vogt, *Can. J. Chem.*, 45, 1519 (1967).
98. J. Mallan and R.L. Bebb, *Chem. Rev.*, 69, 693 (1969).
99. W.E. Truce and M.F. Amos, *J. Amer. Chem. Soc.*, 73, 3013 (1951).
100. W.E. Truce and O.L. Norman, *ibid.*, 75, 6023 (1953).
101. W.E. Truce, W.J. Ray, Jr., O.L. Norman, and D.B. Eickemeyer, *ibid.*, 80, 3625 (1958).
102. F.G. Bordwell, R.H. Imes, and E.C. Steiner, *ibid.*, 89, 3905 (1967).
103. R.G. Pearson and R.L. Dillion, *ibid.*, 75, 2439 (1953).
104. D. Turnbull and S. Maron, *ibid.*, 65, 212 (1943); G.W. Wheland and J. Farr, *ibid.*, 65, 1433 (1943).
105. F.G. Bordwell and G.D. Cooper, *ibid.*, 74, 1058 (1952).
106. C.Y. Meyers, B. Cremonini, and L. Maioli, *ibid.*, 86, 2944 (1964).
107. D.J. Cram, "Fundamentals of Carbanion Chemistry," Academic Press, New York, 1965; M. Gresser, *Mech. React. Sulfur Cmpds.*, 4, 29 (1969); B.S. Thyagarajan, *ibid.*, 4, 115 (1969).
108. D.J. Cram, W.D. Nielsen, and B. Rickborn, *J. Amer. Chem. Soc.*, 82, 6415 (1960).
109. D.J. Cram, D.A. Scott, and W.D. Nielsen, *ibid.*, 83, 3696 (1961).
110. D.J. Cram, R.D. Trepka, and P. St. Janiak, *ibid.*, 88, 2749 (1966).
111. E.J. Corey and T.H. Lowry, *Tetrahedron Lett.*, 793 (1965).

112. E.J. Corey, H. Konig, and T.H. Lowry, *ibid.*, 515 (1962); E.J. Corey and T.H. Lowry, *ibid.*, 803 (1965).
113. F.G. Bordwell, D.D. Phillips, and J.M. Williams, Jr., *J. Amer. Chem. Soc.*, **90**, 426 (1968).
114. F.G. Bordwell, B.B. Jarvis, and P.W.R. Corfield, *ibid.*, **90**, 5298 (1968).
115. S. Wolfe, A. Rauk, and I.G. Csizmadia, *ibid.*, **91**, 1567 (1969).
116. R.R. Fraser and F.J. Schuber, *J. Chem. Soc.*, D, 1474 (1969).
117. R.H. Bavry, Ph.D. Thesis, Purdue University, 1969.
118. L. Field, *J. Amer. Chem. Soc.*, **74**, 2920 (1935).
119. E.M. Kaiser and C.R. Hauser, *Tetrahedron Lett.*, 3341 (1967).
120. W.E. Truce and T.C. Klingler, *J. Org. Chem.*, **35**, 1834 (1970).
121. E.M. Kaiser and C.R. Hauser, *J. Amer. Chem. Soc.*, **89**, 4566 (1967).
122. E.J. Corey and T.H. Lowry, *Tetrahedron Lett.*, 793 (1965).
123. E. Rothstein, *J. Chem. Soc.*, 684 (1934).
124. H.J. Backer, *et al.*, *Rec. Trav. Chim. Pays-Bas.* **70**, 365 (1951).
125. F. Bohlmann and G. Haffer, *Chem. Ber.*, **102**, 4017 (1969).
126. J.W. McFarland and D.N. Buchanan, *J. Org. Chem.*, **30**, 2003 (1965).
127. J.W. McFarland and G.N. Coleman, *ibid.*, **35**, 1194 (1970).
128. V. Baliah and Sp. Shanmuganathan, *ibid.*, **23**, 1233 (1958).
129. E.P. Kohler and M. Tishler, *J. Amer. Chem. Soc.*, **57**, 217 (1935).
130. E.P. Kohler and H.A. Potter, *ibid.*, **58**, 2166 (1936).
131. W.E. Truce and R.H. Knopse, *ibid.*, **77**, 5063 (1955).
132. M.L. Miles and C.R. Hauser, *J. Org. Chem.*, **29**, 2329 (1964).
133. H.O. House and J.K. Larson, *ibid.*, **33**, 61 (1968).
134. G.A. Russell, E.T. Sabourin, and G. Hamprecht, *ibid.*, **34**, 2339 (1969).
135. W.E. Truce, T.C. Klingler, J.E. Parr, H. Feuer, and D.K. Wu, *ibid.*, **34**, 3104 (1969).
136. H. Gilman and F.J. Webb, *J. Amer. Chem. Soc.*, **71**, 4062 (1949).
137. E.A. Lehto and D.A. Schirley, *J. Org. Chem.*, **22**, 989 (1957).
138. H. Nozaki, Y. Yamamoto, and T. Nisimura, *Tetrahedron Lett.*, 4625 (1968).
139. W.M. Ziegler and R. Connor, *J. Amer. Chem. Soc.*, **62**, 2596 (1940).
140. C.Y. Meyers, A.M. Malte, and W.S. Matthews, *ibid.*, **91**, 7510 (1969).
141. B. Zwanenburg and J. ter Wiel, *Tetrahedron Lett.*, 935 (1970).
142. P.F. Vogt and D.F. Tavares, *Can. J. Chem.*, **47**, 2875 (1969).
143. W.E. Truce, unpublished results.
144. H.E. Zimmerman and B.S. Thyagarajan, *J. Amer. Chem. Soc.*, **82**, 2505 (1960).
145. A. Ratajczak, F.A.L. Anet, and D.J. Cram, *ibid.*, **89**, 2072 (1967).
146. W.E. Truce and L.B. Lindy, *J. Org. Chem.*, **26**, 1463 (1961).
147. W.E. Truce and C.T. Goralski, *ibid.*, **34**, 3324 (1969).
148. E. Rothstein, *J. Chem. Soc.*, 1560 (1940).
149. R. Bud and C.J.M. Sterling, *J. Chem. Soc.*, B, 111 (1968).
150. A. Nickon and N.H. Werstillk, *J. Amer. Chem. Soc.*, **89**, 3914 (1967).
151. S.J. Cristol, J.K. Harrington, and M.S. Singer, *ibid.*, **88**, 1529 (1966).
152. W.E. Truce, K.R. Hollister, L.B. Lindy, and J.E. Parr, *J. Org. Chem.*, **33**, 43 (1968).
153. F.G. Bordwell and G.D. Cooper, *J. Amer. Chem. Soc.*, **73**, 5184 (1951).
154. L. Ramberg and B. Bäcklund, *Arkiv. Kemi. Mineral Geol.*, **13A**, No. 27 (1940); *Chem. Abstr.*, **34**, 4725 (1940).
155. L.A. Paquette and J.C. Philips, *Tetrahedron Lett.*, 4645 (1967).
156. L.A. Paquette, *J. Amer. Chem. Soc.*, **86**, 4085 (1964).
157. F.G. Bordwell, in "Organosulfur Chemistry," M.J. Janssen (Editor), Interscience Publishers, New York, 1967, Chapter 16.
158. F.G. Bordwell, J.M. Williams, Jr., and B.B. Jarvis, *J. Org. Chem.*, **33**, 2026 (1968).
159. L.A. Paquette, L.S. Wittenbrook, and V.V. Kane, *J. Amer. Chem. Soc.*, **89**, 4487 (1967).
160. L.A. Paquette and L.S. Wittenbrook, *ibid.*, **90**, 6790 (1968).
161. F.G. Bordwell and J.M. Williams, Jr., *ibid.*, **90**, 435 (1968).
162. F.G. Bordwell and N.P. Neureiter, *ibid.*, **85**, 1209 (1963).
163. (a) N.P. Neureiter, *ibid.*, **88**, 558 (1966), (b) L.A. Paquette and L.S. Wittenbrook, *ibid.*, **90**, 6783 (1968).
164. A. Streitwieser, Jr., "Solvolytic Displacement Reactions," McGraw-Hill Book

Co., Inc., New York, 1962, p. 30.
165. G. Opitz and H. Fischer, *Angew. Chem. Intern. Ed. Engl.,* **4,** 70 (1965).
166. N. Tokura, T. Nagai, and S. Matsumura, *J. Org. Chem.,* **31,** 349 (1966).
167. L.A. Paquette, *Accounts Chem. Res.,* **1,** 209 (1968).
168. R.B. Woodward and R. Hoffmann, *J. Amer. Chem. Soc.,* **87,** 395 (1965).
169. R. Hoffmann and R.B. Woodward, *ibid,* **87,** 2046 (1965).
170. D.C. Dittmer, G.C. Levy, and G E. Kuhlmann, *ibid.,* **91,** 2097 (1969).
171. B.A. Kent and S. Smiles, *J. Chem. Soc.,* 422 (1934).
172. W.J. Evans and S. Smiles, *ibid.,* 181 (1935).
173. W.E. Truce and W.J. Ray, Jr., *J. Amer. Chem. Soc.,* **81,** 481 (1959).
174. T. Takahashi and Y. Maki, *Chem. Pharm. Bull.* (Tokyo), **6,** 369 (1958); *Chem. Abstr.,* **53,** 9228 (1959).
175. R.R. Coats and D.T. Gibson, *J. Chem. Soc.,* 442 (1940).
176. J.D. Loudon, J.R. Robertson, J.N. Watson, and S.D. Aiton, *ibid.,* 55 (1950).
177. K. Florey and A.R. Restivo, *J. Org. Chem.,* **23,** 1018 (1958).
178. A. Levi, L.A. Warren, and S. Smiles, *J. Chem. Soc.,* 1490 (1933).
179. F. Galbraith and S. Smiles, *ibid.,* 1234 (1935).
180. K.G. Kleb, *Angew. Chem.,* **80,** 284 (1968).
181. B.T. Tozer and S. Smiles, *J. Chem. Soc.,* 1897 (1938).
182. B. Karele and O. Neilands, *Zh. Org. Khim.,* **4,** 634 (1968); *Chem. Abstr.,* **69,** 2626 (1968).
183. E. Zbiral, *Monatsh. Chem.,* **95,** 1759 (1964).
184. O.R. Rodig, R.E. Collier, and R.K. Schlatzer, *J. Org. Chem.,* **29,** 2652 (1964).
185. L.A. Warren and S. Smiles, *J. Chem. Soc.,* 1040 (1932).
186. G.E. Bonvicino, L.H. Yogodzinski, and R.A. Hardy, Jr., *J. Org. Chem.,* **27,** 4272 (1962).
187. E.A. Nodiff and M. Hausman, *ibid.,* **29,** 2453 (1964).
188. R. Henriques, *Chem. Ber.,* **27,** 2993 (1894).
189. L.A. Warren and S. Smiles, *J. Chem. Soc.,* 2774 (1932).
190. G. Papalardo, *Gazz. Chim. Ital.,* **90,** 648 (1960).
191. C.S. McClement and S. Smiles, *J. Chem. Soc.,* 1016 (1937).
192. J.F. Bunnett and R.E. Zahler, *Chem. Rev.,* **49,** 362 (1951).
193. M.F. Grundon and B.T. Johnston, *J. Chem. Soc.,* B, 255 (1966).
194. W.E. Truce, C.R. Robbins, and E.M. Kreider, *J. Amer. Chem. Soc.,* **88,** 4027 (1966).
195. W.E. Truce and W.W. Brand, *J. Org. Chem.,* **35,** 1828 (1970).
196. E.M. Kreider, Ph.D. Thesis, Purdue University, 1967.
197. V.N. Drozd, *Dokl. Akad. Nauk SSSR,* **169,** 107 (1966); *Chem. Abstr.,* **65,** 13646 (1966).
198. V.N. Drozd and T. Yu. Frid, *Zh. Org. Khim.,* **3,** 373 (1967); *Chem. Abstr.,* **67,** 2586 (1967).
199. V.N. Drozd and V.I. Sheichenko, *Zh. Org. Khim.* **3,** 554 (1967); *Chem. Abstr.,* **67,** 10965 (1967).
200. F.G. Bordwell and R.J. Kern, *J. Amer. Chem. Soc.,* **77,** 1141 (1955).
201. F.G. Bordwell and P.S. Landis, *ibid.,* **79,** 1593 (1957).
202. Y. Yano and S. Oae, *Tetrahedron,* **26,** 27 (1970).
203. S. Oae, "Elimination Reactions," Tokyo Kagaku Dojin, p. 24, (1965).
204. S.J. Cristol and P. Pappas, *J. Org. Chem.,* **28,** 2066 (1963).
205. F.G. Bordwell and G.D. Cooper, *J. Amer. Chem. Soc.,* **73,** 5184 (1951).
206. F.G. Bordwell and P.S. Landis, *ibid.,* **80,** 2450, 6383 (1958).
207. C.L. Arcus and P.A. Hallgarten, *J. Chem. Soc.,* 4214 (1958).
208. R.C. Krug, J.A. Rigney and G.R. Tichelaar, *J. Org. Chem.,* **27,** 1305 (1962).
209. H.E. Zimmerman and J.H. Munch, *J. Amer. Chem. Soc.,* **90,** 187 (1968).
210. W.L. Parker and R.B. Woodward, *J. Org. Chem.,* **34,** 3085 (1969).
211. A.W. Johnson, *Chem. Ind.* (London), 1119 (1963).
     (a) S.D. Boyd, Ph.D. Thesis, Purdue University, 1971.
212. E.O. Beckmann, *J. Prakt. Chem.,* **17,** 439 (1878).
213. S.M. Kliger, *J. Gen. Chem.* (USSR), **3,** 904 (1933); *Chem. Abstr.,* **28,** 3051 (1934).
214. O. Hinsberg, *Chem. Ber.,* **48B,** 1611 (1915).
215. F.G. Bordwell and W.H. McKellin, *J. Amer. Chem. Soc.,* **73,** 2251 (1951).
216. G.A. Russell, H.D. Becker, and J. Schoeb, *J. Org. Chem.,* **28,** 3584 (1963).

217. W.E. Truce, D.P. Tate, and B.N. Burdge, *J. Amer. Chem. Soc.*, **82**, 2872 (1960).
218. W.E. Truce and F.J. Frank, *J. Org. Chem.*, **32**, 1918 (1967).
219. F.J. Frank, Ph.D. Thesis, Purdue University, 1964.
220. R.E. Dabby, J. Kenyon, and R.F. Mason, *J. Chem. Soc.*, 4881 (1952).
221. D.B. Hope, C.D. Morgan, and M. Wälti, *J. Chem. Soc.*, C, 270 (1970).
222. O. Manousek, O. Exner, and P. Zuman, *Collect. Czech. Chem. Commun.*, **33**, 3988 (1968).
223. L. Horner and R.J. Singer, *Tetrahedron Lett.*, 1545 (1969).
224. R.C. Bowers and H.D. Russell, *Anal. Chem.*, **32**, 405 (1960).
225. H.V. Drushel and J.F. Miller, *ibid*, **30**, 1271 (1958).
226. E.S. Levin and N.A. Osipova, *Zh. Obsch. Khim.* (Eng. tr), **32**, 2060 (1962).
227. L. Horner and H. Neumann, *Chem. Ber.*, **98**, 1715 (1965).
228. W. Bradley, *J. Chem. Soc.*, 458 (1938); T.K. Brotherton and J.F. Bunnett, *Chem. Ind.*, 80 (1957).
229. J.F. Bunnett and T.K. Brotherton, *J. Amer. Chem. Soc.*, **78**, 6265 (1956).
230. C.K. Ingold and J.A. Jessop, *J. Chem. Soc.*, 708 (1930).
231. R. Otto and H. Ostrop, *Annales*, **141**, 96 (1867).
232. R. Otto, *Chem. Ber.*, **18**, 248 (1885).
233. N.J. Leonard and S. Gelfand, *J. Amer. Chem. Soc.*, **77**, 3272 (1955).
234. J.J. Looker, *J. Org. Chem.*, **31**, 2714 (1966).
235. R. Mozingo, D.E. Wolf, S.A. Harris, and K. Folkers, *J. Amer. Chem. Soc.*, **65**, 1013 (1943).
236. N. Kharasch and C.Y. Meyers (Editors), "The Chemistry of Organic Sulfur Compounds," Vol. 2, Pergamon Press, New York, 1966, pp. 62–65.
237. V.E. Cates and C.E. Meloan, *J. Chromatog.*, **11**, 472 (1963).
238. E. Wellisch, E. Gipstein, and O.J. Sweeting, *J. Appl. Polymer Sci.*, **8**, 1623 (1964)
239. E. Gipstein, E. Wellisch, and O.J. Sweeting, *J. Org. Chem.*, **29**, 207 (1964).
240. W.Z. Heldt, *ibid.*, **30**, 3897 (1965).
241. E.M. LaCombe and B. Stewart, *J. Amer. Chem. Soc.*, **83**, 3457 (1961).
242. E.C. Leonard, Jr., *J. Org. Chem.*, **27**, 1921 (1962).
243. E. Fromm and O. Archert, *Chem. Ber.*, **36**, 534 (1903).
244. H. Drews, E.K. Fields, and S. Meyerson, *Chem. Ind.* (London), 1403 (1961).
245. R.M. Dodson and G. Klose, *ibid.*, 450 (1963).
246. W.L. Mock, *J. Amer. Chem. Soc.*, **88**, 2857 (1966).
247. S.D. McGregor and D.M. Lemal, *ibid.*, **88**, 2858 (1966).
248. W.L. Mock, *ibid.*, **91**, 5682 (1969).
249. M.P. Cava and A.A. Deana, *ibid.*, **81**, 4266 (1959).
250. M.P. Cava, R.L. Shirley, and B.W. Erickson, *J. Org. Chem.*, **27**, 755 (1962).
251. M.P. Cava and R.L. Shirley, *J. Amer. Chem. Soc.*, **82**, 654 (1960).
252. F.G. Bordwell, W.H. McKellin, and D. Babcock, *ibid.*, **73**, 5566 (1951).
253. M.A. Naylor and A.W. Anderson, *ibid.*, **76**, 3962 (1954).
254. D.O. Hummel and H.D.R. Schueddemage, *Kolloid. – Z.Z. Polym.*, **210**, 97 (1966).
255. V.S. Foldi and W. Sweeny, *Makromol. Chem.*, **72**, 208 (1964).
256. D.E. O'Connor and W.I. Lyness, *J. Amer. Chem. Soc.*, **86**, 3840 (1964).
257. C.D. Broaddus, *Accounts Chem. Res.*, **1**, 231 (1968).
258. K. Muzika, M. Prochazka, and M. Palecek, *Coll. Czech. Chem. Commun.*, **34**, 635 (1969).
259. H. Mackle, D.V. McNally, and W.V. Steele, *Trans. Faraday Soc.*, **65**, 2060 (1969).
260. H. Mackle, W.V. Steele, *ibid.*, **65**, 2069, 2073 (1969); H. Mackle and D.V. McNally, *ibid.*, **65**, 1738 (1969).
261. R.C. Krug, G.R. Tichelaar, and F.E. Didot, *J. Org. Chem.*, **23**, 212 (1958).
262. A. Kaslová, M. Paleček, and M. Procházka, *Coll. Czech. Chem. Commun.*, **34**, 1826 (1969).
263. E. Stuffer, *Chem. Ber.*, **23**, 3226 (1890).
264. C.J.M. Stirling, *Chem. Ind.*, 933 (1960).
265. D.J. Vrencur, Ph.D. Thesis, Purdue University, 1970.
266. W. Davies, *et al.*, *J. Chem. Soc.*, 2609 (1956).
267. H.R. Snyder, *et al.*, *J. Amer. Chem. Soc.*, **73**, 3258 (1951).

268. V.A. Azovskaya, E.N. Prilezhaeva, and A.U. Stepanyants, *Izv. Akad. Nauk SSSR, Ser. Khim.*, 662 (1969); *Chem. Abstr.*, **71**, 30123 (1969).
269. J. Strating and A.M. van Leusen, *Rec. Trav. Chim. Pays-Bas*, **81**, 966 (1962).
270. J. Strating, J. Heeres, and A.M. van Leusen, *ibid.*, **85**, 1061 (1966).
271. J.B.F.N. Engberts, G. Zuidema, B. Zwanenburg, and J. Strating, *ibid.*, **88**, 641 (1969).
272. A.M. van Leusen and J. Strating, *ibid.*, **84**, 151 (1965).
273. A.M. van Leusen, P.M. Smid, and J. Strating, *Tetrahedron Lett.*, 337 (1965).
274. F. Klages and K. Bott, *Chem. Ber.*, **97**, 735 (1964).
275. D. Hodson, G. Holt, and D.K. Wall, *J. Chem. Soc.*, C, 2201 (1968).
276. J. Diekmann, *J. Org. Chem.*, **28**, 2933 (1963).
277. B. Zwanenburg and J.B.F.N. Engberts, *Rec. Trav. Chim. Pays-Bas*, **84**, 165 (1965).
278. J.B.F.N. Engberts and B. Zwanenburg, *Tetrahedron*, **24**, 1737 (1968).
279. J.B.F.N. Engberts and B. Zwanenburg, *Tetrahedron Lett.*, 831 (1967).
280. A.M. van Leusen, P. Richters, and J. Strating, *Rec. Trav. Chim. Pays-Bas*, **85**, 323 (1966).
281. A.M. van Leusen, R.J. Mulder, and J. Strating, *Tetrahedron Lett.*, 543 (1964).
282. A.M. van Leusen, P.M. Smid, and J. Strating, *ibid.*, 1165 (1967).
283. R.J. Mulder, A.M. van Leusen, and J. Strating, *ibid.*, 3057 (1967).
284. R.G. Laughlin and W. Yellin, *J. Amer. Chem. Soc.*, **89**, 2435 (1967).
285. J.K. Whitehead and H.R. Bentley, *J. Chem. Soc.*, 1572 (1952).
286. M. Barash, *Nature*, **187**, 591 (1960).
287. D.R. Rayner, D.M. von Schriltz, J. Pay, and D.J. Cram, *J. Amer. Chem. Soc.*, **90**, 2721 (1968).
288. O.Z. Sellinger and W.G. Ohlsson, *J. Neurochem.*, **16**, 1193 (1969); R.A. Ronzio, W.B. Rowe, and A. Meister, *Biochemistry*, **8**, 1066 (1969).
289. F. Misani and L. Reiner, *Arch. Biochem.*, **27**, 234 (1950).
290. K. Hayashi, *Chem. Pharm. Bull.* (Tokyo), **8**, 177 (1960).
291. M.A. Sabol, R.W. Davenport, and K.K. Anderson, *Tetrahedron Lett.*, 2159 (1968).
292. H.R. Bentley and J.K. Whitehead, *J. Chem. Soc.*, 2081 (1950).
293. G. Kresze and B. Wustrow, *Chem. Ber.*, **95**, 2652 (1962)
294. L. Horner and A. Christmann, *ibid.*, **96**, 388 (1963).
295. L. Horner, G. Bauer, and J. Doerges, *ibid.*, **98**, 2631 (1965).
296. A. Schönberg and E. Singer, *ibid.*, **102**, 2557 (1969).
297. D. Carr, T.P. Seden, and R.W. Turner, *Tetrahedron Lett.*, 477 (1969).
298. J. Sauer and K.K. Mayer, *ibid.*, 319 (1968).
299. D.J. Anderson, T.L. Gilchrist, D.C. Horwell, and C.W. Rees, *Chem. Commun.*, 146 (1969).
300. F. Misani, T.W. Fair and L. Reiner, *J. Amer. Chem. Soc.*, **73**, 459 (1951).
301. T.W. Rave and T.J. Logan, *J. Org. Chem.*, **32**, 1629 (1967).
302. R. Appel, H.W. Fehlhaber, D. Hänssgen, and R. Schöllhorn, *Chem. Ber.*, **99**, 3108 (1966).
303. R. Wehr, *J. Chem. Soc.*, 3004 (1965).
304. C.R. Johnson, J.J. Rigau, M. Haake, D. McCants, Jr., J.E. Keiser, and A. Gertsema, *Tetrahedron Lett.*, 3719 (1968).
305. C.R. Johnson, E.R. Janiga, and M. Haake, *J. Amer. Chem. Soc.*, **90**, 3890 (1968).
306. C.R. Johnson and G.E. Katekar, *ibid.*, **92**, 5754 (1970).
307. C.R. Johnson, M. Haake, and C.W. Schroeck, *ibid.*, **92**, 6594 (1970).

Chapter 11

# SULFINIC ACIDS AND
# SULFINIC ESTERS

Shigeru Oae* and Norio Kunieda**

*University of Tsukuba* and Osaka City University**

*Ibaraki-ken,* and Sumiyoshi-Ku, Osaka**, Japan*

## 11.1 INTRODUCTION

Sulfinic acids are formed readily either by reduction of sulfonyl chlorides with zinc and alkaline solution or by alkaline hydrolysis of thiolsulfonates. Unlike the sulfonic acid, the sulfinic acid retains a lone electron pair on the sulfur atom and hence often behaves as a nucleophile. At the same time, the sulfinic acid is a strong acid and considered to be a key intermediate in the oxidation of the mercaptan to the corresponding sulfonic acid. Although most sulfinic acids are quite reactive, the chemistry of these acids and their esters are relatively unexplored. However, it will not be long before the chemical behavior of sulfinic acid will be uncovered, like that of sulfoxides.

## 11.2 PREPARATION OF SULFINIC ACIDS

The methods to prepare sulfinic acids cited here are only commonly used ones and reviewed in detail by Truce and Murphy[1] and Stirling.[2]

### 11.2.1 Reduction of Sulfonyl Halides

Sulfonyl chlorides are reduced by treatment with zinc (eq. 11.1)[3] or iron[4] in aqueous caustic alkali solution to the corresponding sulfinic acids in good yields. This is by far the most commonly used procedure. The mechanism of this reduction was studied with $^{18}O$-tracer technique.[5]

$$2RSO_2Cl + 2Zn \longrightarrow (RSO_2)_2Zn + ZnCl$$

$$(RSO_2)_2Zn + Na_2CO_3 \xrightarrow{NaOH} 2RSO_2Na + ZnCO_3 \tag{11.1}$$

Arenesulfonyl chlorides are known to be reduced by sodium sulfide or sodium sulfite[6]:

$$ArSO_2Cl + Na_2SO_3 + H_2O \longrightarrow ArSO_2H + NaCl + NaHSO_4 \tag{11.2}$$

Lithium aluminium hydride also reduces sulfonyl chlorides in good yields to the corresponding sulfinic acids[7a]:

$$RSO_2Cl \xrightarrow{LiAlH_4} RSO_2H + Cl^- \tag{11.3}$$

Electrolytic reduction of sulfonyl chlorides also gives sulfinic acids.[7b]

## 11.2.2 Condensation with Sulfur Dioxide

Arenesulfinic acids are obtained by the direct contact of sulfur dioxide with arenediazonium salts in the presence of copper or cuprous salts (eq. 11.4)[8], with aryl Grignard reagents (eq. 11.5)[9], or with other organometallic compounds (eq. 11.6)[1,10] Alkanesulfinic acids are also prepared similarly.[9,10]

$$ArN_2^+SO_4H^- + 2SO_2 + 2H_2O \xrightarrow{Cu}$$
$$ArSO_2H + N_2 + 2H_2SO_4 \tag{11.4}$$

$$2RMgX + 2SO_2 \longrightarrow (RSO_2)_2Mg \xrightarrow{H^+} 2RSO_2H \tag{11.5}$$

$$RLi + SO_2 \longrightarrow RSO_2^-Li^+ \tag{11.6}$$

Arenesulfinic acids are obtained by the following $AlCl_3$-catalyzed sulfination of aromatic hydrocarbons (eq. 11.7).[11a] The uv irradiation of ethanol in liquid $SO_2$ was reported to give $CH_3CH(OH)SO_2H$. Similar sulfinic acids were also obtained from methanol, benzyl alcohol, diisopropyl ether, methyl isopropyl ether, and diethyl sulfide.[11b]

$$ArH + SO_2 \xrightarrow{AlCl_3} ArSO_2H \tag{11.7}$$

## 11.2.3 Hydrolysis of Sulfenyl and Sulfinyl Halides

Hydrolysis of sulfenyl and sulfinyl chlorides[5,17] is also effective in synthesizing the corresponding sulfinic acids (eq. 11.8), since the chlorides can be prepared readily by treating the corresponding mercaptans or disulfides with chlorine in aqueous media.[18] This procedure is useful for the preparation of the $^{18}O$-labeled sulfinic acids, which are used as starting materials for $^{18}O$-labeled sulfoxides, sulfoximines, and other oxygen-containing organosulfur derivatives.

$$\left.\begin{array}{c} RSCl \\ RSOCl \end{array}\right\} \xrightarrow{H_2O} RSO_2H \tag{11.8}$$

## 11.2.4 Other Methods

Diaryl and alkyl aryl sulfones are found to undergo reductive fission upon treatment with sodium amalgam to the corresponding sulfinic acids and hydrocarbons (eq. 11.9).[12] Cleavage of ethylene disulfones with potassium cyanide also gives the corresponding sulfinic acids (eq. 11.10).[9a,13] Reduction of the toluenesulfonyl derivatives of cysteamine, L-cysteine, and L-homocysteine with metallic sodium in liquid ammonia also gives the corresponding sulfinic acids.[14] Aromatic sulfones containing an $o$-methyl group were found by Truce et al.[15] to rearrange to the $o$-arylmethylated arenesulfinic acid upon treatment with $n$-butyllithium, apparently via a carbanion intermediate. This is illustrated by the rearrangement of mesityl $p$-tolyl sulfone to 2-(4-methylbenzyl)-4,6-dimethylbenzenesulfinic acid (eq. 11.11).

$$ArSO_2R \xrightarrow{\text{Na–Hg}} ArSO_2H + RH \tag{11.9}$$

$$RSO_2CH_2CH_2SO_2R + 2KCN \longrightarrow 2RSO_2K + CNCH_2CH_2CN \tag{11.10}$$

$$\tag{11.11}$$

Formations of sulfinic acids as by-products in such reactions as that of thiolsulfonates with nucleophiles or the elimination of sulfinate ions in the base-catalyzed E2 reaction of sulfones, will be dealt with in the appropriate sections.

## 11.3 STRUCTURES OF SULFINIC ACIDS AND SULFINIC ESTERS

### 11.3.1 Sulfinic Acids

The following two structures, 11.1 and 11.2, are conceivable for sulfinic acids. The choice between the two is still controversial. Dipole moments of several substituted benzenesulfinic acids were measured by Guryanóva et al.[19] and are shown in Table 11.1. These values are all half way between those calculated for 11.1 and 11.2. Thus, it was suggested[19] that these sulfinic acids are mixtures of the two structures 11.1 and 11.2 in both dioxane and benzene.

**11.1**　　　　　　　　**11.2**

TABLE 11.1

Dipole Moments of Benzenesulfinic Acids (D) at 25°C

| Sulfinic Acid | Solvent | |
|---|---|---|
| | Dioxane | Benzene |
| $C_6H_5SO_2H$ | 3.76 | 2.97 |
| $o$-$CH_3$-$C_6H_4SO_2H$ | | 3.02 |
| $p$-$CH_3$-$C_6H_4SO_2H$ | 4.07 | 3.32 |
| $1$-$C_{10}H_7SO_2H$ | 3.88 | |
| $2$-$C_{10}H_7SO_2H$ | 3.98 | |
| $p$-$Cl$-$C_6H_4SO_2H$ | 3.29 | 2.18 |
| $m$-$NO_2$-$C_6H_4SO_2H$ | 4.23 | |
| $o$-$C_2H_5$-$C_6H_4SO_2H$ | 4.40 | 3.99 |

Dipole moments of a few simple sulfinic esters were also determined and are listed in Table 11.2.[20] These data seem to support the pyramidal structure of the sulfinic ester and the semipolar nature of the S-O linkage.

Spectroscopic analyses of sulfinic acids indicate three characteristic absorption bands, i.e., $2790 \sim 2340$ cm$^{-1}$, $1090 \sim 990$ cm$^{-1}$ and $870 \sim 810$ cm$^{-1}$, in the ir spectra of sulfinic acids.[21] There is no strong band in the regions of $1160 \sim 1120$ cm$^{-1}$ and $1320$ cm$^{-1}$, as is characteristic for the sulfone group.[22] The strong band at around $1090$ cm$^{-1}$ is more likely to be that of sulfoxide, which usually appears at ca. $1055$ cm$^{-1}$. Thus, the ir data appear to support the structure **11.1** for the sulfinic acid. Data on the ir spectra of zinc salts of aliphatic and aromatic sulfinates also lead to the same conclusion, since the spectra were found to differ markedly from

TABLE 11.2

Dipole Moments of Sulfinates in Benzene at 25°C

| Sulfinate | $\mu$ (5%) (D) |
|---|---|
| $CH_3SO$-$OCH_3$ | 2.85 |
| $CH_3SO$-$OC_2H_5$ | 2.84 |
| $C_6H_5SO$-$OCH_3$ | 3.43 |
| $C_6H_5SO$-$OC_2H_5$ | 3.51 |
| $4$-$Cl$-$C_6H_4SO$-$OCH_3$ | 2.82 |
| $4$-$Cl$-$C_6H_4SO$-$OC_2H_5$ | 2.74 |
| $4$-$NO_2$-$C_6H_4SO$-$OCH_3$ | 2.84 |
| $4$-$NO_2$-$C_6H_4SO$-$OC_2H_5$ | 3.25 |

those of the corresponding zinc sulfonates.[23] Stretching frequencies of
S-O linkages in ethyl arenesulfinates are shown in Table 11.3.[24] Similar
data on the methyl esters are also available.[25] There is a good linear correlation between the $\nu$-values and the Hammett $\sigma$-values. The relatively
larger $\nu$-values in $CCl_4$ solution than those for the pure compound may be
ascribed to the slightly stronger association in the pure state than in $CCl_4$
solution. The lack of conjugation between the benzene ring and the S-O
bond may be manifested in that the stretching frequency of the S-O bond
in ethyl n-butanesulfinate both in neat and in $CCl_4$ is also 1132 cm$^{-1}$.

Table 11.4 summarizes the uv spectra of sulfinic acids, esters, and
related compounds. There is a clear peak at around 250 nm with intensity,
log $\epsilon$, of 3.4 ~ 3.6. Apparently the uv spectra of sulfinic acids are similar
to those of sulfoxides but not those of sulfones, which have peaks at
260 ~ 270 nm with log $\epsilon$ values of 2.7 ~ 2.8. Accordingly, the uv spectral
data seems also to support the structure 11.1. Other pertinent uv data on
substituted benzenesulfinic acids are listed in Table 11.5.[24] $^1$La is the
K-absorption band (or first absorption band) due to the polarization along
the long axis of substituted benzenesulfinic acid, while $^1$Lb is B band (or
second absorption band) and associated with the transition along the
direction perpendicular to the plane of the molecule, whereas W is considered to be due to the $n$-$\pi*$ excitation of one of the lone pairs of the
sulfur atom.[27] There are only small red-shifts of the $^1$La band in the uv
spectra of substituted benzenesulfinic acids as compared to those of
substituted benzenes, suggesting a rather small electronic interaction
between the benzene ring and the $SO_2$ group.

The nearly identical uv spectrum of $o$-toluenesulfinic acid with that of
the $p$-isomer implies that there is no requirement of coplanarity in the
thorough conjugation between the benzene ring and the sulfinyl group, if
there is any conjugation, since it is not possible to place both the $o$-methyl
group and the sulfinic acid group in the same plane. Ethyl esters of these
acids have nearly identical first absorption bands ($^1$La), however, their
second bands are those produced by the combination of W and $^1$La
bands.[24] The uv spectrum of $p$-chlorobenzenesulfinic acid in water is
shown in Fig. 11.1. In water, the sulfinic acid is completely dissociated;
however, the addition of hydrochloric acid changes the absorption to that
of the undissociated sulfinic acid. From the change of absorption, a p$K_a$
value of 1.14 is obtained.[24] The Raman spectra of sulfinic acids in
dioxane[19] revealed that there is an absorption band corresponding to the
stretching frequency of an S-H bond at around 2550 ~ 2560 cm$^{-1}$. There
is a corresponding band at 2400 ~ 2600 cm$^{-1}$ in the ir spectra of sulfinic

**TABLE 11.3**

Stretching Frequencies of S-O Linkages in Ethyl Substituted
Benzenesulfinates $(cm^{-1})^{24}$

| $R \langle \text{benzene} \rangle \text{-SOEt} \downarrow O$ | H | $p$-CH$_3$ | $p$-CH$_3$O | $p$-Cl | $p$-Br | $p$-NO$_2$ |
|---|---|---|---|---|---|---|
| $\nu$S-O (neat) | 1132 | 1134 | 1126 | 1136 | 1138 | 1138 |
| $\nu$S-O (in CCl$_4$) | 1136 | 1136 | 1135 | 1142 | 1142 | 1149 |

acids in CCl$_4$.[28] These spectroscopic data together with those of the dipole moments may suggest that structure **11.2** is present in tautomeric equilibrium in the sulfinic acid molecule. In nonpolar media such as benzene or dioxane, sulfinic acid could be associated, so that the proton is placed pretty close to the sulfur to behave like an S-H bond.

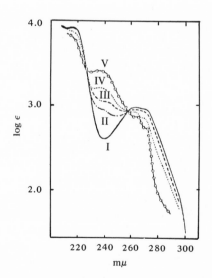

**Fig. 11.1.** UV Spectrum of $p$-chlorobenzenesulfinic acid in:
(I) H$_2$O; (II) 0.01N HCl; (III) 0.02N HCl; (IV) 0.05N HCl;
(V) 1.00N HCl.

**TABLE 11.4**

UV Spectra of Sulfinic Acids, Esters, and Related Compounds

| Compound | $\lambda_{max}$(nm) (log $\epsilon$) |
|---|---|
| $C_6H_5SO_2H$ | 218 (3.99) |
| | 245 (3.40) |
| $p\text{-}CH_3\text{-}C_6H_4SO_2H$ | 223 (4.09) |
| | 242 (3.59) |
| $C_6H_5SO\text{-}OCH_3$ | 223.5 (3.94) |
| | 241.5 (3.59) |
| $C_6H_5SOCH_3$[26] | 250 (3.6) |
| $C_6H_5SO_2CH_3$[26] | 266 (2.9) |
| | 273 (2.8) |

## 11.3.2 Sulfinic Esters

Optically active sulfinic esters have long been known. For example, optically active $\beta$-octyl $p$-toluenesulfinate and menthyl $p$-toluenesulfinate were synthesized nearly fifty years ago.[29] The existence of optically active

**TABLE 11.5**

UV Spectra (nm) of Substituted Benzenesulfinic Acids in Water[24]

| Substituents | In acidic media | | | Sulfinate ions | |
|---|---|---|---|---|---|
| | (undissociated sulfinic acids) | | | | |
| | $^1$La | W | $^1$Lb | $^1$La | $^1$Lb |
| | $\lambda_{max}$ (log$\epsilon$) | $\lambda_{max}$ (log$\epsilon$) | $\lambda_{max}$ (log$\epsilon$) | $\lambda_{max}$ (log$\epsilon$) | $\lambda_{max}$ (log$\epsilon$) |
| H | 215**(3.8) | 237 (3.4) | 265 (2.8) | 215 (3.9) | 265 (2.9) |
| $p\text{-}CH_3$ | 223 (3.8) | 240 (3.5) | 265 (2.8) | 223 (3.9) | 265 (3.0) |
| $p\text{-}Cl$ | 225 (4.3) | 245 (4.0) | 270 (3.2) | 225 (4.2) | 270 (3.2) |
| $p\text{-}Br$ | 230 (4.2) | 250 (3.8) | 275**(2.9) | 230 (4.2) | 275 (3.2) |
| $p\text{-}CH_3O$ | 235* (4.1) | | 275 (3.1) | 232 (4.2) | 275 (3.3) |
| $p\text{-}CH_3CONH$ | 255* (4.2) | | | 243 (4.2) | 280* (~3) |
| $p\text{-}NO_2$ | | 265* | (3.8) | 270* | (4.0) |
| $o\text{-}CH_3$ | 220 (3.8) | 240 (3.5) | 265 (3.0) | 225 (3.9) | 265 (3.5) |
| $o\text{-}CH_3O$ | 222 (3.8) | 245 (3.3) | 282 (3.5) | 225 (3.9) | 280 (3.5) |
| $o\text{-}NO_2$ | | 265* | (3.7) | | 265 (3.7) |
| $o\text{-}COOH$ | 232, (3.7) | | 280 (3.2) | 240**(3.5) | 278 (3.0) |

*Overlapped band.
**Shoulder.

sulfinic esters suggests the esters to have a pyramidal structure similar to that of sulfoxides,[30] as shown in Fig. 11.2. Since sulfinic esters are configurationally stable, they can be resolved into two optically active enantiomers. The absolute configuration around the sulfur atom of (−) menthyl (−) p-iodobenzenesulfinate was determined by X-ray analysis and the sulfur atom in the sulfinate was found to be (S).[31] Physical data of several (-) menthyl sulfinates are listed in Table 11.6

**Fig. 11.2**

**TABLE 11.6**

Physical Data of a Few (−) Menthyl Sulfinates, (−) Menthyl-O-S-R

$$\text{(−) Menthyl-O-S-R}$$
$$\downarrow$$
$$O$$

| R | Absolute config. on S atom | mp or bp (°C) | Optical rotation | Ref. |
|---|---|---|---|---|
| (−) Benzene | S | 49~51 | $[\alpha]_D^{25}$ -205.5° (c;2.0, acetone) | |
| (−) p-Iodobenzene | S | 145.5 | $[\alpha]_D^{25}$ -145.8°(c;0.6, acetone) | a |
| (+) p-Iodobenzene | | 97~8 | $[\alpha]_D^{25}$ +22.7°(c;1.5, acetone) | a |
| (−) p-Toluene | S | 106~7 | $[\alpha]_D^{25}$ −199°(c;2, acetone) | b |
| (−) p-Methoxybenzene | S | 106~8 | $[\alpha]_D^{24}$ −189°(c;1, acetone) | c |
| (−) 1-Naphthalene | S | 118~9 | $[\alpha]_D^{24}$ −433°(c;2.07, acetone) | c |
| (+) Benzyl | | 75.7~76.5 | $[\alpha]_D$ +120°(c;0.25, ethanol) | d |
| (−) Methyl | | 88−90/ 4 mm Hg | $[\alpha]_D^{25}$ −99°(c;2.04, acetone) | e |
| (+) n-Buthyl | | oil | $[\alpha]_D$ +50°(c;2.6, acetone) | f |

(a) H.F. Herbrandson and C.M. Cusano, *J. Amer. Chem. Soc.*, **83**, 2124(1961). (b) H.F. Herbrandson, R.T. Dickerson, Jr., and J. Weinstein, *ibid.*, **78**, 2576(1956). (c) K.K. Andersen, W. Gaffield, N.E. Paranikolaou, J. W. Foley, and P.I. Perkins, *ibid.*, **86**, 5637(1964). (d) M. Axelrod, P. Bickart, J. Jacobus, M.M. Green, and K. Mislow, *ibid.*, **90**, 4835(1968). (e) K.K. Andersen, *J. Org. Chem.*, **29**, 1953(1964). (f) K. Mislow, M.M. Green, P. Laur, J.T. Melillo, T. Simmons, and A.L. Ternay, Jr., *J. Amer. Chem. Soc.*, **87**, 1958(1965). (g) H.F. Herbrandson and R.T. Dickerson, Jr., *ibid.*, **81**, 4102(1959).

On the basis of the pyramidal structure of sulfinate esters, the stable conformation of trimethylene sulfite was at first considered to be of (A)[32]; however, since the observed dipole moment, 3.5 ~ 3.6D, cannot be reconciled with the calculated value, 1.8D, the conformation (B) is now favored.[33] Two conformational isomers of the sulfite ester of pentane-2,4-diol were isolated,[34] and two conformers of thiodan (C), antiinsectide, were also isolated.[35] All these observations support the pyramidal structure of sulfinate esters. The pyramidal structure is also supported by ir and uv spectra, parachor values,[36] and the similarity of chemical behaviors of sulfinate esters with those of analogous sulfoxides.

(A)                    (B)                    (C)

## 11.4 CHEMICAL REACTIVITIES OF SULFINIC ACIDS

### 11.4.1 Acidity[37]

The $pK_a$ values in Table 11.7 reveal that sulfinic acids are quite acidic as compared to carboxylic acids. Benzenesulfinic acid is as acidic as dichloroacetic acid.

The following are the nmr chemical shifts of a few substituted benzenesulfinic acids in dioxane.[38] These results reveal that the electron-withdrawing effect of sulfinic acid group is nearly identical to that of the cyano group.

TABLE 11.7

$pK_a$ Values of Several Sulfinic Acids in Water at $20°C$[37]

| Sulfinic acid | $pK_a$ |
|---|---|
| $C_6H_5SO_2H$ | 1.29 |
| $p$-Cl-$C_6H_4SO_2H$ | 1.14 |
| $C_6H_5CH_2SO_2H$ | 1.45 |
| $C_6H_5CH_2CH_2SO_2H$ | 1.89 |
| $C_6H_5CH_2CH_2CH_2SO_2H$ | 2.03 ~ 2.05 |
| $CH_3CH_2CH_2CH_2SO_2H$ | 2.11 |
| $C_6H_5CH_2CH_2CH_2CH_2SO_2H$ | 2.23 |

$$Cl \quad\quad H \leftarrow 7.36\ ppm$$
$$H \leftarrow 7.50\ ppm$$
$$SO_2H \leftarrow 8.60\ ppm$$

$$CH_3 \quad\quad H \leftarrow 7.16\ ppm$$
$$H \leftarrow 7.42\ ppm$$
$$SO_2H \leftarrow 8.40\ ppm$$

$$CN \quad\quad H \leftarrow 7.88\ ppm$$
$$H \leftarrow 7.88\ ppm$$
$$SO_2H \leftarrow (7.92)$$

## 11.4.2 Oxidation

Sulfinic acids are known to be readily oxidized; for instance, $p$-toluenesulfinic acid undergoes facile autoxidation in acetonitrile at 40°C to give $p$-toluenesulfonic acid and the autoxidation is accelerated by the presence of $Cl^-$ or $Cu$ (II) ion.[39] Lindberg has provided data on polarographic oxidation of several substituted sodium benzenesulfinates with hydrogen peroxide. He pointed out a good linear relationship between the ease of oxidation and the Hammett $\sigma$-values.[40] Half-wave potential values in polarographic oxidation of ethyl $p$-substituted benzenesulfinates were measured in 0.1N HCl solution by Kobayashi and are listed in Table 11.8.[38] These values are also correlated with the Hammett $\sigma$-values. Incidentally the half-wave potential of benzenesulfinate ion is smaller than $-1.9$ v while that of ethyl $n$-butanesulfinate is $-1.22$ v.[38]

Oxidation of sulfinic acids by permanganate, $MnO_4$, does not give the corresponding sulfonic acids but disulfones[41]:

$$2RSO_2H \xrightarrow{MnO_4^-} RSO_2SO_2R \qquad (11.12)$$

### TABLE 11.8

Half-wave Potentials in Polarographic Oxidation of Ethyl $p$-Substituted Benzenesulfinates in 0.1N-HCl Water Solution

| Substituent | H | $CH_3$ | Cl | Br | $CH_3O$ | $NO_2$ |
|---|---|---|---|---|---|---|
| $-E\ 1/2$ (v) | 1.25 | 1.32 | 1.17 | 1.17 | 1.30 | 0.62 |

## 11.4.3 Disproportionation

Sulfinic acids are not very stable and readily undergo disproportionation or decomposition. Aliphatic sulfinic acids are much less stable than aromatic sulfinic acids, and the lower alkanesulfinic acids are known to be the least stable.[42] Disproportionation of sulfinic acids takes place readily in the temperature range of $25 \sim 100°C$ and gives the corresponding thiolsulfonates and sulfonic acids (eq. 11.13).[43] The mechanism of disproportionation of $p$-toluenesulfinic acid in aqueous acetic acid was investigated in detail by Kice and his coworker,[44] and the initial step is considered to involve the formation of the sulfinyl sulfone, **11.3**, and water in a reversible reaction (eq. 11.14). The formation of an intermediate, **11.4**, by intermolecular rearrangement was suggested by Kice *et al.* in the subsequent rate-determining step (eq. 11.15).[44] Once the intermediate is formed, the steps (eqs. 11.16, 11.17) to form the thiol sulfonate and sulfonic acid are believed to be fast. The rate-determining step (eq. 11.15) was suggested to involve the formation of new S-O bond and the simultaneous rupture of the S-S bond of the incipient intermediate **(11.5)**, in which the sulfur atom expands its valence shell beyond the octet (eq. 11.18). Table 11.9 lists the relative magnitudes of the second-order rate constants for the acid-catalyzed disproportionation of several substituted sulfinic acids.[45] Logarithms of the relative magnitudes correlate well with the Hammett $\sigma^+$-values. Further detailed kinetic studies were carried out by Kice and his coworkers[46] who found, by using the sulfinyl sulfone **(11.3)** isolated, that the free energy difference between the sulfinyl sulfone, **11.3**, and the sulfinic acid is about 1.5 kcal. They studied the chemical behavior of the sulfinyl sulfone, **11.3**, amongst which the acid-catalyzed hydrolysis[47] and the reactions with nucleophiles[48] are especially interesting.

$$3ArSO_2H \longrightarrow ArSO_2SAr + ArSO_3H + H_2O \qquad (11.13)$$

$$2ArSO_2H \underset{fast}{\overset{}{\rightleftharpoons}} \underset{\underset{11.3}{\overset{\downarrow}{O}}}{ArS-SO_2Ar} + H_2O \qquad (11.14)$$

$$\underset{\underset{O}{\overset{\downarrow}{}}}{ArS-SO_2Ar} \xrightarrow[r.d.s.]{slow} \underset{11.4}{ArS-O-SO_2Ar} \qquad (11.15)$$

$$11.4 \begin{cases} \xrightarrow{ArSO_2H} ArSO_2SAr + ArSO_3H \\ \xrightarrow{H_2O} ArSO_3H + ArSOH \end{cases} \qquad (11.16)$$

$$ArSOH + ArSO_2H \longrightarrow ArSO_2SAr + H_2O \qquad (11.17)$$

TABLE 11.9

Relative Ratios of Rate of Disproportionations of Arenesulfinic Acids[45]

| $ArSO_2H$ | $k_2 ArSO_2H/k_2 C_6H_5SO_2H$ |
|---|---|
| $p$-$CH_3O$-$C_6H_4$ | 5.4 |
| $p$-$CH_3C_6H_4$ | 2.36 |
| $C_6H_5$ | (1.0) |
| $p$-$Cl$-$C_6H_4$ | 0.55 |
| $p$-$Br$-$C_6H_4$ | 0.53 |
| $p$-$NO_2$-$C_6H_4$ | 0.19 |
| $\beta$-$C_{10}H_7$ | 3.64 |

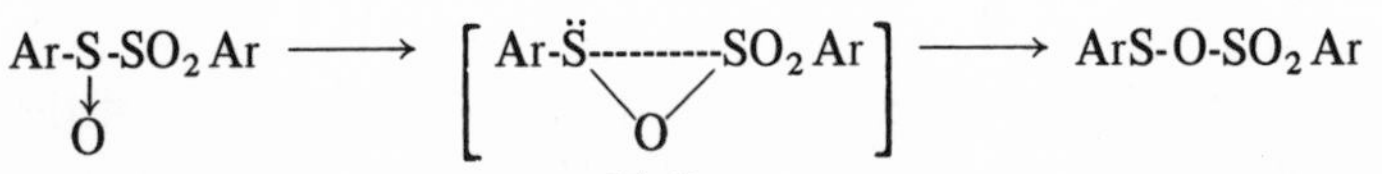

$$\underset{\overset{\displaystyle \downarrow}{O}}{Ar\text{-}S\text{-}SO_2 Ar} \longrightarrow \underset{\mathbf{11.5}}{\left[ Ar\text{-}\ddot{S}\underset{O}{\cdots\cdots}SO_2 Ar \right]} \longrightarrow ArS\text{-}O\text{-}SO_2 Ar \quad (11.18)$$

The sulfinyl sulfones, **11.3**, undergo facile hydrolysis in the presence of nucleophiles.[47,48] Even a trace of chloride ion accelerates the hydrolysis remarkably. Though details of the reaction may differ somewhat with change of the nucleophile, the following scheme illustrates the rate-determining step of the hydrolysis (eq. 11.19), and the relative reactivities of various nucleophiles are shown in Table 11.10.[48]

TABLE 11.10

Relative Reactivities of Various Nucleophiles in the Reactions with Sulfinyl Sulfones (11.3a and 11.3b)

| Nucleophile | $kNu$ ($M^{-1}$, $sec^{-1}$) | | $kNu/k_{Cl^-}$ | |
|---|---|---|---|---|
| | with **11.3a** | with **11.3b** | with **11.3a** | with **11.3b** |
| $F^-$ | 4.0 | | 0.33 | |
| $CH_3CO_2^-$ | 9.0 | | 0.75 | |
| $Cl^-$ | 12.0 | 34 | (1.0) | (1.0) |
| $Br^-$ | 64 | 170 | 5.4 | 5.0 |
| $SCN^-$ | 160 | | 13 | |
| $I^-$ | 1060 | 3050 | 90 | 90 |
| $(NH_2)_2C=O$ | 3350 | | 280 | |

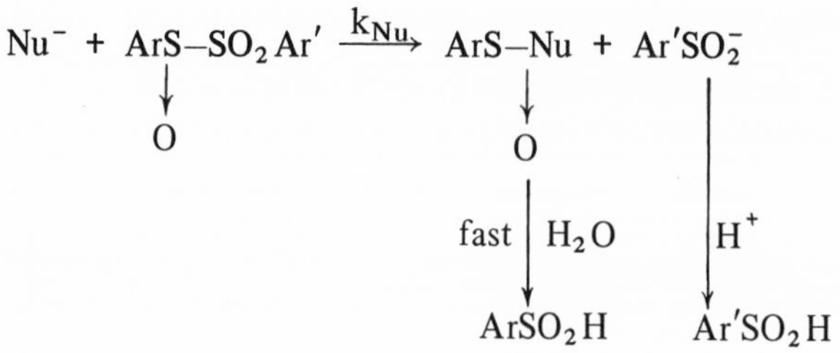

$$Nu^- + ArS-SO_2Ar' \xrightarrow{k_{Nu}} ArS-Nu + Ar'SO_2^-$$

$$
\begin{array}{ccc}
\downarrow & & \downarrow \\
O & & O \\
\end{array}
$$

$$fast \downarrow H_2O \qquad \downarrow H^+ \qquad\qquad (11.19)$$

$$ArSO_2H \qquad Ar'SO_2H$$

$$Ar = p\text{-}CH_3OC_6H_4 \text{ (a)}, \quad p\text{-}CH_3C_6H_4 \text{ (b)}$$

The catalytic action of alkyl sulfides was also studied in detail by Kice et al.[47a] and the following observations were made: (1) the reaction is markedly acid-catalyzed; (2) alkyl sulfides are more reactive than aryl sulfides while sterically nonbulky tetramethylene sulfide is highly reactive, suggesting that the reaction involves the nucleophilic attack of the sulfide on the sulfinyl group of the sulfinyl sulfone 11.3; (3) proton transfer appears to be involved in the departure of the arenesulfinate group at the transition stage, as manifested by the solvent isotope effect ($k_{H_2O}/k_{D_2O} = 1.4$). Thus, the following mechanism was suggested:

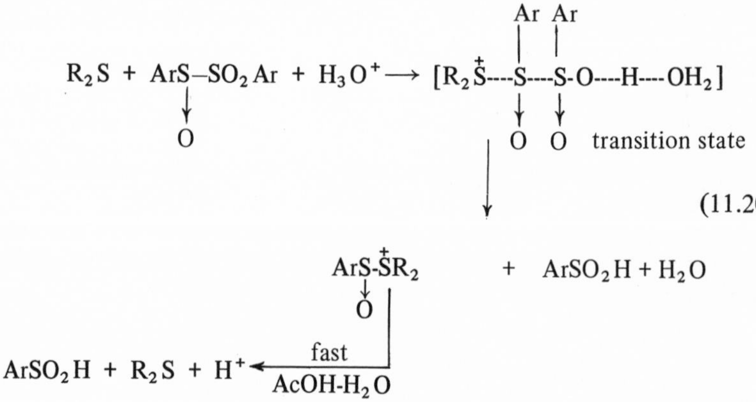

$$
R_2S + ArS-SO_2Ar + H_3O^+ \longrightarrow [R_2\overset{+}{S}\text{---}S\text{----}S\text{-}O\text{----}H\text{----}OH_2]
$$

with structure showing $\overset{Ar}{\underset{O}{|}}$ $\overset{Ar}{\underset{O}{|}}$ and "transition state"

$$\downarrow \qquad (11.20)$$

$$ArS\text{-}\overset{+}{S}R_2 \qquad + \quad ArSO_2H + H_2O$$
$$\downarrow O |$$

$$ArSO_2H + R_2S + H^+ \xleftarrow[AcOH\text{-}H_2O]{fast}$$

### 11.4.4 Reactions with Disulfides, Sulfides, and Other Divalent Sulfur Species

Like disproportionation, reactions of sulfinic acids with disulfides and alkyl sulfides also give the corresponding thiolsulfonates.

11.4.4.1  With Disulfides

The reaction between an arenesulfinic acid and an aryl disulfide gives the corresponding thiolsulfonate (eq. 11.21). The mechanism of this reaction was studied by Kice *et al.* with *p*-toluenesulfinic acid and *p*-tolyl disulfide in aqueous acetic acid.[49] The reaction is first order with respect to sulfinic acid and accelerated noticeably by addition of sulfinic acid, but retarded by the increase of water in the system. Disulfide does participate, but the dependence is not of first order. The mechanism shown in eq. 11.22[49] was suggested to explain all these observations.

$$4ArSO_2H + ArSSAr \longrightarrow 3Ar\text{-}SO_2SAr + 2H_2O$$

$$(11.21)$$

$$ArSO_2H + ArSSAr + H^+ \underset{fast}{\overset{Ke}{\rightleftharpoons}} \text{``Intermediate''}$$
$$\mathbf{11.6}$$

$$\mathbf{11.6} \xrightarrow[slow (r.d.s.)]{k_c} Products$$

$$(11.22)$$

$$\mathbf{11.6} + ArSSAr \xrightarrow[slow (r.d.s.)]{k_o} Products + ArSSAr$$

The involvement of disulfide may be expressed by eq. 11.23. The detailed course of this reaction involves the following sequence. The initial step involves the formation of an intermediate, **11.7**, in an equilibrium (eq. 11.24). The intermediate, **11.7**, either undergoes disproportionation (eq. 11.25) or undergoes nucleophilic attack by disulfide (eq. 11.26) in the rate-determining step. The subsequent steps to form thiosulfonate (eqs. 11.27, 11.28 and 11.29) are anticipated to be fast.

$$k_1 = K_c[k_o + k_c(ArSSAr)] (ArSSAr) \qquad (11.23)$$

$$ArSO_2H + ArSSAr + H^+ \underset{fast}{\rightleftharpoons} \underset{\substack{|\ \ | \\ O\ Ar}}{ArS\text{-}\overset{+}{S}\text{-}SAr} + H_2O \qquad (11.24)$$
$$\mathbf{11.7}$$

$$\text{r.d.s.} \begin{cases} \textbf{11.7} \xrightarrow[\text{slow}]{} \underset{O}{Ar\overset{\downarrow}{S}\text{-}SAr} + ArS^+ & (11.25) \\[3em] \textbf{11.7} + ArSSAr \xrightarrow[\text{slow}]{} \underset{\substack{Ar \\ \textbf{11.8}}}{ArS\text{-}\overset{+}{S}\text{-}SAr} + \underset{O}{Ar\overset{\downarrow}{S}\text{-}SAr} & (11.26) \end{cases}$$

$$ArS^+ + ArSO_2H \longrightarrow ArSO_2SAr + H^+ \qquad (11.27)$$

$$\textbf{11.8} + ArSO_2H \longrightarrow ArSSAr + ArSO_2SAr + H^+ \qquad (11.28)$$

$$2Ar\underset{O}{\overset{\downarrow}{S}}\text{-}SAr \longrightarrow ArSO_2SAr + ArSSAr \qquad (11.29)$$

## 11.4.4.2 With Dialkyl Sulfides

Like disulfides, dialkyl sulfides also react with arenesulfinic acids to afford the corresponding thiolsulfonates. An interesting feature of this reaction is that a S-C bond of dialkyl sulfide is cleaved and one alkyl group is oxidized. Thus, primary alkyl sulfide gives aldehyde, while secondary alkyl derivatives form ketones. The following mechanisms for primary and secondary alkyl sulfides, respectively, were postulated by Kice and his coworker[50]:

$$5ArSO_2H + (RCH_2)_2S \longrightarrow 2ArSO_2SAr + ArSO_2SCH_2R \\ + RCHO + 3H_2O \qquad (11.30)$$

$$5ArSO_2H + (RR'CH)_2S \longrightarrow 2ArSO_2SAr + ArSO_2SCHRR' \\ + \underset{R'}{\overset{R}{\diagdown}}C{=}O + 3H_2O \qquad (11.31)$$

The reaction was found to be of first-order dependence with respect to both dialkyl sulfide and sulfinic acid, and the effects of sulfuric acid and water on the rate are also nearly the same as in the case of disulfide. Thus, the mechanism shown in eqs. 11.32 and 11.33 was suggested for the reaction with primary alkyl sulfide. The initial step (11.32) is a reversible reaction to form an intermediate, **11.9**, which undergoes decomposition in the subsequent rate-determining step (eq. 11.33) to yield disulfide and aldehyde. A similar mechanism can be postulated for the reaction of sulfinic acid with secondary dialkyl sulfide.

An anomalous reaction takes place with 2-octyl sulfide, as shown below (eq. 11.34). In this case, 2-octyl sulfide is evidently oxidized to the sulfoxide by sulfinic acid. Table 11.11 summarizes the rates of the reaction of $p$-toluenesulfinic acid with a few alkyl sulfides. Rates for secondary alkyl sulfides are smaller than those for primary alkyl sulfides, presumably because the equilibrium in eq. 11.32 is less favorable for secondary alkyl sulfides than primary sulfides due to larger steric hindrance of secondary alkyl groups. Diphenyl sulfide does not react at all. The relatively small nucleophilicity of sulfur and the strong C-S bond in diphenyl sulfide seem to be responsible for the lack of reactivity, since both steps, eqs. 11.32, 11.33, become unfavorable. A detailed mechanism of the rate-determining decomposition of the intermediate, **11.9**, in eq. 11.33 was presented, as shown in eq. 11.35, on the basis of further kinetic study.[52]

$$ArSO_2H + (RCH_2)_2S + H^+ \rightleftharpoons ArS \overset{+}{\underset{\underset{O}{\downarrow}}{-}} S \Big\langle \overset{CH_2R}{\underset{CH_2R}{}} + H_2O \qquad (11.32)$$

$$\mathbf{11.9}$$

$$\mathbf{11.9} \xrightarrow{\text{slow}} ArS\text{-}S\text{-}CH_2R + RCHO + H^+ \qquad (11.33)$$

$$2ArSO_2H + \left( \underset{CH_3}{\overset{C_6H_{13}}{}} CH \right)_2 S \longrightarrow ArSO_2SAr$$

$$+ \left( \underset{CH_3}{\overset{C_6H_{13}}{}} CH \right)_2 S{\rightarrow}O + H_2O \qquad (11.34)$$

**TABLE 11.11**

Rate Constants of the Reaction between $p$-Toluenesulfinic Acid and Various Alkyl
Sulfides in Aqueous Acetic Acid at 70°C[51]

$(CH_3COOH-0.56M\ H_2O, 0.6M\ H_2SO_4)$

| Alkyl sulfide | $k \times 10^3\ M^{-1}sec^{-1}$ |
|---|---|
| $i$-Propyl | 2.6 |
| 2-Butyl | 2.8 |
| 2-Octyl | 2.4 |
| $n$-Butyl | 17.5 |
| Ethyl | 10.2 |

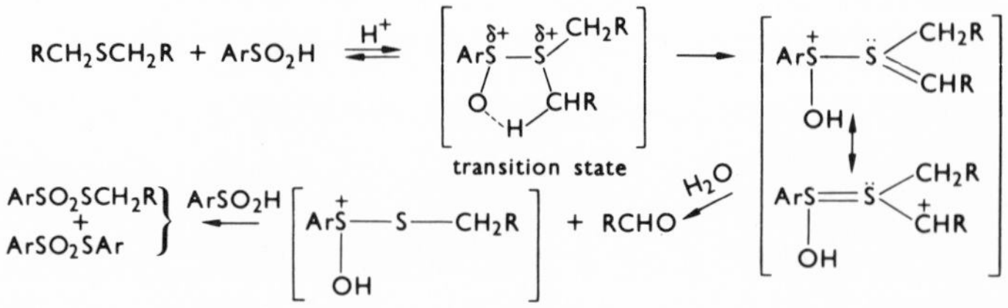

(11.35)

## 11.4.4.3 With Thiolsulfinates

Thiolsulfinates are relatively unstable and hence very reactive. Under
acidic conditions, thiolsulfinates undergo disproportionation to give the
corresponding disulfides and thiolsulfonates[53]:

$$2ArS\text{-}SAr \longrightarrow ArSSAr + Ar\text{-}SO_2S\text{-}Ar \qquad (11.36)$$

In acetic acid containing small amounts of water and sulfuric acid,
diaryl thiolsulfinates react with sulfinic acids, and the reaction is far faster
than the acid-catalyzed disproportionation of the thiolsulfinate itself. The
stoichiometry of the reaction is as shown in eq. 11.37. The reaction was
studied in detail by Kice *et al.*[54] and found to be accelerated by the
addition of alkyl or aryl sulfides. A thorough kinetic study revealed the
following features of the reaction: (1) the reaction is of first-order dependence with respect to both thiolsulfinate and sulfide, but not sulfinic
acid; (2) the rate of reaction varies markedly with the structure of the
alkyl or aryl sulfide, i.e., being higher with a more nucleophilic sulfide
as shown in Table 11.12, while the logarithms of the rates log $k_s$, can be

## TABLE 11.12

Relative Reactivities of Various Sulfides in the Acid-Catalyzed Reaction of Phenyl Benzenethiolsulfinate with Benzenesulfinic Acid, at $39.4°C$, $CH_3COOH$ $(0.56M\ H_2O, 0.1M\ H_2SO_4)$, Medium

| Sulfide | $k_s(M^{-1}, sec^{-1})$ |
|---------|------------------------|
| $(CH_2)_4S$ | 370 |
| $C_2H_5S$ | 200 |
| $n\text{-}Bu_2S$ | 170 |
| $(C_6H_5CH_2)_2S$ | 21 |
| $(HOOCCH_2CH_2)_2S$ | 3.1 |
| $C_6H_5CH_2SC_6H_5$ | 0.42 |
| $(C_6H_5)_2S$ | 0.0087 |
| $(HOOCCH_2)_2S$ | 0.0055 |

nicely correlated with the Taft $\sigma$-values, with a $\rho$-value of $-2.0$; (3) the reaction is strongly acid-catalyzed, and there is a linear relationship between the Hammett HO functions and the log $k_s$ values; (4) the solvent isotope effect ($k_{HOAc}/k_{DOAc}$) is 0.75. Based on these observations, the scheme shown in eq. 11.38[54] was postulated as the most reasonable mechanism for the reaction.

$$2ArSO_2H + PhS\text{-}SPh \longrightarrow 2ArSO_2SPh + H_2O \qquad (11.37)$$
$$\underset{O}{\downarrow}$$

$$PhS\text{-}SPh + H^+ \rightleftharpoons Ph\overset{+}{S}\text{-}SPh$$
$$\underset{O}{|} \qquad \underset{OH}{|}$$

$$R_2S + PhS\text{-}\overset{+}{S}Ph \longrightarrow \left[ R_2\overset{\delta+}{S}\text{---}S\text{---}\overset{\delta+}{S}Ph \right] \longrightarrow R_2\overset{+}{S}\text{-}SPh + PhSOH$$
$$\underset{OH}{|} \qquad \underset{Ph\ OH}{|\ |}$$
$$\text{transition state}$$

$$\hspace{10cm} (11.38)$$
$$R_2\overset{+}{S}\text{-}SPh + ArSO_2H \xrightarrow{\text{fast}} ArSO_2SPh + R_2S + H^+$$

$$PhSOH + R_2S + H^+ \rightleftharpoons R_2\overset{+}{S}\text{-}SPh + H_2O$$

$$PhSOH + ArSO_2H \longrightarrow ArSO_2SPh + H_2O$$

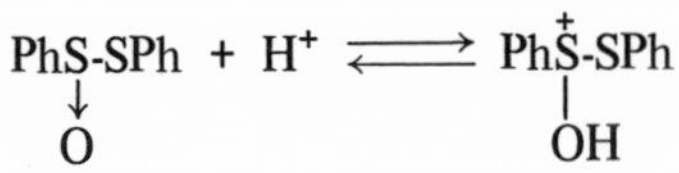

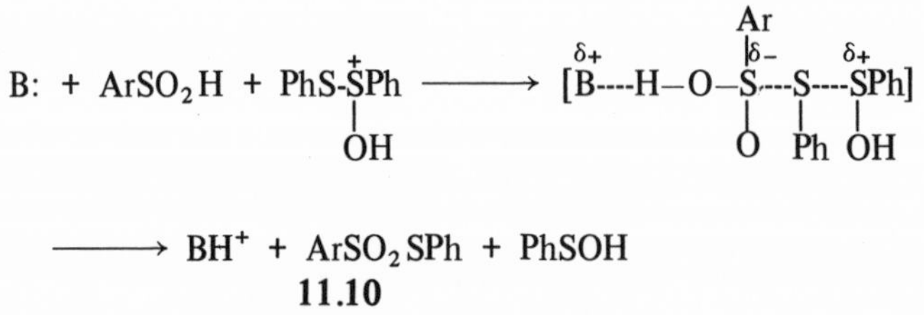

$$\text{PhS-SPh} + \text{H}^+ \rightleftharpoons \overset{+}{\text{PhS}}\text{-SPh}$$
$$\underset{\text{O}}{\downarrow} \qquad\qquad\qquad\qquad \underset{\text{OH}}{|}$$

$$\text{B:} + \text{ArSO}_2\text{H} + \text{PhS-}\overset{+}{\underset{\overset{|}{\text{OH}}}{\text{SPh}}} \longrightarrow [\text{B----H--O--}\overset{\overset{\displaystyle\text{Ar}}{|}}{\underset{\overset{|}{\text{O}}}{\overset{\delta-}{\text{S}}}}\text{---}\overset{|}{\underset{\text{Ph}}{\text{S}}}\text{----}\overset{\delta+}{\underset{\text{OH}}{\text{SPh}}}] \qquad (11.39)$$

$$\longrightarrow \text{BH}^+ + \text{ArSO}_2\text{SPh} + \text{PhSOH}$$
$$\textbf{11.10}$$

The normal uncatalyzed reaction between thiolsulfinate and sulfinic acid is of first-order dependence on both substrates, and the reaction is general-acid catalyzed, while the solvent isotope effect, $k_{\text{HOAc}}/k_{\text{DOAc}}$, is known to be 0.8 at 39.4°C. Since sulfinic acids of higher acidities are known to react faster, the reaction undoubtedly involves the nucleophilic attack of thiolsulfinate on the electron-deficient sulfur atom of sulfinic acid, as shown in eq.11.39, in which B could be a general base such as thiolsulfinate or water. As for the fate of the benzenesulfenic acid, **11.10**, the following path is a possibility (eq. 11.40). The other alternative is the

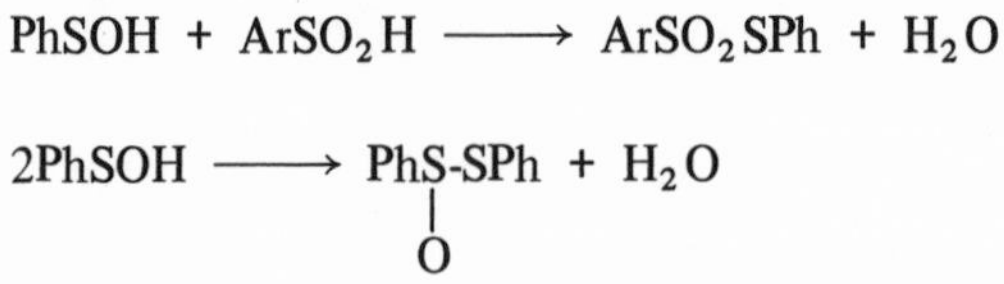

$$\text{PhSOH} + \text{ArSO}_2\text{H} \longrightarrow \text{ArSO}_2\text{SPh} + \text{H}_2\text{O}$$
$$\qquad\qquad\qquad\qquad\qquad\qquad\qquad\qquad\qquad (11.40)$$
$$2\text{PhSOH} \longrightarrow \text{PhS-SPh} + \text{H}_2\text{O}$$
$$\qquad\qquad\qquad\qquad\quad \underset{\text{O}}{|}$$

following scheme (eq. 11.41). This is in keeping with the actually observed reaction[55] shown in eq. 11.42.

$$\text{PhSOH} + \text{Ph}\underset{\overset{\downarrow}{\text{O}}}{\text{S}}\text{-SPh} \longrightarrow \text{PhSO}_2\text{H} + \text{PhSSPh} \qquad (11.41)$$

$$\text{CH}_3\text{SOR} + \text{CH}_3\underset{\overset{\downarrow}{\text{O}}}{\text{S}}\text{-SCH}_3 \longrightarrow \text{CH}_3\text{SSCH}_3 + \text{CH}_3\underset{\overset{\downarrow}{\text{O}}}{\text{SOR}} \qquad (11.42)$$

$$(11.43a)$$

The thiolsulfinate undergoes facile thermal decomposition and disproportionates into disulfide and thiolsulfonates. [56] Hydrolysis of $o$- and $p$-nitrobenzenesulfenyl chlorides is considered to involve the formation of the thiolsulfinates, which upon disproportionation yield the sulfides and the thiolsulfonates (eq. 11.43a).[57] Previously, hydrolysis of $o$- or $p$-nitrobenzenesulfenyl chloride was suggested to give the corresponding sulfinic anhydride,[58] **11.11**, which was not found to be the case.

The reaction between aryl disulfides and pyridine oxide might also involve the formation of a thiolsulfinate, which eventually gives the corresponding disulfide and thiolsulfonate, as shown in eq. 11.43b.[59] While the incipient formation of a sulfenate radical is conceivable in the thermal disproportionation of thiolsulfinate, involvement of a similar incipient sulfenate radical, **11.12**, was suggested for the reaction of $p$-nitrobenzenesulfenyl chloride with pyridine N-oxide (eq. 11.44).[60] Another example may be the thermal decomposition of arenesulfenyl nitrate shown in eq. 11.45.[61]

$$(11.43b)$$

$$O_2N-\text{⟨⟩}-SCl \; + \; \text{(pyridine N-oxide, R)} \; \longrightarrow \; \left[ \text{(complex)} \right] Cl^-$$

$$\longrightarrow \; O_2N-\text{⟨⟩}-SO\cdot \; + \; \text{(pyridinium, R)} \quad Cl^- \qquad (11.44)$$

**11.12**

$$(2) \; \mathbf{11.12} \; \longrightarrow \; O_2N-\text{⟨⟩}-SO_2S-\text{⟨⟩}-NO_2$$

$$ArS\text{-}O\text{-}NO_2 \; \xrightarrow{\Delta} \; ArS\text{-}O\cdot \; + \; NO_2$$

$$\qquad\qquad\qquad\qquad\qquad\qquad\qquad (11.45)$$

$$2\,ArSO\cdot \; \longrightarrow \; ArSO_2SAr$$

The reaction of $p$-nitrobenzenesulfinyl chloride and pyridine N-oxide is also considered to involve the homolytic cleavage of the N–O bond, to result in the formation of the disulfide, thiolsulfonate, sulfonic acid, and pyridine.[62]

### 11.4.4.4 With Thiols

Arenesulfinic acids react readily with acidic thiols, such as acylthiol to afford diacyl disulfides and acyl aryl disulfides, **11.13**, nearly quantitatively[63]:

$$p\text{-Tol-SO}_2\text{H} \; + \; 3\,Ar\underset{\overset{\|}{O}}{C}SH \; \longrightarrow \; (Ar\underset{\overset{\|}{O}}{C}\text{-S})_2 \; + \; Ar\underset{\overset{\|}{O}}{C}\text{-SS-Tol-}p \qquad (11.46)$$

**11.13**

A detailed kinetic study revealed that the reaction is of first-order dependence with respect to both acylthiol and arenesulfinic acid, while the solvent isotope effect, $k_{H_2O}/k_{D_2O}$, is 3.0 at 50°C. The effects of substituents were measured with substituted benzoylthiols in the reaction of $p$-toluenesulfinic acid in acetonitrile and the results are listed in Table 11.13. The rates correlate well with the Hammett $\sigma$-values with a $\rho$ value of +0.7. Based on these observations, a mechanism involving the rate-determining attack of benzoylthiolate on $p$-toluenesulfinic acid was suggested[63]:

**TABLE 11.13**

Observed Rate Constants for the Reaction between $p$-Toluenesulfinic Acid and $p$-Substituted Benzoylthiol in $CH_3CN$ at 50°C

| $p$-R-$C_6H_4SO_2H$<br>R: | $k_{obs}$. $min^{-1}$ |
|---|---|
| Cl | 0.0121 |
| H | 0.0080 |
| $CH_3$ | 0.0060 |
| $CH_3O$ | 0.0052 |

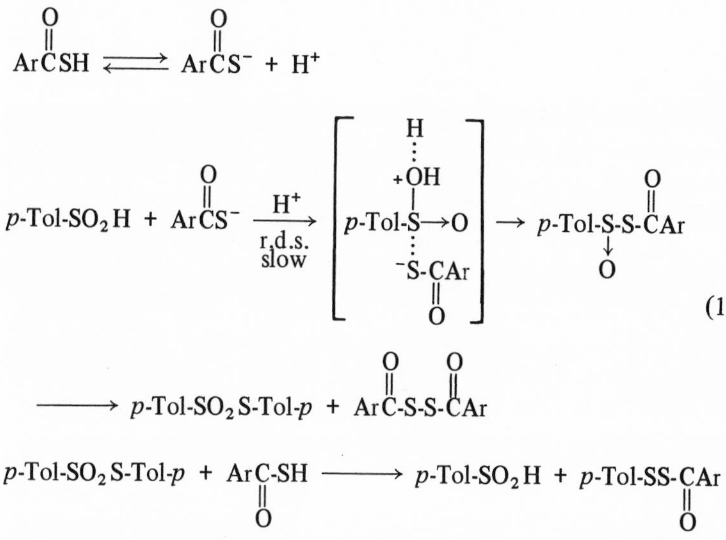

$$ArCSH \; \underset{\longleftarrow}{\rightleftharpoons} \; ArCS^- + H^+$$
(with carbonyl O above ArC in each)

$$p\text{-Tol-SO}_2H + ArCS^- \xrightarrow[\substack{r,d.s.\\ slow}]{H^+} \left[ \begin{array}{c} H \\ \vdots \\ +OH \\ | \\ p\text{-Tol-S}\rightarrow O \\ \vdots \\ {}^-S\text{-CAr} \\ \| \\ O \end{array} \right] \rightarrow p\text{-Tol-S-S-CAr}$$

$$(11.47)$$

$$\longrightarrow p\text{-Tol-SO}_2\text{S-Tol-}p + Ar\overset{O}{\underset{\|}{C}}\text{-S-S-}\overset{O}{\underset{\|}{C}}Ar$$

$$p\text{-Tol-SO}_2\text{S-Tol-}p + Ar\overset{}{\underset{\underset{\|}{O}}{C}}\text{-SH} \longrightarrow p\text{-Tol-SO}_2H + p\text{-Tol-SS-}\overset{}{\underset{\underset{\|}{O}}{C}}Ar$$

## 11.4.5 Formation of Sulfones

### 11.4.5.1 Reaction with RX

Sulfinic acids readily react with alkyl and aralkyl halides to afford the corresponding sulfones as shown below (eqs. 11.48, 11.49 and 11.50). The

$$RSO_2Na + IC_2H_5 \longrightarrow RSO_2C_2H_5 + NaI \qquad (11.48)^{64}$$

$$p\text{-}CH_3C_6H_4SO_2H + PhCH_2Cl \longrightarrow p\text{-}CH_3C_6H_4SO_2CH_2Ph \; (11.49)^{65}$$

$$CH_3SO_2H + RSO_2CCl{=}CHCl \longrightarrow RSO_2CCl{=}CHSO_2CH_3 \; (11.50)^{65}$$

TABLE 11.14

Rate Constants of the Reactions of Sodium Benzenesulfinates with Sodium
Bromoacetate, (eq. 11.51), and with Sodium Bromoacetamide, (eq. 11.52)

a) $\quad$ R⟨⟩-$SO_2$Na + BrCH$_2$$\overset{\overset{O}{\|}}{C}$-ONa $\longrightarrow$ R⟨⟩-$SO_2$CH$_2$$\overset{\overset{O}{\|}}{C}$-ONa $\qquad$ (11.51)

b) $\quad$ R⟨⟩-$SO_2$Na + BrCH$_2$$\overset{\overset{O}{\|}}{C}$-NH$_2$ $\longrightarrow$ R⟨⟩-$SO_2$CH$_2$$\overset{\overset{O}{\|}}{C}$-NH$_2$ $\qquad$ (11.52)

| Substituent | $k_2$ $(1, \text{mol}^{-1}, \text{min}^{-1})$ | | Substituent | $k_2$ $(1, \text{mol}^{-1}, \text{min}^{-1})$ | |
|---|---|---|---|---|---|
| | Reaction (a) | Reaction (b) | | Reaction (a) | Reaction (b) |
| H | 0.144 | 0.336 | $m$-Cl | 0.0789 | 0.137 |
| $o$-CH$_3$ | 0.0567 | 0.098 | $p$-Cl | 0.109 | 0.184 |
| $m$-CH$_3$O | 0.130 | 0.254 | $o$-NO$_2$ | 0.00609 | 0.0102 |
| $p$-CH$_3$O | 0.238 | 0.547 | $m$-NO$_2$ | 0.0478 | 0.0673 |
| $o$-Cl | 0.0183 | 0.0399 | $p$-NO$_2$ | 0.0401 | 0.0616 |

rates of the reactions of both sodium bromoacetate and bromoacetamide
with substituted sodium benzenesulfinates were measured and are
summarized in Table 11.14.[6 6a]

There is a good linear relationship between the log $k_2$ values of $m$- and
$p$-substituted benzenesulfinates and the Hammett $\sigma$-values. Other examples
are also known.[6 6 b,c]

Sulfinic acids also react readily with sulfenyl halides to yield
thiolsulfonates (eq. 11.53).[67,68] Sulfinate ions are ambient ions (11.14)
and the following canonical structure can be drawn. Therefore, both sulfur

$$\text{RSO}_2\text{H} + \text{R}'\text{SCl} \longrightarrow \text{RSO}_2\text{SR}' \qquad (11.53)$$

$$(\text{R}, \text{R}': \text{Ph}, p\text{-CH}_3\text{C}_6\text{H}_4-, etc.)$$

$$\underset{\underset{-O}{|}}{\overset{+}{R\text{-}\overset{..}{S}}\rightarrow O^-} \longleftrightarrow \underset{\underset{O}{\|}}{R\text{-}\overset{..}{S}\text{-}O^-} \longleftrightarrow \underset{\underset{-O}{|}}{R\text{-}\overset{..}{S}=O}$$

**11.14**

and terminal oxygen can be the nucleophilic centers. Since most reactions
are carried out in protic media containing either water or alcohol or both
and the negatively-charged terminal oxygen is tightly hydrogen bonded,
the central sulfur becomes the reaction center as in the reactions shown
earlier, i.e., eqs. 11.48, 11.49, 11.50 and 11.53, are sulfones are formed.

Like sulfenyl halides, sulfinyl chlorides also react with sodium sulfinate to give the corresponding sulfinyl sulfones[69] (eq. 11.54), while the reaction with sulfonyl chlorides yields the corresponding disulfones (eq. 11.55).[70] In all these reactions, the nucleophilic center of the sulfinate ion is the sulfur atom. However, in undissociated sodium sulfinate, the negative charge is concentrated on the terminal oxygen, which then becomes the reacting center in such a reaction as, for example, with a Meerwein reagent, a powerful alkylating agent, in methylene chloride as shown below (eq. 11.56).[71] The same reaction takes place with free

$$RSO_2Na + R'SOCl \longrightarrow RSO_2SOR' \qquad (11.54)$$

$$RSO_2Na + R'SO_2Cl \longrightarrow RSO_2SO_2R' \qquad (11.55)$$

$$\underset{\underset{O}{|}}{RS\text{-}O}\ Na^+ + (C_2H_5)_3O^+BF_4^- \longrightarrow \underset{\underset{O}{|}}{RS\text{-}OC_2H_5} \qquad (11.56)$$

$$R = C_6H_5-, p\text{-}NO_2C_6H_4-$$

sulfinic acids when pyridine is added.[71] Formation of sulfinyl esters is observed also with other reactive organic halides, such as $ClCOOC_2H_5$ and $ClSOOC_2H_5$ as shown below[72] (eqs. 11.57, 11.58 and 11.59). In these reactions, $^{18}O$-labeled sodium sulfinate gives $^{18}O$-labeled $CO_2$ and $SO_2$. In the former reaction, e.g., eq. 11.58, an anhydride (11.15) formed in the

$$RSO_2Na + \underset{\underset{O}{\|}}{Cl\text{-}COR'} \longrightarrow \underset{\underset{O}{\downarrow}}{RS\text{-}OR'} + CO_2 + NaCl \qquad (11.57)$$

$$RS^{18}O_2Na + \underset{\underset{O}{\|}}{Cl\text{-}COR'} + R''OH \longrightarrow$$

$$\underset{\underset{^{18}O}{\downarrow}}{R\text{-}S\text{-}R''} + NaCl + C^{18}O_2 + R'OH \qquad (11.58)$$

$$RS^{18}O_2Na + \underset{\underset{O}{\downarrow}}{Cl\text{-}SOR'} \longrightarrow \underset{\underset{^{18}O}{\downarrow}}{RS\text{-}OR'} + NaCl + S^{18}O_2 \qquad (11.59)$$

initial step apparently undergoes facile alcohol exchange. Another anhydride, **11.16**, thus formed, would undergo $S_N i$ type reaction via either path (a) or path (b), as shown below (eq. 11.60). The tracer experiment revealed that the $^{18}O$ originally incorporated in the sulfinic acid is equally divided into the $CO_2$ and the ester formed.[73] A similar scheme involving an anhydride, **11.18**, similar to the former one, **11.17**, is conceivable for the reaction shown in eq. 11.59,[74] though the general-base-catalyzed decomposition of **11.18** can also explain the $^{18}O$ data.

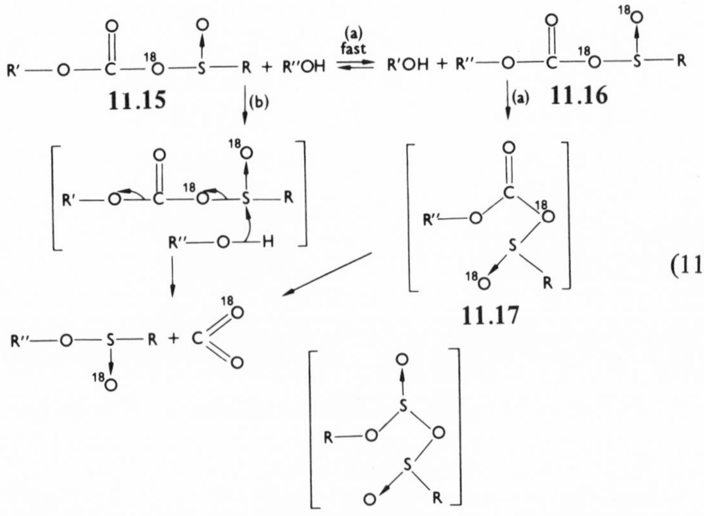

$$(11.60)$$

The reaction between sodium $p$-toluenesulfinate and acyl chloride has long been known, and the stoichiometry of the reaction is shown below (eq. 11.61).[75] The reaction also involves the formation of an acid anhydride intermediate (**11.19**),[76] which in the subsequent reaction gives carboxylic anhydride and sulfinyl sulfone or ethyl sulfinate and carboxylic acid in the presence of ethanol, as shown in eq. 11.62.

$$RSO_2Na + R'COCl \longrightarrow 1/2(R'CO)_2O + 1/4RSO_2SR$$

$$(11.61)$$

$$+ 1/4RSO_3Na + 1/4RSOCl + 3/4NaCl$$

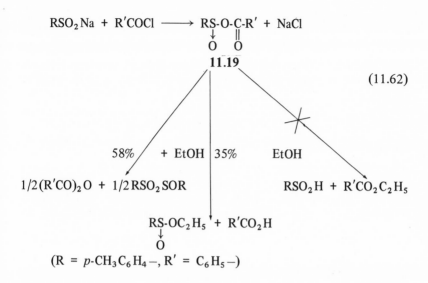

$$(R = p\text{-}CH_3C_6H_4-, R' = C_6H_5-)$$

There is a similar reaction between sodium carboxylate and alkyl chlorosulfite as shown below (eq. 11.63). The reaction proceeds via the formation of an acid anhydride, **11.20**. In fact $n$-butyl benzoyloxysulfinate

$$RCO_2Na + ClSOR' \longrightarrow RC\text{-}O\text{-}S\text{-}OR' + NaCl \qquad (11.63)$$

**11.20**

$(R = C_6H_5, R' = n\text{-}C_4H_9)$ was actually isolated. Thermal decomposition of the anhydride, **11.20**, gives a minute amount (2%) of $n$-butyl benzoate while the major path, 98%, of the reaction proceeds according to the following scheme, yielding benzoic anhydride and $n$-butyl sulfide (eq. 11.64).[77] This disproportionation is catalyzed by a base such as

$$C_6H_5C\text{-}O\text{-}S\text{-}OC_4H_9\text{-}n \longrightarrow 1/2(PhCO)_2O + 1/2(n\text{-}C_4H_9O)_2SO$$

$$+ 1/2SO_2 \qquad (11.64)$$

triethanolamine; the following mechanism was postulated (eq. 11.65). The

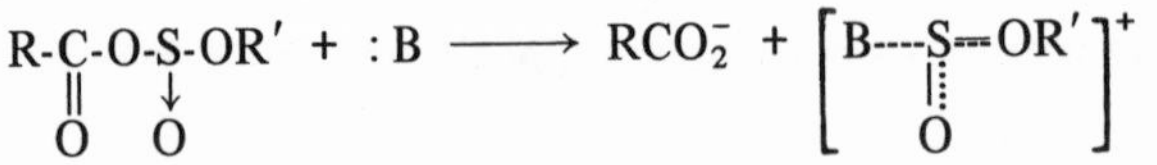

$$\text{R-}\underset{\underset{O}{\|}}{\underset{O}{C}}\text{-O-}\underset{\downarrow}{\underset{O}{S}}\text{-OR}' \; + \; :\text{B} \longrightarrow \text{RCO}_2^- \; + \; \left[ \text{B---}\underset{O}{\overset{}{\underset{}{S}}}\text{=OR}' \right]^+$$

**11.20**

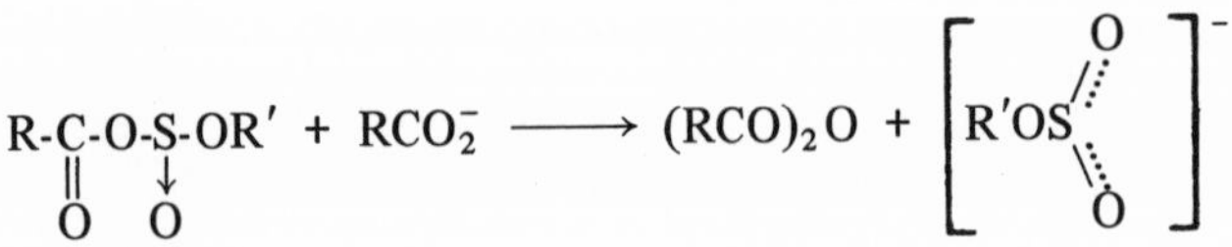

$$\text{R-}\underset{\underset{O}{\|}}{\underset{O}{C}}\text{-O-}\underset{\downarrow}{\underset{O}{S}}\text{-OR}' \; + \; \text{RCO}_2^- \longrightarrow (\text{RCO})_2\text{O} \; + \; \left[ \text{R'OS} \underset{\cdot\cdot}{\overset{O}{\diagup}} \right]^- \qquad (11.65)$$

**11.20**

anhydride, **11.20**, is known to react readily with alcohol, resulting in the formation of a mixed alkyl sulfite as shown below (eq. 11.66), indicating that the sulfinyl group is the preferential attacking site for nucleophiles. In the reaction between sodium benzoate and methyl chlorosulfite, methyl benzoate is obtained in only 5% yield. The formation of methyl benzoate is also considered to proceed *via* an anhydride, **11.20**, since half the [18]O in the original [18]O-labeled benzoic acid is found to remain in the ester,[78] as illustrated in eq. 11.67.

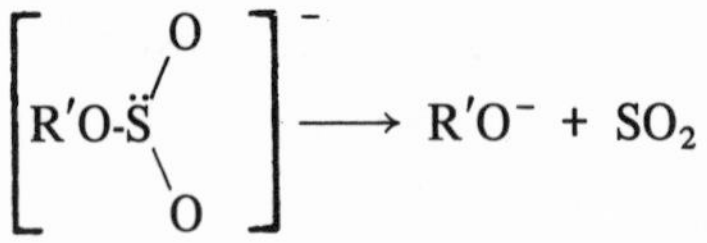

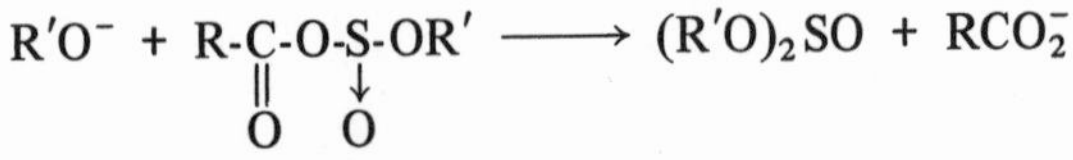

$$\left[ \text{R'O-}\underset{O}{\overset{O}{\underset{\diagdown}{\diagup}}}\underset{}{S} \right]^- \longrightarrow \text{R'O}^- \; + \; \text{SO}_2$$

$$\text{R'O}^- \; + \; \text{R-}\underset{\underset{O}{\|}}{\underset{O}{C}}\text{-O-}\underset{\downarrow}{\underset{O}{S}}\text{-OR}' \longrightarrow (\text{R'O})_2\text{SO} \; + \; \text{RCO}_2^-$$

**11.20**

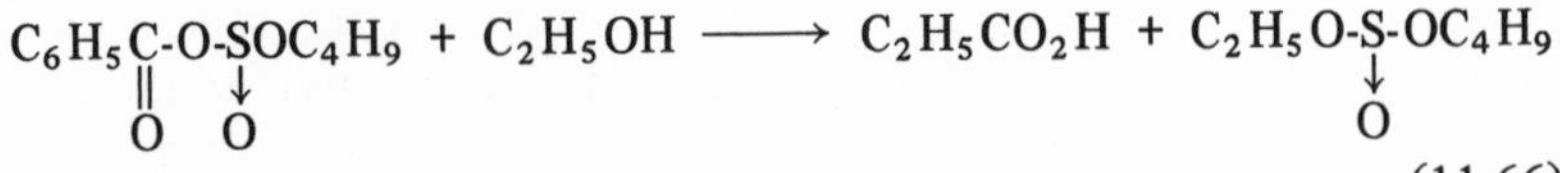

$$\text{C}_6\text{H}_5\underset{\underset{O}{\|}}{\underset{O}{C}}\text{-O-}\underset{\downarrow}{\underset{O}{S}}\text{OC}_4\text{H}_9 \; + \; \text{C}_2\text{H}_5\text{OH} \longrightarrow \text{C}_2\text{H}_5\text{CO}_2\text{H} \; + \; \text{C}_2\text{H}_5\underset{\downarrow}{\underset{O}{O\text{-S}}}\text{-OC}_4\text{H}_9$$

$$\qquad\qquad\qquad\qquad\qquad\qquad\qquad\qquad\qquad\qquad\qquad\qquad\qquad\qquad (11.66)$$

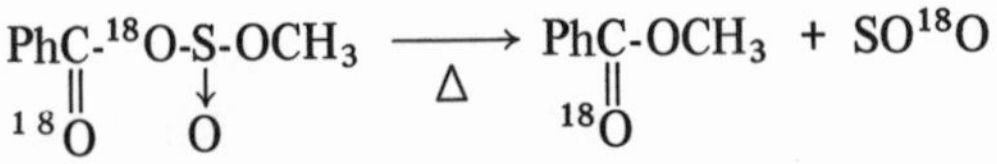

$$\text{PhC-}^{18}\text{O-}\underset{\downarrow}{\underset{O}{S}}\text{-OCH}_3 \xrightarrow{\;\Delta\;} \text{PhC-OCH}_3 \; + \; \text{SO}^{18}\text{O} \qquad\qquad (11.67)$$

$$\underset{^{18}O}{\overset{\|}{}} \qquad\qquad\qquad\qquad\qquad \underset{^{18}O}{\overset{\|}{}}$$

**11.20**

Such trithiocarbonate S,S-dioxides as $PhSO_2CS_2Me$ and $p\text{-}CH_3C_6H_4SO_2CS_2Ph$ were prepared from $PhSO_2Na$ and $ClCS_2Me$ and by the reaction of $p\text{-}CH_3C_6H_4SO_2Na$ with $ClCS_2Ph$, respectively. In the reaction between $MeSO_2Na$ and $ClCS_2Ph$, $(PhS)_2CS$ and a small amount of $MeSO_2CH(SPh)SSPh$ were obtained.[79]

### 11.4.5.2 Addition to Olefinic Double Bonds

Sulfinic acids add to $\alpha,\beta$-unsaturated Michael-type olefins to yield the corresponding sulfones.[80] The reaction is catalyzed either by a buffer solution or weak acids. The following are but a few examples (eqs. 11.67, 11.68, 11.69 and 11.70). In all these Michael-type addition reactions, the nucleophilic center of sulfinic acid appears to be the sulfur atom, presumably due both to protonation shielding the terminal oxygen and also the markedly high thermodynamic stability of the resulting sulfone.

The following methods of disulfone formation are also known (eqs. 11.71a and 11.71b).[80e]

$$p\text{-}CH_3C_6H_4SO_2Na + CH{=}CHCN \xrightarrow{H_3BO_3} p\text{-}CH_3C_6H_4SO_2CH_2CH_2CN$$

$$(11.68a)$$

$$PhSO_2H + CH_2{=}CHCOOH \xrightarrow{H_3BO_3} PhSO_2CH_2CH_2COOH \quad (11.68b)$$

$$p\text{-}CH_3C_6H_4SO_2Na + CH_2{=}CHCOOCH_3 \xrightarrow{CH_3COOH}$$

$$p\text{-}CH_3C_6H_4SO_2CH_2CH_2COOCH_3 \quad (11.69)$$

$$p\text{-}CH_3C_6H_4SO_2Na + CH_2{=}CClCN \underset{CH_3COOH}{\overset{H_3BO_3}{\rightleftharpoons}} \begin{array}{l} p\text{-}CH_3C_6H_4SO_2CH{=}CHCN \\ p\text{-}CH_3C_6H_4SO_2CH_2{-}CHClCN \end{array}$$

$$(11.70)$$

$$HO_2SCH_2CH_2SO_2H + CH_2{=}CHCONH_2 \longrightarrow$$

$$(NH_2COCH_2CH_2SO_2CH_2)_2 \quad (11.71a)$$

$$HO_2S(CH_2)_4SO_2H + CH_2{=}CHCN \longrightarrow (NCCH_2CH_2SO_2CH_2CH_2)_2$$

$$(11.71b)$$

The addition of substituted benzenesulfinic acids to acrylonitrile in aqueous buffers was studied kinetically by Ogata *et al.*,[81] who found that the reaction follows second-order kinetics and the rate is independent of pH while the effects of substituents correlate well with their Hammett $\sigma$-values.

### 11.4.5.3 Reactions with Aldehydes

Sulfinic acids add to the carbonyl group of aldehydes to yield the corresponding $\alpha$-oxysulfones,[82,83,84] **11.21**; compare the similar addition of $H_2SO_3$ to aldehydes and ketones shown below:

$$RSO_2H + R'CHO \longrightarrow \underset{\underset{\textbf{11.21}}{\overset{|}{R'}}}{RSO_2CHOH} \qquad (11.71c)$$

$$p\text{-}CH_3C_6H_4SO_2H \begin{array}{l} \xrightarrow{+\ HCHO} \quad p\text{-}CH_3C_6H_4SO_2CH_2OH \\ \hspace{3.5cm} (11.72)^{85} \\ \xrightarrow{+\ CH_3CHO} \quad p\text{-}CH_3C_6H_4SO_2\underset{\underset{CH_3}{|}}{CHOH} \\ \hspace{3.5cm} (11.73) \\ \xrightarrow{+\ CH_3CH=CHCHO} \quad p\text{-}CH_3C_6H_4SO_2\underset{\underset{CH_3CH=CH}{|}}{CHOH} \\ \hspace{3.5cm} (11.74) \end{array}$$

The following Mannich-type reactions are also known[66,86,87] (eqs. 11.75, 11.76 and 11.77). One pathway is that the $\alpha$-oxysulfone, **11.21**, formed in the initial step, undergoes dehydrative condensation with sulfinic acid to give the final product, while the other may involve the prior acid-catalyzed condensation of acyl amide with formaldehyde, as shown in eq. 11.78. Both routes are equally conceivable.

$$CH_3SO_2H + CH_2O + H_2N\overset{\overset{O}{\|}}{C}R \longrightarrow CH_3SO_2CH_2NH\overset{\overset{O}{\|}}{C}R \qquad (11.75)$$

$$CH_3SO_2H + CH_2O + H_2NSO_2CH_3 \longrightarrow CH_3SO_2CH_2NHSO_2CH_3$$

$$(11.76)$$

$$RSO_2H + CH_2O + NH_2CO_2Et \longrightarrow RSO_2CH_2NHCO_2Et + H_2O$$

$$(11.77)^{88a}$$

$$\underset{\substack{\|\\ R\text{-}C\text{-}NH_2}}{\overset{O}{}} + CH_2O \xrightarrow{H^+} \underset{\substack{\|\\ R\text{-}C\text{-}NH\text{-}CH_2OH}}{\overset{O}{}} \xrightarrow{-H_2O}$$

$$(11.78)$$

$$[\underset{\substack{\|\\ R\text{-}C\text{-}N=CH_2}}{\overset{O}{}}] \xrightarrow{CH_3SO_2H} \underset{\substack{\|\\ R\text{-}C\text{-}NHCH_2SO_2CH_3}}{\overset{O}{}}$$

### 11.4.5.4 Reactions with Lactones

Since sulfinic acids are nucleophiles, they open up lactone rings to afford sulfonyl carboxylic acids:

$$\underset{\substack{|\quad\ |\\ O\text{---}C=O}}{R\text{-}CH(CH_2)_n} + R'SO_2H(\text{or Na}) \longrightarrow \underset{\substack{|\\ R}}{R'SO_2CH(CH_2)_nCOOH(\text{or Na})}$$

$$(11.79)$$

$$n = 1, \beta\text{-lactone}; n = 2, \gamma\text{-lactone}$$

In this case too, the sulfur is the nucleophilic center for attack on $C_\beta$ or $C_\gamma$, resulting in alkyl-O cleavage forming a $\beta$- or $\gamma$-sulfonyl carboxylic acid:

$$\underset{\substack{|\quad\ |\\ O\text{---}C=O}}{CH_2CH_2} + p\text{-}CH_3C_6H_4SO_2Na \longrightarrow p\text{-}CH_3C_6H_4SO_2CH_2CH_2COONa$$

$$(11.80)^{88b}$$

$$\underset{\substack{|\qquad\quad\ |\\ O\text{------}C=O}}{CH_2CH_2CH_2} + PhSO_2Na \longrightarrow PhSO_2CH_2CH_2CH_2COONa$$

$$(11.81)^{89}$$

### 11.4.5.5 Reactions with Quinones

The addition of sulfinic acids to quinones has been known to yield dihydroxy-diaryl sulfones[90], for example eq. 11.82. The addition of $\omega$-chloroalkane sulfinic acids to $p$-benzoquinones is followed by subsequent intramolecular nucleophilic displacement (eq. 11.83).[92] These additions are typical 1,4-addition to conjugated unsaturated ketones by another sulfur nucleophile. A kinetic study of this reaction was conducted by Ogata $et$ $al.$,[93] who found that the rates of addition of substituted benzenesulfinic acids to $p$-benzoquinone can be nicely correlated with the Hammett $\sigma$-values.

Addition of arenesulfinates to N,N-dialkyl-$p$-quinondiimines gives the four products shown in eq. 11.84.[94] The variation in the products formed was investigated under several conditions at pH 5~9.

$$p-\text{AcNHC}_6\text{H}_4\text{SO}_2\text{H} \ + \quad \text{(quinone)} \quad \longrightarrow \quad p-\text{AcNHC}_6\text{H}_4\text{SO}_2-\text{(dihydroxyphenyl)} \qquad (11.82)^{91}$$

$$\text{Cl(CH}_2)\,\text{SO}_2\text{H} \ + \quad \text{(quinone)} \quad \longrightarrow \quad \text{(sulfone product)} \qquad (11.83)^{92}$$

$$(11.84)$$

## 11.4.5.6  Reactions with Azo, Azine and Diazonium Compounds

Refluxing azobenzenes or phenazine with excess arenesulfinic acid in alcohol gives $p$-arenesulfonylated azobenzenes[95a] or $p$-arenesulfonyl phenazine[95b] (eqs. 11.85 and 11.86). In this reaction, sulfinic acids are reduced to the corresponding thiolsulfonates. At room temperature, sulfinic acids add to either the $-N=N-$ double bond or the azine group of phenazine to form isolable addition intermediates, **11.22, 11.23**; however, these addition intermediates cannot be involved in the rearrangement, since the yields of rearrangement products are so low when these intermediates, **11.22, 11.23**, are heated alone, as compared to those in the normal refluxing of the mixtures.[96] The normal rearrangement reaction of benzenesulfinic acid with azobenzene was studied with unsymmetrically substituted azobenzene, and the results shown in eqs. 11.87 and 18.88 were obtained. In this reaction, sulfinic acids are not only nucleophilic acids adding to $-N=N-$ double bond, but oxidizing agents oxidizing hydride in order to complete the reaction, and in turn being reduced to the corresponding thiolsulfonates. Although no detailed information is available for interpretation of the mechanism of the reaction, the orientation of the rearrangement of the sulfonyl group in this reaction is exactly opposite to that in the reaction between asymmetrically substituted azoxybenzenes and arenesulfinyl chlorides yielding the corresponding arenesulfonyl azobenzenes. Equation 11.89 is a good example to illustrate this.[96]

$$(11.85)^{95a}$$

(Ar; $p-CH_3C_6H_4$, $p-ClC_6H_4$)

$$(11.86)^{95b}$$

**11.22**                              **11.23**

$$(11.87)$$

$$(11.88)$$

$$(11.89)$$

Addition of aromatic sulfinic acids to diazonium ions has also been known to occur, but the products are arenesulfonylhydrazines by Michael-type addition products to N≡N bonds.[97]

### 11.4.5.7 Reactions with N,N-Dimethylaniline[98]

When N,N-dimethylaniline is refluxed with excess arenesulfinic acid in ethanol, p-dimethylaminobenzyl aryl sulfone, **11.24**, and p-methyl-aminobenzyl aryl sulfone, **11.25**, are obtained along with p-dimethyl-aminophenyl aryl sulfoxide, **11.26** , diaryl disulfide, **11.27**, diaryl thiol-sulfonate, **11.28**, and arenesulfonic acid (eq. 11.90). The yield of the demethylation product, **11.25**, varies with the arenesulfinic acid used, while the demethylation appears to be limited solely to N,N-dimethyl; N-methyl aromatic amines and other alkylated aromatic amines do not undergo a similar reaction. Based on a detailed survey of intermediates in-volved in the reaction, the scheme shown in eq. 11.91 was suggested as the major path of the reaction.[98]

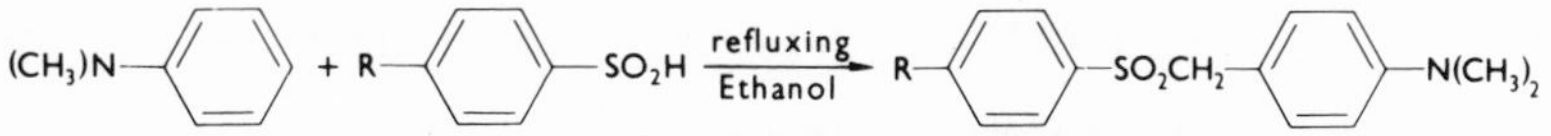

**11.24** major

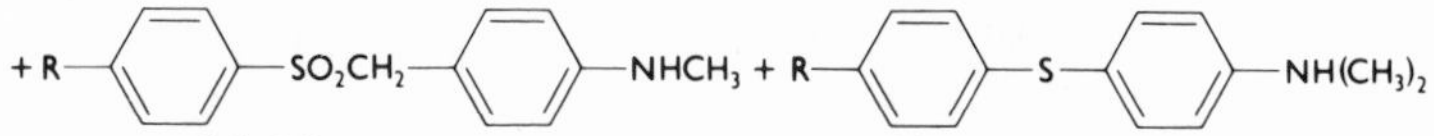

**11.25** major **11.26** minor

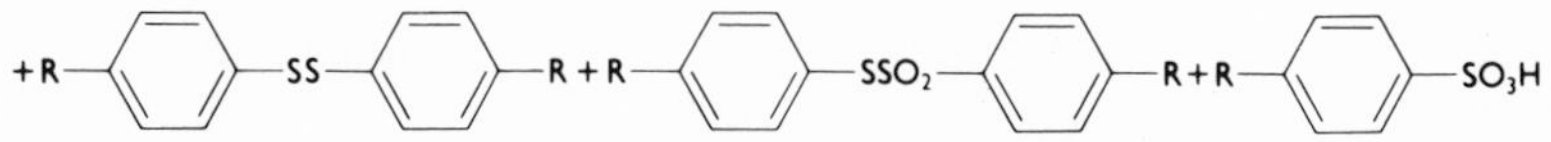

**11.27** major **11.28** major minor

(11.90)

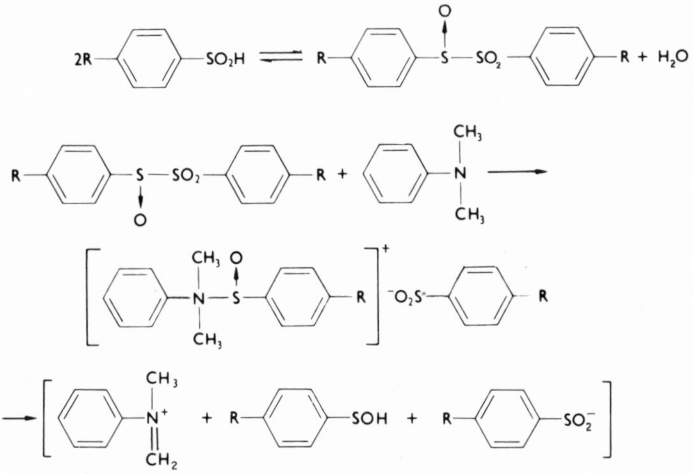

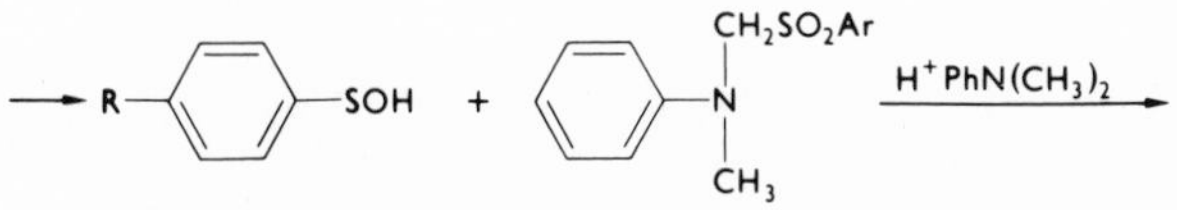

**11.29** **11.30**

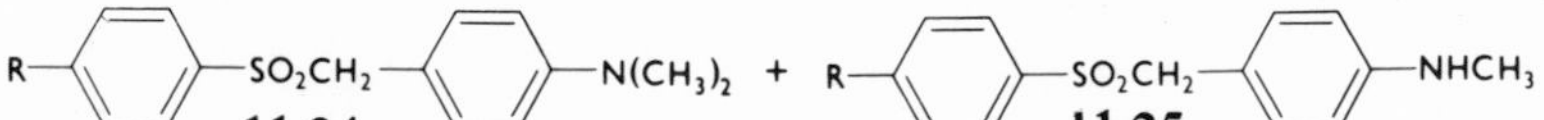

**11.24** **11.25**

**11.29** 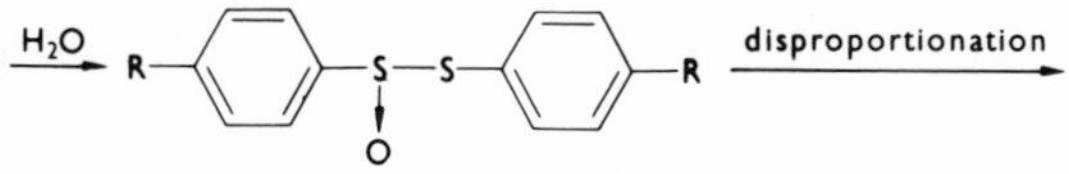

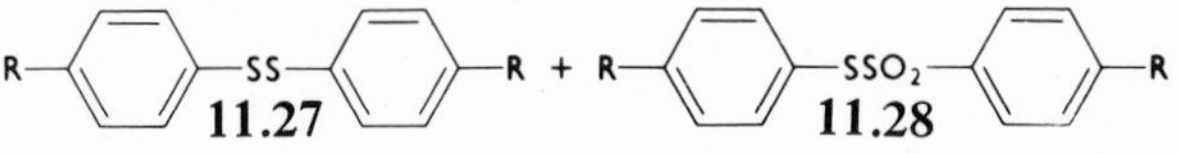

**11.27** **11.28**

(11.91)

## 11.5 CHEMICAL REACTIVITIES OF SULFINIC ESTERS

### 11.5.1 Reactions with Grignard Reagents

Reactions of sulfinic esters with Grignard reagents give the corresponding sulfoxides (eq. 11.92). Although this is a normal Grignard reaction, it is extremely significant, since it results in complete inversion of the configuration around the sulfur and gives optically highly pure sulfoxides from optically active sulfinates. By treating $l$-menthyl $p$-toluenesulfinate with alkyl or aryl Grignard reagent, $d$-ethyl $p$-tolyl sulfoxide of very high optical activity is obtained in a high yield (eq. 11.93).[99] Stereochemistry of this reaction and absolute configurations of the resulting sulfoxides have been extensively investigated and well clarified by Mislow and his coworkers.[100] An [18]O-tracer experiment with [18]O-labeled methyl sulfinate was conducted by Kobayashi *et al.*, who found the sulfinyl oxygen to be retained throughout the reaction, as shown below (eq. 11.94).[101] This is in keeping with the $S_N2$-type inversion process that has been suggested earlier.

$$\underset{\overset{\uparrow}{\text{O}}}{\text{R-S-OR}'} \;+\; \text{R}''\text{MgX} \;\longrightarrow\; \underset{\overset{\uparrow}{\text{O}}}{\text{R-S-R}''} \;+\; \text{R}'\text{OMgX} \qquad (11.92)$$

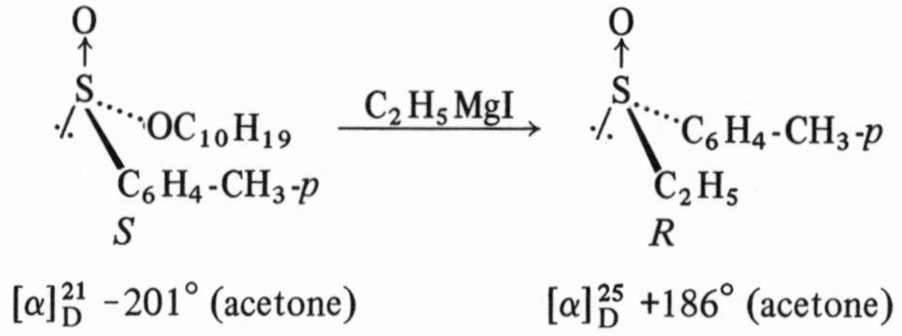

$$
\underset{S}{\overset{\overset{\overset{\uparrow}{\text{O}}}{\text{S}}}{\underset{\text{C}_6\text{H}_4\text{-CH}_3\text{-}p}{\cdots\text{OC}_{10}\text{H}_{19}}}}
\;\xrightarrow{\text{C}_2\text{H}_5\text{MgI}}\;
\underset{R}{\overset{\overset{\overset{\uparrow}{\text{O}}}{\text{S}}}{\underset{\text{C}_2\text{H}_5}{\cdots\text{C}_6\text{H}_4\text{-CH}_3\text{-}p}}}
\qquad (11.93)
$$

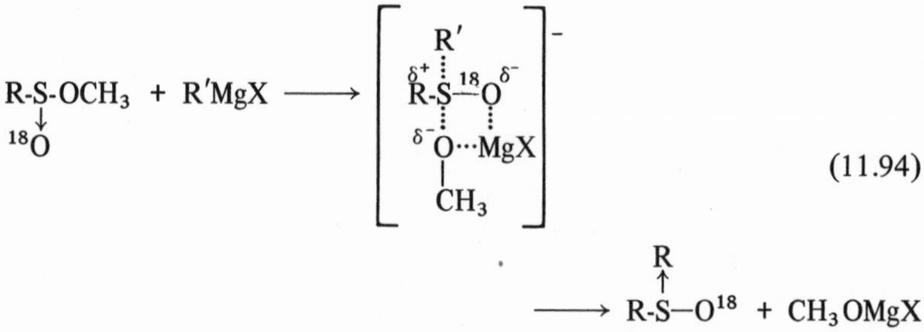

$$[\alpha]_D^{21} -201° \text{ (acetone)} \qquad\qquad [\alpha]_D^{25} +186° \text{ (acetone)}$$

$$
\underset{^{18}\text{O}}{\overset{\downarrow}{\text{R-S-OCH}_3}} \;+\; \text{R}'\text{MgX} \;\longrightarrow\;
\left[\;
\begin{array}{c}
\text{R}' \\
\overset{\delta^+}{\text{R-S}}\!\!-\!\!^{18}\!\overset{\delta^-}{\text{O}} \\
\delta^-\text{O}\cdots\text{MgX} \\
\text{CH}_3
\end{array}
\;\right]^{-}
$$

$$\qquad (11.94)$$

$$\longrightarrow\; \underset{}{\overset{\overset{\uparrow}{R}}{\text{R-S}}}\!\!-\!\text{O}^{18} \;+\; \text{CH}_3\text{OMgX}$$

## 11.5.2  Rearrangements

Certain sulfinic esters rearrange to the corresponding sulfones upon heating. For example, α-methylbenzyl p-toluenesulfinate rearranges to α-methylbenzyl p-tolyl sulfone upon heating.[102] However, the rearrangement appears to be limited solely to those sulfinic esters of either benzylic or allylic alcohols that give stable carbonium ions upon heterolysis, as shown in Table 11.15.[103] Ethyl benzenesulfinate does not give the corresponding sulfone, but decomposes upon pyrolysis.[104] Since the rearrangement is known to proceed fast in polar media such as acetic acid and nitromethane and slowly in nonpolar media, the ionic mechanism shown in eq. 11.95 involving the formation of carbonium ions has been postulated.

$$\underset{\underset{O}{\downarrow}}{R\text{-}S\text{-}OR'} \longrightarrow [RSO_2^- + {}^+R'] \longrightarrow \underset{\underset{O}{\downarrow}}{R\overset{\overset{O}{\|}}{\text{-}S\text{-}}R'} \qquad (11.95)$$

**TABLE 11.15**

Yields of Rearrangement Products in the Following Pyrolysis in Either Nonpolar Media or Neat

$$\underset{\underset{O}{\downarrow}}{R\text{-}S\text{-}OR'} \longrightarrow \underset{\underset{O}{\downarrow}}{R\overset{\overset{O}{\|}}{\text{-}S\text{-}}R'}$$

| R | R' | Yield of sulfone (%) |
|---|---|---|
| p-Tolyl | Methyl | 0 |
| | Phenyl | 0 |
| | Benzyl | 0 |
| | o-Nitrobenzyl | 0 |
| | α-Methylbenzyl | 20~30 |
| | α-Dimethylbenzyl | 87 |
| | Diphenylmethyl | 100 |
| | p-Chloro-diphenylmethyl | 95 |
| | Phenyl-o-tolylmethyl | 95 |
| | Phenyl-p-tolylmethyl | 80 |
| p-Chlorophenyl | Diphenylmethyl | 100 |

Trityl $o$-toluenesulfinate undergoes rearrangement to give the corresponding sulfone on refluxing for 1.5 hr in chloroform (eq. 11.96).[105] The rate of the rearrangement increases in polar solvents in the following order: $CCl_4 <$ $CHCl_3 <$ nitrobenzene, while addition of $Bu_4NN_3$ into the reaction system in acetonitrile results in the formation of trytyl azide in yields of up to 45%. Based on these observations, the scheme shown in eq. 11.97 was proposed for the mechanism of the rearrangement.

$$\text{(11.96)}$$

$$R\text{-}S\text{-}OR' \rightleftharpoons [R\text{-}S\overset{\delta^-}{\text{-}}\overset{\delta^+}{O}\text{----}R'] \rightleftharpoons [R\text{-}SO_2^- + {}^+R'] \xrightarrow{Bu_4NN_3} R'N_3$$

$$\text{(11.97)}$$

In A-1 type solvolysis of alkyl 2,6-dimethylbenzenesulfinate, both rearrangement and substitution take place concurrently, and the rate appears to increase with the increase of solvent polarity while the yield of rearrangement product, the sulfone, also increases according to the increase in carbonium ion stability of the alkyl group.[106]

Among rearrangements of allylic esters of benzenesulfinate, those of crotyl, 11.29, and $\alpha$-methylallyl benzenesulfinate, 11.30, are interesting since both give crotyl phenyl sulfone, 11.31, in nearly identical yields.[104] This reaction apparently involves the resonance-stabilized carbonium ion intermediate, which eventually collapses to the thermodynamically more stable crotyl sulfone, 11.31 (eq. 11.98). Sulfinic esters of propargyl alcohols are also known to rearrange to the corresponding sulfones (eq. 11.99). For example $(+)-R$ 1-methyl 2-propynyl $p$-toluenesulfinate was found to rearrange to $(-)$ 1,2-butadienyl $p$-tolyl sulfone, apparently via the five-membered cyclic mechanism shown in eq. 11.99.[107]

Ph-S-CH$_2$-CH=CH-CH$_3$
$\downarrow$
O
**11.29**

[PhSO$_2^-$ +

Ph-S-CH-CH=CH$_2$
$\downarrow$ |
O CH$_3$        ($^+$CH$_2$-CH=CH-CH$_3$ $\longleftrightarrow$ CH$_2$=CH-$^+$CH-CH$_3$)]
**11.30**                                              (11.98)

$\longrightarrow$ Ph-SO$_2$-CH$_2$-CH=CH-CH$_3$
**11.31**

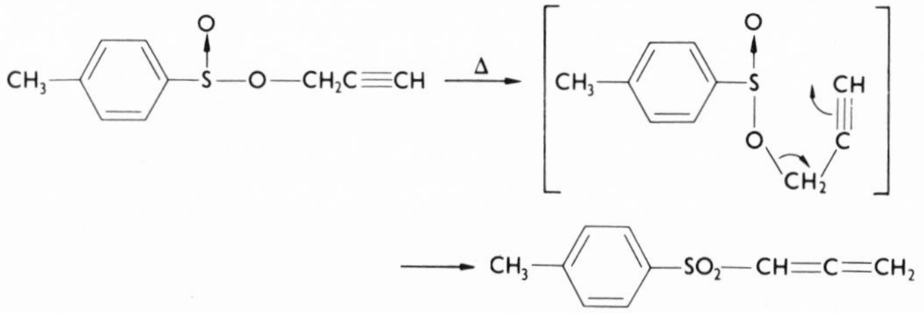

$$\text{(11.99)}$$

### 11.5.3 Hydrolysis of Sulfinic Esters and Acids

Alkyl $p$-toluenesulfonates undergo hydrolysis $via$ alkyl-O fission in both acidic and alkaline conditions.[108] However, this is not the case with sulfinic esters. For example, hydrolysis of methyl $p$-toluenesulfinate with H$_2$$^{18}$O–dioxane mixture gives methanol in both acidic and alkaline media with no $^{18}$O incorporation[109] (eq. 11.100); that is, acyl-O fission takes place. The reason for this difference is not at all clear; however, the weaker S–O bond and smaller repulsive field effect due to the terminal negatively charged O$^-$ atom in sulfinates compared to those in the corresponding sulfonates may be partly or jointly responsible for the different mode of hydrolysis.

In the acid hydrolysis of sulfinic esters with H$_2$ $^{18}$O, usually more than one equivalent $^{18}$O is found in the resulting sulfinic acids.[110a] In a separate experiment, sulfinic acids were found to undergo oxygen exchange in the presence of acid. The fast equilibrium suggested by Kice et al.[44] (eq. 11.101) may be responsible. Recently, Kobayashi and his coworker have conducted a kinetic study of the oxygen-exchange reaction of substituted benzenesulfinic acids in H$_2$ $^{18}$O, and they propose a mechanism for the reaction involving the formation of the sulfinylsulfone as an intermediate.[111a]

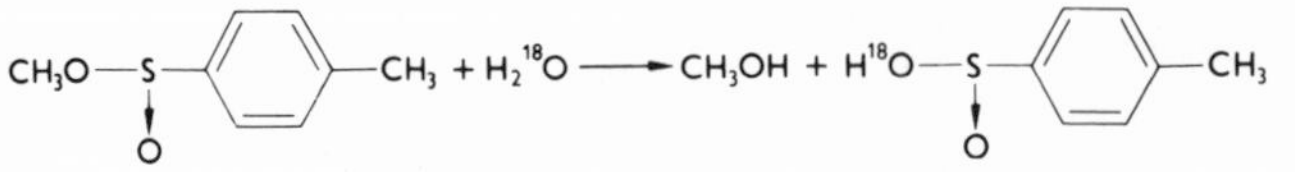

$$CH_3O-\underset{O}{\overset{|}{S}}-\underset{}{\bigcirc}-CH_3 + H_2{}^{18}O \longrightarrow CH_3OH + H^{18}O-\underset{O}{\overset{|}{S}}-\underset{}{\bigcirc}-CH_3 \qquad (11.100)$$

$$2ArSO_2H \underset{fast}{\overset{fast}{\rightleftarrows}} Ar\text{-}\underset{O}{\overset{|}{S}}\text{-}SO_2\text{-}Ar + H_2O \qquad (11.101)$$

The hydrolysis of $o$-phenylene sulfite proceeds $10^3$ times faster than the open-chain analogue.[111b] Similar five-membered cyclic sulfonates and phosphates are also known to undergo fast hydrolysis. Steric strain is considered to be responsible for the rapid hydrolysis of these five-membered esters of hetero atoms. In this case too, acyl-O fission takes place by the attack of water on the sulfur atom. Halide ions such as $Cl^-$ and $F^-$ are known to accelerate the hydrolysis, and the two schemes, (a) and (b), shown in eq. 11.102 are suggested.[112]

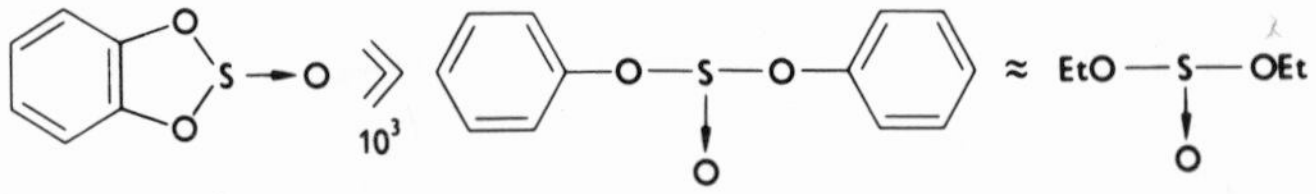

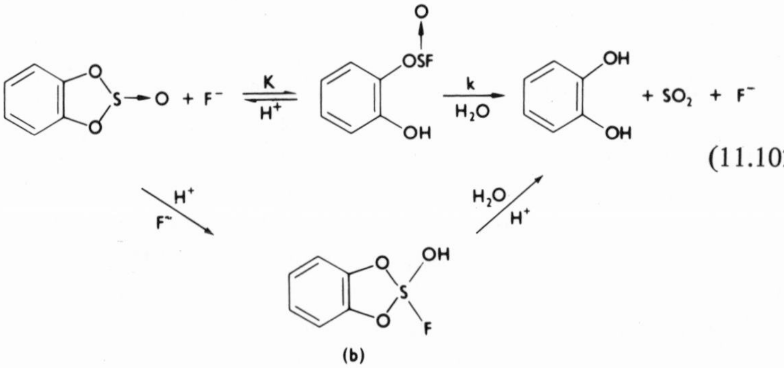

$$(11.102)$$

Another interesting observation is the following reaction of cyclic sulfites with benzaldehyde (eq. 11.103). With the use of $^{18}O$-labeled $p$-nitrobenzaldehyde, formation of the unlabeled compound was obtained as shown in eq. 11.104, and the mechanism shown in eq. 11.105 was suggested.[112]

$$\underset{\substack{CH_3 \\ | \\ C-O \\ | \ H \\ CH_3 | \quad S{\to}O \\ | \\ C-O \\ | \\ H}}{} + PhCHO \longrightarrow \underset{\substack{CH_3 \\ | \\ C-O \\ | \ H \\ CH_3 | \quad CHPh \\ | \\ C-O \\ | \\ H}}{} + SO_2 \qquad (11.103)$$

$$\underset{\substack{CH_2-O \\ | \qquad S{\to}O \\ CH_2-O}}{} + {}^{18}O{=}\underset{\substack{| \\ H}}{C}\text{-}C_6H_4NO_2 \longrightarrow \underset{\substack{CH_2-O \\ | \qquad CHC_6H_4NO_2 \\ CH_2-O}}{} + SO^{18}O \qquad (11.104)$$

$$\underset{\substack{RCH-O \\ | \qquad S{\to}O \\ RCH-O}}{} + H^+ \longrightarrow \underset{\substack{RCH-O \\ | \qquad S^+ \\ RCH-O \quad OH}}{} \xrightarrow{+\ R'CHO} \underset{\substack{RCH-O \quad O{=}CHR' \\ | \qquad S \\ RCH-O \quad OH}}{}$$

$$(11.105)$$

$$\underset{\substack{RCH-O \quad {}^+CHR' \\ | \qquad S \\ RCH-O \quad OH}}{} \xrightarrow{-H^+} \underset{\substack{RCH-O \quad CHR' \\ | \qquad O \\ RCH-O{-}S}}{} \longrightarrow SO_2 + \underset{\substack{RCH-O \\ | \qquad CHR' \\ RCH-O}}{}$$

Sulfinic esters undergo the reaction with hydrazine (eq. 11.106).[113] The reaction appears to involve nucleophilic attack of hydrazine on sulfur to form sulfinyl hydrazide, which undergoes disproportionation to give the final products. Sulfinic acids also react with hydrazoic acid to give sulfonamides, presumably *via* the scheme shown in eq. 11.107.[114] However no Schmidt-type reaction is observed.

$$4\text{R-S-OR}' + 3\text{N}_2\text{H}_4\text{H}_2\text{O} \longrightarrow 2(\text{R-S})_2 + 4\text{R}'\text{OH} + 7\text{H}_2\text{O} + 3\text{N}_2$$
$$\downarrow$$
$$\text{O}$$

(11.106)

$$\text{R-S-OH} + \text{H}^+ \longrightarrow \text{R-}\overset{+}{\text{S}}\text{-OH} + \text{N}_3^- \longrightarrow \text{R-S-OH}$$

with substituents: O (below first), OH (below second), N$_3$/OH (on third)

(11.107)

$$\xrightarrow{-\text{N}^2} \quad \text{R-S-OH} \longrightarrow \text{R-S-OH} \longrightarrow \cdot\text{R-S}{\rightarrow}\text{O}$$

with substituents: Ṅ:/OH, NH(double bond)/O, NH$_2$/O

## 11.6 REFERENCES

1. W.E. Truce and A.M. Murphy, *Chem. Revs.*, **48**, 69 (1951).
2. C.J.M. Stirling, "The Chemistry of Organic Sulfur Compounds," Vol. 3. (1966).
3. a) F.C. Whitmore and F.H. Hamilton, *Org. Syn.*, Coll Vol. **1**, 492 (1941).
   b) H. Gilman, E.W. Smith, and H.J. Oatfield, *J. Amer. Chem. Soc.*, **56**, 1413 (1934).
4. Ger. P. 912, 091 May. 24 (1954).
5. N. Kunieda, K. Sakai, and S. Oae, *Bull. Chem. Soc. Japan*, **41**, 3015 (1968).
6. a) S. Smiles and C.M. Bere, *Org. Syn.*, Coll. Vol. 1, 7 (1941).
   b) S. Krishna and H. Singh, *J. Amer. Chem. Soc.*, **50**, 792 (1928).
   c) M. Kulka, *ibid.*, **72**, 1216 (1950).
7. a) L. Field and F.A. Geunwald, *J. Org. Chem.*, **16**, 964 (1951).
   b) S.G. Mairanovski and M.B. Neiman, *Dokl. Akad. Nauk SSSR*, **79**, 85 (1951).
8. a) H.R. Todd and R.L. Shriner, *J. Amer. Chem. Soc.*, **56**, 1382 (1934).
   b) J. Thomas, *J. Chem. Soc.*, **95**, 342 (1909).
   c) W.A. Silvester and W.P. Wynne, *ibid.*, **693** (1936).
9. a) C.S. Marval and R.S. Johnson, *J. Org. Chem.*, **13**, 822 (1948).
   b) E. Rothstein, *J. Chem. Soc.*, 309 (1937).
   c) J.V. Braun, *Chem. Ber.*, **63**, 2839 (1930).
   d) H.G. Haulton and H.V. Tartar, *J. Amer. Chem. Soc.*, **60**, 544 (1938).
10. W.E. Truce and J.F. Lyons, *ibid.*, **73**, 126 (1951).
11. a) R.M. Hann, *ibid.*, **57**, 2166 (1936).
    b) J.R. Nooi, P.C. Van-der Hoeven and W.P. Haslinghuis, *Tetrahedron Lett.*, 2531 (1970).
12. R.E. Dabby, J. Kenyon, and R.E. Mason, *J. Chem. Soc.*, 4881 (1952).
13. W.M. Ziegler and R. Connor, *J. Amer. Chem. Soc.*, **62**, 2596 (1940).
14. D.B. Hope, C.D. Morgan, and M. Walti, *J. Chem. Soc.*, 270 (1970).
15. a) W.E. Truce, W.J. Ray, Jr., O.L. Norman, and D.B. Eickemeyer, *J. Amer. Chem. Soc.*, **80**, 3625 (1958).
    b) W.E. Truce and W. Brand, *ibid.*, **35**, 1828 (1970).
16. a) R.N. Haszeldine and J.M. Kidd, *J. Chem. Soc.*, 2091 (1955).
    b) K.R. Brower and I.B. Dauglass, *J. Amer. Chem. Soc.*, **73**, 5787 (1951).

17. For details see "Sulfur Bonding" by C.C. Price and S. Oae, Ronald Press, New York (1962).
18. I.B. Douglass and D.R. Poole, *J. Org. Chem.*, **23**, 330 (1958); S. Oae and K. Ikura, *Bull. Chem. Soc. Japan*, **39**, 1309 (1966).
19. E.N. Guryanova and Ya. K. Syrkin, *Zhur. Fiz. Khim.*, **23**, 105 (1949).
20. O. Exner, P. Dembech, and P. Vivarelli, *J. Chem. Soc.*, (B), 278 (1970).
21. S. Detoni and D. Hodži, *J. Chem. Soc.*, 3163 (1955).
22. D. Barnard, J.M. Fabian and H.P. Koch, *ibid.*, 2442 (1949). E.S. Waight, *ibid.*, 2440 (1952).
23. J.P. Weidner and S.S. Block, *Appl. Spectr.*, **23**, 337 (1969).
24. M. Kobayashi and N. Naga, *Bull. Chem. Soc. Japan*, **39**, 1788 (1966).
25. G. Ghersetti and G. Modena, *Ann. Chem.* (Rome), **53**, 1083 (1963); B. Bonini, S. Ghersetti, and G. Modena, *Gazz. Chem. Ital.*, **93**, 1222 (1963).
26. C.C. Price and J.J. Hydock, *J. Amer. Chem. Soc.*, **74**, 1943 (1952).
27. a) Y. Hirota, "Interpretation of UV and Visible Absorption Spectra," Chapter 4, Kyoritsu Shuppan (1965).
    b) H.H. Jaffé and M. Orchin, "Theory and Applications of Ultraviolet Spectroscopy," p. 466, John Wiley & Sons, N.Y. (1962).
28. H. Bredereck, G. Brod, and G. Höshele, *Chem. Ber.*, **88**, 438 (1955).
29. H. Phillips, *J. Chem. Soc.*, **127**, 2552 (1925).
30. O. Bastiasen and H. Viervoll, *Acta Chem. Second.*, **2**, 702 (1948).
31. E.B. Fleischer, M. Axelrod, M. Green and K. Mislow, *J. Amer. Chem. Soc.*, **86**, 3395 (1964).
32. D.G. Hellier, J.G. Tilett, H.F. van Woerden, and R.F.M. White, *Chem. Ind.*, **1963**, (1956).
33. a) G. Wood and M. Miskow, *Tetrahedron Lett.*, 4433 (1966).
    b) H.F. van Woerden and E. Havinga, *Rec. Trav. Chim. Pays-Bas*, **86**, 341, 353 (1967).
    c) H.F. van Woerden, *Chem. Revs.*, **63**, 557 (1960).
34. P.C. Lauterbru, J.G. Pritchard, and R.L. Vollmer, *J. Chem. Soc.*, 5307 (1963).
35. S.E. Forman, A.J. Durbetaki, M.V. Cohen, and R.A. Olofson, *J. Org. Chem.*, **30**, 169 (1954).
36. S. Sugden, J.B. Reed, and H. Wilkins, *J. Chem. Soc.*, 1525 (1925).
37. P. Rumpf and J. Sadet, *Bull. Soc. Chem. France*, No. 4, 447 (1958); R.R. Coats and D.T. Gibson, *J. Chem. Soc.*, 442 (1940).
38. M. Kobayashi, private communication.
39. H. Bredereck, A. Wagner, R. Blashke, G. Demetriades, and K.G. Kottenhahn, *Chem. Ber.*, **92**, 2628 (1959).
40. B.J. Lindberg, *Acta Chem. Scand.*, **20**, 1843 (1966).
41. a) C.M. Suter, "Organic Chemistry of Sulfur," Wiley, N.Y. (1948).
    b) P. Allen, L.S. Karger, J.D. Haygood, and J. Strensel, *J. Org. Chem.*, **16**, 767 (1951).
42. C.S. Marvel and R.S. Johnson, *ibid.*, **13**, 822 (1948).
43. R. Otto, *Annales*, **145**, 13, 317 (1868).
44. a) J.L. Kice and K.W. Bowers, *J. Amer. Chem. Soc.*, **84**, 605 (1962).
    b) J.L. Kice and K.W. Bowers, *J. Org. Chem.*, **29**, 1162 (1963).
45. J.L. Kice, D.C. Hampton, and A. Fitzgerald, *ibid.*, **20**, 882 (1965).
46. J.L. Kice, G. Guaraldi, and C.G. Venier, *ibid.*, **31**, 3561 (1966).
47. a) J.L. Kice and G. Guaraldi, *J. Amer. Chem. Soc.*, **88**, 5236 (1966).
    b) J.L. Kice and G. Guaraldi, *ibid.*, **89**, 4113 (1967).
48. J.L. Kice and G. Guaraldi, *Tetrahedron Lett.*, 6135 (1966).
49. a) J.L. Kice and J.W. Bowers, *J. Amer. Chem. Soc.*, **84** 2384 (1962).
    b) J.L. Kice and E.H. Morkved, *ibid.*, **86**, 2270 (1964).
50. J.L. Kice and K.W. Bowers, *ibid.*, **84**, 2390 (1962).
51. J.L. Kice and E.H. Morkved, *J. Org. Chem.*, **29**, 1942 (1964).
52. J.L. Kice, B.R. Toth, D.C. Hampton, and J.F. Barbow, *ibid.*, **31**, 848 (1966).
53. a) E. Vinkler and F. Klivenyi, *Acta Chim. Acad, Sci. Hung.*, **19**, 15 (1957); *ibid.*, **22**, 345 (1960).
    b) D. Barnard, *J. Chem. Soc.*, 4675 (1957).
54. J.L. Kice, C.G. Venier and L. Heasley, *J. Amer. Chem. Soc.*, **89**, 3557 (1967).
55. T.L. Moore and D.E. O'Connor, *J. Org. Chem.*, **31**, 3587 (1966).
56. H.J. Backer and H. Kloosterziel, *Rec. Trav. Chim.*, **73**, 129 (1954).

57. S. Oae and S. Kawamura, *Bull. Chem. Soc. Japan*, **35**, 1156 (1962).
58. a) T. Zincke and S. Lenhardt, *Annales*, **400**, 1 (1913).
    b) T. Zincke and F. Farr, *Annales*, **291**, 55 (1912).
    c) N. Kharasch, S.J. Potempa, and H.L. Wehrmeister, *Chem. Revs.*, **39**, 269 (1946).
59. K. Ikura and S. Oae, *Tetrahedron Lett.*, 3791 (1968).
60. a) S. Oae and K. Ikura, *Bull. Chem. Soc. Japan*, **38**, 58 (1965).
    b) S. Oae and K. Ikura, *ibid.*, **40**, 1421 (1967).
61. R.M. Topping and N. Kharasch, *Chem. Ind.*, 178 (1961).
62. S. Oae and K. Ikura, *Bull. Chem. Soc. Japan*, **39**, 1306 (1966).
63. S. Tamagaki, R. Ichihara and S. Oae, unpublished data.
64. R. Otto, *Chem. Ber.*, **13**, 1272 (1880).
65. H.J. Backer, J.S. Strating, and J.F.A. Fazenberg, *Rec. Trav. Chim. Pays-Bas*, **72**, 813 (1953).
66. a) B. Lindberg, *Acta Chem. Scand.*, **17**, 393 (1963).
    b) I.B. Douglass, F.J. Ward, and R.V. Norton, *J. Org. Chem.*, **32**, 324 (1967).
    c) R.J. Mulder, A.M. von Lausen, and J. Strating, *Tetrahedron Lett.*, 3061(1967).
67. F. Klivényi, *Magyar, Kém. Falyoirot*, **64**, 121 (1958).
68. C.J.M. Stirling, *J. Chem. Soc.*, 3597 (1957).
69. H. Bredereck, A. Wagner, H. Beck, and R.J. Klein, *Chem. Ber.*, **93**, 2736 (1960).
70. C.M. Suter, "Organic Chemistry of Sulfur," John Wiley & Sons, N.Y. (1948).
71. M. Kobayashi, *Bull. Chem. Soc. Japan*, **39**, 1296 (1966).
72. a) R. Otto and A. Rosing, *Chem. Ber.*, 2493 (1885).
    b) R. Otto, *ibid.*, 2504 (1885).
73. M. Kobayashi and M. Terao, *Bull. Chem. Soc. Japan*, **39**, 1292 (1966).
74. M. Kobayashi, M. Terao, and A. Yamamoto, *ibid.*, **39**, 802 (1966).
75. a) E.P. Kohler and M.B. MacDonald, *Amer. Chem. J.*, **22**, 219 (1899).
    b) H.T. Hookway, *J. Amer. Chem. Soc.*, **71**, 3240 (1949).
    c) K. Schank, *Annales*, **702**, 75 (1967).
76. M. Kobayashi, *Bull. Chem. Soc. Japan*, **39**, 967 (1966).
77. M. Kobayashi and A. Yamamoto, *ibid.*, **39**, 961 (1966).
78. M. Kobayashi and R. Kiritani, *ibid.*, **39**, 1782 (1966).
79. N.H. Nilsson, C. Jacobsen and A. Senning, *J. Chem. Soc.*, (D), 314 (1971).
80. a) O. Achmatowicz and J. Michalski, *Roczniki. Chem.*, **30**, 243 (1956).
    b) L. Kh. Fel'Dmann and V.N. Mikhailova, *Zh. Obshch. Khim.*, **32**, 944 (1962).
    c) I. Kh. Fel'Dmann and V.N. Mikhailova, *ibid*, **33**, 2111 (1963).
    d) V.N. Mikhailova, N. Borisova, and D. Stankevich, *Zh. Organ. Khim.*, **2**, 1437 (1966).
    e) M.T. Beachem and J.T. Shaw, U.S.P., 3,040,088, June 19 (1962).
81. Y. Ogata, Y. Sawaki and M. Isono, *Tetrahedron*, **26**, 3045 (1970).
82. H. Bredereck and E. Bäder, *Chem. Ber.*, **87**, 129 (1954).
83. L. Field and P.H. Settlage, *J. Amer. Chem. Soc.*, **73**, 5870 (1951).
84. H. Bredereck, E. Bäder, and G. Hoshele, *Chem. Ber.*, **87**, 784 (1954).
85. V. Mayer, *J. Prakt. Chem.*, **63**, 167 (1901).
86. G. Rewson and J.B.F.N. Engerts, *Tetrahedron*, **26**, 5653 (1970).
87. E. Bäder and H.D. Hermann, *Chem. Ber.*, **88**, 41 (1955).
88. a) J.B.F.N. Engberts and J. Strating, *Rec. Trav. Chim. Pays-Bas*, **83**, 733 (1964).
    b) T.T. Gregory, U.S.P., 2,659, 752, Nov. 17 (1953).
89. Ger.P., 832,158, Feb. 21 (1952).
90. W.B. Price and S. Miles, *J. Chem. Soc.*, 3154 (1928).
91. S. Pickholz, *ibid.*, 685 (1946).
92. Ger.P., 913, 179, June 10 (1954).
93. Y. Ogata, Y. Sawaki, and M. Isono, *Tetrahedron*, **26**, 1731 (1970).
94. K.T. Finley, R.S. Kaiser, R.L. Reeves, and G. Werimont, *J. Org. Chem.*, **34**, 2083 (1969).
95. a) W. Bradley and J.D. Hannon, *J. Chem. Soc.*, 2713 (1962).
    b) W. Bradley and J.D. Hannon, *ibid.*, 4438 (1962).
96. S. Oae, T. Maeda and O. Yamada, unpublished data.
97. J.A. Sprung and W.A. Schmidt, U.S.P., 2, 513, 826.
98. S. Oae, O. Yamada, and T. Maeda, unpublished data.
99. K.K. Andersen, *Tetrahedron Lett.*, 93,(1962).

Sulfinic Acids and Sulfinic Esters                                                    647

100. a) K. Mislow, M.M. Green, P. Laur, J.T. Melillo, T. Simmons, and A.L. Terney,
          Jr., J. Amer. Chem. Soc., 87, 1958 (1965).
      b) M. Axelrod, P. Bickart, J. Jacobus, M.M. Green, and K. Mislow, ibid., 90,
          4835 (1968).
      c) K.K. Andersen, W. Gaffield, N.E. Papanikoiaou, J.W. Foley, and R.I. Perkins,
          ibid., 86, 5637 (1964).
101. M. Kobayashi and M. Terao, Bull. Chem. Soc. Japan, 39, 1343 (1966).
102. C.C. Arcus, M.P. Balfe, and J. Kenyon, J. Chem. Soc., 485 (1938).
103. A.H. Wragg, J.S. McFadyen, and T.S. Stevens, ibid., 3603 (1958).
104. A.C. Cope, D.E. Morrison, and L. Field, J. Amer. Chem. Soc., 72, 59 (1950).
105. D. Darwish and E.A. Preston, Tetrahedron Lett., 113 (1964).
106. D. Darwish and R. McLaren, ibid., 1231 (1962).
107. G. Smith and C.J.M. Stirling, J. Chem. Soc., (C), 1530 (1971).
108. A. Streitwiser, Jr., Chem. Revs., 56, 571 (1956).
109. C.A. Bunton and B.N. Henry, Chem. Ind., 466 (1960).
110. a) M. Kobayashi, private communication.
      b) S. Oae and N. Kunieda, unpublished data.
111. a) M. Kobayashi, H. Minato, and Y. Ogi, Bull. Chem. Soc. Japan, 45, 1224
          (1972).
      b) C.A. Bunton and G. Schwerin, J. Org. Chem., 31, 842 (1966).
112. M. Kobayashi, A. Yabe, and R. Kiritani, Bull. Chem. Soc. Japan, 39, 1785
          1966.
113. M. Kobayashi and A. Yamamoto, ibid., 39, 2736 (1966).
114. M. Kobayashi and A. Yamamoto, ibid., 39, 2733 (1966).

*Chapter 12*

# REACTIONS OF SULFONATE AND SULFATE ESTERS

Emil Thomas Kaiser

*Department of Chemistry, University of Chicago*
*Chicago, Illinois 60637, USA*

## 12.1 INTRODUCTION

This chapter is organized on the basis of a classification of sulfonate and sulfate esters according to the nature and number of the alcohols from which they are derived. In the case of the sulfate esters, the reason for this mode of organization is obvious. However, in treating the chemistry of the sulfonate esters, we could have used the nature of the parent sulfonic acids as the basis for the organization of our discussion. We did not do this for two reasons. One is that we wished to present a reasonably parallel treatment of the chemistry of sulfonate and sulfate esters. The other is that we felt that, from a mechanistic standpoint, reactivity patterns emerge most clearly when the reactions of sulfonate esters are examined in terms of the kinds of alcohol groups which can be displaced.

While numerous examples will be presented of reactions which proceed by way of displacement at carbon, the ensuing discussion will focus most strongly on the mechanisms by which displacement occurs at sulfur in the reactions of sulfonate and sulfate esters. Our justification for this approach is that this text is concerned primarily with reactions which sulfur undergoes in its various oxidation states in organic compounds. We will not attempt to provide exhaustive coverage of the syntheses or reactions of sulfonate and sulfate esters. Rather, we will present selected syntheses in each section of this chapter, and the reactions to be discussed will be ones for which some quantitative data exist, allowing at least a tentative formulation of reasonable reaction mechanisms.

## 12.2  SULFONATE ESTERS OF ALIPHATIC ALCOHOLS

The reaction of a sulfonyl halide with the appropriate alcohol is the most common and convenient method used for the preparation of these compounds. Typical examples of this reaction are given in eqs. 12.1 and 12.2.

$$(12.1)^1$$

$$(12.2)^2$$

An enormous literature exists for the reactions of sulfonate esters of aliphatic alcohols.[3] Most of the displacement reactions seen occur with C-O rather than S-O bond cleavage. The mechanisms involved can be of the $S_N1$ or $S_N2$ type, depending on the alkyl group at which substitution takes place. In addition to products from direct displacement at the alkyl carbons, elimination or rearrangement reactions are often observed. To illustrate these points we will now present some examples that indicate the diversity of the reactions which sulfonate esters of aliphatic alcohols undergo.

$$\text{(norbornyl)}-\overset{\underset{H}{|}}{C}-O-SO_2-\underset{}{\langle \rangle}-Br$$

$$\Big\downarrow \text{AcOH} \qquad\qquad (12.3)^4$$

$$AcO-\text{(norbornyl)} \quad + \quad Br-\langle \rangle-SO_3H$$

$$\text{(norbornyl)}-\overset{\underset{}{|}-H}{\underset{O-SO_2-\langle\rangle-Br}{}}$$

$$\Big\downarrow \begin{array}{l}75\%\ \text{aqueous}\\ \text{acetone}\end{array} \qquad\qquad (12.4)^5$$

$$HO-\text{(norbornyl)} \quad + \quad Br-\langle\rangle-SO_3H$$

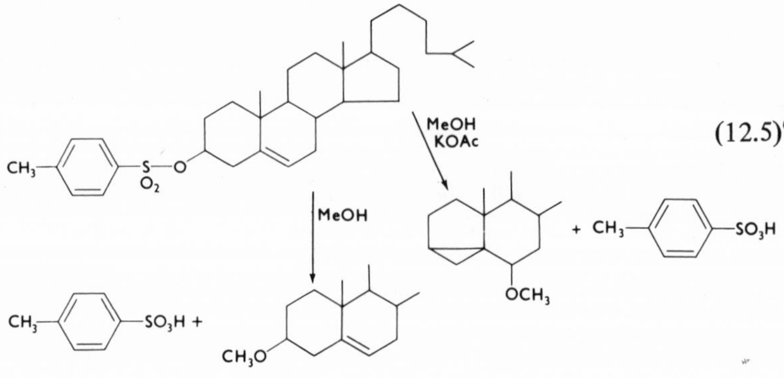

$(12.5)^6$

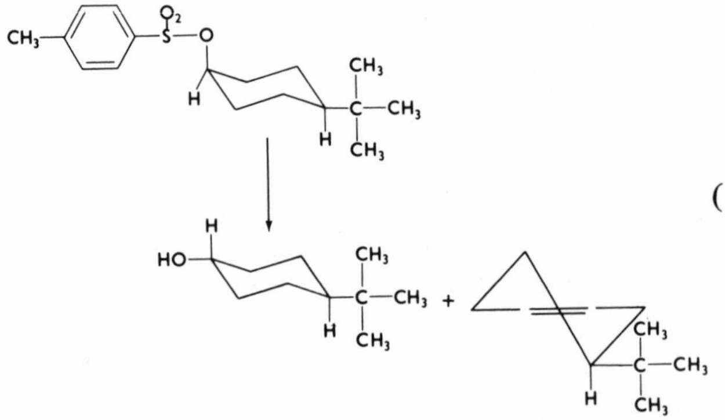

$(12.6)^7$

An interesting departure from the usual pattern of C-O bond cleavage in the reactions of sulfonate esters of aliphatic alcohols was observed in the reaction of $3\beta$-cholestanyl methane sulfonate[8] with potassium $t$-butoxide (see eq.12.7) in an anhydrous DMSO-benzene mixture. The principal product of this reaction, $3\beta$-cholestanol, arose apparently from S-O bond cleavage caused by nucleophilic attack on the sulfonate sulfur by the $t$-butoxide anion. In contrast, when $3\beta$-cholestanol $p$-toluenesulfonate was subjected to the same reaction conditions, C-O bond cleavage occurred and olefin was the major product.

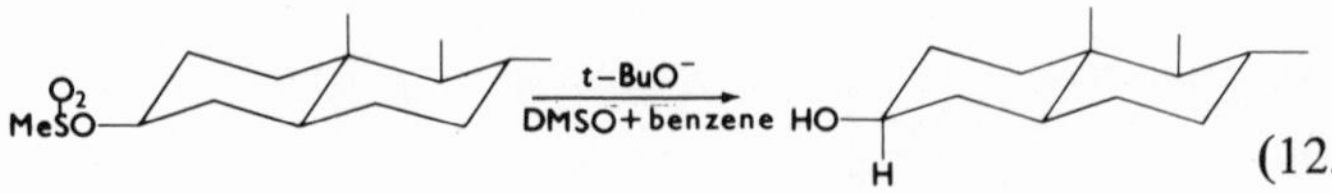

$(12.7)$

In the reactions of neopentyl $p$-toluenesulfonate Bordwell *et al.*[9] found that the type of bond cleavage which occurred showed a strong dependence on the nature of the attacking nucleophile. While iodide, morpholine and several mercaptide ions reacted with the sulfonate ester by way of C-O bond scission, displacement by methoxide resulted in S-O bond cleavage, as shown: eq. (12.8).

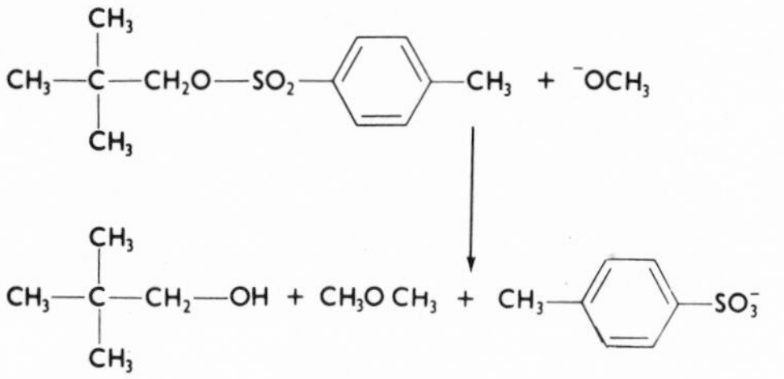

$$
\begin{array}{c}
\text{CH}_3 \\
\text{CH}_3{-}\overset{|}{\underset{|}{\text{C}}}{-}\text{CH}_2\text{O}{-}\text{SO}_2{-}\!\!\bigcirc\!\!{-}\text{CH}_3 \;+\; {}^{-}\text{OCH}_3 \\
\text{CH}_3 \\
\downarrow \\
\text{CH}_3 \\
\text{CH}_3{-}\overset{|}{\underset{|}{\text{C}}}{-}\text{CH}_2{-}\text{OH} \;+\; \text{CH}_3\text{OCH}_3 \;+\; \text{CH}_3{-}\!\!\bigcirc\!\!{-}\text{SO}_3^{-} \\
\text{CH}_3
\end{array}
$$

(12.8)

## 12.3 SULFONATE ESTERS OF AROMATIC ALCOHOLS

An illustrative synthesis of a sulfonate ester of an aromatic alcohol is shown below[10]:

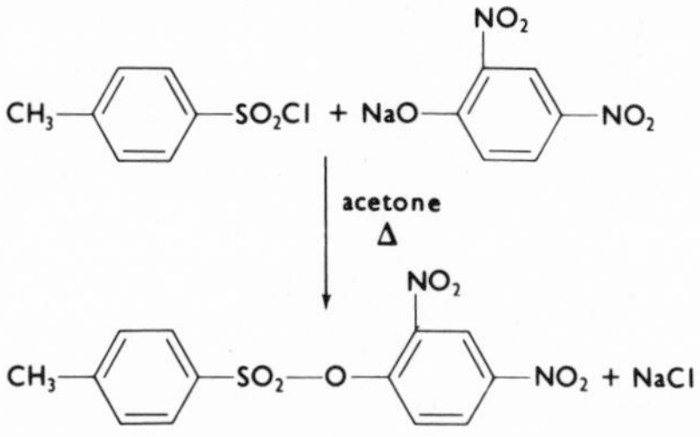

$$
\begin{array}{c}
\text{CH}_3{-}\!\!\bigcirc\!\!{-}\text{SO}_2\text{Cl} \;+\; \text{NaO}{-}\!\!\bigcirc\!\!{-}\text{NO}_2 \;\;(\text{NO}_2) \\
\downarrow \; \overset{\text{acetone}}{\Delta} \\
\text{CH}_3{-}\!\!\bigcirc\!\!{-}\text{SO}_2{-}\text{O}{-}\!\!\bigcirc\!\!{-}\text{NO}_2 \;+\; \text{NaCl} \;\;(\text{NO}_2)
\end{array}
$$

(12.9)

Whereas, as discussed above, sulfonate esters of aliphatic alcohols usually react with C-O fission, the sulfonate esters of aromatic alcohols show a tendency for S-O fission. Thus, when phenyl $p$-toluenesulfonate and phenyl methane sulfonate were hydrolyzed in oxygen-18-enriched alkaline aqueous dioxane, no oxygen-18 enrichment was observed in the product phenol (eqs. 12.10 and 12.11).[11,12] Reduction of phenyl $p$-toluenesulfonate with lithium aluminium hydride was also found to occur with S-O cleavage, yielding phenol.[13]

$$CH_3\text{-}\underset{}{\langle\bigcirc\rangle}\text{-}SO_2\text{-}O\text{-}\underset{}{\langle\bigcirc\rangle} \quad + \quad ^{18}OH^-$$

$$\downarrow \qquad\qquad\qquad (12.10)$$

$$CH_3\text{-}\underset{}{\langle\bigcirc\rangle}\text{-}SO_2{}^{18}O^- \quad + \quad \underset{}{\langle\bigcirc\rangle}\text{-}OH$$

$$CH_3\text{-}SO_2\text{-}O\text{-}\underset{}{\langle\bigcirc\rangle} \quad + \quad ^{18}OH^-$$

$$\downarrow \qquad\qquad\qquad (12.11)$$

$$CH_3SO_2{}^{18}O^- \quad + \quad \underset{}{\langle\bigcirc\rangle}\text{-}OH$$

The possibility exists that in nucleophilic substitution reactions like those shown in eqs. 12.10 and 12.11 pentacoordinate sulfur intermediates are formed along the reaction pathway as illustrated by eq. 12.12. Christ- man and Oae[14] found that phenyl $p$-toluenesulfonate recovered after partial hydrolysis of this compound in an oxygen-18-enriched alkaline medium showed no incorporation of excess oxygen-18. While this ex- periment is consistent with the postulation of an $S_N2$ direct displace- ment mechanism in the hydrolysis of the sulfonate ester, it does not conclusively rule out the intervention of a pentacoordinate species like that shown in eq. 12.12. For example, it is conceivable that pentaco- ordinate species are formed irreversibly. Alternatively, the equilibration in the pentacoordinate intermediate of the oxygens originally present in the sulfonyl group with those introduced from the solvent might be very slow relative to the reversal of the intermediate to the starting materials.

$$CH_3\text{-}\underset{}{\langle\bigcirc\rangle}\text{-}\underset{O_2}{\overset{}{S}}\text{-}O\text{-}\underset{}{\langle\bigcirc\rangle} \quad + \quad OH^-$$

$$\updownarrow$$

$$CH_3\text{-}\underset{}{\langle\bigcirc\rangle}\text{-}\underset{O\;\;O^-}{\overset{OH}{S}}\text{-}O\text{-}\underset{}{\langle\bigcirc\rangle} \quad\longrightarrow\quad CH_3\text{-}\underset{}{\langle\bigcirc\rangle}\text{-}SO_3H \quad + \quad \underset{}{\langle\bigcirc\rangle}\text{-}O^- \qquad (12.12)$$

In a number of instances, both C-O and S-O fission have been observed in reactions of sulfonate esters of aromatic alcohols. An example in which both types of bond cleavage have occurred is shown in eq. 12.13.[15] It has been suggested that the proportion of C-O to S-O cleavage has a marked dependence on the polarizability of the attacking nucleophile.[15-17] If steric conditions are favorable, highly polarizable nucleophiles like thiolate anions attack at carbon, causing C-O cleavage, whereas the less polarizable ones favor S-O cleavage.

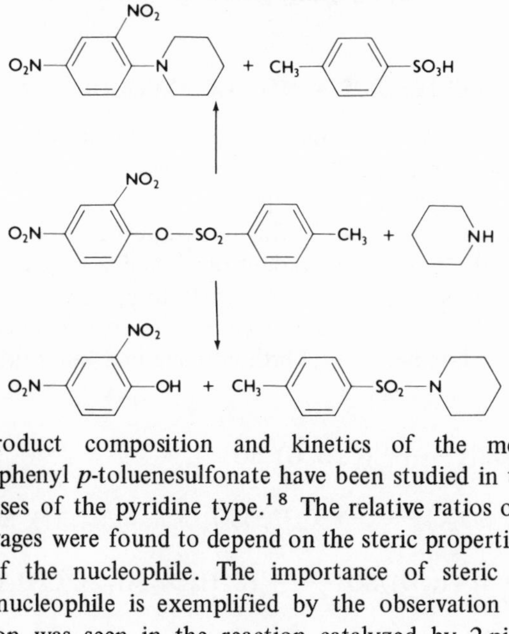

(12.13)

The product composition and kinetics of the methanolysis of 2,4-dinitrophenyl p-toluenesulfonate have been studied in the presence of tertiary bases of the pyridine type.[18] The relative ratios of C-O and S-O bond cleavages were found to depend on the steric properties and the base strength of the nucleophile. The importance of steric effects in the attacking nucleophile is exemplified by the observation that only S-O bond fission was seen in the reaction catalyzed by 2-picoline (see eq. 12.14), while 80% S-O bond cleavage was found in the case where 4-picoline was the catalyst (see eqs. 12.14 and 12.15). For bases which were approximately equally hindered sterically, the catalytic rate constants for S-O fission were found to obey the Brønsted relationship, with $\beta$ values of approximately 0.7.

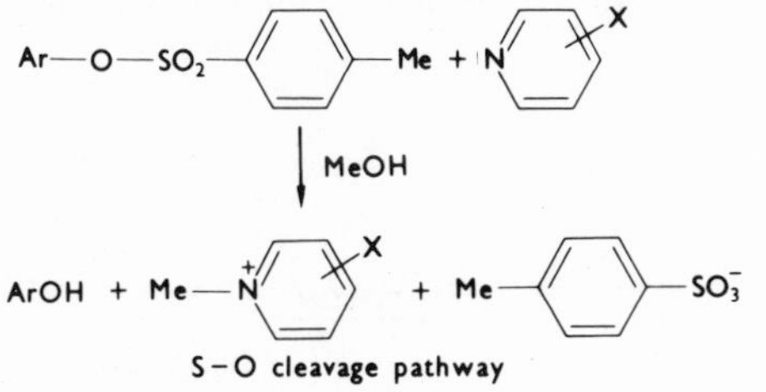

(12.14)

S-O cleavage pathway

$$\text{Ar}-\text{O}-\text{SO}_2-\text{C}_6\text{H}_4-\text{Me} + \text{N-pyridinium-X}$$

↓ MeOH                                                                    (12.15)

$$\text{Ar}-\overset{+}{\text{N}}\text{-pyridinium-X} + {}^-\text{O}_3\text{S}-\text{C}_6\text{H}_4-\text{Me}$$

C – O cleavage pathway

## 12.4 SULFATE DIESTERS OF ALIPHATIC ALCOHOLS

Numerous syntheses of sulfate diesters of aliphatic alcohols have been reported. The older literature on this topic has been summarized by Suter.[19] A synthetic route for the preparation of di-*n*-butyl sulfate in high yield is illustrated in eq. 12.16.[20] This procedure has been reported to be applicable to the synthesis of various other similar sulfates, e.g., di-*n*-propyl sulfate. Other, somewhat less satisfactory, procedures for the preparation of di-*n*-butyl sulfate include the reaction of *n*-butyl chlorosulfonate with either *n*-butyl orthoformate or *n*-butyl sulfite.[21,22]

$$2C_4H_9OH + SOCl_2 \longrightarrow (C_4H_9O)_2SO$$

$$(C_4H_9O)_2SO + SO_2Cl_2 \longrightarrow C_4H_9OSO_2Cl + C_4H_9Cl + SO_2$$
(12.16)
$$C_4H_9OSO_2Cl + (C_4H_9O)_2SO \longrightarrow (C_4H_9O)_2SO_2 + C_4H_9Cl + SO_2$$

As in the case of sulfonate esters discussed in Sect .12.2 of this chapter, there is a strong tendency for sulfate diesters of aliphatic alcohols to react with C-O bond cleavage. Dialkyl sulfates are well-known alkylating agents, and some typical alkylation reactions for dimethyl sulfate are illustrated in eqs. 12.17–12.19. An alkylation reaction involving diethyl sulfate is shown in eq. 12.20.

$$\text{(aryl-COCH}_3\text{, OH, HO-)} + \text{NaOH} + (CH_3O)_2SO_2 \xrightarrow{C_2H_5OH} \text{(aryl-COCH}_3\text{, OCH}_3\text{, CH}_3O\text{-)}$$  (12.17)[23]

$$(12.18)^{24}$$

$$(12.19)^{25}$$

$$(12.20)^{26}$$

Experiments in oxygen-18-enriched solvent have demonstrated that dimethyl sulfate hydrolyzes with C-O bond cleavage in neutral and alkaline solutions, as illustrated for hydroxide attack in eq. 12.21.[27-29] No evidence was found for the exchange of oxygen between water and the sulfonyl group of dimethyl sulfate.[29] Measurements of the entropy and heat capacity of activation for the attack of water on dimethyl and diethyl sulfate have been performed, and the results appear to be consistent with the hypothesis that reaction occurs by an $S_N2$ mechanism[30] (see eq. 12.22).

$$\text{HO}^-\text{CH}_3\text{-O-SO}_2\text{-OCH}_3 \longrightarrow \text{H-O-CH}_3 + {}^-\text{OSO}_2\text{OCH}_3 \qquad (12.21)$$

$$(12.22)$$

## 12.5 SULFATE DIESTERS OF AROMATIC ALCOHOLS

The synthesis of a representative sulfate diester, diphenyl sulfate, is illustrated in eq. 12.23. Although several synthetic procedures have been reported for the preparation of diaryl sulfates and a considerable number of these compounds have been described in the literature,[33-35] very little has been reported concerning their chemistry. The kinetics of the alkaline hydrolysis of diphenyl sulfate have been studied, and it was found that phenol and monophenyl sulfate were produced[31] (see eq. 12.24). The monoaryl sulfate esters are, in general, much more stable to attack by hydroxide ion than the corresponding diaryl esters, so that the latter compounds can usually be hydrolyzed cleanly to the monoester stage without the occurrence of further decomposition. To the best of the author's knowledge, the position of bond cleavage in the alkaline hydrolysis of diphenyl sulfate has not been established experimentally, but on the basis of studies on related compounds (to be discussed in Sect. 12.9), it seems likely that S-O bond fission occurs predominantly.

$$2 \, C_6H_5OH + SO_2Cl_2 \xrightarrow{\text{pyridine}} (C_6H_5O)_2SO_2 \qquad (12.23)^{31,32}$$

$$(C_6H_5O)_2SO_2 + OH^- \longrightarrow C_6H_5OSO_3^- + C_6H_5OH \qquad (12.24)$$

## 12.6 SULFATE MONOESTERS OF ALIPHATIC ALCOHOLS

Monoalkyl sulfates are produced when the corresponding dialkyl sulfates are hydrolyzed in alkaline media in which the monoesters are relatively stable (see eq. 12.21). The sulfate monoesters are formed also by the reaction of alcohols with sulfuric acid, pyrosulfate or chlorosulfonic acid. One of the numerous synthetic procedures employed is illustrated in the case of ethyl hydrogen sulfate in eq. 12.25. This compound was not obtained in a pure state in the acid form because the equilibrium shown in eq. 12.26 was rapidly attained. Frequently, monoalkyl sulfates are isolated as their barium salts, from which the free acids can be obtained by the addition of the appropriate amounts of sulfuric acid.

On alkaline hydrolysis, which usually requires vigorous treatment, monoalkyl sulfates tend to undergo C-O bond cleavage, as shown for monomethyl sulfate in eq. 12.27.[27]

$$C_2H_5OH + SO_3 \xrightarrow{\text{liquid } SO_2} C_2H_5OSO_3H \qquad (12.25)[36]$$

$$2C_2H_5OSO_3H \rightleftharpoons (C_2H_5O)_2SO_2 + H_2SO_4 \qquad (12.26)[36]$$

$$CH_3\text{-}O\text{-}SO_3^- + {}^{18}OH^- \longrightarrow H^{18}OCH_3 + SO_4^{2-} \qquad (12.27)$$

In acidic solution monoalkyl sulfates generally react with S-O bond fission as illustrated in eqs. 12.28 and 12.29 for 3-hydroxypropyl monosulfate[27] and isopropyl monosulfate.[37]

$$HO\text{-}CH_2CH_2CH_2\text{-}OSO_3H \xrightarrow{H_2{}^{18}O} HO\text{-}CH_2CH_2CH_2\text{-}OH + H_2SO_3{}^{18}O$$
$$(12.28)$$

$$CH_3\text{-}CHOSO_3H \xrightarrow{H_2{}^{18}O^+} CH_3\text{-}CHOH + H_2SO_3{}^{18}O \qquad (12.29)$$
$$\quad\ \ |\qquad\qquad\qquad\qquad\quad |$$
$$\quad\ CH_3 \qquad\qquad\qquad\qquad CH_3$$

Since the hydrolysis in acidic media of the monoalkyl sulfates is probably just the reverse of the esterification of alcohols by sulfuric acid,[38,39] the mechanism of the hydrolysis reaction has been suggested to occur as shown in eq. 12.30. Considerable evidence exists that the sulfonating species in the esterification reactions is $SO_3$.[40-42] Support for the mechanism of eq. 12.30 has come from studies on the solvent- and acidity-function dependence and on the activation parameters for the hydrolysis of alkyl hydrogen sulfates.[37,43]

$$ROSO_3H \rightleftharpoons R\text{-}\overset{+}{\underset{H}{O}}\text{-}SO_3^- \xrightarrow{\text{slow}} ROH + SO_3 \qquad (12.30)$$

## 12.7 SULFATE MONOESTERS OF AROMATIC ALCOHOLS

As in the case of the sulfate monoesters of aliphatic alcohols, the mono-esters of aromatic alcohols are frequently prepared by the alkaline hydro-lysis of the corresponding diesters. However, it is also often convenient to prepare the monoesters directly from their parent alcohols, as shown for 2-[4(5)-imidazolyl] phenyl sulfate in eq. 12.31.[44]

$$(12.31)$$

Since the nucleophilic displacement reactions of $p$-nitrophenyl sulfate have been studied especially thoroughly,[45] they will be discussed here. The measured first-order rate constants for the hydrolytic breakdown of $p$-nitrophenyl sulfate to $p$-nitrophenol and inorganic sulfate are virtually independent of pH in the range 7–12, but show a first-order dependence on the hydroxide ion concentration at higher pH values.[45] The hydroxide-ion-dependent reaction seems to involve attack of hydroxide ion on both sulfur, causing S-O bond fission, and on aromatic carbon, causing C-O bond cleavage. Thus, a study of the alkaline hydrolysis of $p$-nitrophenyl sulfate in an oxygen-18-enriched medium showed that about 33% of the reaction occurred by C-O bond cleavage and the remainder by S-O fission.[46]

The hydroxide-ion-independent    decomposition    of $p$-nitrophenyl sulfate seen in the pH 7–12 region occurs primarily by way of S-O bond cleavage[45] and might in principle proceed *via* a unimolecular elimination pathway or a bimolecular reaction involving water as the attacking nucleophile. Various mechanistic criteria, including the activation entropy, the effects of added nucleophiles like fluoride and sulfite, the effect of adding potassium iodide, and the results of changes in solvent composition, have been considered in an attempt to distinguish between these alternative hypotheses.[45] It was concluded that the experimental results obtained do not clearly indicate which of the two mechanisms applies to the pH-independent hydrolysis of $p$-nitrophenyl sulfate; however, if the water is indeed participating as a nucleophile, the data imply that, in the transition state for reaction, there is a small degree of bond formation be-tween solvent and ester. In the case of the attack of hydroxide ion at sul-fur in $p$-nitrophenyl sulfate (the high-pH reaction), it is not known whether or not a transient pentacoordinate sulfur species is formed.

The reactions of amines with $p$-nitrophenyl sulfate have also been investigated.[45] As illustrated for primary amines in eq. 12.32, reaction can occur either *via* C-O bond cleavage, leading to anilide formation, or S-O bond fission, resulting in amine sulfamate formation.

$$O_2N-\!\!\langle \quad \rangle\!-\!O-SO_3^- \; + \; RNH_2$$

$$O_2N-\!\!\langle \quad \rangle\!-OH + RNHSO_3^- \qquad O_2N-\!\!\langle \quad \rangle\!-NHR + HSO_4^-$$

(12.32)

The S-O bond fission process was examined most closely,[45] and it was found that the reaction-rate constants for the attack of amines at sulfur were not very sensitive to changes in the basicity of the incoming nucleophile ($\beta = 0.20$).[47] This result is consistent with the hypothesis that, in the transition states for these reactions, bond formation and the development of positive charge on the nitrogen of the attacking amine have occurred only to a limited extent. On the other hand, from the low reactivity of phenyl sulfate to attack by dimethylamine, for instance, it has been argued that the reactions of the amines at sulfur are quite sensitive to the nature of the leaving group ($\rho = 0.4$, minimum). The picture that emerges from the study of the S-O bond cleavage reactions of $p$-nitrophenyl sulfate with amines is very similar to that of nucleophilic reactions at phosphorus in phosphate monoester dianions[48] and indicates that the transition states for the reactions at sulfur may have similar characteristics to those suggested in the phosphate cases. That is, in the transition states for the reaction of amines at the sulfur of $p$-nitrophenyl sulfate, there appears to be a small degree of bond formation by the attacking nucleophile and considerable cleavage of the bond to the leaving group. The principal driving force for these reactions is probably due to electron donation from the negatively charged oxygen atom of the sulfate moiety, resulting in the expulsion of the $p$-nitrophenolate group.

Besides the hydroxide dependent and independent hydrolysis reactions of $p$-nitrophenyl sulfate discussed above, the hydrolysis of monoaryl sulfates in acidic media has been examined. From measurements of the kinetic solvent isotope effect, the dependence of rate on acidity, and the effects of changes in the nature of the substituents in the aryl residue and in the solvent composition on the hydrolysis of aryl sulfates in acidic solution,[49-51] it has been suggested that the mechanism of the acidic hydrolysis reaction is that shown in eq. 12.33, a process similar to the one indicated earlier in eq. 12.30 for the hydrolysis of alkyl hydrogen sulfates.

$$\text{ArOSO}_3\text{H} \;\rightleftharpoons\; \text{Ar-}\overset{+}{\underset{\text{H}}{\text{O}}}\text{-SO}_3^- \;\xrightarrow{\text{slow}}\; \text{ArOH} + \text{SO}_3 \qquad (12.33)$$

Recently, a number of cases of intramolecular catalysis of the hydrolysis of monoaryl sulfates have been described. Studies on salicyl sulfate have shown that the catalytic effect of the carboxyl group present in the substrates is to increase the rate of hydrolysis about 200-fold over that due to hydronium ion at pH 3-4.[50] On the basis of the kinetic solvent isotope effect ($k_{\text{H}_2\text{O}}/k_{\text{D}_2\text{O}} = 2$) and the observation of a positive entropy of activation, either of the two mechanistic hypotheses shown in eqs. 12.34 and 12.35 appear to be reasonable. In the mechanism of eq. 12.34, the carboxyl group acts as an intramolecular general acid catalyst. In that of eq. 12.35, a combination of intramolecular nucleophilic catalysis by the carboxylate group with specific acid catalysis is postulated to occur.

$$(12.34)$$

$$(12.35)$$

In the pH range 4–7, intramolecular catalysis is important in the hydrolysis of 2-[4(5)-imidazolyl] phenyl sulfate.[44] The possible mechanisms considered fell into two general categories. As shown in eq. 12.36, the imidazole moiety can function as an intramolecular general acid catalyst. Alternatively, a combination of acid catalysis with intramolecular nucleophilic catalysis by the imidazole group is possible (eq. 12.37). To investigate the latter possibility, 4(5)-(2'-hydroxyphenyl) imidazole N-sulfate was prepared, and its hydrolysis in the appropriate pH region was found to proceed about 70 times faster than that of the

O-sulfate. Therefore, it was not surprising that attempts to detect accumulation of the N-sulfate during the hydrolysis of 2-[4(5)-imidazolyl] phenyl sulfate failed. On the basis of trapping experiments with fluoride ion and several other mechanistic arguments, the mechanism of eq. 12.36 has been favored as the principal pathway for the intramolecular hydrolysis reaction, although it is not possible to completely exclude the possibility that reaction occurs by the pathway of eq. 12.37.[44]

The mechanisms of the hydrolysis reactions of mixed sulfuric-carboxylic acid anhydrides and sulfuric-phosphoric acid anhydrides are closely related to the reactions of monoalkyl and monoaryl sulfates.[52] As the pathways of biologically important reactions involving sulfates[53-55] are elucidated, it will be interesting to see their relationship to the mechanisms discussed above.

## 12.8   CYCLIC SULFONATE AND SULFATE ESTERS OF ALIPHATIC ALCOHOLS

The reason for separately considering the chemistry of the cyclic sulfonate and sulfate esters (as will be done in this section and the following one) is that the five-membered cyclic esters show a greatly enhanced reactivity at sulfur as compared to their acyclic analogues.[56] Indeed, we shall consider principally the chemistry of the five-membered cyclic species because the reactions of the larger ring compounds are quite similar to those of the acyclic esters which have been reviewed already.

The methods of preparation of 3-methyl-3-hydroxy-1-butanesulfonic acid sultone, an aliphatic five-membered cyclic sulfonate, 4-hydroxy-1-butanesulfonic acid sultone, an aliphatic six-membered cyclic sulfonate, and ethylene sulfate, an aliphatic five-membered cyclic sulfate, are illustrated in eqs. 12.38–12.40, respectively.

$$(12.38)^{57}$$

$$(12.39)^{58}$$

$$(12.40)^{59,\,60}$$

At pH 7, the relative rate constants for the attack of water (see eq. 12.41) on the five-membered cyclic sulfonate, 3-hydroxy-1-propane-sulfonic acid sultone, the six-membered cyclic sulfonate, 4-hydroxy-1-butanesulfonic acid sultone, and the acyclic ester, ethyl ethanesulfonate, are in the ratio 37:1:7.[61] The small differences seen in the reactivity of these compounds can be accounted for primarily on the basis of the small differences measured in the entropies of activation for their reactions.

The attack of hydroxide ion on the five-membered sultone also proceeds somewhat faster than that on the six-membered cyclic ester.[61] It is especially interesting that in the case of the five-membered cyclic ester, the use of oxygen-18-labeled solvent has revealed that there is 14% S-O bond fission (eq. 12.42), in addition to the predominant pathway involving C-O bond cleavage (eq. 12.43).

$$\underset{\underset{CH_2}{\overset{CH_2}{\big|}}}{\overset{CH_2}{\diagdown}}\!\!\!\diagup\!\!\!\overset{}{\underset{O}{\diagdown}}SO_2 \;+\; H_2O \;\longrightarrow\; HO\text{-}CH_2\text{-}CH_2\text{-}CH_2\text{-}SO_3H \qquad (12.41)$$

$$H^{18}O^- \;+\; \underset{\underset{CH_2}{\overset{CH_2}{\big|}}}{\overset{CH_2}{\diagdown}}\!\!\!\diagup\!\!\!\overset{}{\underset{O}{\diagdown}}SO_2 \;\longrightarrow\; HO\text{-}CH_2\text{-}CH_2CH_2\text{-}SO_2{}^{18}O^- \qquad (12.42)$$

$$H^{18}O^- \;+\; \underset{\underset{CH_2}{\overset{CH_2}{\big|}}}{\overset{CH_2}{\diagdown}}\!\!\!\diagup\!\!\!\overset{}{\underset{O}{\diagdown}}SO_2 \;\longrightarrow\; H^{18}O\text{-}CH_2\text{-}CH_2\text{-}CH_2\text{-}SO_3^- \qquad (12.43)$$

In contrast, the six-membered cyclic ester reacts exclusively with C-O fission (eq. 12.44), within the limits of experimental measurement. Therefore, it is clear that the rate of attack of hydroxide ion at sulfur is considerably enhanced in the five-membered cyclic ester system as compared to the six-membered sulftone, but it has not been feasible to measure this enhancement quantitatively.

$$\underset{\underset{\underset{CH_2}{\diagdown}}{\overset{CH_2}{\big|}}}{\overset{CH_2}{\diagup}}\!\!\!\overset{CH_2}{\diagdown}\underset{O}{\overset{SO_2}{\big|}} \;+\; H^{18}O^- \;\longrightarrow\; H^{18}O\text{-}(CH_2)_4\text{-}SO_3^- \qquad (12.44)$$

The hydrolytic behavior of the aliphatic cyclic sulfate esters[27] has strong similarities to that found for the cyclic sulfonates. In the pH range 2-9, ethylene sulfate, trimethylene sulfate and dimethyl sulfate undergo first-order hydrolysis to the corresponding monoesters with relative rates of $12:1:6$. In more alkaline solutions, the hydroxide-ion-catalyzed reactions become important, and the second-order rate constants for attack by $OH^-$ are in the ratio of $103:1:5.5$.[27] Solvolysis studies in oxygen-18 solvent have shown that the cyclic sulfate esters react predominantly with C-O bond cleavage in either neutral or alkaline solution. As in the case of the five-membered cyclic sulfonate, 3-hydroxy-1-propanesulfonic acid sultone, the five-membered cyclic sulfate undergoes 14% S-O bond fission in strongly alkaline solution (see eqs. 12.45 and 12.46). Only C-O bond fission was detected in the alkaline hydrolysis of either trimethylene sulfate or, as discussed in Sect. 12.4, dimethyl sulfate.

$$\text{H}^{18}\text{O}^- + \quad \begin{array}{c} \text{CH}_2\text{-O} \\ | \qquad\quad \diagdown \\ \qquad\quad\; \text{SO}_2 \\ | \qquad\quad \diagup \\ \text{CH}_2\text{-O} \end{array} \quad\longrightarrow\quad \text{H}^{18}\text{O-CH}_2\text{-CH}_2\text{-OSO}_3^- \qquad (12.45)$$

$$\text{H}^{18}\text{O}^- + \quad \begin{array}{c} \text{CH}_2\text{-O} \\ | \qquad\quad \diagdown \\ \qquad\quad\; \text{SO}_2 \\ | \qquad\quad \diagup \\ \text{CH}_2\text{-O} \end{array} \quad\longrightarrow\quad \text{HO-CH}_2\text{-CH}_2\text{-O-SO}_2{}^{18}\text{O}^- \qquad (12.46)$$

From a reasonable estimate of the experimental error in the experiment with labeled solvent where the extent of S-O bond fission was determined, it appears that the rate of attack of hydroxide ion at sulfur is at least 5000 times greater for the five-membered than for the six-membered cyclic ester.

The five-membered cyclic ester, cis-cyclohexane-1, 2-diol cyclic sulfate, reacts with C-O bond fission (at least predominantly) in acidic solution, but with S-O cleavage in alkali, as demonstrated by experiments with oxygen-18-labeled solvent and by the stereochemistry of the products formed[62, 63] (eqs. 12.47 and 12.48). Thus, this example shows again that despite the usual behavior of sulfate diesters of aliphatic alcohols which generally react with hydroxide ion by way of C-O cleavage, the five-membered cyclic esters show a definite tendency to undergo at least some S-O bond fission in alkali. Similar findings were made with trans-cyclohexane-1, 2-diol cyclic sulfate[63] and with the meso and optically active butane-2, 3-diol cyclic sulfates.[64]

$$(12.47)$$

$$(12.48)$$

Thermochemical measurements have shown that the heat of saponification of ethylene sulfate exceeds that for dimethyl sulfate by about 5.7 kcal/mole.[27] This difference is approximately the same as the difference between the heat of hydrolysis of an aliphatic five-membered cyclic phosphate ester, methyl ethylene phosphate, and its acyclic analogue, methyl hydroxyethyl phosphate.[27,65] In the phosphate series, the rate enhancement for attack of hydroxide ion at the phosphoryl group is enormous (a factor of at least $10^7$) when five-membered cyclic phosphate esters are compared to the corresponding open-chain esters. The heat of hydrolysis measurements on the sulfate and phosphate esters indicate that there is considerable ring strain in the five-membered cyclic species, and relief of this strain may provide the driving force for the extremely rapid hydrolysis of the five-membered cyclic phosphates. In the sulfate series, where there is a strong preference for C-O cleavage on hydrolysis, little relief of the ring strain probably occurs when hydroxide ion attacks at carbon in the five-membered cyclic ester, ethylene sulfate, and therefore, the rate of reaction of this compound by this pathway is not especially enhanced. On the other hand, attack of hydroxide ion at sulfur may cause relief of the ring strain in the transition state for this pathway and that could be the reason why a substantial amount of S-O cleavage is seen in the alkaline hydrolysis of ethylene sulfates while it is not observed for either acyclic or larger ring sulfate esters.

The structures of ethylene sulfate[60] and methyl ethylene phosphate[66] have been determined by X-ray crystallography, and several of the pertinent features of these structures are shown in Figs. 12.2 and 12.2. One feature common to both of these structures is the small value of the ring angle around the sulfur or phosphorus, which reflects the strain present in

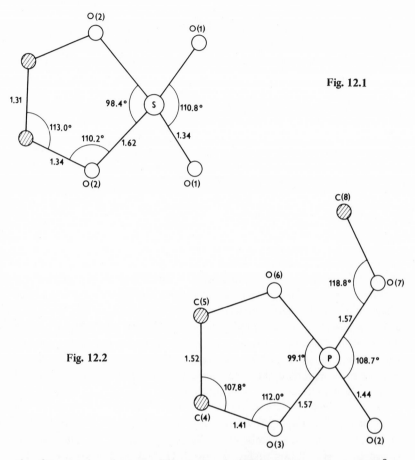

Fig. 12.1

Fig. 12.2

the five-membered rings. The internal O-S-O bond angle is 98.4° in ethylene sulfate,[60] and the corresponding O-P-O bond angle in methyl ethylene phosphate has the same value. There is strong evidence for the postulation of pentacoordinate intermediates in the hydrolysis of five-membered cyclic phosphates,[67] but similar evidence is lacking for the cyclic sulfates. Nevertheless, the transition state formed in the course of the attack of hydroxide ion at the sulfur atom of ethylene sulfate may have a structure with approximately trigonal-bipyramidal geometry, in which the ring angle at sulfur is close to 90° and the five-membered ring spans one apical and one equatorial position. Since relatively little perturbation of the ring angle at sulfur would be required to achieve such a transition state geometry, it is easy to see why nucleophilic displacement reactions at sulfur should be facilitated for such an ester.

## 12.9 CYCLIC SULFONATE AND SULFATE ESTERS OF AROMATIC ALCOHOLS

For historical reasons the chemistry of the sulfate esters will be discussed first in this section. To obtain a direct comparison of the rates of hydroxide ion attack at the sulfur atom in a five-membered cyclic sulfate and an open-chain analogue, studies were performed on the alkaline hydrolysis of the aromatic sulfates catechol cyclic sulfate and diphenyl sulfate.[68] It was anticipated that nucleophilic attack of hydroxide ion at the aromatic carbon atoms in these compounds should be extremely unlikely. The synthesis of catechol cyclic sulfate is illustrated:

$$
\text{catechol} + SO_2Cl_2 \xrightarrow[\text{pyridine}]{\text{petroleum ether}} \text{catechol cyclic sulfate} \qquad (12.49)
$$

A comparison of the second-order rate constants for the hydroxide-ion-catalyzed reaction of these compounds showed that catechol cyclic sulfate hydrolyzes $2 \times 10^7$ times faster than diphenyl sulfate. This was the first observation of such an enormous rate enhancement for the hydrolysis of a five-membered cyclic ester containing a heteroatom other than phosphorus. Later, hydrolytic studies in oxygen-18-enriched solvents substantiated the assumption that hydroxide ion attack occurs at the sulfur atoms in these compounds:[69]

$$
\text{catechol cyclic sulfate} + H^{18}O^- \longrightarrow \text{o-hydroxyphenyl } OSO_2{}^{18}O^- \qquad (12.50)
$$

$$
\text{diphenyl } O{-}SO_2{-}O \text{ diphenyl} + H^{18}O^- \longrightarrow O{-}SO_2{-}^{18}O^- + \text{phenol } OH \qquad (12.51)
$$

Additional studies have demonstrated that the hydrolytic reactions of five-membered cyclic sulfonates also show a large rate acceleration (ca. a factor of $10^5$–$10^6$) relative to those of the acyclic or six-membered analogues. For instance, the alkaline hydrolysis of 2-hydroxy-α-toluene-sulfonic acid sultone is $8 \times 10^5$ faster than that of phenyl α-toluene-sulfonate, its open-chain analogue.[70] Although the rate constant for the alkaline hydrolysis of the six-membered cyclic sulfonate, β-2-hydroxy-

phenylethanesulfonic acid sultone, was found to be somewhat larger than
that for the reaction of phenyl α-toluenesulfonate, it was much smaller
than that for 2-hydroxy-α-toluenesulfonic acid sultone.[71] Thus, it appears
that the five-membered cyclic ester is uniquely labile.

By analogy to the behavior of catechol cyclic sulfate, it is reasonable to
propose that the hydrolysis of 2-hydroxy-α-toluenesulfonic acid sultone
occurs *via* attack of hydroxide ion at sulfur with concomitant S-O bond
cleavage. However, the hydrolytic mechanisms outlined in eqs. 12.52 and
12.53, which do not involve direct attack of hydroxide ion at sulfur, must
be considered.

In the mechanisms of eqs. 12.52 and 12.53, a carbanion and/or a sulfene are proposed to be reactive intermediates in the hydrolysis of 2-
hydroxy-α-toluenesulfonic acid sultone. A variety of experiments have
been performed to test these mechanisms. If the concerted elimination reaction shown in eq. 12.52 occurs, it can be predicted that, if the
reaction were run in $D_2O$, the resultant sulfonic acid should have one
deuterium atom in the methylene group. The hydrolytic reaction can be
examined under conditions where the hydrogens of the methylene group
of the product sulfonic acid do not exchange with the deuterium of the
solvent $D_2O$. When the hydrolysis of 2-hydroxy-α-toluenesulfonic acid
sultone was conducted in a $D_2O$-$OD^-$ solution in which the sultone was
in excess over $OD^-$, the sultone recovered, after all the $OD^-$ was consumed,
was found to have undergone extensive exchange of deuterium into its
methylene group.[56] This observation indicated that a carbanion is formed
rapidly and reversibly from the sultone in basic solution, and this leaves
the question open whether or not carbanion and/or sulfene intermediates
lie along the reaction pathway for the alkaline hydrolysis of
2-hydroxy-α-toluenesulfonic acid sultone.

$$(12.52)$$

$$(12.53)$$

In the case of 5-nitrocoumaranone, a lactone with $\alpha$ protons which undergo ionization with a $pK_a$ of 9.8, an experimental approach involving the measurement of the solvent isotope effect for the decomposition of the fully ionized ester was used to determine if a mechanism analogous to that in eq. 12.53 was operating.[72] This approach is limited to cases in which the carbon acid is fully ionized at an alkalinity accessible in aqueous media. The $pK_a$ for the ionization of the labile $\alpha$ protons of 2-hydroxy-5-nitro-$\alpha$-toluenesulfonic acid sultone is greater than 14.[73] Several general methods involving comparative isotope exchange and hydrolysis rate measurements have been developed to determine if carbanions are intermediates in the alkaline hydrolysis of an ester.[74] From experiments using these methods, it appears that mechanisms involving carbanion and/or sulfene intermediates do not provide the predominant pathways by which 2-hydroxy-$\alpha$-toluenesulfonic acid sultone hydrolyzes.

1-Naphthol-8-sulfonic acid sultone hydrolyzes in alkali at a rate comparable to that for 2-hydroxy-$\alpha$-toluenesulfonic acid sultone.[71] Since the former compound does not have a methylene group, it cannot be reacting *via* carbanion and/or sulfene intermediates, and presumably, hydroxide ion attacks directly at the sulfur atom in the hydrolysis reaction (eq. 12.54). While the finding that 1-naphthol-8-sulfonic acid sultone is very labile to alkali does not bear upon the occurrence or nonoccurrence of such intermediates in the hydrolysis of 2-hydroxy-$\alpha$-toluenesulfonic acid sultone, it demonstrates at the very least that a sulfene pathway is not obligatory for the rapid hydrolysis of a five-membered sultone. From all of the data presently in hand, it seems reasonable to conclude that the high reactivity seen in the alkaline hydrolysis of aromatic five-membered cyclic sulfonates reflects the ease with which hydroxide ions attack at the sulfur atoms in these cyclic systems.

As in the cases of the aliphatic cyclic sulfonate and sulfate esters, there is no evidence which requires the postulation of pentacoordinate intermediates in which the attacking hydroxide ion is covalently bound to sulfur in the hydrolysis of the aromatic cyclic sulfonate and sulfate esters. For example, when incomplete hydrolyses of catechol cyclic sulfate, 2-hydroxy-$\alpha$-toluenesulfonic acid sultone and $\beta$-2-hydroxyphenylethanesulfonic acid sultone were carried out in alkaline solutions containing excess oxygen-18, no significant exchange was observed when the starting esters were reisolated.[69] Thus, by this technique, the reversible formation of pentacoordinate intermediates like that shown in eq. 12.55 has not been detected. This does not mean that the mechanism of eq. 12.55 can be ruled out. For example, pentacoordinate intermediates such as the one shown in eq. 12.55 might be reversibly formed in the hydrolysis of the cyclic esters, but the oxygens external to the ring might not equilibrate during the lifetimes of these intermediates. Alternatively, the mechanisms of reaction of the cyclic esters might involve irreversible formation of pentacoordinate intermediates.

$$(12.54)$$

$$(12.55)$$

The uncertainty which exists concerning the intermediacy of pentacoordinate sulfur species in the hydrolysis of the aromatic cyclic esters affects the interpretation of data which have been obtained on substituent effects in the reaction of 5-substituted 2-hydroxy-α-toluene-sulfonic acid sultones.[75] The preparation of one of these compounds, 5-methoxy-2-hydroxy-α-toluenesulfonic acid sultone, is illustrated in eq. 12.56. A linear relationship exists between the logarithms of the rate constants for the alkaline hydrolysis of these compounds and the appropriate Hammett para-substituents constants, $\sigma_p$. A positive $\rho$-value of +1.23 was obtained, indicating that electron-withdrawing substituents in the 5-position of the aromatic ring have an accelerating effect on the reaction illustrated by eq. 12.57. If the mechanism of eq. 12.58—which involves a concerted nucleophilic displacement reaction by hydroxide ion at sulfur—holds, then the $\rho$-value measured suggests that the sulfur-oxygen bond is significantly cleaved in the transition states for the hydrolysis of the sultones. Thus the transition states proposed in this mechanism would appear to have considerable polar character, according to the substituent-effect results.

However, the possibility that pentacoordinate intermediates do lie along the pathway for the alkaline hydrolysis of sultones introduces considerable ambiguity in the interpretation of the $\rho$-value measured for the decomposition of the 5-substituted 2-hydroxy-α-toluenesulfonic acid sultones. It does seem likely that the $\rho$-value for the first step of the mechanism shown in eq. 12.59 would not be large and that the observed

$p$-value should reflect primarily the effects of substituents on the reaction of step 2, in which the pentacoordinate intermediate undergoes ring opening to give the product sulfonic acid. According to this hypothesis, the transition state for the ring-opening reaction (step 2) would be quite polar. Because it is unclear which of the mechanisms illustrated in eqs. 12.58 and 12.59 applies to the hydrolysis of the aromatic five-membered sultones, the most definite statement that can be made is that the transition state in at least one step in the hydrolysis of these compounds must be very polar.

$$\text{(12.56)}$$

$$\text{(12.57)}$$

$$\text{(12.58)}$$

$$\text{(12.59)}$$

As in the cases of the ethylene sulfate and methyl ethylene sulfate (Sect. 12.8), X-ray crystallographic structure determinations have shown that ring angles around sulfur or phosphorus in 2-hydroxy-α-toluene-sulfonic acid sultone[76] (Fig. 12.3), catechol cyclic sulfate[77] (Fig. 12.4) and catechol cyclic phosphate[78] are small, reflecting the strain presumably present in the five-membered rings. While it is not clear if penta-coordinate intermediates are involved in the alkaline hydrolysis of catechol cyclic sulfate and 2-hydroxy-α-toluenesulfonic acid sultone, the transition states in these reactions may have structures approximating trigonal-bipyramidal geometry, as considered earlier in this chapter in the case of ethylene sulfate.

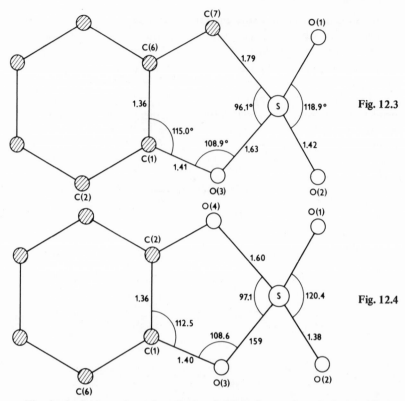

Fig. 12.3

Fig. 12.4

The hydrolysis and methanolysis of β-2-hydroxyphenylethanesulfonic acid sultone, γ-2-hydroxyphenylpropanesulfonic acid sultone and biphenylene cyclic sulfate have been studied recently in strongly basic media.[79] An important objective of this work was to obtain kinetic evidence testing the existence of ionizable pentacoordinate sulfur species in the alkaline hydrolysis of these cyclic esters. For example, in strongly alkaline solutions, it is possible that, as shown in eq. 12.60, a penta-coordinate intermediate formed by the reaction of hydroxide ion with β-2-hydroxyphenylethanesulfonic acid sultone can undergo ionization to give another pentacoordinate species which has a double negative charge. If both pentacoordinate species decompose to the product sulfonic acid, the kinetics of reaction of the sultone could show a mixed first-order and second-order dependence on hydroxide ion. In the basic methanolysis of β-2-hydroxyphenylethanesulfonic acid sultone, the pentacoordinate intermediate which might be formed (eq. 12.61) would not be capable of ionizing further. The kinetics of the methanolysis reaction would not be expected therefore to show a second-order dependence on methoxide ion.

$$(12.60)$$

$$(12.61)$$

The rate laws observed for a number of reactions in strongly basic media have been correlated with the acidity functions $H_-$ when the solvent was water, and $H_M$ when the solvent was methanol.[80] While detailed mechanistic interpretations based on acidity functions may be questionable, their use in determining the order of a reaction with respect to hydroxide[81] or methoxide[80] ion seems to be valid. The hydrolysis and methanolysis rate constants measured for $\beta$-2-hydroxyphenylethane-sulfonic acid sultone, $\gamma$-2-hydroxyphenylpropanesulfonic acid sultone and biphenylene cyclic sulfate showed a first-order dependence on the functions $a_W 10^{H_-}$ and $a_M 10^{H_M}$, respectively.[79] Hence, the kinetic results obtained in very basic aqueous solution do not require the postulation of ionizable pentacoordinate intermediates. These findings demonstrate that the $H_-$ and $H_M$ functions are useful mechanistic tools in the study of displacement reactions at sulfonyl sulfur, even though the structures of the indicators employed in establishing these acidity functions are very different[80,82] from those of the sulfonate and sulfate esters investigated.

Imidazole and N-methylimidazole have been found to act as catalysts in the hydrolysis of cyclic sulfate and sulfonate esters.[83] Catalysis by these bases was studied in water and deuterium oxide. The ratios of the second-order rate constants for attack by imidazole, $k_{Im}^{H_2O}/k_{Im}^{D_2O}$, in water and deuterium oxide, respectively, were found to be 3.6 for catechol cyclic sulfate and 4.2 for 2-hydroxy-5-nitro-$\alpha$-toluenesulfonic acid sultone. The ratio of $k_{MeIm}^{H_2O}/k_{MeIm}^{D_2O}$ in the case of N-methylimidazole was 3.5 for the reaction of both cyclic esters.

Nucleophilic and general base catalysis of the hydrolysis of esters by imidazole are kinetically indistinguishable processes.[84] However, either the magnitude of the deuterium kinetic solvent isotope effect or the detection of an acylimidazole in the case of nucleophilic catalysis can be used to distinguish between the two types of processes. From the kinetic solvent isotope effect values, it appears that imidazole and N-methylimidazole function as general basic rather than nucleophilic catalysts in the hydrolysis of the aromatic five-membered cyclic sulfate and sulfonate esters. The hydrolysis of catechol cyclic sulfate and 2-hydroxy-5-nitro-$\alpha$-toluenesulfonic acid sultone catalyzed by imidazole can be pictured to proceed as shown below.

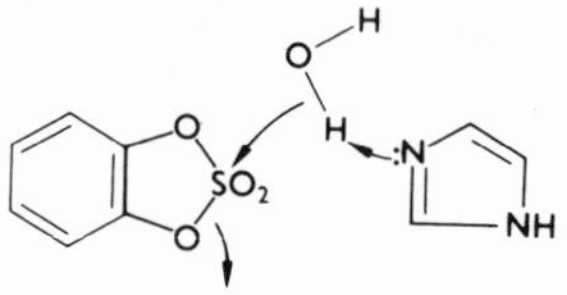

$$(12.62)$$

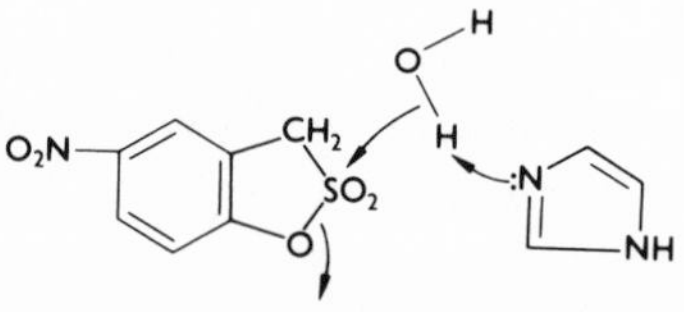

$$(12.63)$$

The question can be raised why the imidazole and N-imidazole-catalyzed reactions of the aromatic five-membered cyclic esters do not proceed by nucleophilic attack of the imidazole species on the sulfur atom as shown in eq. 12.64, followed by hydrolysis of the sulfonylimidazoles formed. The excellence of the leaving group formed when the sulfate or sulfonate rings are broken might suggest that nucleophilic catalysis of the hydrolysis of the cyclic sulfates and sulfonates rather than general base catalysis should occur. However, it is possible that nucleophilic attack by imidazole to give sulfonylimidazoles, as illustrated in eq. 12.64, may occur, but this attack may not be detected. The reason why the formation of sulfonylimidazoles like that involved in the pathway given by eq. 12.64 may not be detected is that they could recyclize to form the starting cyclic esters much faster than they hydrolyze to the corresponding sulfuric or sulfonic acids. In other words, catechol cyclic sulfate and 2-hydroxy-5-nitro-$\alpha$-toluenesulfonic acid sultone could be attacked by imidazole and N-methylimidazole by both nucleophilic and general base-catalyzed pathways, but only the latter pathway would contribute significantly to the hydrolysis of the cyclic esters.[85]

$$(12.64)$$

## 12.10 CONCLUSIONS

From the discussion presented in this chapter, it should be evident that reactions at the sulfur atoms in sulfonate and sulfate esters have not been as extensively explored as those at the carbon atoms of carbonyl groups. One of the most basic aspects of these kinds of processes, the detection and characterization of reactive intermediates, has been thoroughly studied in numerous reactions which carbonyl groups undergo, but there is no reaction at the sulfonyl groups of sulfonate and sulfate esters for which compelling evidence for corresponding intermediates exists at the present time.

To keep this chapter to a reasonable length, the biological reactions which sulfonate and sulfate esters undergo have not been covered in any detail. It should be mentioned that the chemistry of these reactions is being actively investigated, and in several instances, strong mechanistic parallels have been demonstrated to exist between related organic and biological reactions.[44,52,56]

In conclusion, much interesting chemistry undoubtedly remains to be discovered in the reactions of sulfonate and sulfate esters. It is the hope of the author that this chapter may provide aid to those future investigators who will contribute to progress in this area of research.

## 12.11 REFERENCES

1. A.T. Roos, H. Gilman and N.J. Beaber, *Org. Syn.*, Coll. Vol. 1, 145 (1941).
2. S. Marvel and V.C. Sekera, *Org. Syn.*, Coll. Vol. 3, 366 (1955).
3. See, for example, the discussion of displacement reactions in "Physical Organic Chemistry" by L.P. Hammett, 2nd Edition, McGraw-Hill Book Co., New York (1970), pp. 147–185.
4. S. Winstein and D. Trifan, *J. Amer. Chem. Soc.*, 74, 1154 (1952).
5. S. Winstein and D. Trifan, *ibid.*, 74, 1147 (1952).
6. S. Winstein and R. Adams, *ibid.*, 70, 838 (1948).
7. S. Winstein and N.J. Holness, *ibid.*, 77 5562 (1955).
8. F.C. Chang, *Tetrahedron Lett.*, 305 (1964).
9. F.G. Bordwell, B.M. Pitt and M. Knell, *J. Amer. Chem. Soc.*, 73, 5004 (1951).

10. J.F. Bunnett and J.Y. Bassett, Jr., *J. Amer. Chem. Soc.*, **81**, 2104 (1959).
11. C.A. Bunton and Y. Frei, *J. Chem. Soc.*, 1872 (1951).
12. C.A. Bunton and V.A. Welch, *ibid.*, 3240 (1956).
13. H. Schmid and P. Karrer, *Helv. Chim. Acta*, **32**, 1371 (1949).
14. D.R. Christman and S. Oae, *Chem. Ind.*, 125 (1959).
15. J.F. Bunnett and J.Y. Bassett, Jr., *J. Org. Chem.*, **27**, 2345 (1962).
16. J.F. Bunnett and J.Y. Bassett, Jr., *J. Amer. Chem. Soc.*, **81**, 2104 (1959).
17. J.F. Bunnett and J.Y. Bassett, Jr., *J. Org. Chem.*, **27**, 1887 (1962).
18. A. Kirkien-Konasiewicz, G.M. Sammy and A. Maccoll, *J. Chem. Soc.* (B), 1364 (1968).
19. C.M. Suter, "The Organic Chemistry of Sulfur," John Wiley and Sons, Inc., New York (1944), pp. 48-74.
20. C.M. Suter and H.L. Gerhart, *Org. Syn.*, Coll. Vol. **2**, 111-113 (1943).
21. R. Levaillant, *Compt. Rend.*, **197**, 648 (1933).
22. C. Barkenbus and J.J. Owen, *J. Amer. Chem. Soc.*, **56**, 1204 (1934).
23. G.N. Vyas and N.M. Shah, *Org. Syn.*, Coll. Vol. **4**, 837 (1963).
24. R.E. Benson and T.L. Cairns, *ibid.*, Vol. **4**, 588 (1963).
25. L. Field and R.D. Clark, *ibid.*, Vol. **4**, 674 (1963).
26. H. Gilman and W.E. Catlin, *ibid.*, Vol. **1**, 471 (1941).
27. E.T. Kaiser, M. Panar and F.H. Westheimer, *J. Amer. Chem. Soc.*, **85**, 602 (1963).
28. D.N. Kursanov and R.V. Kudryavtsev., *J. Gen. Chem. U.S.S.R.*, **26**, 3323 (1956), English translation, Consultants Bureau, Inc., N.Y.
29. I. Lauder, I.R. Wilson and B. Zerner, *Australian J. Chem.*, **14**, 41 (1961).
30. R.E. Robertson and S.E. Sugamori, *Can. J. Chem.*, **44**, 1728 (1966).
31. E.T. Kaiser, I.R. Katz and T.F. Wulfers, *J. Amer. Chem. Soc.*, **87**, 3781 (1965).
32. L. Denivelle, *Compt. Rend.*, **199**, 211 (1934).
33. L.J. Bollinger, *Bull. Soc. Chim. France*, 156 (1948).
34. H. Geis and E. Pfeil, *Annales*, **578**, 11 (1952).
35. R. Cramer and D.D. Coffman, *J. Org. Chem.*, **26**, 4164 (1961).
36. D.S. Breslow, R.R. Hough and J.T. Fainlough, *J. Amer. Chem. Soc.*, **76**, 5361 (1954).
37. B.D. Batts, *J. Chem. Soc.* (B), 551 (1966).
38. G. Williams and D.J. Clark, *ibid.*, 1304 (1956).
39. N.C. Deno and M.S. Newman, *J. Amer. Chem. Soc.*, **72**, 3852 (1950); **73**, 1920 (1951).
40. R.L. Burwell, Jr. *ibid.*, **74**, 1462 (1952).
41. V. Gold and D.P.N. Satchell, *J. Chem. Soc.*, 1635 (1956).
42. W.A. Cowdrey and D.S. Davies, *ibid.*, 1871 (1949).
43. B.D. Batts, *ibid.*, (B), 547 (1966).
44. S.J. Benkovic and L.K. Dunikoski, Jr., *Biochemistry*, **9**, 1390 (1970).
45. S.J. Benkovic and P.A. Benkovic, *J. Amer. Chem. Soc.*, **88**, 5504 (1966).
46. B. Spencer, *Biochem. J.*, **69**, 155 (1968).
47. See W.P. Jencks, "Catalysis in Chemistry and Enzymology," McGraw-Hill, New York, (1969), p. 81, for a treatment of β-values and other topics relevant to the present discussion.
48. W.P. Jencks and M. Gilchrist, *J. Amer. Chem. Soc.*, **86**, 1410 (1964).
49. G.N. Burkhardt, W.G.K. Ford and E. Singleton, *J. Chem. Soc.*, 17 (1936); G.N. Burkhardt, A.G. Evans and E. Warhurst, *ibid.*, 25 (1936); G.N. Burkhardt, C. Horrex and D.I. Jenkins, *ibid.*, 1649, 1654 (1936).
50. S.J. Benkovic, *J. Amer. Chem. Soc.*, **88**, 5511 (1966).
51. J.L. Kice and J.M. Anderson, *ibid.*, **88**, 5242 (1966).
52. See, for instance, S.J. Benkovic and R.C. Hevey, *ibid.*, **92**, 4971 (1970).
53. J.D. Gregory and F. Lipmann, *J. Biol. Chem.*, **229**, 1081 (1957); P.W. Robbins and F. Lipmann, *J. Amer. Chem. Soc.*, **78**, 2652 (1956); F. Lipmann, *Science*, **128**, 575 (1958).
54. E. Meezan and E.A. Davidson, *J. Biol. Chem.*, **242** 1685 (1967).
55. A.S. Balasubramanian and B.K. Backhawat, *Indian J. Exp. Biol.*, **1**, 179 (1963).
56. E.T. Kaiser, *Acc. Chem. Res.*, **3**, 45 (1970).
57. F.G. Bordwell, R.D. Chapman and C.E. Osborne, *J. Amer. Chem. Soc.*, **81**, 2002 (1959).

58. W.E. Truce and F.D. Hoerger, *ibid.*, **76**, 5357 (1954).
59. J. Brunken and E.J. Poppe, Ger. P. 1,049,870 (1959); *Chem. Abstr.*, **55**, P2488c (1961).
60. F.P. Boer, J.J. Flynn, E.T. Kaiser, O.R. Zaborsky, D.A. Tomalia, A.E. Young and Y.C. Tong, *J. Amer. Chem. Soc.*, **90**, 2970 (1968).
61. A. Mori, M. Nagayama and H. Mandai, *Bull. Chem. Soc. Japan*, **44**, 1669 (1971).
62. J.S. Brimacombe, A.B. Foster and M. Stacey, *Chem. Ind.*, 262 (1959).
63. J.S. Brimacombe, A.B. Foster, E.B. Hancock, W.G. Overend and M. Stacey, *J. Chem. Soc.*, 201 (1960).
64. H.K. Garner and H.J. Lucas, *J. Amer. Chem. Soc.*, **72**, 5497 (1950).
65. J.R. Cox, Jr., R.E. Wall and F.H. Westheimer, *Chem. Ind.*, 929 (1959).
66. T.A. Steitz and W.N. Lipscomb, *J. Amer. Chem. Soc.*, **87**, 2488 (1965).
67. F.H. Westheimer, *Acc. Chem. Res.*, **1**, 70 (1968).
68. E.T. Kaiser, I.R. Katz and T.F. Wulfers, *J. Amer. Chem. Soc.*, **87**, 3781 (1965).
69. E.T. Kaiser and O.R. Zaborsky, *ibid.*, **90**, 4626 (1968).
70. O.R. Zaborsky and E.T. Kaiser, *ibid.*, **88**, 3084 (1966).
71. E.T. Kaiser, K. Kudo and O.R. Zaborsky, *ibid.*, **89**, 1393 (1967).
72. P.S. Tobias and F.J. Kézdy, *ibid.*, **91**, 5171 (1969).
73. In DMSO, a $pK_a$ of 15.6 was found for the ionization of the methylene group in the five-membered sulfone. This value is significantly lower than that measured for the corresponding ionization in phenyl α-toluenesulfonate. D.F. Mayers, Ph.D. Thesis, University of Chicago (1973).
74. P. Müller, D.F. Mayers, O.R. Zaborsky and E.T. Kaiser, *J. Amer. Chem. Soc.*, **91**, 6732 (1969).
75. O.R. Zaborsky and E.T. Kaiser, *ibid.*, **92**, 860 (1970). *Chem. Commun.*, 197 (1967).
77. F.P. Boer and J.J. Flynn, *J. Amer. Chem. Soc.*, **91**, 6604 (1969).
78. E.T. Kaiser, T.W.S. Lee and F.P. Boer, *ibid.*, **93**, 2351 (1971).
79. J.H. Smith, T. Inoue and E.T. Kaiser, *J. Amer. Chem. Soc.*, **94**, 3098 (1972).
80. C.H. Rochester, "Acidity Functions," Academic Press, New York (1970).
81. C.H. Rochester, *Trans. Faraday Soc.*, **59**, 2826 (1963).
82. G. Yagil, *J. Phys. Chem.*, **71**, 1034 (1967).
83. E.T. Kaiser, K.-W. Lo, K. Kudo and W. Berg, *Bioorg. Chem.*, **1**, 32 (1971).
84. T.C. Bruice and S. Benkovic, "Bioorganic Mechanisms," Vol. 1, W.A. Benjamin, Inc., New York (1966), pp. 46–66.
85. Added in proof: Nucleophilic catalysis by imidazde has been found in the solvolysis of the six-membered cyclic sulfonate β-(2-hydroxy-3,5-dinitrophenyl) sulfonic acid sulfone. W. Berg, P. Campbell and E.T. Kaiser, *J. Amer. Chem. Soc.*, **94**, 7933 (1972).

# Subject Index